MEDICAL RADIOLOGY

Diagnostic Imaging and Radiation Oncology

Springer

Berlin
Heidelberg
New York
Barcelona
Budapest
Hong Kong
London
Milan
Paris
Santa Clara
Singapore
Tokyo

A.M. Davies · H. Pettersson (Eds.)

Orthopedic Imaging
Techniques and Applications

With Contributions by

T.H. Berquist · H. Bonél · J.A. Bouffard · M. Breitenseher · V.N. Cassar-Pullicino
N. Chemla · A. Chevrot · A.M. Davies · J.L. Drape · A.M. Dupont · N. Egund
H.K. Genant · F. Gires · D. Godefroy · J. Haller · J. Hodler · H. Imhof · K. Jonsson
F. Kainberger · J.J. Kaye · M.V. Maffey · C. Masciocchi · I.W. McCall · E.G. McNally
A. Minoui · J. Moutounet · W.R. Obermann · E. Pessis · H. Pettersson · M. Reiser
L. Sarazin · C. Schiepers · E.R. Tjin A Ton · S. Trattnig · D. Vanel · M. van Holsbeeck
C. van Kujik · I. Watt

Foreword by

A.L. Baert

Preface by

A.M. Davies and H. Pettersson

With 366 Figures in 598 Separate Illustrations

Springer

Dr. A. Mark Davies
MRI Centre
Royal Orthopaedic Hospital
Bristol Road South
Birmingham B31 2AP
United Kingdom

Professor Dr. Holger Pettersson
Department of Radiology
University Hospital
University of Lund
S-22185 Lund
Sweden

MEDICAL RADIOLOGY · Diagnostic Imaging and Radiation Oncology

Continuation of
Handbuch der medizinischen Radiologie
Encyclopedia of Medical Radiology

ISBN-13: 978-3-642-64341-5   e-ISBN-13: 978-3-642-60295-5

DOI: 10.1007/978-3-642-60295-5

Library of Congress Cataloging-in-Publication Data. Orthopedic imaging: techniques and applications/A.M. Davies, H. Pettersson (eds.); with contributions by T.H. Berquist . . . [et al.].; foreword by A.L. Baert. p. cm. – (Medical radiology) Includes bibliographical references and index. ISBN 3-540-63187-9 (alk. paper) 1. Orthopedics – Diagnosis. 2. Musculoskeletal system – Imaging. I. Davies, A.M. (Arthur Mark), 1954- . II. Pettersson, Holger, 1942- . III. Berquist, Thomas H. (Thomas Henry), 1945- . IV. Series. [DNLM: 1. Musculoskeletal Diseases–diagnosis. 2. Diagnostic Imaging – methods. 3. Orthopedics. WE 141 0767 1998] RD734.078 1998 616.7'0754 – dc21 DNLM/DLC for Library of Congress 97-49118 CIP

Cover design: de'blik, Berlin
Typesetting: Best-set Typesetter Ltd., Hong Kong

SPIN: 10546210   21/3134 – 5 4 3 2 1 0 – Printed on acid-free paper

Foreword

Musculoskeletal studies account for an important proportion of the daily clinical practice of most radiologists. For many years following Röntgen's discovery of x-rays in 1895, these studies were confined to plain films and conventional tomography, which substantially limited the contribution of radiology in achieving better diagnosis and treatment of orthopedic pathologic conditions. The advent of digital radiography, ultrasound, computer tomography and especially magnetic resonance imaging has greatly enhanced the potential of radiologic imaging in this field. Among the benefits to accrue from these techniques are the detailed visualization of soft tissue anatomy and pathologic changes, progress in the noninvasive study of joint pathology, and improved staging of primary bone tumors. The need for an update of our knowledge in orthopedic imaging is therefore immense.

The editors, Dr. A.M. Davies and Prof. H. Pettersson, have been able to acquire the collaboration of a number of international leaders in bone and soft tissue imaging for the production of this book, which took an exceptionally brief period from the date of conception to printing. Accordingly, readers will find in this work the latest developments in techniques and radiologic interpretation. The up-to-date nature of the information provided, and the expertise which it embodies, will undoubtedly be of great help in daily clinical practice, not only to general radiologists and orthopedic surgeons but also to musculoskeletal radiologists working in subspecialties.

I am very grateful to the editors and to the authors for their excellent contributions.

Leuven   ALBERT L. BAERT

Preface

Few can have envisaged the dramatic developments that have occurred in all aspects of imaging in the past 20 years. The greatest impact has arguably been in the subject of musculoskeletal imaging. There is therefore a continuous need to update radiologists, orthopaedic surgeons and others working in this field. To this end the purpose of this book is twofold. First, to acquaint the reader with the full range of techniques available for imaging musculoskeletal problems, describing how they work and emphasising indications and contraindications. Amongst the nine chapters in this first section are contributions on computer tomography, magnetic resonance imaging, scintigraphy and bone densitometry. The remaining ten chapters discuss the optimal application of these techniques to specific clinical problems. These chapters are divided by either the anatomy involved or the underlying pathological process and highlight practical solutions to everyday clinical problems. The editors are grateful to all the authors for their contributions to this book, which aims to offer a comprehensive overview of current musculoskeletal imaging applicable to all specialties involved in this area of clinical practice.

Birmingham
Lund
A.M. DAVIES
H. PETTERSSON

Contents

Imaging Techniques and Procedures

1 Radiography

H. Pettersson and K. Jonsson

CONTENTS

1.1
Introduction

For the evaluation of musculoskeletal lesions, any combination of the different diagnostic imaging modalities may be appropriate, as will be discussed in the following chapters. However, the classical x-ray examination, which has now been used in medical practice for 100 years, is still important. Indeed, for the diagnostic imaging workup of most musculoskeletal lesions, it is the primary examination.

The advantage of radiography is that it provides an overview of the bony anatomy. Thus, all major pathology concerning trauma, joint disease, and structural changes within bone is well diagnosed with radiography. However, subtle changes may need complementary examination with computed tomography (CT), magnetic resonance imaging (MRI), or isotope scans. The choice of one or more complementary modalities is then dependent on the clinical status and the findings at the radiographic examination.

H. Pettersson, MD, Professor, Department of Radiology, University Hospital, University of Lund, S-22185 Lund, Sweden
K. Jonsson, MD, Department of Radiology, University Hospital, University of Lund, S-22185 Lund, Sweden

When using x-rays as a diagnostic tool, it should always be borne in mind that sending a patient for x-ray means that the patient will be exposed to ionizing radiation, with its potential hazards. The benefits of the examination should therefore always be weighed against the possible damage caused by the radiation. However, using modern equipment, with highly sensitive film-screen combinations, or using digital radiography, the radiation doses are very small for all routine examinations.

1.2
Physics

The x-rays are generated in an x-ray tube, which fundamentally consists of a vacuum tube, with a cathode and an anode (minus- and plus-poles, respectively) (Fig. 1.1). The cathode is an electrically heated filament from which electrons are emitted when a high potential (voltage) is applied between the poles. Because of the high potential, the electrons are drawn to the anode, made up of a metal plate – usually molybdenum, tungsten, or wolfram, to resist the high temperature. The accelerated electrons are concentrated to a narrow beam, which hits a small area of the anode plate, the so-called focus. Upon hitting the focus, most of the energy of the electrons is transformed into heat, but approximately 1% is instead transformed into x-rays. These x-rays are directed through a window out of the tube, which for radiation protection is encapsulated in a lead cover.

X-rays are electromagnetic waves, part of the nonvisible spectrum, with a very short wavelength and a high frequency (Kiuru 1995). They have a very high energy and can penetrate most types of tissue in the human body. The penetration ability depends on the potential (voltage) between the anode and the cathode. The higher the potential, the shorter the wavelength of the x-rays, and the stronger the penetration. If the voltage is lower, the x-rays have a longer wavelength with less penetration ability. When the primary x-rays hit the body, the rays inter-

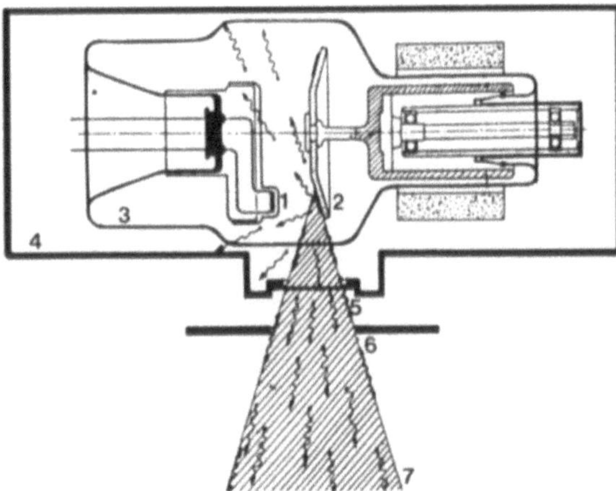

Fig. 1.1. Schematic drawing of an x-ray tube in its lead cover. *1*, Cathode with the electrically heated filament; *2*, anode (on a rotating disc for cooling); *3*, the glass tube; *4*, the tube housing (lead cover); *5*, window with aluminum filter; *6*, lead window for primary focusing; *7*, x-ray radiation

act with the materia, which gives rise to secondary radiation, scattered in all directions. Only a minor part of the primary radiation passes the body unaffected, and it is this radiation that is used for the production of the images. The detector medium of these x-rays originally was photographic films, because x-rays will blacken a photographic film similar to visible light. Later, different kinds of screens and film-screen combinations were used to optimize the image on the film, depending on the type of examination and the organ examined. Today, other types of detector media are also available, with computers being used for image production instead of photochemical techniques.

1.3
Radiographic Techniques

To obtain a radiograph, principally two techniques are used today: the analogue and the digital.

With the *direct analogue technique* the radiograph is created directly on the detector medium, that is by direct exposure of the x-rays on the radiographic film (film-screen combination) or a fluorescent screen. With the *indirect analogue technique*, the primary image obtained by a fluoroscopic technique is not observed directly, but is instead transferred to an image intensifier, which enhances the brightness of the primary image. The intensified information may then be registered by a TV camera or shown on a monitor.

With the *digital technique*, the radiation that has passed the examined body is sampled by a detector medium, which conveys this information to computer systems for further management (see below).

1.4
Positioning of the Patient

Irrespective of the radiographic technique used, it is imperative to obtain adequate projections of the area examined. To acquire a valid, three-dimensional impression of a lesion and the anatomical landmarks related to the lesion, it is necessary to obtain images in two perpendicular planes (Fig. 1.2). For almost all routine examinations, these planes should be identical with the anatomical frontal and sagittal planes, i.e., the radiographic AP/PA and lateral views. The axial plane will not be studied with a radiographic technique, but rather with, for instance, CT or MRI, as will be described in the following chapters.

The most adequate projections vary not only with the examined anatomical region, but also with the indication for the examination. Details concerning positioning in radiography, according to anatomical area and lesion, are available in special textbooks (BALLINGER 1991; BERNAU 1995; CLARK and SWALLOW 1991).

To avoid disturbing distortion in the image it is also mandatory that the x-rays should hit the examined bone or joint perpendicular to the surface of the structure, and that the structure should be parallel with the plane of the detector medium (Fig. 1.3). In the normal patient this is no problem, but in the diseased or impaired patient difficulties may arise. In such situations, either the x-ray tube may be tilted or the limb may be supported in the position desired. For example, in a locked elbow one has to take two AP films, one with the x-rays perpendicular to the forearm and one with the x-rays perpendicular to the humerus (Fig. 1.4). In a patient with extensive outward rotation of the femur (e.g., caused by femoral neck fracture), the femoral neck is difficult to evaluate on an AP film. Then, the tube must be tilted accordingly, or pillows must be placed under the pelvis to tilt the patient in such a way that the x-rays penetrate the femoral head and neck perpendicular to the neck (Fig. 1.5). In a patient with scoliosis it may be difficult to evaluate the intervertebral discs, but often one can straighten out the scoliosis if the patient is lying on the contralateral side for the lateral view and it is then often possible to evaluate the discs. It is also possible to tilt the tube so that the x-rays run parallel to the intervertebral disc in question.

1.5
Fluoroscopy

If the patient is severely ill or hurt, it may be inappropriate or hazardous to put too much emphasis on positioning the patient to obtain the standard views. Rather it may be more efficient and safer to use fluoroscopy to find the adequate projection, tilting only the tube.

Fluoroscopy may also be of help to evaluate joint motion and instability in a joint, for instance in analyzing wrist movement. The complex motion of a wrist may, however, be difficult to evaluate just by fluoroscopy, and the latter should be combined with cineradiography or recording on videotape to make a more detailed analysis possible. Fluoroscopy may also be of value to analyze the presence and movement of intra-articular bodies or osteochondral fragments (RESNICK 1995a).

1.6
Magnification Radiography

Magnification radiography may improve the quality of an examination. There are two principle methods of magnification. In *optical magnification*, the radiograph is taken on a fine-grain industrial film, the focal spot is of conventional size, and the object is placed close to the film to obtain the highest possible spatial resolution. The image is then viewed with optical magnification. The other method is *radiographic magnification*, where the focal spot is small, 100–150 µm, and the film is placed at a distance from the object with an air gap between the object and the film (Fig. 1.6).

In all situations where subtle changes are expected, magnification radiography may be of help. Thus, it may be used for examination of patients with rheumatoid arthritis to reveal erosions that are not

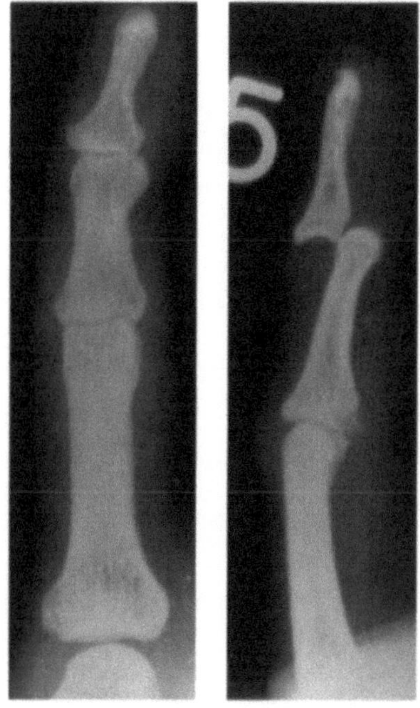

Fig. 1.2 a,b. Images must be obtained in two perpendicular planes, to give a valid three-dimensional impression of a lesion. a Frontal and b lateral view of the index finger. In a, very little is seen of the total dislocation, which is obvious in b

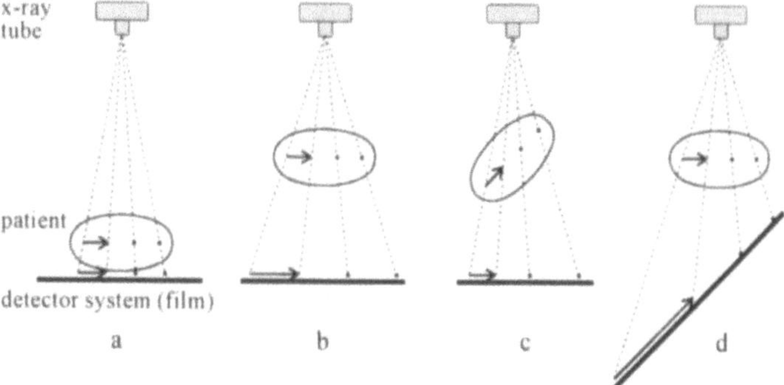

Fig. 1.3 a–d. Distortion and magnification in the image. Because of the imaging geometry, there will always be some distortion and magnification of the image. a It is only when the patient is close to the film, and the examined part of the body is parallel to the film and perpendicular to the central beam, that the magnification and distortion are minimized. b If the distance between the examined body part and the film increases, there is an increased magnification. c,d If there is an angle between the examined body part and the film, the image is distorted

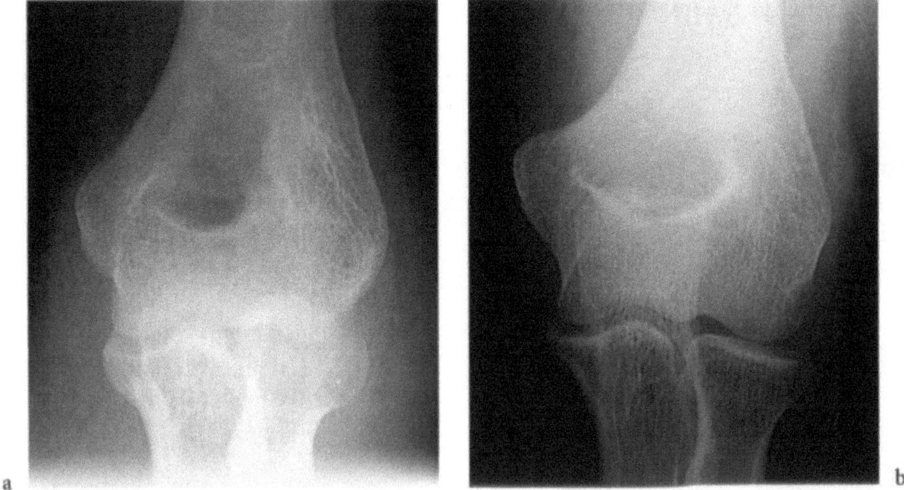

Fig. 1.4 a,b. Examination of an elbow, fixed in flexion after trauma; AP views. **a** Perpendicular to the humerus; **b** perpendicular to the forearm. It is only in **b** that the joint is clearly visible, and the fracture in the radial head is seen

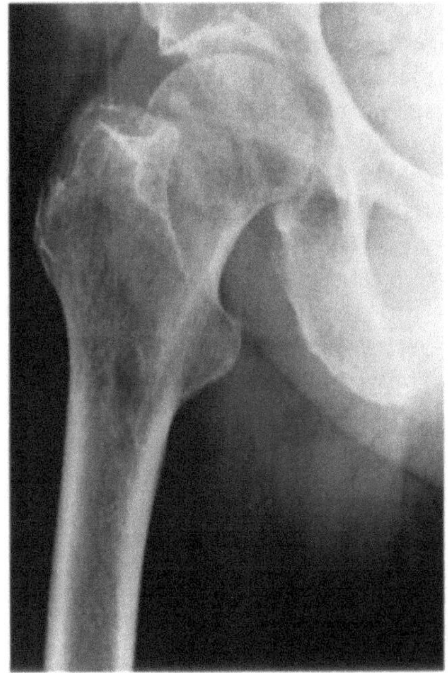

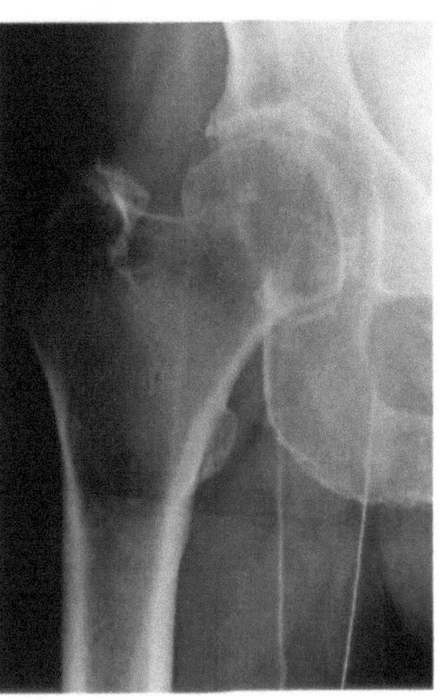

Fig. 1.5 a,b. Hip; femoral neck fracture. In **a** the patient is examined in the "resting" position. Because of the lesion, the leg is heavily rotated outwards, and the fracture is hardly seen. In **b**, the tube and the patient have been tilted, so that the femoral neck is more parallel to the film and perpendicular to the x-ray beam. The fracture is now obvious

seen with the conventional technique. Also in hyperparathyroidism, subperiosteal bone resorption is better evaluated with magnification techniques (GENANT and RESNICK 1995).

There are two limitations to the use of magnification radiography. One is that the field of view is reduced and only a limited part of the object is shown. The other is that there may be an increased

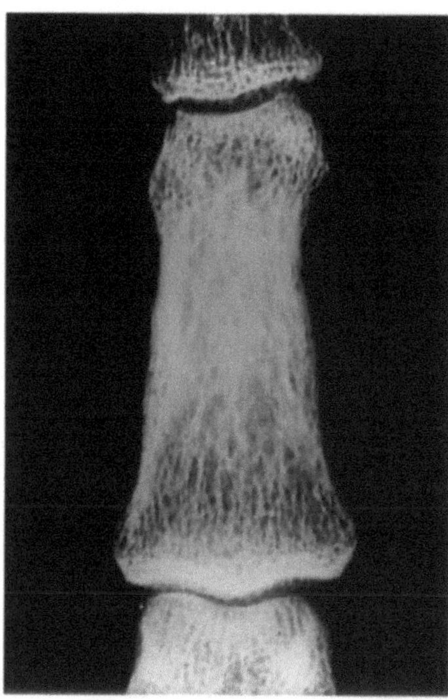

Fig. 1.6. Magnification radiography. Using magnification radiography, details of the skeletal structure are better evaluated, as here in a patient with hyperparathyroidism

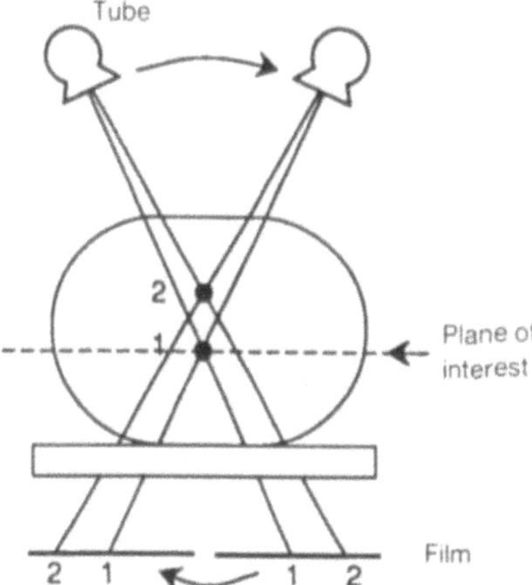

Fig. 1.7. Conventional tomography. The x-ray tube and the film move together, in such a way that the projections of all points in the plane of interest remain stationary on the film. In the figure, point *1* is located in the plane of interest, and is imaged sharply, while point *2* is located outside this plane, and its image on the film is blurred due to gross movement unsharpness. (From SMITH 1995)

radiation dose to the patient, even though film-screen systems are now available that help to reduce the radiation.

1.7
Low Kilovolt Technique

The low kilovolt, or mammography technique, uses a molybdenum target and filter and low kilovoltage, between 28 and 35 kV, for generation of the x-rays (FISCHER 1995). The technique results in greater contrast between fat and water-equivalent tissue and between water-equivalent tissue and bone. It allows a more specific soft tissue diagnosis and a precise analysis of the margins of thinner portions of the bone. With this technique thickened soft tissues are easily diagnosed, such as thickening of tendons and joint capsules in arthritides. It is also of value for the diagnosis of gouty tophi and rheumatoid nodules. Calcifications in soft tissues or in joints and vascular calcifications are well seen with this technique. The fact that the x-ray spectrum is of low energy with reduced penetration and increased radiation dose reduces the application to thin body parts, such as the hands and feet.

1.8
Conventional Tomography

In conventional tomography (for computed tomography, see Chap. 3), the conventional x-ray tube and the film-screen combination are used to define a predetermined plane of the examined body, while the structures above and below this plane are eliminated or blurred (Fig. 1.7) (RESNICK 1995b). This is achieved by moving the x-ray tube and the film-screen combination in a defined relation to each other, while the examined part of the body remains stationary. The motion of the x-ray tube and the film may be either unidirectional (linear tomography) or pluridirectional (circular, elliptical, spiral, or hypocycloidal tomography). The more complex the movement, the better the quality of the image with less longitudinal streaking (which may be seen in unidirectional tomography). The disadvantages with tomography are the long examination time and the high radiation exposure to the patient (depending on the number of sections obtained).

The advent of CT and MRI has obviated the need for conventional tomography in most cases, but the method may still be of great help in some cases where conventional radiography is equivocal or not diagnostic. Conventional tomography may provide valuable information on lesions of a number of joints, for instance the sternoclavicular joint, temporomandibular joints, and sacroiliac joints. It is

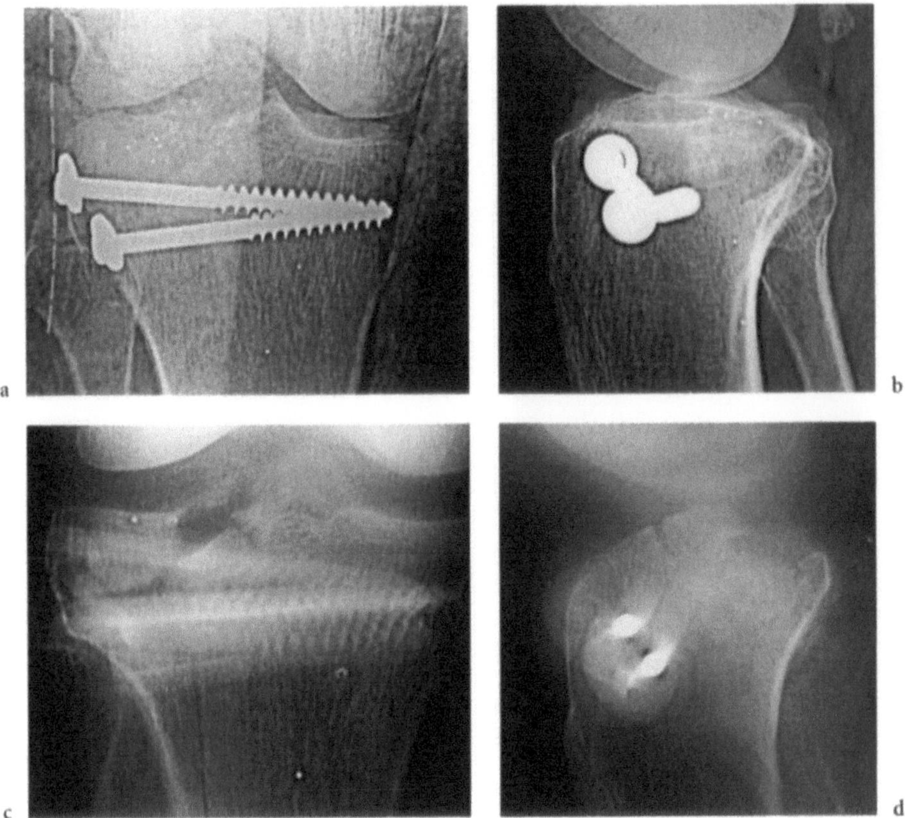

Fig. 1.8 a–d. Knee with a tibial plateau fracture, after surgery. **a,b** Plain x-ray, AP and lateral; **c,d** conventional tomography, AP and lateral. On the plain films, the exact position of the fragments cannot be evaluated, while on the tomograms, the position of the different fragments can easily be defined

also very valuable for evaluation of fractures of the tibial plateau of the knee. Fragments of the joint surface may be surprisingly dislocated and deep; such fragments may be clearly visible using tomography but hardly discernible on plain x-rays (Fig. 1.8).

1.9
Digital Radiography

As described above, in the classical analogue radiographic examination, a photographic film or a film-screen combination is used to detect the radiation that has passed through the patient, and the image is displayed as an x-ray film or on the viewbox. In *digital radiography* the radiation that has passed the patient is instead caught by some type of detector system, and the energy carried by the radiation is conveyed to signals that are processed in computers. The resulting images may be displayed on monitors or possibly on hard copies. Today the most fre-

quently used detector system is the so-called imaging plate, based on phosphor or selenium (SCHMIDT and DEININGER 1990; PETTERSSON 1992). In the routine setting, it is common to have a dual display of the images, using two different computer programs: one that gives an image similar to that obtained with a conventional film-screen combination, and one with low global contrast and increased unsharp masking which increases the contrast step interfaces (Fig. 1.9).

1.9.1
Potential Advantages and Limitations

The potential advantages of digital radiography may be summarized as the broad exposure range, the free choice of data processing, and possibilities for reporting at work stations, while the limitations are the limited spatial resolution, the risk of contrast artifacts, and difficulties in visual evaluation of bone density.

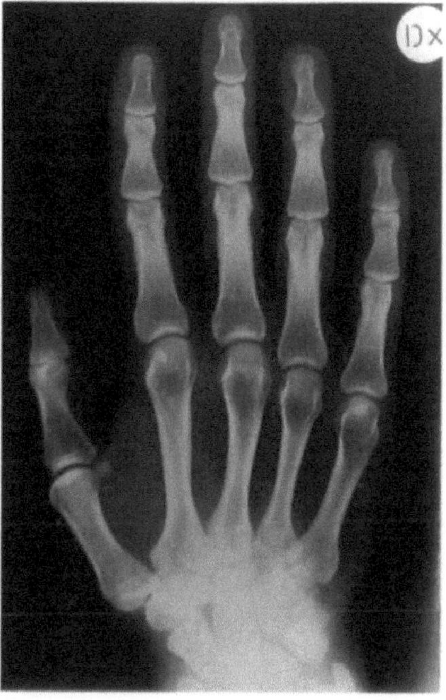

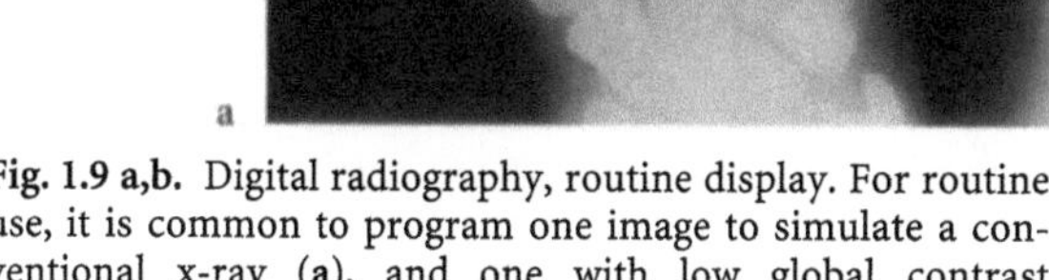

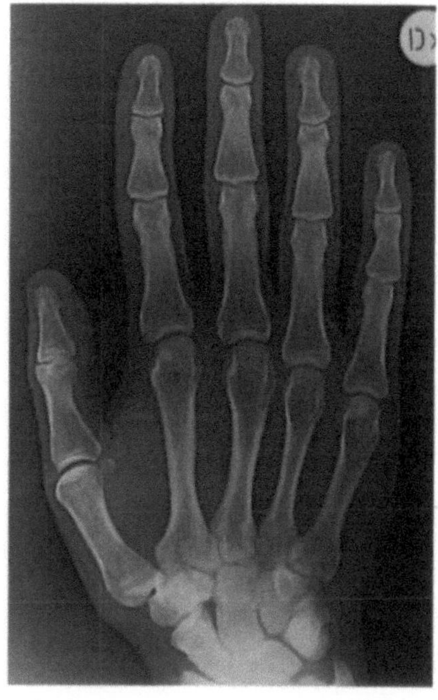

Fig. 1.9 a,b. Digital radiography, routine display. For routine use, it is common to program one image to simulate a conventional x-ray (a), and one with low global contrast and increased unsharp masking, better revealing both the skeletal structure and the soft tissues in one and the same image. (b)

The *broad exposure* range is based on the fact that for digital radiography, the dose-response curve is linear over 4 or 5 orders of magnitude, as compared with 1–2 orders of magnitude for conventional radiography. This means that for practical reasons it is almost impossible to make bad exposures, as is exemplified by Fig. 1.10. It also means that the radiation dose can be decreased in the individual case without disturbing the diagnostic accuracy (JÓNSSON et al. 1996).

The free choice of *data processing* means that from one and the same exposure, it is possible to produce an unlimited number of images with varying gray scale, global contrast, etc. Hence, from a theoretical point of view it is now possible to obtain the image that is best suited for displaying the pathology looked for in each individual case.

Reporting at work stations instead of at viewboxes means that all image information may be handled using computers, opening up the possibility of image communication within the department and between departments and hospitals – a situation that in the past could only be dreamt of.

The *limited spatial resolution* has until now been regarded as a limiting factor, but with modern systems this is no longer the case (JÓNSSON et al. 1995; SCOTT et al. 1993). *Contrast artifacts* may appear as an illusion of a radiolucent zone at high-contrast steps, for instance between metal, cement and bone in a joint prosthesis. However, using appropriate programs, or using the routine of two different programs for display of each examination, this problem may easily be avoided.

The *apparent bone density* is highly dependent on which computer program is used to produce the image, and hence difficulties may arise. Again, this may be controlled if the examiner is aware of the problem, using two different programs as a routine for each examination.

1.9.2
Clinical Use

Given the potential advantages described above, digital radiography is now successfully used in many large departments, and has proven superior to the conventional x-ray examination, for instance in intensive care wards, in operating rooms, and for all patients undergoing examination of areas where high-contrast differences appear or soft tissue examination. Based on a broad general clinical impression, as well as on several prospective scientific studies, it may today be said that digital radiography

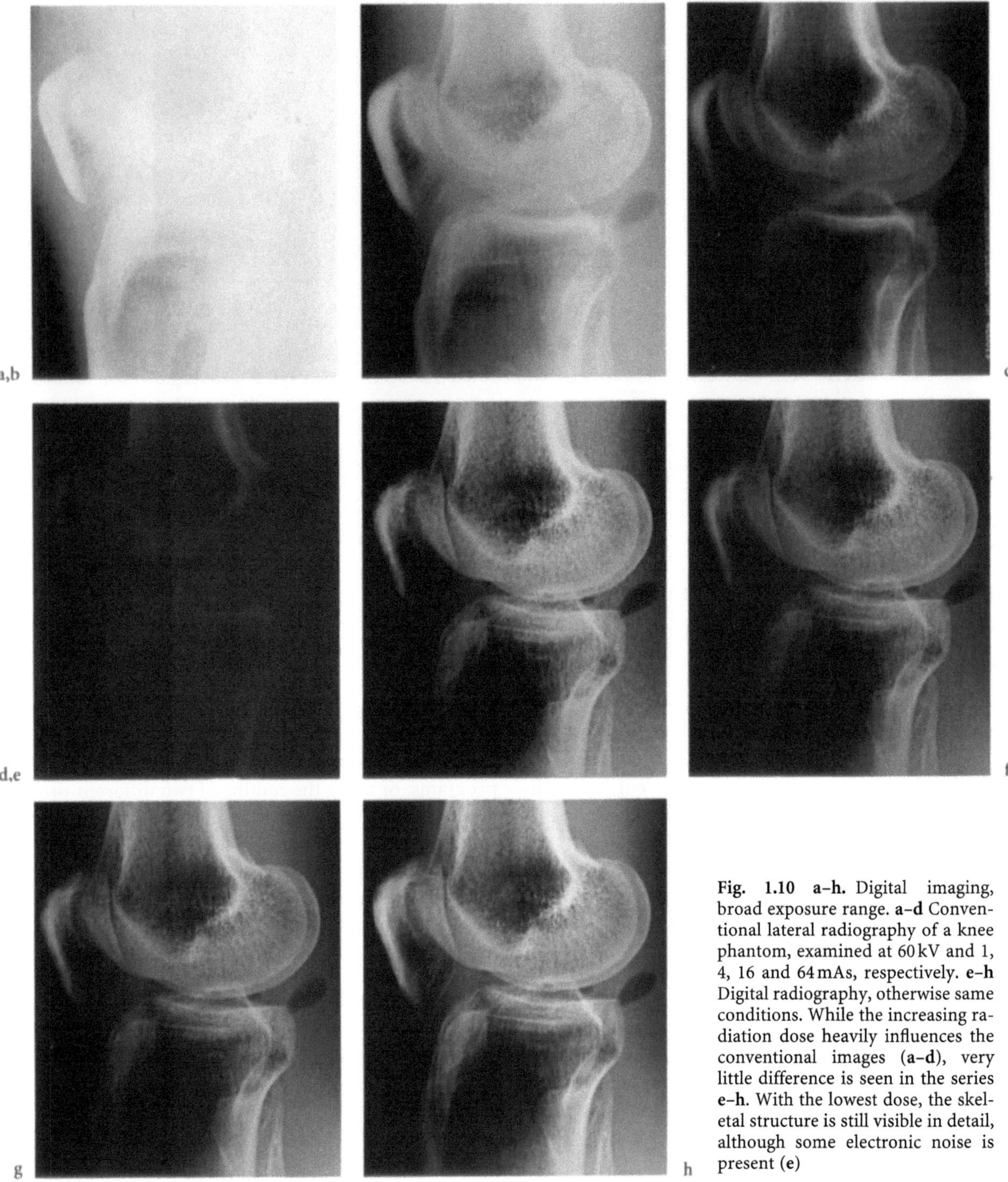

Fig. 1.10 a–h. Digital imaging, broad exposure range. a–d Conventional lateral radiography of a knee phantom, examined at 60 kV and 1, 4, 16 and 64 mAs, respectively. e–h Digital radiography, otherwise same conditions. While the increasing radiation dose heavily influences the conventional images (a–d), very little difference is seen in the series e–h. With the lowest dose, the skeletal structure is still visible in detail, although some electronic noise is present (e)

for routine musculoskeletal examinations is equal or superior to conventional radiography (WILSON et al. 1994; KREIPKE et al. 1990). Also, the digital technique opens up new possibilities both for computed measurements of distances, angles, volumes, etc. and for bone density measurements. Given the wealth of information handled by the computer for each image, there is also potential for computer-aided diagnosis and for adjustment of the image in each individual case to fit the human possibilities for perception and cognition, there by serving as a basis for maximization of the diagnostic accuracy.

References

Ballinger PW (1991) Merrill's atlas of radiographic positions and radiologic procedures, 7th edn. Mosby, St. Louis

Bernau A (1995) Orthopädische Röntgendiagnostik. Einstelltechnik. Urban & Schwarzenberg, München

Clark KC, Swallow RA (1991) Positioning in radiography, 11th edn. Butterworths-Heinemann, London

Fischer E (1995) Low kilovolt radiography. In: Resnick D (ed) Diagnosis of bone and joint disorders, 3rd edn. Saunders, Philadelphia, pp 89–107

Genant HK, Resnick D (1995) Magnification radiography. In: Resnick D (ed) Diagnosis of bone and joint disorders, 3rd edn. Saunders, Philadelphia, pp 72–88

Jónsson A, Hannesson P, Herrlin K, et al. (1995) Computed vs. film-screen magnification radiography of fingers in hyperparathyroidism. Acta Radiol 36:290–294

Jónsson A, Herrlin K, Jonsson K, et al. (1996) Radiation dose reduction in computed skeletal radiography: the effect of image quality. Acta Radiol 37:128–131

Kiuru A (1995) Radiophysics. In: Pettersson H (ed) A global textbook of radiology. The NICER Institute, Oslo

Kreipke DL, Silver DI, Tarer RD, Braunstein EM (1990) Readability of cervical spine imaging: digital versus film/screen radiographs. Comput Med Imaging Graph 14:119–125

Pettersson H (1992) Digital skeletal radiography. In: Resnick D, Pettersson H (eds) Skeletal radiology. NICER series on diagnostic imaging. Merit Communications, London, pp 1–8

Resnick D (1995a) Fluoroscopy. In: Resnick D (ed) Diagnosis of bone and joint disorders, 3rd edn. Saunders, Philadelphia, pp 68–71

Resnick D (1995b) Conventional tomography. In: Resnick D (ed) Diagnosis of bone and joint disorders, 3rd edn. Saunders, Philadelphia, pp 108–147

Schmidt CH, Deininger HK (1990) Die digitale Bildverstärkerradiographie: ein neues Konzept für die traumatologische Röntgendiagnostik. Fortschr Röntgenstr 152: 51–55

Scott WW Jr, Rosenbaum JE, Ackerman SJ, et al. (1993) Subtle orthopaedic fractures: teleradiology workstation versus film interpretation. Radiology 187:811–815

Smith H-J (1995) Modalities and methods. In: Pettersson H (ed) A global textbook of radiology. The NICER Institute, Oslo

Wilson AJ, Mann FA, West OC, et al. (1994) Evaluation of the injured cervical spine: comparison of conventional and storage phosphor radiography with a hybrid cassette. Radiology 193:419–422

2 Arthrography

J.J. KAYE

CONTENTS

2.1
Introduction

Arthrography of the knee was a very early application in the study of joints. Gas arthrography was used early in the 1930s, but there was poor contrast resolution between tissues (BIRCHER 1931). Later in the 1930s, iodinated contrast materials were utilized for arthrography, but the toxicity of the early contrast materials precluded their widespread use. In the 1940s, water-soluble contrast materials were utilized for contrast arthrography and allowed for the development of arthrography as a means to evaluate

J.J. KAYE, MD, Professor of Radiology, New York Medical College, Chairman, Department of Radiology, Saint Vincents Hospital and Medical Center, 153 West 11th Street, New York, NY 10011, USA

joints. Virtually all joints in the human body have been studied by arthrography, and the technique became extremely popular during the late 1960s, 1970s, and 1980s. Several textbooks have been devoted to the subject (FREIBERGER and KAYE 1979; RICKLIN et al. 1979; THIJN 1979; DALINKA 1980; STOKER 1980; ARNDT et al. 1981; PAVLOV et al. 1983a). In many locations where magnetic resonance imaging (MRI) is available, this technique has largely replaced arthrography of joints because of the excellent contrast in soft tissues and the multiplanar capabilities of this technique (KAYE 1994). However, when the cost of the procedure is an issue and where MRI is not readily available, arthrography remains an excellent technique to study most joints.

2.2
Knee Arthrography

2.2.1
Technique and Normal Findings

The extensive study by LINDBLOM in 1948 probably did more than any other to establish the safety and reliability of this technique. Lindblom used single positive contrast arthrography of the knee. Subsequent advances included the double-contrast method initially described using a horizontal x-ray beam (ANDREN and WEHLIN 1960). Further advances related to the fluoroscopic and spot film method of filming arthrograms described by several authors (BUTT and McINTYRE 1969; RICKLIN et al. 1979). The technique of double-contrast arthrography of the knee using the fluoroscopic method has now been adopted by most experienced arthrographers.

Meticulous attention to the details of technique is essential. There are a number of detailed descriptions of the fluoroscopic method of knee arthrography (BUTT and McINTYRE 1969; DALINKA 1980; RICKLIN et al. 1979; FREIBERGER 1979b). Following complete aspiration of any synovial fluid which may

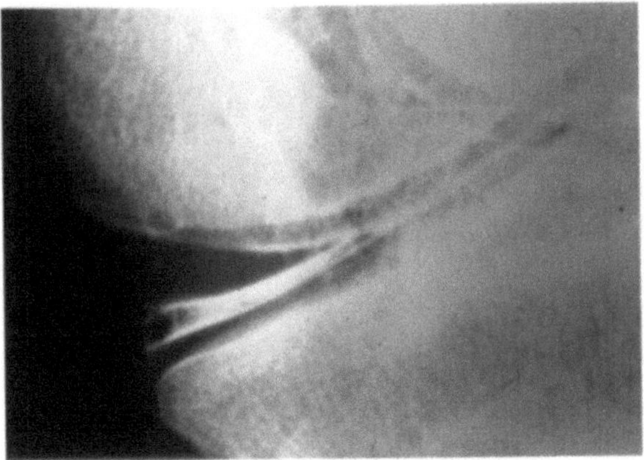

Fig. 2.1. Normal double-contrast knee arthrogram of the medial meniscus. The meniscus is coated by a thin layer of positive contrast material and partially surrounded by air. No contrast material is seen within the meniscus or at its periphery

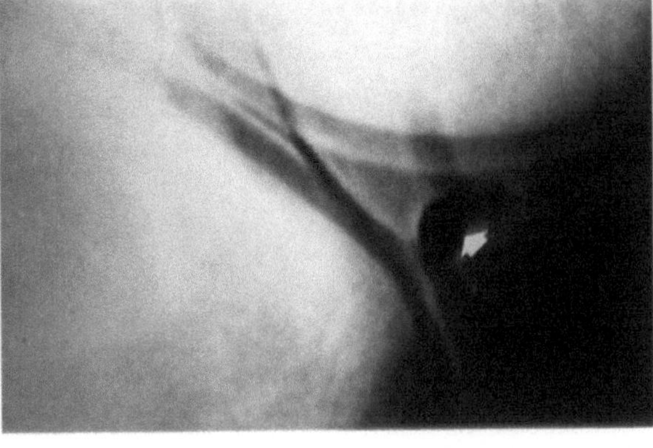

Fig. 2.2. Normal double-contrast knee arthrogram of the lateral meniscus. The popliteus tendon sleeve (*arrow*) is filled with air and contrast material. The normal meniscus is coated by a thin layer of positive contrast material. No air or contrast material is seen within the substance of the meniscus

be present, a small amount (2–5 cc) of water-soluble positive contrast material is injected via either a medial or a lateral approach in the parapatellar region. This is followed by approximately 20 cc of room air. The knee is then exercised and filming begins. In general, 9–18 well-collimated exposures of each meniscus are made with the knee being turned slightly so that the examination proceeds in a fluoroscopically controlled fashion from the back to the front of the meniscus. Stress is placed upon the knee to open up the compartment being examined and to insure that the meniscus is coated by a layer of positive contrast material and surrounded or partially surrounded by air. Recumbent lateral radiographs and sitting lateral radiographs of the knee after injection allow evaluation of the cruciate ligaments (PAVLOV and TORG 1978; PAVLOV 1979; PAVLOV et al. 1983b).

The normal medial meniscus is a roughly triangular structure which is firmly attached to the capsule throughout its anterior to posterior extent; it is roughly C-shaped (KAYE 1979a). The normal medial meniscus will be coated by a thin layer of positive contrast material and surrounded by air, and no contrast material will be seen within its substance or at its periphery (Fig. 2.1). The normal lateral meniscus is more nearly circular in shape, and is of about the same width anteriorly as it is posteriorly. The anterior portion of the lateral meniscus is firmly attached to the capsule, similar to the medial meniscus. Posteriorly, however, the popliteus tendon becomes partially intra-articular and crosses the meniscus, partially surrounded by the popliteus tendon sheath. There are defects in the normal attachments of the

lateral meniscus posteriorly (MCINTYRE 1972; JELASKO 1975; WICKSTROM et al. 1975; HARLEY 1977; KAYE 1979a). Contrast material and air instilled into the knee in the course of arthrography will fill the tendon sheath of the popliteus and pass across the periphery of the posterior portion of the lateral meniscus, making the demonstration of tears more difficult (Fig. 2.2).

2.2.2
Meniscal Abnormalities

Injuries frequently cause tears or disruptions of the meniscal cartilage, and these are readily seen on double-contrast arthrography. Tears can be vertical, oblique, horizontal, or radial, or they may be complex, combining several types (FREIBERGER 1979a). The configuration of the tear depends upon the degree of separation of the fragments and in part is influenced by the amount of distraction applied to the knee in the course of arthrography.

Vertical concentric tears split the meniscus into inner and outer portions (Figs. 2.3, 2.4). If the anterior and posterior portions of the inner fragment remain attached to the outer portion but the medial fragment is displaced, this is referred to as a bucket-handle tear. The central displaced fragment may be displaced into the intercondylar notch, where it may be difficult to identify. If the fragments remain in close apposition, then positive contrast agent alone or with a small amount of air will seep into the tear. If a tear is oblique and the inner fragment is displaced into the intercondylar region, the appearance

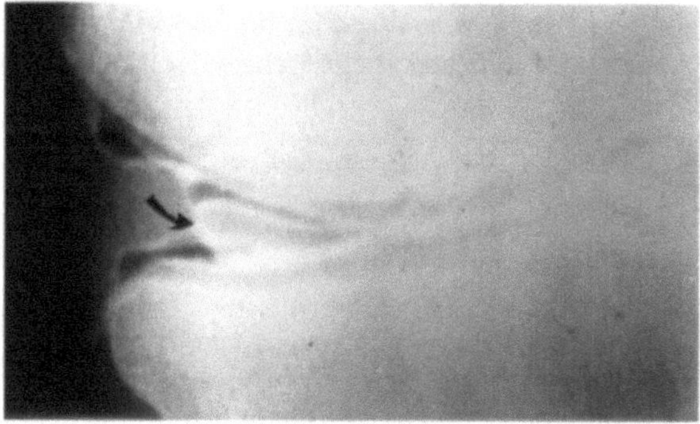

Fig. 2.3. Torn medial meniscus. A vertical concentric tear of the medial meniscus is present (*arrow*), with slight inferior displacement of the central meniscal fragment

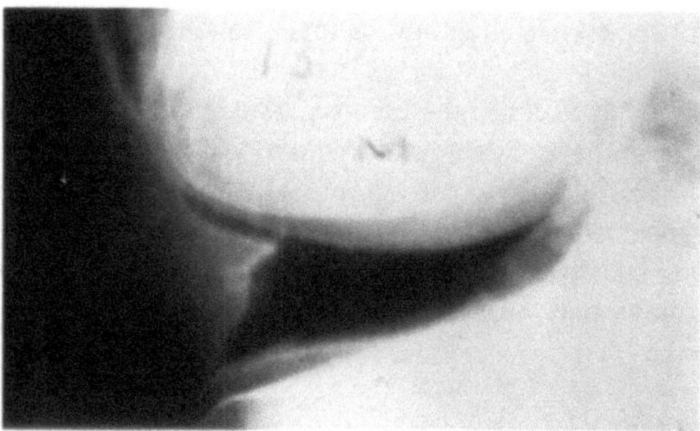

Fig. 2.4. Bucket handle tear. The medial meniscal contour is truncated by a tear

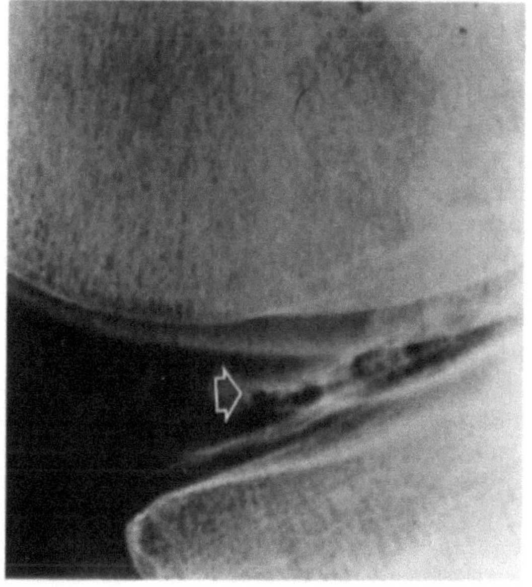

Fig. 2.5. Horizontal tear of the medial meniscus. Positive contrast material extends into a horizontal tear (*arrow*)

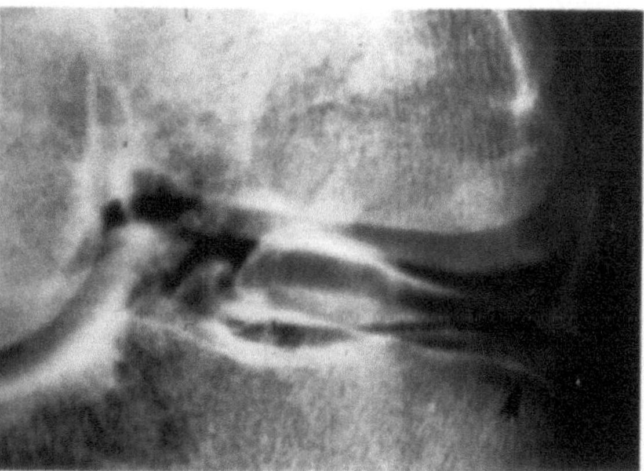

Fig. 2.6. Torn discoid lateral meniscus. The lateral meniscus extends too far into the joint with a bulbous central portion. At the periphery of this discoid meniscus, a tear is filled with contrast material (*arrow*)

may be that of a meniscus that is too small with an abnormal contour (Fig. 2.4). Small vertical tears beginning at the apex of the meniscus may be present, and the fragments are typically close together. In this case, a small area of opacification at the inner edge of the triangular meniscus is seen on the arthrogram. This is referred to as a parrot beak tear. Vertical radial tears occur roughly at right angles to the tangent of the meniscus, and are seen on arthrography as small areas of central opacification of the meniscus. Horizontal tears typically occur in the older individual and are seen as linear areas of contrast material extending into the substance of the meniscus from near its apex (Fig. 2.5). Complex tears may assume a variety of configurations with air and positive contrast material seen within the menisci. When the posterior segment of the lateral meniscus is torn, there may be compression of the popliteus tendon sheath (Pavlov and Goldman 1980).

A discoid meniscus is completely or partially disc shaped rather than crescent shaped. Discoid menisci are more common on the lateral side of the knee. They are frequently torn because their abnormal shape causes them to be unusually stressed. Discoid menisci are seen on arthrography as extending too far into the knee, reaching or nearly reaching the intercondylar notch, and projecting between the articular cartilage of the femoral condyle and that of the tibial plateau (Fig. 2.6). When a discoid meniscus is demonstrated, a careful examination of the meniscus should be made for the presence of a tear.

Cysts occur in the menisci, usually at the periphery of the meniscus, and frequently associated with a horizontal tear. Arthrographically, these are seen as linear areas of contrast material extending into the substance of the meniscus and ending at a contrast-filled cavity near the margin of the joint.

2.2.3
Ligamentous Abnormalities

The cruciate ligaments are evaluated on the lateral radiograph with the tibia stressed anteriorly with respect to the femur (PAVLOV 1979; PAVLOV et al. 1983b). The anterior surface of the anterior cruciate ligament and the posterior surface of the posterior cruciate ligament are then well visualized (Fig. 2.7). When these ligaments are normal, these synovial surfaces are straight, although there might be slight posterior bowing of the posterior cruciate ligament. The anterior cruciate ligament is considered to be torn when there is pooling of contrast material in the location of the anterior cruciate ligament or when there is deformity of its surface (Fig. 2.8). Similar criteria are applied to the posterior aspect of the posterior cruciate ligament to determine tears.

When there is a tear of the medial collateral ligament or the lateral capsular ligament, and the patient is examined within 48 h of the tear, there will be leakage of contrast material from the joint through the tear. This is visualized on the spot films of the menisci and indicates the presence of a ligamentous tear.

2.2.4
Other Abnormalities

A number of other abnormalities can be studied by knee arthrography (STAPLE 1972; SCHNEIDER and FREIBERGER 1979). A considerable portion of the articular cartilage is visualized by knee arthrography coated by positive contrast material and partially surrounded by air. However, only those portions of articular surface which are in tangent to the x-ray beam are accurately evaluated. The most common abnormality in the articular cartilage is osteoarthritis with fissuring or erosion of articular cartilage.

Popliteal cysts represent distended gastrocnemius-semimembranosus bursae and communicate with the knee joint through a small opening or channel. On arthrography, these fill with air and positive contrast material, allowing the diagnosis to be readily made. When a popliteal cyst is present, this may be projected over the posterior horn of the medial meniscus, interfering with the arthrographic examination of this area. Popliteal cysts may rupture, particularly in patients with rheumatoid arthritis, and the leakage of synovial fluid may mimic thrombophlebitis.

Inflammatory conditions, including rheumatoid arthritis, will cause multiple synovial villi to project

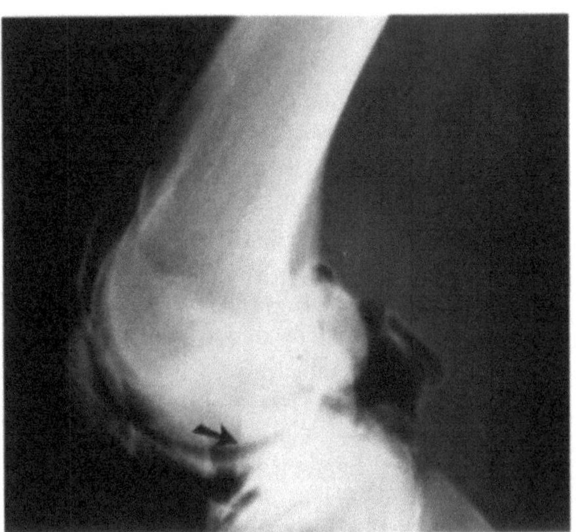

Fig. 2.7. Normal cruciate ligament. The anterior cruciate ligament is seen as a straight line of contrast material (*arrow*)

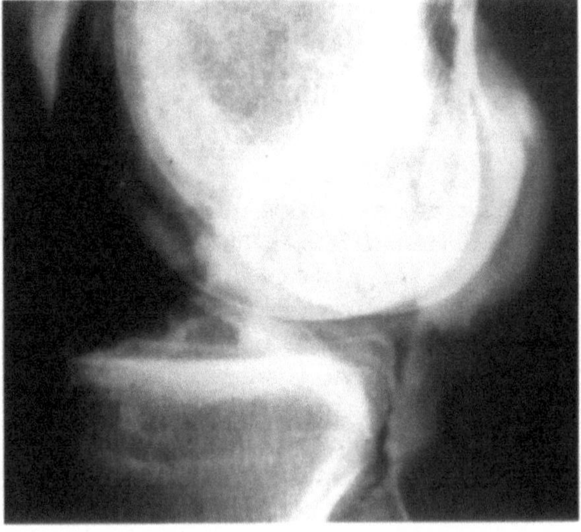

Fig. 2.8. Torn anterior cruciate ligament. No anterior cruciate ligament is demonstrated. The tibia is anteriorly displaced (an anterior drawer sign)

into the air and contrast material present within the joint during arthrography. Similar appearances may be seen in pigmented villonodular synovitis, lipoma aborescens, and synovial hemangiomas. Multiple loose bodies may be detected within the joint cavity in synovial chondromatosis.

2.2.5
Accuracy of Knee Arthrography

Most experienced examiners achieve an accuracy of greater than 90% with respect to meniscal tears (FREIBERGER et al. 1966; NICHOLAS et al. 1970). Accuracy for anterior cruciate ligament abnormalities has been reported to be approximately 90% (PAVLOV and TORG 1978; PAVLOV et al. 1983a). Knee arthrography is a simple, safe, and reliable method to evaluate the menisci, ligaments, and other intra-articular structures, and it still has a role in the evaluation of patients with internal derangements of the knee or other suspected abnormalities.

2.3
Shoulder Arthrography

Shoulder arthrography is usually utilized in the evaluation of patients with persistent pain or weakness in the absence of abnormal conventional radio-graphs. The most commonly demonstrated abnormalities are partial and complete rotator cuff tears, adhesive capsulitis, post-dislocation capsular deformities, and abnormalities of the bicipital tendon at its sleeve.

2.3.1
Technique and Normal Findings

The injection is performed with the patient supine and the shoulder in slight external rotation (KILLORAN et al. 1968; SCHNEIDER et al. 1975). Under fluoroscopic control, a lead marker is placed over the center of the joint and this point is marked on the skin. Following administration of local anesthesia, a 22 gauge spinal needle is directed straight downward into the joint. For a single-contrast study, 10–12 cc of water-soluble contrast medium is injected into the joint space. For routine double-contrast study, 2–4 cc of positive contrast material is injected followed by approximately 10 cc of room air (GHELMAN and GOLDMAN 1977; GOLDMAN and GHELMAN 1978; GOLDMAN 1979a). Following removal of the needle, the shoulder is exercised slightly and then routine radiographs are obtained in internal and external rotation with an axillary view and a bicipital groove view. For double-contrast studies, the internal and external rotation views are taken in the erect position with slight cranial tube angulation. If the initial

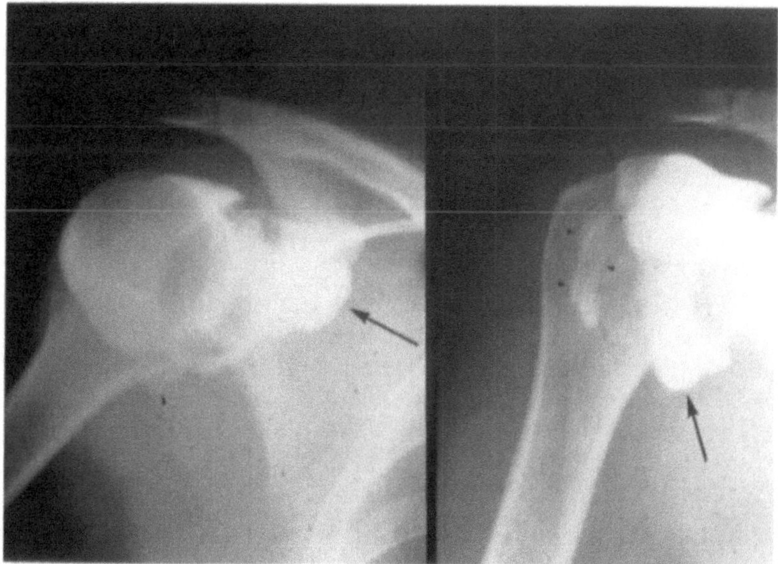

Fig. 2.9. Normal positive contrast shoulder arthrogram. On the left, in internal rotation, the subscapularis bursa fills (*arrow*). On the right, in external rotation, the axillary recess fills (*arrow*). No contrast material is seen extending into the rotator cuff or filling the subacromial-subdeltoid bursa. (*Arrowheads* indicate biceps tendon sleeve)

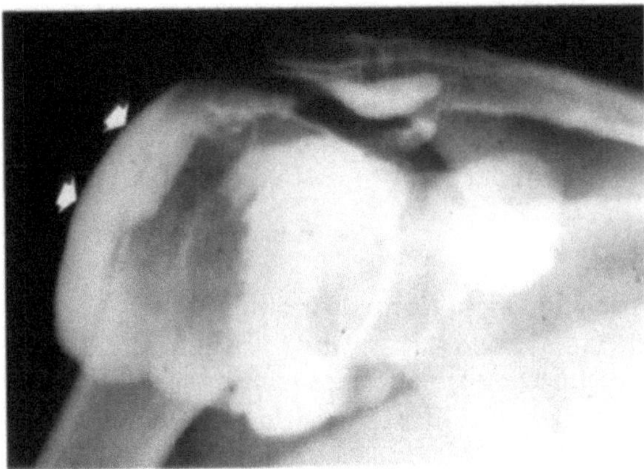

Fig. 2.10. Complete tear of the rotator cuff. Contrast material injected into the shoulder joint fills the subacromial-subdeltoid bursa (*arrows*) through a tear in the rotator cuff

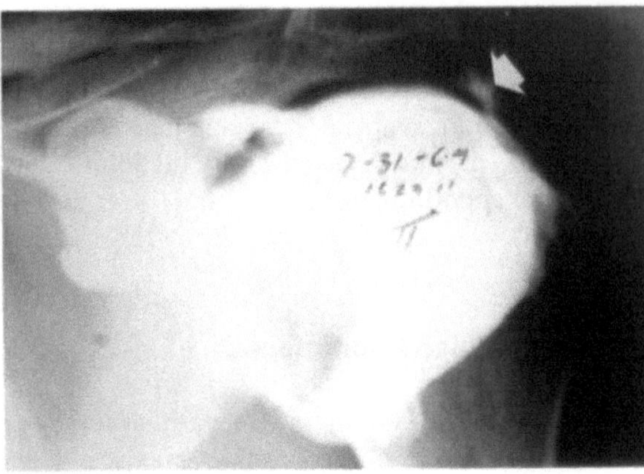

Fig. 2.11. Partial rotator cuff tear. Contrast material injected into the joint extends into a tear in the substance of the rotator cuff (*arrow*)

studies show no evidence of a rotator cuff tear, then a second set of radiographs is obtained after the exercise. Contrast material injected into a normal shoulder outlines the joint recesses and the humeral head, but does not fill the subacromial-subdeltoid bursa (Fig. 2.9).

2.3.2
Rotator Cuff Tears

The arthrographic criterion of a complete rotator cuff tear is visualization of contrast material in the subacromial-subdeltoid bursa. Contrast material injected into the joint in the course of arthrography flows through the tear in the rotator cuff and fills the bursa. On internal and external rotation views, contrast material is seen extending below the acromion and lateral to the humeral head distal to the greater tuberosity (Fig. 2.10). On axillary projections, contrast material extends across the humeral shaft. Partial rotator cuff tears are identified as abnormal collections of contrast above the articular cartilage but not extending through the full thickness of the rotation cuff (Fig. 2.11). Only those partial rotator cuff tears that begin on the undersurface of the rotator cuff can be diagnosed by shoulder arthrography.

2.3.3
Post-dislocation Capsular Deformity

Documentation of prior anterior dislocation of the shoulder depends upon the presence of capsular and

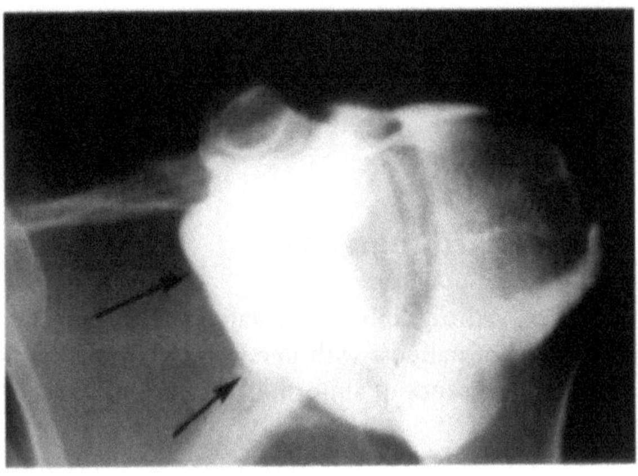

Fig. 2.12. Post-dislocation capsular deformity. The shoulder arthrogram demonstrates loss of definition between the subscapularis and axillary recesses after anterior dislocation of the shoulder. A bag-like structure anteriorly (*arrows*) is due to prior stripping of the capsule

osteocartilaginous deformities. A Hill-Sachs defect may contain pooled contrast material on the view taken in internal rotation. A Bankart deformity is usually seen on conventional radiographs. Cartilaginous Bankart deformities are best diagnosed on CT following double-contrast arthrography (Deutsch et al. 1984; Schuman et al. 1983). Following anterior dislocation of the shoulder, the capsule is stripped away from the glenoid and a characteristic post-dislocation deformity is noted with a bag-like collection of contrast obliterating the normal distinction between the axillary and subscapularis recesses (Fig. 2.12).

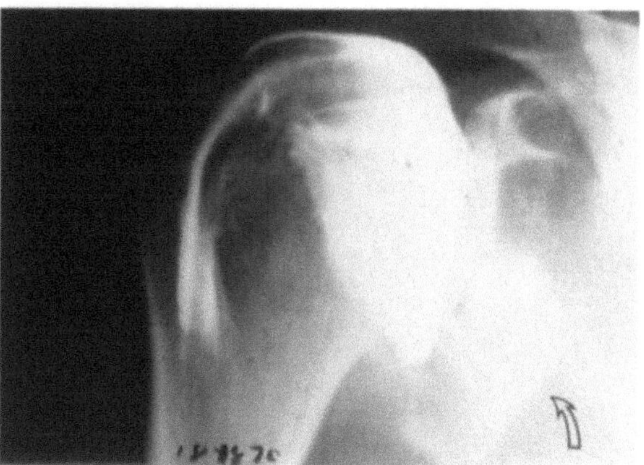

Fig. 2.13. Adhesive capsulitis. Contrast material was injected into the shoulder with difficulty, and outlines a retracted and small joint cavity. Note that contrast material has leaked out of the shoulder through the injection site (*arrow*)

2.3.4
Adhesive Capsulitis

With adhesive capsulitis, considerable difficulty is found in injection of even small amounts of contrast material into the shoulder. If adhesive capsulitis is suspected, single-contrast arthrography is probably the better choice. Synovial irregularities at the margins of the contracted and retracted capsule can be appreciated on shoulder arthrography (NEVAISER 1962; KAYE and SCHNEIDER 1979) (Fig. 2.13).

2.3.5
Other Abnormalities

Other abnormalities can be evaluated, including tears in the long head of the biceps and dislocations of the biceps tendon (KAYE and SCHNEIDER 1979).

2.4
Hip Arthrography

Even since the introduction of MRI, hip arthrography has continued to play an important role in the evaluation of both the child and the adult.

2.4.1
Technique

Whether in the child or in the adult, hip arthrography is best performed utilizing fluoroscopic control

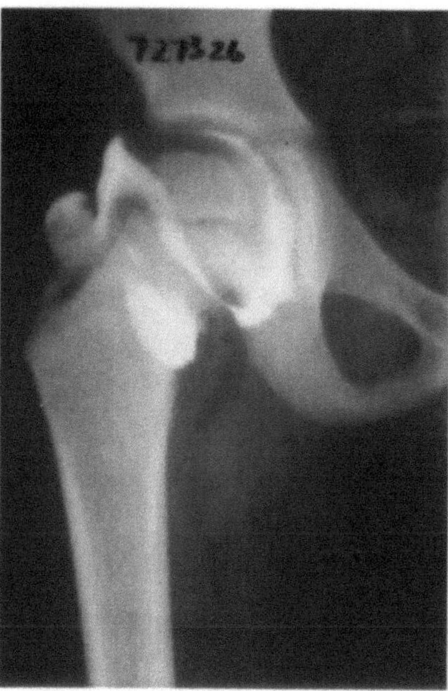

Fig. 2.14. Normal hip arthrogram. Contrast material injected into the joint outlines the joint capsule and the articular cartilage of the femoral head in this child

(GHELMAN and FREIBERGER 1979; GOLDMAN 1979b). The most commonly used approach is the anterolateral approach, with the needle directed downwards onto the femoral neck; it is not directed towards the joint space itself, but rather towards the femoral neck, which is intracapsular. In contrast to knee and shoulder arthrography, hip arthrography is almost always done as a single positive contrast study. More detailed descriptions of technique are beyond the scope of this chapter, but can be found in standard textbooks of arthrography. The normal hip arthrogram outlines the cartilaginous femoral head and the fibrocartilaginous acetabular labrum (Fig. 2.14).

2.4.2
Developmental Dysplasia of the Hip

Hip arthrography is not used in the diagnosis of developmental dysplasia of the hip, which should be established on the basis of physical examination. Hip arthrography does reveal the degree of coverage of the cartilaginous femoral head by the fibrocartilaginous labrum of the acetabulum (SEVERIN 1939, 1941; FREIBERGER 1973; GOLDMAN 1979b) (Fig. 2.15). It can also be used to determine whether the

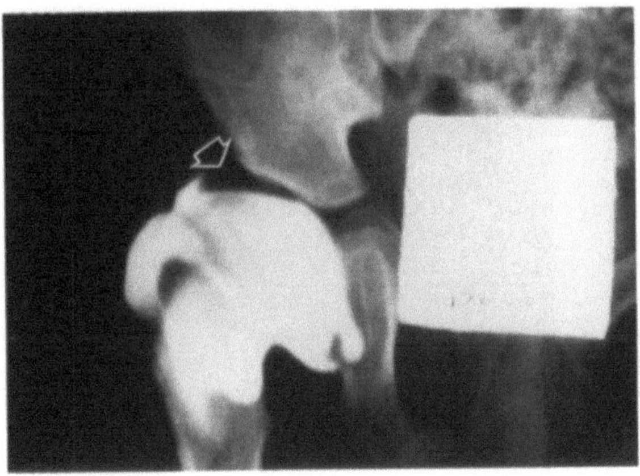

Fig. 2.15. Mild acetabular dysplasia due to developmental dysplasia of the hip. The hip is seen to be located on this arthrogram and the fibrocartilaginous labrum (*arrow*) is not infolded. There is mild acetabular dysplasia

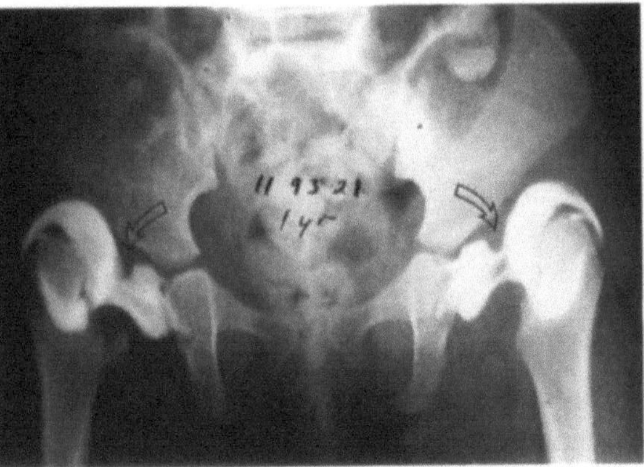

Fig. 2.16. Developmental dysplasia of the hip. This child, with bilateral dislocation of hips, shows the characteristic hourglass configuration of the joint capsule. There is infolding of the fibrocartilaginous labrum bilaterally (*arrows*)

fibrocartilaginous labrum has become infolded (Fig. 2.16). When arthrography is done using fluoroscopy, the position of best reduction can also be determined.

In many locations, ultrasound has assumed a very important role in the evaluation of infants with developmental dysplasia of the hip, and similar anatomic information can be obtained by ultrasound without joint puncture or radiation dose.

2.4.3
Septic Arthritis

Arthrography plays an important, although secondary role, in the evaluation of those infants and children suspected of having a septic hip (GLASSBURG and OZONOFF 1978; KAYE 1973). The diagnosis of septic arthritis must be excluded whenever this is a differential diagnostic consideration, since the damage done by septic arthritis is so great and happens so rapidly. The most important part of the procedure is then the aspiration of the hip, to establish or exclude the diagnosis of infection. Arthrography is performed at the end of the aspiration procedure to confirm the intra-articular location of the aspirating needle; anatomic information about the hip is secondary.

While the presence of fluid within the joint can be determined by ultrasound, only an aspiration will tell whether any joint fluid present is due to infection, and only aspiration will allow culture of joint fluid, which may establish the offending organism. While aspirations may be performed under ultrasound control, they are often more readily done by fluoroscopic guidance. Arthrograms done at the end of fluoroscopically guided aspirations exclude the possibility of false-negative aspirations.

2.4.4
Legg-Calvé-Perthes Disease

Arthrography has been used in patients with Legg-Calvé-Perthes disease, not to establish the diagnosis but to evaluate any deformity of the femoral head which has occurred, since the shape of the ossific nucleus does not accurately mirror the shape of the articular cartilage (KATZ 1968; GOLDMAN 1979b). In addition, arthrography can be helpful to determine the degree of coverage of the epiphysis, the amount of incongruity which may be present, and the position of best fit of the femoral head in the acetabulum.

In many locations, MRI has been used to evaluate many of these features of the femoral head and acetabulum in patients with Legg-Calvé-Perthes disease. Arthrography may still be useful in those patients in whom fluoroscopic manipulation and choice of position of best fit is desirable.

2.4.5
Adult Hips Without Prostheses

In adult patients with unexplained hip pain, hip arthrography may demonstrate the presence of intra-articular loose bodies, such as may be seen in patients with synovial chondromatosis or osteochondromatosis (GHELMAN and FREIBERGER 1979). Intra-articular fragments, which may be cartilaginous or osteocartilaginous, may lodge within the hip after dislocation and be outlined by the contrast material used during arthrography.

A villous synovitis may be present in patients with rheumatoid arthritis or other inflammatory arthritides and also in patients with pigmented villonodular synovitis. Villi and nodular masses may erode the femoral head and neck.

In addition, in patients who present with inguinal masses, hip arthrography may demonstrate that the mass is due to the presence of a communicating and enlarged iliopsoas bursa.

Still another use for hip arthrography in adults without prior surgery occurs when abnormality, and usually some degree of osteoarthritis, is demonstrated on radiographs, but symptoms are disproportionately severe. In these patients intra-articular injections of local anesthetics and corticosteroids may be used as a therapeutic trial to determine whether the patient might benefit from reconstructive surgery. In addition, in patients who have hip and back disease, both usually degenerative in nature, injection of local anesthetics and corticosteroids may help to determine the cause of the patient's symptoms. In both of these situations, the arthrogram is performed only to ascertain that the injecting needle is in an intra-articular location; anatomic information is purely secondary.

2.4.6
Arthrography of Hip Prostheses

Considerable metal is implanted in most hip prostheses. Since this severely degrades any anatomic information which may be gained from MRI, hip arthrography is still frequently used to evaluate loosening of prostheses, with or without infection. In this situation, the most important information to be gained comes from culture of the aspirated joint fluid. The arthrogram serves primarily to confirm the intra-articular location of the aspirating needle (GHELMAN and FREIBERGER 1979). Loosening of prosthetic components is shown by tracking of contrast material at the cement–bone interface (SALVATI et al. 1971; DUSSAULT et al. 1977; GHELMAN and FREIBERGER 1979; WEISSMAN 1997) (Fig. 2.17). Minor modifications of technique are helpful in performing hip aspirations after prostheses; it is usually most helpful to choose the anterolateral approach in such a way that the needle

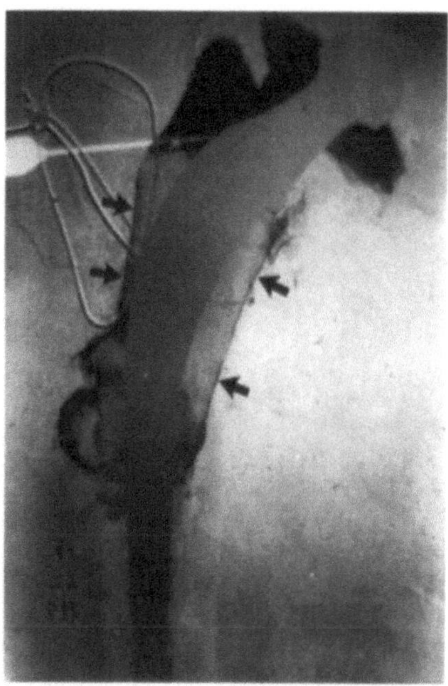

Fig. 2.17. Subtraction hip arthrogram after total hip arthroplasty shows contrast material extending into the cement bone interface (*arrows*), indicating a loose femoral component

can be visualized throughout the course to the neck of the prosthesis. Subtraction techniques may be helpful (SALVATI et al. 1974; ANDERSON and STAPLE 1979). Many hip aspirations and arthrograms are performed in patients with prostheses that are deemed to be loose based upon conventional radiographs. In this situation, the aspiration and arthrogram are done to exclude infection. Note should be made that negative cultures from aspirated fluid from a loose prosthesis do not absolutely exclude the diagnosis of infection, which may be walled off, particularly about the femoral component and lodged within the medullary cavity of the femoral shaft. In addition, it should be noted that the failure to demonstrate contrast material extending around the cement–bone interfaces of prosthetic components is not an absolute indication that the prosthesis is not loose. Again, granulation tissue or fibrous tissue may preclude the tracking of contrast material about loose prosthetic components. Bursae or abscess cavities can also be demonstrated (Fig. 2.18).

2.5
Wrist Arthrography

Wrist arthrography is used to evaluate structures not visualized on conventional radiographs. These in-

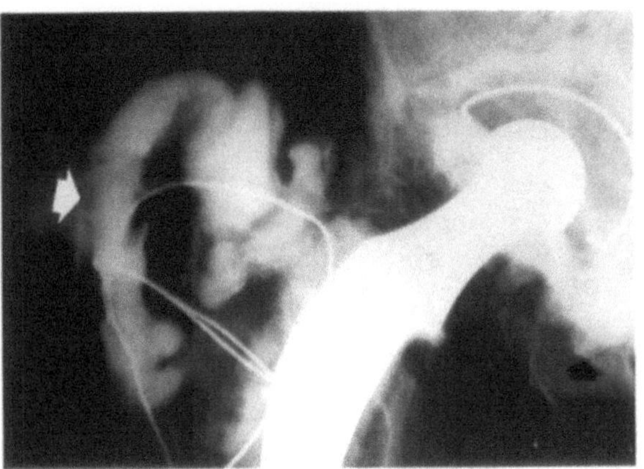

Fig. 2.18. Infected total hip arthroplasty with communicating bursa. Contrast material injected into the hip fills a bursa around the greater trochanter (*arrow*). Purulent material was aspirated from the hip at the time of the arthrogram and cultures were positive

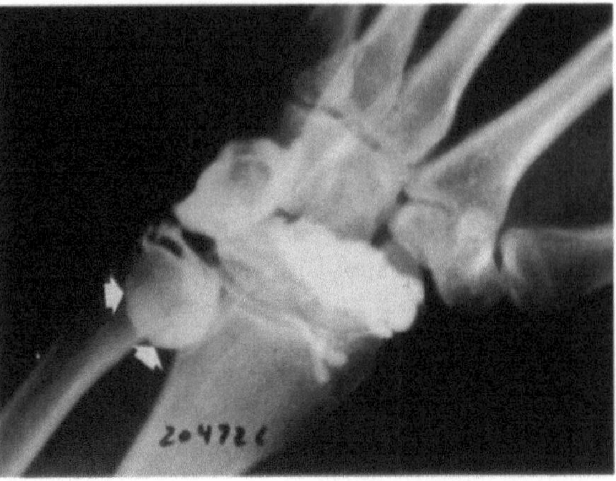

Fig. 2.19. Wrist arthrogram demonstrating triangular fibrocartilage tear. Contrast material injected into the radiocarpal joint fills the distal radioulnar joint (*arrows*) through a tear in the triangular fibrocartilage

clude the intra-articular ligaments and the articular cartilages, especially the triangular fibrocartilage (GOLDMAN 1979c; DALINKA et al. 1981; GILULA et al. 1983; LEVINSON and PALMER 1983; LEVINSON et al. 1987; MANASTER 1991). In the past, wrist arthrography has also been used to evaluate inflammatory arthritides, such as rheumatoid arthritis.

2.5.1
Technique

Wrist arthrograms are generally performed under fluoroscopic control using very small needles and positive contrast material alone. Injection of the radiocarpal joint may be combined with injection, at a later time, of the midcarpal and distal radioulnar joints. Triple-compartment arthrograms are felt to be more reliable in detecting small perforations of intraosseous ligaments and the triangular fibrocartilage.

All compartments to be injected are approached under fluoroscopic control using a dorsal approach. More detailed descriptions of the technique can be found in textbooks of arthrography.

2.5.2
Abnormal Wrist Arthrograms

Most tears of the triangular fibrocartilage fill by injection of the radiocarpal joint and are noted when the radiocarpal joint is injected and the distal radioulnar joint fills (Fig. 2.19). Partial tears of the

triangular fibrocartilage are visualized when that compartment of the joint which is injected is in contact with the tear, which does not go through the full thickness of the triangular fibrocartilage.

Intraosseous intercarpal ligamentous injuries also occur, and can be demonstrated by wrist arthrography. The most common of these is a tear of the scapholunar ligament, which may result in scapholunar dissociation. In this setting, injection of one compartment, such as the radiocarpal, will cause filling of another intercarpal compartment, the midcarpal joint. Such filling may also occur through disruptions that have occurred in the ligament between the lunate and the triquetrum.

Magnetic resonance imaging of the wrist has been performed in a number of centers to evaluate not only these structures but also the extrinsic ligaments about the wrist.

2.6
Other Joints

Arthrography of the elbow has been performed to evaluate the integrity of articular cartilage in patients with osteochondral injuries or Panner's disease of the capitellum. In addition, arthrography may be useful to determine the presence or absence of loose bodies within the joint (ETO et al. 1975; HUDSON 1981; PAVLOV et al. 1979) (Fig. 2.20). MRI may also be used for both of these applications.

Ankle arthrography has been utilized in patients with suspected ligamentous injuries (BROSTROM et al. 1965; FUSSELL and GODLEY 1973; KAYE 1979b)

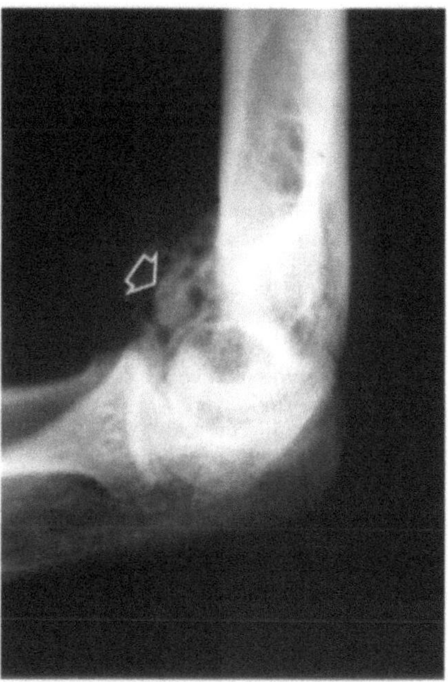

Fig. 2.20. A double-contrast elbow arthrogram demonstrates a loose body (*arrow*) anteriorly within the joint capsule

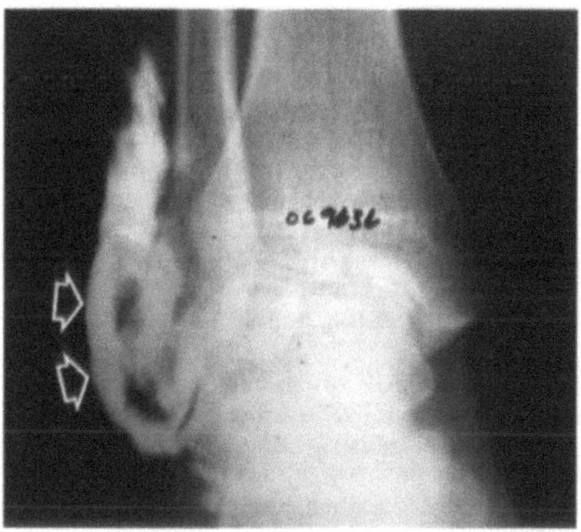

Fig. 2.21. Ankle arthrogram demonstrating a tear of the anterior talofibular ligament. Contrast material injected into the ankle joint leaks from the joint through the tear and is seen lateral to the fibula (*arrows*)

(Fig. 2.21) and also to evaluate the integrity of the articular cartilage in patients with osteochondritis dissecans or osteochondral factors of the talus. MRI is also quite useful in both of these areas.

2.7
Summary

Arthrography has long provided a useful tool in the evaluation of a variety of joints. The applications, as described above, are many. MRI has had an impact on the frequency with which these studies are requested, but arthrography remains simple, safe, and reliable as a means for studying joint disorders.

References

Anderson LS, Staple TC (1979) Arthrography of total hip replacements using subtraction technique. Radiology 109:470–472

Andren L, Wehlin L (1960) Double contrast arthrography of the knee with horizontal roentgen ray beam. Acta Orthop Scand 29:307–314

Arndt RD, Horns JW, Gold RH (1981) Clinical arthrography. Williams and Wilkins, Baltimore

Bircher E (1931) Pneumographic des Knies und der anderen Gelenke. Schweiz med Wochenschr 61:1210–1211

Brostrom L, Liljedahl SO, Lindvall H (1965) Sprained ankles. II. Arthrographic diagnosis of recent ligament ruptures. Acta Chir Scand 129:485–499

Butt WP, McIntyre JL (1969) Double contrast arthrography of the knee. Radiology 92:487–499

Dalinka MK (1980) Arthrography. Springer, Berlin Heidelberg New York

Dalinka MK, Turner ML, Osterman AL, Batra P (1981) Wrist arthrography. Radiol Clin North Am 19:217–226

Deutsch AL, Resnick D, Mink JH, et al. (1984) Computed and conventional arthrotomography of the glenohumeral joint: normal anatomy and clinical experience. Radiology 140:603–609

Dussault RG, Goldman AB, Ghelman B (1977) Roentgenographic diagnosis of loosening and/or infection in hip prostheses. Correlation between roentgen and surgical findings. J Can Assoc Radiol 28:119–123

Eto RT, Anderson PW, Harley JD (1975) Elbow arthrography with the application of tomography. Radiology 115:283–288

Freiberger RH (1973) Congenital dislocation of the hip. Curr Prob Radiol 5:4–16

Freiberger RH (1979a) Meniscal abnormalities. In: Freiberger RH, Kaye JJ (eds) Arthrography. Appleton-Century-Crofts, New York, pp 55–91

Freiberger RH (1979b) Technique of knee arthrography. In: Freiberger RH, Kaye JJ (eds) Arthrography. Appleton-Century-Crofts, New York, pp 5–30

Freiberger RH, Kaye JJ (1979) Arthrography. Appleton-Century-Crofts, New York

Freiberger RH, Killoran PJ, Cardona G (1966) Arthrography of the knee by double contrast method. AJR 97:736–747

Fussell ME, Godley DR (1973) Ankle arthrography in acute sprains. Clin Orthop 93:278–290

Ghelman B, Freiberger RH (1979) The adult hip. In: Freiberger RH, Kaye JJ (eds) Arthrography. Appleton-Century-Crofts, New York, pp 189–256

Ghelman B, Goldman AB (1977) The double contrast shoulder arthrogram: evaluation of rotator cuff tears. Radiology 124:251–254

Gilula LA, Totty WG, Weeks PM (1983) Wrist arthrography: the value of fluoroscopic spot viewing. Radiology 146:555–556

Glassburg GB, Ozonoff MB (1978) Arthrographic findings in septic arthritis of the hip in infants. Radiology 128:151–155

Goldman AB (1979a) Double contrast shoulder arthrography. In: Freiberger RH, Kaye JJ (eds) Arthrography. Appleton-Century-Crofts, New York, pp 165–188

Goldman AB (1979b) Hip arthrography in infants and children. In: Freiberger RH, Kaye JJ (eds) Arthrography. Appleton-Century-Crofts, New York, pp 217–235

Goldman AB (1979c) The wrist. In: Freiberger RH, Kaye JJ (eds) Arthrography. Appleton-Century-Crofts, New York, pp 277–289

Goldman AB, Ghelman B (1978) The double contrast shoulder arthrogram: a review of 158 studies. Radiology 127:655–664

Harley JD (1977) An anatomic-arthrographic study of the relationship of the lateral meniscus and the popliteus tendon. AJR 128:181–187

Hudson TM (1981) Elbow arthrography. Radiol Clin North Am 19:227–241

Jelasco DV (1975) The fascicles of the lateral meniscus. An anatomic-arthrographic correlation. Radiology 114:335–339

Katz JF (1968) Arthrography in Legg-Calvé-Perthes disease. J Bone Joint Surg [Am] 50:467–472

Kaye JJ (1973) Bacterial infections of the hips in infancy and childhood. Curr Probl Radiol 5:11–29

Kaye JJ (1979a) Anatomy and arthrography of the normal menisci. In: Freiberger RH, Kaye JJ (eds) Arthrography. Appleton-Century-Crofts, New York, pp 31–53

Kaye JJ (1979b) The ankle. In: Freiberger RH, Kaye JJ (eds) Arthrography. Appleton-Century-Crofts, New York, pp 237–256

Kaye JJ (1994) Magnetic resonance imaging of the knee. A senior musculoskeletal radiologist's perspective. MRI Clin North Am 3:497–500

Kaye JJ, Schneider R (1979) Positive contrast shoulder arthrography. In: Freiberger RH, Kaye JJ (eds) Arthrography. Appleton-Century-Crofts, New York, pp 137–163

Killoran PJ, Marcove RC, Freiberger RH (1968) Shoulder arthrography. AJR 103:658–668

Levinsohn EM, Palmer AK (1983) Arthrography of the traumatized wrist. Radiology 146:647–651

Levisohn EM, Palmer AK, Coren AB, Zlaberg E (1987) Wrist arthrography: the value of the three compartment injection technique. Skeletal Radiol 16:539–544

Lindblom K (1948) Arthography of the knee, roentgenographic and anatomic study. Acta Radiol (Suppl) 74:1–112

Manaster BJ (1991) The clinical efficacy of triple-injection wrist arthrography. Radiology 178:267–270

McIntyre JL (1972) Arthrography of the lateral meniscus. Radiology 105:531–536

Nevaiser JS (1962) Arthrography of the shoulder joint. Study of the findings in adhesive capsulitis of the shoulder. J Bone Joint Surg [Am] 44:1321–1329

Nicholas JA, Freiberger RH, Killoran PJ (1970) Double contrast arthrography of the knee: its value in the management of two hundred and twenty-five knee derangements. J Bone Joint Surg [Am] 52:203–220

Pavlov H (1979) Cruciate ligaments. In: Freiberger RH, Kaye JJ (eds) Arthrography. Appleton-Century-Crofts, New York, pp 93–107

Pavlov H, Goldman AB (1980) The popliteus bursa: an indicator of subtle pathology. AJR 134:313–321

Pavlov H, Torg JS (1978) Double contrast arthrographic evaluation of the anterior cruciate ligament. Radiology 126:661–665

Pavlov H, Ghelman B, Warren RF (1979) Double-contrast arthrography of the elbow. Radiology 130:87–95

Pavlov H, Ghelman B, Vigority VJ (1983a) Atlas of knee menisci: an arthrographic-pathologic correlation. Appleton-Century-Crofts, Norwalk

Pavlov H, Warren RF, Sherman MF, Cayea PD (1983b) The accuracy of the double contrast arthrographic evaluation of the anterior cruciate ligament. A retrospective review of 163 surgically confirmed cases. J Bone Joint Surg [Am] 65:175–183

Ricklin P, Ruttimann A, Del Buono MS (1979) Meniscus lesions. Grune and Stratton, New York

Salvati EA, Freiberger RH, Wilson PD Jr (1971) Arthrography for complications of total hip replacement. J Bone Joint Surg [Am] 53:701–709

Salvati EA, Ghelman B, McLaren T, Wilson PD Jr (1974) Subtraction technique in arthrography for loosening of total hip replacement fixed with radiopaque cement. Clin Orthop 101:105–109

Schneider R, Freiberger RH (1979) Extrameniscal abnormalities. In: Freiberger RH, Kaye JJ (eds) Arthrography. Appleton-Century-Crofts, New York, pp 109–135

Schneider R, Ghelman B, Kaye JJ (1975) A simplified injection technique for shoulder arthrography. Radiology 114:738–739

Severin E (1939) Arthrography in congenital dislocation of the hip. J Bone Joint Surg 21:304–313

Severin E (1941) Arthrograms of hip joints of children. Surg Gynecol Obstet 72:601–604

Shuman WP, Kilcoyne RF, Matsen FA, Rogers JV, Mack LA (1983) Double-contrast computed tomography of the glenoid labrum. AJR 141:581–587

Spiegel PK, Staples SH (1975) Arthrography of the ankle joint: problems in diagnosis of acute lateral ligament injuries. Radiology 114:587–590

Staple TW (1972) Extrameniscal lesions demonstrated by double contrast arthrography of the knee. Radiology 102:311–319

Stoker DJ (1980) Knee arthrography. Chapman and Hall, London

Thijn CJP (1979) Arthography of the knee joint. Springer, Berlin Heidelberg New York

Weissman BN (1997) Imaging of total hip replacement. Radiology 202:611–623

Wickstrom KT, Spitzer RM, Olsson HE (1975) Roentgen anatomy of the posterior horn of the lateral meniscus. Radiology 116:617–619

3 Computed Tomography

V.N. Cassar-Pullicino

CONTENTS

V.N. Cassar-Pullicino, MD, Department of Diagnostic Imaging, The Institute of Orthopaedics, The Robert Jones & Agnes Hunt Orthopaedic & District Hospital, Oswestry, Shropshire SY10 7AG, UK

3.1 Introduction

In 1935, Grossman coined the term "tomography" from the Greek *"tomos"*, meaning section or cut, to denote the method using conventional radiography to depict specific layers within the human body. In conventional tomography the image of the section is orientated parallel to the film. Computed tomography (CT) refers to the method of obtaining information from the passage of x-ray beams through a selected area of the body which in turn is processed with the aid of a sophisticated computer. The computer then produces clear sharp images of the internal structure of the body in the axial plane.

Roentgen's discovery of x-rays in 1895 had a profound impact on the understanding, diagnosis and management of orthopaedic disorders. Similarly, the invention of CT produced another giant step towards the refinement of the diagnosis and management of orthopaedic conditions by generating a dramatic increase in diagnostic information. Advances in technology, physics, computer science (hardware and software), mathematics, engineering, and image processing have been harnessed and applied to medicine. Apart from increasing the opportunity for an early and accurate diagnosis, CT also helped to reduce the number of unpleasant investigative interventional procedures. It paved the way for the introduction of magnetic resonance imaging (MRI). In some instances CT has been superseded by MRI as the preferred method of investigation, e.g. spinal disorders, but it still enjoys an unparalleled superiority in the depiction of mineralised tissue (bone and calcification) (Cassar-Pullicino et al. 1992). More often than not, CT complements MRI in the understanding of the pathological processes involving the musculoskeletal system (Fig. 3.1).

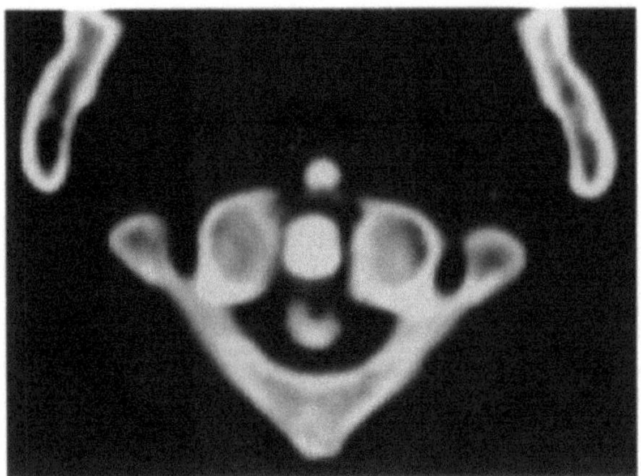

Fig. 3.1. Axial image of atlas showing ossification of the anterior and posterior longitudinal ligaments in front and behind the odontoid peg

3.2
Historical Perspective

The desire to image the body in the axial plane dates as far back as 1937, when Watson in England developed a tomographic technique referred to as transverse axial tomography. The clinical application of this technique, however, was severely restricted due to the poor quality of the images produced. The underlying principle of CT, image reconstruction from multiple projections (GABOR 1980), stems from the work of Radon, who in 1917 proved that it was possible to reconstruct an image of an object from various projections taken from different directions (RADON 1917). Image reconstruction from projections found clinical application in medicine, when Dr. A.M. Cormack in the early 1960s applied the reconstruction techniques to nuclear medicine. In 1967 Dr. Godfrey Hounsfield, while investigating pattern recognition with the help of computerised reconstruction techniques, deduced that measurements of x-ray transmission through an object from all directions could provide information about the internal structure of the body. This led to the production of the world's first clinically useful CT scanner manufactured by EMI for imaging the brain (HOUNSFIELD 1973). The proven value of this new technique in the diagnosis of intracranial disorders led to a rapid development of CT with applications to the rest of the body. The development was so rapid that in the space of just 7 years EMI, having introduced CT in 1973, dropped out of the increasingly competitive CT market in 1980. By then various "generations" of CT scanner design had evolved.

Improvements in generator technology, computer power and reconstruction algorithms helped to further optimise the image information quality, and quantitative CT analysis of bone mineral content was introduced in 1986. Further milestones included the production of ultra-fast CT scanners based on electron beam technology without use of an x-ray tube or mechanical motion of the components of the CT scanner. More recently, in 1990 spiral (helical) CT was introduced by Dr. W. Kalender utilising a continuous mode of CT imaging (KALENDER et al. 1990).

Dr. Godfrey N. Hounsfield and Dr. A.M. Cormack were winners of the 1979 Nobel prize for medicine and physiology for their contribution to the development of CT (HOUNSFIELD 1980; CORMACK 1980). The CT scanner resulted in Hounsfield gaining other awards, distinctions and prizes including the prestigious McRobert engineering award in 1972, which was accompanied by the referee's comments that no comparable discovery had been made in the field of x-ray techniques since Roentgen discovered x-rays in 1895.

3.3
Digitisation

In general terms, depending on the form or method of generation, there are two basic forms of images, *analogue* and *digital*. Analogue images are continuous images, e.g. a black and white photograph, or radiograph, and the signals generated by electronic devices such as the detectors used in CT. Digital images are numerical (discrete) representations of objects which require a digital computer for their formation; only digital images can be processed by a computer. The conversion of an analogue image into digital data for input to a computer is known as *"digitisation"*. *Reconstruction* is the process that takes a digital image and changes it into a visible physical image.

In the process of digitisation there are three distinct steps: scanning, sampling and quantisation (BAXES 1984; SEERAM 1994). Scanning commences the process by dividing the picture (photograph, radiograph etc.) into small regions, each with an x and a y co-ordinate called picture elements or *pixels*, resulting in a grid of rows and columns which form what is known as the *matrix* of the picture. Every individual pixel therefore has a separate location within the matrix. The second step employed in digitisation is sampling, which is the measurement of the brightness of each pixel within the entire image. Lastly, in quantisation, the brightness value of each pixel sampled is assigned a grey level which is either positive or negative in number. The pixels making up the matrix of the original picture are now transformed into a range of numbers or grey levels, each with a precise location on the matrix grid. The total number of grey levels in turn compose the grey scale of the image and at the end of digitisation, the information has been transformed into an array of numbers representing the analogue image, which is in turn sent to the computer for further processing.

3.4
Principles of CT

Similar steps in the digitisation of an image are found in the CT process, which employs a digital image processing system. In CT a three-dimensional slice of information is digitised into a two-dimensional image display. The slice is divided into small regions called *voxels* (volume element) because the dimension of depth dependent on the *slice thickness* (z) is added to the pixel (Fig. 3.2). In CT this is done by acquiring the data utilising an x-ray tube which moves around the patient resulting in the transmission and attenuation of x-rays through the voxels making up the slice to be examined. The voxels are sampled by the transmitted x-ray beam which are picked up by the detectors of the scanner. In the final step, the analogue signal of CT produced by the detectors is quantised and transformed into a digital array for input into the computer. In turn, the digital data in CT are subjected to several imaging processing algorithms so that the output image can be displayed for viewing. CT involves the sequential digitisation of patient slices. In the process of digitisation, transfer of analogue information to digital information for computer processing takes place requiring the help of analogue to digital converters, while in image display digital to analogue converters transfer the data to analogue format.

The goals of CT are: (a) to minimise the problem of superimposition, (b) to improve the contrast of the image and (c) to record the very small differences in tissue contrast (HOUNSFIELD 1973).

3.5
Technical Aspects

X-ray tubes produce the energy that creates the CT image. The x-ray beam is attenuated after passing

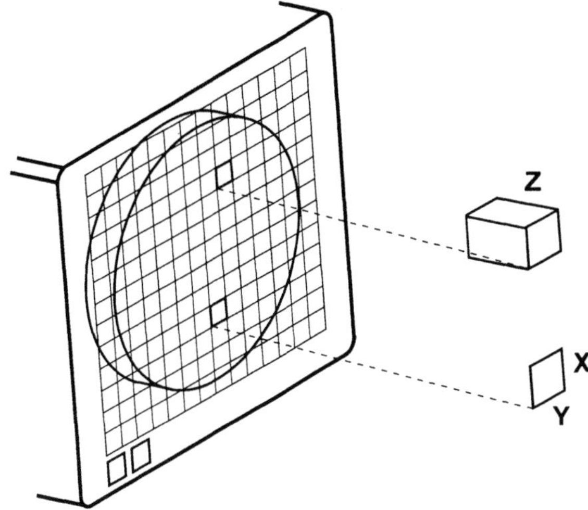

Fig. 3.2. Three-dimensional patient information (voxels) is transferred to a two-dimensional representation (pixels) in the matrix of the monitor (z = slice thickness)

through the patient and the transmitted x-rays are measured by the detectors. The x-ray tube with the detectors is hidden out of sight in the gantry of the scanner and rotates around the patient during scanning. Irrespective of detector material, each detector cell is sampled many times and it is the function of the detectors in turn to convert the x-ray photons into electrical signals (analogue) which must in turn be converted into digital (numerical) information for input into the computer. Analogue-to-digital converters in the data acquisition system convert the electrical signal to a digital format. The computer then reconstructs the CT image utilising numerous mathematical complex techniques referred to as reconstruction algorithms. The image information in its digitised state is translated into a matrix by assigning each pixel within the matrix a specific value or CT density number. The digitised data are then in turn sent to the display processor, which converts them into the various shades of the grey scale. A reconversion of digital (numerical) to analogue (electrical) information is required with the help of a digital-to-analogue converter, which enables the resultant image to be displayed on the cathode-ray tube of the television monitor. The processed image is a tomographic image which can also be stored on magnetic tape or optical discs, or recorded on film for permanent archiving.

Computed tomography is a digital imaging system using computers to process images. The process encompasses three essential components: (a) data acquisition in the form of analogue signals, (b) digital image processing of the analogue information into digital data, which allows it to be processed by the computer, and (c) image display, which requires the reconversion of the reconstructed digital information into analogue information because the display device (television monitor) works only with analogue signals (Fig. 3.3) (SEERAM 1994; ROMANS 1995).

3.5.1
Data Acquisition

Data acquisition requires three essential components: the x-ray generator, the gantry and the patient table. The generator produces high voltage which is transmitted to the x-ray tube. The x-ray tube and the detectors are housed in the gantry, the most recognisable component of the CT scanner.

There are two methods of data acquisition: (a) conventional slice by slice acquisition and (b) spiral continuous (volume) data acquisition.

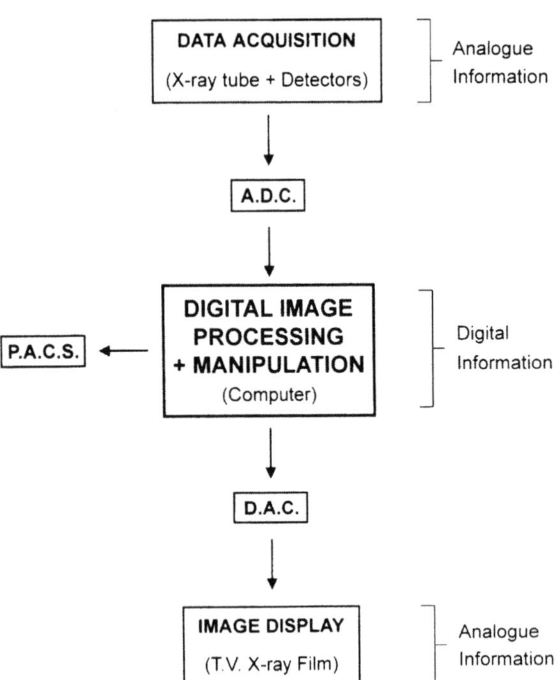

Fig. 3.3. Basic infrastructure of CT's imaging system. *A.D.C.,* Analogue-to-digital converter; *D.A.C.,* digital-to-analogue converter; *P.A.C.S.,* picture archiving and communications system

3.5.1.1
Conventional Slice by Slice Acquisition

In conventional slice by slice acquisition the x-ray tube rotates around the patient and collects data from each slice in turn. As the tube travels along this path, x-ray energy is emitted and passes through the patient, who is placed within the opening of the gantry; the transmitted x-ray energy is then retrieved by the detectors. The configuration of the x-ray tube and detectors describes the data acquisition geometry of the CT scanner and determines the scanner generation. The first two generations are no longer in use, while the third generation of conventional CT design is the most widely used configuration in which the x-ray tube and detectors are coupled and rotate 360° around the patient to collect transmission measurements using a fan beam of radiation. As both the x-ray tube and the detectors move in a circle within the gantry, these scanners are sometimes referred to as rotate-rotate scanners (Fig. 3.4). In the fourth generation type of scanner a different configuration exists whereby the x-ray tube rotates 360° around the patient and the detector array is fixed in a 360° circle within the gantry. As the tube rotates within the fixed detector array it produces a fan-shaped beam whose width determines the number of detectors in use at any one time. This type of data acquisition geometry is often referred to as a rotate-only CT system (Fig. 3.5). In conventional slice by slice acquisition (third or fourth generation)

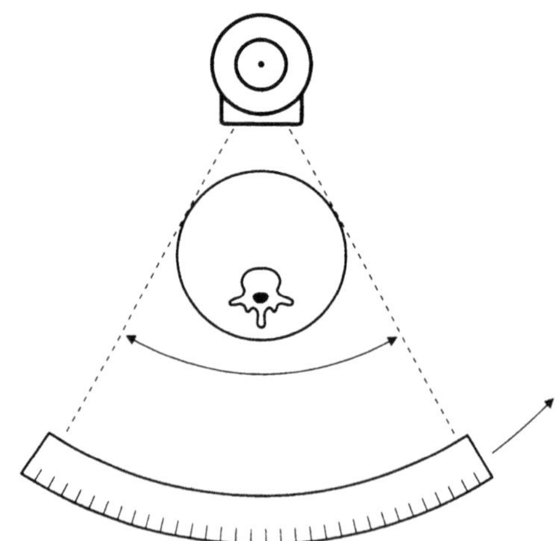

Fig. 3.4. Fan beam geometry: third generation rotate/rotate design. The tube and detectors move as a couple within the gantry

the x-ray tube rotates around the patient and after collecting data comes to a halt. The patient is then moved (fed) further into the gantry into a position to allow the next slice to be scanned. This step by step or incremental process continues until all the slices have been obtained.

The gantry and patient couch are often referred to as the scanner. The gantry is the framework around the patient, which houses the hardware imaging components such as the slip-rings, x-ray tube, high-tension generator, collimators, detectors, and detector electronics referred to as the data acquisition system (DAS). CT detectors are of two types, scintillation detectors and gas ionisation detectors. Scintillation detectors comprise a crystal coupled to a photodiode whereas gas ionisation detectors are xenon gas chambers that produce electrical signal as a result of ionisation. The signal then goes to the DAS, which acts as a translator between the detectors and the computers. A key component of the DAS is the analogue-to-digital converter (ADC), which changes transmission measurements from the patient (analogue data) into digital signals which are then transmitted to the computer.

The gantry aperture is the opening in which the patient is positioned during the scanning procedure. It is usually 70 cm in aperture and allows access to the patient from both the front and the back of the gantry, which is important when interventional techniques are done under CT control and when general anaesthetic is required for the CT scan. The gantry on modern scanners is capable of being tilted to accommodate virtually all types of patient and clinical examinations, with a variable tilt of usually + or −25–30°. Inherent within the gantry is a laser beam which allows optimal patient positioning and serves as a reference point for the commencement and end of the procedure and as a guide in interventional procedures.

There are three primary types of acquisition geometry, namely parallel beam geometry (first generation), fan beam geometry (second, third and fourth generation scanners), and spiral geometry, found in spiral/helical CT (ROMANS 1995). The detector and computer designs are virtually optimal. However, substantial improvements in data acquisition due to the introduction of slip-ring technology have been applied to both the third and the fourth generation CT designs, allowing the x-ray tube to rotate continuously on a slip-ring within the gantry. Electrical connections are made by sliding contact from the stationary gantry to the rotating ring. These systems also have large data acquisition memories so that many scans can be performed in rapid succession.

3.5.1.2
Spiral Continuous (Volume) Data Acquisition

Volume data acquisition utilises the special beam geometry referred to as spiral (helical) scanning, which produces data from a volume of tissue rather than one slice at a time. The patient is fed into the scanner while the x-ray tube rotates continuously and as a result traces a spiralled path which scans an entire volume of tissue during a single breath-hold (Fig. 3.6).

In both types of scanning methods, source collimators are located in the x-ray tube; these limit the amount of x-ray beam emerging, with variation from 1 to 10 mm. The collimators are used by the operator in selecting the slice thickness on the grounds that

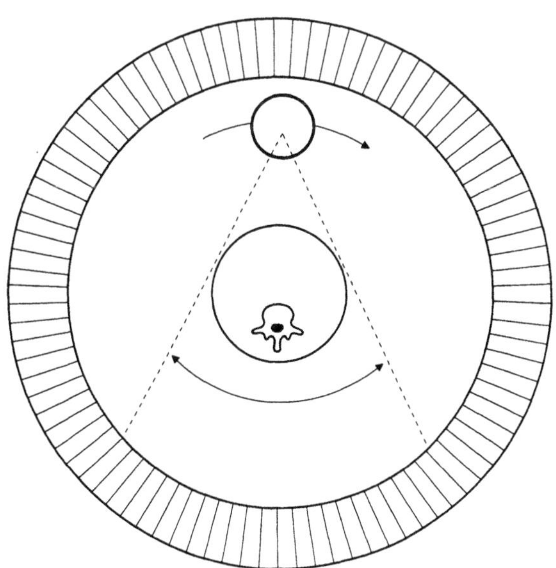

Fig. 3.5. Fan beam geometry: fourth generation rotate-only design. Only the tube rotates in the gantry

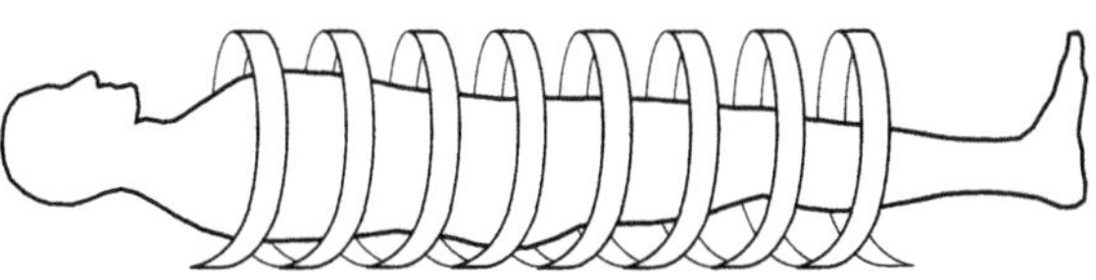

Fig. 3.6. Spiral geometry: helical/volumetric/continuous acquisition of data

the finer the collimation, the thinner the slice, leading to improvement in contrast resolution. Also in both systems the detectors measure the transmitted radiation that passes through the patient from all the different locations and allows measurement of the relative transmission values, which in turn are sent to the computer and stored as raw data.

The patient lies on the table in readiness for the commencement of the CT programme, which is operator dependent. To start with the CT scanner is used to make a digital radiograph of the patient as this is useful in identifying the exact region of interest that requires to be included in the CT scanning protocol. This digital radiograph (scannogram, scout film, topogram) can be used in assessing the various portions of the body such as the spine in scoliosis, or the legs in leg length discrepancy, but it is more commonly employed in helping to identify the beginning and end of the CT investigative protocol and also in helping to identify the exact location of each CT slice. The digital image is produced by fixing the x-ray tube at one position in the gantry, turning on the beam, and passing the patient through it as the table is advanced (Fig. 3.7). The process of moving the table by a specified measure in an incremental fashion into the gantry yields patient information slice by slice. A numeric read-out of the table location relative to the gantry is displayed. It is also important to employ anatomical landmarks which allow the level of the table to be referenced; this will ensure consistency in follow-up examinations as well as before and after contrast enhancement. CT images are usually acquired in the transverse plane and on most scanners it is possible to place the patient either head first or feet first and in either the supine or the prone position within the gantry. The pedestal of the patient couch has in-built mechanical and electrical components that facilitate movement of the table top. The table top is moved in the vertical plane to allow easy transfer of geriatric patients, trauma and paediatric patients and it also moves in the horizontal plane, which allows the patient to be scanned and fed incrementally without the need of repositioning. There is usually, however, a weight limit of about 20 stones (ca. 127 kg) beyond which the couch does not allow the patient to be scanned. Obviously the other limiting factor is the girth dimensions of the patient in relation to the 70-cm aperture of the CT gantry. As a rule, however, compared with MRI there are no problems associated with claustrophobia.

Every time the x-ray tube is activated, information is gathered from the detectors and fed into the systems computer. The data are saved and stored on the

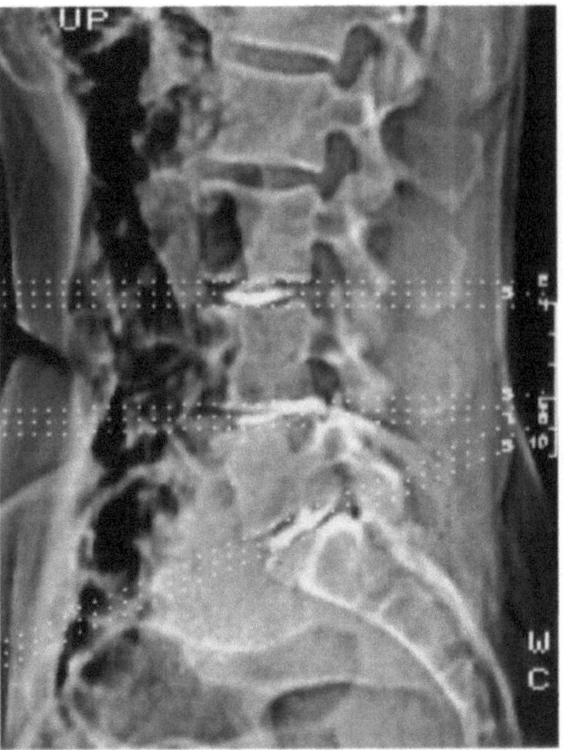

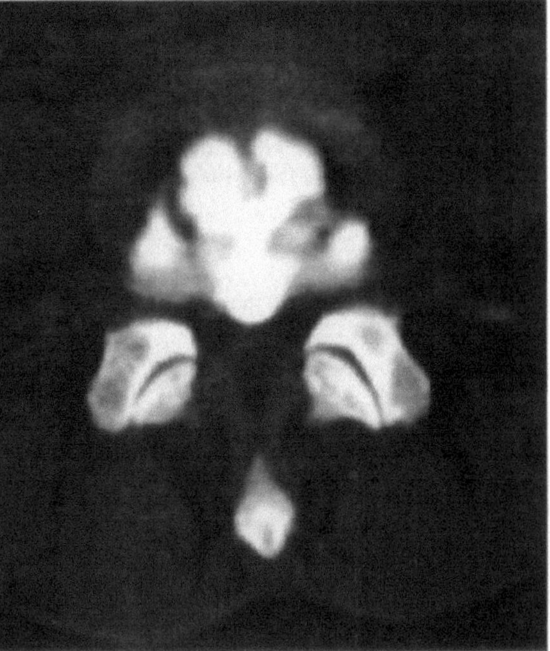

Fig. 3.7 a,b. Digital lateral radiograph of the lumbar spine (topogram, scout view). **a** Note the chosen areas of interest in the plane of the distal three lumbar discs, which have intradiscal contrast medium. **b** Axial post-discogram image depicting a large posterior annular tear

hard disc and the thousands of bits of data are collectively termed "*raw data*". These data can be saved for long-term usage utilising floppy discs, optical discs or magnetic type.

3.5.2
Digital Image Processing

Inherent in this is the processing of raw data from the hard disc to reconstruct the digital image. The computer assigns one value (CT number or Hounsfield number) to each pixel of the matrix. This value is the average of all measurements within that particular pixel. The two-dimensional pixel which houses the information in fact represents a three-dimensional portion (voxel) of patient tissue within the slice that was scanned. In the image reconstruction process highly complex mathematical reconstruction algorithms are used incorporating Fourier transformation, convolution and interpolation. Once the data are averaged, each pixel has one associated number which collectively constitute image data. In addition, the computer system performs image manipulation and a wide range of processing operations including windowing, image enhancement, measurement, multiplanar reformat, three-dimensinal imaging and quantitative measurements (BAXES 1984; SEERAM 1994).

3.5.3
Image Display

The CT images are displayed by the monitor's cathode-ray tube, recorded on x-ray film, or stored on magnetic tape or optical or floppy discs. The television monitor can only display about 256 shades of the grey scale. There are over 4000 different Hounsfield units, but the human eye can only differentiate approximately 20–30 shades of grey. Due to these limitations, a grey scale is employed in image display.

The tissues within the CT image are displayed by varying shades of grey based on basic radiation principles employed in plain radiography. X-ray energy passes through or is attenuated by given structures within the body in varying amounts depending on the density and anatomic number of the structure. It is the amount of x-ray beam that passes through the body which determines the shade of grey on the image in both conventional radiography and CT imaging. By convention, x-ray beams that pass through objects unimpeded are represented by a black area on the image while those completely stopped by an object cannot be detected and appear white on the image. Metal has a very high capacity for beam attenuation and so do surgical clips, orthopaedic implants etc., which appear white on the CT image. Air or gas, on the other hand, has a very low inherent density with little attenuation capacity and produce a black area on the CT image. The degree of beam attenuation on a CT image can be quantified and its measurement expressed in CT numbers or Hounsfield units. CT numbers are established on a relative basis with the attenuation of water used as a reference with a CT number of 0, while those for bone and air are +1000 and −1000 respectively on the Hounsfield scale. Tissues with a beam attenuation less than that of water are designated negative while those with a beam attenuation greater than that of water have a proportionately positive Hounsfield unit. A proportional conversion is carried out between CT numbers and grey-scale brightness levels using the Hounsfield scale where the upper (+1000) and lower (−1000) limits of the scale represents white and black respectively. This enables the continuation of the convention employed in conventional radiography where bone appears white and air appears black. All the other values between these two limits represent varying shades of grey. This relationship between the CT numbers and the grey scale is referred to as "windowing".

3.5.4
CT Numbers

The transmission values measured by the detector array depend on the degree of beam attenuation by the various components of the tissues within the body slice included in the scan. The computer calculates the CT numbers and a numerical image is produced of all the pixels. The system enables the measurement of an unknown structure that appears on a CT image to be calculated by comparison with measurements of known substances, helping to approximate and determine the composition of the unknown tissue structure. Knowledge of CT numbers for various tissues is quite useful and underlies the basic principles of CT interpretation (Fig. 3.8). Volume averaging, however, needs to be borne in mind and this has diagnostic implications (see Sect. 3.10). Strict adherence and application of CT numbers in the clinical setting will inevitably result in diagnostic errors (LEVI 1982).

3.6
Principles of Interpretation

A thorough knowledge and understanding of the anatomical structures present in the region that is being scanned is a fundamental prerequisite to the

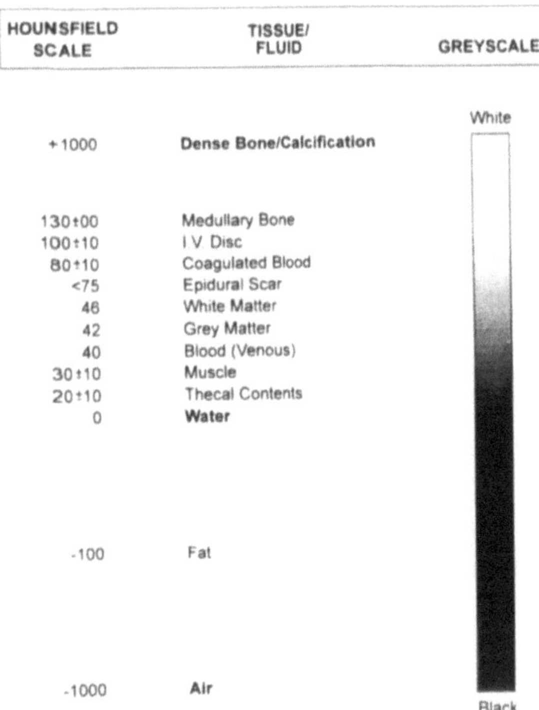

Fig. 3.8. Relationship between CT numbers, tissue/fluid and brightness level on the grey scale

correct interpretation of CT images. Similarly, a thorough understanding of the pathological processes that alter the characteristics of the musculoskeletal tissues is very important. In the distinction of normal from abnormal tissues, detection of disease requires correct interpretation of the CT density measurements of the image display.

Prior to MRI, CT, when compared with plain film radiography, provided numerous advantages in orthopaedic disorders, including the cross-sectional display, excellent contrast resolution, accurate measurement of tissue attenuation coefficients, noninvasiveness, reformatting and three-dimensional imaging. CT identifies undetectable and partially definable lesions on conventional plain films. However, its spatial resolution is notably poorer. Furthermore, the dose is generally higher for similar anatomical regions when compared with plain film radiography, and the imaging is limited to the axial plane, although the gantry can be tilted up to 30° to the transverse section. Other disadvantages include partial volume averaging, especially where soft tissues are surrounded by large amounts of bone in a confined space, e.g. the spinal canal. A series of artefacts can also plague the image quality, reducing detail and perceptibility. Nevertheless, the capability of sub-

millimetre resolving power for high-contrast objects such as bone, coupled with the capability of high-resolution CT mode utilising thin sections (1–3 mm), usually ensures high-quality images of fine bony structures that help to produce excellent contrast resolution, which is approximately 0.3%. The use of extended window scales, −1000 to +3000 HU, permits visualisation of the entire range of densities in the musculoskeletal system. Radiation dose is still a very important consideration and with the introduction and rapid advance in MRI, the role of CT in the assessment of musculoskeletal disorders has been adjusted.

3.6.1
Soft Tissues

The density value of soft tissues is determined overwhelmingly by the contents of protein, water and fat and the relative proportions found in specific areas. As expected, fat-containing lesions, e.g. lipoma, tend to produce negative CT numbers (−80 to −100 HU). This is, however, not synonymous with the conclusion that fat-containing lesions are always lipomas because well-differentiated liposarcomas do have a great amount of fat within them which can be very hard to distinguish from benign lipomas (Fig. 3.9). Alternatively, conventional lipomas can also have within them fibrotic as well as calcific components which in turn alter the attenuation co-efficient characteristics (ANDRE and RESNICK 1995). The amount of interposing adipose tissue determines the extent to which soft tissue anatomical planes and neurovascular structures can be demarcated. The low CT number of this intervening fat allows depiction of the epidural space, muscle planes, muscle compartments and para-osseous outlines and the distinction of artery, vein and nerve. Soft tissue calcification in necrotic material, tumour or denatured protein, as well as phleboliths in haemangiomas, produces a high CT number and indeed CT is more sensitive than plain x-rays and MRI in this regard (Fig. 3.10). However, liquefying necrosis, commonly found in abscesses, rapidly expanding tumours, and following chemotherapy, leads to a reduction in density such that it approximates that of water. In the assessment of intraspinal soft tissues CT is quite useful. In postoperative states two causes need to be differentiated, namely hypertrophic extradural fibrosis from recurrent intervertebral disc herniation. The attenuation value of epidural fibrosis (40–75 HU) is typically less than that of a

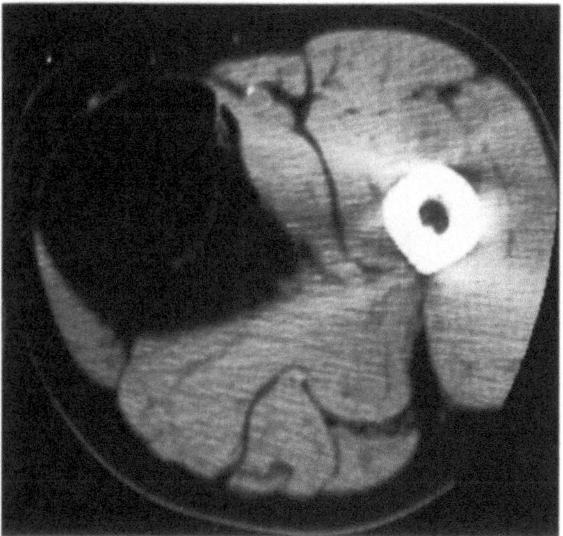

Fig. 3.9. Axial image of a well-differentiated liposarcoma in the thigh. Note the low CT density and the presence of septa

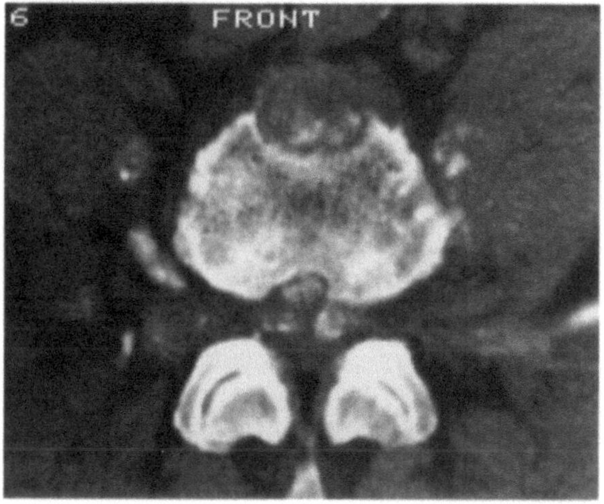

Fig. 3.10. Tuberculosis of the spine with areas of calcification within para-vertebral, epidural, and anterior and posterior subligamentous abscesses

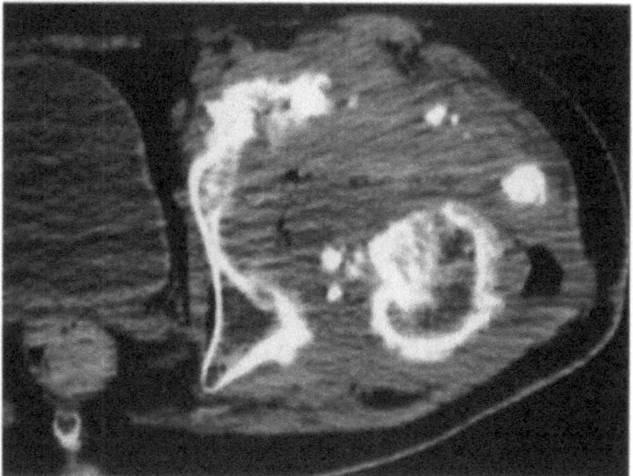

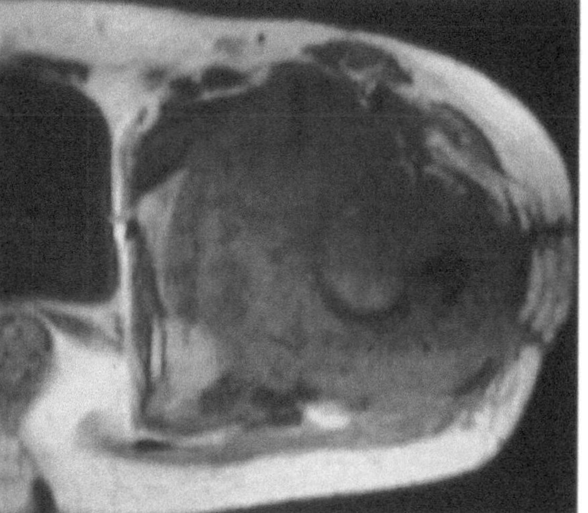

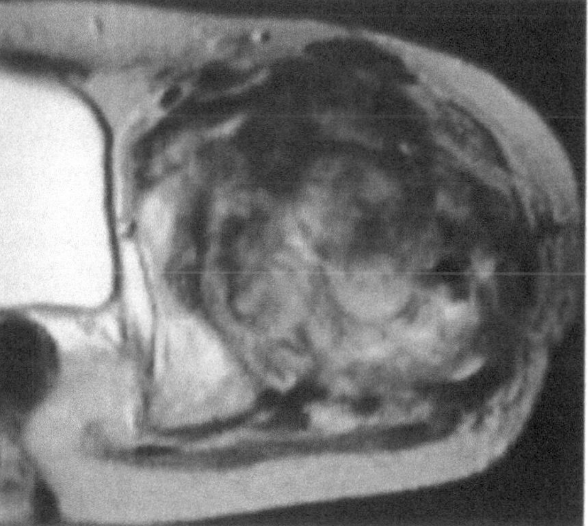

Fig. 3.11. Septic arthritis in a paraplegic patient showing joint destruction, bone debris and gas within the soft tissues on CT (a). The diagnosis is not as obvious on T1- and T2-weighted axial MR images (b and c, respectively)

disc herniation (90–120 HU) (SCHUBIGER and VALAVANIS 1982; WEISZ 1986). However, there are common exceptions to this rule that limit its diagnostic value, which is sometimes improved with intravenous enhancement with CT as well as with MRI and MR enhanced imaging. Gas can be depicted by its low CT number very easily when present within bone or soft tissue and is a sign of ischaemic necrosis, osteomyelitis or subchondral or para-articular cysts (SILVER et al. 1992), and is also seen within intervertebral disc and gas-containing disc fragments within the spinal canal (Fig. 3.11).

3.6.2
Bone

Changes in attenuation values within bone are easily discernible if they occur in the medulla or if a bone normally enjoys an abundance of marrow and trabecular bone, e.g. vertebral body, proximal femur and proximal humerus. In narrow bones, e.g. the ribs, fibula and phalanges, the use of attenuation values is significantly restricted due to the error secondary to partial volume averaging. The normal fatty marrow exhibits a negative CT number and in the presence of disease which generates oedema, through either an inflammatory or a tumoral cause, a high attenuation value is seen in the diseased marrow which is useful in assessing the extent of the offending process.

3.6.3
Fluids

The CT density value of water-filled contents within cystic lesions will vary depending on the protein contents, electrolytes and measurement inaccuracies. Exudates with a protein content greater than 30 g per litre have CT values of 20–30 HU, whereas transudates have CT values of <20 HU. Cystic structures, being avascular, do not enhance after intravenous contrast enhancement (MEANEY et al. 1992). The density value of blood (55 ± 5 HU) is largely determined (40 HU) by the haemoglobin content within the blood corpuscles, with 15 HU contributed by the CT density of plasma. When blood coagulates, haemoconcentration results in an increased CT density compared with venous blood which lasts up to 7 days after the onset of haemorrhage. Later, decomposition of blood products and protein absorption lead to a reduction in CT density, which can approximate that of water depending on the residual protein content. Furthermore, blood-containing lesions either in soft tissues or intra-osseously can also produce fluid-fluid levels (DAVIES et al. 1992) depicted because of different attenuation coefficients of the sediment from the supernatant (Fig. 3.12).

3.7
Image Manipulation

Image manipulation incorporates digital image processing techniques which modify the image data to enhance the visibility of information while suppressing non-useful information, thereby allowing enhancement, transformation and analysis. Image

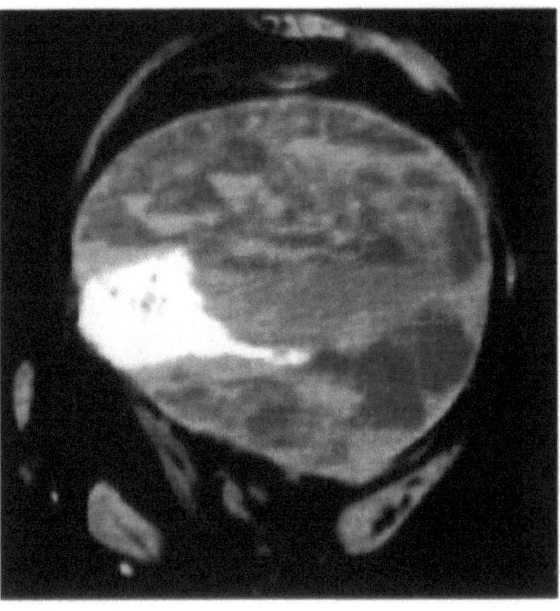

Fig. 3.12. Fluid-fluid levels in a telangiectatic osteosarcoma of the distal femur

manipulation does not produce any additional information, and the processed image is either less or at best equal in its information content to that of the original image. Windowing (grey level mapping) is the most commonly used point processing technique in CT. Other major programmes for CT image manipulation include region of interest analysis, statistical highlighting, multiplanar reformatting, 3D imaging, quantitative CT, etc. Although the dynamic range of CT is very large, ranging from −1000 (air) to +1000 (dense bone), the monitors have in relative terms a very limited grey scale and are unable to portray all of the available CT numbers. Using windowing capabilities, the image is displayed within these limitations so that the available shades of grey are assigned to a selectable range of CT numbers. If the monitor allows only 32 grey shades for display in a single width of CT numbers ranging from 0 to 320 HU for example, each grey shade would represent 10 Hounsfield units. To detect more subtle tissue differences than 10 HU, the operator must employ a narrower window width.

3.7.1
Windowing

The window controls found on the operating console include the window width and window level, which are used to alter picture contrast. The window width refers to the range of CT numbers while the window level is the centre of that range. By this technique the CT image grey scale can be manipulated using the CT numbers that make up the image. These numbers are altered by the operator to produce the optimum demonstration of the different structures of interest present on the image. The picture can therefore be changed to concentrate on soft tissues or dense structures, such as bone.

3.7.1.1
Window Width

The absorption measurement range in CT is expressed in Hounsfield units (HU) and is referred to as the window width. The number of Hounsfield units assigned to each level of grey is determined by the window width. The grey scale assigns higher Hounsfield values as lighter shades of grey towards white, while the lower CT numbers are represented by the darker shades towards black. The window width determines the range of Hounsfield units represented on a specific image and the maximum number of shades of grey that can be displayed on the CT monitor. The CT numbers that fall within the window width range selected are therefore assigned various shades of grey. All values higher than the selected range appear white, while values lower than the window width range appear black. Widening (increasing) the window width assigns more numbers to each shade of grey. Wide window levels (400–2000 HU) are best for imaging tissue types that vary greatly, allowing the inclusion of a large number of anatomical structures with different inherent densities. On the other hand, a narrow (decreased) window width (50–400 HU) allows greater density discrimination, which is particularly useful in structures with small differences in CT numbers. Wide window settings decrease contrast while narrow window widths enhance contrast discrimination, allowing differentiation, for example, of white from grey matter in the brain (Fig. 3.13). Wide window settings also suppress the display of inherent noise of the image, which is particularly useful in the presence of metal artefacts.

3.7.1.2
Window Level

The window level selects the centre CT value of the window width and therefore determines which Hounsfield numbers are displayed on the image. The window levels should be centred close to the average

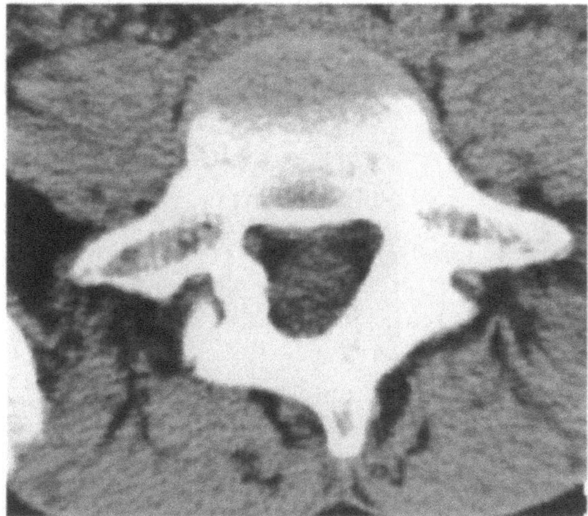
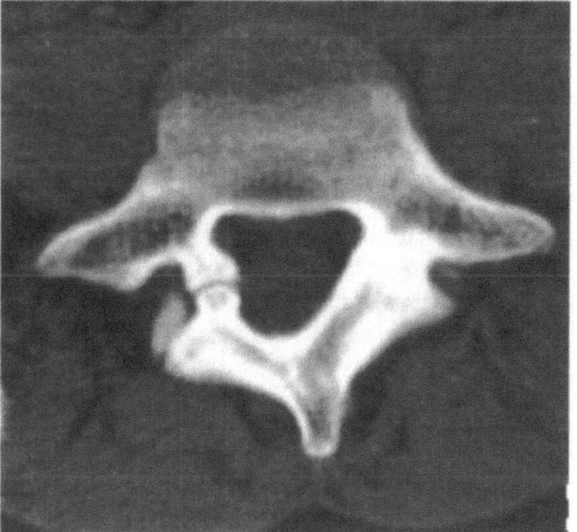

Fig. 3.13. Axial image of L5 vertebra on soft tissue (W400 L40) (a) and bone (W1600 L300) settings (b). The spondylolysis is easily missed in **a**

attenuation of the tissues of interest. If, for example, a window width of 300 HU is chosen with a window level of 0, then the Hounsfield values ranging from +150 to −150 will appear on that image. Any value lower than −150 would appear black while values higher than +150 would appear white. To summarise, therefore, changing the window width essentially alters image contrast whereas the window level is intended to optimise image display (SEERAM 1994).

3.7.1.3
Statistical Highlight Windowing

Statistical highlight windowing is quite useful and paints in white a user-defined range of CT densities

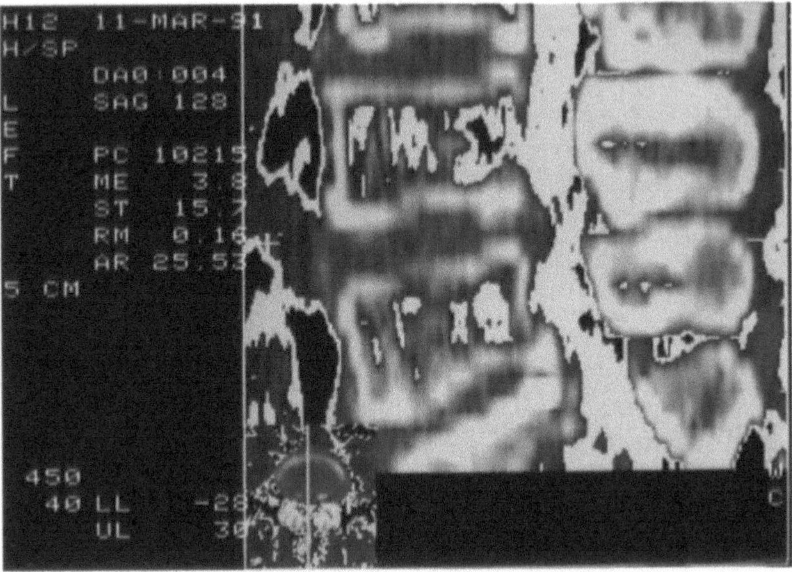

Fig. 3.14. 2D reformatted image in the sagittal plane of the lumbar spine with statistical highlighting (blink mode) of the theca (upper level 30, lower level −28) showing L4/5 disc prolapse

which is an electronic means of aiding interpretation and diagnosis (Fig. 3.14). All the CT images have a scale placed alongside the image for size reference which is used in measuring the size of lesions, calculating the placement of a biopsy needle, leg lengthening (AITKEN et al. 1985), calculating the distance of tumour from joint etc. Simultaneously the degree of angulation of the measurement line from the horizontal or vertical plane is also provided, which is quite useful in determining the angle of femoral anteversion (HERNANDEZ et al. 1981), tibial torsion (LAASONEN et al. 1984) and patellofemoral geometry (WALKER et al. 1993) (Fig. 3.15).

3.7.2
Image Reformatting

In image reformatting (2D and 3D), the software programme creates coronal, sagittal and para-axial images from the transverse axial scans obtained contiguously. This is helpful in determining the extent of lesions, in assessing the degree of malalignment in fractures and in localising lesions and intra-articular bone fragments or foreign bodies. The quality of the reformatted image depends quite crucially on the quality of the transverse axial images obtained, the slice thickness and the absence of any movement of the patient.

3.7.2.1
2D Reformatting

The fundamental prerequisite in optimising this software programme is that common features apply to all the transaxial slices obtained in an identical display field of view, image centre, gantry tilt and contiguous acquisition (Fig. 3.16). The reformation software also provides the possibility to create additional reformatted images from the topogram. Transdiscal lumbar images can be obtained by reformatting instructions from the original images which were obtained without any gantry tilt. The thinner the transaxial slice, the better the quality of the reformatted image, and as a rule employing overlapping slices also improves its quality (SEERAM 1994; ROMANS 1995). Spiral CT data can be retrospectively reconstructed to create these overlapping images, which optimise reformations. Although useful, the reformatted images produced never enjoy the same high quality in terms of resolution as those actually scanned in the axial plane. Also, when compared to direct acquisition in the same plane they are of an inferior quality.

3.7.2.2
3D Reformatting

Special software computer graphics produce 3D images from the contiguous sets of slices which re-

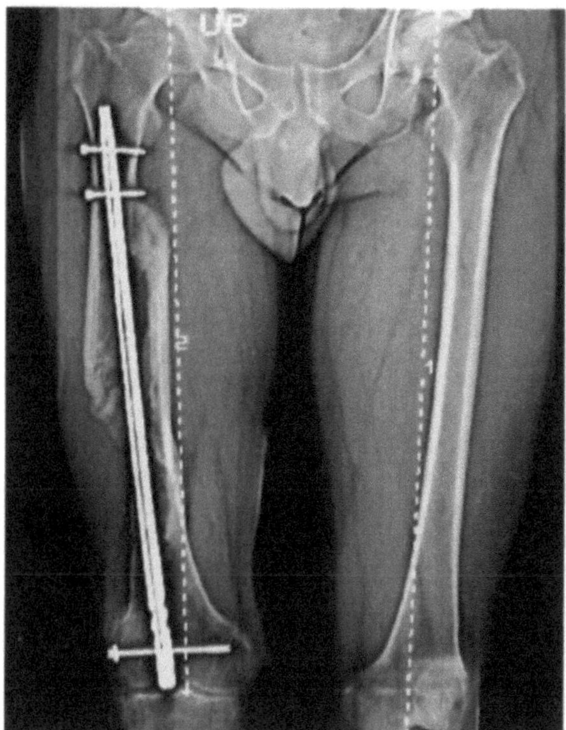

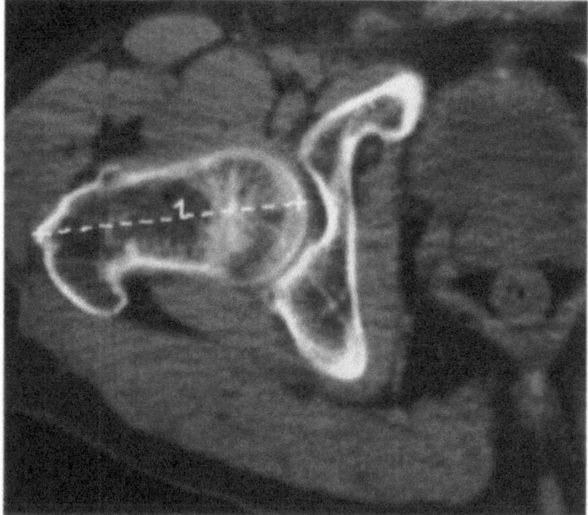

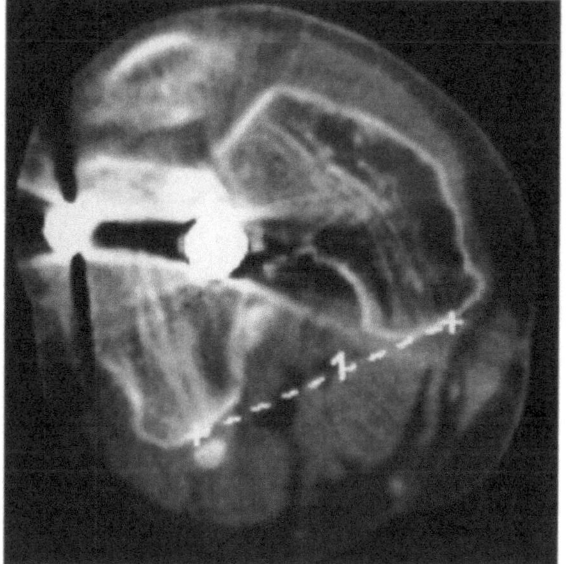

Fig. 3.15. CT measurements of post-traumatic femoral deformity showing malalignment, shortening (a) and femoral retroversion (b, c). Note that the internal fixation produces a streak artefact in the axial images but not on the digital radiograph (a)

semble the intact patient structure. The most common indicators are trauma, tumour and congenital abnormalities (TOTTY and VANNIER 1984). There is increasing interest in its potential role in orthopaedic and reconstructive surgical planning preoperatively, but assessment can also take place postoperatively to determine the surgical outcome. The 3D images can be rotated to allow visualisation of the abnormality from all aspects, and in addition isolation of a structure preventing crucial information can also be obtained, as for example in disarticulating the femoral head when one is interested in the acetabulum (ANDRE and RESNICK 1995).

The rendering techniques that transform conventional serial transaxial CT image data into simulated 3D images are of two types: (a) thresholding- or surface-based techniques and (b) percentage or semi-transparent volume-based techniques. In thresholding-based imaging the operator selects the orientation of the view, the threshold and the lighting characteristics for the surface. A low threshold setting (−450 HU) delineates a soft tissue surface while a higher one (+450 HU) isolates the bones. If bone is the tissue to be imaged the two appropriate Hounsfield units are chosen (threshold) so that all attenuation values below the lower threshold (150 HU) will not be included in the scan and similarly, values above the 3000 HU will also not be included. This surface technique is particularly satisfactory in studying the skeleton as in orthopaedics or craniofacial surgery. Volumetric rendering, however, refers to the use of the entire data set for the generation of 3D images. The major advantage of volumetric rendering is the ability to display all the

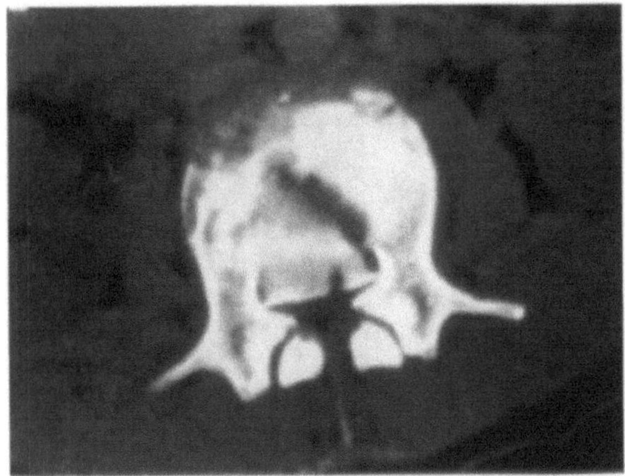

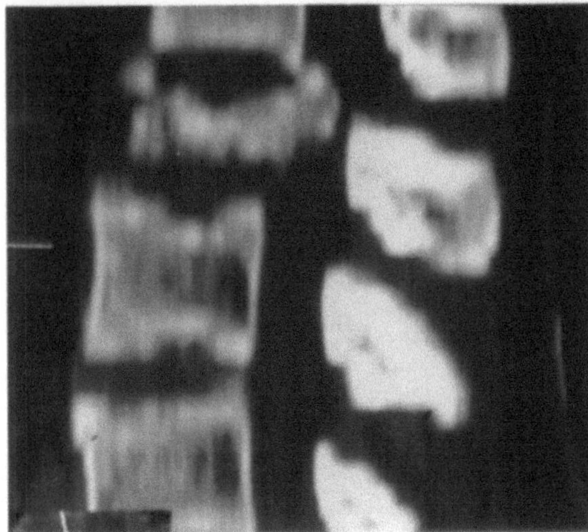

Fig. 3.16. Axial (a) and 2D sagittal reformation (b) of burst fracture of L2 with a large retropulsed bone fragment in the spinal canal

information from the original set of CT slices (FISHMAN 1991). With the advent of spiral CT, many of the original problems associated with 3D reformatting have been eliminated because spiral CT allows the collection of any number of slices with very narrow collimation with practically no risk of patient movement. The generation of custom-made orthopaedic implants and life-size models of various body parts from CT scan data has been employed for many preoperative surgical planning operations and applications.

3.8
Image Quality

Many factors affect image quality, including x-ray beam characteristics, dose, transmissivity of the sub-

ject, slice thickness, CT geometry, motion, computer processing, pixel size, reconstruction algorithm and the display resolution. Factors referred to as scanning parameters which can be regulated by the operator and influence image quality include the mA, scan time, slice thickness, field of view, scan algorithm and the kV if this is not fixed. As in conventional radiography, the thicker and denser the part being examined, the more mAs are required to produce an adequate image. Other factors, however, such as patient size and specification of hardware and software are beyond the control of the operator. For example, an important physical parameter of the grey scale display monitor is its resolution. This is related to the size of the display pixel matrix, which can range from 64×64 to 124×124, but high-performance monitors can also display an image with a 2048×2048 matrix.

Two commonly used terms in defining image quality are "*spatial resolution*" and "*contrast resolution*". Spatial resolution is defined as the ability to present small objects and differentiate between closely spaced objects, while contrast resolution is the ability to differentiate small density differences on the image. As a general rule, thinner slices produce sharper images. This is due to the previously described digitisation process that takes place in CT whereby to create an image, the information in a voxel is displayed in two dimensions in pixel format. Therefore, the thicker the slice, the more pronounced the inaccuracies in the averaging of CT attenuation that is displayed in the pixel. This is the so-called volume averaging or partial volume effect. High-resolution CT, introduced in the mid 1980s, optimises the spatial resolution of conventional scanners by employing a narrow beam collimation to ensure that thin slices are obtained. Slice thicknesses of 1–2 mm are utilised, thereby reducing artefacts caused by partial volume averaging. Utilising a smaller field of view will also reduce the pixel size, which will further increase the spatial resolution.

Compared with conventional radiography, CT has a much higher contrast resolution. CT can image tissues that vary only slightly in density and anatomic number. Whereas radiography can discriminate density differences of about 10%, CT has the advantage of detecting density differences ranging from 0.25% to 0.5%. Modern CT scanners can resolve high-contrast objects as small as 0.25 or 0.5 mm, which corresponds to a maximum resolving power of 0.5–0.75 lines per mm. Image intensifiers and fluoroscopy provide a somewhat better resolution of 1–2 lines per mm whereas screen film systems under normal circumstances resolve 2.5–4 lines per mm. The contrast resolution of CT scanners (0.3%), however, is much better than that of conventional radiography and it is indeed this fact that has made CT so prominent despite its lower spatial resolution. Contrast resolution is enhanced further by the intracavitary or intravenous application of contrast material. The spatial resolution can be increased by a higher patient dose. As a rule, to double the resolution by reducing noise, a fourfold increase in patient dose is required. In the assessment of the lumbar spine, where it is necessary to detect small soft tissue structures such as nerve roots, epidural veins, bone morphology and disc herniation, a high requirement for spatial and contrast resolution necessitates an increase in patient dose.

3.9
Contrast Media Enhancement

The iodine atoms in the injected contrast material are responsible for the increase in attenuation. Contrast administration followed by CT can be intravenous, intrathecal, intradiscal or intra-articular. In the intra-articular administration of contrast material both positive contrast (contrast material, dye) and air can be used in demonstrating intra-articular pathology. Furthermore, contrast injection in sinography can also be used with CT to demonstrate the sinus tract and the relationships to sequestra, bone cavities and areas of soft tissue abnormality.

3.9.1
Intravenous Contrast Medium

Intravenous contrast material increases the ability of the enhanced structure to attenuate the x-ray beam. In so doing it can help identify suspected soft tissue masses when the unenhanced initial CT scan is unremarkable. Furthermore, injected contrast medium assesses the vascularity of the soft tissue or osseous lesion when this feature has diagnostic and therapeutic implications by defining the anatomical plane and, in particular, delineating the relationship of the neurovascular structures to the soft tissue or osseous lesion (Fig. 3.17). Two different phases separated in time occur following intravenous injection of contrast medium. Intravascular opacification is assessed with fast scanning techniques during the first passage of the contrast medium. Hypervascular areas, e.g. neoplastic vascular neogenesis, are easily differ-

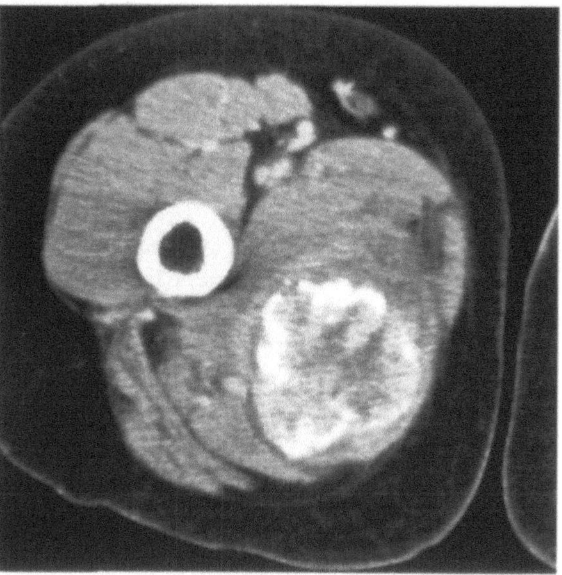

Fig. 3.17. Myositis ossificans of the thigh showing peripheral ossification with peripheral contrast enhancement following intravenous contrast medium

entiated from normal surrounding parenchyma, while avascular areas, e.g. cysts, necrosis and haematomas, do not obviously enhance. Later, parenchymal opacification results, which is also dependent on the vascularity of the tissues and occurs secondary to the diffusion of contrast medium in the interstitial spaces of normal and pathological tissues. It is important to realise at the very outset that hypervascularity is not synonymous with malignancy and this feature is not a reliable means of differentiating between benign and malignant aetiology. Fibrous tissue, haemangiomas and myosistis ossificans, for example, quite commonly enhance following intravenous contrast injection. Furthermore, despite the use of intravascular enhancement, it is sometimes quite difficult to differentiate displacement by benign lesions from invasion by malignant lesions of soft tissue and bone. Intravascular contrast injection has some use in the assessment of the spine, especially postoperatively in helping to differentiate epidural scar, which enhances significantly, from recurrent lumbar herniated intervertebral disc, which is usually associated with a thin peripheral rim of enhancing tissue (SCHUBIGER and VALAVANIS 1982; DE SANTIS et al. 1984). There are, however, some exceptions to this and, furthermore, contrast enhancement is not specific as it can also occur in intraspinal tumours. Prior to MRI, intravenous enhancement was used in the assessment of cervical disc prolapse in the unoperated state with a satisfactory diagnostic yield.

3.9.2
Intrathecal Contrast Medium

The advent of MRI has caused a dramatic reduction in the use of myelography and CT myelography (TEPLICK and HASKIN 1983). CT myelography, however, still has applications particularly in the assessment of the postoperative spine and when MRI is not available (MODIC 1991). A delay of 2–4 h between the myelogram and the CT scan allows the contrast medium to dilute; scans done earlier, when the contrast material is too dense, causes masking of intradural detail (Fig. 3.18). Post-myelographic CT should ideally be done by rolling the patient over before transfer to the CT table to prevent layering of contrast medium and CSF. The cause of encroachment on the neural tissue as well as the exact location of this compression can be easily identified and differenti-

ated between soft tissue, e.g. disc prolapse, and osteophyte.

3.9.3
CT Discography

CT discography is employed in some centres to help identify the exact location of annular disease with and without associated disc herniation (Fig. 3.7). CT does not differentiate the nucleus pulposus from the annulus fibrosis and therefore requires the presence of contrast medium within the nucleus to help separate the two components of the intervertebral disc. However, MRI with and without intravenous gadolinium is increasingly becoming the investigation of choice in patients with problems in the spine.

3.9.4
CT Arthrography

CT arthrography with air or radiopaque contrast medium or both within the joint is extremely useful in defining the intra-articular structures with particular reference to osteo-cartilaginous loose bodies

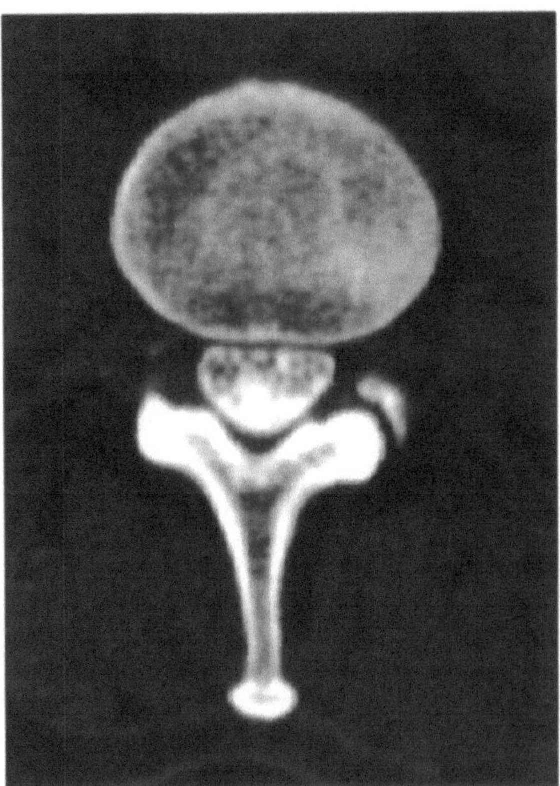

Fig. 3.18. Post-myelogram CT image of L2 showing the nerve roots, which appear as filling defects, as well as posterior layering of the contrast medium

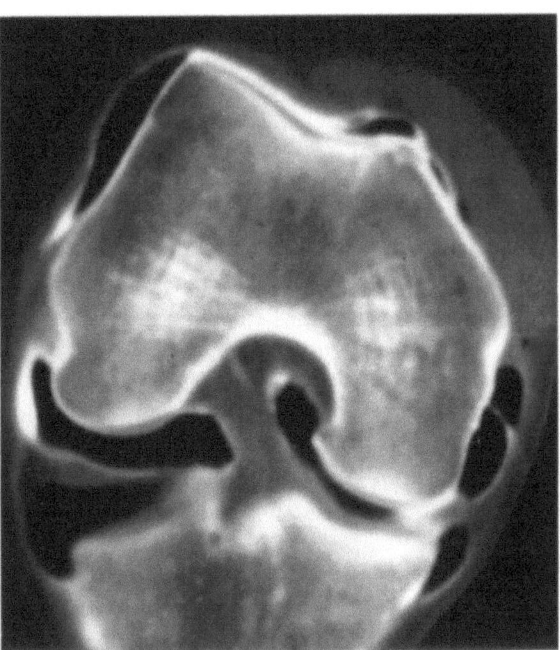

Fig. 3.19. Double-contrast CT arthrography of the knee in the coronal oblique plane to show anterior cruciate ligament, menisci and articular cartilage

within the knee, hip, ankle (DAVIES and CASSAR-PULLICINO 1989) and elbow (HOLLAND et al. 1994) (Fig. 3.19). It is also particularly valuable in defining the presence of a communication between the joint and associated peri-articular soft tissue masses such as synovial cysts or ganglia (MEANEY et al. 1992). The glenoid labrum, patellar cartilage (REISER et al. 1982a), synovial plicae (BOVEN et al. 1983) and

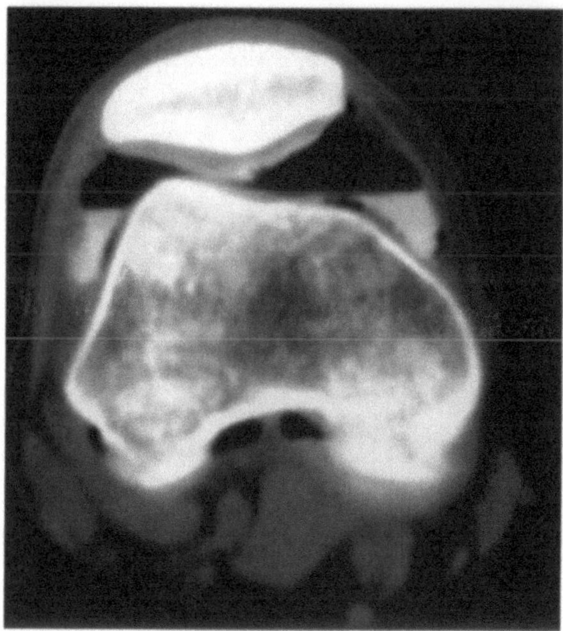

Fig. 3.20. Contrast inbibition due to chondromalacia of the patellar cartilage at double-contrast CT arthrography of the knee in the axial plane (a), while in another patient ulceration and fissuring is clearly evident (b)

osteochondral abnormalities are well identified and more importantly the subchondral bone is also well assessed (Fig. 3.20).

3.10
Artefacts

Systematic errors caused mainly by motion, x-ray beam hardening and image reconstruction inaccuracies lead to falsification of CT density values. The factors responsible are collectively known as artefacts. Artefacts cause errors in the images that are unrelated to the information obtained from the subject being studied. They degrade image quality and therefore affect the diagnostic yield. The main artefacts one encounters in CT are those due to volume averaging, blooming, ring artefact, beam-hardening artefact, streak artefact and motion artefact (BRANT 1986). Furthermore, artefacts specific to spiral CT, namely "break up" and "stair-step" artefacts, are also encountered.

3.10.1
Volume Averaging

When CT numbers are computed the calculations are based on the x-ray attenuation co-efficient of small volumes of patient tissue. Each CT image produced is a two-dimensional representation of a three-dimensional volume (slice) of patient tissue. Each two-dimensional pixel therefore represents a three-dimensional volume (voxel) of patient tissue. Each pixel has assigned to it the CT number which represents the calculated (computed) volume average attenuation of each voxel. If the voxel is entirely homogeneous, containing tissue of the same density throughout, then the computer accurately assigns the correct CT number. However, when objects of widely differing x-ray attenuation occupy the same voxel of patient tissue, the calculated attenuation value (CT number) is directly proportional to their average value, and therefore is characteristic of neither. For example, if a voxel contains three similar tissue types in which the CT numbers are closely approximated, e.g. blood (CT = 40), grey matter (CT = 43) and white matter (CT = 46), then the CT number for that voxel is based on the average of the three tissues, i.e. the CT number is 43. This is known as partial volume averaging and gives rise to the partial volume effect and artefact. The volume average effect, however, is present in every CT image and must always be considered in image interpretation. The partial volume effect is always present in structures with a diameter smaller than the slice thickness and therefore this effect can be minimised by using thin slices and small pixel sizes. Partial volume artefacts appear as bands and streaks and are usually located at the interface of structures with widely differing CT densities, e.g. bone and air.

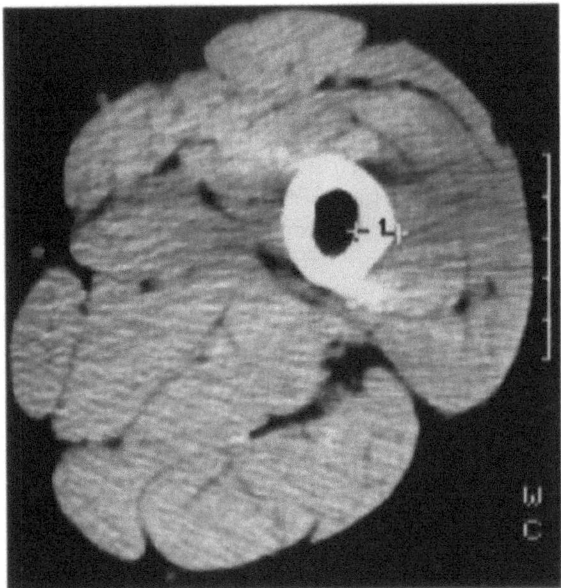

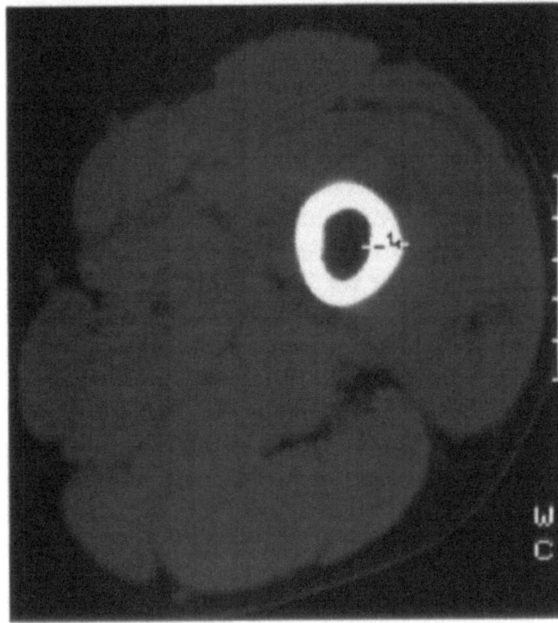

Fig. 3.21. Axial image of femur on soft tissue (a) and bone (b) settings to demonstrate a "blooming" artefact whereby the dimensions/sizes of bone and soft tissue structures alter in the same image depending on the window settings chosen

3.10.2
Blooming (Point Spread Effect)

In 1981 BAXTER and SORENSON demonstrated that size measurements are most accurate in CT when the window level is centred mid-way between the background density and the object density. Under optimal conditions, size measurements in CT are accurate to approximately 1 mm. However, inappropriate manipulation and choice of window width and window level will have a significant influence on the measurement of the size of structures from CT images (Fig. 3.21).

Manipulation of the window controls results in object boundaries which appear to shift position in different viewing settings (BAXTER and SORENSON 1981). The clinical significance and the measurement errors are particularly important in orthopaedics when one measures the cortical thickness and intramedullary canal dimensions preoperatively, as well as in the sequential follow-up of the size of lesions within the lung such as metastases.

3.10.3
Ring Artefact

A high-density artefactual ring may arise when using CT scanners with rotating detectors, indicating that the detectors are not appropriately and optimally calibrated. Each detector is responsible for a ring of data as the gantry rotates. The calibration error is therefore projected in the data ring of the faulty detector. Such an artefact is usually found with the third-generation type of scanners and is seen on multiple scans in different patients obtained consecutively.

3.10.4
Beam-Hardening Artefact

As the polychromatic x-ray beam traverses the patient, the low-energy photons are attenuated to a much larger extent than the high-energy photons. As a result, attenuation is less at the end of an x-ray beam than at its beginning. This results in the main energy of the beam becoming progressively increased and produces the phenomenon called beam hardening (YOUNG et al. 1983). The CT reconstruction algorithms work on the premise that any change in beam intensity is due to a change in tissue attenuation, and this inevitably leads to errors because of the beam-hardening effect. The errors appear as streaks of low density extending from structures of high x-ray attenuation. This is well seen in the posterior fossa at the petrous edges and the internal occipital protuberance, in the lumbar spine at the level of the facet joints, and also in bone outside the field of view, e.g. the shoulders when one images the spine (Fig. 3.22).

3.10.5
Streak Artefact

High-density objects, particularly those made of metal, are prone to produce streak artefacts. Streaks in the CT image reflect errors in radiation detection along the projected beam. In the data acquisition during the scanning protocol the metal object absorbs radiation and results in incomplete projection profiles. The loss of information leads to the appearance of a typical star-shaped artefact (Fig. 3.15). Computer programs designed to reduce the metal artefact in essence complete the incomplete profile by interpolation techniques (metal reduction program) (MORIN and RAESIDE 1981).

3.10.6
Motion Artefact

Errors in image reconstruction are produced when structures move voluntarily or involuntarily during image acquisition. Motion is demonstrated in the

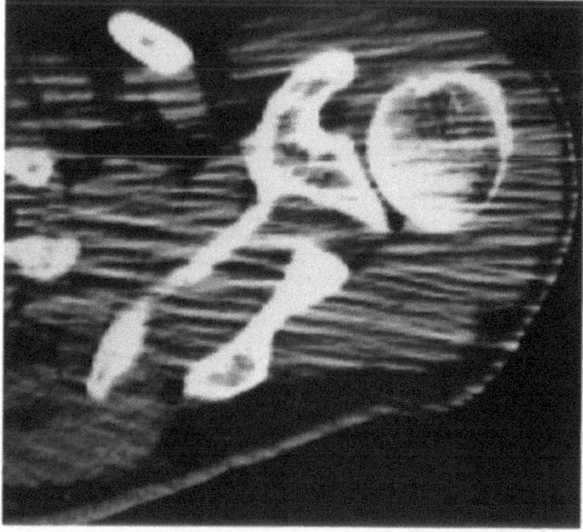

Fig. 3.22. Streaking degrading the axial image of the shoulder due to motion and a beam-hardening effect from the contralateral shoulder and trunk in a large patient

image as prominent streaks from high- to low-density interfaces or as a blurred duplicated image. This error is likely to occur in patients who are uncooperative due to cerebral irritation, pain or age and with long scanning times, since these factors increase the risk of motion. Respiration also can have profound effects, for example in CT of the spine in the prone position; in such a case the effect may be eliminated by placing the patient in the supine position.

3.11
Radiation Considerations

The prerequisite that the potential benefits of CT outweigh the risks from ionising radiation must be met and considered prior to requesting a CT investigation. It has been shown relatively recently that in the United Kingdom 20% of the radiation dose from medical investigations is attributable to CT. Every effort therefore should be employed to minimise the radiation dose and risk to the patient.

If there were no scatter radiation, the dose of an entire CT study would equal that of a single slice, assuming that each section is exposed only once. Unfortunately, scatter radiation from adjacent slices results in a higher dose than that used for a single slice. The effective dose equivalent (gonadal dose) can, however, be kept to low levels by prudent collimation when one is examining particularly the peripheral skeleton. Clearly this would not apply in investigating the pelvic structures and viscera. Furthermore, in conventional radiography, it is the skin that receives the most radiation as the dose decreases quickly after the x-ray beam penetrates the body due to absorption. In CT, because the x-ray tube rotates around the patient, the centre of the patient receives nearly as much radiation as the periphery. The dose delivered from a CT study usually is significantly higher than that received from a conventional x-ray study. There are many factors which influence the dose received by the patient, including the type of scanner, the rotational angle, the exposure factors utilised, filtration, collimation, detector efficiency, scan field diameter, slice thickness, spacing, overlapping of slices, matrix size and use of repeat scans with and without intravenous contrast enhancement (ROMANS 1995). With the advent of spiral (helical) CT geometry, significant information in respect of a large quantity of patient tissue can be obtained with a marked reduction in the radiation dose to the patient.

3.12
Clinical Applications

Requesters of CT investigations are encouraged to visit the CT scanning suite to familiarise themselves with the layout and distribution of hardware. The CT radiographer or technician drives the scanner at the operating console. The console has function buttons to commence data acquisition, which takes place in a separate adjacent room where the scanner proper (gantry and patient table) is located. The data are then transferred to the computers and by separate function buttons, again located on the operating console, image manipulation can take place. Some CT suites also have an additional satellite console which can be used for archiving and image manipulation.

The CT scans are usually acquired in the axial plane because of the obvious limitations in patient positioning within the gantry, which precludes direct scanning of certain anatomical regions in the sagittal or coronal plane. Clearly, with the help of patient co-operation and gantry tilt, the peripheral skeleton can be examined with some success in the sagittal and coronal planes. Data obtained from contiguous transaxial scans, however, can be reformatted by pixel rearrangement in any of the desired planes.

In musculoskeletal disease, CT needs to be regarded as a contrast-enhancing technique and as a general rule, preliminary plain radiographs of the musculoskeletal area require examination. Based on a combination of the clinical and plain film radiological information, a CT strategy for a particular region of interest may be developed. For example, in the presence of a very large soft tissue or osseous lesion, thick non-contiguous CT cuts may be all that are required, while if the lesion is small and clinically undetectable with normal plain films but the bone scan is abnormal (as is often the case in osteoid osteoma), then thin section slices will be required. In the spine the existence of abnormal spinal curves may require alteration of the patient position, and angulation of the gantry is desirable in the exclusion of a spondylolysis of the posterior neural arches in the lumbar spine. Although CT can be operated on a menu basis, optimum yield only results when specific instructions and understanding of the clinical problems as well as the surgeon's requirements are incorporated prior to obtaining the CT scan.

The choice of the prone, supine or sometimes lateral position of the patient in CT depends on patient comfort, the structure to be examined and the clinical status of the patient. The lumbar spine, for example, is best examined with the patient in the supine position with the knees flexed to eliminate the normal lumbar lordosis. In double-contrast CT arthrography, patient position influences the visualisation of structures, e.g. as seen in the assessment of the glenoid labrum in the shoulder. In the acutely injured patient obviously the supine position is the only one adopted, but in interventional procedures the prone or the lateral position can be preferred as these positions allow proper placement and safe advance of the needle.

3.12.1
Complex Anatomy

Although desirable, the principle in skeletal radiology which requires the acquisition of at least two views in plain film radiography at right angles to each other cannot always be achieved. This is especially the case in complex anatomy, such as the shoulder and pelvic girdles, the hip, the sternoclavicular joints, the sacrum and the sacroiliac joints. CT can provide one of the views at 90° to a conventional plain film by imaging the region of interest in the axial plane, thereby helping to define osseous status and articular congruity (Fig. 3.23)

3.12.2
Trauma

Computed tomography is very sensitive in confirming or excluding fractures and in defining the extent of fracture, joint involvement, instability and osteo-

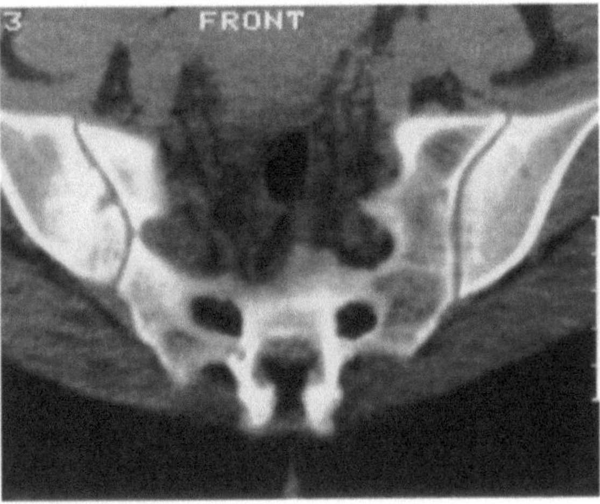

Fig. 3.23. Unilateral septic sacro-iliitis shown by erosions and surrounding sclerosis in the right ilium

cartilaginous damage. The presence of casts or plaster of Paris does not alter the image quality of CT. In the traumatised spine axial images of the area of abnormality or possible abnormality are supplemented by coronal, sagittal and 3D reformations which show horizontally directed fractures (odontoid and Chance fractures and dislocations) optimally (HANDEL and LEE 1981). The spinal canal dimensions and the presence of bone fragments within the spinal canal are also assessed. CT is superb in the study of acute (DALINKA et al. 1985) and insufficiency fractures of the pelvis. Angling the CT gantry further enhances information concerning the sacrum and sacroiliac joints. Trauma to the shoulder girdle is optimally assessed by CT with particular reference to excluding the presence and sequelae of dislocation (DEUTSCH et al. 1984). CT arthrography in the chronic stage identifies osseous and cartilaginous sequelae of dislocation. In the hand and feet (MARTINEZ et al. 1985), CT assesses specific injuries to the hook of hamate (NORMAN et al. 1985), scaphoid, mid-foot, and hind-foot in helping to identify the presence and extent of injury (Fig. 3.24). In any articular location which has been traumatised, CT arthrography is the method of choice, rather than unenhanced CT, in delineating osteochrondral fractures and excluding intra-articular loose bodies (PASSARIELLO et al. 1983; REISER et al. 1982b).

3.12.3
Infection

Although CT plays virtually no role in the acute diagnosis of osteomyelitis, it is very useful in chronic osteomyelitis, helping to identify the presence, num-ber and location of sequestra (Fig. 3.25). Intra-osseous and para-osseous abscesses as well as sinus tracts are well established by CT, allowing CT-guided biopsy with or without aspiration (WING et al. 1985). Depending on the therapeutic regimen employed (antibiotic treatment, surgery etc.), CT can also be used to monitor the response to treatment of abscesses, especially in the axial skeleton.

3.12.4
Tumours

Plain film radiography still remains the most effective means of detection and preliminary diagnosis of primary bone tumours. CT is effective, especially in complex anatomical sites such as the pelvis, shoulder and spine, in delineating and determining the cause of the tumour (Fig. 3.26). It delineates matrix calcification, cortical and cancellous bone involvement, and extension into soft tissue or muscle, optimal visualisation being achieved by a properly timed contrast injection (LUKENS et al. 1982). CT-guided biopsy can be used with pretreatment assessment of the lungs to exclude pulmonary metastases. Although both MRI and CT can demonstrate equally well the cartilaginous cap in an osteochondroma, MRI is superior in demonstrating soft tissue and medullary definition of spread. However, neither modality can yet distinguish with precision the exact boundary between tumour margin and reactive soft tissue oedema intra-osseously and within the soft tissues (ANDRE and RESNICK 1995). The superior sensitivity of MRI in detecting the extensive reactive

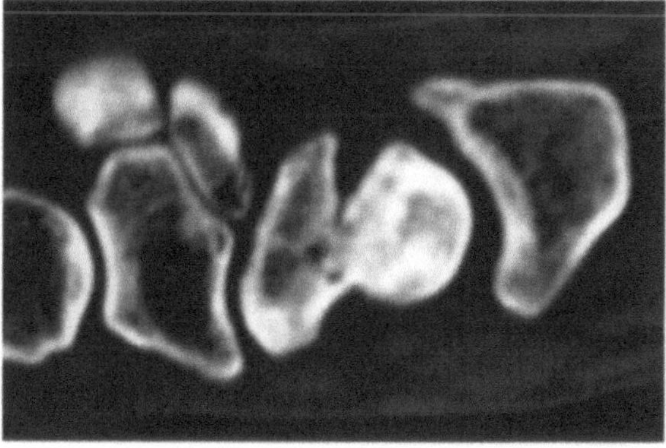

Fig. 3.24. Sagittal oblique high-resolution CT of the scaphoid following fracture healing. Note dorsal defect at fracture site which suggested non-union on the plain radiographs

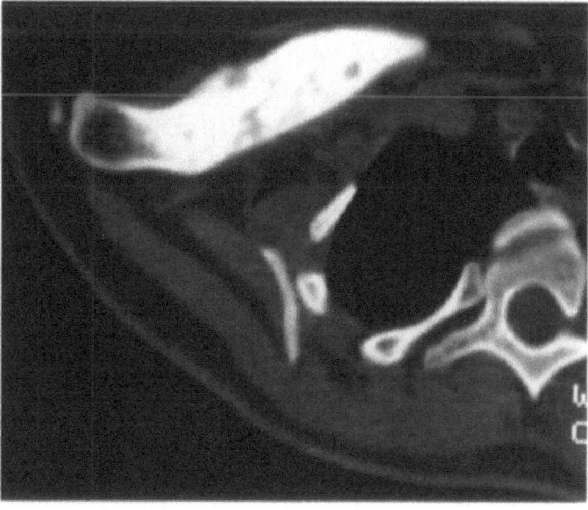

Fig. 3.25. Chronic recurrent sclerosing osteomyelitis of the right clavicle

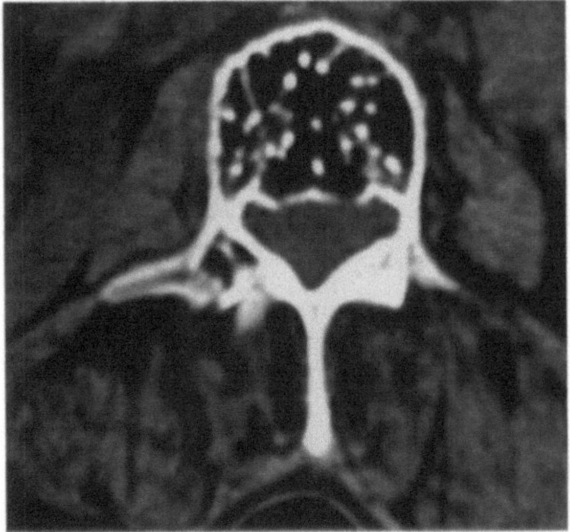

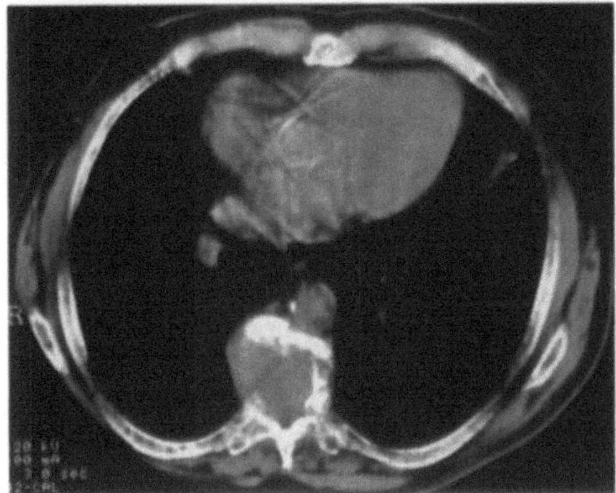

Fig. 3.26. Benign (a) and malignant (b) tumours of the vertebrae. The thickened dense trabeculae surrounded by vascular and fatty tissue are characteristic of a haemangioma (a) while the total bone destruction, soft tissue mass and multiple lytic defects in the ribs and sternum are characteristic of multiple myeloma (b)

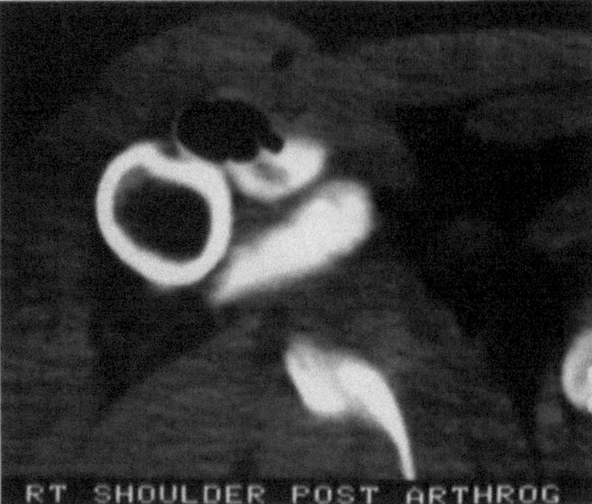

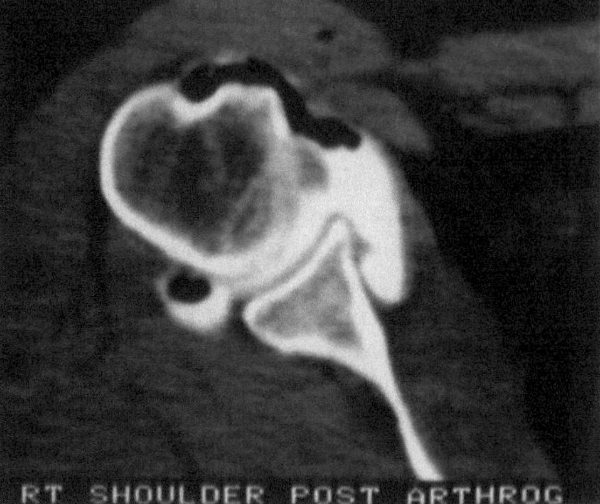

Fig. 3.27. Double-contrast CT arthrogram of the shoulder showing a medially located long head of biceps surrounded by contrast medium (a) due to dislocation from its normal location in the bicipital groove (b). Note the "filling defect" lying anteromedial to the humeral head

neighbouring marrow physicochemical alterations (oedema) in osteoid osteoma and some cases of chondroblastoma risks the loss of perception of the underlying problem, especially when it is small.

In the assessment of soft tissue tumours, CT in general has been replaced by MRI. The identification of adipose tissue within a mass is equally well accomplished by CT and MRI, but further characterisation of fat-containing masses is best accomplished by MRI. CT, however, is superior in detecting calcification or ossification within a soft tissue mass, e.g. phleboliths in haemangioma and peripheral calcification in myositis ossificans. The zonal pattern of mineralisation essential to the radiological diagnosis of early myositis ossificans can be seen by CT when the plain films are non-specific (Fig 3.28).

3.12.5
Joint Disease

Mono-articular involvement by pigmented villonodular synovitis and synovial osteochondromatosis usually gives rise to plain film abnormalities, but CT can define the extent of osseous involvement in such disorders, especially when they occur in the hip and shoulder (CASSAR-PULLICINO et al. 1992). The advantage of CT over MRI in this context is its ability

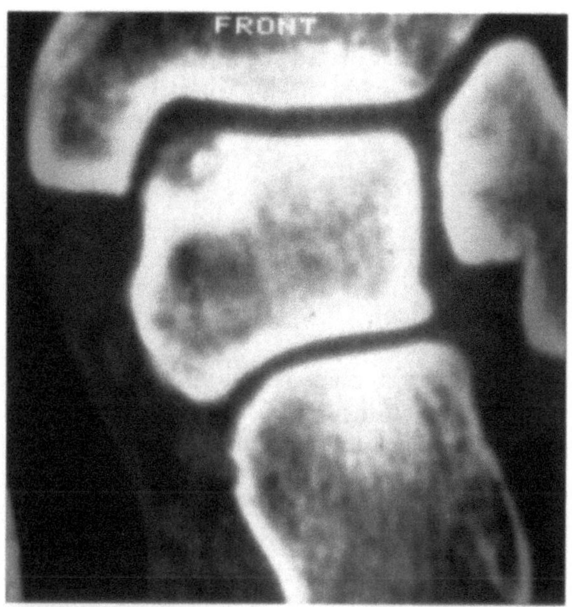

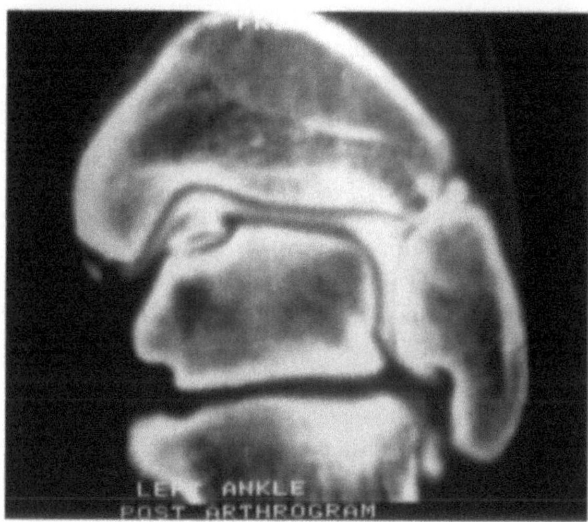

Fig. 3.28. Pre (a) and post CT arthrogram (b) coronal images of osteochondritis dissecans of the talus showing loss of articular cartilage overlying the defect

to distinguish clearly between the calcification and the haemosiderin, which tend to give similar low signals on MRI. CT is especially useful in defining accurately the presence of disease with particular reference to the sacroiliac joints, facet joints, costo-vertebral joints, sterno-clavicular joints, temporo-mandibular joints and the ankle joint (Figs. 3.27, 3.28), for which purpose plain films are notoriously suspect. CT arthrography may demonstrate communication with the joint in confirming the presence of synovial cysts, ganglia or bursa (MEANEY et al. 1992).

3.12.6
Ischaemic Necrosis

Computed tomography is not the method of choice in confirming or excluding early avascular necrosis (DIHLMANN 1982). Scintigraphy and MRI are excellent methods for detection of avascular necrosis of the femoral head, but CT with multiplanar reformatting is valuable in staging the disease and helping to plan surgical intervention. The cross-sectional display of the femoral head, especially when combined with multiplanar reconstructions, determines the surface area involvement as well as verifying collapse of the subchondral bone and delineating loose intra-articular osseous fragments.

3.12.7
Paediatric Disorders

There is a vast gamut of applications of CT in various paediatric disorders, ranging from the delineation of concentricity following closed or open reduction in developmental dysplasia of the hip (HERNANDEZ and POZNANSKI 1985) to various measurements which are important in assessment of deformity, such as the femoral angle of anteversion, tibial torsion, vertebral rotation in scoliosis and leg length discrepancy. Congenital coalition of the tarsus, especially in the hind-foot, is also well demonstrated by CT. CT myelography optimally delineates diastematomyelia, the tethered conus and anomalies of the posterior osseous structures of the vertebrae, as well as excluding the presence of associated intraspinal tumours such as lipomas and dermoids. It is also invaluable preoperatively in the assessment of slipped upper femoral epiphysis by defining the degree of rotation of the slip in relation to the femoral neck prior to surgical fixation. With increased availability MRI should replace most of these CT applications.

3.12.8
Metabolic Bone Disease

Quantitative techniques have been developed which help in the assessment of bone mineral content in specific anatomical osseous sites including the calcaneum, the femoral neck and vertebral bodies. Information derived from the one site does not necessarily correspond to that obtained from another site. The trabecular bone of the vertebral body is a good site for assessing and analysing bone mineral content as it is highly responsive to metabolic stimuli and it can be separated spatially from the less responsive cortical bone. Quantitative CT of the spine is a proven useful non-invasive means of determining bone mineral content and can be serially monitored with time, but there is no direct relationship between the bone mineral content assessed by CT and the risk factor for fracture.

3.12.9
Spinal Disorders

The advent of CT had a major impact on the diagnosis and management of disorders of the spine in both the traumatised and the untraumatised state (TEPLICK and HASKIN 1983). High-resolution CT was the first effective non-invasive method prior to MRI for the evaluation of disc herniations (Fig. 3.29). It enjoys an accuracy rate approaching 93% in this respect, and the disc material usually measures 70–100 HU, which is approximately twice the density of the thecal sac. This difference, as well as the contrast provided by epidural fat, allows accurate delineation of disc margins. Ten percent of disc herniations show areas of calcification while a smaller percentage may contain gas secondary to a vacuum phenomenon. CT differentiates these features from each other while MRI demonstrates both as low signal. Calcified intervertebral disc prolapses and ligament ossification are best assessed by CT rather than by MRI (Figs. 3.1, 3.29). In the cervical spine CT is more sensitive than MRI in detecting and differentiating osteophytes from soft disc lesions. MRI is the pri-

mary modality of choice in the assessment of the spine, the spinal canal and neural foraminae (MODIC 1991). CT is useful in the characterisation and localisation of certain primary neoplastic processes involving the vertebrae, e.g. osteoid osteoma (Fig. 3.30).

3.13
Spiral (Helical) CT

Also known as continuous acquisition or volumetric scanning, spiral (helical) CT is the most important innovation since the invention of CT in the early 1970s (KALENDER et al. 1990; KALENDER and POLACIN 1991). The spiral method uses a continually rotating x-ray gantry with constant x-ray output and uninterrupted table movement. The slip-rings which rotate on a sliding contact or brushes are stationary, allowing the source detector assembly to rotate continuously while maintaining electrical contact. In this technique the patient is transported continuously through the gantry while data are acquired continuously during several 360° scans. This type of scanning produces scan times shorter than 1 s and as the x-ray tube rotates continuously around the patient, it produces a geometric path similar to a spiral or helix, winding around the patient. Spiral CT provides a volume or block of data of information which in turn allows images to be manipulated in ways not available in conventional CT. It is important to note, however, that the information is also acquired in thin slices and not as one block of

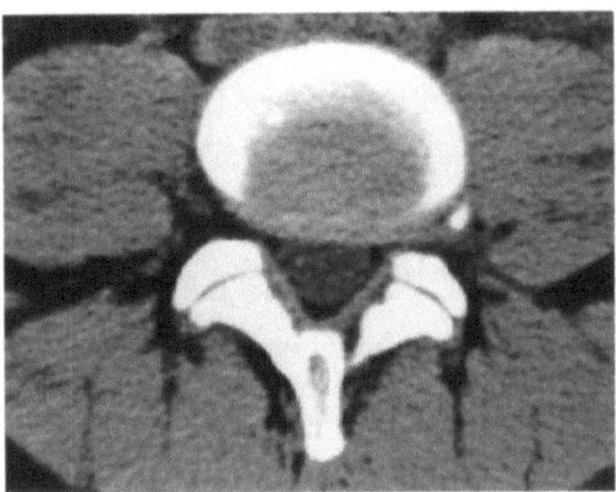

Fig. 3.29. Axial image of L4/5 intervertebral disc showing a "far out" lateral disc prolapse compressing the nerve root

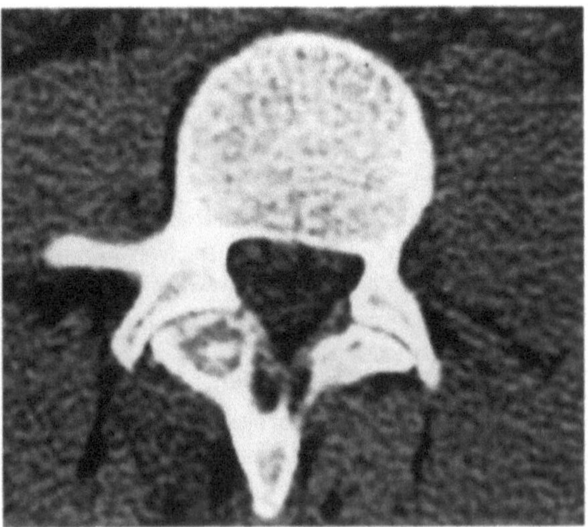

Fig. 3.30. Osteoid osteoma with calcified nidus causing expansion of the right lamina of L3

information data at any one time. Spiral CT allows one to position the slice where one wants, but not to select the slice width as well. The slice thickness cannot be made any thinner than the slice width with which the data were acquired in the first place. In the same amount of time required for conventional scanners, spiral CT can acquire a volume of data 4–9 times larger. The images produced, although virtually indistinguishable from conventional CT images, are not exactly axial. Each slice is slightly at an angle.

In the early 1990s KALENDAR and his colleagues described the fundamental principles and practical implementation of spiral CT (KALENDER et al. 1990). One of the major advantages of spiral scanning is the capability of changing the reconstructed slice incrementation. Due to the continuous acquisition of raw data, reconstructions can be carried out retrospectively at any point. Staggered slices can be created retrospectively with as little as 1 mm difference. This allows a more accurate assessment of the Hounsfield unit of a lesion, as well as its detection by reducing further the partial volume effect. Furthermore, the additional images allow the acquisition of a series of overlapping images which are very useful in producing superior three-dimensional and multiplanar reformatted images. All this is obtained without any increased radiation exposure to the patient. Although spiral CT has the capability of altering slice incrementation retrospectively from the raw data, it should be reiterated that the actual slice thickness cannot be altered as this is determined by the x-ray beam collimation utilised during the scanning process (SILVERMAN 1995).

It is important that the patient moves at a constant speed into the gantry and usually a table speed of about 10 mm per second is chosen during a continuous 1-s scan. The relation of table speed to slice thickness is referred to as *"pitch"*. With a pitch of 1 to 1 the table moves at a speed that allows all the anatomical areas to be covered. If a 24-s scan is taken, then the anatomical volume scan is 24 cm. Adjusting the pitch which is the relation of table speed to slice thickness stretches or compresses the spiral. Pitch therefore affects both slice thickness and image resolution. One of the problems that results from acquiring data using spiral CT is that there is no defined slice, which means that localisation of a particular slice is difficult and, in turn, the projection data can be inconsistent. Utilising highly sophisticated interpolation mathematical techniques, a dedicated reconstruction algorithm synthesises a perfectly planar slice from the original spiral data as well as affording the capability of selecting a slice anywhere

between the start and end positions besides the spacing and numbering of slices.

The advantages of spiral scanning are increased speed (spiral scanning usually reduces total examination time by more than 50%), less requirement for contrast medium in contrast-enhanced investigations, a significant reduction in slice misregistration, reduced motion artefacts, the ability to change slice incrementation retrospectively, and a vast improvement in the three-dimensional and multiplanar reformations. These advantages are particularly useful in the assessment of the traumatised patient, enabling multiple areas to be examined simultaneously with a significant reduction in time, and also in paediatric disorders. Spiral CT is also useful in imaging fractures because of its speed, longitudinal axis resolution, ease of use and rapid multiplanar capabilities (FISHMAN et al. 1993). In conventional CT, the scanning time usually is in the range of 4–6 min if the scanner performs between 8 and 12 scans per minute, whereas with spiral CT data acquisition requires only 24–40 s. In addition to the marked reduction in total examination time, the rapidity of data acquisition is important because it helps to minimise and prevent inadvertent motion. In fact, all the data can be acquired in a single breath-hold time. After the data have been acquired, the ability to reconstruct at any preselected interval becomes very important, especially when the clinician is scanning a small area of interest in the spine or peripheral skeleton. The benefits of spiral CT are seen in all the clinical applications of CT described above (SILVERMAN 1995; FISHMAN et al. 1993).

Spiral CT also assesses optimally the two pulmonary problems frequently encountered in musculoskeletal disorders, namely pulmonary embolism and metastatic disease. Spiral CT, by eliminating respiratory motion and minimising partial volume errors, results in a high rate of detection of nodules of a smaller size than are detected with conventional CT. Angiography is among the top applications for spiral CT and computer algorithms produce images that resemble conventional arteriograms; this technique is increasingly being applied for the exclusion or confirmation of clot within the pulmonary vasculature.

3.14
Interventional Procedures

Computed tomography is a valuable tool for use in interventional procedures such as biopsies and abscess drainage. CT-guided percutaneous procedures can be done because of precise three-dimensional localisation of lesions by CT, which also produces an access route by showing the relationship of the lesion to surrounding structures (Fig. 3.31). The tip of the needle within the structure is visualised and therefore procedures can be performed in small lesions. Improving the accuracy of the procedure diminishes the associated risks and patients can be placed in a variety of positions to allow easier access to the lesion or anatomical location. CT can be used quite effectively in helping to place the proposed injection medium in the correct place, as, for example, in nerve root sleeve injection (Fig. 3.32), chemonucleolysis and tumor ablation. Interventional techniques in neoplasms employ CT to ensure accurate biopsy localisation. This is especially useful when the lesion is not seen on a plain film and is small and deeply located, for example in the pedicles. CT-guided biopsy ensures sampling from the correct location and is especially useful when the lesion is in the spine or pelvis. The percutaneous therapy of malignant tumour, especially a recurrence, by local tumour ablation with alcohol is possible. Percutaneous therapy is also increasingly being employed under CT guidance for treatment of

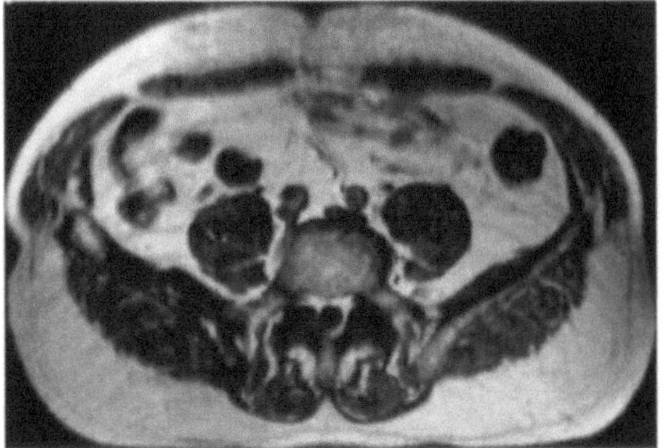

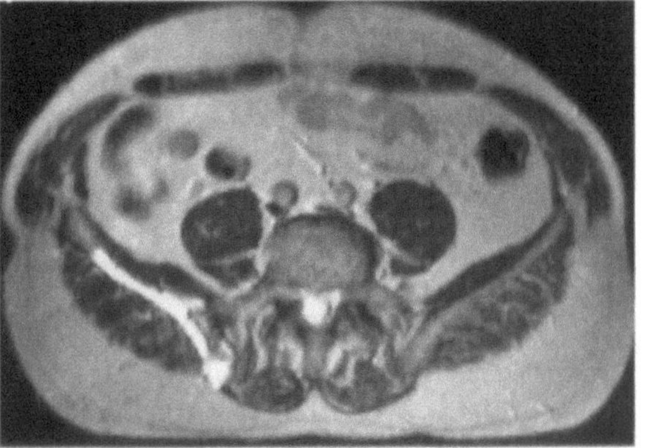

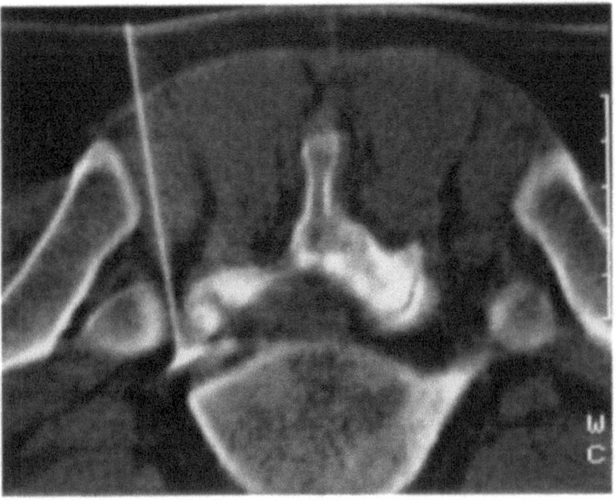

Fig. 3.31. Vertebral bone biopsy under CT guidance

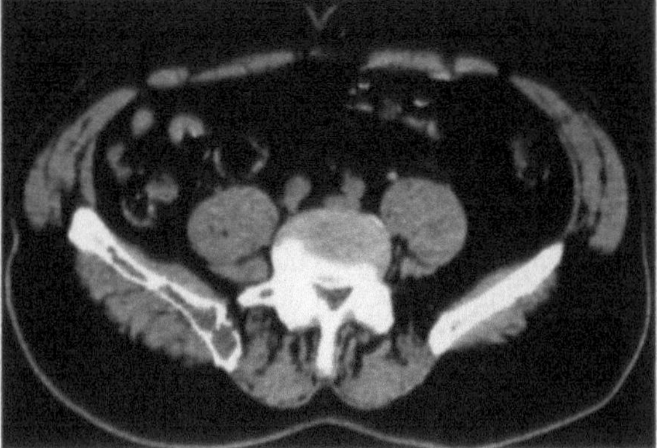

Fig. 3.32. CT-guided nerve root block of L5 with injected contrast medium in the root sleeve

Fig. 3.33. The presence of an intramedullary lesion within the right ilium is well seen by T1-weighted (a) and T2-weighted (b) MRI images but the intact nature of the cortex is best appreciated on the CT image (c)

osteoid osteoma (DOYLE and KING 1989; ROGER et al. 1996). Spiral CT can also be used to monitor the placement of drainage tubes, for example in abscesses (especially in the pelvis and around the spine), and to monitor regression with time.

3.15
Conclusion

It is essential to regard CT as simply a tool in the vast armamentarium which is now available for the investigation of patients with musculoskeletal disorders. Reliance on a single technique is not recommended as one technique very rarely provides all the answers to the questions being posed by the attending physician or surgeon. MRI and ultrasound do not employ ionising radiation and in this respect have a significant advantage over CT. However, paradoxically, because of the exquisite sensitivity of MRI to chemical changes within bone and soft tissue as a result of pathological states, the MRI features can sometimes be quite confusing (Fig. 3.33). This statement is not intended to undermine the invaluable role of MRI in orthopaedic disorders, but merely to put it into the correct context. Prudence is necessary at all times when assessing the role of these high technology modes of imaging.

References

Aitken AGF, Flodmark O, Newman DE, et al. (1985) Leg length determination by CT digital radiography. Am J Roentgenol 144:613–616

Andre M, Resnick D (1995) Computed tomography. Diagnosis of bone and joint disorders, 3rd, vol 1. Saunders, Philadelphia, pp 118–170

Baxes GA (1984) Digital image processing. A practical primer. Prentice-Hall, Englewood Cliffs, NJ

Baxter BS, Sorenson JA (1981) Factors affecting the measurement of size and CT numbers in computed tomography. Invest Radiol 16:337–341

Boven F, De Boeck M, Potvliege R (1983) Synovial plicae of the knee on computed tomography. Radiology 147:805–809

Brant WE (1986) Physics and artifacts. In: Vogler JB, Helms CA, Callen PW (eds) Normal variants and pitfalls in imaging. Saunders, Philadelphia, pp 1–3

Cassar-Pullicino VN, McCall IW, Wan S (1992) Intra-articular osteoid osteoma. Clin Radiol 45:153–160

Cormack AM (1980) Early two-dimensional reconstruction and recent topics stemming from it. Nobel award address. Med Phys 7:277–282

Dalinka MK, Arger P, Coleman V (1985) CT in pelvic trauma. Orthop Clin North Am 16:471–480

Davies AM, Cassar-Pullicino VN (1989) Demonstration of osteochondritis dissecans of the talus by coronal computed tomographic arthrography. Br J Radiol 62:1050–1055

Davies AM, Cassar-Pullicino VN, Grimer RJ (1992) The incidence and significance of fluid levels on computed tomography of oseous lesions. Br J Radiol 65:193–198

De Santis M, Crisi G, Vici FF (1984) Late contrast enhancement in the CT diagnosis of herniated lumbar disk. Neuroradiology 26:303–307

Deutsch AL, Resnick D, Berman JL, et al. (1984) Computerized and conventional arthrotomography of the glenohumeral joint: normal anatomy and clinical experience. Radiology 153:603–609

Dihlmann W (1982) CT analysis of the upper end of the femur. The asterisk sign and ischaemic bone necrosis of the femoral head. Skeletal Radiol 8:251–258

Doyle T, King K (1989) Percutaneous removal of osteoid osteomas using CT control. Clin Radiol 40:514–517

Fishman EK (1991) Three dimensional imaging. Radiology 181:321–327

Fishman EK, Wyatt SH, Bluemke DA, et al. (1993) Spiral CT of musculoskeletal pathology: preliminary observations. Skeletal Radiol 22:253–256

Gabor HT (1980) Image reconstruction from projections. Academic Press, New York

Handel SF, Lee Y-Y (1981) Computed tomography of spinal fractures. Radiol Clin North Am 19:69–89

Hernandez RJ, Poznanski AK (1985) CT evaluation of pediatric hip disorders. Orthop Clin North Am 16:513–541

Hernandez RJ, Tachdjian MO, Poznanski AK, et al. (1981) CT determination of femoral torsion. Am J Roentgenol 137:97–101

Holland P, Davies AM, Cassar-Pullicino VN (1994) Computed tomographic arthrography in the assessment of osteochondritis dissecans of the elbow. Clin Radiol 49:231–235

Hounsfield GN (1973) Computerized transverse axial scanning (tomography). Part I. Description of the system. Br J Radiol 46:1016–1022

Hounsfield GN (1980) Computed medical imaging. Nobel award address. Med Phys 7:283–290

Kalender WA, Polacin A (1991) Physical characteristics of spiral CT scanning. Med Phys 18:910–915

Kalender WA, Seissler W, Klotz E, Vock P (1990) Spiral volumetric CT with single breath-hold technique, continuous transport, and continuous scanner rotation. Radiology 176:181–183

Laasonen EM, Jokie P, Lindholm TS (1984) Tibial torsion measured by computed tomography. Acta Radiol 25:325–329

Levi C, Gray JE, McCullough EC, et al. (1982) The unreliability of CT numbers as absolute values. Am J Roentgenol 139:443–447

Lukens JA, McLeod RA, Sim FH (1982) Computed tomographic evaluation of primary osseous malignant neoplasms. Am J Roentgenol 139:45–48

Martinez S, Herzenberg JE, Apple JS (1985) Computed tomography of the hindfoot. Orthop Clin North Am 16:481–496

Meaney JF, Cassar-Pullicino VN, Etherington R, et al. (1992) Ilio-psoas bursa enlargement. Clin Radiol 45:161–168

Modic MT (ed) (1991) Imaging of the spine. Radiol Clin North Am 29, no. 4

Morin RL, Raeside DE (1981) A pattern recognition method for the removal of streaking artifact in computed tomography. Radiology 141:229–233

Norman A, Nelson J, Green S (1985) Fractures of the hook of hamate: radiographic signs. Radiology 154:49–54

Passariello R, Trecco F, DePaulis F, et al. (1983) Computed tomography of the knee joint: clinical results. J Computed Assist Tomogr 7:1043–1049

Radon J (1917) On the determination of functions from their integrals along certain manifolds. Ber Saech Akad Wiss Leipzig Math Phys Kl 69:262

Reiser M, Karpf PM, Bernett P (1982a) Diagnosis of chondromalacia patellae using CT arthrography. Eur J Radiol 2:181–185

Reiser M, Rupp N, Karpf PM, et al. (1982b) Erfahrungen mit der CT-Arthrographie der Kreuzbänder des Kniegelenkes. ROFO 137:372

Roger B, Bellin MF, Wioland M, et al. (1996) osteoid osteoma: CT guided percutaneous excision confirmed with immediate follow-up scintigraphy in 16 outpatients. Radiology 210:239–242

Romans LE (1995) Introduction to computed tomography. Williams & Wilkins, Baltimore

Schubiger O, Valavanis A (1982) CT differentiation between recurrent disc herniation and postoperative scar formation: the value of contrast enhancement. Neuroradiology 22:251–254

Seeram E (1994) Computed tomography: physical principles, clinical applications and quality control. Saunders, Philadelphia

Silver DA, Cassar-Pullicino VN, Morrissey BM, et al. (1992) Gas-containing ganglia of the hip. Clin Radiol 46:257–260

Silverman PM (ed) (1995) Helical (spiral) computed tomography. Radiol Clin North Am 33, no. 5

Teplick JG, Haskin ME (eds) (1983) CT of the lumbar spine. Radiol Clin North Am 21, no. 2

Totty WG, Vannier MW (1984) Complex musculoskeletal anatomy: analysis using three dimensional surface reconstruction. Radiology 150:173–177

Walker C, Cassar-Pullicino VN, Vaisha R, et al. (1993) The patello-femoral joint – a critical appraisal of its geometric assessment utilising conventional axial radiography and computed arthro-tomography. Br J Radiol 66:755–761

Weisz GM (1986) The value of CT in diagnosing postoperative lumbar conditions. Spine 11:164–166

Wing VW, Jeffrey RB Jr, Federle MP, et al. (1985) Chronic osteomyelitis examined by CT. Radiology 154:171–174

Young SW, Muller HH, Marshall WH (1983) Computed tomography: beam hardening and environmental density artifact. Radiology 148:279–283

4 Magnetic Resonance Imaging

H. Bonél and M. Reiser

CONTENTS

4.1
Concepts of Magnetic Resonance Physics

This chapter focuses on the most important basic principles of musculoskeletal magnetic resonance imaging (MRI). The aim is to make the phenomenon of MRI comprehensible; a more detailed presentation of MRI physics is not intended.

4.1.1
The Magnetic Resonance Process

The magnetic resonance (MR) process can be considered to be a simple reemission phenomenon. Energy is applied to a patient to be reemitted, detected, and processed.

The MR process is based on the interaction between a strong, external magnetic field (B_0) and the magnetic spin of nuclei of the tissue of interest inside the gantry. The tissue nuclei themselves act as very small magnets. When the tissue of interest is placed in the strong external magnetic field, the nuclei of the tissue are aligned along this very powerful magnetic field, producing an equilibrium magnetization of the tissue (Fig. 4.1). This tissue magnetization is then disrupted by properly tuned radiofrequency (RF) pulses. When the RF pulse is turned off, the nuclei recover ("relax") to equilibrium in the main magnetic field, and by relaxing they produce RF signals (Fig. 4.2).

The RF signals produced by tissue relaxation are proportional to the magnitude of the initial alignment, to the proton density of the tissue, and to the different rates at which nuclei of a distinct chemistry

H. Bonél, MD, Ludwig-Maximilian-Universität, Klinikum Großhadern, Institut für Radiologische Diagnostik, Marchioninistrasse 15, D-81377 München, Germany
M. Reiser, MD, Professor, Ludwig-Maximilian-Universität, Klinikum Großhadern, Institut für Radiologische Diagnostik, Marchioninistrasse 15, D-81377 München, Germany

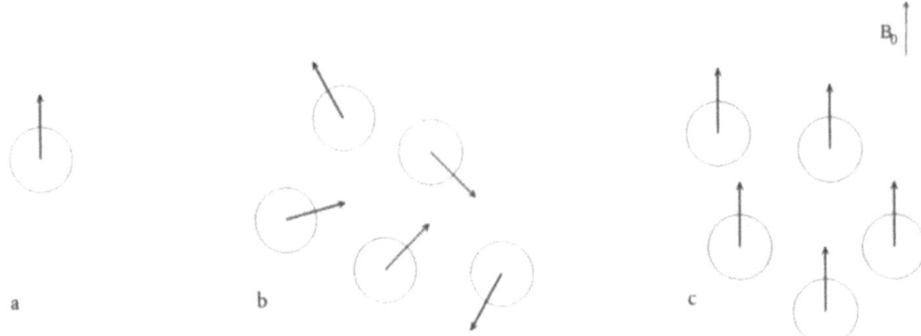

Fig. 4.1. Proton (**a**). In the absence of a strong magnetic field, the magnetic vectors of the nuclei are randomly oriented and produce no net magnetic effect (**b**). When tissue is placed in the magnetic field, some nuclei align with the strong magnetic field B_0 (**c**)

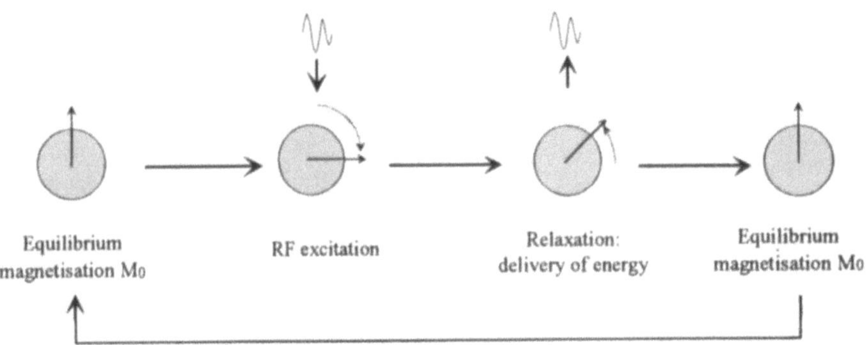

Fig. 4.2. The magnetic resonance cycle

and chemical surrounding relax. The differences in the RF signals measured can be used to calculate a gray scale for image presentation ("tissue contrast").

In order to obtain a significant difference in tissue contrast, the signals are measured, or read out, after a user-defined time has elapsed from the initial RF excitation. This time span is called the relaxation time (TR) and is – like all time measurements in MRI sequences – measured in milliseconds. The image is calculated from the signal using a mathematical process, which is called Fourier transformation.

In theory, many elements could be imaged by MR. For musculoskeletal imaging, MR is primarily applied to hydrogen (^{1}H, "protons"). As water is most prevalent in living systems, hydrogen is most abundant. Also, hydrogen produces the highest signal per nucleus. Therefore, MRI referring to hydrogen is most effective for medical purposes.

Magnetic Vectors. Before the RF vector is applied, some of the many hydrogen nuclei are aligned paral-lel to the main magnetic field (B_0). This equilibrium state is often referred to as *longitudinal* magnetization, and the tissue net magnetization is then named M_0 (Figs. 4.2, 4.3a).

When an RF pulse is applied, this longitudinal alignment is disturbed (Fig. 4.3b) and *transverse* magnetization results. The transverse magnetization can only be measured for a short period of time. This transverse relaxation time, often referred to as T2 or the spin-spin relaxation time, is dependent on the homogeneity of B_0 and the tissue composition. Spin-spin relaxation is a process that rapidly reduces after the excitation pulse. In many sequences, further RF pulses are applied during this period of transverse magnetization. After the T2 relaxation time, trans-verse magnetization has returned to 37% of its original strength (Fig. 4.4).

Different tissues vary in the time span that they need to return to complete longitudinal magnetization. The time span until net magnetization of the tissue has reached 63% of M_0 is called T1, or the spin-lattice relaxation time, because energy from

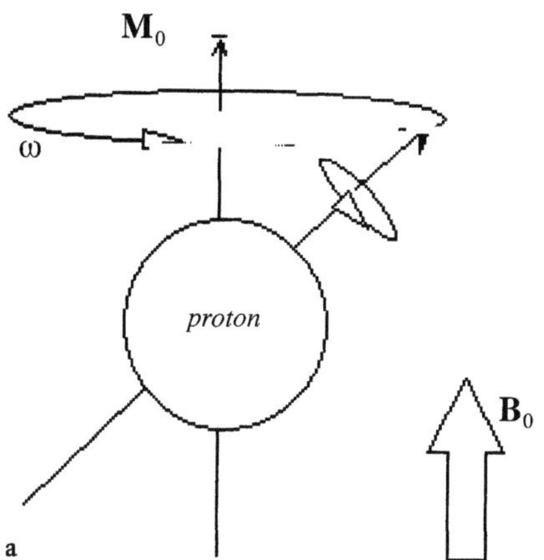

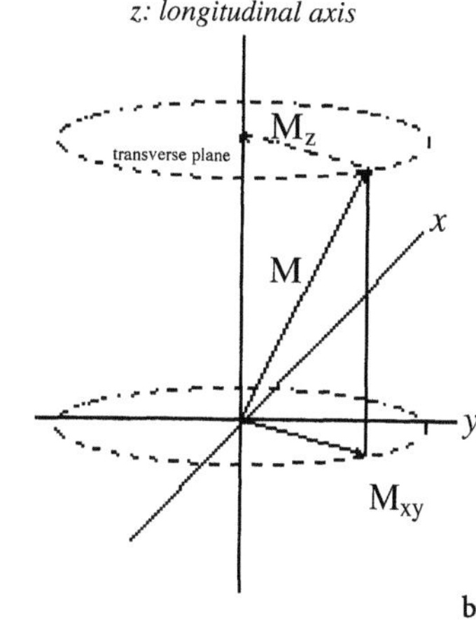

Fig. 4.3. a Magnetization under the influence of a strong magnetic field $\mathbf{B}_0$ (z-axis). The magnetic vector of the proton oscillates ("precession ω"), but aligns along the external field $\mathbf{B}_0$. The resultant magnetic vector of the proton spin in the stationary magnetic field $\mathbf{B}_0$ is called $\mathbf{M}_0$. **b** After excitation using a radiofrequency pulse, the net magnetization $\mathbf{M}$ of a proton spin can be seen as the sum of two vector components: $\mathbf{M}_{xy}$ and $\mathbf{M}_z$. $\mathbf{M}_{xy}$ represents the transverse magnetization in the XY direction, $\mathbf{M}_z$ the magnetization along the main magnetic field $\mathbf{B}_0$. $\mathbf{M}_{xy}$ is the detectable magnetization of the proton spin. During excitation $\mathbf{M}_z$ is smaller the $\mathbf{M}_0$, but it increases during RF relaxation

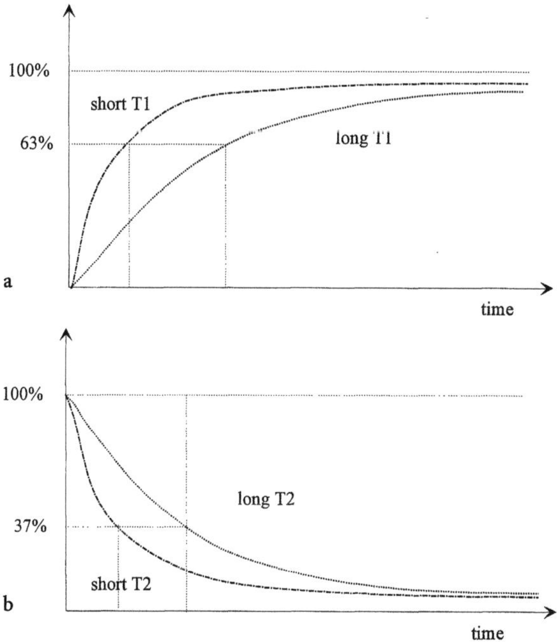

Fig. 4.4. Longitudinal (**a**) and transverse (**b**) magnetization for different tissues

RF-excited protons dissipates to their molecular environment ("lattice") (Fig. 4.1).

4.1.2
Hardware

The main components necessary for MRI imaging are the main magnet, the transmitter coils, and the receiver coils. Figure 4.5 gives an overview of these components, which are described in detail below.

4.1.2.1
Main Magnets

The purpose of the main magnet is to create a very homogeneous magnetic field. The region of the patient to be imaged has to be in the center of homogeneity of the main magnetic field. Special coils are used for adjusting the main magnetic field once the patient is inside (active "shimming" with shim coils), in order to compensate for the magnetic field of the patient and to ensure maximal homogeneity.

The magnets in routine use for musculoskeletal imaging have a field strength ($\mathbf{B}_0$) of 0.07–2.0 T. By comparison, the magnetic field of a 1.0-T system is 20 000 times stronger than that of the earth or 20 times stronger than a magnet used to hold notes, e.g., on refrigerator doors. Table 4.1 provides information on the magnets used in imaging, and examples of the three types of magnet are shown in Fig. 4.6.

Imaging systems with a field strength of 0.5 T or more are often referred to as "high-field" systems. Both high- and low-field systems have certain advantages (Table 4.2). There is a consensus that high-field systems produce images that look subjectively better. However, although multiple studies have been performed to compare images from different systems using various field strengths (LEE et al. 1995;

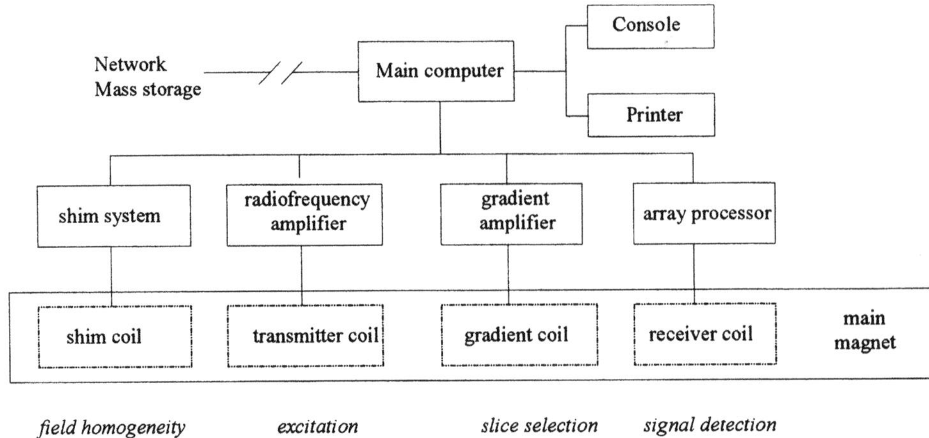

Fig. 4.5. Main components of an MRI system: the shimming, transmitter, gradient, and receiver coils form the major components along with the main magnet

Table 4.1. A comparison of magnets used in commercially available MRI systems

Type	Magnetic field strength	Important imaging features
Permanent magnet	Up to 0.3 T	Quite massive and heavy construction or small maximal field of view More susceptible to temperature changes Uses least energy of all systems Increased patient comfort (e.g., reduction of claustrophobia)
Resistive magnet	Up to 0.4 T	Electricity input rapidly increases with field strength Easy containment of stray fields Increased patient comfort (e.g., claustrophobia)
Superconductive magnet	Used for more than 0.5 T	Large stray fields, resulting in siting problems Superconductive coil must be cooled near to absolute zero (4 K) at all times High cost, including installment and maintenance

STEINBERG et al. 1990), high-field systems have not proved to be substantially better in lesion detection.

4.1.2.2
Transmitter Coils

An RF pulse perpendicular to the main magnet vector $\mathbf{B}_0$ is used to excite the protons in the tissue in question. The coil used to deliver the RF pulse is called the transmitter coil.

4.1.2.3
Receiver Coils

Receiver coils are designed to detect the MRI signal in the body part being imaged. As both noise and signal are measured with the receiver coil, it is important to match the size of the receiver coil, which determines the sensitive volume of the coil, to the volume of the body that is imaged. In order to improve the signal-to-noise ratio, the coil with the smaller diameter is usually the better choice, providing the better coil load or "filling factor." For example, a surface coil is a better choice for knee imaging than the body coil, because it picks up less noise.

Therefore, in musculoskeletal imaging, local coils or surface coils are usually preferred if applicable (Fig. 4.7). A surface coil is a receiver coil which can be placed very close to the region of interest, and therefore picks up less noise from the outside and more signal from the region being examined. Flat surface coils, for example, are best for imaging of the spine. Shoulder, elbow, wrist, knee, and ankle are frequently imaged with flexible coils, which can be

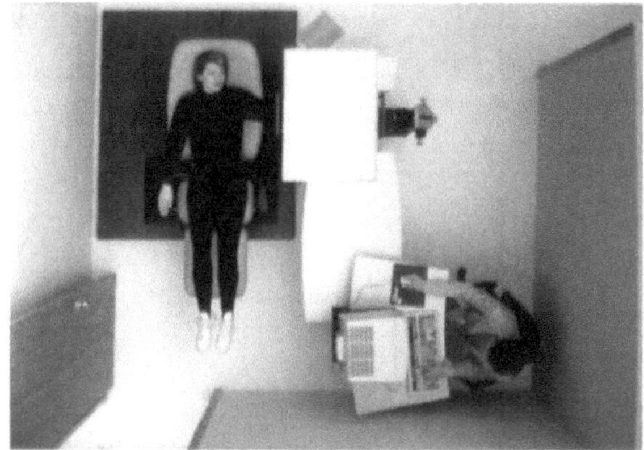
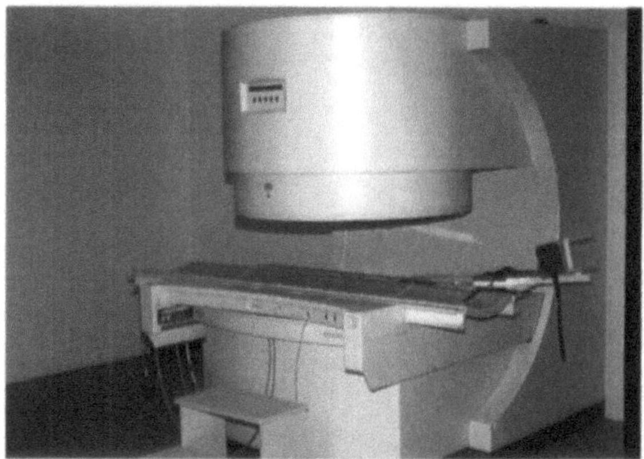
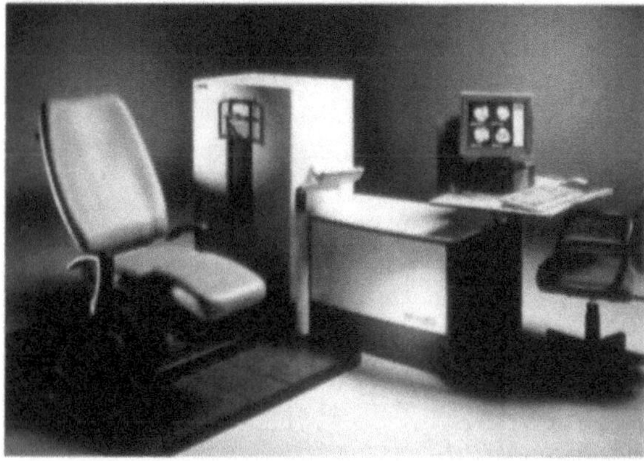
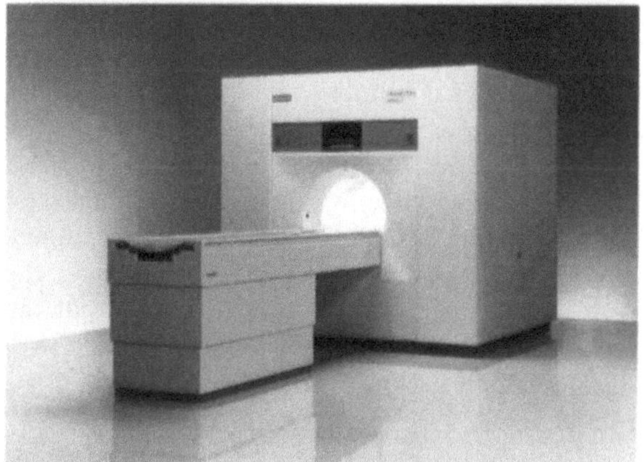

Fig. 4.6 a–d. Examples of three MRI systems used in musculoskeletal imaging. **a** Permanent magnet with a field strength of 0.2 T (Esaote Artoscan). The gantry leaves little more space than for the receiver coil and the joint being imaged. **b** Side view of the Artoscan. The orientation of the field of the 900-kg permanent magnet is transverse. **c** Resistive magnet with a field strength of 0.2 T (Siemens Magnetom Open). The magnetic field is directed transversely from the upper to the lower Helmholtz coil, and this more "open" design leaves more space for kinematic studies and avoids off-center imaging, as the stretcher can be moved in two directions between the coils. **d** Superconductive magnet with a field strength of 1.0 T (Siemens Impact). The magnetic field is oriented along the longitudinal axis of the patient (parallel to the spine)

Table 4.2. A comparison of low- and high-field systems

Low-field systems	High-field systems
Better tissue contrast	Better signal-to-noise ratio
Images appear "noisy"	Subjectively more impressive images
Small magnetic footprint	Large magnetic footprint (large magnetic stray field)
Long examination time	Shorter acquisition time
More sensitive to magnetic susceptibility	Increased readout bandwidth, decreasing susceptibility artifacts
No frequency-selective fat suppression	Larger metal artifacts
Smaller metal artifacts	High installment costs
Fewer siting problems	High maintenance costs, e.g., loss of cryogens
Lower cost	Less comfort due to noise and increased likelihood of claustrophobia
More comfortable positioning	Spectroscopy

wrapped around the body part of interest and in this way provide an optimal filling factor.

Numerous receiver coil constructions have been developed to improve the signal-to-noise ratio. *Circularly polarized (quadrature)* coils are designed to improve the signal-to-noise ratio of the detected protons. The protons of the body region being imaged have a preferred rotation direction. Coils consisting

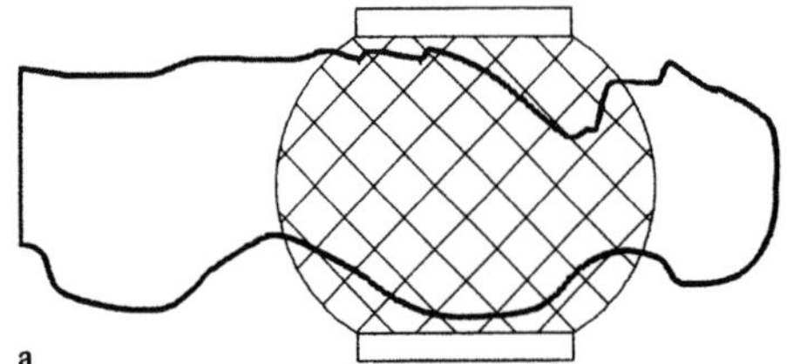
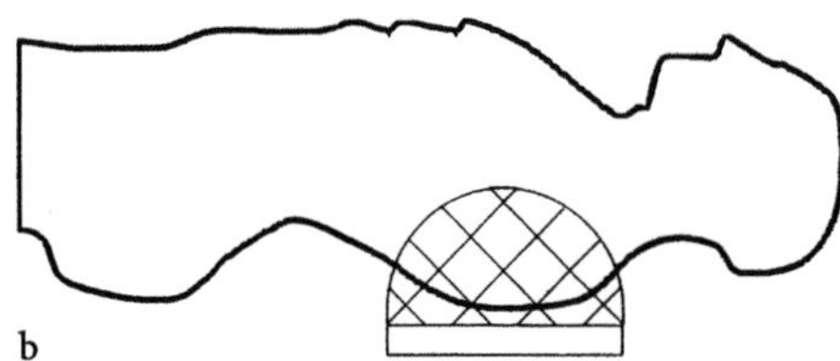

Fig. 4.7. Impact of receiver coil configuration in a clinical setting: When the thoracic spine is imaged, the body coil picks up noise from a large volume (**a**), while a surface coil picks up noise from a very small volume containing mainly the spine (**b**)

of simple wire loops, so-called linear coils, pick up the signal only from this direction, but the noise from both directions. Quadrature coils are designed to pick up the signal and noise from this direction and ignore the noise from the other direction. *Phased array or multicoil arrays* are a combined form of several coils, which only slightly overlap. Each of these coils is connected to its own receiver channel, providing the same signal-to-noise ratio over the body volume that it is able to measure. As a result, the higher signal-to-noise ratio typically obtained with surface coils is combined with the added advantage of imaging of a larger body volume. For example, images of the spine or pelvis can be obtained in minutes. However, phased array coils are more expensive and require more time for image calculation, as data of four to six channels have to be combined.

4.1.2.4
Gradient Coils

The three pairs of orthogonally arranged gradient coils are able to produce a linear variation of the main magnetic field in any orthogonal or, if coupled, oblique direction. Maximal gradient amplitudes, ranging from 10 to 30 mT/m, and gradient ramp times, reflecting the time needed to reach a stable plateau value of the gradient field, are vital to the performance of the MR system.

Loud noise is audible if the gradients are operated at their maximal capacity, which can be quite disturbing to the patient.

4.1.3
Space Encoding

Space encoding in MR systems is achieved using slice selection gradients (along the x-axis) and phase encoding gradients (along the y-axis).

Only proton vectors precessing at a certain frequency can be "flipped" by an RF pulse with this frequency. This frequency is dependent on the strength of the magnetic field that the protons are in at this moment.

The *slice selection gradient* G_z is only turned on during the application of the RF pulse. By variation of the gradient in the measured object, only a certain slice of the tissue being imaged is affected by the RF pulse (Fig. 4.8a). If the corresponding readout frequency is used, the receiver will only measure the signal from this region. As the gradient and readout frequency can be planned and are known in detail, the localization of the region which is contributing to the signal is recognized in the MRI system.

When the slice selection gradient has been applied with the corresponding RF excitation, all protons in the selected slice precess in phase at the same frequency. If a *phaseencoding gradient* G_y is switched on, the precession frequencies of the protons will be changed according to their location along the y-axis (Fig. 4.8b). When G_y is turned off, the precession along the y-axis returns to the same nominal frequency; however, the phases of the proton spin are still altered according to the relative position along the y-axis. This principle of phase encoding is repeated for the same plane: if a higher read-out matrix is desired, e.g., 512 instead of 256, twice as many *phase encoding steps* have to be measured, and imaging time is prolonged accordingly.

The *read-out gradient* G_x is switched on to vary the nuclei along the selected slice in frequency along the x-axis (Fig. 4.8c). The nuclei in the selected slice now vary in frequency along the x-axis, and in phase along the y-axis. The differences in phase are used to add the second dimension to the image information.

The two-dimensional (2D) data set obtained, often named k-space, is the basis for the Fourier transform. The Fourier transform uses the raw data matrix to produce a gray scale image, in which

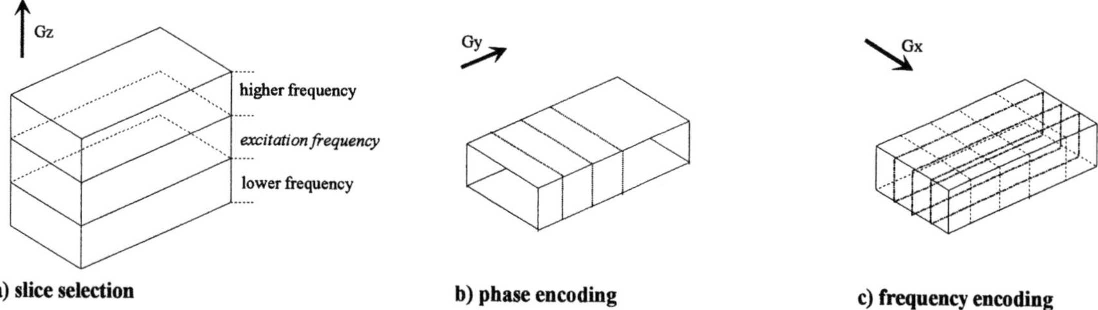

Fig. 4.8. a When gradient G_z is applied to the tissue, a slice is selected. **b** The phase encoding gradient is switched on to create different precession frequencies depending on their location. **c** During read-out, the frequency encoding gradient is applied. The protons precess at the same frequency; the phase variation, however, is retained. Many frequency encoding steps are needed

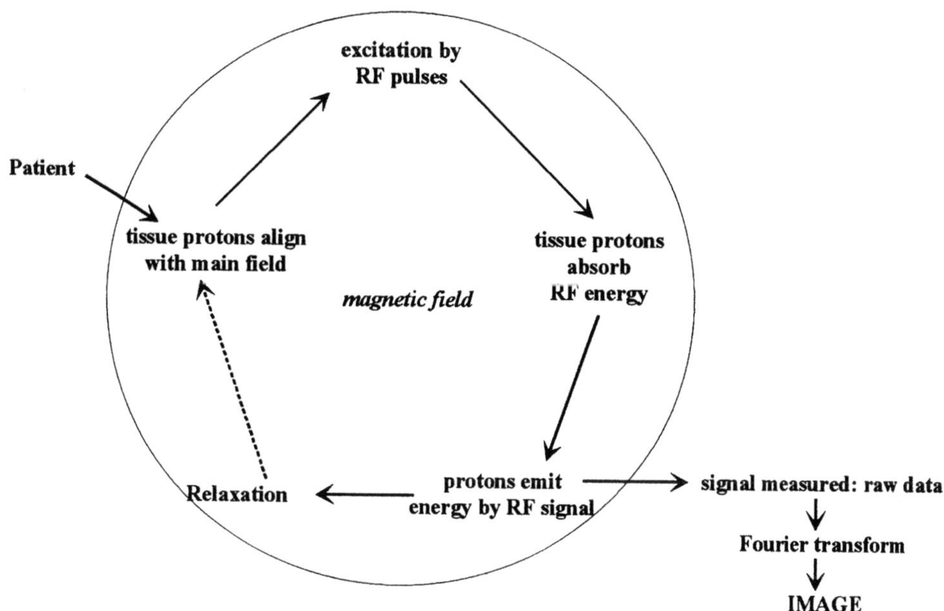

Fig. 4.9. Flowchart summarizing the steps involved in MR image formation

voxels representing higher signal intensities are depicted brighter.

In most MR systems, slices are acquired in a *sequential order*. Signal from adjacent slices is measured. Because the profile of the RF excitation pulse is not uniform, a small layer of the neighboring tissue is also excited. This overlap tissue between the slices is saturated by RF pulses from both sides, causing reduced signal from this region. This phenomenon decreases image quality and is called *cross-talk*. In order to compensate for the imperfect RF pulses, a *slice gap*, which is the space between the slices, is not measured. In most sequences, a slice gap of 10%

of the slice thickness measured will quite effectively minimize cross-talk effects. Another method to reduce cross-talk effects is to change the excitation order from sequential to interleaved.

Instead of measuring 2D slices in the described way, a larger volume can be measured. In volume or *3D imaging* an entire volume is excited by RF pulses. The sequence can be set up to have the same spatial resolution in nearly all dimensions, which means that an isotropic (cubic) voxel can be presented. Isotropic 3D sequences therefore can be reconstructed in virtually any direction in space and allow a slice thickness as small as 1 mm or even less. Signal intensity is high in comparison to 2D sequences, and a slice gap is not required.

Figure 4.9 shows a flowchart summarizing the steps involved in MR image formation.

4.2
Pulse Sequences

4.2.1
Basic Concepts

Pulse sequences can be considered measurement programs that are designed to be adjusted to the size of the musculoskeletal region being imaged, measure its tissue composition, and compose an image impression that reflects this composition and the anatomic localization.

In order to be shown in an MR image, a tissue must contain a sufficient amount of protons. For example, the proton density of the cortical bone is very low. Cortical bone therefore has very low signal in all sequences compared to other tissues and appears very dark or black on all MR images independent of the sequence used. Similarly, a healthy tendon or ligament has very low signal intensity.

If the proton density of the tissue is sufficient, it can be depicted according to tissue composition and to the signal intensity it delivers. MR sequences are designed to provide information on the chemical surrounding of the protons being measured. Numerous sequences have been developed to show fat, water, cartilage, or contrast agent at a high signal intensity in comparison with other adjacent anatomic structures.

If a sequence is sensitive to the different T1 relaxation times of the different tissues, it is named *T1-weighted*. Images using T1 contrast show body fat and paramagnetic contrast agents at high signal intensity. In *T2-weighted* sequences, water or tissues containing more water are emphasized, while fat appears of intermediate signal intensity (Figs. 4.10, 4.11). In fat suppression techniques, fat is depicted as very dark, while free water is detected easily. *Proton density* (PD = density weighted, DW) images reflect the local concentration of protons in the specific tissue, whereas water and fat are not specifically contrasted (Tables 4.3, 4.4).

Before evaluating a diagnostic image, the radiologist has to decide which tissue content is depicted at which gray level by the sequence used. Because a standard of gray levels representing signal intensities has not been defined, it is reasonable to begin with

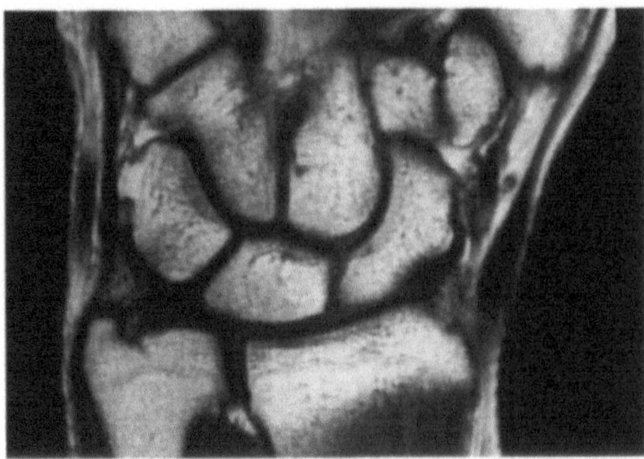

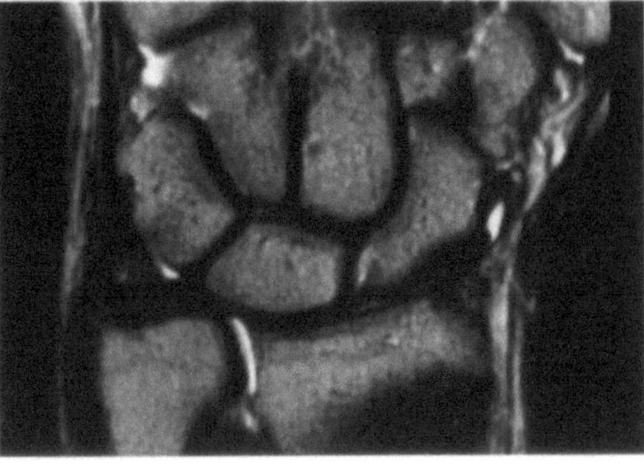

Fig. 4.10 a,b. Wrist of a healthy 27-year-old male. **a** T1-weighted image contrast depicts body fat, such as the bone marrow or subcutis, with a very high signal intensity. **b** Joint fluid becomes bright (reflecting a high signal intensity) in a T2-weighted sequence, while body fat is depicted with intermediate signal intensity

the interpretation of MR images by checking the appearance of fatty bone marrow and the subcutis, which usually contain fat, and structures containing water-like fluid, e.g., joint fluid or cerebrospinal fluid. The sequence type and parameters, and the window center and width used for printing are the main factors responsible for tissue appearance in the MR image and therefore should always be noted – and, if necessary, corrected.

4.2.2
Spin Echo

The spin-echo sequence (SE) is a standard sequence in musculoskeletal imaging. SE imaging is used to provide T1, T2, and PD contrast.

T1-weighted SE sequences are reasonably fast and provide a good tissue signal. Fatty structures, e.g., the bone marrow, are depicted with a particularly

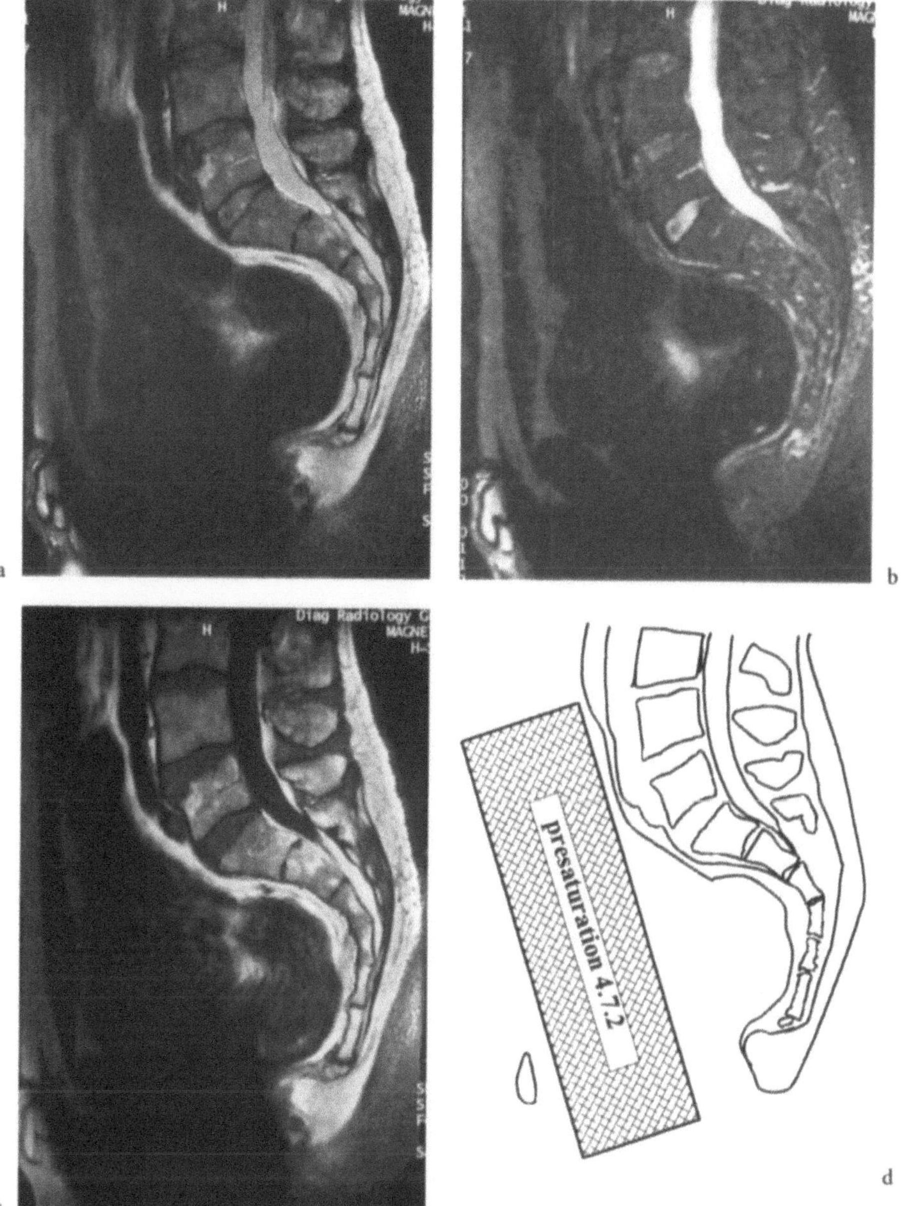

Fig. 4.11 a–d. Lower lumbar spine and sacral bone of a 55-year-old male. **a** T2-weighted sequence in which CSF appears bright, while fat is shown as intermediate gray. **b** STIR image of the same slice: Suppressed fat of the bone marrow of the vertebral bodies is displayed dark, whereas CSF again appears bright. **c** T1-weighted image for comparison: dark CSF, bright fat

fine contrast in comparison to the surrounding tissue. Excellent anatomic orientation and easy identification of hematoma and intravenous contrast are the main reasons why at least one SE sequence should be part of all musculoskeletal examinations. Relative insensitivity to artifacts is another advantage of SE imaging. Both a short TR (less than 700 ms at 1.5 T, less than 550 ms at 0.2 T) and a short TE (about 20 ms or less) are used to achieve T1 contrast in SE sequences.

If a long TR (>1000 ms) and a long TE (80–120 ms or even longer) are utilized, a *T2-weighted* contrast is obtained (Table 4.5). Tissues containing water are depicted brighter in comparison to fat or muscle. Therefore, pathologic structures often appear brighter in T2-weighted spin-echo images. For

Table 4.3. Examples of tissue contrast in MR images

	Signal characteristics	High signal intensity	Low signal intensity
T1-weighted	↑ fat ↔ water	Subcutis Fatty bone marrow	Cortical bone Ligaments Tendons
T2-weighted	↑ water ↔ fat	Joint fluid CSF	Cortical bone Ligaments Tendons
Fat suppression	↑ water ↓ fat	Edema Joint fluid	Fatty bone marrow Subcutis
Proton density/density weighted	↔ water ↔ fat		Cortical bone Ligaments Tendons

Table 4.4. Tissue relaxation times and relation to image contrast

	Appearing dark in the image	Bright
T1-weighted image	Long T1	Short T1 (**fat**, contrast agent)
T2-weighted image	Short T2	Long T2 (**water**)

Fat: short T1 and T2 relaxation times; water: long T1 and T2 relaxation times.

Table 4.5. Image contrast and sequence parameters in SE imaging

	Short TE (15–30 ms)	Long TE (80–120 ms)
Short TR (400–800 ms)	**T1 contrast**	(mixed contrast)
Long TR (1000–4000 ms)	**Proton density weighted**	**T2 contrast**

example, in recent tendinous lesions or meniscal lesions water is found in the region of the pathology. Because a high TR is necessary, T2-weighted SE sequences take a long time. Motion or pulsation artifacts are common. Since faster sequences have been able to provide T2-weighted images of the same quality in a more reasonable examination time, SE sequences are used less frequently than the newer T2-weighted fast-spin-echo or gradient-recalled echo sequences, which are described below.

If a short TE (20 ms) is combined with a long TR (1800–3000 ms), *proton density* (PD) contrast results (Table 4.5). In musculoskeletal imaging, PD-weighted images can be useful in tumor diagnosis: the tumor is appreciated well and the surrounding soft tissue edema is depicted at high signal intensity. Although SE PD imaging is still in common use in neuroradiology, in musculoskeletal imaging newer sequences are usually preferred.

4.2.3
Fast Spin Echo

Turbo-spin-echo (TSE) sequences, also called fast-spin-echo (FSE) sequences, differ from SE sequences in echo measurement: many echoes are measured in each TR cycle ("echo train"). The longer the echo train, the shorter the sequence. In medical imaging, echo trains of 3–16 have been approved.

Because the examination time can be drastically reduced, especially in T2-weighted imaging, T2-weighted TSE sequences are commonly utilized, particularly in high-field MR systems. Image contrast differs from T2-weighted SE images: fat appears considerably brighter in TSE sequences. Pathologic processes adjacent to fat are more difficult to identify. Nevertheless, TSE imaging provides sufficient contrast for musculoskeletal imaging (VAHLENSIECK et al. 1994). In some MRI systems, PD and T1-weighted images can also be obtained using a TSE sequence, and these sequences are usually referred to as "multi-echo" sequences.

4.2.4
Gradient Echo

In gradient-recalled echo (GRE) sequences, the flip angle (FA) is used in addition to TR and TE to determine T1- or T2-weighted image contrast. SE sequences use a pre-set FA of 90°. In GRE sequences, an FA different from 90° can be chosen (Table 4.6). As a rule, a small FA (e.g., 20°) results in T2 contrast, and a larger FA (e.g., 75° or more) in T1 contrast. If the FA is smaller than 90°, the return to complete longitudinal magnetization takes less time. This permits shorter acquisition cycles. GRE sequences are usually shorter than SE acquisitions.

4.2.4.1
Susceptibility

All materials have a tendency to counteract the presence of a magnetic field. The "magnetizability" of a material is determined by this counteraction and is referred to as *magnetic susceptibility*, that is the tendency of the material to distort an applied magnetic field.

In musculoskeletal MRI, susceptibility is detected at air-tissue or fat-water interfaces. The accompanying field inhomogeneities produce dephasing, which results in T2* signal loss. In SE imaging, rephasing pulses are used to produce T2 signal. In GRE sequences, the rephasing pulses are not applied, and due to the energy loss because of tissue susceptibility only a smaller signal, named T2*, can be measured (Fig. 4.12).

At low field strengths, effects of magnetic inhomogeneities are less significant than in high-field imaging. Therefore, there is a smaller difference

between T2 and T2* relaxation. At low field strengths, long TE gradient-echo images show similar contrast to T2-weighted SE sequences (PETERFY et al. 1997), and especially bone marrow pathologies are detected more easily.

4.2.4.2
More Advanced GRE Sequences

To reduce acquisition time, more advanced GRE sequences have been developed. Table 4.7 summarizes the GRE sequences and gives examples of acronyms frequently used of these sequences.

4.2.4.2.1
STEADY-STATE GRE
Steady-state GRE sequences use a rewinder-gradient to establish an equilibrium between longitudinal and transverse magnetization. T1 and T2 contrast is determined by the ratio of the T1 and T2 relaxation times. Mixed-weighted images result if short repeti-

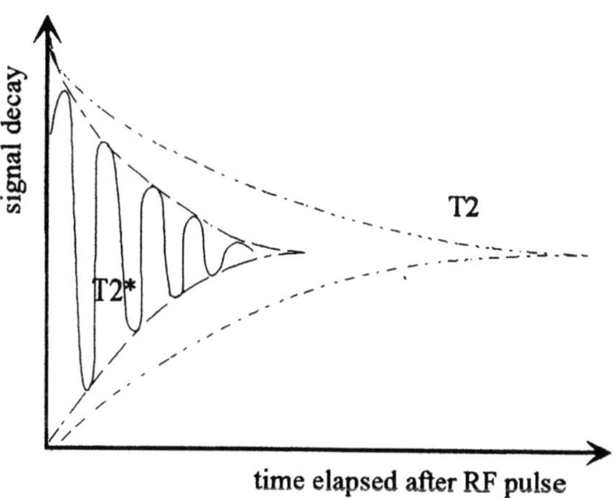

Fig. 4.12. Signal loss in T2* (GRE) is much faster than in T2 (SE) owing to large molecule interactions increasing transverse spin dephasing

Table 4.6. Image contrast in GRE sequences: most important criteria shown in italics

Contrast	Preferred GRE type	Examples of flip angle	Examples of TE
Proton density		*Small FA: 5–15°*	Short TE: 8–15 ms
T1		*Large FA: 45–90°*	8–15 ms
T2	*Contrast-enhanced GRE*		
Mixed T2/T1	*Steady-state GRE using a short TE*		
T2* (effective T2)		5–20°	*Long TE: 30–60 ms*

Table 4.7. Acronyms for GRE imaging as used by producers

	Basic GRE	Spoiled GRE	Steady-state GRE	Contrast-enhanced GRE
Siemens	–	FLASH	FISP	PSIF
General Electric	MPGR	SPGR	GRASS	SSFP
Philips	–	CE-FFE T1	FFE	CE-FFE T2
Picker	FE	PSR	FAST	CE-FAST
Elscint	–	SHORT	F-SHORT	E-SHORT
Toshiba	PFI	–	FE	–

tion (<250 ms) and echo times are combined with intermediate flip angles (15–40°). Using very small flip angles (<5°), proton density contrast is obtained. Longer echo times produce T2-weighted contrast, while large flip angles (>40°) produce T1-weighted contrast.

4.2.4.2.2
SPOILED GRE

An RF pulse or a gradient is used to nullify, or "spoil," residual transverse magnetization. Spoiled GRE sequences therefore are deficient of a T2 equilibrium state that can be used for T2-weighted imaging, and they are commonly used with large flip angles and short echo times to obtain T1-weighted image contrast.

4.2.4.2.3
CONTRAST-ENHANCED GRE

This sequence produces a relatively strong T2-weighted contrast and is not in widespread use because of its relatively low overall signal-to-noise ratio.

4.2.5
Inversion Recovery
and Short Tau Inversion Recovery

Inversion recovery imaging uses a prepulse in conjunction with either a GRE or an SE sequence. The *inversion prepulse* is applied primarily to influence T1 relaxation effects. Structures depending primarily on T1 relaxation can be emphasized (T1-weighted inversion recovery) or suppressed (short tau inversion recovery).

An inversion pulse inverts net magnetization: a magnetization along the positive z-axis is brought to the negative z-axis. When this inversion pulse is switched off, the net magnetization increases along the z-axis, starting from maximum negative value, passing through zero, and then increasing along the positive direction.

While the longitudinal magnetization is increasing from $-M_0$ to M_0, a signal generation scheme is applied. The combination with an SE acquisition is most commonly used and is referred to as an "inversion recovery" sequence (IR). Especially in low-field systems, however, a GRE acquisition might be used instead. The time between the cessation of the inversion pulse and the start of the image acquisition is named the inversion time (synonyms: tau, TI). TI determines tissue contrast (Fig. 4.13).

By proper choice of the inversion time, fat signal can be nullified. Because of the relatively short inversion time, this sequence has been named *STIR*, or short inversion time recovery sequence. Fat has relatively short T1 and T2 relaxation times. When a short TI is used, most tissues still have negative magnetization, because their T1 relaxation time is longer. As these tissues also have a longer T2 relaxation time, their transverse magnetization contributes more signal strength (Fig. 4.13). This results in attractive contrast. The applications of STIR imaging in musculoskeletal diagnosis are therefore multiple: for example, the very high signal from fatty bone marrow can be nullified to show the accompanying bone edema in stress fractures, inflammation, or tumors (VAHLENSIECK et al. 1993).

4.2.6
Chemical Shift Selective Saturation

Chemical shift selective saturation (CHESS) techniques use a saturation pulse shortly before the signal measurement. Protons in fatty tissue are saturated by crusher impulses and therefore do not contribute to the signal measured. CHESS techniques can be part of any pulse sequence, but are difficult to establish in the presence of magnetic inhomogeneity or at a low field strength.

CHESS techniques show cartilage with good contrast against suppressed fatty bone marrow. Furthermore, fatty tumors can be differentiated from hematoma or blood. CHESS imaging has not

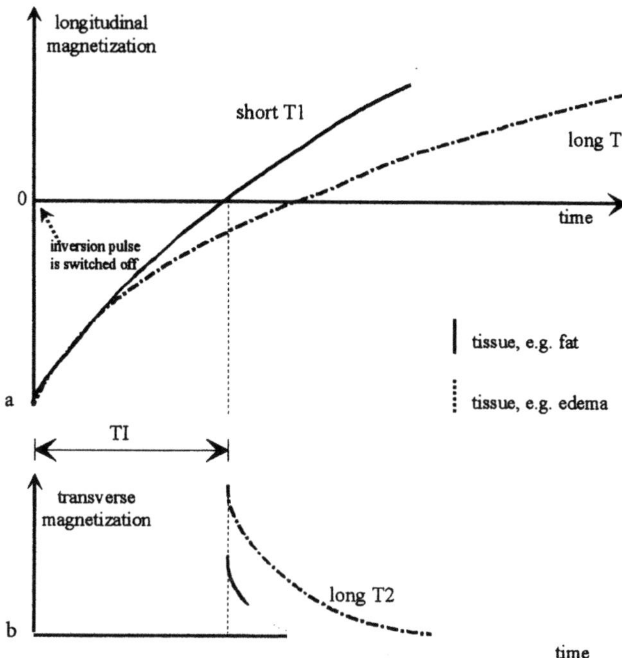

Fig. 4.13 a,b. Inversion recovery imaging: T1 and T2 relaxation curves. **a** Longitudinal magnetization of two different tissues. During the inversion pulse, longitudinal magnetization is M_0. When switched off, longitudinal magnetization increases at different rates for different tissues. **b** Transverse magnetization of the two tissues: relevant T2 relaxation starts after the inversion time (T1)

yet become established for routine musculoskeletal imaging.

4.2.7
Magnetization Transfer Imaging

In musculoskeletal tissues, only spins from free water contribute to measurable MR signal: spins from bound water do not contribute to the MR signal. Different tissues contain different amounts of water in the free (contributing) and bound pool.

In magnetization transfer (MTC) imaging, high-intensity, off-resonance pulses with a narrow bandwidth are utilized to saturate the bound water pool with little effect on the free water pool (Fig. 4.14). A new equilibrium is created due to cross-relaxation between the free and bound water pool. This decreases longitudinal magnetization in the tissue examined, and the T1 relaxation time is shortened while the T2 relaxation time remains unchanged. After the MTC pulse, a standard sequence, for example GRE or SE, is used for signal measurement.

In musculoskeletal imaging, MTC imaging exerts its influence mainly on muscle and cartilage; tissues like fat or blood are hardly changed in contrast.

Table 4.8. Fast and ultrafast sequences

Sequence type	Acronyms	Acquisition time per slice
Fast gradient echoes	Turbo-Flash Snapshot-GRE	1–5 s
Combined gradient and spin-echo	GRASE	100–300 ms
Echo-planar imaging	EPI	50–100 ms

4.2.8
Ultrafast Sequences

Ultrafast sequences are rarely used in musculoskeletal imaging. The very short acquisition times of these sequences allow, for example, real-time kinematic examinations in joint pathologies or contrast dynamic examinations in tumor patients. Table 4.8 shows the typical acquisition times.

In fact routine examinations require anatomic detail more than kinematic impression, and few high-field scanners are equipped with EPI or GRASE sequences. Ultrafast sequences are therefore rarely used in musculoskeletal imaging.

4.2.9
Suggested Sequences

Articular or periarticular abnormalities (Table 4.9) can be identified using double-echo T2-weighted sequences in two planes. Axial plus, either coronal

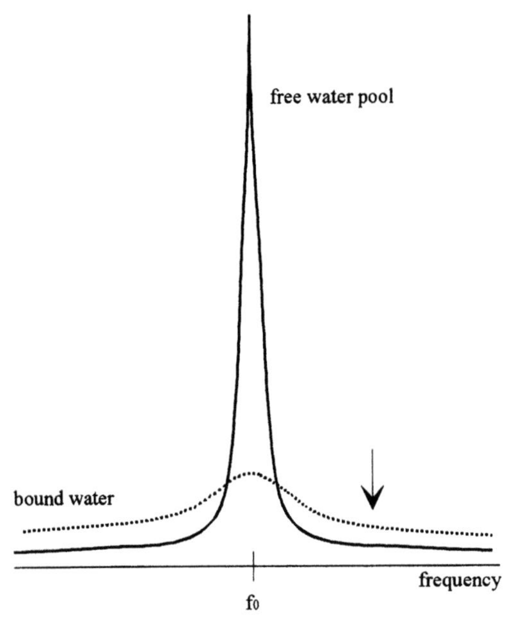

Fig. 4.14. Magnetization transfer imaging. The spectral lines of both the free and the bound pool are centered at the same Lamour frequency f_0. High-intensity, off-resonance pulses with a narrow bandwidth selectively destroy signal from the bound pool by using differences in line width (*arrow*)

or sagittal planes (depending on the region and suspected pathology) are suggested. Radial or oblique planes may be needed in addition. Also, T1-weighted SE sequences give a very good anatomic impression. A joint pathology, however, is usually more easily detected on T2-weighted sequences, as abnormal areas often contain more water and appear of high signal intensity. In addition, joint fluid has a high signal intensity on T2-weighted images.

Standard T1- and T2-weighted SE and TSE sequences are often inconclusive for *osseous abnormalities*. So, in addition, fat-suppressed images are used. STIR sequences produce additive effects of T1- and T2-weighted sequences. By suppressing fat signal, they provide better contrast between normal, suppressed (low signal), and pathologic bone marrow. Fat suppression techniques providing a T2-weighted effect are also very useful.

Fat-suppressed T1- and T2-weighted sequences are preferred for early diagnosis of ischemic bone disease. Intravenous contrast provides additional information in such cases.

For the examination of *cartilage*, GRE sequences are suitable. Special steady-state GRE sequences have been designed for cartilage evaluation. Whenever patient motion is a problem or kinematic examinations have to be performed, GRE sequences are beneficial.

Three-dimensional sequences have the added advantage of allowing volume reconstruction, very thin slices, and, depending on the voxel symmetry, free reconstruction in virtually any direction. Combined

Table 4.9. Suggested pulse sequences for examination of articular or periarticular disorders (BERQUIST 1993)

Indication	Pulse sequence	Contrast
Articular cartilage	GRE: FA 55–60° Fat-suppressed 3D GRE	
Bone marrow	T1-weighted spin-echo: TR 500 TE 20 STIR T2-weighted fat-suppressed sequences	✓
Dynamic studies	GRE: FA 90°	✓
Kinematic studies	GRE: FA 20°	
Menisci Labral abnormality	PD-weighted spin-echo: TR 2000 TE 20	
Metal artifacts	T1-weighted SE: low TR and TE	
Muscle Ligament Tendon	PD-weighted spin-echo: TR 2000 TE 20 T2-weighted spin-echo: TR 2000 TE 80 T2-weighted fat-suppressed sequences STIR	
Synovial disorder	T1-weighted spin-echo: TR 500 TE 20	✓

with fat saturation, 3D GRE steady-state sequences are state of the art for cartilage evaluation [3D spoiled GRASS (General Electric) or DESS (Siemens)] in high-field systems. Three-dimensional reconstruction of the menisci, cruciate ligaments, rotator cuffs, and wrist ligaments are used for superior depiction (DISLER et al. 1993; TOTTERMAN et al. 1996).

Synovial pathologies are more obvious with intravenous contrast. Both synovial proliferation and enhancement are depicted using contrast. Dynamic gadolinium-enhanced studies are useful in the evaluation of *musculoskeletal neoplasms.*

4.3
Kinematic MRI

Various joint pathologies cause pain only in certain positions of the joint. Kinematic fluoroscopy, providing projection radiography images, is a proven method for the clarification of these pathologies. However, soft tissue, providing the major support to the joint, is not depicted.

Kinematic MRI combines the advantages of better soft tissue presentation, functional aspects, and the opportunity to voluntarily position multiple slices. The major drawbacks of kinematic MRI are the long acquisition times, which result in long examination times and do not allow the same real-time motion as fluoroscopic examinations, and the design of most MR systems, which allows only a limited range of motion.

Numerous positioning devices have been developed to allow reproducible *static images* acquired in different angles. Combination of images obtained in the same frame but at different angles to yield a sequential loop (*"cine mode"*) is the basis for the evaluation of these functional images. For this type of examination, GRE or SE sequences are utilized. For all major peripheral joints and for the mandibular joints and the spine, first experiences have been published (MUHLE et al. 1995). Use of open or half-open MR systems allows a complete range of motion for most joints; however, the lower field strength of these systems prolongs the examination time (MINAMI et al. 1991).

Motion-triggered MRI examinations use a sensor that allows motion-free images during continuous active (slow) joint movement. Standard or fast GRE techniques are most commonly employed (MELCHERT et al. 1992). Examination time is shortened, and stress examinations are possible.

Ultrafast sequences could in the future allow near real-time examinations. These sequences are, unfortunately, confined to high-field systems and therefore have the disadvantage of a limited range of motion.

4.4
Contrast Agents

Ions or molecules with at least one unpaired electron generate, when placed in a magnetic field, a magnetic momentum which is 4–8 times stronger than that of the proton nucleus and aligns with the main magnetic field. Depending on their composition, these molecules are called ferromagnetic or paramagnetic. In an MR setting, these substances can be used as components of relaxation-enhancing contrast agents.

4.4.1
Agents and MR Properties

In medical MRI, the paramagnetic compounds gadolinium (Gd^{3+}), iron (Fe^{2+}, Fe^{3+}), and manganese are of greatest interest. All of the above are effective in accelerating T1 relaxation. Gadolinium has gained the most widespread acceptance. As a free ion, gadolinium is acutely toxic at clinically relevant doses.

Chelation to molecules both reduces toxicity and determines the pharmacologic properties of the contrast agent. Diethylene triamine penta-acetic acid combines with gadolinium to form a complex of strongly reduced toxicity (Gd-DTPA) which permits intravenous injection in routine clinical examinations. For most organ systems, including the musculoskeletal, it has very similar distributional properties to the common iodinated contrast agents used in radiography. Table 4.10 lists the most important contrast agents used in musculoskeletal MRI.

Administration in the recommended dose according to body weight will shorten the T1 relaxation in tissues, increasing signal intensity where paramagnetic contrast is accumulated (Fig. 4.15). Therefore, a T1-weighted sequence best demonstrates contrast enhancement. In a typical examination protocol, a T1-weighted SE sequence is used to acquire pre- and postcontrast images using the same imaging parameters (e.g., TR 500 ms, TE 20 ms). In comparison, T2-weighted sequences show, little contrast effect and are therefore rarely used after contrast. The effect of contrast enhancement increases with field

Table 4.10. Approved contrast agents used in musculoskeletal MRI

Chemical name	Abbreviation	Example of trade name
Gadopentetate dimeglumine	Gd-DTPA	Magnevist
Gadodiamide	Gd-DTPA-BMA	Omniscan
Gadoteridol	$Gd-HP-DO_3A$	ProHance
Gadoterate meglumine	Gd-DOTA	Dotarem

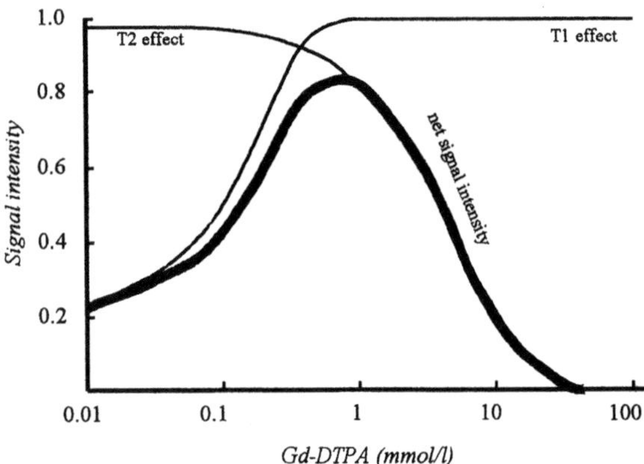

Fig. 4.15. Effect of Gd-DTPA concentration on signal intensity. The *thin lines* show T1 and T2 shortening by varying Gd-DTPA concentrations; the *thick line* shows the calculated net signal intensity. Maximal enhancement occurs at approximately 1 mmol/l. (Spin-echo TR = 250 ms, TE 28 ms, frequency = 10 MHz). (Adapted from ENGEL et al. 1990)

strength (CHANG et al. 1994). As in STIR imaging the inversion time is adjusted to tissue properties before contrast, STIR sequences must always be used before administration of a contrast agent.

Inversion recovery sequences are most sensitive for the detection of contrast at medium inversion time between the lesion T1 before and after contrast.

4.4.2
Pharmacology

After intravenous administration, the contrast agent rapidly equilibrates in the intravascular and interstitial compartments, which together form that part of the extracellular space which is important for contrast media distribution in musculoskeletal imaging. Passive diffusion or even specific uptake processes are the basis for intracellular distribution and

depend on the molecular structure of the contrast agent.

The excretion pathway is also determined by the chemical structure of the molecule complex (chelate). Hydrophilic low-molecular-weight chelates do not bind to plasma proteins and are eliminated from the body by nonspecific glomerular filtration. Molecules with both hydrophilic and hydrophobic properties are partly taken up by the liver and excreted into the bile. The greater the lipophilicity, the greater the distribution into fat storage sites, cell membranes, and reticuloendothelial cells in liver and spleen and the slower the clearance. Complete and quick clearance, however, is most desirable to minimize toxicity. Therefore, in musculoskeletal imaging hydrophilic agents (Table 4.10) are the agents of choice. Fast extracellular distribution and efficient renal excretion are the major characteristics of this important class of MR contrast agents. In their distribution properties, hydrophilic MR contrast agents resemble iodinated contrast agents used in computed tomography (CT) as well as more analogous DTPA complexes of technetium-99m or indium-113 used in scintigraphic studies (HEINDEL et al. 1978). Tissue perfusion, capillary properties, and kidney function are the major determinants of MR contrast enhancement.

4.4.3
Administration

4.4.3.1
Intravenous Injection

All MR contrast agents shown in Table 4.10 are approved for intravenous use. Body weight is used to calculate the correct dose, e.g., 0.2 ml Magnevist per kilogram body weight approximates the preferred concentration of 0.1 mmol/l in the extracellular space. After intravenous administration, a time lag before renal elimination suffices to repeat precontrast T1-weighted sequences.

In contrast dynamic imaging, a short sequence (usually 30 s to 2 min) is performed once before and repeatedly after intravenous contrast injection for a certain period of time (e.g., 8 min). A signal-versus-time curve is obtained and is useful in certain indications.

4.4.3.2
Direct and Indirect Arthrography

Direct injection of diluted contrast medium in the joint space produces a positive contrast of the joint space. The whole joint compartment appears bright on T1-weighted sequences. Intra-articular structures forming the surface of the joint space, such as the rotator cuff, cartilage, or intra-articular ligaments, are visualized with lower signal intensity and are depicted darker.

After intravenous injection in the standard dose, only a very small proportion of the MR contrast medium distributes to the joint space. The intra-articular enhancement reaches its climax after 1 h and can be augmented by physical activity. The intra articular accumulation of contrast agent is detected by T1-weighted sequences, especially if fat-suppressed sequences are used. In *indirect arthrography*, too, intra-articular surfaces are contrasted, and the joint fluid mixed with the small amount of the contrast agent creates a positive contrast.

4.4.4
Indications

In musculoskeletal imaging, intravenous contrast is not routinely indicated. Established indications for *intravenous contrast* are:

1. Chronic inflammatory processes
2. Differentiation of liquid versus solid and edematous versus infiltrative tissues, e.g., in malignant tumors
3. Assessment of bone vitality, e.g., in scaphoid fractures

Clinical trials on *dynamic* MR examinations have been performed using fast GRE sequences in patients with musculoskeletal tumors (ERLEMANN et al. 1989). Dynamic sequences can be of use in the differential diagnosis of malignant from benign tumors; the specificity, however, is low (MIROWITH et al. 1992).

The *intra-articular* use of gadopentetate dimeglumide is still being explored (HAJEK et al. 1987; WINALSKI et al. 1993). Increased sensitivity has been shown for lesions of the rotator cuff and glenoid labrum or cartilage lesions of the knee joint (HODLER et al. 1992; PALMER et al. 1993). Indications have to be restricted because of the invasiveness of the procedure, and superiority over conventional arthrography has yet to be proven. Few studies have so far been completed on *indirect arthrography*. Superiority over conventional MR examinations has been shown for the shoulder, knee, and ankle (HODLER et al. 1992; VAHLENSIECK et al. 1995, 1996; DRAPÉ et al. 1993a,b). As there is no direct injection of fluid in the joint, distention of the joint capsule is minimal in indirect arthrography. But because this distention is an important contributing factor for arthrographic joint evaluation, it could be the limiting factor of indirect arthrography, e.g., in rotator cuff examination.

4.5
Image Quality

An MR image is composed of pixels, which are the smallest elements of the image. The signal intensity measured in the voxel, the smallest volume unit considered by MR measurement in the tissue, determines the brightness of the corresponding pixel presented. The greater the signal intensity emitted by the voxel, the brighter the pixel.

The ideal MR image would have high signal, high spatial resolution, high contrast, low noise, and no artifacts. These ideal criteria have to be balanced against a reasonable examination time and the image quality required for a diagnostically valuable image which shows the anatomic features and depicts the pathology with detectable contrast and spatial resolution.

4.5.1
Spatial Resolution

The size of the voxel is adjusted by setting the field of view, matrix size, and slice thickness. Spatial resolution determines the ability to resolve anatomic detail.

Signal intensities in a voxel are averaged, and details within a voxel are lost. The smaller the voxel, the better fine detail is resolved; however, image contrast decreases when a smaller voxel volume is used.

4.5.2
Signal-to-Noise Ratio

The brightness of each pixel is determined by the absolute signal intensity measured in the corresponding voxel and the noise detected by the receiver coil. The background noise is the limiting factor in MRI. For quantification, average signal is divided by the standard deviation of background noise, resulting in a value named the signal-to-noise ratio (SNR). Data processing procedures, such as filtering, have comparatively little effect in noise reduction.

Signal averaging is one of the methods used to increase SNR. Voxels are measured more than once and the signals from these successive measurements are summed. The total signal intensity increases approximately linearly with the number of measurements. If the number of excitations (NEX) is increased from 1 to n, SNR increases by $\sqrt{n}$ and scan time increases by n.

Signal intensity increases as the square of the magnetic field strength, whereas noise only increases linearly. In a similar way, other hardware parameters (such as RF coils and receiver coils), pulse sequence timing (TE, TR, flip angle), and the voxel size exert a considerable influence on SNR.

Tables 4.11 and 4.12 give an impression of the influence of the aforementioned factors and the magnitude of this influence.

In practice, once pulse sequence timing has been optimized in an MRI system, overall examination time, number of repetitions, SNR, and spatial resolution have to be balanced.

4.5.3
Image Contrast

Contrast is defined as the signal intensity difference between two tissues. In addition to spatial resolution and SNR, contrast is vital for clinical MRI. Apart from the intrinsic composition of the sample, image contrast is influenced by adjustable pulse sequence these parameters. As the influence of these parameters has already been discussed (see Sect. 4.2), tissue factors are focused on here.

Apart from the T1 and T2 relaxation times of the specific tissue, proton density is probably the most important factor influencing image contrast. As the net magnetization present in a voxel depends on the number of "mobile protons" in this voxel, the number of protons determines its maximal possible signal intensity. Mobile protons, present in water, fat, and the hydration layer of proteins, contribute measurable signal intensity. Immobile protons, found in solid materials such as cortical bone and some fibrous tissues, contribute little.

Table 4.11. Pulse sequence parameters and their influence on SNR

Pulse sequence parameter	Effect on SNR of a decrease in the parameter	Effect on SNR of an increase in the parameter	Other effects
Relaxation time (TR)	SNR $\downarrow$	SNR $\uparrow$	Imaging time $\uparrow$ and number of sections $\uparrow$ and less T1 if $\uparrow$
Echo time (TE)	SNR $\uparrow$	SNR $\downarrow$	More T2 if $\uparrow$
Number of excitations (NEX)	SNR $\downarrow$	SNR $\uparrow$	Imaging time
Slice thickness	SNR $\downarrow$	SNR $\uparrow$	Image detail
Matrix size	SNR $\uparrow$	SNR $\downarrow$	Spatial resolution

Table 4.12. Parameters influencing SNR: approximate quantitative influence

Parameter	Effect on SNR	Acquisition time
Number of excitations doubled	141%	Doubled
Bandwidth in frequency-encoding direction decreased to one-half	141%	–
Matrix size doubled: increased from 128 × 128 to 256 × 256	35%	Doubled
Slice thickness doubled	141%	Decreased by a smaller number of excitations

T1 and T2 relaxation times act as modulating factors and determine whether tissues with a high density of protons are measured with higher or lower signal intensity. If a tissue produces very little signal intensity, it can be contrasted with surrounding structures richer in protons.

4.5.4
Coil Centering

In most systems, the area of highest homogeneity is situated in the center of the main magnet. Maximal signal intensity, minimal image distortion, and the preferred region for the effect of RF and gradient coils are the reasons why coil centering is essential in MRI. Few high-field systems allow far off-center imaging.

4.6
Fast MRI

4.6.1
Reduced Number of Excitations

Single-excitation imaging can be routinely used in high-field systems, whereas in low-field systems the intrinsically lower signal-to-noise ratio usually requires more than one excitation. However, the loss of signal-to-noise caused by fewer acquisitions can often be compensated for by using a slightly higher TR, especially if FSE sequences are used. Therefore, frequently more PD- weighted images, are acquired instead of true T1-weighted images, if no intravenous contrast is needed in a clear-cut clinical setting.

In addition to the low signal-to-noise ratio, single-excitation imaging is much more influenced by undesirable artifacts from repetitive motion, for example ghost artifacts from respiration and pulsatile flow, and from thermal noise. These undesirable features tend to be averaged out by multiple acquisitions.

4.6.2
Reduced Matrix Size

As spatial resolution is decreased using a reduced matrix, the image appears blurred and fine detail is lost. However, image contrast decreases if a high-resolution matrix is utilized, and scan time increases. Therefore, a suitable matrix has to be chosen according to the detail and size of the region being imaged.

If matrix size is increased along the phase encoding direction, more phase encoding steps are needed, and the scan time increases proportionally. Changing the matrix size along the frequency encoding direction has no effect on the scan time. Therefore, if the body part being imaged is elliptic or rectangular, the field of view and read-out direction can be adjusted accordingly.

4.6.3
Rectangular Field of View

Use of a rectangular field of view (FOV) is especially suitable for sagittal images of the spine, because it accommodates the shape of the area of interest.

Using a rectangular FOV, the number of phase encoding steps is reduced, while the spatial resolution is preserved. If only 50% of the lines are measured (Fig. 4.16), scan time is cut by half and the FOV is also reduced to half of its original size along the preparation direction. Pixel size is unchanged; however, signal-to-noise ratio is decreased proportionally to the square root of the fraction of lines measured.

4.6.4
Number of Slices

Often, the number of slices can be reduced without loss of diagnostic information. For example, if a knee joint is examined in both the sagittal and the coronal orientation, the very medial and lateral slices can be spared in the sagittal views, as these soft tissue structures are already available in coronal slices (Fig. 4.17).

4.6.5
Sequences

In addition to spin-echo imaging, which, because of its realistic anatomic detail, should be part of every musculoskeletal study, faster GRE and TSE techniques can be used for further clarification (cf. Sect. 4.2).

4.6.6
Half-Fourier Scan

Half-Fourier imaging, sometimes referred to as half-scan, uses the symmetry of the data set obtained

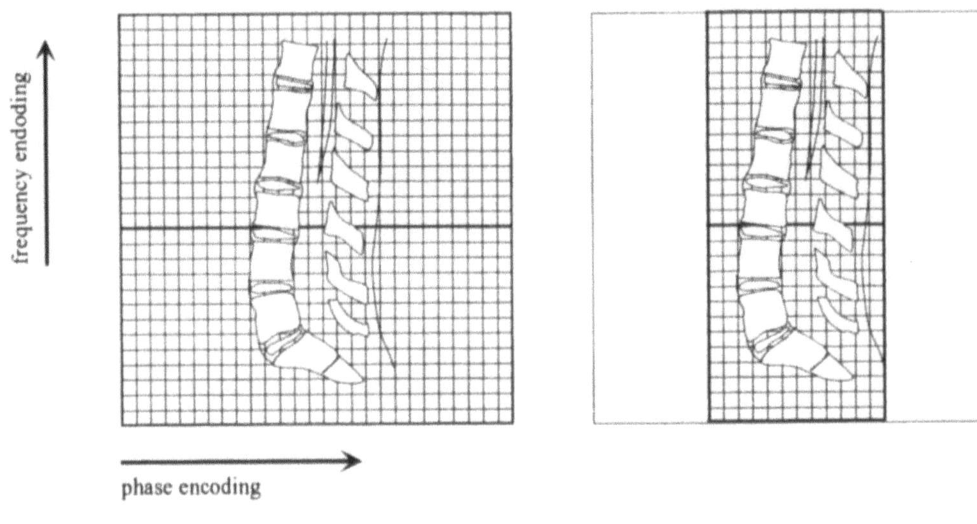

Fig. 4.16. Rectangular field of view (*FOV*). If the FOV is reduced to half its size and the matrix is adjusted accordingly, pixel size is unchanged

Fig. 4.17. Axial scout views of the knee joint showing coronal slice positioning (**a**), and sagittal slice positioning with unnecessarily many (*,**b**) or an adequate number (**c**) of slices

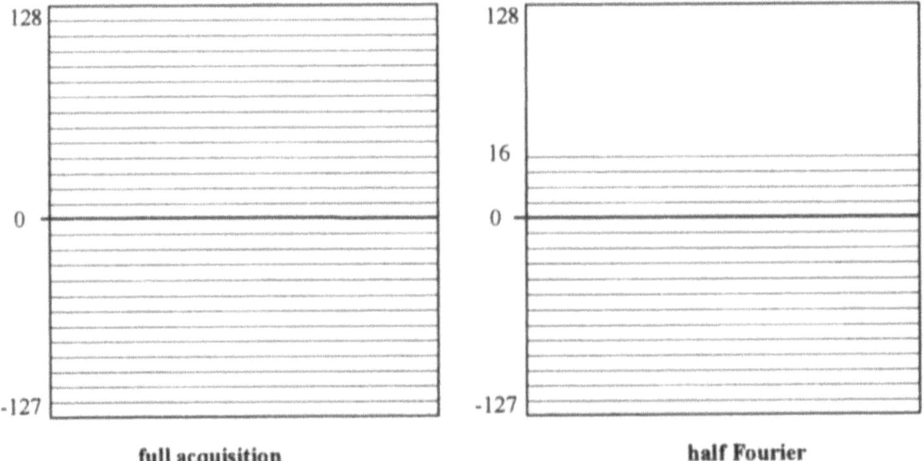

Fig. 4.18. Half-Fourier scan. Using a half-Fourier scan, only just over half the data are acquired in this example of a 256 pixel matrix

during an MR measurement. An asymmetric fraction of the data set, representing just over half of the data samples, is acquired, and the remaining part of the image is synthesized on the basis of these data (Fig. 4.18). Spatial resolution is preserved, but reduction of overall signal to noise and sensitivity to field inhomogeneities are major drawbacks of this technique.

When full spatial resolution is necessary and the expected signal-to-noise ratio is high, long scan times can be reduced. Therefore, excellent results are achieved using 3D scans of a large FOV combined with a large slice thickness.

4.7
Artifacts

4.7.1
Motion Artifacts

A variety of causes are responsible for patient motion, especially voluntary movement (mainly in pediatric imaging), respiratory and cardiac movement, peristaltic movements of the bowels, swallowing, or cerebrospinal fluid flow. The consequence is blurring of the image in the direction of the motion. Ghost images are independent of the direction of motion and are discovered along the direction of the phase encoding gradient.

Signal averaging is the most commonly applied method to reduce motion artifacts. Similar to background noise, the artifacts are averaged out. Total imaging time is increased. Cardiac and respiratory triggering is based on the use of electronic sensors, such as ECG, to synchronize the acquisition sequence. Blurring and ghost artifacts are eliminated, but longer measurement times and limited imaging strategies are major disadvantages. These techniques are rarely employed in musculoskeletal imaging.

Regional Presaturation. Regional presaturation techniques (REST) are most commonly employed in musculoskeletal imaging. Additional RF impulses presaturate tissues that are not of interest for the clinical question and still are sources of motion artifacts (Fig. 4.19). As these tissues cannot contribute any net magnetization during signal measurement, presaturated areas appear dark in the image. Regional presaturation is most effective in eliminating artifacts due to motion or blood flow. In order to eliminate artifacts caused by blood flow, the

presaturation pulses are usually applied perpendicular to the blood vessels and in the direction of inflow.

4.7.2
Flow Artifacts

Blood flow in both arteries and veins or CSF flow can cause typical signal alterations. Signal may decrease, often referred to as "flow void," or increase, mimicking intra- or extraluminal pathologies.

Flow enhancement may be reduced using regional saturation techniques (Fig. 4.20). Also, the use of a higher TR may contribute to the reduction of flow enhancement. Flow void is more difficult to suppress. The basic problem is that inflowing fluid has not been influenced by preparation pulses. Partial "refreshing" of the flowing blood in the selected slice

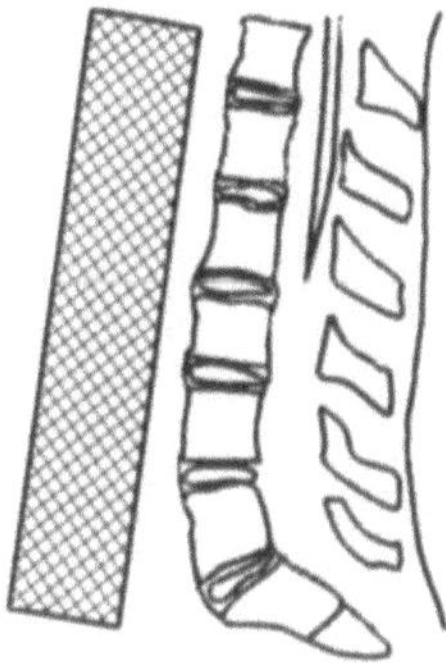

Fig. 4.19. Presaturation to eliminate motion artifacts from peristaltic movement. The presaturation impulse, which saturates the signal from the prevertebral bowels, is shown on the *left* (sagittal scout view of the lumbar spine used for planning of axial slices). Also compare Fig. 4.11

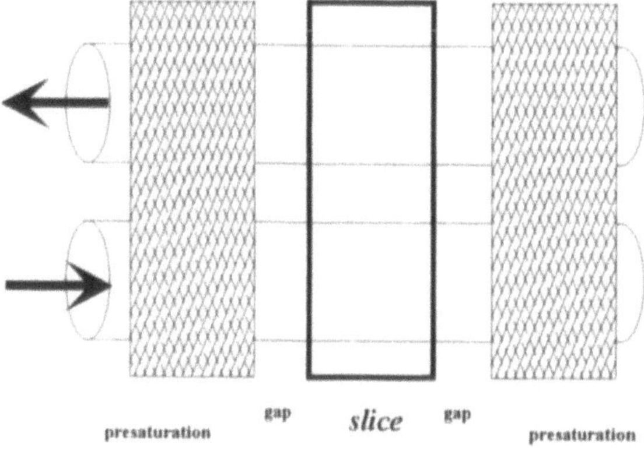

Fig. 4.20. Parallel regional presaturation technique used to eliminate flow enhancement. Inflowing blood or CSF is saturated and does not contribute signal to image formation

is the most common technique to reduce related artifacts. Changing the echo time may also be worth trying.

4.7.3
Static Field

Even small inhomogeneities in the magnetic field can cause distortions or signal loss. Ferromagnetic implants can locally distort the magnetic field, but nonferromagnetic objects can also condense magnetic field lines. Switching the gradients induces eddy currents in the metallic objects, resulting in local field inhomogeneity. Often the signal loss and the distortion exceed by far the size of the metallic object.

As a rule, GRE and STIR sequences show a larger area of signal loss, whereas in SE and TSE sequences a smaller area is affected by image distortion (Fig. 4.21).

4.7.4
Partial Volume Averaging

In 2D sequences, anisotropic voxels are usually used. These voxels often measure more than 3 mm along one axis; as a consequence more than one anatomic structure, e.g., fat and tendon, may contribute to the voxel, and average signal intensity results. This average signal intensity, however, is inconclusive with regard to the true tissue composition of a particular structure in a particular voxel of the imaged volume. Therefore, partial volume effects should be avoided. As a rule, the interpretation of a specific structure should only be done if the structure is contained in parallel slices or in perpendicular slices. For example, the Achilles tendon may be evaluated in an axial or sagittal, but not in a coronal orientation.

4.7.5
Magic Angle

Increased signal intensity can sometimes be observed in normal tendons in the MR image. The appearance of this increase in signal intensity depends on the angle of the tendon in the image and the direction of the main magnetic field. For example, magic angle artifacts are often found in tendons (ankle, shoulder) and menisci (lateral posterior horn) at about 50–60° in correlation with the constant magnetic induction field B_0.

If a short TE is utilized in an SE sequence, this effect is more significant. To test whether the observed increase in signal intensity is due to magic angle effects or pathology, a T2-weighted sequence

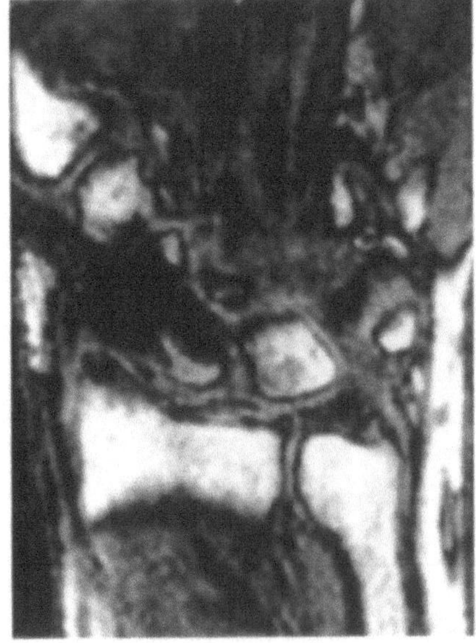
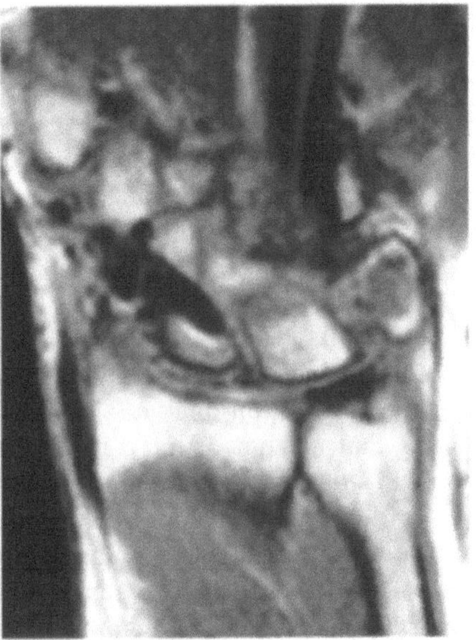

Fig. 4.21 a,b. Scaphoid fracture stabilized by Herbert screw. Using the GRE sequence, there is major signal loss in a larger area (**a**). If an SE sequence is used, the image is distorted but the scaphoid bone is depicted to a much greater extent (**b**)

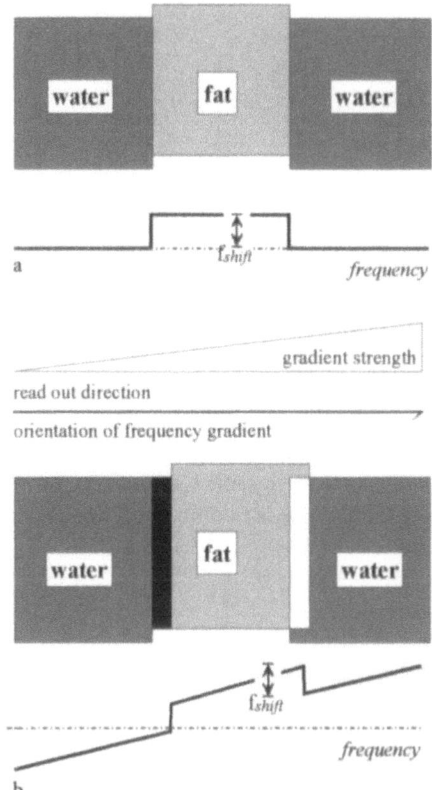

Fig. 4.22 a,b. Chemical shift artifact. **a** Before the application of the frequency gradient, the frequency graph is horizontal with a 220 Hz (f_{shift}) jump over the fat tissue at 1.5 T. **b** Situation after application of the frequency gradient. The frequency gradient is perpendicular to the white and black bands and increases from left to right. The fatty tissue is shifted to the right, and a band of low signal intensity remains. Signal intensities are additive at the right fat/water interface, and a bright band is produced

can be used to image the same slice, or, as an even more reliable alternative, the patient can be repositioned.

4.7.6
Chemical Shift

Depending on the main magnetic field strength, the resonance frequencies of water and fat are separated by about 3.5 parts per million. For a 1.0-T field, the absolute difference in frequency is therefore about 150 Hz, and for 1.5 T it is about 220 Hz. The chemical shift phenomenon occurs at all field strengths, but is more conspicuous at higher field strengths.

A "zebra banding" with a white or dark band at the fat/water junction is the typical appearance of the chemical shift artifact (Fig. 4.22b). At 0.5 T, the dis-

placement at the fat/water interface measures about 1 or 2 pixels, and at 1.5 T about 5 pixels.

4.7.7
Wrap-around, Aliasing, or Backfolding

If the dimension along the preparation direction of the object being imaged is larger than the field of view, wrap-around or aliasing can occur. Those parts of the object which are outside of the field of view are folded back to the opposite side of the image presented. Aliasing is caused by the reduction of the field of view, because the variety of precessional frequencies is increased in the image but the number of phase encoding steps and sampling is not adjusted accordingly.

Increasing the field of view along the preparation direction (oversampling) suppresses the wrap-around artifacts. Using oversampling, the unwanted area is contained in the preparation direction, but ignored during image reconstruction. Regional presaturation is another method to decrease backfolding.

4.8
Patient Management

4.8.1
Safety

Attention must be paid to all metallic objects, especially ferromagnetic ones, and to all mechanical or electrical appliances worn by the patient. Ferromagnetic objects, particularly if loosely attached to the patient or freed inside the examination room, become projectile masses. Mechanical appliances such as watches or insulin pumps are often magnetized and come to a standstill. Electrical appliances may malfunction: especially in cardiac pacemakers, inducted currents are potentially hazardous and may be life-threatening. Wires or similarly shaped metallic objects can be subject to HF or magnetic induction and may cause burns to the patient.

The magnetic attraction of a supposed ferromagnetic object can be tested with a bar magnet. For implanted materials, however, information should be obtained from the producer of the implanted material, or, if such information is unavailable, an extracorporeal sample provided by the producer should be tested first.

Table 4.13. Magnetic field strength typically tolerated by hospital equipment

1 gauss	10 gauss	20 gauss
Normal unshielded color monitor	Shielded color monitor	Watches
CT scanner	Magnetic data carriers	Computers
Nuclear scanner	Operator console	
	Shielded CT scanner	

The easiest and most efficient way to remove most hazardous objects is to ask the patient to change to a gown from personal clothing. The patient's medical history should be screened for implants and surgery following a standardized scheme.

Table 4.13 gives an impression as to which clinical devices usually tolerate a certain magnetic field strength.

4.8.2
Patient Selection and Scheduling

From the referring physician, a small but very important amount of information has to be obtained to schedule an examination appropriately. A thorough knowledge of the patient's history, the findings on clinical examination, and the suspected abnormality is essential to determine whether MRI is the best modality to detect and analyze the problem. Conventional x-rays in standard projections provide the most comprehensive and most cost-effective radiological basis to decide upon further imaging studies and, with few exceptions, should have been performed prior to an MRI examination.

The anatomic region to be examined and the presence of electrical implants such as cardiac pacemakers or metallic implants such as prosthesis or even surgical clips have to be considered. The radiographs of the patient should be reviewed and the local relationship between metal implants should be carefully thought about. Other patient status factors include clinical factors like pain, psychological factors such as claustrophobia or very young age, and high body weight.

A standard examination protocol should be determined to estimate examination time. As patients not arriving in time leave an open slot of 30 min to 1 h, it is advisable to contact outpatients personally to confirm the appointment shortly before the examination date. This also provides a good opportunity to remind the patient about the length of the examination and appropriate dress. Also, if possible, scheduling should allow the immediate scan of another "back-up" patient, e.g., an inpatient, if an outpatient does not attend punctually.

When the patient arrives, an experienced receptionist can take the chance of the first personal contact to assess whether the patient's fears might be relieved by answering some questions or whether further intervention is required on the part of the radiologist.

4.8.3
Special Considerations

4.8.3.1
Biologic Effects, Pregnancy

As there is no ionizing radiation (as in radiography or computed tomography), MRI is estimated to be a safe imaging modality. However, MRI is a relatively new method, and long-term effects resulting from the considerable energy applied to patients remain to be determined. The static and changing magnetic fields, the radiofrequency pulse, and the administration of contrast media are potentially dangerous to the health of patient.

To date, long-lasting negative side-effects of MRI have not been observed at field strengths up to 2.0 T. Nevertheless, there is no conclusive evidence that medical MRI scanning is completely safe. Therefore as a rule MRI scanning should not be performed if not clearly indicated, and this rule should certainly be observed in pregnant patients.

4.8.3.2
Claustrophobia and Very Young Patients

Significant claustrophobia is found in less than 5% of patients. Also, children of less than 6 years of age may not cooperate sufficiently for the time span necessary to obtain an optimal study.

New low field systems provide larger openings of the gantry and easier access and are less noisy. A close relative or friend can sit at the side of the gantry. For patients with mild claustrophobia, this works well. If necessary, mild sedation can be administered prior to the examination. For children and most adults, oral sedation is generally sufficient. General or intravenous anesthetics are necessary in older children and a small number of adult patients.

4.8.3.3
Metal Implants

It is not advisable to examine patients with intracranial aneurysm clips.

Most surgical clips used today are not ferromagnetic. As a rule, if an object has been in place for 6 months or longer, the risk of motion is reduced because of surrounding scar tissue holding the metallic object in place.

Unsuspected foreign bodies are a potentially more difficult problem. If there is any concern, a plain radiograph prior to the MRI examination will reveal any major metallic foreign body. However, the patient may be unaware of shrapnel or other foreign bodies, so that they are only detected during the examination. If the object is not close to a nerve or another region where it might cause damage, examination can be continued with close monitoring.

Orthopedic appliances usually contain ferromagnetic impurities. The extent of the artifact in the MRI image is influenced by the composition, size, and configuration of the metal implant. Artifacts, however, may also result from minute metal parts that remain after any invasive procedure, including surgery or even arthroscopy. In this case, the risk is considerable smaller, and little can be done until the artifact is discovered during the examination.

4.8.3.4
Electric Implants

Patients with cardiac pacemakers are generally not candidates for an MRI study. Insulin pumps, however, can be removed for the duration of the examination.

4.8.3.5
Other Considerations

Most body coils leave an opening of only about 0.5 m, which can make *large patient size* a limiting factor.

Also, positioning of patients weighing more than 130 kg (285 lbs) can be a problem in many standard systems. New low-field to middle-field systems (0.1–0.5 T) have shorter magnets or larger openings, providing better access for patient positioning.

4.8.4
Patient Preparation

Electric and mechanical devices have to be left outside the electric field. Insulin pumps, for example, can be substituted by direct injection prior to the examination. The surveillance necessary in polytrauma or intensive care patients has to be provided in the examination room and often requires the presence of anesthesiologists.

Patients who have more than mild pain in the examination position will find it difficult to stay immobile for the duration of the examination. Appropriate premedication with diazepam or pain medication can be administered in advance on the ward.

4.8.5
General Aspects of the Examination

The anatomic position and the size of the structures of interest have to be considered in the selection of the pulse sequence, the size of the field of view, the slice thickness, and the orientation. Kinematic studies provide additional information for evaluation of problems such as subluxation of the peroneal tendon and patellofemoral pain syndromes. Clear-cut indications and the posing of specific questions by the consulting physician result in a considerable reduction in examination time. Standard protocols, however, have to be designed for screening whenever clinical data are not suggestive of a certain pathology.

4.8.6
Patient Positioning and Coil Selection

Patient size, the body part to be examined, and the examination time have to be considered when deciding upon patient positioning and coil selection. To achieve maximum signal-to-noise ratio and resolution, the smallest possible coil that covers the anatomic site and can be positioned most closely should be used.

Patient positioning for examination of the *lower extremities* is not difficult. Structures of interest can be positioned near the midline. This allows the use of surface coils with a reduced field of view. Generally, circumferential coils are preferred. However, for kinematic studies, flat or coupled flat coils allow a larger range of motion.

Positioning for examination of the *upper extremities* is more difficult. This is especially true for large patients and MRI systems that do not allow a small, off-center field of view. For the hand, wrist, and elbow, this problem can be partially solved by rotation of the joint to the center of the gantry. If the patient is too large, rotation above the head may be necessary. This position, however, is uncomfortable and cannot be maintained for a long time. As a consequence in about one-quarter of patients, motion artifacts reduce the quality of the images (BERQUIST 1991). Positioning of the shoulder is strenuous. Most systems are too small to allow a centered position of the shoulder. Also, motion artifacts caused by respiration have to be excluded from measurement by the use of presaturation pulses.

Positioning of the *spine and pelvis* is not difficult, as these regions are easy to center and are not prone to motion artifacts. For the cervical spine, special coils are available. The thoracic and lumbar spine and the pelvis with hip joints can be examined with body coils or phased array coils. Motion artifacts from respiration, the heart and blood flow, and peristaltic movement have to be excluded by the use of presaturation.

4.8.7
Printing

Signal inhomogeneity, ample tissue contrast, and the many steps in image acquisition and processing are the reasons why the printing of MR images is quite difficult. Maximization of tissue contrast over the region of interest and minimization of the conspicuity of background noise are the main goals in image printing. In some instances, two window settings are necessary because the structure components vary too much in composition, e.g., in knee studies two different settings may be used for evaluation of menisci and bones.

Because of the large number of images in preset sequence packages, such as 3D sequences, images should be preselected for printing. Prints should always include patient data, examination time, sequence details, window settings, and a scout image to show slice orientation.

More advanced laser printers offer internal quality controls of the printer console, the laser imager, and the film processor. These quality controls should be included in the routine, and in addition performed if an image appears of low contrast ("gray and flat") in comparison to the viewing monitor.

4.8.8
Quality Control

Quality control is more complicated in MRI than in most other imaging modalities. Numerous artifacts have to be recognized. Imaging parameters have to be adjusted appropriately, and ineffective parameter settings have to be corrected. Humidity and temperature levels have to be checked in accordance with the instructions of the producer.

A log of signal-to-noise ratios should be maintained for each receiver coil using standard phantoms. In this way, hardware errors are recognized more easily. Downward trends in the signal-to-noise ratio should be noticed, and structured noise patterns should be perceived. In this case, raw data should be saved for the service team. When service personnel are not available immediately and the patient scan has to be continued, the application of thicker slices in combination with larger fields of view and more excitations improves image quality. In any case, a call should be made to the service team to check for immediately available solutions and on-site tests.

References

Berquist TH (1991) Magnetic resonance techniques in musculoskeletal disorders. Rheum Dis Clin North Am 17:599–615

Berquist TH (1993) Optimizing MR imaging techniques for articular disorders. In: Weissman BN (ed) Syllabus: a categorical course in musculoskeletal radiology: advanced imaging of joints: theory and practice. RSNA

Chang KH, Ra DG, Han MH, Cha SH, Kim HD, Han MC (1994) Contrast enhancement of brain tumors at different MR field strengths: comparison of 0.5T and 2.0T. AJNR 15:1413–1419; discussion 1420–1423

Disler DG, Kappaturam SV, Chew FS, Rosnethal DI, Patel D (1993) Menical tears of the knee: preliminary comparison of three dimensional reconstruction with two dimensional MR imaging and arthroscopy. AJR 160:343–345

Drapé JL, Thelen P, Gay-Depassier P, Silbermann O, Benacerraf R (1993a) Intraarticular diffusion of Gd-DOTA after intravenous injection in the knee: MR evaluation. Radiology 188:227–234

Drapé JL, Thelen P, Gay-Depassier P, Silbermann O, Benacerraf R (1993b) Intraarticular diffusion of Gd-DOTA after intravenous injection in the knee: MR imaging evaluation. Radiology 188:227–234

Engel A, Hajek P, Kramer J (1990) Magnetic resonance arthrography: enhanced contrast by gadolinium contrast in the rabbit and humans. Acta Orthop Scand 61(Suppl):1–57

Erlemann R, Reiser M, Peters PE, et al. (1989) Musculoskeletal neoplasms: static and dynamic Gd-DTPA-enhanced MR imaging. Radiology 171:767–773

Hajek PC, Sartoris DJ, Neumann CH, et al. (1987) Potential contrast agents for MR arthrography: in vitro evaluation and practical observations. AJR 149:97–104

Heindel ND, Burns HD, Honda T, Brady LW (1978) The chemistry of radiopharmaceuticals. Masson, New York

Hodler J, Kursunoglu-Brahme S, Snyder SJ, et al. (1992) Rotator cuff disease: assessment with MR arthrography versus standard MR imaging in 36 patients with arthroscopic comfirmation. Radiology 182:431–436

Lee DH, Vellet AD, Eliasziw M (1995) MR imaging field strength: prospective evaluation of the diagnostic accuracy of MR for diagnosis of multiple sclerosis at 0.5 and 1.5 Tesla. Radiology 194:257–262

Melchert UH, Schröder C, Brossmann J, Muhle C (1992) Motion triggered cine MR imaging of active joint movement. Magn Reson Imaging 10:457–460

Minami M, Yoshikawa K, Matsuoka Y, Itai Y, Kokubo T, Iio M (1991) MR study of normal joint function using a low field strength system. J Comput Assist Tomogr 15:1017–1023

Mirowith SA, Totty WG, Lee JKT (1992) Characterization of musculoskeletal masses using a dynamic Gd-DTPA enhanced spin echo MRI. J Comput Assist Tomogr 16:120–125

Muhle C, Melchert UH, Brossmann J, Schröder C, Wiskirchen J, Heller M (1995) Positionsgestell zur kinematischen MRT der Halswirbelsäule. Fortschr Röntgenstr 162:252–254

Palmer WE, Brown JH, Rosenthal DI (1993) Rotator cuff: evaluation with fat suppressed MR arthrography. Radiology 188:683–687

Peterfy CG, Roberts T, Genant HK (1997) Dedicated MR imaging. Radiol Clin North Am 35:1–20

Steinberg HV, Alarcon JJ, Bernadino ME (1990) Focal hepatic lesions: comparative MR imaging at 0.5 and 1.5 T. Radiology 174:153–156

Totterman SM, Miller RJ, McCance SE, Meyers SP (1996) Lesions of the triangular fibrocartilage complex: MR findings with a three-dimensional gradient-recalled-echo sequence. Radiology 199:227–232

Vahlensieck M, Seelos K, Träber F, Gieseke J, Reiser M (1993) Magnetresonanztomographie mit schneller STIR-Technik: Optimierung und Vergleich mit anderen Sequenzen an einem 0,5 Tesla System. Fortschr Röntgenstr 159:288–294

Vahlensieck M, Lang P, Seelos K, Yang-Ho-Sze D, Grampp S, Reiser M (1994) Musculoskeletal MR imaging: turbo (fast) spin echo versus conventional spin-echo and gradient echo imaging at 0.5 Tesla. Skeletal Radiol 23:607–610

Vahlensieck M, Wischer T, Schmidt A, et al. (1995) Indirekte MR Arthrographie: Optimierung der Methode und erste klinische Erfahrung bei frühen degenerativen Gelenkschäden am oberen Sprunggelenk. Fortschr Röntgenstr 162:338–341

Vahlensieck M, Peterfy CG, Wischer T, et al. (1996) Indirect MR arthrography: optimization and clinical applications. Radiology 200:249–254

Winalski S, Aliabadi P, Wright RJ, Shortkroff S, Sledge CB, Weisman BN (1993) Enhancement of joint fluid with intravenously administered gadopentetate dimeglumide: technique, rationale, and implications. Radiology 187:179–185

5 Scintigraphy

C. Schiepers

CONTENTS

5.1
Introduction and Historical Perspective

Bone scintigraphy is one of the common procedures in a nuclear medicine service. The procedure is relatively straightforward to carry out and is largely standardized throughout diagnostic imaging departments. Bone scans account for 30%–60% of the routine work load. Modern equipment has greatly enhanced the ease of operation and permits imaging in planar and tomographic as well as whole body mode (vide infra).

Scintigraphy is an extremely sensitive procedure for evaluating a variety of skeletal disorders, and also has applicability in soft tissue evaluations. The main indications for referral are screening of patients with malignancy, trauma, orthopedic problems, sports injuries, endocrine, and rheumatologic disorders.

Bone is a specialized form of connective tissue, with hardness as its characterizing feature. Within the soft organic matrix of cells and intercellular substance, minerals such as calcium, phosphate, carbonate, and citrate are deposited. Bone is a dynamic tissue which is continually formed, remodeled, and reabsorbed. Basically, this is performed by three different cell types: (a) osteoblasts, which produce the organic bone matrix, (b) osteocytes, which form the inorganic matrix, and (c) osteoclasts, which are active in bone resorption. These processes can be followed with radioactive elements in tiny amounts, which are appropriately called (radio)tracers or currently radiopharmaceuticals. Bone is a metabolically active structure and osteogenesis and resorption occur continually in the normal skeleton. Metabolic rates are affected by disease processes, and can be greatly enhanced, as in Paget's disease, or decreased, as in involutional osteoporosis.

Bone seekers are elements or substances that mainly localize in the skeleton, the most important ones being Ca, Sr, Ra, P, and S. Two series of natural radioactive elements have biological importance since they have been introduced in humans in the past: (a) the thorium series (^{232}Th to ^{208}Pb) and (b) the uranium-radium series (^{238}U to ^{206}Pb). Whether it was accidental or by design, both series have caused damage to the human organism (McLean and Budy 1964).

Other used radioactive elements are all artificial. Beta emitters ^{32}P and ^{45}Ca were initially used to measure bone mineralization. Later on, gamma-emitting radionuclides were introduced (^{47}Ca, ^{85}Sr) because they allowed external counting.

In 1962 ^{18}F-fluoride was introduced as a bone imaging agent by Blau and collaborators. The annihilation radiation of this positron emitter is relatively high with 511 keV and suited for rectilinear scanners. Van Dyke et al. reported in 1965 the use of

C. Schiepers, MD, PhD, Department of Radiological Sciences, Olive View – UCLA Medical Center, 14445 Olive View Drive, Sylmar, CA 91342, USA

^{18}F with a gamma camera. Since the advent of positron emission tomography (PET), this radiopharmaceutical has been revived and allows for true regional quantification of bone blood flow and fluoride influx rate (SCHIEPERS et al. 1990).

The next major breakthrough was the development of ^{99m}Tc-labeled polyphosphate complexes by SUBRAMANIAN and MCAFEE in 1971. This made bone scanning possible on routine gamma cameras or Anger scintillation systems. The photopeak of 140 keV of ^{99m}Tc is ideal for the sodium iodide detector (of gamma cameras) and allows for a high dose activity that may be administered (700–900 MBq of ^{99m}Tc). These developments have led to the present place of bone scintigraphy in clinical practice.

5.2
Radiopharmaceuticals

5.2.1
Bone Seekers

For the natural radioactive materials, there never was a place in diagnosis or treatment. The accidents with radium in the 1920s and plutonium experiments before World War II still cast their dark shadow. Together with the actual use of atomic bombs on Japan, these events are responsible for the strong suspicion with which nuclear medicine is viewed. However, the medical use of radionuclides in tracer amounts for diagnosis and ablative doses for therapy is an adjunct to medicine and greatly benefits patients.

Today, the role of ^{45}Ca and ^{85}Sr or ^{87}Sr in diagnosis is obsolete. Interestingly, ^{89}Sr has become available for the palliative treatment of painful bone metastases and is approved by the FDA (Food and Drug Administration of the United States).

5.2.2
Fluoride

Radioactive fluoride, ^{18}F$^-$, on the other hand, is again used in clinical practice following the introduction of PET systems. The skeletal uptake of approximately 70% is quite high, and 25% is excreted in the urine by 6 h. The half-life of 109.8 min is relatively short but permits transportation and reasonable imaging times. Thus, ^{18}F$^-$ forms an excellent tracer to study the fluoride kinetics in the skeleton, and provides absolute quantification of regional blood flow. The

small solutes leave the capillaries in bone by passive free diffusion, and traverse through the fluid spaces to reach the osseous tissues. The uptake mechanism of fluoride is adsorption in the water shell around newly formed bone crystals and the exchange with hydroxyl ions of the hydroxyapatite in the bone matrix. The adsorption of fluoride is a process requiring minutes to hours; the actual incorporation in the bone matrix takes days and, therefore, cannot be measured accurately with this tracer.

5.2.3
Technetium Complexes

Presently, labeled diphosphonates are the radiopharmaceuticals of choice for skeletal scintigraphy, e.g., methylene diphosphonate (MDP), hydroxymethylene diphosphonate (HMDP), or hydroxyethylene diphosphonate (HEDP). Pyrophosphate (PYP) can also be used, having a somewhat higher uptake in the soft tissues, and its indication is now limited to myocardial infarct imaging. In order to obtain stable chelated complexes, reducing agents (SnCl$_2$) are needed. These keep technetium in a low valence state so that binding occurs, i.e., reduce Tc(7+) from the generator eluate to Tc(4+) in the complexes.

Generally, the clearance from the vascular compartment is fast, with half-times of 2–4 min. Peak uptake varies for the different agents, but is usually around 1 h. However, the bone to background ratio also varies, due to the different clearance and uptake rates of the organs, and the maximum ratio occurs much later at 4–6 h. Patient convenience is an important factor, as well. The combination of contrast, peak uptake, radionuclide decay, and practical issues results in imaging 2–4 h after tracer administration. At this time about one-third of the administered dose is bound to bone, one-third is excreted in the urine, and the remainder is associated with other tissues, about 10% of which is bound to blood proteins.

The aforementioned diphosphonates are biologically active compounds, which can be followed in the body along their metabolic pathway because of their specific affinity to a certain organ or organ system. This is achieved by attaching a radionuclide (^{99m}Tc) to the compound, which can be detected by an external imaging device. Therefore, trace amounts can be monitored and only minimal doses have to be administered, contrary to the pharmacological doses that are needed with contrast agents such as iodine (see Chaps. 2, 3, 7), or gadolinium (see Chap. 4). The radiopharmaceutical is administered intravenously, and strict precautions to ensure sterility and antipyrogenicity need to be followed. The routine procedure is to add an eluate of pertechnetate solution from the generator to a kit of diphosphonates under sterile conditions. Quality control is performed to check the radiochemical purity, i.e., percentage of desired radiopharmaceutical present in the syringe, usually more than 95%.

Adverse reactions to the injection of the radiopharmaceutical are virtually nonexistent. The reported incidents are usually related to other agents in the kits that are necessary for stabilization, e.g., pH buffers, reducing agents to keep technetium in a low valence state, and/or metabolites.

5.3
Methods

5.3.1
Imaging Equipment

The image acquisition equipment in nuclear medicine is based on scintillation detection. Gamma cameras have a special crystal comprised of thallium-doped sodium iodide (NaI monocrystal with small Tl impurities), which converts the imparting photon (or gamma ray) into a light flash. This process is called scintillation and routine gamma cameras are most sensitive for energies of 100–150 keV. For a positron camera the imaging principle is similar. Positrons are unstable and almost immediately combine with an electron. This process is called annihilation and results in the emission of two gamma photons of 511 keV which are detected with crystals designed for this higher energy. The light photons or scintillations are amplified with a photomultiplier tube and converted to an electronic signal, analog or digital, which can be processed by a computer and displayed. Hence, the terms "scintigraphy" and "scintigram" were introduced to describe this imaging technique. The name "single-photon detection" is used to distinguish conventional gamma decay, e.g., ^{99m}Tc, with one gamma photon per disintegration, from positron decay with two gamma photons per annihilation.

A collimator is mounted on the face of the camera to ensure that gamma rays perpendicular to the crystal are detected and other angles absorbed. The collimator is a mechanical device that affects both the sensitivity and the resolution. PET systems do not have collimators, but utilize electronic collimation; hence, their hundredfold increase in sensitivity over gamma cameras.

Several geometric configurations have been designed for nuclear imaging equipment. The standard gamma camera has one head which can be tilted, angled, and moved to image patients in the supine, sitting, or standing position. In addition, gantries have been developed to rotate the camera head around the patient. Thus, tomographic imaging is possible analogous to CT where the x-ray tube rotates around the patient. Due to the significantly lower photon flux in nuclear imaging, acquisition duration is prolonged.

Presently, gantries with two or three heads are available to shorten the acquisition duration. Varying angles of 60°, 90°, and 180° between the camera heads are possible to execute specific protocols and accelerate the acquisition. Systems with detectors over the full 360° are the standard in PET, but not in conventional single-photon imaging. A feature of all tomographic systems is the simultaneous acquisition of multiple image planes.

5.3.2
Positron Emission Tomography

The annihilation radiation of ^{18}F is readily detected with a positron camera. PET systems are optimized for 511 keV and allow correction for attenuation effects. For a more detailed description of this methodology the reader is referred to the literature (PHELPS et al. 1986). The applicability of ^{18}F-fluoride PET in clinical practice has been dealt with elsewhere (SCHIEPERS 1993). A detailed description of the various blood flow determination methods with ^{18}F⁻ can be found in the literature: quantitative with a gamma camera (CHARKES 1980), based on whole body clearance (WOOTTON et al. 1976, 1981), and PET (HAWKINS et al. 1992).

Previous work has shown the applicability of ^{18}F to study kinetics of normal vertebrae (SCHIEPERS et al. 1990; HAWKINS et al. 1992), metabolic bone disease (SCHIEPERS et al. 1991, 1997; RYAN and FOGELMAN 1995), and osteonecrosis (SCHIEPERS et al. 1994). Bone remodeling is closely related to bone blood flow as shown by tetracycline labeling (REEVE et al. 1988). The evaluation of bone graft viability with fluoride PET has also been published (BERDING et al. 1995).

The feasibility of whole body imaging with ^{18}F-fluoride in normal bone and oncological disorders

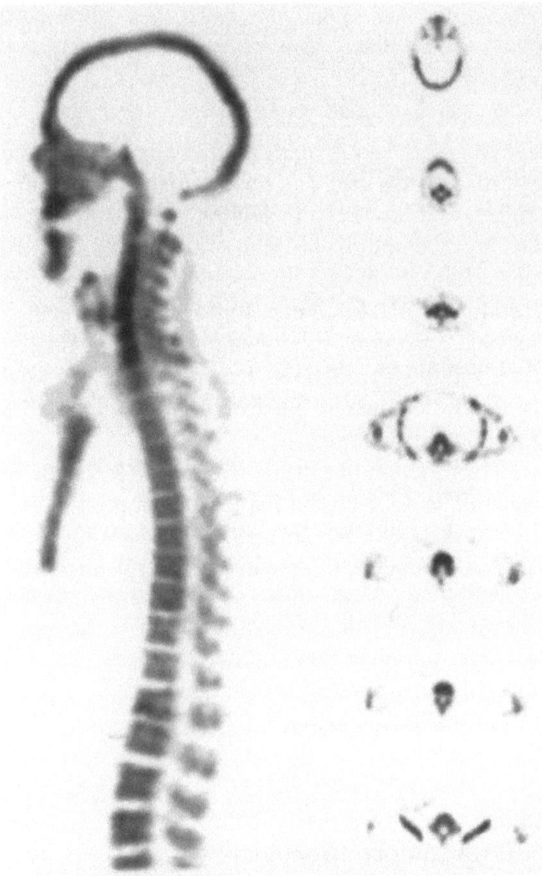

Fig. 5.1. Fluoride PET scan of a 28-year-old male with juvenile osteoporosis. Sagittal plane of the torso of a volumetric dataset acquired with a 2D PET system 1 h after administration of 300 MBq of ^{18}F-fluoride. Corresponding axial slices of the head, jaw, neck, thoracic and lumbar spine, and pelvis are shown on the right. Note the increased uptake in the superior and inferior aspects of thoracic and lumbar vertebrae

was reported by HoH et al. in 1993. An example of a whole body bone scan with fluoride is shown in Fig. 5.1 with a volumetric display of the various slices.

5.3.3
Single-Photon Imaging

In the remainder of this chapter we will discuss routine bone scanning with ^{99m}Tc complexes.

Current gamma cameras have been optimized for the 140-keV photopeak of ^{99m}Tc, since this is the most often used radionuclide in nuclear medicine. As was discussed above, this radionuclide is attached to a biologic compound that has a specific affinity to bone. The compound is followed through the body by detecting the decaying atoms of ^{99m}Tc.

There is no special patient preparation for a bone scan. After the tracer administration, the patient is advised to drink plenty of fluids and to void frequently. Thus, excretion of tracer is enhanced and the radiation dose to the bladder may be minimized. Before scanning, the patient is asked to empty the bladder. Self-evidently, patients need to be instructed about possible contamination because of tracer in the urine. Various imaging protocols are available and will be discussed briefly.

5.3.4
Planar Static Imaging

Images are acquired during a "steady state" of the tracer distribution throughout the body. As has been mentioned before, this refers to the delayed phase 2–4 h after the tracer administration. Images are acquired in both anterior and posterior views of the skeleton. Dependent upon the size of the camera, multiple views are obtained to encompass the entire body. The termination of an image can be accomplished in several ways:

1. By time, e.g., a fixed time of 1–3 min per view.
2. By counts, e.g., 500–1000 kcounts per image: Obviously, the lower uptake in the distal extremities would call for very long acquisition times. A preset number for the various parts of the body is normally present based on prior experience with the available equipment. In this way the left and right side can be compared within an image.
3. Iso-time, i.e., the time to acquire a high-quality image of, for instance, the posterior chest is recorded, and all subsequent images are acquired for the same duration. The advantage of this protocol is that uptake between images can be compared directly, in addition to the left/right evaluations.

The standard available options of zooming and acquisition of spot views under specific angles, e.g., anterior or posterior oblique, can be attempted if a certain body area needs to be inspected in detail. A commonly applied view is the TOD or tail-on-detector, which is used to evaluate the pelvis in an attempt to exclude overlying activity from the bladder. Most institutions will mount low-energy, high-resolution collimators, and preferably ultra-high resolution for tomography, since physicians like high-resolution images.

Previously, pinhole images were recommended in case high magnification was needed, e.g., for evaluation of the caput femoris in osteonecrosis (see Chap. 12). With the currently available equipment, this is no longer necessary. Camera sensitivity and resolution have been improved and a zoomed image (1.5–4 times) of the area of interest, with appropriately increased acquisition time, will suffice. It is important to note that the information density is the relevant parameter here. In other words, if the zoom is 2, the imaged area of the object is only a quarter of the original matrix (both x and y dimensions are cut by half). Therefore, the acquisition duration needs to be increased by 4 in order to maintain the information density, i.e., acquire the same number of counts per pixel.

5.3.5
Dynamic or Multiphase Imaging

The movement of tracer immediately after the injection can be followed with flow imaging. Subsequently, the blood pool phase is acquired. After an interval of 2–3 h the delayed phase of bone scintigraphy is performed. Therefore, this protocol has been named "three-phase bone imaging." Certain groups have advocated four-phase imaging, in which case an additional 24-h view of the area under investigation is acquired. The camera size determines the body area that can be studied during the first or flow phase. For the other phases larger body areas can be evaluated, by acquiring multiple views.

For the flow phase, images of 2–4 s duration are acquired for a total time of 60–90 s. A matrix of 64 × 64 is sufficient. For the blood pool 500-kcount images in a 128 × 128 matrix are recommended. According to FOGELMAN, the blood pool phase needs to be completed within 10 min in order to limit the contribution of bony uptake (FOGELMAN et al. 1993; RYAN and FOGELMAN 1995).

The delayed images are usually recorded with high resolution, i.e., pixel size of 3–4 mm.

5.3.6
Whole Body Imaging

Whole body imaging is routinely used in most nuclear medicine clinics. The patient is scanned in posterior and anterior views. This can be accomplished by passing the patient through the camera gantry or by moving the detector over the patient on the stationary bed. Special dual-headed camera systems have been developed to image both sides in a single pass. This protocol is ideal for screening purposes and additional spot views may be acquired of suspicious areas. A 1024 × 256 matrix is needed with a scan time of 15–30 min per head.

5.3.7
Tomographic Imaging

Tomographic sections of a certain body part may be obtained with Single-photon emission computerized tomography (SPECT). This is only possible for the delayed phase, since tomography assumes an equilibrium distribution of the radioactivity in the body. An additional requirement is patient compliance with immobility. Special gantries have been developed to permit the camera to rotate around the patient. Currently, single-, dual-, and triple-headed systems are available. With more camera heads, shorter acquisitions may be achieved, greatly enhancing patient convenience and increase of throughput.

Best results are obtained with a 360° acquisition, a 128 × 128 matrix for high resolution, 3–6° angular steps, and 20–30 s per view. This results in a total acquisition time of 30–45 min, which is tolerable for most patients. Multiheaded systems are preferable to decrease the acquisition time to clinically acceptable proportions.

5.3.8
Equipment Quality Assurance

Obviously, high-quality imaging assumes properly functioning equipment, which meets all the criteria concerning homogeneity, resolution, linearity, etc. The demands on a tomographic system are higher than those on a conventional planar imaging system. Since more of the equipment is digitized and controlled by computers, rigorous testing routines are implemented. To this end, quality control procedures need to be performed on a frequent basis (daily, weekly, and monthly). It is beyond the scope of this book to go into any detail and the reader is referred to standard textbooks.

5.4
Image Interpretation

Knowledge of normal uptake in the skeleton is mandatory. This experience is usually gained through

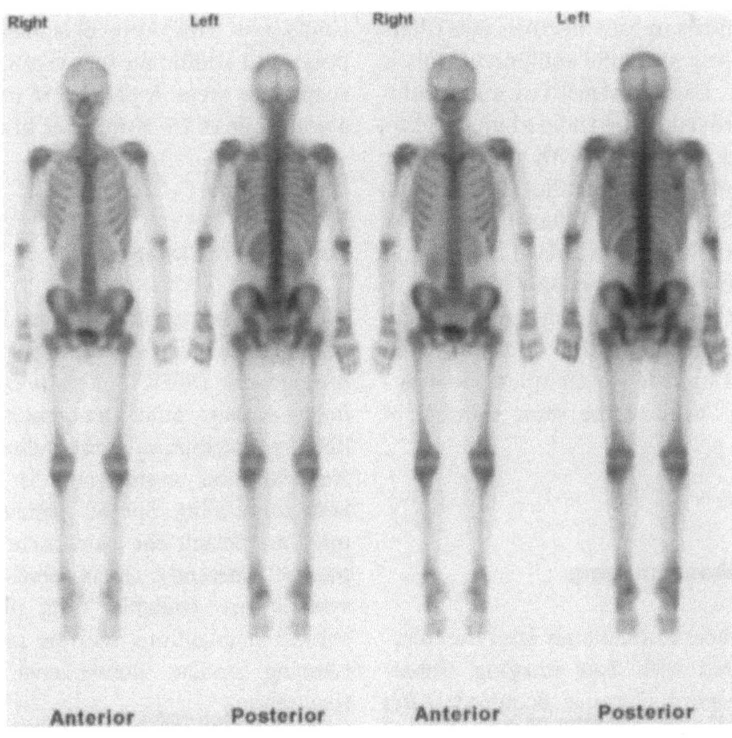

Fig. 5.2. Normal whole body bone scan of a 34-year-old white female. The images were taken 3 h after i.v. administration of 900 MBq ^{99m}Tc-MDP. The format is the so-called dual-intensity display, i.e., scaled to 75% and 100% of the maximum count, for both the anterior and the posterior view

exposure to training and interpreting sessions with experts. Fortunately, bone scintigraphy is a routine and common procedure, so that each radiology or nuclear medicine resident should easily get acquainted and become proficient during his or her training period. Normal variants, however, can be tricky, and many an atlas is devoted to these. An example of a normal whole body bone scan is given in Fig. 5.2.

The first step is to check for focal or diffuse abnormalities, i.e., areas of increased and/or decreased uptake. Since the human body is full of symmetries, the next step is to compare uptake on the left versus the right. Another hallmark is comparison of uptake between body parts. This is easily accomplished with the whole body mode. In the multiple spot-view mode of static imaging, an iso-time acquisition is necessary to compare uptake between images. In pediatric patients, the growth plates are active, which translates into increased uptake. Additional information may be retrieved from the different phases, e.g., increased uptake during the flow phase, indicating hyperemia. Multiphase imaging is important to differentiate increased uptake in the soft tissues from truly increased bone uptake (as seen in the third or fourth phase). Tomography greatly enhances contrast and eliminates superimposed activity by providing three-dimensional images, i.e., in axial, coronal, and sagittal planes.

Common pitfalls that may lead to false-positive results are patient rotation, obscuring the symmetry; genitourinary contamination; external artifacts, e.g., belt buckle, neck lace, earrings, breast prosthesis; dental procedures or disease; degenerative changes; and radiopharmaceutical problems.

A distinctive feature of bone scintigraphy is its high sensitivity for the detection of abnormalities such as fractures, infection, degenerative changes, metabolic bone disorders, and metastases, but the test is notoriously nonspecific. Many disease entities present with abnormal uptake on the bone scan. However, certain patterns may favor one diagnosis over another. For instance, a linear array of hot spots in the rib cage suggests fractures. Multiple scattered areas of focally increased uptake are highly suspicious for metastatic disease. Slight to moderately increased uptake in a diffuse pattern in joints suggests degenerative changes, especially when it is also seen in neighboring joints.

Needless to say, the clinical context is important, since it may focus the possibilities and limit the number of differential diagnoses. In the present dis-

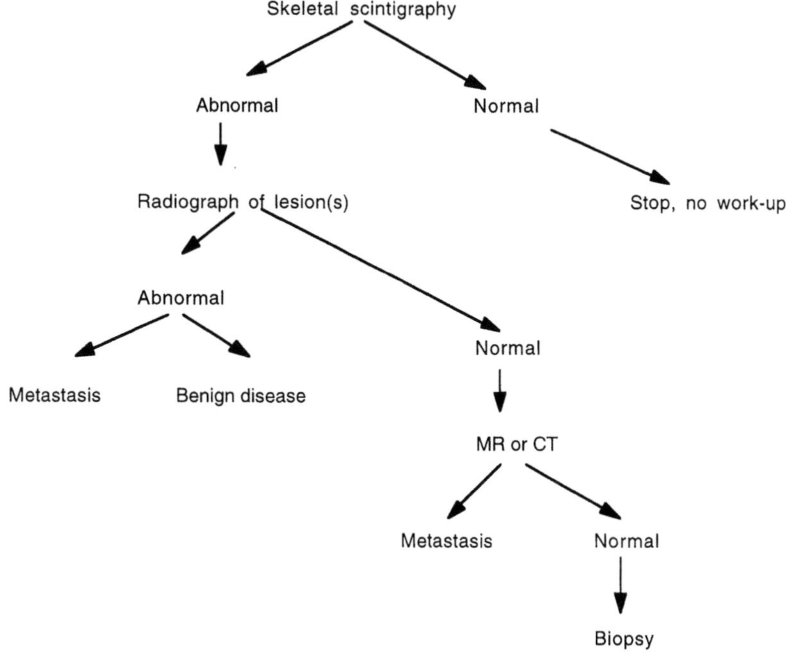

Fig. 5.3. Workup scheme for abnormalities seen on a bone scan of an oncologic patient

cussion, the image interpretation was purposely described first. It is our policy to read the films "blind or blank" to gather all available information. Secondly, the clinical history and signs and symptoms (e.g., through patient interview) are added and a final report dictated. This sequence prevents omissions and increases the likelihood that the majority of differential diagnoses are included.

The last but not least important step is correlative imaging (POMERANZ et al. 1994; RYAN and FOGELMAN 1995). In my view, it is impossible to provide the referring physician with adequate information if the bone scan is not interpreted in conjunction with other image modalities, such as conventional radiography, computed tomography (CT), magnetic resonance imaging (MRI), or ultrasonography (US). More sophisticated procedures such as CT (see Chap. 3), MRI (see Chap. 4), and angiography (see Chap. 7) are usually done after the bone scan, and are perhaps prompted by abnormalities detected on scintigraphy and plain films. The combination of findings of all imaging modalities and correlative interpretation should provide the diagnosis in most cases.

5.5
Clinical Applications

Skeletal scintigraphy is indicated in a whole set of situations. In the following sections we will address the most common ones.

5.5.1
Oncology

The intent here is to evaluate the skeleton for the presence and extent of malignant disease. By surveying the entire skeleton, this is a main referral indication for this highly sensitive study. It is superior to conventional radiography. Since the study is not very specific and in some malignancies the number of false-positives exceed that of true-positives, a combination of scintigraphy and radiography is necessary. For the spine, especially vertebrae, MR imaging is recommended to confirm presence or absence of bone metastases (see Chap. 19). A logical decision tree would be one as given in Fig. 5.3; the sequence may be modified as befits the individual laboratory and/or health system (POMERANZ et al. 1994).

In general, bone metastases reveal increased uptake (BROWN et al. 1993). Since the metastases are usually located in the bone marrow, it is not the metastasis itself that is seen on the bone scan, but the reaction of the bone to the expanding malignant bone marrow. In highly aggressive and fast expanding tumors, therefore, the lesions are cold, since there is not enough time for the bone to respond and the regional bone blood flow may be jeopardized to such an extent that the tracer cannot be delivered. Cold lesions have been reported for leiomyosarcoma, ductal breast cancer, multiple myeloma, etc.

Primary bone tumors generally show a very high uptake. Bone scintigraphy is indicated to evaluate the extent of disease and screening for metastases. Monitoring of therapy response is no indication since the bone scan remains positive for a long time. ^{201}Tl-chloride and ^{18}F-fluorodeoxyglucose (FDG) are better radiopharmaceuticals for this purpose.

Skeletal scintigraphy is extremely useful in the diagnosis and screening of osteogenic sarcoma, Ewing's sarcoma, and chondrosarcoma (see also Chap. 19).

An interesting finding is the so-called flare phenomenon, an increasing uptake in lesions and skeleton after initiation of chemotherapy, hemibody radiation or high-dose radionuclide therapy. In general, this is related to the response of affected bone to the therapeutic agents and is usually associated with a therapeutic effect.

5.5.2
Infection

For the evaluation of osteomyelitis a three-phase bone scan is performed in which there is usually increased flow to the affected area. The blood pool is also increased and the delayed images (third phase) show abnormal uptake in the bone, which further increases in the fourth phase. If the initial increased uptake decreases in time and appears not to affect the bones, a diagnosis of soft tissue disease such as cellulitis may be established. In dubious cases an infection survey with labeled white blood cells (WBCs) is recommended to check for localized infection (BROWN et al. 1993). Alternately, MR imaging of the affected area may be performed to check for bone marrow edema.

5.5.3
Trauma, Fractures

In general, the bone scan will be positive 1–2 days after the traumatic event. Fractures will show in-

creased uptake up to 1 year in about two-thirds of cases (COLLIER et al. 1993).

Sports injuries are an emerging field and bone scintigraphy is indicated to differentiate stress fractures from shin splints or periostitis, in athletes. This is very important because the therapy is so different for these entities.

5.5.4
Metabolic Bone Disease

The main disorder in this field is Paget's disease of bone. Osteoporosis is also a common referral indication, not to visualize the disease, but to assess effects of the disease such as compression and pathologic fractures. In the past, the 24-h retention index was popular as an indirect measure to assess bone mass. The retention index of diphosphonates was popularized by FOGELMAN in the early 1980s (FOGELMAN et al. 1993; RYAN and FOGELMAN 1995) and appeared useful in hyperparathyroidism, renal osteodystrophy, and osteomalacia. Currently, dual energy X-ray absorptiometry has replaced this application and fulfills this role of measuring bone density and bone mass. The method is precise and reproducible (see Chap. 9).

5.5.5
Benign Bone Disease

Osteoid osteoma is an extremely painful benign tumor, especially at night. If the radiographs are negative, scintigraphy is extremely useful, not only in diagnosis but also in evaluating the surgery. Recurrent pain with persistently increased uptake suggests a remaining nidus and need for reoperation. Fibrous dysplasia is an entity that can be confirmed with scintigraphy. In the case of exostoses, the activity may be related to the intensity of uptake and solitary or multiple presence may be assessed.

A routine referral is low back pain with normal radiographs. When planar scintigraphy is negative, SPECT needs to be performed to exclude facet joint disease, occult fracture, spondylolysis, or spondylolisthesis.

5.5.6
Vascular Bone Disorders

Decreased blood flow, either congenital, traumatic, or postsurgical, may lead to necrosis. In stage I (1–5

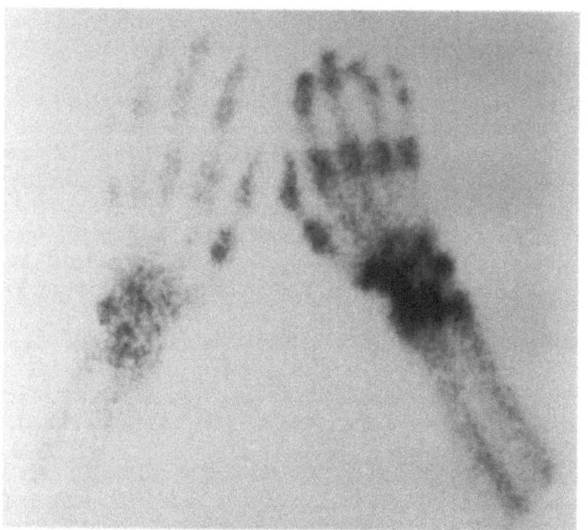

Fig. 5.4. Bone scan of a patient with RSD of the left arm and hand. Delayed or third phase image in palmar view. Note the typical periarticular uptake in the left phalangeal, metatarsal, and carpal bones, which is increased when compared to the normal hand/wrist on the right

weeks), this can be diagnosed with the bone scan as a cold lesion. Gradually, repair mechanisms are activated, leading to increased uptake around the site of injury. SPECT is very helpful to delineate the lesion, e.g., in avascular necrosis (see Chap. 12).

Common referral indications in this field are: osteonecrosis, Legg-Calvé-Perthes disease, slipped epiphysis, and ischemic injuries like frostbite and burns. Reflex sympathetic dystrophy (RSD) is a special syndrome, characterized by increased flow in stage I (3–6 months), and the typical increased periarticular uptake on the delayed scan. RSD is a complex entity with widely varying signs and symptoms. The contribution of scintigraphy in the diagnosis and therapy of upper extremity RSD has been reported (SCHIEPERS 1997). In Fig. 5.4 a scan is shown which reveals the increased uptake in joints of the affected limb.

5.5.7
Orthopedic Prostheses

Uptake is increased during the first year after a prosthesis (hip, knee, shoulder, or elbow implant). The time that the delayed scan is positive is somewhat longer for noncemented than cemented prostheses, limiting the usefulness of scintigraphy during the first months after surgery (RAHMY et al. 1994). Increased uptake around the stem and tip usually heralds loosening. The differential diagnosis with

infection has to be made by performing an infection survey with ^{67}Ga-citrate or labeled WBC and plain films. If the imaging findings are still inconclusive, addition of a colloid scan may be indicated to assess the presence and location of normal but displaced bone marrow (see Chap. 18).

Another complication is heterotopic ossification in the soft tissues. This is often encountered in people with orthopedic implants and also in paraplegics. The increased tracer uptake is clearly localized in the soft tissue. Another name for this abnormality is myositis ossificans.

5.5.8
Soft Tissue Abnormalities

Increased diphosphonate uptake can also be seen in the soft tissues. The main causes for these abnormal tracer localizations are:

1. Trauma or contusion, leading to cell necrosis and exposure of microcalcification sites, e.g., cardioversion, electric burns, infiltrated injection, infarcts (myocardium, brain), rhabdomyolysis, fat necrosis, muscle trauma, radiotherapy
2. Hematoma, vascular calcification, calcinosis, heterotopic bone formation
3. Infection and inflammation, e.g., cellulitis, surgical scar, tendinitis, dermatomyositis, polymyositis

Most institutions will use pyrophosphate as the tracer, since its avidity for soft tissue is the highest. Extensive experience with this agent is available for the evaluation of myocardial infarct 1–3 weeks after the event.

5.6
Conclusion

Bone scintigraphy is an extremely sensitive test to evaluate a large spectrum of abnormalities related to the skeleton. The study is nonspecific and plain radiographs are usually necessary to reduce the number of diagnostic possibilities. The addition of sophisticated imaging modalities as CT, MRI and angiography provides the opportunity for correlative imaging, which will yield the final diagnosis in the vast majority of patients.

References

Berding G, Burchert W, van den Hoff J, et al. (1995) Evaluation of the incorporation of bone grafts used in maxillofacial surgery with [^{18}F]fluoride ion and dynamic positron emission tomography. Eur J Nucl Med 22:1133–1140

Blau M, Nagler W, Bender MA (1962) Fluorine-18: a new isotope for bone scanning. J Nucl Med 3:332–334

Brown ML, Collier BD, Fogelman I (1993) Bone scintigraphy. Part 1. Oncology and infection. J Nucl Med 34:2236–2240

Charkes ND (1980) Skeletal blood flow: implications for bone scan interpretation. J Nucl Med 21:91–98

Collier BD, Fogelman I, Brown ML (1993) Bone scintigraphy. Part 2. Orthopedic bone scanning. J Nucl Med 34:2241–2246

Fogelman I, Collier BD, Brown ML (1993) Bone scintigraphy. Part 3. Bone scanning in metabolic bone disease. J Nucl Med 34:2247–2252

Hawkins RA, Choi Y, Huang SC, et al. (1992) Evaluation of the skeletal kinetics of ^{18}F-fluoride ion with PET. J Nucl Med 33:633–642

Hoh CK, Hawkins RA, Dahlbom M, et al. (1993) Whole body skeletal imaging with [^{18}F]fluoride ion and PET. J Comput Assist Tomogr 17:34–41

McLean FC, Budy AM (1964) Radiation, isotopes, and bone. Academic Press, New York

Phelps ME, Mazziotta JC, Schelbert HR (eds) (1986) Positron emission tomography and autoradiography: principal applications for the brain and the heart. Raven Press, New York

Pomeranz SR, Pretorius HT, Ramsingh PS (1994) Bone scintigraphy and multi-modality imaging in bone neoplasia: strategies for imaging in the new health care climate. Semin Nucl Med 24:188–207

Rahmy AI, Tonino AJ, Tan WD (1994) Quantitative analysis of technetium-99m-methylene diphosphonate uptake in unilateral hydroxy-apatite-coated total hip prostheses: first year of follow-up. J Nucl Med 35:1788–1791

Reeve J, Arlot M, Wootton R, et al. (1988) Skeletal blood flow, iliac histomorphometry, and strontium kinetics in osteoporosis: a relationship between blood flow and corrected apposition rate. J Clin Endocrinol Metab 66:1124–1131

Ryan PJ, Fogelman I (1995) The bone scan: where are we now? Semin Nucl Med 25:76–91

Schiepers C (1993) Skeletal fluoride kinetics of ^{18}F$^-$ and positron emission tomography (PET): in vivo estimation of regional bone blood flow and influx rate in humans. In: Schoutens A, Arlet J, Gardeniers JWM, Hughes SPF (eds) Bone circulation and vascularization in normal and pathological conditions. Plenum Press, New York, pp 95–101

Schiepers C (1997) Clinical value of dynamic bone and vascular scintigraphy in diagnosing reflex sympathetic dystrophy of the upper limb. In: Cooney WP (ed) Hand clinics, post-traumatic upper extremity RSD. Saunders, Philadelphia, pp 423–429

Schiepers CWJ, Hawkins RA, Choi Y, et al. (1990) Kinetics of bone metabolism assessed with ^{18}F$^-$ and PET. Eur J Nucl Med 16:450

Schiepers C, Geusens P, Vleugels S, et al. (1991) Positron emission tomography (PET) with ^{18}F$^-$ to evaluate metabolic rate in bone disorders. J Min Bone Res 6:S243

Schiepers C, Broos P, Nuyts J, et al. (1994) Positron emission tomography with F-18 fluoride in the high risk femoral head for osteonecrosis. J Nucl Med 35:35P

Schiepers C, Nuyts J, Bormans G, et al. (1997) Fluoride kinetics of the axial skeleton measured in-vivo with positron emission tomography (^{18}F$^-$ – PET): initial experience in metabolic bone disease. J Nucl Med 38:1970–1976

Subramanian G, McAfee JF (1971) A new complex of ^{99m}Tc for skeletal imaging. Radiology 99:192–198

Van Dyke D, Anger HO, Yano Y, Bozzini C (1965) Bone blood flow shown with ^{18}F and the positron camera. Am J Physiol 209:65–70

Wootton R, Reeve J, Veall N (1976) The clinical measurement of skeletal blood flow. Clin Sci Mol Med 50:261–268

Wootton R, Tellez M, Green JR, Reeve J (1981) Skeletal blood flow in Paget's disease of bone. Metab Bone Dis Rel Res 4 & 5:263–270

6 Ultrasound

J.A. Bouffard and M. van Holsbeeck

CONTENTS

6.1
Introduction

Ultrasound, which was introduced into medical imaging in the 1970s, is fast becoming an important method of diagnosis in musculoskeletal radiology. The most practical use of ultrasound is in the evaluation of soft tissue pathology. The other imaging modalities used to evaluate soft tissue structures include magnetic resonance imaging (MRI), computed tomography (CT), soft tissue radiography, xeroradiography, thermography, and nuclear medicine scans. MRI and ultrasound are currently the modalities of choice for imaging soft tissue structures and lesions.

The ever-continuing improvement of ultrasound equipment is benefitting bone radiologists. Machines, software and transducers now have technical refinements suited for musculoskeletal detail. The resourcefulness of the musculoskeletal radiologist lies in uncovering uses for ultrasound technology developed for other purposes such as breast imaging and intraoperative scanning. The number of applications of musculoskeletal ultrasound may surpass the number for abdominal ultrasound.

Some musculoskeletal radiologists are now developing newer maneuvers and techniques for the examination of diverse joints and muscle groups. Even joints such as the temporomandibular joints and the spine (JACOBSON et al. 1997) are now being investigated with ultrasound. The positioning of patients (CRASS et al. 1987), different stress maneuvers (MACK et al. 1988) and use of topographic landmarks are actually extensions of the clinical examination. The unselfish exchange of ideas between musculoskeletal imagers on how to perform ultrasound examinations in patients helps in establishing standardized protocols for ultrasound scanning.

The medical literature is now replete with information about the sensitivity and specificity of ultrasound in many areas of the musculoskeletal system. The effort to show that ultrasound has the accuracy to diagnose several lesions, such as rotator cuff tears or median nerve enlargement (LEE et al. 1995), has provoked a comparison with MRI, its closest competitor in soft tissue imaging. It is to the benefit of the bone radiologists that ultrasound continues to be proven accurate in the diagnosis of many diseases, especially in tendon pathology. Eventually, ultrasound and MRI will be considered complementary modalities.

Ultrasound also has the capability of delineating the different structures according to their echotextures and, thereby, of giving an excellent pictorial representation. Ultra-high-frequency energy shows multiple interfaces within the tendon as a template of internal structure (MARTINOLI et al. 1993). This imaging principle based on physical changes in composition is quite different from the imaging with MRI, which is based on changes in chemical composition. The small field of view may appear constricting to many physicians, who are accustomed to having a panoramic view of the structures they wish to visualize. This may not be true for

J.A. BOUFFARD, MD, Division of musculoskeletal Radiology, Henry Ford Hospital, 2799 West Grand Blvd., Detroit, MI 48202, USA
M. VAN HOLSBEECK, MD, Division of musculoskeletal Radiology, Henry Ford Hospital, 2799 West Grand Blvd., Detroit, MI 48202, USA

specialists who focus on exclusive areas such as joints in the case of rheumatologists or tendons in the case of hand surgeons. Extended field of view imaging has been developed for ultrasound, which may be appealing to physicians who wish to see a larger perspective (WENG et al. 1997).

Ultrasound is a modality that should be part of the armamentarium of the musculoskeletal imager. It is an extension of soft tissue radiography and a complementary technique to MRI. The ability of ultrasound to visualize minute structures such as the fascicular pattern of tendons or thickened synovium only enhances the ability of the radiologist to diagnose musculoskeletal disease. The mobility of the ultrasound equipment (and, soon, its portability) is essential for those patients who cannot be moved. This mobile modality is becoming ubiquitous in the hospital or clinical setting.

6.2
Equipment

High-frequency linear array transducers have improved musculoskeletal scanning dramatically. A complete range of transducers is available to the musculoskeletal sonographer, from 5 to 15 MHz. The 7.5-MHz linear array probe is used most often. However, it appears that transducers with frequencies centered above 10 MHz, in particular 13 through 15 MHz, will open newer applications in musculoskeletal sonography.

The 5-MHz curved linear, 10-MHz linear array, and 12-MHz linear array transducers complete the necessary number of scanheads to visualize musculoskeletal anatomy. The 5-MHz curved linear transducer is used for deeper structures such as the hip and popliteal fossa. The 10-MHz linear array probe can focus on small lesions like partial defects in tendons. The 12-MHz and higher frequency linear transducers may replace the 10-MHz probe for the exquisite definition of small structures such as ligaments and nerves in hand and foot pathology (SILVESTRI et al. 1995).

Assigning a curve to the face of a transducer has enabled the musculoskeletal sonographer to snugly coaptate the probe into certain areas such as the antecubical and popliteal fossae. Present-day curved linear array transducers have resolution and penetration comparable to the linear array probes. Footprints of the transducers vary from 2 to 4 cm. The latter is a more universal format; however, the smaller footprint has the advantage of insinuating

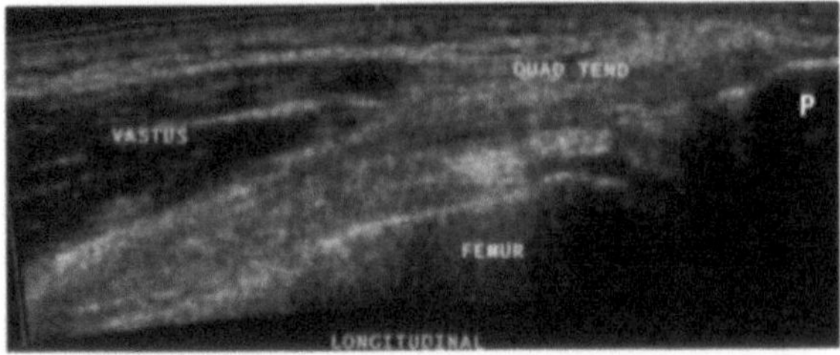

Fig. 6.1. Ultrasound extended field of view in the longitudinal plane of the quadriceps segment of the extensor mechanism, rivalling the perspective offered by MRI. *P*, Patella

into narrow spaces between bony elements. Not only the length of the footprint but also the curvature of the scanhead will be critical in addressing artifact caused by anisotropy.

Another recent advance in ultrasound is the extended field of view (WENG et al. 1997). This is a carryover technique of the articulated B-mode ultrasound that now strings an uninterrupted collage of spot views, thereby forming a single image of an entire segment such as the heel-cord complex of the Achilles tendon or elongated lesions such as a dissecting Baker's cyst. The ability to zipper together images to show an entire section of a limb now rivals the perspective of segmental anatomy offered by MRI (Fig. 6.1). An advantage of this new ultrasound technique is that the imaging plane can be adjusted to the type of pathology and it can be correctly formatted during the examination. Unlike MRI, image reconstruction in ultrasound is instantaneous and can be repeated as many times as necessary during the study.

6.3
Shoulder

Shoulder ultrasound is the most requested musculoskeletal sonographic study in our hospital. Age-related changes and overuse syndrome force many patients to seek medical attention for a painful shoulder. It is a convenient joint to study because the patient sits upright and is only partially undressed during the examination. Rotator cuff disease is best investigated with shoulder ultrasound. Lesions associated with rotator cuff disease, such as long biceps tenosynovitis and subacromiodeltoid bursitis, are within the field of exploration and are readily visualized. Ultrasound can therefore demonstrate both direct and contributing findings of rotator cuff pathology.

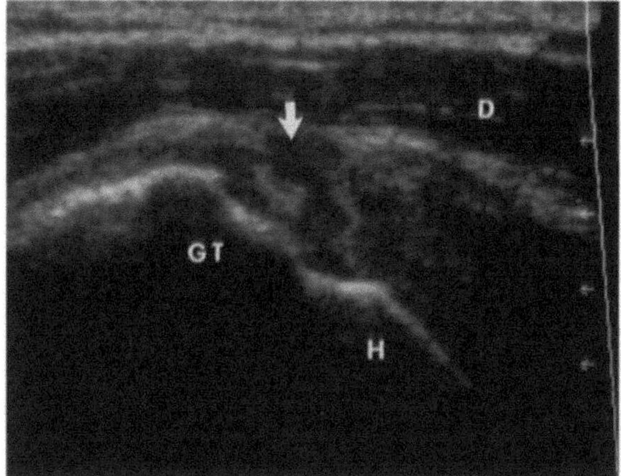

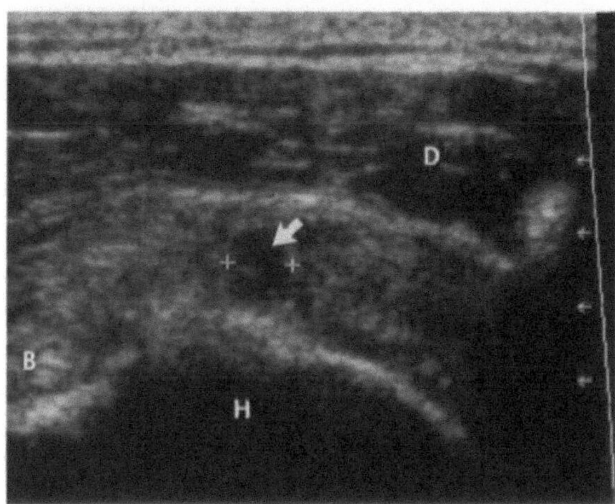

Fig. 6.2. Longitudinal (a) and transverse (b) scans of a full-thickness tear (*arrow*) of the supraspinatus in orthogonal planes. *H*, Humerus; *GT*, greater tuberosity; *D*, deltoid; *B*, long biceps tendon

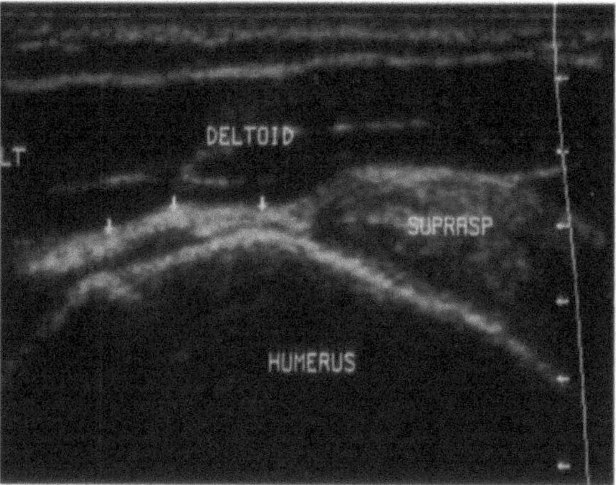

Fig. 6.3. Longitudinal view of a retracted supraspinatus often poorly termed "atrophy" (*arrows*) because of the hyperreflectivity of dipping subacromiodeltoid bursa and granulation tissue

The diagnosis of rotator cuff tears is the most important indication for shoulder sonography (VAN HOLSBEECK et al. 1996b). The three classical criteria of ultrasound for full-thickness tear are: nonvisualization of the rotator cuff, atrophy or retraction of the cuff tendon, and a sharply marginated hypoechoic tendon defect extending from the bursal to the articular tendon surfaces. The hypoechoic defect is the most common sign (Fig. 6.2). Very rarely, a hyperechoic defect may represent a full-thickness tear. It is presumed that this hyperreflective cleft in the rotator cuff may represent an acute tear with blood, or an interface artifact due to overlapping torn fascicles of the tendon. "Atrophy" of the rotator cuff as a sign of full-thickness tear is a misleading term (Fig. 6.3). The tendon volume loss is visualized as the tapering or dipping of the supraspinatus towards the ledge of the greater tuberosity. The ta-

pered defect has the same echotexture as the main tendon substance of the supraspinatus, thereby confusing physicians that there is merely "thinning" of the cuff. The hyperechogenicity of the tapered portion is from the hyperreflective peribursal fat/fascia and scar tissue. It is better to use the term "retraction" rather than "atrophy."

Rotator cuff tears commonly occur in the "critical zone" of the supraspinatus, which is the anterior first centimeter of the tendon adjacent to the bicipital groove. Cuff tears are usually classified as vertical or horizontal (VAN HOLSBEECK and INTROCASO 1993). Confirmation of tears is based on visualization in orthogonal planes and the presence of supporting signs such as tenosynovitis of the long biceps tendon or subacromiodeltoid bursitis. Tears isolated to the infraspinatus and subscapularis are less common.

Newer criteria for full-thickness rotator cuff tears have been recognized and include: naked tuberosity, cartilage interface, and deltoid herniation signs. The "naked tuberosity" sign is seen as cortical irregularity of the greater tuberosity as it is uncovered by the receding torn cuff. Three-quarters of patients with rotator cuff tears will have bony changes of the greater tuberosity; we have substantiated this in a cadaver study. The "cartilage interface" sign appears as a curvilinear high-level echo of the hyaline cartilage of the proximal humerus because the overlying rotator cuff defect offers no impedance to sound waves. Therefore, the size of the tear can be estimated measuring the length of the accentuated interface. The "deltoid herniation" sign appears as the overlying deltoid muscle forms a smooth "dimple"

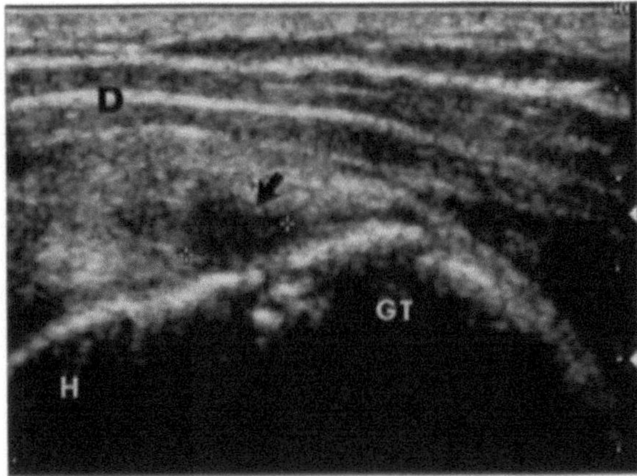

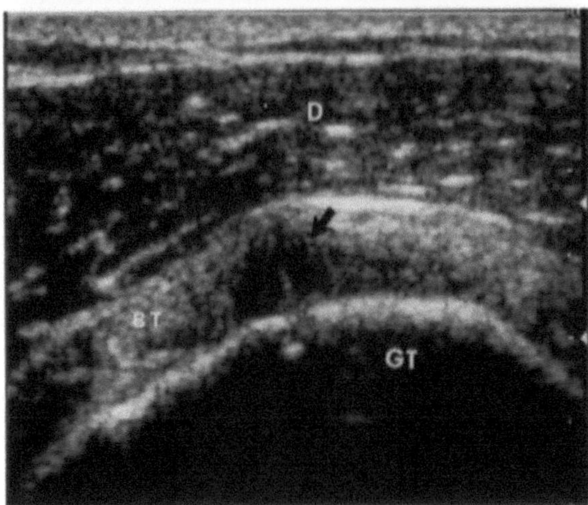

Fig. 6.4. Longitudinal (**a**) and transverse (**b**) scans of mixed-echogenic partial-thickness tear (*arrows*) of the supraspinatus. Note the subjacent cortical irregularity of the greater tuberosity. *H*, Humerus; *GT*, greater tuberosity; *D*, deltoid; *BT*, long biceps tendon

indenting the bursal surface of the rotator cuff and, when larger, gives the appearance of a pinched or flat tire against the rim of the convex humeral head (VAN HOLSBEECK and INTROCASO 1993).

The associated signs of full-thickness rotator cuff tears include tenosynovitis of the long biceps tendon, subacromiodeltoid bursitis, glenohumeral effusion and, sometimes, the "geyser" sign. Subacromiodeltoid bursitis may be seen as fluid distending the sac over the rotator cuff, elongated bursa below the ledge of the greater tuberosity, or focal/diffuse bursal thickening (VAN HOLSBEECK and STROUSE 1993; HOLLISTER et al. 1995). Glenohumeral effusion is seen as a hypoechoic "halo" forming around the hyperreflective triangular posterosuperior glenoid labrum (VAN HOLSBEECK et al. 1995). The "geyser" sign is a clinical finding adapted into musculoskel-

etal ultrasound. This sign represents fluid from the glenohumeral joint tracking across the torn rotator cuff and into the acromioclavicular joint, forming a mushroom-like fluid distention of the dorsal acromioclavicular capsule. The fluid appearance can be hypoechoic or anechoic.

Partial-thickness tears are more challenging to diagnose with ultrasound than are full-thickness tears (VAN HOLSBEECK et al. 1995). The former may appear as sharply marginated focal abnormalities but, more often, as mixed hypo- and hyperechogenic defects (Fig. 6.4). These focal lesions may be on the bursal aspect, intrasubstance, or at the articular surface of the cuff. The most common site is at the articulating surface in the proximity of the anatomic neck. A bursal aspect partial-thickness tear over the greater tuberosity may mimic the "atrophy" of the full-thickness tears. The incipient or earliest sign of cuff tears is the rim rent (CODMAN 1934), which is a very small partial-thickness tear. It appears as a "bull's-eye" or "halo" lesion at the cuff insertion into the greater tuberosity, often with subjacent cortical pitting (VAN HOLSBEECK et al. 1995). This defect should not be confused with the anisotropic artifacts seen at the distal anchor of the supraspinatus as it inserts into the greater tuberosity.

Ultrasound has also been helpful in the evaluation of other pathology of the shoulder. Lesions of the long biceps tendon including subluxation, tenosynovitis, and tears can be detected. Suprascapular ganglion is detected as a hypoechoic or cystic-appearing lesion within the spinoglenoid groove medial to the glenohumeral joint when investigating the infraspinatus recess. Diseases of the acromioclavicular joint such as separation or osteolysis may cause pain mimicking rotator cuff symptoms. These diagnoses can be made sonographically as well.

Undisplaced greater tuberosity fractures may be diagnosed easily, demonstrating interrupted cortices at the anatomic neck and metaphysis of the proximal humerus (PATTEN et al. 1992). After an episode of trauma, patients with radiographically occult greater tuberosity fracture are usually investigated with ultrasound because their symptoms simulate rotator cuff tears.

6.4
Elbow

The elbow is usually examined with ultrasound for joint effusion whether inflammatory or infectious, and for loose bodies. In the coronoid fossa, increased

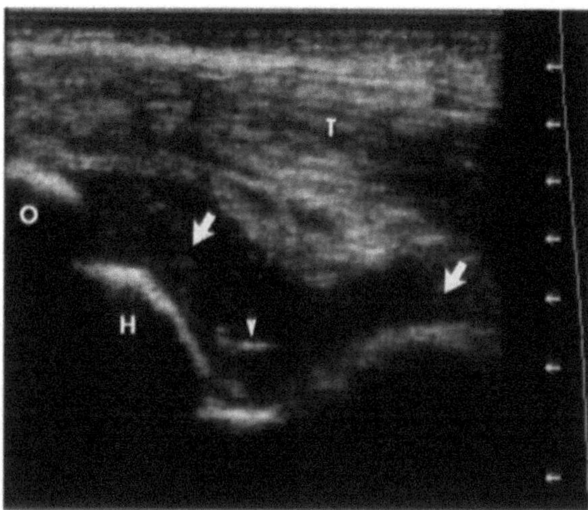

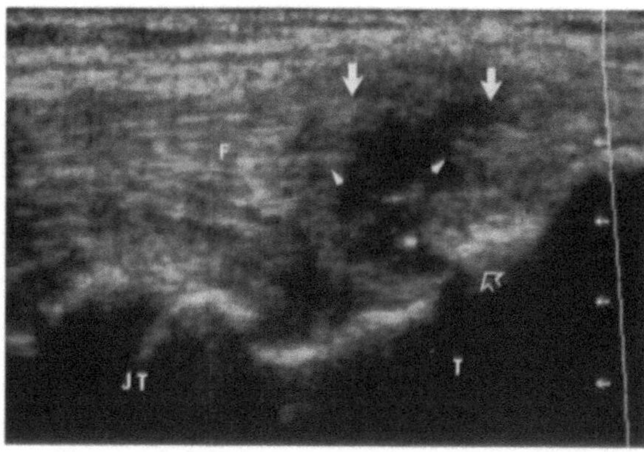

Fig. 6.6. Longitudinal scan of the medial elbow of a sportsman with golfer's elbow with hypoechoic fusiform enlargement of the common flexor tendon (*arrows*) with interrupted fibers (*arrowheads*) and irregular subjacent cortex (*open arrow*) of the medial epicondyle. *T*, Trochlea; *JT*, trochleoulnar joint; *F*, common flexor tendon

Fig. 6.5. Longitudinal scan of the posterior elbow with an effusive olecranon fossa (*arrows*) and a loose body (*arrowhead*) appearing as a solitary high-level echo. *O*, Olecranon process; *H*, distal humerus; *T*, triceps tendon

joint fluid causes the ultrasound equivalent of the "anterior fat pad" sign, and in the olecranon fossa, the equivalent of the "posterior fat pad" sign. Subacute hemarthrosis may appear hyperechoic and therefore silhouette out the uplifted anterior or posterior fat pads. It is not uncommon for the joint fluid in the elbow to be loculated in either the medial or the lateral recess. Ultrasound is helpful in determining where the largest pool of fluid is prior to ultrasound-guided aspiration.

Loose bodies in the elbow are best detected with ultrasound (Fig. 6.5) (VAN HOLSBEECK and INTROCASO 1991). The most common niche is the olecranon fossa. Loose bodies appear as high-level, usually mobile, echoes with posterior acoustic shadowing. Knowledge of the different recesses of the elbow enables the musculoskeletal sonographer to pinpoint loose bodies. If joint effusion is minimal or absent, saline arthrosonography with a sterile introduction of physiological saline and epinephrine helps distend the elbow capsule and enables the sonographer to confirm the floating or mobile loose body. Ultrasound is valuable not only in detecting loose bodies but also in localizing the fragments. Elbow arthroscopy is used for treatment. This technique needs accurate preoperative localization prior to successful surgery.

Tennis elbow or radial epicondylitis is seen as fusiform enlargement and segmental hypoechogenicity of the common origin of the extensor tendons as they insert into the epicondyle, along with some cortical irregularity of the subjacent bone (VAN HOLSBEECK and INTROCASO 1991). Similar tendon changes affect the common flexor tendons in golfer's elbow (Fig. 6.6). In another medial elbow pathology, acute or subacute ulnar collateral ligament tears can be seen as transverse hypoechoic clefts interrupting the ligaments. In some cases, these ulnar collateral ligament tears will be associated with hyperechoic avulsion factures. When the tears are chronic, association with intrasubstance heterotopic ossification is not uncommon. Transchondral injury of the capitulum is quite common in the "little league elbow." Ultrasound detects a combination of ulnar collateral ligament damage and osteochondritis dissecans in these patients. Loose bodies can form in these joints as well.

Tears of the distal long biceps tendon are challenging to diagnose with ultrasound but, nonetheless, readily evaluated with this modality. An acute tear is demonstrated as a measurable hypoechoic gap between the retracted portion of the tendon and the radial tubercle. This acute defect is seen at the level of the antecubital fossa, and better appreciated on the transverse views as a hypoechoic abnormality perifocal to the pulsating radial artery. The distal tuberosity insertion of the tendon is best seen with the forearm in pronation. All distal biceps tendons avulse from this radial tubercle.

Ultrasound is practical in investigating the posterior elbow, demonstrating tendonitis or tears of the triceps. Tears of this tendon are usually of partial thickness and often accompanied by a small avulsed bony fragment from the olecranon process. Ultrasound can distinguish fluid-filled olecranon bursitis from homogeneous chronic fibrosis of this elbow pad, which is helpful for physicians who have tried unsuccessfully to aspirate an unmistakable clinically enlarged bursa. The ulnar nerve or the "funny bone" at the level of the elbow can be investigated for tumors, neuritis, or subluxation.

Pediatric radiology is benefitting from musculoskeletal ultrasound which can visualize cartilage not seen on radiograph. In children who sustain trauma to the elbow, transchondral defects or occult epiphyseal abnormalities can be detected. In addition, radiocapitellar subluxation can be diagnosed because ultrasound clearly shows the hypoechoic cartilaginous articulating surfaces and their joint congruency. Ultrasound has helped us distinguish traumatic and congenital dislocation. Similarly, at birth, it has helped us distinguish epiphysiolysis from dislocation (VAN HOLSBEECK and INTROCASO 1991).

6.5
Hand/Wrist

Carpal tunnel syndrome and ganglion cysts are the most common indications for ultrasound in the hand and wrist. The course, morphology and neighboring structures of the median nerve are clearly visualized. Deflection of the median nerve from its usual course by masses, tenosynovitis, or encroaching lumbrical muscles of the clenched hand can be responsible for carpal tunnel syndrome (LEE et al. 1995). Ganglion cysts, granulomata, or myofibromata impinging on the median nerve cause similar symptoms. Median nerve enlargement at the proximal carpal tunnel can be measured with ultrasound, the results showing a good correlation to those of electromyography. At the level of the flexor retinaculum on the palmar aspect of the wrist, the median nerve is normal when it measures $12\,mm^2$ or less. The median nerve is abnormal in size if it measures $15\,mm^2$ or greater (Fig. 6.7) (LEE et al. 1995). Contour changes of the median nerve are important, and may be appreciated in both transverse and longitudinal views. A bilobed appearance or an existing "waistline" defect of the median nerve may be seen in symptomatic patients. A "buckle" deformity in the median nerve in the longitudinal view signifies an abnormality and can be correlated to the patient's symptomatology. A median nerve should be investi-

gated starting a couple of inches proximal to the medial elbow epicondyle and ending at the distal carpal tunnel. Occupational overuse can cause nerve edema at the pronator or carpal tunnel level.

Ganglion cysts, like any other cystic lesion, are readily investigated with ultrasound. Uncomplicated cystic masses appear completely anechoic with imperceptible walls. In musculoskeletal imaging, the usual underlying bony structures prevent the characteristic "through-transmission" of classical cysts. The most common location for ganglion cysts is over the scapholunate joint (Fig. 6.8). With proper positioning of the wrist and transducer compression, a communicating neck into the articular space of the scapholunate or radiocarpal joint may be uncovered. Ganglia of the wrist are often dorsal lesions, and can be accentuated by asking the patient to partially

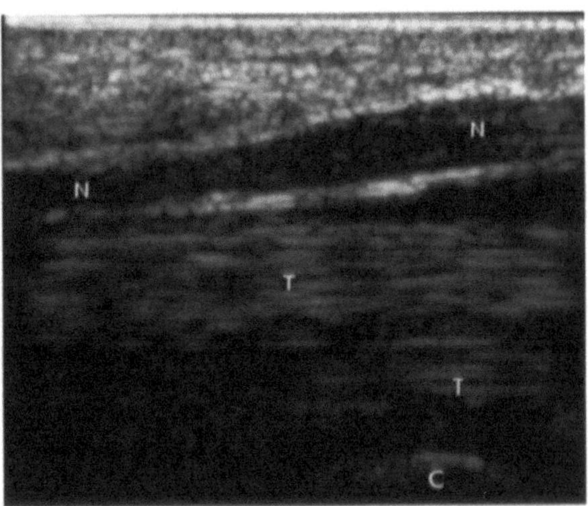
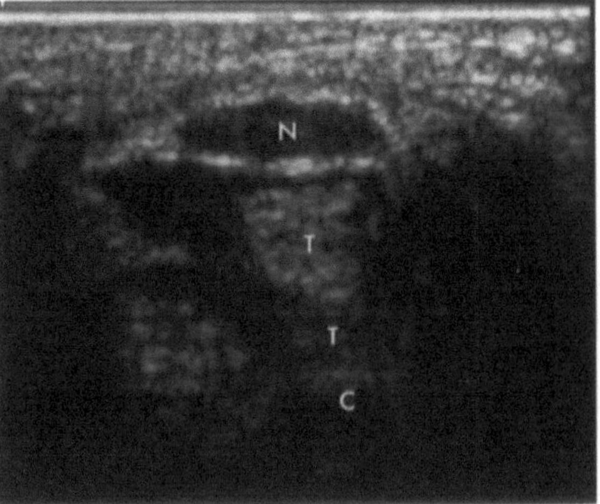

Fig. 6.7. Longitudinal (**a**) and transverse (**b**) scans of the enlarged median nerve (*N*), larger than the flexor tendons (*T*), in a patient with carpal tunnel syndrome. *T*, Superficial and deep tendons of the index finger; *C*, capitate

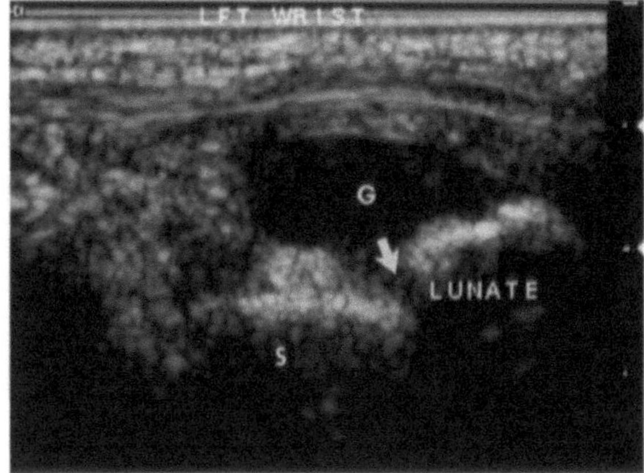

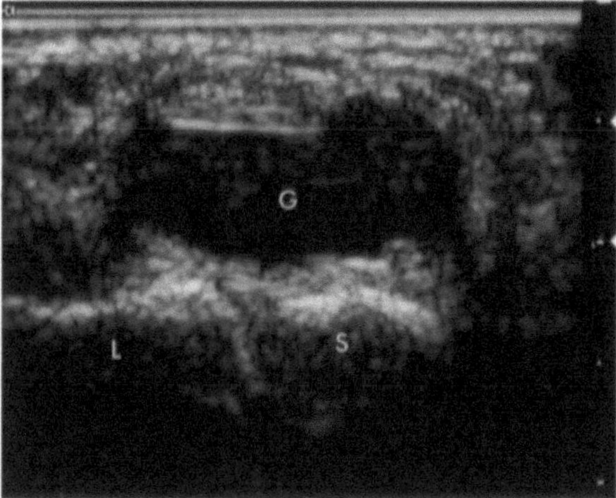

Fig. 6.8. Longitudinal (**a**) and transverse (**b**) scans of the dorsal wrist showing a scapholunate ganglion cyst (*G*) with intra-articular communications (*arrow*). *S*, Scaphoid; *L*, lunate

suspend his supinated hand and wrist in the air by grasping any available pole or handle of the ultrasound machine. Internal debris within the cyst indicates complications and may represent loose bodies, microbubbles from infection, cellular debris, or cholesterol crystals.

Ultrasound is important both in the diagnosis of ganglia and in accurate localization of the mass. Masses underneath tendons are difficult to diagnose and clinically appear as tenosynovitis, effusion, or tumor. Ganglion cysts perifocal to tendons may be mistaken for tenosynovitis, foreign bodies, or fibromata. The localization of the cyst in relation to the tendons and carpal bones is helpful in the operative planning. The cause of the ganglia can often be shown, such as a rent in the capsule of the joint, a focal degeneration of the tendon sheath, or an underlying bony protuberance. Ultrasound can clearly delineate any bony irregularities as a cause or effect

of the ganglion (VAN HOLSBEECK and INTROCASO 1991).

The other abnormalities of the hand and wrist for which ultrasound may be of value are lesions affecting the palmar aponeurosis and ligament tears in between or around carpal bones, which are usually post-traumatic. It is practical to use ultrasound in the detection and localization of retained foreign bodies. Acutely, these appear as short high-level echoes with posterior acoustic shadowing. Metallic foreign bodies cast a "comet tail" artifact. A chronic foreign body granuloma appears as a linear or punctate hyperreflective echo surrounded by a hypoechoic halo of granulation tissue. The diagnosis can be made sonographically but more importantly, the foreign body can be localized with great accuracy. The location relative to neurovascular structures and synovial spaces will expose potential complications. Ultrasound-guided removal can be considered in acute cases.

6.6
Thumb/Fingers

Musculoskeletal ultrasound has been very helpful in diagnosing traumatic lesions of the digits, mostly tendon lesions. Many clinicians can readily diagnose tendon and ligament lesions of the hand, but ultrasound helps to identify the extent and severity of the lesion and also helps classify the type of defect, establishing whether it is going to be a simple strain or tear of the tendon and identifying the presence of an accompanying bony avulsion fragment.

Gamekeeper's thumb is readily visualized as the unilateral enlargement and decreased echogenicity of the ulnar collateral ligament, without having to subject the thumb metacarpophalangeal joint to stress (BRONSTEIN et al. 1994). In addition, it is not uncommon to see a measurable difference in the distracted joint space of the affected thumb in comparison to the contralateral side (Fig. 6.9). The practical use of ultrasound is in detecting a complication of Gamekeeper's thumb, the Stener lesion. This is the displaced part of the proximal stump of the ulnar collateral ligament which retracts and rests above the adductor fascia rather than staying underneath. This malalignment prevents spontaneous healing and bridging with the distal stump. The Stener lesion is visualized in the longitudinal view as a hypoechoic round lesion resting on the condyle of the thumb metacarpal proximal to the metacarpophalangeal joint. The transverse view is ultrasonographically

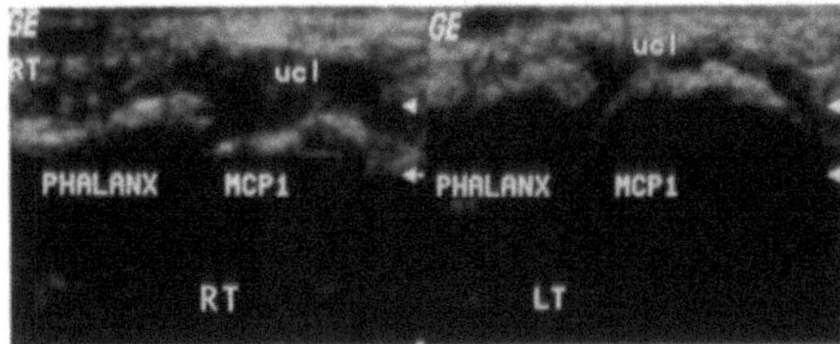

Fig. 6.9. Split-screen image of right (*RT*) and left (*LT*) thumb metacarpophalangeal joints (MCP 1 and phalanx) showing a right gamekeeper's thumb with an enlarged hypoechoic and torn ulnar collateral ligament (*ucl*)

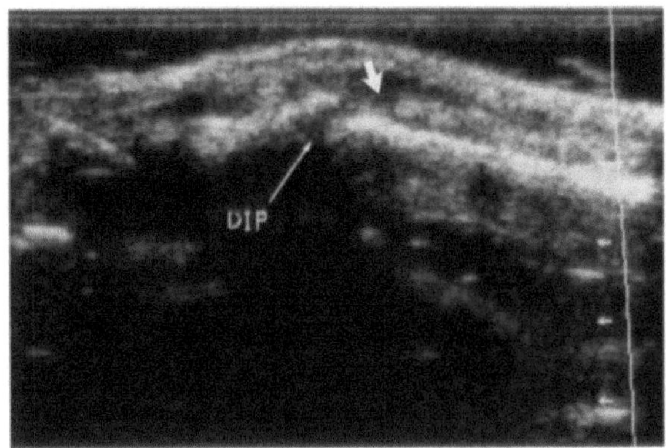

Fig. 6.10. Longitudinal scan of the dorsal distal fifth finger with mallet finger of tendon origin (*arrow*) seen as a hypoechoic tendon insertion. *DIP*, Distal interphalangeal joint

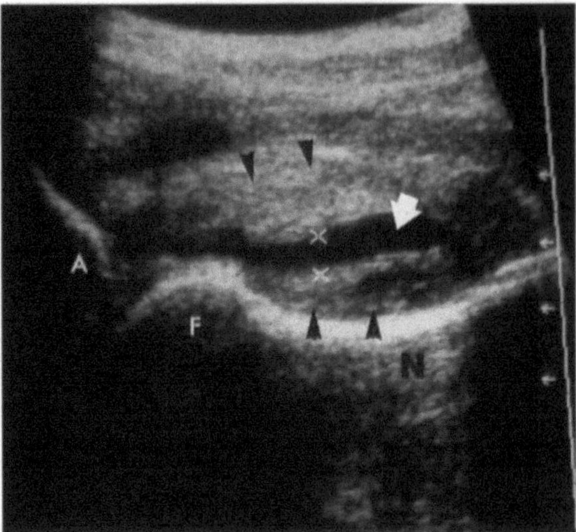

Fig. 6.11. Longitudinal scan of pediatric hip with chronic effusion (*arrow*) and resultant synovitis (*arrowheads*). *A*, Acetabulum; *F*, femoral head; *N*, femoral neck

pathognomonic and appears as a "bull's-eye" or "halo" lesion resting on the convexity of the metatarsal condyle but perched above the adductor fascia and adjacent to the extensor tendon.

Chronic tenosynovitis or DeQuervain's syndrome poses as a hyperechoic "boggy" distended tendon sheath around the extensor pollicis brevis and abducter pollicis longus at the level of the anatomic "snuff box" (VAN HOLSBEECK and INTROCASO 1991). Ultrasound is helpful in determining whether the tendons share a common sheath or two separate envelopes, in the latter case warning the hand surgeon that he/she should infiltrate both tendon sheaths to achieve the desired therapy.

Dorsal or extensor hood injury is commonly seen in rheumatoid patients and in individuals who sustain direct trauma to their knuckles. The extensor tendon at the level of the knuckles is held centrally over the condyle of the metacarpal by sagittal bands. A tear commonly on the radial aspect of the sagittal band causes the extensor tendon to sublux towards the ulna. With the real-time capability of ultrasound,

the extensor tendon can be seen to sublux as the patient clenches his fist (VAN HOLSBEECK and INTROCASO 1998). The other tendinous injuries of the hand that may be investigated with ultrasound are the boutonnière deformity and mallet finger (VAN HOLSBEECK and INTROCASO 1998). The former is seen as a defect of the extensor tendon over the proximal interphalangeal joint and as a hypoechoic segment or interrupted fibrillar pattern. The mallet finger deformity (Fig 6.10) of tendon origin can be diagnosed with ultrasound as a strain with a hypoechoic segment over the distal interphalangeal joint of the finger or a retracted defect of the tendon. A small avulsion fracture of a mallet finger of bony origin may be appreciated by ultrasound and yet be occult on x-ray.

A tear of the pulley can result in abnormal bowstring deformity of the flexor tendons. Ultrasound can confirm this diagnosis dynamically by showing how the tendon pulls away from the proximal phalanx during resisted finger flexion (VAN HOLSBEECK and INTROCASO 1998).

6.7
Hip

Hip effusion is difficult to recognize on radiographs. Detection can be achieved with MRI or ultrasound. Ultrasound is the most practical way to detect effusive hips. A right-left hip comparison is helpful to establish the presence of extra fluid in the hip. Fluid usually appears hypoechoic and pools over the femoral neck on the longitudinal view (Fig. 6.11). Greater amounts of fluid distend the capsule further and form a "halo" around the femoral head (MARCHAL et al. 1987). Ultrasound can characterize accompanying synovitis of the hip without the use of contrast such as gadolinium in MRI. Synovial ostochondromatosis, loose bodies, or pigmented villonodular synovitis are lesions of the hip demonstrable by ultrasound.

A common and often forgotten lesion of the hip is bursitis, predominantly affecting the greater trochanter bursa. Bursitis appears as a curvilinear hypoechoic mass draped around the greater trochanter. Synovial thickening accompanying bursitis is indicative of inflammatory or infectious origin. Other bursae, around the iliopsoas and lesser trochanter, can also be visualized (VAN HOLSBEECK and INTROCASO 1998).

Ultrasound is useful in the workup of postoperative hips that are suspected to have infection or loosening. Infection shows markedly increased fluid in the joint. The fluid separating the capsule from the native femoral neck at the level of the calcar is the best place to measure. The presence of extra-articular fluid collection enables the examiner to aspirate separate pools of fluid, whether extra-articular or intra-articular. It has been found that extra- or peri-articular fluid is often secondary to infection (VAN HOLSBEECK et al. 1994).

6.8
Knee

The knee is the most commonly injured joint. While ultrasound should not compete with MRI in the detection of intra-articular lesions of the knee, it is more practical than MRI for visualizing extra-articular structures. The extensor mechanism, ligaments, tendons, and capsule of the knee are better investigated with ultrasound. The popliteal fossa is within reach of ultrasound as well (VAN HOLSBEECK and INTROCASO 1991).

Tears of the tendons of the extensor mechanism appear as discrete hypoechoic defects interrupting the fibrillar pattern of the quadriceps or patellar tendons. Quadriceps tendon tears usually occur 1–2 cm above the base of the patella. Musculotendinous junction tears are also common. The eccentric full-thickness tears of the quadriceps tendon usually affect the musculotendinous junction of either the

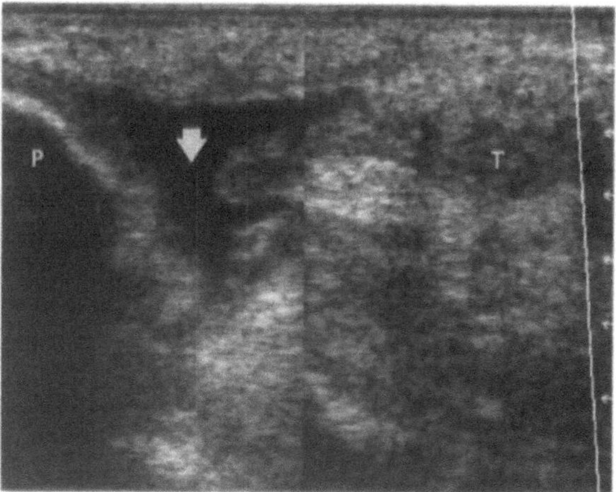

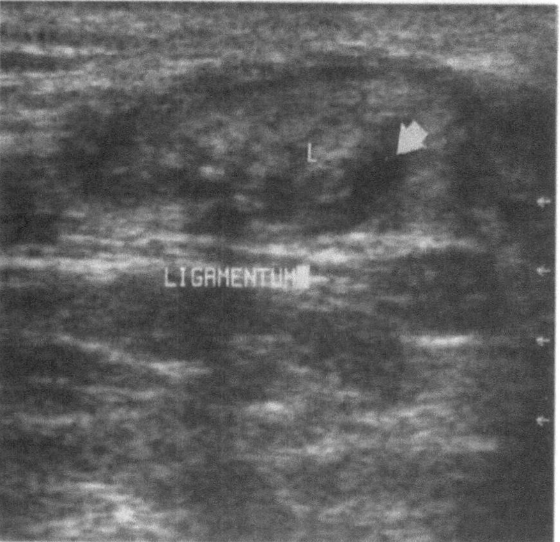

Fig. 6.12. Longitudinal (a) and transverse (b) scans of the proximal patellar tendon with an acute anechoic full-thickness tear (*arrow*) subjacent to the patellor apex (*P*). *T*, Slackened torn distal stump of the patellar tendon; *L*, torn ligamentum patella in transverse view

vastus lateralis or the vastus medialis. Total discontinuity of the quadriceps tendon can be seen with ultrasound, but can confound the clinical examination because of the marked overlying soft tissue swelling. Focal or segmental tears of the quadricep tendons can be distinguished from tendon discontinuity which necessitates surgery. Tears of the distal quadriceps tendon can rip into the prepatellar bursa, in what may appear to be extra-articular prepatellar bursitis (STROME et al. 1995).

Ultrasound is excellent for the evaluation of patellar tendon disease. Tendinitis or jumper's knee of the patellar tendon appears as segmental hypoechoic enlargement of the tendon subjacent to the patellar apex. In the transverse view, this is often a focal and nodular hypoechoic lesion defect most commonly in the central third of the patellar tendon. Diffuse patellar tendinitis shows fusiform enlargement of the proximal patellar tendon and hypoechoic edema of the tendon substance, and comparison to the asymptomatic knee can confirm this. Patellar tendon tears (Fig. 6.12) most often occur subjacent to the patellar apex as a discrete sharply marginal abnormality usually transverse or obliquely oriented. Calcific tendinitis usually appears as a focal curvilinear high-level echo with posterior acoustic shadowing within the edematous patellar tendon. Tendon tears of the extensor mechanism may be bilateral, easily investigated by ultrasound with right-left comparisons.

Osgood-Schlatter's disease is a clinical diagnosis. With ultrasound, it is confirmed as fusiform hypoechoic enlargement of the distal patellar tendon at the tibial tuberosity insertion associated with irregularity or fragmentation of the bone–cartilage apophysis of the tibial tuberosity, distention or inflammation of the deep infrapatellar bursa, and often synchronous superficial infrapatellar bursitis. The mirror lesion of the proximal patellar tendon, Sinding-Larsen-Johanssen disease, appears as the pediatric equivalent of proximal jumper's knee but also in association with osteochondral fragmentation of the patellar apex and calcification (BOUFFARD et al. 1993).

The suprapatellar bursa serves as the "window" with a view on the intra-articular status of the knee. Ultrasound can detect acuity with anechoic and uncomplicated fluid distending the suprapatellar pouch, or chronicity with irregular synovial thickening readily differentiated from hypoechoic fluid containing floating debris and, sometimes, lipoma arborescens (VAN HOLSBEECK and POWELL 1995).

Cartilaginous defects of the knee, especially of the femoral condyles, can be observed with ultrasound with the proper flexion and extension of the knee. In transverse or longitudinal views, normal hyaline appears as a smooth thick black stripe atop the intact subchondral plate of the convex anterior and posterior femoral condyles. Osteochondral defects cause contour deformities of the cartilage and subchondral plate fragmentation. In addition, the posterior patellar cartilage can be investigated with ultrasound by laterally pushing the patella in an extended knee and placing the transducer obliquely and longitudinally along the lateral facet. A medial push on the patella will allow one to look at the medial facet. This imaging of patellar cartilage is useful in chondromalacia

and osteochondritic diseases (VAN HOLSBEECK and POWELL 1995).

Medial collateral ligament injury on ultrasound appears as hypoechoic swelling of the ligament, often at the femoral insertion. Strains demonstrate no discrete interruption of the medial collateral ligament, while acute or subacute tears appear as jagged hypoechoic defects clearly interrupting the fibrillar pattern of the collateral ligament again at the femoral condylar level. Lateral collateral ligament strain or tears usually demonstrate contour change of the normally straight oblique hypoechoic ligament. A tear appears as a truncated bulbous hypoechoic stump of the lateral collateral ligament at its insertion into the head of the fibula (VAN HOLSBEECK and INTROCASO 1991).

Meniscal cysts are easily seen on MRI. Sometimes the intrasubstance meniscal tear cannot be appreciated because of close coaptation. With ultrasound, this is readily detected as a hypoechoic cyst sitting at the base of the usually triangular hyperechoic meniscus. The often associated meniscal tear can be distracted with graded valgus or varus stress. Ultrasound may be more practical than MRI when differentiating between meniscocapsular separation or meniscal cyst. Ultrasound has the benefit of real-time and stress imaging. In traumatic meniscocapsular separation, stress views create a total and wider plate-like separation between the outer margin of the meniscus and the detached capsule, while a meniscal cyst remains as a hypoechoic round focus at the base of the meniscus (VAN HOLSBEECK and INTROCASO 1991).

The pes anserinus is a small structure that can be overlooked by the relatively large field of view used in MRI. Bursitis and/or tendinitis may both appear as amorphic increased signal on T2-weighted images but a merely hypointense widening on T1-weighted images. Ultrasound can discriminate between the hypoechoic enlarged tendinitis and a well-defined hypoechoic cystic lesion representing bursitis or ganglion cyst.

Baker's cyst should be evaluated with ultrasound. This appears in the transverse view as a hypoechoic "boomerang lesion," with the apex towards the hyperreflective semimembranosus tendon at the level of the posterior femoral condyle (Fig. 6.13). On longitudinal scanning, a Baker's cyst appears like a bag sagging from the crossing of the semimembranosus and medial gastrocnemius tendons with its cul-de-sac pointing caudad towards the calf of the leg. The practicality of ultrasound is in its ability to detect not only the presence or size of a Baker's cyst,

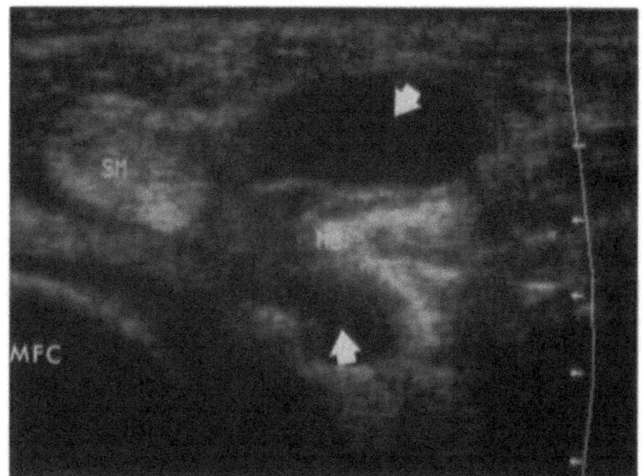

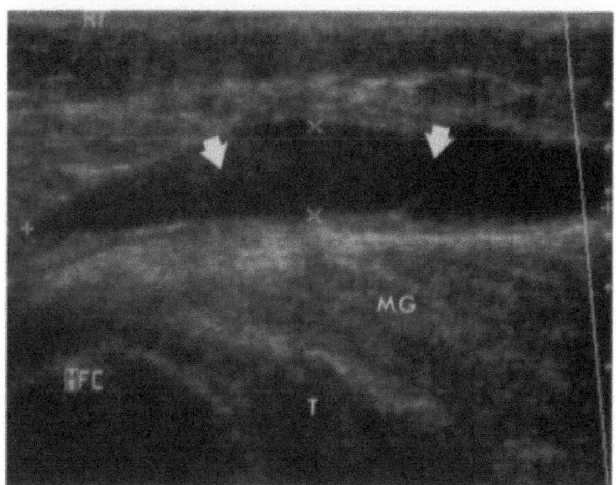

Fig. 6.13. Transverse (a) and longitudinal (b) scans of the popliteal fossa with an uncomplicated Baker's cyst (*arrows*). *SM*, Semimembranosus tendon; *MG*, gastrocnemius; *MFC*, posterior medial femoral condyle; *T*, tibial condyle

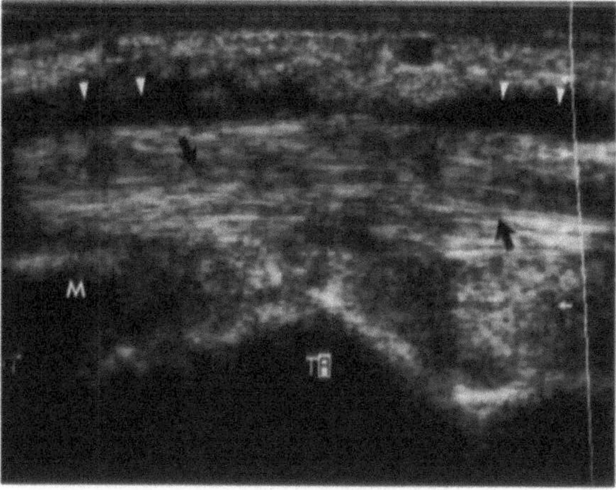

Fig. 6.14. Longitudinal scan along the medial ankle with diffuse hypoechoic posterior tibial tendinitis (*arrows*) and attendant tenosynovitis (*arrowheads*). M, Medial malleolus; *TA*, talus

but also accompanying complications such as synovitis, loose bodies, rupture, or leaking. Without the benefit of radiographic contrast, synovitis and chronic or inflammatory disease of the Baker's cyst appear as irregular undulating thickening of the wall, while loose bodies appear as curvilinear high-level echoes with posterior acoustic shadowing that are usually mobile and change in position between prone and decubitus views. Rupture or leaking is diagnosed when the usually round cul-de-sac of the Baker's cyst in its most caudad portion converts into a pointed or "stiletto" tip with fluid tracking down between the subcutaneous layer and the adjacent muscle fascia. Active leaking or rupture of the Baker's cyst will show calf muscular edema, often with distended intramuscular veins. Ultrasound has the advantage of being able to track dissecting Baker's cysts all the way down to the level of the ankle (VAN HOLSBEECK and INTROCASO 1991).

Ultrasound can be complementary to MRI or CT in the characterization of popliteal masses, whether vascular, neurogenic, or lipomatous. Color or power Doppler capability of ultrasound is instrumental in distinguishing vascular lesions and neurogenic tumors (VAN HOLSBEECK and INTROCASO 1991).

6.9
Ankle

The ankle is a joint investigated frequently with ultrasound, possibly second only to the shoulder. The imaging of the tendons of the ankle should be done with ultrasound, and the examination includes the posterior tibial, peroneal, and Achilles tendons. These three main tendons of the ankle are visualized throughout their entirety. The second practical use of ultrasound of the ankle would be in the investigation of the tibiotalar joint for effusion or loose bodies. Lesions affecting the ligaments of the ankle would be a third indication (VAN HOLSBEECK and INTROCASO 1991).

Chronic tendinitis or tendinosis appears as hypoechoic intrasubstance vacuoles which may coalesce, forming larger hypoechoic geographic defects. The tendon shows fusiform or diffuse enlargement. Partial tears may arise from these intrasubstance defects may arise, extending only towards one surface of the tendon (RESNICK and KANG 1997).

The posterior tibial tendon is a commonly afflicted structure as it is one of the main supports for the plantar arch of the foot. Tendinitis is visualized

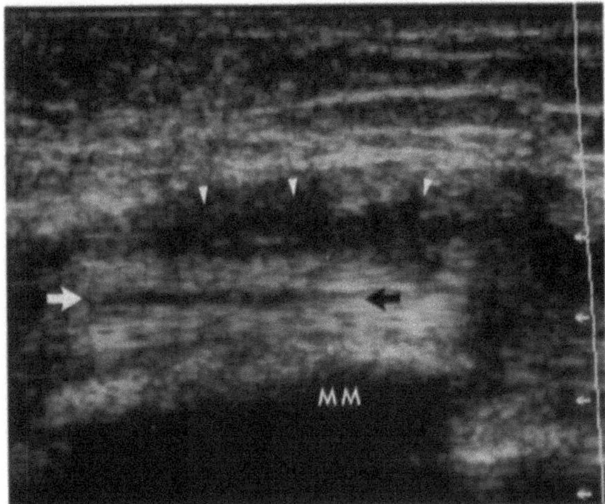

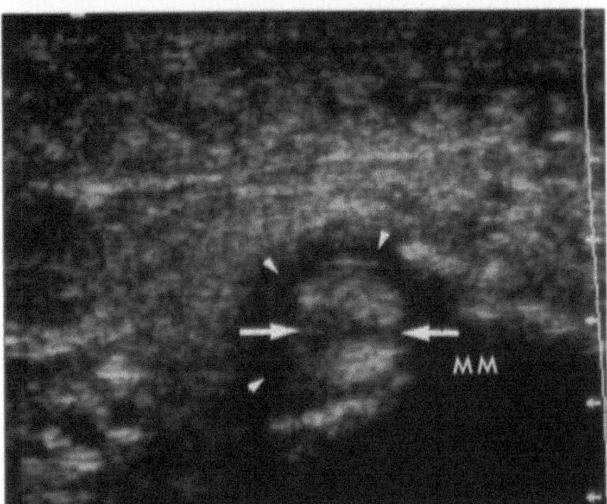

Fig. 6.15. Longitudinal (a) and transverse (b) scans of the medial ankle along the inframalleolar segment of the posterior tibial tendon with a longitudinal tear (*arrows*) and a "coffee bean" appearance on the transverse view (b). Note resultant tenosynovitis (*arrowheads*). *MM*, Medial malleolus

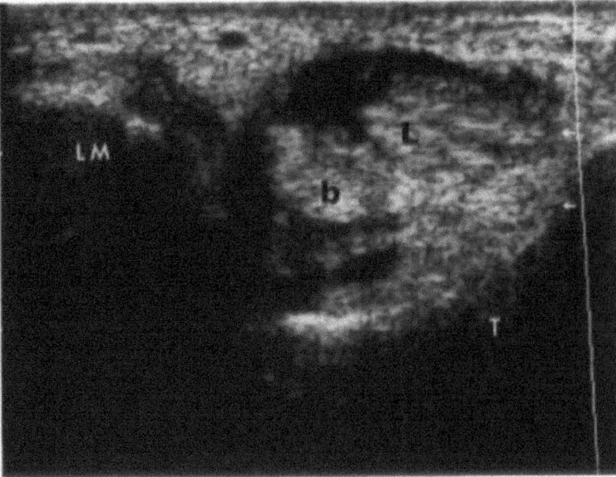

Fig. 6.16. Transverse view of the lateral ankle with peroneal tenosynovitis appearing as a "halo" of fluid in the common tendon sheath of both tendons. *L*, Longus; *b*, brevis; *LM*, lateral malleolus; *T*, talus

as hypoechogenicity of the tendon with increased interfibrillar distances (Fig. 6.14). This may be focal or diffuse. Tenosynovitis appears as a hypoechoic "halo" surrounding the tendon, and may also be focal or diffuse. Tears of the posterior tendon most often occur at the inframalleolar level, sometimes at the malleolar level, and, rarely, at the supramalleolar level. Tendon discontinuity is often seen as an "hourglass" appearance between the hypoechoic proximal and distal stumps of the tendon. The residual bridge between the distracted tendon ends, which represent the strung-out tendon sheath, may appear as slackened or serpiginous. A unique tear of the posterior tibial tendon, difficult to diagnosis with other modalities, is the intrasubstance longitudinal split (Resnick and Kang 1997). This is clearly seen

as a sharply marginated longitudinal hypoechoic cleft along the long axis of the tendon on the longitudinal view and as a transverse hypoechoic cleft in the transverse view, giving the posterior tibial tendon a "coffee bean" appearance (Fig. 6.15) (van Holsbeeck et al. 1996a).

Insertion tendinopathy of the posterior tibial tendon is visualized as fan-shaped hypoechoic enlargement of the tendon as it inserts into the navicular. This is seen usually in individuals with accessory navicular and less commonly in those with os externum tibiale, another ossicle in the distal posterior tibial tendon.

Peroneal tendons suffer the same fate as posterior tibial tendons, but subluxation is more common. Real-time imaging of subluxing peroneal tendons, because of torn retinaculum, can be visualized by applying stress upon the ankle. The foot is dorsiflexed and the hind foot is everted, forcing the peroneal longus tendon to roll externally over the brevis or both tendons to barrel anteriorly over the lateral malleolus. Peroneal tendons share a common sheath, and tenosynovitis is represented as a typical "halo" surrounding both tendons (Fig. 6.16) (Resnick and Kang 1997).

Tears may be selective, affecting either the peroneal brevis or the peroneal longus or, sometimes, both. Tears of the peroneal tendons do not necessarily occur at the same levels. A particular type of tear is seen in the peroneal tendons. The tendency of these tendons to sublux over the fibula causes the peroneal brevis to be sandwiched between the subjacent fibula and the overlying peroneal longus.

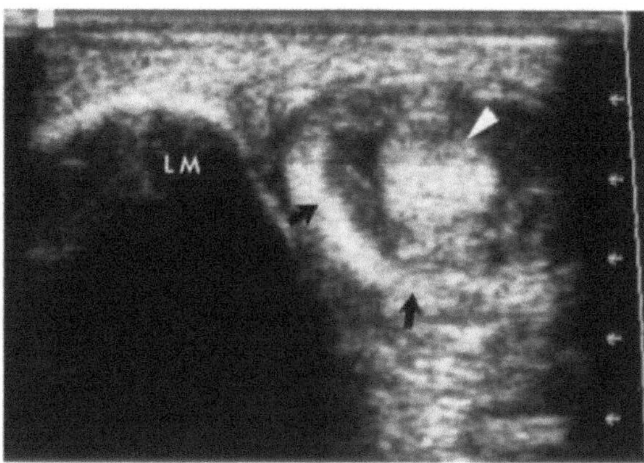

Fig. 6.17. Transverse view of the lateral ankle with the "rosette" appearance of a chronic longitudinal split of the peroneal brevis (*arrows*) and peroneal longus tendinitis (*arrowhead*). *LM,* Lateral malleolus

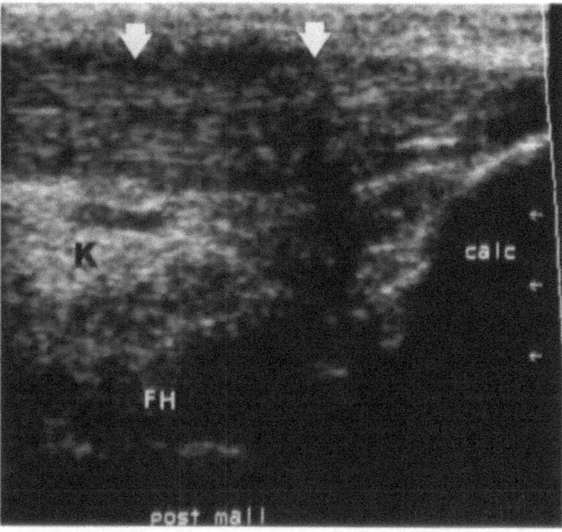

Fig. 6.18. Longitudinal scan of the posterior heel showing focal tendinitis (*arrows*) of the swollen hypoechoic distal Achilles tendon as it inserts into the calcaneus (*calc*). *FH,* Flexor hallucis longus; *K,* Kager's triangle

With the longus crowding the brevis against a usually hypertrophic spur of the fibula, there is chafing of the brevis, and a predominantly longitudinal tear often causes the brevis to split into two parts. The defect has a "rosette" appearance in a transverse view. The peroneal longus represents the usually hypoechoic inhomogeneous center while the split brevis forms the ring around the cluster (Fig. 6.17). Typically, this defect occurs at the malleolar level (RESNICK and KANG 1997).

Disease of the Achilles tendon covers the entire spectrum of tendon pathology, ranging from paratendonitis to complete tendon discontinuity. Investigation of this tendon by ultrasound is usually focused at the level of the posterior malleolus. The defects of the Achilles tendon, whether tendonitis or tears, most often occur at this level. Tendonitis appears as lengthy fusiform hypoechogenicity of the tendon (Fig. 6.18). This often affects the entire tendon width, but may appear focally eccentric towards the medial or lateral aspect of the tendon. In the transverse view, the normal Achilles tendon appears reniform with its concavity directed medially. In tendinitis, the Achilles tendon assumes a round configuration. Right and left comparison helps in characterizing the severity and extent of the tendinitis (RESNICK and KANG 1997).

Tears of the Achilles tendon appear as a measurable gap, again at the level of the posterior malleolus. Unique to the evaluation of Achilles tendon tears is the degree of herniation by the anterior fat pad of Kager's triangle (Fig. 6.19). Insinuating Kager's fat into the Achilles tendon defect inhibits spontaneous healing. Tears of the Achilles tendon are evaluated in

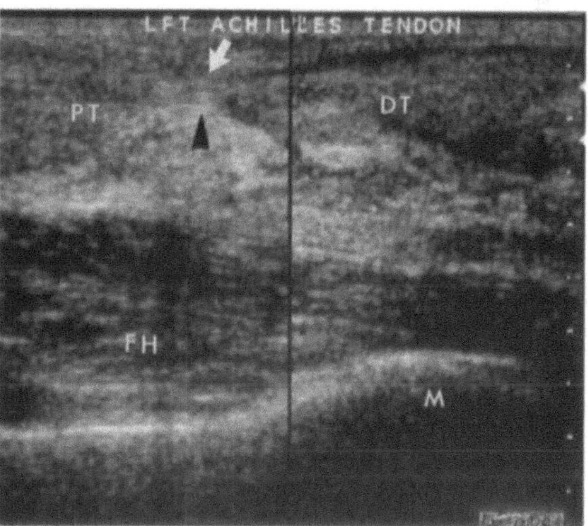

Fig. 6.19. Longitudinal scan of the posterior ankle with a sub-acute tear (*arrow*) of the Achilles tendon at the level of the posterior malleolus (*M*). Kager's fat has herniated into the interrupted Achilles tendon substance (*arrowhead*). *FH,* Flexor hallucis longus; *PT,* proximal Achilles stump; *DT,* distal Achilles stump

the neutral, plantar flexion, and dorsiflexion positions. The amount of cephalocaudal translation of the separated tendon stumps determines the complexity of surgical repair and implies chronicity of the tear. Evaluation of these tendon tears through ultrasound is very helpful in individuals who exhibit severe overlying soft tissue swelling, when clinicians

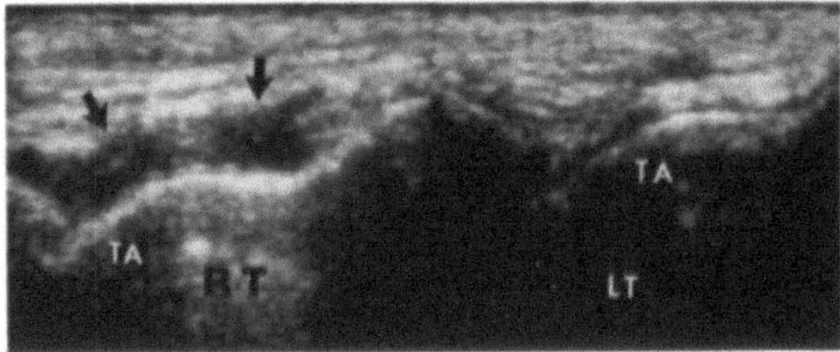

Fig. 6.20. Dual-screen image of the anterior ankles in the longitudinal plane. The capsule is distended (*arrow*) in the effusive right (*RT*) tibiotalar joint in comparison to the asymptomatic left joint (*LT*). *TA*, Talar dome

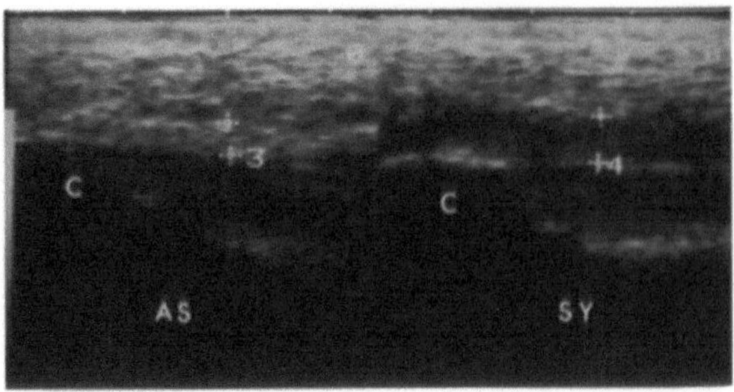

Fig. 6.21. Split-screen image of the plantar heels in the longitudinal view. The symptomatic (*SY*) heel shows an enlarged hypoechoic plantar fasciitis in comparison to the asymptomatic (*AS*) heel. *C*, Calcaneus

cannot palpitate the defect (RESNICK and KANG 1997).

The plantaris runs alongside the medial aspect of the Achilles tendon. It may appear relatively hypoechoic and mimic focal Achilles tendinitis, especially at the musculotendinous junction or proximal Achilles tendon. The plantaris itself may also exhibit isolated tendinitis and/or tears. The plantaris is often intact in full-thickness Achilles tendon tears. False-negative clinical tests are attributed to the preserved pull of this small tendon.

Ankle sprains can imply partial or complete tears of the ligaments. The most commonly torn ankle ligament is the anterior talofibular, giving rise to the "classical" sprain. This defect is seen as a strain when there is hypoechogenicity of the ligament and as a tear when there is discontinuity of this ligament, usually associated with an appreciable distraction of the talofibular joint. Fluid in the lateral recess would be seen splaying the talofibular ligament outwardly. Similar ultrasound appearance of the calcaneofibular and deltoid ligaments can be seen especially in the acute or subacute phases of ankle sprains. Ultrasound has been helpful in evaluating the syndesmosis of the ankle. A "high-ankle" sprain involves a tear of the anterior tibiofibular ligament which may extend through the distal interosseous ligament and up the interosseous membrane for a variable distance. Strains or tears are noted as hypoechogenicity of the tibiofibular ligaments, often accompanied by a more hypoechoic vertical cleft through the ligament. Stress maneuvers such as eversion of the foot distract and confirm these defects (BOUFFARD et al. 1996).

The tibiotalar joint is readily examined with ultrasound. Anechoic fluid may be seen distending the anterior capsule all the way to its distal talar neck insertion (Fig. 6.20). Loose bodies can be detected as freely moving fragments in joint effusion (BARGIELA et al. 1996a). Passive movement of the joint can prove their mobility. Loculated effusion may occur towards either the lateral or the medial aspect of the ankle. A posterior recess loculated effusion is also possible, outlining the common os trigonum, which should not be mistaken for a loose body. Chronic effusion may elicit synovitis. Further characterization of the synovitis depending on its echotexture can be made. Inflammatory synovitis usually is of an intermediate-level echo forming an undulating margin with increased vascular flow on power Doppler angiography. Infiltrative or chronic synovitis appears more jagged and causes further distention and thickening of the tibiotalar capsule with little vascular flow.

Ganglion cysts of the foot are often uncovered when examining the ankle because they occur on the dorsum talonavicular joint and are commonly accompanied by hypertrophic spurring. Further investigation of such hypoechoic lesions may detect some internal debris such as floating cellular particles.

6.10
Foot

Ultrasound has been helpful in the evaluation of foot lesions such as plantar fasciitis, plantar fascial tears, fibromata of the plantar aponeurosis, and Morton's neuroma. Right-left comparison of the plantar fascia at its calcaneal insertion demonstrates fasciitis (Fig. 6.21) as hypoechoic fusiform enlargement. Plantar fasciitis is a soft tissue pathology and the finding of a bone spur on radiographs is irrelevant. Tears of the plantar fascia at its calcaneal insertion demonstrate more marked hypoechogenicity, often extending into the deeper layers of the foot. The torn fascia

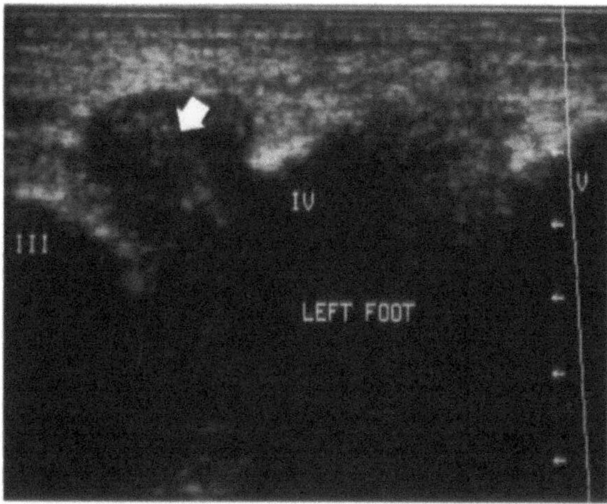

Fig. 6.22. Transverse view of the plantar forefoot with an enlarged hypoechoic mass in the 3rd interdigital space compatible with a Morton's neuroma (*arrow*). *III, IV, V* = 3rd, 4th and 5th metatarsal condyles

assumes a "buckle" deformity rather than a fusiform contour (VAN HOLSBEECK and POWELL 1995).

Plantar fibromata are seen on ultrasound as marginated hypo-hyperechoic lesions. These masses may be situated in the subcutaneous layer immediately adjacent to the aponeurosis or may be located in the substance of the aponeurosis. Bilaterality and multiplicity is the rule for plantar fibromata. They are most often located in the medial aspect of the plantar aponeurosis. Extension of the fibromata more distally can affect the ball of the foot (VAN HOLSBEECK and POWELL 1995).

Most ankle and foot orthopedic surgeons and podiatrists can clinically diagnose Morton's neuroma. Ultrasound has, however, been of value in determining the number of tumors and in differentiating the disease from tendinitis, bursa de novo, tenosynovitis, metatarsophalangeal arthritis, or stress fractures of the metatarsals (BARGIELA et al. 1996b). The examiner presses the interdigital space of the foot between the transducer, which is on the plantar aspect of the foot, and his finger of the opposite hand. Morton's neuroma appears as a predominantly hypoechoic round lesion with significant through-transmission at the level of the metatarsophalangeal joint in the transverse view (Fig. 6.22). More characteristically on the longitudinal view, the neuroma demonstrates a relatively hypoechoic nerve tract ingressing into the mass. "Clinical palpation," which is the ultrasound equivalent of eliciting tenderness, will be noted during the examination. Morton's

neuroma often occurs in the third and second interdigital spaces. Accurate anatomic localization is possible sonographically but may be difficult clinically because of anatomic variations in nerve distribution (VAN HOLSBEECK and POWELL 1995).

6.11
Extra-Articular Disease

Musculoskeletal ultrasound allows the examination of tendon, bursal, and ligamentous disease. A great number of lesions are located extrasynovially and extracapsularly. Ultrasound has been helpful in investigating tears of the pectoralis muscles and determining whether these are strictly muscular or tendinous. Tendon tears need surgery. The clearly retracted or disrupted tendon of the pectoralis major often leaves a stump at its humeral insertion and distention of the residual tendon sheath may be seen. Instead of CT, ultrasound may be used in the investigation of ventral hernias or rectus hematoma of the abdomen. Investigation of the pelvis may show ischial tuberosity bursitis and insertion tendon tears of the abductors or rectus femoris.

Muscle disease and lesions can be visualized with ultrasound with great accuracy. Muscle structures in the longitudinal view have a pennate configuration. Disruption of this pattern is typical of tears ranging from focal tears to crush-type injuries which demonstrate a large geographic hypoechoic defect in the muscle. Myositis ossificans is seen as regional hypoechogenicity of the muscle with high-level oblique echoes representing calcification in a pennate distribution. Pyomyositis, necrotizing myopathy, and muscle infarct in diabetics and in people who are immune suppressed can be detected as focal hypoechoic poorly defined lesions often punctuated with high-level echoes possibly representing abscess or gas. Ultrasound has been very helpful in identifying these sites for ultrasound-guided aspirations (VAN HOLSBEECK and INTROCASO 1991).

Muscle tears are common in the quadriceps and hamstring groups. Sequelae of muscle injury can result in loculated walled-off hematoma or seroma. Both appear as predominantly hypoechoic defects, with hematoma possessing a thicker and more defined wall in comparison to the seroma. Hematomata have internal debris representing the blood particulates while seromas have usually completely anechoic fluid. Calf tears most often occur around the fascial layer in between the gastrocnemius and soleus or at the Achilles musculotendinous junction

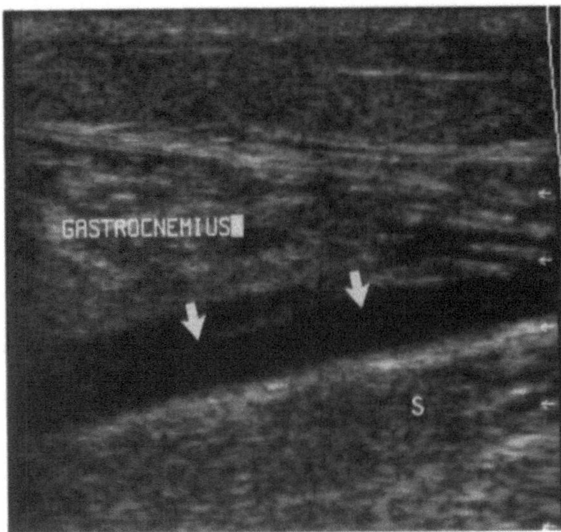

Fig. 6.23. Longitudinal view of the calf of a tennis player demonstrating an acute or subacute tear of the distal medial gastrocnemius from the soleus (*S*) separated by bloody fluid (*arrows*) in the fascia

(Fig. 6.23). Compressible hypoechoic fluid or hematoma can clearly be seen tracking along the fascial plane. A follow-up study of any muscle injury can be carried out with ultrasound and show an involuting mass.

6.12
Conclusion

Soft tissue evaluation is an indication for musculoskeletal ultrasonography. The most practical use of ultrasound is in the evaluation of tendon structures and pathology. There is no better choice than ultrasound in the imaging of tendons. With proper patient positioning and maneuvers and with present-day transducers, all tendons can be investigated. Attendant lesions such as tenosynovitis and perifocal bony defects are simultaneously addressed. The second practical use of ultrasound is in articular structures and diseases. Bursal disease along with accompanying synovitis can be detected. Fluid within the bursa and joint can be measured and further characterized because of internal debris such as blood, crystals, or loose bodies. With current ultrasound equipment giving better contrast resolution, muscular disease can be investigated.

More specifically, ultrasound should be performed when investigating rotator cuff tears, tennis or golfer's elbow, carpal tunnel syndrome, cysts of the wrist or foot, tendons of the hands, retained foreign bodies, joint effusion, extra-articular ligamentous or tendon diseases of the knee, meniscal cysts, Baker's cysts and its complications, acute muscle trauma, posterior tibial or peroneal tendon disease, Achilles tendon afflictions, plantar fasciitis, and Morton's neuroma.

Ultrasound offers real-time imaging and right-left side comparison. This modality and equipment are universal, therefore offering easier access for both patients and physicians. The initial cost of ultrasound machines is relatively low compared to MRI and CT. Throughout the world, the reimbursement for ultrasound is lower than that for the other modalities, except for x-rays. The proximity of the patient to the radiologist and imager brings home the point for which we all trained in medical school: communication skills and patient contact. A musculoskeletal imager must be conversant with ultrasound in order to give his patients the most practical approach to the diagnosis and treatment of musculoskeletal diseases.

References

Bargiela A, Frankel DA, Bouffard JA, Craig JG, Shirazi KK, van Holsbeeck M (1996a) Ultrasound depiction of loose bodies in synovial joints. Radiology 201(P):295

Bargiela A, Frankel DA, Craig JG, Shirazi KK, Bouffard JA, van Holsbeeck M (1996b) Ultrasound evaluation of interdigital nerve and Morton's neuroma. Radiology 201(P):295

Bouffard JA, Eyler WR, Introcaso JH, van Holsbeeck M (1993) Sonography of tendons. Ultrasound Q II:259

Bouffard JA, Goitz HT, van Holsbeeck M (1996) Sonographic evaluation of high ankle sprain. Radiology 201(P):399

Bronstein AJ, Koniuch MP, van Holsbeeck M (1994) Ultrasonographic detection of thumb ulnar collateral ligament injuries: a cadaveric study. J. Hand Surg 19A:304

Codman EA (1934) The shoulder. Thomas Todd, Boston

Crass JM, Craig EF, Feinberg SB (1987) The hyperextended internal rotation view in rotator cuff ultrasonography. J Clin Ultrasound 15:416

Hollister MS, Mack LA, Pattern RM, et al. (1995) Association of sonographically detected subacromial/subdeltoid bursal effusion and intraarticular fluid with rotator cuff tear. AJR 165:605

Jacobson JA, Starok M, Pathria MN, Garfin SR (1997) Pseudarthrosis: US evaluation after posterolateral spinal fusion. Radiology 204:853–858

Lee D, van Holsbeeck M, Janevski P, Ganos D, Ditmars D, Darian V (1995) Ultrasound of the median nerve in carpal tunnel syndrome: correlation with electromyography. Radiology 193 (P):337

Mack LA, Nyberg DA, Matsen FA (1988) Sonographic evaluation of the rotator cuff. Radiol Clin of North Am 26: 161

Marchal G, van Holsbeeck M, Raes L, et al. (1987) Ultrasonography in transient synovitis of the hip in children. Radiology 162:825

Martinoli C, Derch LE, Pastorino C, Bertolotto M, Silvestri E (1993) Analysis of echotexture of tendons with ultrasound. Radiology 186:839

Patten RM, Mack LA, Wang KY, Lingel J (1992) Nondisplaced fractures of the greater tuberosity of the humerus; sonographic detection. Radiology 182:201

Resnick D, Kang HS (1997) Internal derangements of joints: emphasis on MR imaging, 1st edn. Saunders, Philadelphia

Silvestri E, Martinoli C, Derchi L, Bertolotto M, Chiaramondia M, Rosenberg I (1995) Echotexture of peripheral nerves: correlation between ultrasound and histologic findings and criteria to differentiate tendons. Radiology 197:291

Strome GM, Bouffard JA, van Holsbeeck M (1995) Knee. In: Fornage B (ed) Musculoskeletal ultrasound. Churchill-Livingstone, New York, pp 201–219

van Holsbeeck M, Introcaso J (1991) Musculoskeletal ultrasound. Mosby-Year Book, St. Louis

van Holsbeeck M, Sherman L (1991) Sonographic detection of septic hip arthroplasty. March, Diagn Radiol

van Holsbeeck M, Introcaso J (1993) Ultrasound of tendons. Patterns of disease. Instruction Course Lectures 47:475

van Holsbeeck M, Strouse PJ (1993) Sonography of the shoulder evaluation of the subacromial-subdeltoid bursa. AJR 160:561

van Holsbeeck M, Powell A (1995) Ankle and foot. In: Fornage B (ed) Musculoskeletal ultrasound. Churchill-Livingstone, New York, pp 221–237

van Holsbeeck M, Introcaso J (1998) Musculoskeletal ultrasound (2nd edn.) Mosby-Year Book, St. Louis (in print)

van Holsbeeck M, Eyler WR, Sherman LS, et al. (1994) Detection of infection in loosened hip prostheses: efficacy of sonography. AJR 163:381

van Holsbeeck M, Kolowich PA, Eyler WR, et al. (1995) Ultrasound detection of partial-thickness tear of the rotator cuff. Radiology; 197:443

van Holsbeeck M, Boruta PA, Miller SD, Wu KK, Katcherian DA (1996a) Ultrasound in the diagnosis of posterior tibial tendon pathology. Foot and Ankle International 17:555

van Holsbeeck M, Craig JG, Bouffard, JA, Shirazi KK (1996b) Shoulder pain. RSNA Special Course in Ultrasound (P):117

Weng L, Trimulai AP, Lowery CM, Nock LF, Gustafson DE, Von Behren PL, Kim JH (1997) Ultrasound extended-field-of-view imaging technology. Radiology 203:877

7 Interventional Radiological Techniques

A. Chevrot, J.L. Drape, D. Godefroy, A.M. Dupont, F. Gires,
N. Chemla, E. Pessis, L. Sarazin, A. Minoui, and J. Moutounet

CONTENTS

7.1
Introduction

A number of interventional techniques are used under guidance by fluoroscopy, computed tomography (CT), or ultrasound (US). Their indications are decided upon by the entire medical team (physician, radiologist, and surgeon) working together (EL-KHOURY et al. 1994).

A. Chevrot, MD, Head; J.L. Drape, MD; D. Godefroy, MD;
A.M. Dupont, MD; F. Gires, MD; N. Chemla, MD; E. Pessis,
MD; L. Sarazin, MD; A. Minoui, MD; J. Moutounet, MD,
Department of Radiology B, Hôpital Cochin, 27, rue du
Faubourg-Saint Jacques, F-75675 Paris, France

This chapter will consider in particular the following topics:

1. General problems of interventional procedures
2. Bone biopsies
3. Special indications for therapeutic arthrography
4. Percutaneous treatment of periarticular and soft tissue calcifications
5. Treatment of adhesive capsulitis
6. Other techniques such as injections for the treatment of nerve entrapment, vertebroplasty, and percutaneous ablation of small bone lesions (osteoid osteoma).

7.2
General Problems

7.2.1
Needle Guidance

7.2.1.1
Fluoroscopic Guidance

Needle guidance can easily be achieved using an ordinary fluoroscopic unit which allows continuous control of the needle's position. Using one's anatomical knowledge, the point of skin puncture is chosen, and a metallic indicator is placed immediately in front of the target area. The needle is introduced and directly centered on the x-ray beam until bone contact is felt, which indicates the depth of the target. This method also permits an oblique approach since the x-ray beam can be manipulated at various angles (Fig. 7.1). A small amount of contrast medium is then usually injected in order to ensure the proper positioning of the needle.

It is also possible to guide the needle by means of CT scan or even US, but the latter technique is more difficult and less safe due to difficulty in maintaining asepsis, and is used only in the case of soft tissue lesions (BAZZOCCHI et al. 1988; FORNAGE et al. 1991; KONERMANN et al. 1995). A brief description of CT guidance follows.

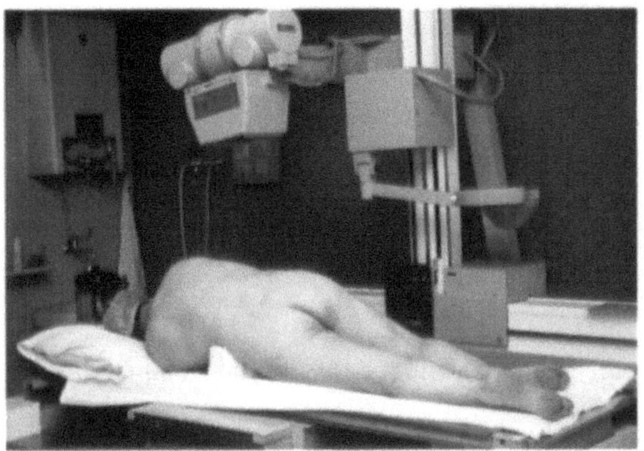

Fig. 7.1. Fluoroscopic unit. The pillow placed under the patient not only facilitates the lumbar approach, but also reduces the irradiation

7.2.1.2
CT Guidance

With the patient lying on the CT unit table, the radiologist places a number of small metallic indicators on the skin in front of the deep bone lesion. The CT slice corresponding to the lesion is chosen (Fig. 7.2). On the basis of the anatomic details of this slice and the positions of the metallic indicators, it is possible to calculate the correct point of puncture, the angle of the needle, and the corresponding depth. The patient is then removed from the gantry, and the needle is introduced centimeter by centimeter. CT controls of the correct direction of the needle may be made from time to time, by reintroducing the patient into the gantry. These procedures are followed until the complete success of the operation is ensured. The overall intervention takes somewhat longer than is the case with fluoroscopic guidance (BABU et al. 1994). Fast scanners capable of rapid image reconstruction allow immediate location of the needle tip in relation to the lesion being biopsied and to the vital organs (REUTHER 1994).

7.2.2
Asepsis

Asepsis is of fundamental importance to the radiologist. One must take particular care with regard to the patient's skin, the needles, and the sterility of the interventional x-ray room, especially in the case of steroid injections. It is most important for the radiologist to wash his or her hands, wrists, fingers, and nails very thoroughly using soap. Although not sufficient without a thorough hand wash, it is also useful

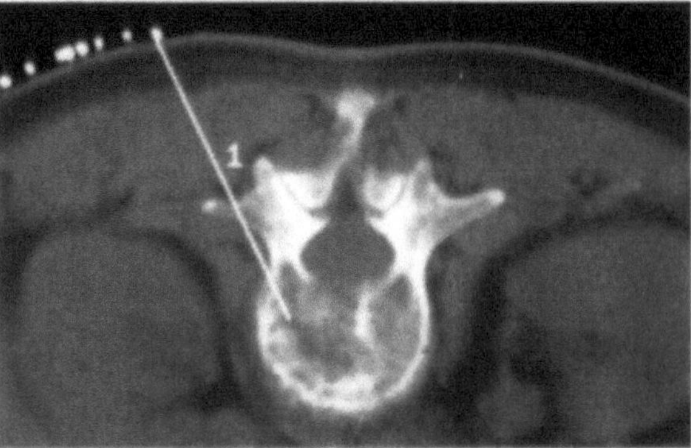

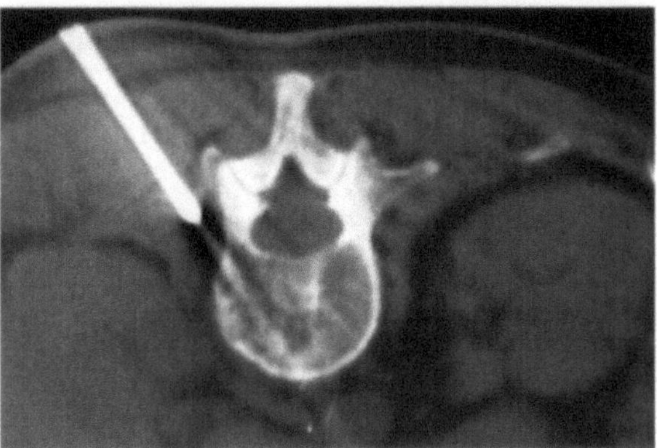

Fig. 7.2 a,b. CT guidance. **a** A selected slice is chosen crossing the target (here the L3 vertebral body). **b** Under local anesthesia the trephine is introduced, following the preselected direction

to wear surgical clothes and sterile rubber gloves. Simple washing of the hands with alcohol is not a safety procedure. The patient's skin must be cleansed with iodine alcohol or mercuric solution (it is not recommended that both be used at the same time because of the danger of skin irritation). The use of sterile windowed drape is a further precaution. Needles and syringes must be sterile; we recommend single-use disposable syringes.

The x-ray room must be as sterile as an operating room. It must also be furnished with intensive care supplies (oxygen, saline solution, water-based steroids, ETC) to assure readiness in the case of allergic reactions to iodine or anesthetic injections, or hemorrhage. Thorough preparation of the radiological operating room is a prerequisite of any interventional procedure.

7.2.3
Postprocedure Infection

The danger of iatrogenic infection should always be kept in mind: such infection is very serious and may sometimes put the patient's life at risk. A rapid diagnosis is of the utmost importance. The usual pattern of presentation is as follows:

- Symptoms appear after a 48-h delay and comprise general discomfort and fever.
- Swelling, redness, and even pus are observed at the point of puncture.
- The usual biological findings of infection are present, i.e., a high sedimentation rate and increase in the white blood cell count.

When infection is suspected the patient should be transferred immediately to a specialized medical unit in order to search for the causal bacteria, and to administer appropriate emergency treatment.

The person responsible for postprocedure infection is the operator, and the preparation should therefore never be delegated to another person.

7.2.4
Patient Care and Local Anesthesia

Premedication is not always necessary but sometimes it is useful to administer some light drug, such as valium per os.

For local anesthesia a O.5% solution of lidocaine is used. The maximum dose is 7 mg/kg; thus several bottles of 20 ml can be used. Local anesthesia is better than general anesthesia since it enables the patient to communicate radiating pain due to possible nerve impingement.

The injection is performed preoperatively into the chosen point of puncture using an appropriate syringe and a thin needle. Since the nerve endings are located on the skin and the periosteum, it is only necessary to inject these areas, and unnecessary to use further anesthesia for the transfixed muscles.

In certain cases, one has to take into consideration the risk of hemorrhage during or after the procedure, especially if the use of a large needle or a trephine is planned. When these devices are to be used, a prior check of the patient's blood coagulation is necessary. In the event of hypocoagulation, the procedure must be delayed until the problem is solved.

Hospitalization is not required in the majority of cases, an exception being patients undergoing deep lesion biopsies, in whom there is a risk of late occurring, insidious deep hemorrhage.

7.2.5
Contrast Media

Contrast media should always be on hand. Only iodinated water-soluble agents are to be used. For spinal interventions, special neural agents are necessary.

7.3
Bone Biopsies

The advantages of various biopsy procedures for musculoskeletal tumors have been extensively discussed elsewhere (LOGAN et al. 1996; SIMON and BIERMANN 1993; SKRZYNSKI et al. 1996).

Radiological evidence often provides a clear indication of the growth rate of a lesion and therefore of latency or aggressivity. Accordingly, the radiologist can confidently identify the latent or slow-growing lesion which does not require biopsy.

A percutaneous needle biopsy undertaken by the radiologist using information and guidance from imaging investigations offers many advantages, and it can usually be performed under sedation and local anesthesia.

7.3.1
Choice of Needle

A variety of tissues may be assessed by aspiration or trephining techniques.

With a lytic lesion and particularly myeloma or a lytic metastasis, it is adequate to obtain aspirated material with a fine-bore (18–20 gauge) needle of appropriate length (BENNETT et al. 1990; CIVARDI et al. 1994). A biopsy gun facilitates the sampling (SCHWEITZER and DEELY 1993; SCHWEITZER et al. 1995).

When the lesion is mainly composed of a soft tissue mass, a cutting needle of the Trucut type is satisfactory (HAUENSTEIN et al. 1995) (Figs. 7.3, 7.4).

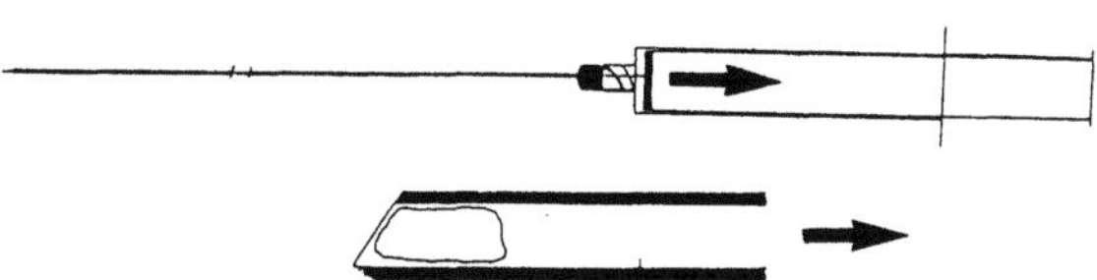

Fig. 7.3. The trucut type needle creates negative pressure in the tip of the needle, to facilitate the sampling

For bone biopsy a strong trephine is necessary (ASTROM et al. 1995; LANGER-CHERBIT et al. 1994) (Fig. 7.5). A large variety of trephines are available on the various medical markets of the world. All consist of an external cannula measuring 1–3 mm in external diameter and 10–20 cm in length Another longer (by 1–2 cm) serrated or cutting cannula is introduced through the external cannula for lesion sampling. Multiple and safe sampling is possible whilst maintaining the external cannula in contact with the lesion.

Hand-held drilling is usually sufficient. A pneumatic or electric drill is useful for the trephination of sclerotic lesions when used to move the cutting needle. Occasionally a small lightweight sterilized hammer may be used in place of a drill (VORWERK et al. 1989).

Biplane radiographs are taken at the end of the procedure with the device still in place to confirm that the biopsy samples are from the correct location.

7.3.2
Ideal Requirements for Skeletal Biopsy

Ideally skeletal biopsy (STOKER et al. 1991) should be executed after radiological staging procedures. It should be remembered that even a thin-needle biopsy may initiate fresh hemorrhage within the neoplasm and that this may alter the magnetic resonance signal or even the size of the tumors.

The radiologist must choose the appropriate imaging technique. Most needle biopsies are better performed by an experienced radiologist in the radiology department, where there is an available choice of imaging methods for accurate direc-

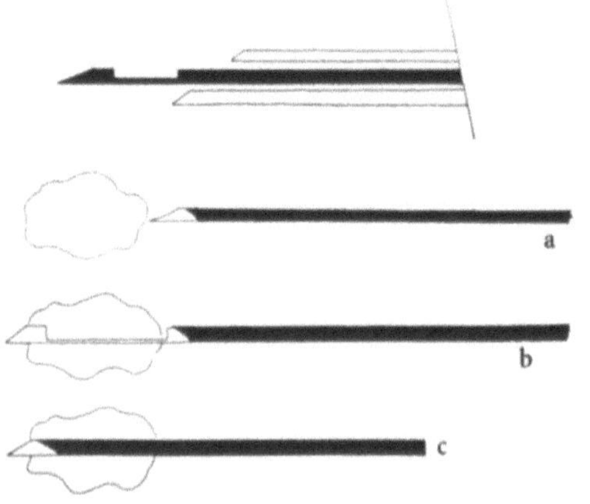

Fig. 7.4. Cutting needle (Trucut type)

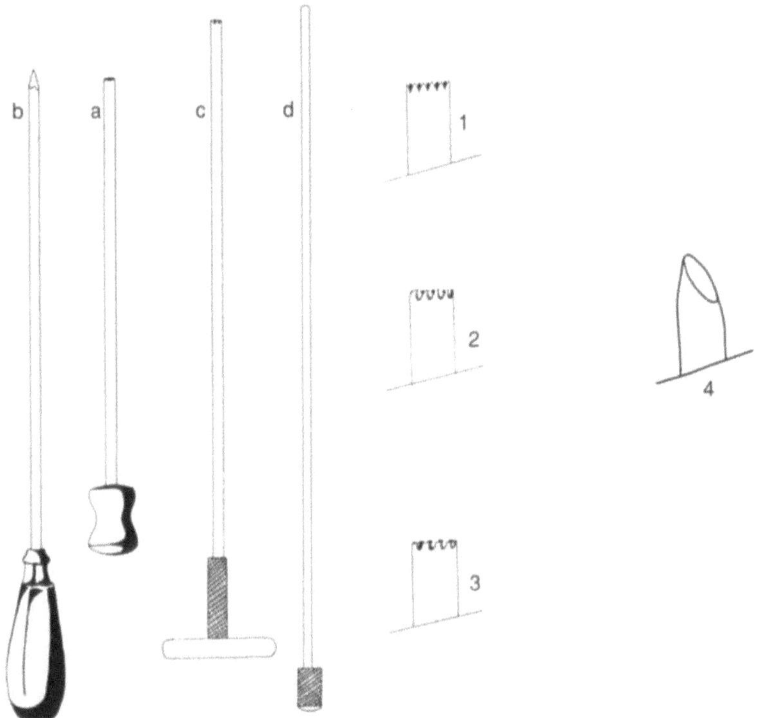

Fig. 7.5. Solid bone trephine (Mazabraud's type). *a–d*, Different parts of the device; *1–4*, different types of cutting end

tion of the needle. As part of the oncological team, the radiologist must be consulted at an early stage. The use of CT as a primary or subsidiary technique for biopsy is a personal decision which may be modified by the immediate availability of the imaging equipment. It is possible to limit this technique to the few difficult regions encountered.

The biopsy should be carried out at the treatment center rather than at the referring hospital if there is any chance that the tumor(s) might prove resectable and suitable for limb salvage. The location of the biopsy needle track may be of importance for subsequent limb salvage operations.

Local anesthesia is usually quite adequate except in children and is even desirable in the case of vertebral lesions. Use of lidocaine for biopsy does not interfere with the diagnosis of microorganisms (SCHWEITZER and DEELY 1993; SCHWEITZER et al. 1995).

The biopsy specimen should be examined both by a histopathologist experienced in osteoarticular pathology and in the bacteriological department. It is not always possible to differentiate neoplasm and infection radiologically and it is wise on most occasions to send some material to both the histopathological and the microbiological department (HOWARD et al. 1994). In cases of suspected infection after the biopsy, it is recommended that one or two hemocultures be obtained since the causal bacteria are often found in the blood after the procedure. The overall complication rate with percutaneous biopsies is only 0.2%, and even in the spine it is only 2.2%.

Although a positive diagnosis on the basis of needle biopsy is considered definitive, a negative result must be regarded with suspicion and may eventually necessitate open surgical biopsy.

7.3.3
Vertebral Biopsy

After sedation, the use of a large volume (20 ml) of 0.5% lidocaine will not only produce good analgesia but also displace the soft tissue away from the needle track (i.e., the pleura in the dorsal approach).

The approach has to be posterolateral (Fig. 7.6). The skin is punctured approximately 8 cm from the median plane in the lumbar region and approximately 6 cm from the median plane in the thoracic region, so as to avoid puncturing the pleura (BENDER

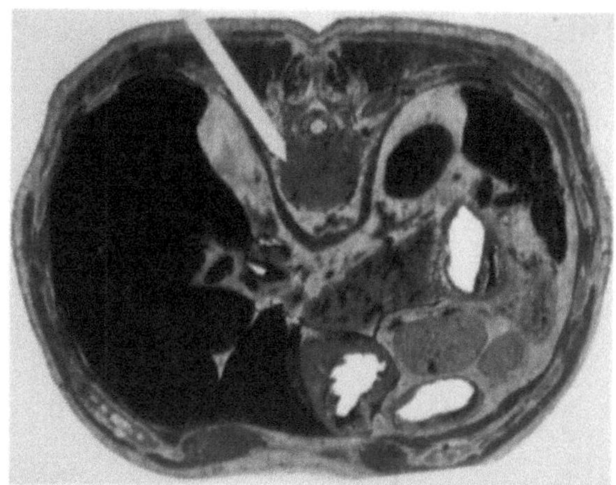

Fig. 7.6. Vertebral biopsy: posterolateral approach

et al. 1986). Under fluoroscopy, the position of the needle must be checked in both the AP and the lateral planes (Fig. 7.7).

Some radiologists prefer to direct the trephine into the vertebral body through the axis of the pedicle (Fig. 7.8); this has the advantage of direction control but severely restricts the biopsy region if the whole of the centrum is not involved. One of the more difficult regions for fluoroscopically controlled biopsy is the upper thoracic spine (GHELMAN et al. 1991).

In cases of suspected infection, it is recommended that contrast medium be injected through the needle in order to visualize cavities (Fig. 7.9).

Cervical biopsies can be performed by the anterolateral approach (Fig. 7.10).

7.3.4
Nonvertebral Biopsy: Limbs and Limb Girdles

When reconstructive surgery is proposed, the needle track should be in a location which is acceptable to the surgeon. Scars of earlier biopsies may modify the surgeon's strategy. Whenever possible the needle should not cross a synovial space, thereby running the risk of converting an intracompartmental lesion into an extracompartmental one (Fig. 7.11).

7.3.5
Management of the Specimen

All the material obtained should be sent for microbiological and histopathological examination since

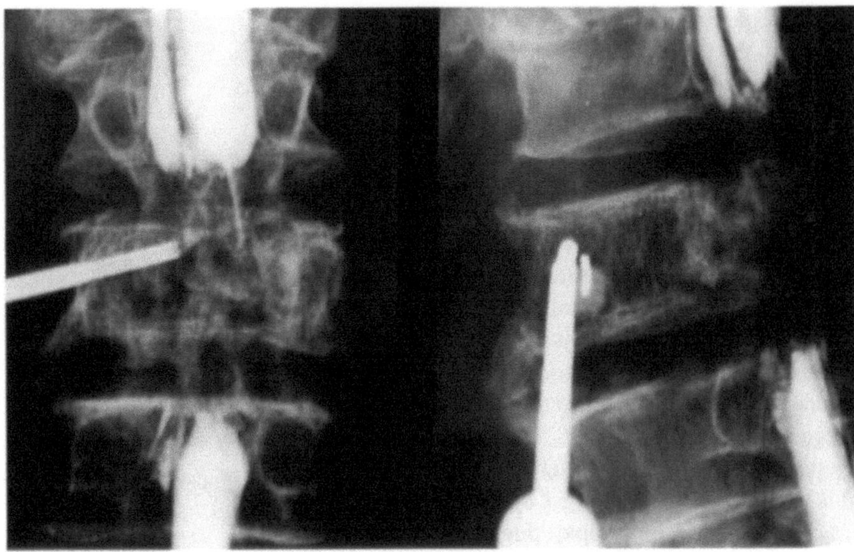

Fig. 7.7. Biopsy of L3 (Paget's disease with compression of the dural sac)

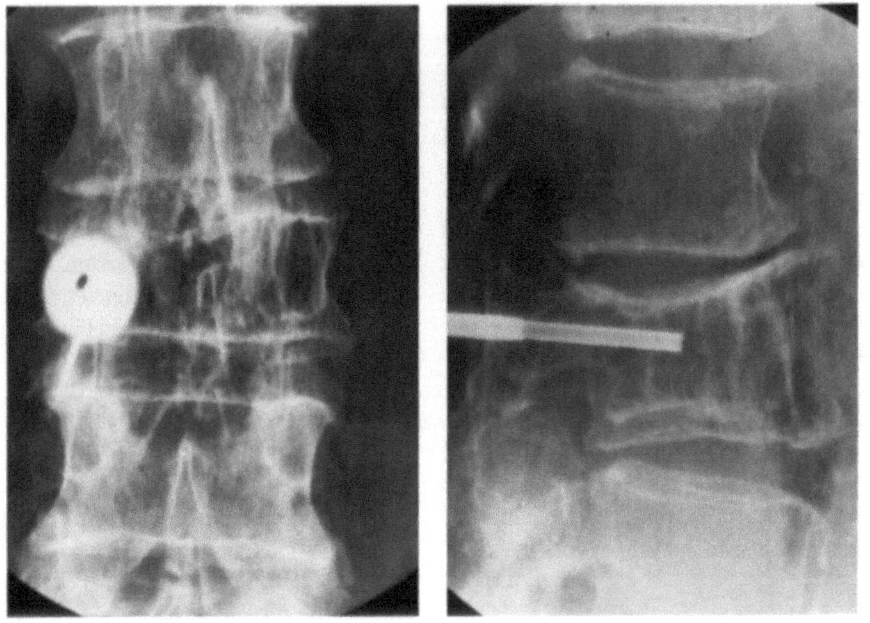

Fig. 7.8 a,b. Transpedicular approach. **a** AP view; **b** lateral view (solitary myeloma)

tumor and infection can be confused. Even a blood clot may contain cells or organisms; it is important to flush out the needle before completing the operation. It is in any case impossible to detect what is in the blood clot by the naked eye and sometimes the proportion of malignant cells may be greater in this material than in the bony core. Imprint or smear preparations are made by the operator; these can be used for instant staining or for immunohistological examination. A part of the specimen is then placed in a sterile container and sent for culture. The remainder of the core is fixed in formol-saline; afterwards it will usually require decalcification.

7.3.6
Aftercare

Apart from surveillance during the period of sedation, in most cases close nursing care is not required.

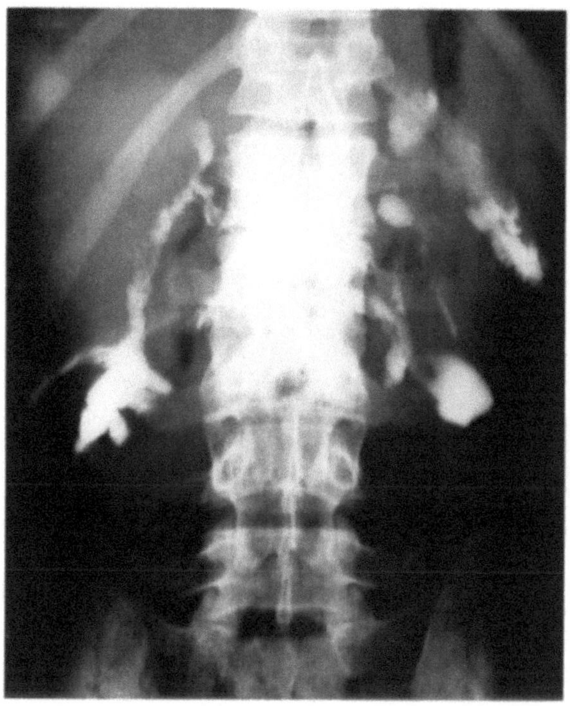

Fig. 7.9. Postbiopsy injection of contrast medium to check the abscessed cavities

The majority of patients are accommodated as day cases for just a few hours until the sedative effects have worn off and they are able to go home.

7.3.7
Results

Tissue adequate for diagnosis is obtained in ca.95% of cases. In 80%–90% of cases an accurate positive or negative result is obtained. We have recorded no false-positive results.

7.4
Percutaneous Injections

Percutaneous therapies require needles and medical injections.

7.4.1
Choice of Needle

It is best to use a single-use disposable needle. The needle itself must be as long as the depth of the target

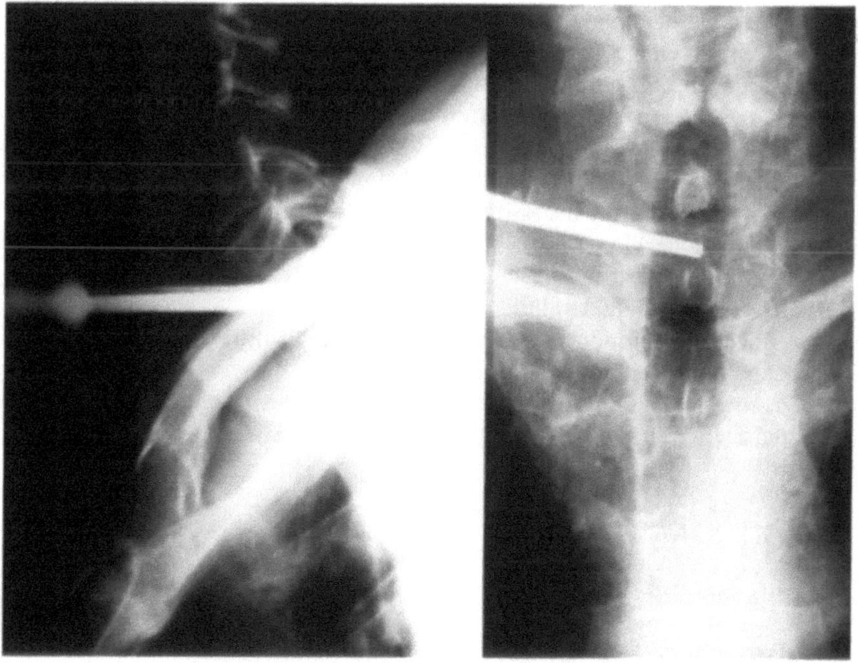

Fig. 7.10. Cervical biopsy

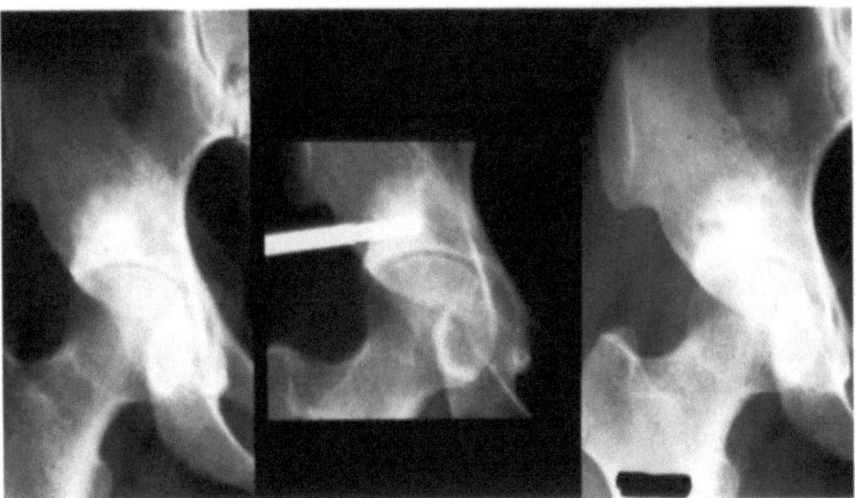

Fig. 7.11. Iliac biopsy by the lateral approach (prostate metastasis)

(shorter for the wrist or a finger joint but longer for the shoulder or the knee and longer still for the facet joints or the hip). Thin needles are safer, especially if there is a risk of vascular or pleural puncture. It is recommended, however, that a sufficiently rigid (i.e., not too thin) needle be used, since such a needle is easier to control at depth.

7.4.2
Choice of Contrast Medium

It is necessary to use water-soluble iodinated agents like Hexabrix for the peripheral joints or other well-tolerated agents for central joints. If the contrast medium is injected directly into the joint, the patient will not feel anything. By contrast, slight discomfort is common when a small amount of the contrast is accidently injected outside the joint into the soft tissues. This indicates to the operator the necessity of a change in the position of the needle.

Occasionally it is possible to use sterilized air as contrast medium. However, an iodinated lipid solution should never be injected as it produces extensive and destructive articular "foreign body" reaction.

The quantity of contrast medium is chosen according to the injected cavities; it is usually less than 5 ml in total.

7.4.3
Choice of Late-Acting Steroids

Late-acting steroids (LAS, Table 7.1) are the most useful locally injected drugs employed in these pro-

cedures. The joint is injected in order to decrease inflammatory phenomena in degenerative diseases or in the case of true inflammatory disease (rheumatoid arthritis). Although LAS act locally because of their crystalline constitution, they also have general effects. For example large doses must be avoided in patients with diabetes, gastric disease, or chronic infections (e.g., tuberculosis).

The action life of these drugs corresponds to the size of the crystals: the bigger the crystal, the longer the action life. However, some large crystals can produce or induce calcium deposition (by means of soft tissue necrosis). It appears safer to use very small amounts of the drug or to inject saline solution or anesthetic fluid at the same time in order to dilute the crystals.

Details of the appropriate technique in respect of individual (peripheral or central) joints are given below. These are sometimes presented in list form for ease of reference.

7.4.4
Peripheral Joints

Favorite sites of puncture of peripheral joints are shown in Fig. 7.12.

7.4.4.1
Wrist

- Dorsal approach
- Needle: 22 gauge, 3 cm long
- Point of puncture: between scaphoid and radius

Table 7.1. Crystal sizes and prednisone dose equivalents corresponding to 1 ml of the injecting solution

Name	Crystal size	Prednisone dose equivalent
Cortivazol (Altim)	+++	36 mg
Betamethasone (Betnesa depot)	++	50 mg
Betamethasone acetate and disodium phosphate (Celestene chrondose)	++	60 mg
Dexamethasone *tert*-Butyl acetate (Decadron TBA suspension)	+	25 mg
Dexamethasone acetate (Dectancyl suspension)	++	30 mg
Methylprednisolone (Depo-Medrol suspension)	+++	50 mg
Paramethasone (Dilar suspension)	++	50 mg
Betamethasone dipropionate and disodium phosphate (Diprostene)	++	60 mg
Triamcinolone hexacetonide (Hexatrione)	++++	25 mg
Prednisolone Acetate (Hydrocortancyl suspension)	++	25 mg
Hydrocortisone sodium succinate (Hydrocortisone suspension)	+	6 mg
Triamcinolone acetonide (Kenacort 80 suspension retard)	+++	50 mg

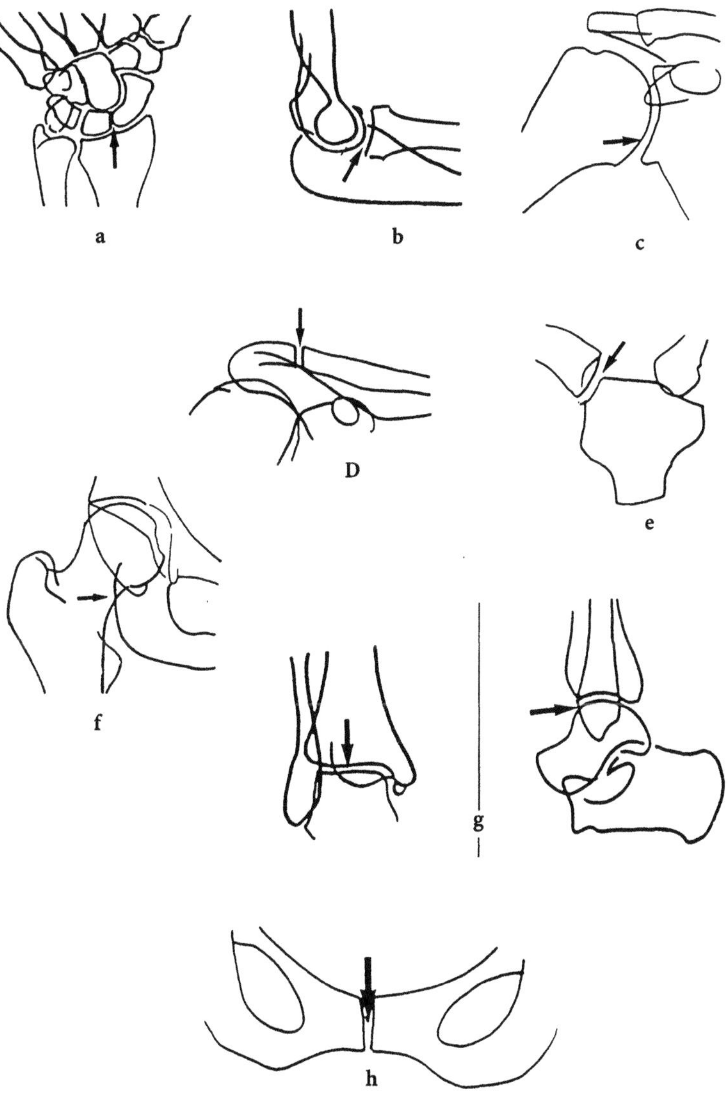

Fig. 7.12 a–h. Favorite sites of puncture: **a** wrist; **b** elbow; **c** shoulder; **d** acromioclavicular joint; **e** sternoclavicular joint; **f** hip; **g** ankle; **h** pubic joint

- 1 ml contrast
- 1 ml LAS

7.4.4.2
Metacarpophalangeal Interphalangeal Joints

- Dorsal approach
- Needle: 22 or 26 gauge
- 1 ml LAS

7.4.4.3
Elbow

- External approach
- Point of puncture: between the head of the radius and the capitulum of the humerus
- 1–3 ml of contrast
- 1 ml LAS

7.4.4.4
Shoulder (Glenohumeral Joint)

- Anterior approach, with patient supine
- Needle: 18 to 20 gauge; 3 or 4 cm long
- Point of puncture: inferior aspect of the joint between the head of the humerus and the glenoid surface (4–6 ml contrast)
- 1 ml LAS

7.4.4.5
Shoulder (Acromioclavicular Joint)

- Superior approach
- 1 ml LAS

7.4.4.6
Sternoclavicular Joint

- Anterior approach
- 1 ml LAS

7.4.4.7
Hip

- Anteroinferior approach
- Point of puncture: inferior aspect of the joint
- Needle: 18–20 gauge; spinal needle

- 3–5 ml contrast
- 1 ml LAS

7.4.4.8
Knee

Usually this joint is punctured without fluoroscopic control, by a lateral approach under the patella. But it is advisable to check the correct placement of the injection under fluoroscopy because it is very easy to inject outside the joint in the prefemoral fat.

In the case of large popliteal cysts more LAS (2–3 ml) can be injected into the joint. The cyst can be directly punctured and drained.

7.4.4.9
Ankle and Foot

- Anterior approach
- Needle: 20–22 gauge
- 1 ml LAS

7.4.4.10
Pubic Joint

- Anterosuperior approach
- Long needle
- 1 ml LAS

7.4.5
Central Joints

7.4.5.1
Lumbar Facet Joints

Sciatica or low back pain can be due to the lumbar zygapophyseal facet (BOUGH et al. 1990; MAHESHWARAN et al. 1995; TOURNADE et al. 1992). According to LYNCH and TAYLOR (1986) only intra-articular injections are effective. The injection of a facet joint is quite easy since the joint has a large recess below the inferior aspect of the posterior facet (SELLIER et al. 1987) (Figs. 7.13, 7.14). Details of the technique are as follows:

- Patient in strict prone position, on the fluoroscopy table
- Point of puncture: below the posterior facet against the inferior lamina
- 1 ml contrast

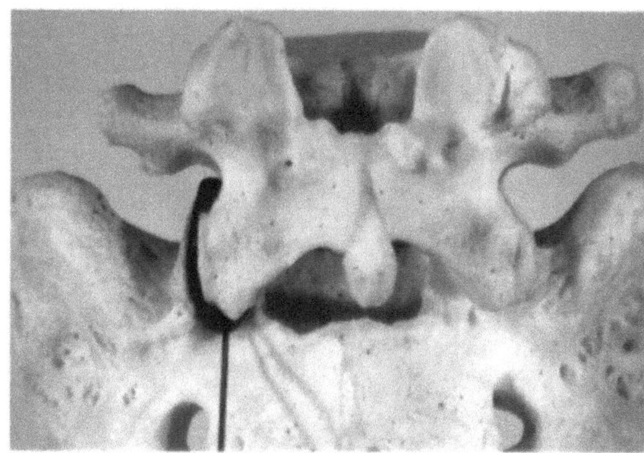

Fig. 7.13. Facet joint approach: point of puncture for L5–S1 facet joint

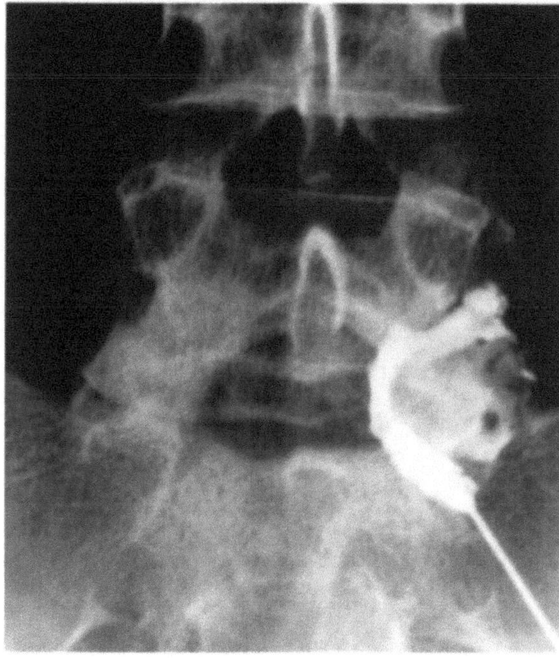

Fig. 7.14. L5–S1 arthrography

– 1 ml LAS
– CT can be used as a guide (SCHLEIFER et al. 1995)

Lumbar intraspinal facet cysts associated with significantly degenerated facet joints can be treated by this facet injection (HSU et al. 1995; VALLEE et al. 1987) (Fig. 7.15).

7.4.5.2
Cervical Facet Joints

Cervical facet joints are less easy to inject. A posterolateral approach is commonly used (HOVE and GYLDENSTED 1990) (Fig. 7.16).

7.4.5.3
Costovertebral Joints

A posterior approach is used to inject costovertebral joints, but the injection sometimes misses the joint cavity, which is very small.

7.4.5.4
C1–C2 Joint

We perform C1–C2 arthrography by means of the posterolateral approach; this is a painstaking procedure but very safe (CHEVROT et al. 1995) (Figs. 7.17–7.19).

7.4.5.5
Discography and Nucleolysis

It is possible to inject the lumbar or the cervical disc.

7.4.5.5.1
LUMBAR DISC
Techniques employed include intradiscal steroid injection and chemonucleolysis, as discussed below.

Intradiscal Steroid Injection. Lumbar discography is a well-known technique (BOGDUK and MODIC 1996; GUYER and OHNMEISS 1995; CODERC et al. 1990) in which contrast medium is injected into the nucleus of an intervertebral disc with the purpose of ascertaining whether that disc is responsible for the patient's pain.

Using a standard fluoroscopic unit and with the patient in the prone position, injection is performed via the posterolateral extradural approach. An 18-gauge spinal (guide) needle and a 22-gauge 15-cm disc needle are used for this purpose. The needle is driven through the posterior muscles. For L5–S1 the approach is via the "access triangle," i.e., the iliac crest, the transverse process of the upper vertebra, and the upper face of the lower vertebra (Fig. 7.20). It may also be necessary to bend the tip of the needle to reach the L5–S1 intervertebral space (BONAFE et al. 1993). The injection of 1 ml contrast medium enables one to check the intervertebral space, posterior seepage of the contrast medium proving that there is a tear of the annulus fibrosus (Fig. 7.21). If the pain provoked is of the same nature as that usually suffered by the patient, it can be concluded that the painful disc has been identified. The operator then injects a very small amount (0.5 ml) of LAS, choosing

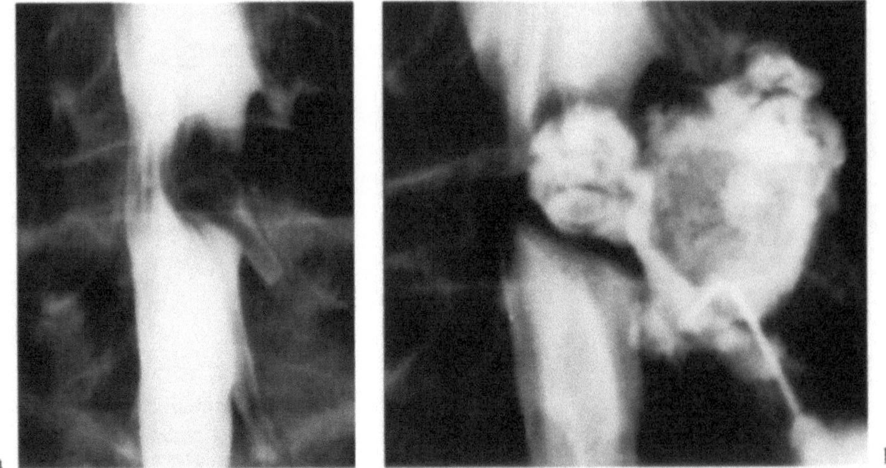

Fig. 7.15 a,b. L4–L5 facet joint arthrography in a case of intravertebral compressive ganglion cyst. **a** Myelography with the left compression of the dural sac. **b** Facet injection filling the joint and the cyst

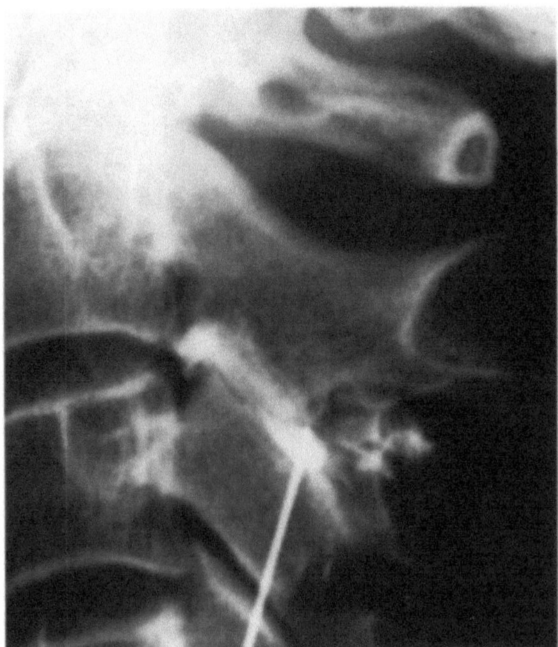

Fig. 7.16. C3–C4 facet joint arthrography

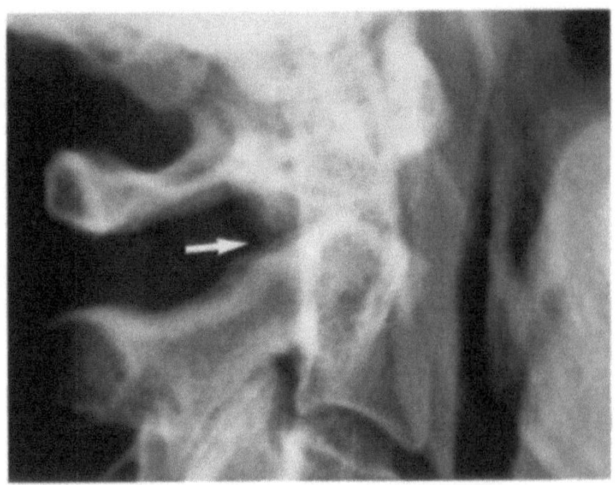

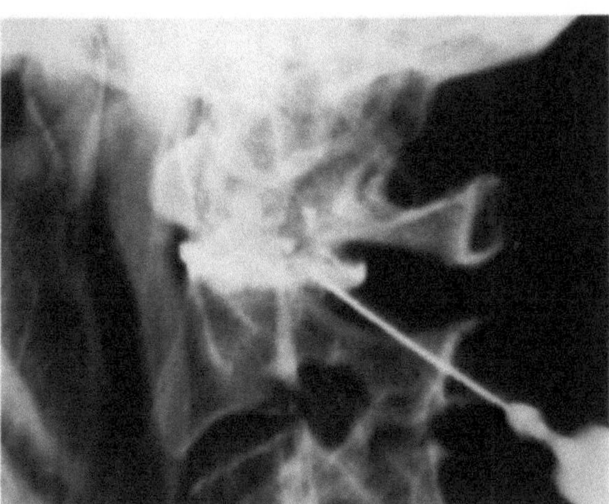

Fig. 7.18. C1–C2 joint arthrography

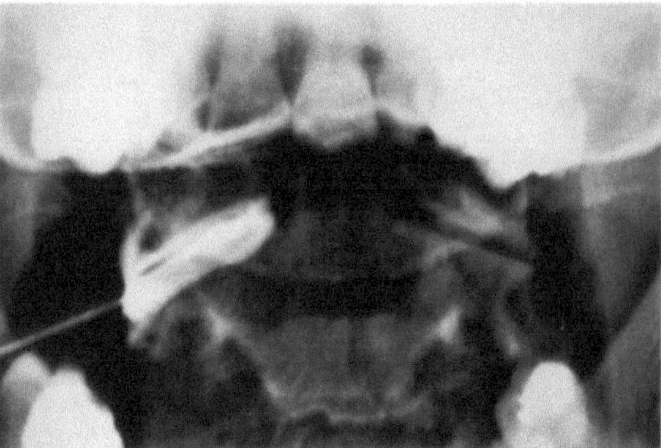

Fig. 7.19. C1–C2 joint arthrography

Fig. 7.17. C1–C2 joint arthrography: point of puncture (*arrow*)

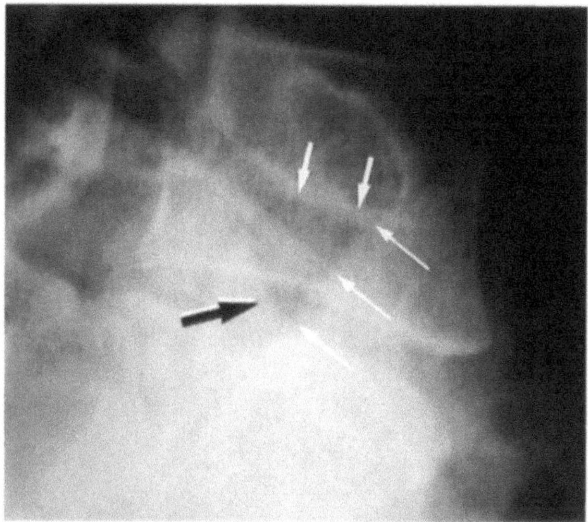

Fig. 7.20. Posterolateral approach for L5–S1 discography. Note the "access triangle," i.e., the iliac crest (*thin white arrow*), the transverse process of the upper vertebra (*two medium white arrows*), and the upper face of the lower vertebra (*black arrow*)

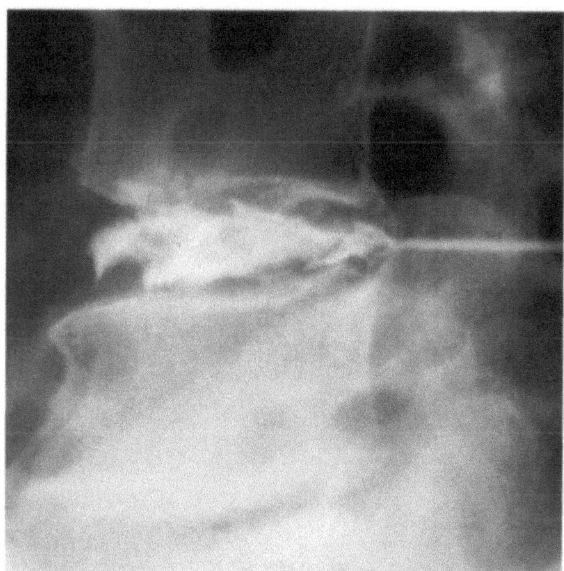

Fig. 7.21. L4–L5. discography: degenerative space

one with very small crystals. In this respect it should be remembered that the intervertebral cavity is very small. Soft tissue calcifications surrounding the annulus have been reported following this procedure, and sometimes severe narrowing of the intervertebral space is observed several months later. These manifestations appear when a significant amount of steroid has been injected. Thus the use of water-soluble steroids or LAS with very small crystals is compulsory (L'HUILLIER et al. 1988).

The procedure is performed on an outpatient basis. The patient is advised to wear a rigid girdle for 6 weeks after the injection. This immobilization helps

the medication to take effect. The success rate is ca. 70%.

Chemonucleolysis. Intradiscal injection (BOUILLET 1990) of chemopapain (TOURNADE et al. 1992) is also used in cases of sciatica. Due to the risk of anaphylactic shock, the patient requires intensive care surveillance both during and after the injection, the hospitalization period usually lasting 4–5 days. The efficiency rate is almost 80% (BENOIST 1996; LOUWAEQE et al. 1996; GOSAL and HARRISON 1995).

Nucleotomy. Another form of treatment is nucleotomy. Although we have tried this technique, we no longer use it due to the high rate of unsatisfactory results (CASTRO et al. 1992; MOCHIDA and ARIMA 1993). Some authors use other procedures like "laser nucleotomy" (CHOY et al. 1987).

Discitis. Discitis after discography is due to bacterial penetration into the intervertebral disc via a contaminated needle and has an incidence of 1%–4% (GUYER et al. 1988). Adding an antibiotic to the intradiscal suspension or giving it intravenously 30 min prior to injection prevents any radiographic, macroscopic, or histological signs of discitis (OSTI et al. 1990).

7.4.5.5.2
CERVICAL DISC
(ZEIDMAN et al. 1995; SCHELLHAS et al. 1996).

- Anterolateral approach; the radiologist presses the cervical vessels and pharynx with his finger (Fig. 7.22)
- 18–20 gauge spinal needle
- 1 ml contrast
- 1 ml LAS

7.4.5.6
Sacroiliac Joint (Fortin et al. 1994)

- Patient in prone position; posterior approach
- Point of puncture: inferior aspect of the joint
- Needle: 18 to 20 gauge; 3 cm long
- 1 ml LAS

7.4.6
Lumbar Epidurogaphy

Lumbar epidurography has been proposed as a means of diagnosing painful epidural fibrosis

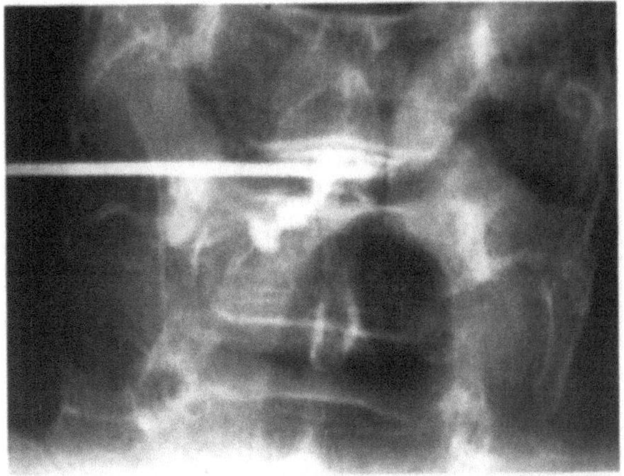

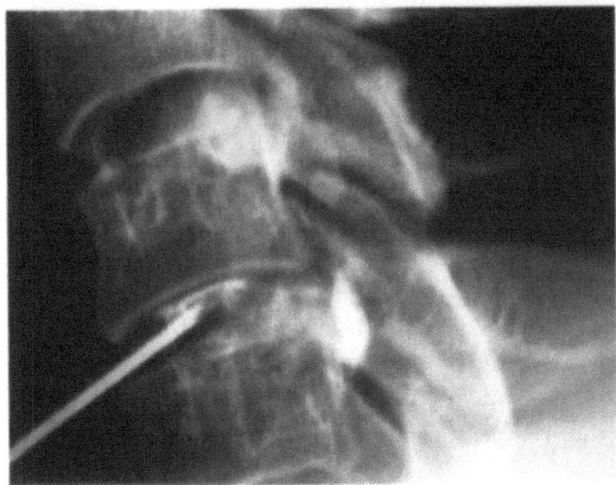

Fig. 7.22 a,b. C5-C6 discography: **a** frontal view; **b** lateral view

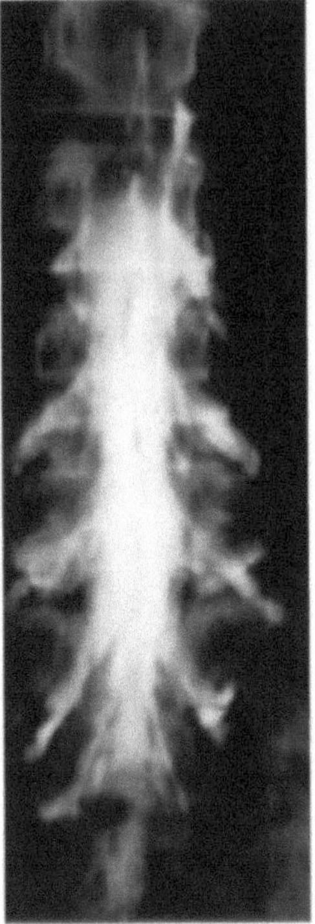

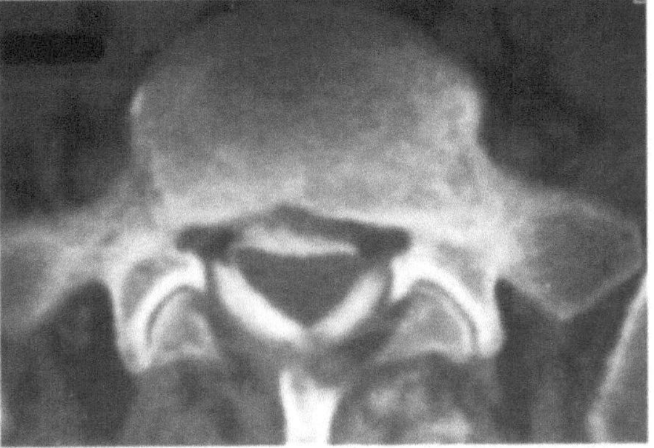

Fig. 7.23 a,b. Epidurography: **a** AP view; **b** CT (normal pattern)

(REVEL et al. 1988). Injection of LAS can improve the patient's condition. (DEVULDER et al. 1995; STEWART et al. 1987). Injection can be made into any interlaminal space or the sacral hiatus (Fig. 7.23).

7.4.7
Bursa and Tendon Sheath Injections

7.4.7.1
Bursae

There are a number of bursae in the body, and each of them can be the site of effusion, inflammation (microtrauma, gout or other crystal deposition diseases, or inflammatory disease involvement) or infection (infectious bursitis). Direct puncture allows samples to be acquired if infection is suspected. LAS injections are used in order to reduce the inflammatory phenomena. It is also possible to inject synovial cysts, either by direct puncture or by injecting the neighboring joint.

7.4.7.2
Bursa of Olecranon

The bursa of olecranon (bursa subcutanea olecrani) can be injected behind the olecranon.

7.4.7.3
Subacromial Bursa

Injection of the subacromial bursa is possible by positioning a needle just under the acromion.

7.4.7.4
Bursa of Gluteus Maximus Muscle

The bursa of gluteus maximus muscle is often involved in cases of tendinitis of the gluteus maximus. MRI or US helps to establish the diagnosis. Of course, it would be a mistake to inject a greater trochanter that is affected by tuberculosis (TB); this classic TB location should always be taken into consideration, even though the condition is rarely encountered.

7.4.7.5
Bursae of the Foot

There are many bursae in the foot including the bursa tendinis calcanei (bursa of Achilles tendon).

7.4.7.6
Interspinous Bursa

In Baastrup's disease (also known as "kissing spine"), the abnormal contact between spinous processes can produce bursae and sometimes bursitis. It is possible to diagnose this disease with plain films or MRI. Direct puncture of the interspinous processes, injection of contrast medium, and LAS injection can produce positive results in the treatment of some cases of associated lower back pain.

7.4.7.7
Tendon Sheaths

Contrast opacification of tendon sheaths around the wrist or the ankle allows for injection. The puncture is performed directly into the tendon. Injection of a small amount of contrast medium confirms the proper positioning of the tip of the needle, after which 1 ml of LAS can be injected.

7.4.8
Treatment of Calcifying Tendonitis

The natural behavior of a tendinous calcification is to cause acute temporary pain and to heal spontaneously. One month later, even without treatment, the calcification no longer appears on the plain film control (HAYES and CONWAY 1990).

In cases of painful calcifying tendonitis, the puncture of a calcification (under fluoroscopic control) allows the withdrawal of a small amount of calcified material. The resulting pain is controlled by steroid injection. This is a very good treatment of the painful calcifying shoulder periarthritis, which is healed in more than 90% of cases (MOUTOUNET et al. 1992). With the small number of failures, a new trial can be performed without hesitation. Some cases of subsequent tear of the rotator cuff have been reported, but without real inconvenience.

Most other sites of painful soft tissue calcification can be treated in the same way (GALVEZ et al. 1995; YOSIPOVITCH and YOSIPOVITCH 1993). The radiologist can pinpoint the calcification using his equipment and treat the affected area with precision.

7.4.9
Treatment of Adhesive Capsulitis

Adhesive capsulitis mainly involves the shoulder (frozen shoulder), which is painful and immobile due to adhesions in the synovial membrane and the fibrous capsule itself. It may occur following a number of shoulder diseases, principally SUDECK'S disease and transient osteoporosis, or any surgical procedure involving the shoulder (MURNAGHAN 1988; HULSTYN and WEISS 1993). The aim of the transcutaneous treatment is to break the adhesions under local anesthesia (GAVANT et al. 1994).

The radiological procedure includes an injection of anesthetic and LAS into the scapulohumeral joint. A capsular tear occurs during arthrography. The same injection is performed into the subacromial bursa. Strong active and passive motions of the shoulder, assisted by the doctor, are performed to try to improve the range of motion of the joint. Subsequent mechanical treatment is recommended for 1–2 weeks. The result of this procedure is often very satisfactory.

Adhesive capsulitis in other locations can be treated in the same way: as well as in the shoulder, it may occur in the hip, the ankle (PALLADINO and CHAN 1987), the knee, and even the wrist joint (BRANDSER et al. 1995).

7.4.10
Treatment of Nerve Entrapment

It is possible to infiltrate various deep sites of the body in order to treat nerve entrapment, e.g.:

- Carpal canal
- Subscapular groove
- Greater sciatic foramen
- Alcock's canal
- Tarsal sinus
- Tarsal canal

CT or fluoroscopic control allows a precise canal approach for the injection of anesthetic or LAS. The aim is to achieve a decompressive effect through the anti-inflammatory properties of the steroids. Such treatment is used in patients with carpal tunnel syndrome (BONINGER et al. 1996; PAPAIOANNOU et al. 1992; ROSENBAUM 1993) or Guyon's canal sydrome (ANTUNA et al. 1995).

Suprascapular nerve entrapment is an acquired neuropathy secondary to compression of the nerve in the bony suprascapular notch (CALLAHAN et al. 1991; FEHRMAN et al. 1995; FRITZ et al. 1992; HASHIMOTO et al. 1994).

In Alcock's canal syndrome, perineal neuralgia arises due to compression of the pudendal nerve in the ischiorectal fossa (OBERPENNING et al. 1994). Treatment is by infiltration of cortisone derivatives into the pudendal nerve canal.

ROMANOFF et al. (1989) have described saphenous nerve entrapment at the adductor canal which may be treated by a saphenous nerve block at the adductor canal.

Lumbar lateral canal entrapment can be treated by intraforaminal injections (STOCKLEY et al. 1988).

7.5
Vertebroplasty

Injection of acrylic cement into a vertebral lesion is proposed to obtain pain relief (CHIRAS et al. 1995) and to improve the strength of the vertebral body. The main indications are vertebral metastases (WEILL et al. 1996), malignant vertebral tumors, painful hemangioma, and osteoporotic, acutely painful vertebral collapse (CARDON et al. 1994). Vertebroplasty can also be used for vascular embolization if vertebral surgery for a tumor is planned.

A special single-use needle is introduced into the vertebral body, using a small hammer. A posterolateral or transpedicular approach is used under fluoroscopic guidance, with the patient under sedation. Acrylic cement is prepared by mixing 20 g of cement powder and 10 ml of hardener liquid. To render the cement radio-opaque, 1 g tantalium powder is mixed (i.e., Sulfix 60: 48.4 g powder; 19.23 g liquid). Once prepared, the viscosity of the mixture progressively increases. After 2 min the cement reaches the appropriate viscosity for injection using an appropriate syringe. No more than 2 ml to 4 ml is injected, under fluoroscopic control using a lateral view (GANGI et al. 1994). The injection is stopped if the cement overflows behind the vertebral body line. The procedure can be performed with the usual fluoroscopic unit: a biplane fluoroscopic unit is unnecessary. It is possible to turn the patient after positioning the needle. Proper distribution of the cement can be controlled by means of CT. Pain relief can be achieved even when lesion filling is incomplete (COTTEN et al. 1996).

Nerve impingement can be produced if the cement penetrates into the vertebral canal or the foramen (Fig. 7.24).

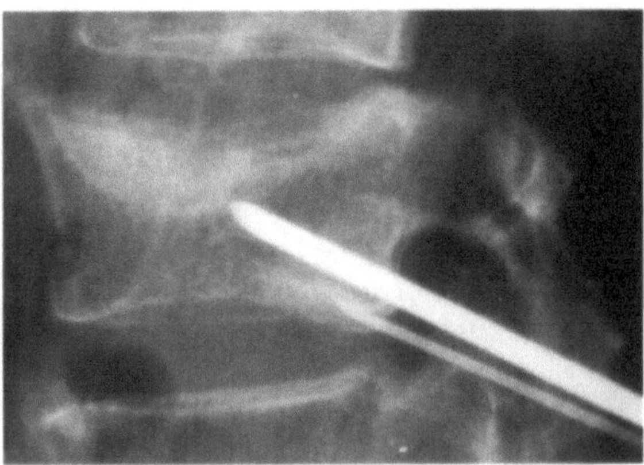

a

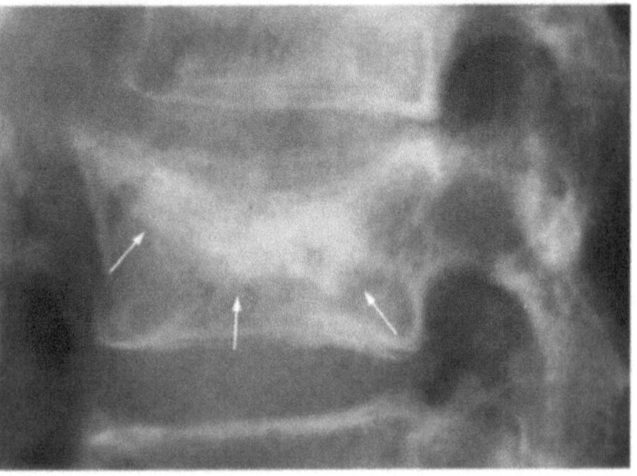

b

Fig. 7.24 a,b. Vertebroplasty. a L3 compression fracture with the needle in place. b Cement after injection (*arrows*)

7.6
Percutaneous Removal
of Small Bone Lesions

Small lesions less than 1 cm in diameter, such as osteoid osteoma, can be removed by means of percutaneous trephining. Other authors propose destruction of the nidus using electrocoagulation or laser treatment (ADAM et al. 1995; ASSOUN et al. 1993; D'ERME et al. 1995; DE BERG et al. 1995; GALLOY et al. 1996; KOHLET et al. 1995; ROSENTHAL et al. 1992, 1995; TOWBIN et al. 1995) or even alcoholization (SANHAJI et al. 1996). In all these procedures, the device is guided to the appropriate site under fluoroscopic or CT guidance (Fig. 7.25). General anesthesia is recommended since the procedure is painful. Thus the procedure is frequently performed by a team comprising radiologist, surgeon, and anesthesist.

On the other hand, KNEISL and SIMON (1992) have reported that long-term administration of nonsteroidal anti-inflammatory drugs can often be as effective as excision for the treatment of osteoid osteoma, without the morbidity that is associated with surgery. Such treatment might be of particular value in patients in whom operative treatment would be complex or might lead to disability.

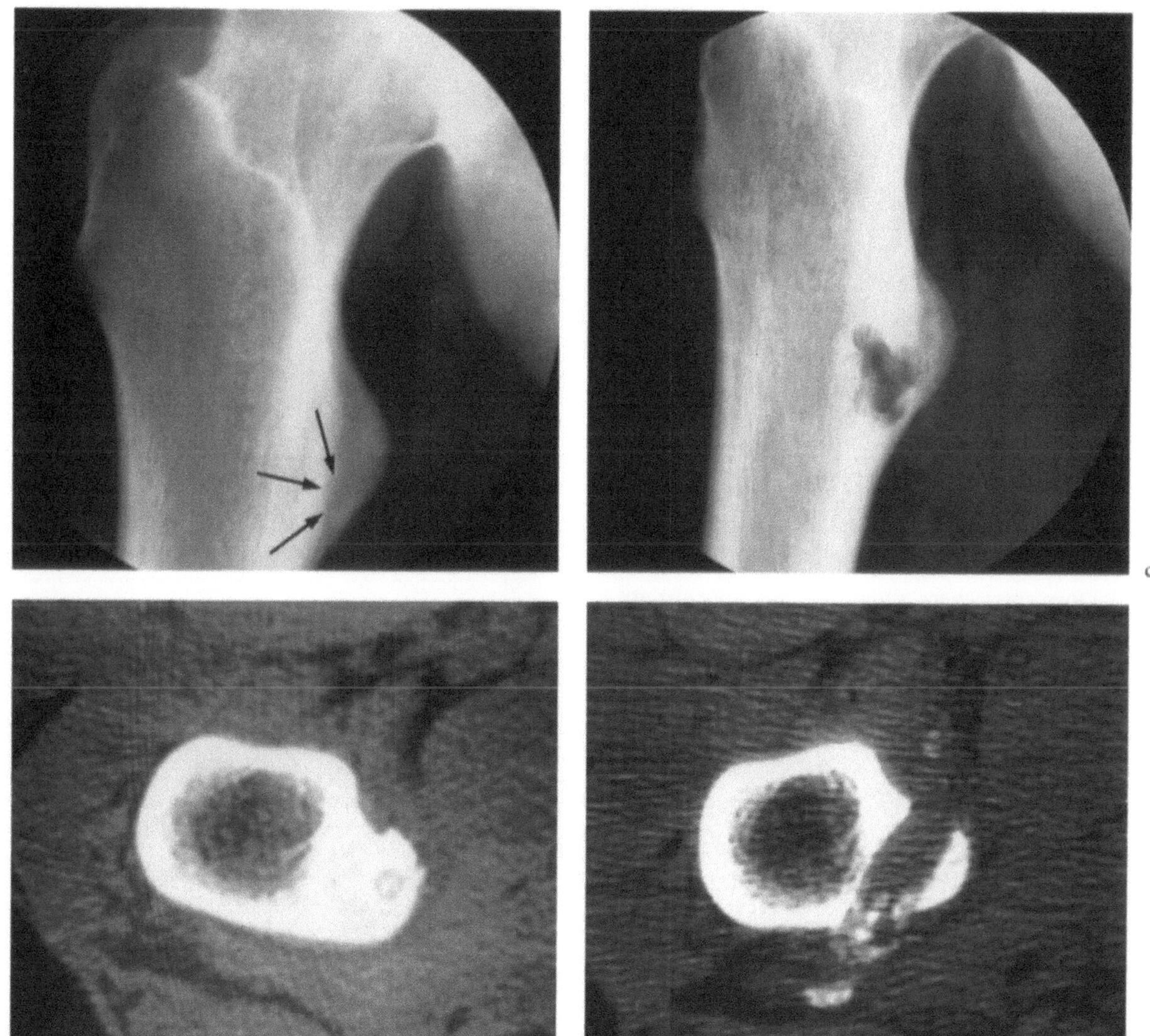

Fig. 7.25 a–d. Percutaneous treatment of an osteoid osteoma. **a** Plain film (lesion indicated by *arrows*); **b** CT before the procedure; **c** plain film; **d** CT after the procedure

7.7
Conclusion

Safety is the key point in all of the above-discussed interventional radiological techniques. The major problem remains the possibility of iatrogenic infection, which justifies the precautions taken by the radiologist. The risk of bleeding is not to be feared owing to the sole use of needles.

Percutaneous needle biopsy of lesions affecting the musculoskeletal system should be considered a routine radiological procedure. Although relatively safe, the procedure requires expertise. An experienced radiologist and the cooperation of a skilled pathologist are essential. Consultation with the orthopedic surgeon is also important, especially when resection of the lesion is contemplated. Recent advances in imaging techniques and the availability of various cutting and trephine needles have made it easier to perform biopsies safely and accurately, even in difficult locations. The biopsy procedure obviates surgery in many instances and facilitates appropriate surgical planning in others. This review offers a pragmatic approach to percutaneous needle biopsy of skeletal lesions. It is emphazised that biopsy for suspected musculoskeletal tumors should not be delegated to junior residents and should be performed in centers that have experience in the management of such tumors.

The discussed techniques are not only very safe but provide great relief to the patient. They can be performed quickly, usually as a day procedure, and are inexpensive. There are numerous indications for the techniques, especially in the elderly, in whom the quality of life can be dramatically improved. Thus doctors everywhere should receive appropriate instruction to enable them to safely perform these procedures, and it is to be hoped that more radiologists will be encouraged to undertake them.

References

Adam G, Keulers P, Vorwerk D, Heller KD, Fuzesi L, Gunther RW (1995) The percutaneous CT-guided treatment of osteoid osteomas: a combined procedure with a biopsy drill and subsequent ethanol injection. Rofo Fortschr Geb Rontgenstr Neuen Bildgeb Verfahr 162:232–235

Antuna SA, Gutierrez CF, Paz Jimenez J (1995) Ulnar nerve compression in Guyon's canal caused by a pseudotumor of the pisiform. Acta Orthop Belg 61:245–248

Assoun J, Railhac JJ, Bonnevialle P, et al. (1993) Osteoid osteoma: percutaneous resection with CT guidance. Radiology 188:541–547

Astrom KG, Sundstrom JC, Lindgren PG, Ahlstrom KH (1995) Automatic biopsy instruments used through a coaxial bone biopsy system with an eccentric drill tip. Acta Radiol 36:237–242

Babu NV, Titus VT, Chittaranjan S, Abraham G, Prem H, Korula RJ (1994) Computed tomographically guided biopsy of the spine. Spine 19:2436–2442

Bazzocchi M, Gozzi G, Zuiani C, Pozzi Mucelli RS (1988) Ultrasonic-guided fine-needle biopsy of osteolytic lesions. Radiol Med (Torino) 76:23–27

Bender CE, Berquist TH, Wold LE (1986) Imaging-assisted percutaneous biopsy of the thoracic spine. Mayo Clin Proc 61:942–950

Bennett JD, Yacyshyn BJ, Haddad RC, Lefcoe MS (1990) Fine-needle aspiration of bone lesions. Can Assoc Radiol J 41:65–68

Benoist M (1996) Vingt ans de chymonucleolyse lombaire. Presse Med 25:743–745

Bogduk N, Modic MT (1996) Lumbar discography. Spine 21:402–404

Bonafe A, Tremoulet M, Manelfe C (1993) Traitements percutanes des hernies discales lombaires. Criteres radiologiques de decision therapeutique. Neurochirurgie 39:105–109

Boninger ML, Robertson RN, Wolff M, Cooper RA (1996) Upper limb nerve entrapments in elite wheelchair racers. Am J Phys Med Rehabil 75:170–176

Bough B, Thakore J, Davies M, Dowling F (1990) Degeneration of the lumbar facet joints. Arthrography and pathology. J Bone Joint Surg [Br] 72:275

Bouillet R (1990) Treatment of sciatica. A comparative survey of complications of surgical treatment and nucleolysis with chymopapain. Clin Orthop 251:144–152

Brandser EA, Renfrew DL, Schenck RR (1995) Adhesive capsulitis of the wrist. Can Assoc Radiol J 46:137–138

Callahan JD, Scully TB, Shapiro SA, Worth RM (1991) Suprascapular nerve entrapment. A series of 27 cases – comment. J Neurosurg 75:1001, and 74:893–896

Cardon T, Hachulla E, Flipo RM, et al. (1994) Percutaneous vertebroplasty with acrylic cement in the treatment of a Langerhans cell vertebral histiocytosis. Clin Rheumatol 13:518–521

Castro WH, Jerosch J, Hepp R, Schulitz KP (1992) Restriction of indication for automated percutaneous lumbar discectomy based on computed tomographic discography. Spine 17:1239–1243

Chevrot A, Cermakova E, Vallee C, Chancelier MD, Chemla N, Rousselin B, Langer-Cherbit A (1995) C1–2 arthrography. Skeletal Radiol 24:425–429

Chiras J, Sola-Martinez MT, Weill A, Rose M, Cognard C, Martin-Duverneuil N (1995) Vertebroplasties percutanees. Rev Med Interne 16:854–859

Choy DS, Case RB, Fielding W, Hughes J, Liebler W, Ascher P (1987) Percutaneous laser nucleolysis of lumbar disks (letter). N Engl J Med 317:771–772

Civardi G, Livraghi T, Colombo P, Fornari F, Cavanna L, Buscarini L (1994) Lytic bone lesions suspected for metastasis: ultrasonically guided fine-needle aspiration biopsy. J Clin Ultrasound 22:307–311

Coderc E, Chevrot A, Vallee C, et al. (1990) L'annulographie. J Radiol 71:331–337

Cotten A, Dewatre F, Cortet B, et al. (1996) Percutaneous vertebroplasty for osteolytic metastases and myeloma: effects of the percentage of lesion filling and the leakage of methyl methacrylate at clinical follow-up. Radiology 200:525–530

D'Erme M, Del Popolo P, Diotallevi R, Pasquali-Lasagni M (1995) Trattamento percutaneo dell'osteoma osteoide sotto guida con Tomografia Computerizzata. Radiol Med (Torino) 90:84–87

de Berg JC, Pattynama PM, Obermann WR, Bode PJ, Vielvoye GJ, Taminiau AH (1995) Percutaneous computed-tomography-guided thermocoagulation for osteoid osteomas. Lancet 346:350

Devulder J, Bogaert L, Castille F, Moerman A, Rolly G (1995) Relevance of epidurography and epidural adhesiolysis in chronic failed back surgery patients. Clin J Pain 11:147–150

Ekelund AL, Rydell N (1992) Combination treatment for adhesive capsulitis of the shoulder. Clin Orthop 282:105–109

el-Khoury GY, Renfrew DL, Walker CW (1994) Interventional musculoskeletal radiology. Curr Probl Diagn Radiol 23:161–203

Fehrman DA, Orwin JF, Jennings RM (1995) Suprascapular nerve entrapment by ganglion cysts: a report of six cases with arthroscopic findings and review of the literature. Arthroscopy 11:727–734

Fornage BD, Richli WR, Chuapetcharasopon C (1991) Calcaneal bone cyst: sonographic findings and ultrasound-guided aspiration biopsy. J Clin Ultrasound 19:360–362

Fortin JD, Aprill CN, Ponthieux B, Pier J (1994) Sacroiliac joint: pain referral maps upon applying a new injection/arthrography technique. II. Clinical evaluation. Spine 19:1483–1489

Fritz RC, Helms CA, Steinbach LS, Genant HK (1992) Suprascapular nerve entrapment: evaluation with MR imaging. Radiology 182:437–444

Galloy MA, Routy A, Gerbier R, Lascombes P, Hoeffel JC (1996) La resection percutanee des osteomes osteoides. J Chir (Paris) 133:37–42

Galvez J, Linares LF, Villalon M, Pagan E, Marras C, Castellon P (1995) Acute calcific periarthritis of the fingers. Rev Rhum Engl Ed 62:602–604

Gangi A, Kastler BA, Dietemann JL (1994) Percutaneous vertebroplasty guided by a combination of CT and fluoroscopy. AJNR 15:83–86

Gavant ML, Rizk TE, Gold RE, Flick PA (1994) Distention arthrography in the treatment of adhesive capsulitis of the shoulder. J Vasc Intervent Radiol 5:305–308

Ghelman B, Lospinuso MF, Levine DB, O'Leary PF, Burke SW (1991) Percutaneous computed-tomography-guided biopsy of the thoracic and lumbar spine. Spine 16:736–739

Gosal HS, Harrison DJ (1995) Magnetic resonance imaging before chemonucleolysis for lumbar disc prolapse. Eur Spine J 4:206–209

Guyer RD, Ohnmeiss DD (1995) Lumbar discography. Position statement from the North American Spine Society Diagnostic and Therapeutic Committee. Spine 20:2048–2059

Guyer RD, Collier R, Stith WJ, Ohnmeiss DD, Hochschuler SH, Rashbaum RF, Regan JJ (1988) Discitis after discography. Spine 13:1352–1354

Hashimoto BE, Hayes AS, Ager JD (1994) Sonographic diagnosis and treatment of ganglion cysts causing suprascapular nerve entrapment. J Ultrasound Med 13:671

Hauenstein KH, Vinee P, Adler CP (1995) Percutaneous needle biopsy in skeletal metastases. Indications, technique, value and results. Radiologe 35:39–46

Hayes CW, Conway WF (1990) Calcium hydroxyapatite deposition disease. Radiographics 10:1031–1034

Hove B, Gyldensted C (1990) Cervical analgesic facet joint arthrography. Neuroradiology 32:456–459

Howard CB, Einhorn M, Dagan R, Yagupski P, Porat S (1994) Fine-needle bone biopsy to diagnose osteomyelitis. J Bone Joint Surg [Br] 76:311–314

Hsu KY, Zucherman JF, Shea WJ, Jeffrey RA (1995) Lumbar intraspinal synovial and ganglion cysts (facet cysts). Ten-year experience in evaluation and treatment. Spine 20:80–89

Hulstyn MJ, Weiss AP (1993) Adhesive capsulitis of the shoulder. Orthop Rev 22:425–433

Kneisl JS, Simon MA (1992) Medical management compared with operative treatment for osteoid-osteoma. J Bone Joint Surg [Am] 74:179–185

Kohler R, Rubini J, Postec F, Canterino I, Archimbaud F (1995) Traitement de l'osteome osteoide par forage resection percutane sous controle tomodensitometrique (F.R.O.P.). A propos de 27 cas. Rev Chir Orthop 81:317–325

Konermann W, Wuisman P, Hillmann A, Rossner A, Blasius S (1995) Value of sonographically guided biopsy in the histological diagnosis of benign and malignant soft-tissue and bone tumors. Z Orthop Ihre Grenzgeb 133:411

Langer-Cherbit A, Chemla N, Vacherot B, Dupont AM, Godefroy D, Chevrot A (1994) Interet et resultats de la biopsie osseuse profonde rachidienne radioguidee. J Radiol 75:603–608

L'Huillier F, Chevrot A, Vallee C, Gires F, Wybier M, Pallardy G (1988) Lomboradiculalgie et hernie discale calcifiee. J Radiol 69:763–766

Logan PM, Connell DG, O'Connell JX, Munk PL, Janzen DL (1996) Image-guided percutaneous biopsy of musculoskeletal tumors: an algorithm for selection of specific biopsy techniques. AJR 166:137–141

Louwaege A, Goubau J, Deldycke H, et al. (1996) Efficiency of discography followed by chemonucleolysis in the treatment of sciatica. J Belge Radiol 79:68–71

Lynch MC, Taylor JF (1986) Facet joint injection for low back pain. A clinical study. J Bone Joint Surg [Br] 68:138–141

Maheshwaran S, Davies AM, Evans N, Broadley P, Cassar-Pullicino VN (1995) Sciatica in degenerative spondylolisthesis of the lumbar spine. Ann Rheum Dis 54:539–543

Mochida J, Arima T (1993) Percutaneous nucleotomy in lumbar disc herniation. A prospective study. Spine 18:2063–2068

Moutounet J, Chevrot A, Wybier M, Godefroy D (1992) Ponction-infiltration radio-guidee des calcifications des periarthrites rebelles de l'epaule. Ann Radiol (Paris) 35:156–159

Murnaghan JP (1988) Adhesive capsulitis of the shoulder: current concepts and treatment. Orthopedics 11:153–158

Oberpenning F, Roth S, Leusmann DB, van Ahlen H, Hertle L (1994) The Alcock syndrome: temporary penile insensitivity due to compression of the pudendal nerve within the Alcock canal. J Urol 151:423–425

Osti OL, Fraser RD, Vernon-Roberts B (1990) Discitis after discography. The role of prophylactic antibiotics. J Bone Joint Surg [Br] 72:271–274

Palladino SJ, Chan R (1987) Adhesive capsulitis of the ankle. J Foot Surg 26:484–492

Papaioannou T, Rushworth G, Atar D, Dekel S (1992) Carpal canal stenosis in men with idiopathic carpal tunnel syndrome. Clin Orthop 285:210–213

Reuther G (1994) CT-guided biopsies of the axial skeleton. The approaches and results. Rofo Fortschr Geb Rontgenstr Neuen Bildgeb Verfahr 160:78–83

Revel M, Amor B, Mathieu A, Wybier M, Vallee C, Chevrot A (1988) Sciatica induced by primary epidural adhesions. Lancet 1:527–528

Romanoff ME, Cory PC Jr, Kalenak A, Keyser GC, Marshall WK (1989) Saphenous nerve entrapment at the adductor canal. Am J Sports Med 17:478–481

Rosenbaum RB (1993) The role of imaging in the diagnosis of carpal tunnel syndrome. Invest Radiol 28:1059–1062

Rosenthal DI, Alexander A, Rosenberg AE, Springfield D (1992) Ablation of osteoid osteomas with a percutaneously placed electrode: a new procedure. Radiology 183:29–33

Rosenthal DI, Springfield DS, Gebhardt MC, Rosenberg AE, Mankin HJ (1995) Osteoid osteoma: percutaneous radiofrequency ablation. Radiology 197:451-454

Sanhaji L, Gharbaoui IS, Hassani RE, Chakir N, Jiddane M, Boukhrissi N (1996) Un nouveau traitement de l'osteome osteoide: la sclerose percutanee a l'ethanol sous guidage scanographique. J Radiol 77:37–40

Schellhas KP, Smith MD, Gundry CR, Pollei SR (1996) Cervical discogenic pain. Prospective correlation of magnetic resonance imaging and discography in asymptomatic subjects and pain sufferers. Spine 21:300–311

Schleifer J, Fenzl G, Wolf A, Diehl K (1994) Treatment of lumbar facet joint syndrome by CT-guided infiltration of the intervertebral joints. Radiologe 34:666–670

Schleifer J, Kiefer M, Hagen T (1995) Lumbar facet syndrome. Recommendation for staging before and after intra-articular injection treatment. Radiologe 35:844–847

Schweitzer ME, Deely DM (1993) Percutaneous biopsy of osteolytic lesions: use of a biopsy gun. Radiology 189:615–616

Schweitzer ME, Deely DM, Beavis K, Gannon F (1995) Does the use of lidocaine affect the culture of percutaneous bone biopsy specimens obtained to diagnose osteomyelitis? An in vitro and in vivo study. AJR 164:1201–1203

Sellier N, Vallee C, Chevrot A, et al. (1987) La sciatique par kystes synoviaux et diverticules articulaires lombaires a developpement intra-rachidien. Etude saccoradiculographique, tomodensitometrique et arthrographique. Rev Rhum Mal Osteoartic 54:297–301

Simon MA, Biermann JS (1993) Biopsy of bone and soft-tissue lesions. J Bone Joint Surg [Am] 75:616–621

Skrzynski MC, Biermann JS, Montag A, Simon MA (1996) Diagnostic accuracy and charge-savings of outpatient core needle biopsy compared with open biopsy of musculoskeletal tumors. J Bone Joint Surg [Am] 78:639–643, and 644–649

Stewart HD, Quinnell RC, Dann N (1987) Epidurography in the management of sciatica. Br J Rheumatol 26:424–429

Stockley I, Getty CJ, Dixon AK, Glaves I, Euinton HA, Barrington NA (1988) Lumbar lateral canal entrapment: clinical, radiculographic and computed tomographic findings. Clin Radiol 39:144–149

Stoker DJ, Cobb JP, Pringle JA (1991) Needle biopsy of musculoskeletal lesions. A review of 208 procedures. J Bone Joint Surg [Br] 73:498–500

Tournade A, Patay Z, Krupa P, Tajahmady T, Million S, Braun M (1992) A comparative study of the anatomical, radiological and therapeutic features of the lumbar facet joints. Neuroradiology 34:257–261

Towbin R, Kaye R, Meza MP, Pollock AN, Yaw K, Moreland M (1995) Osteoid osteoma: percutaneous excision using a CT-guided coaxial technique. AJR 164:945–949

Vallee C, Chevrot A, Benhamouda M, Gires F, Wybier M, Sellier N, Pallardy G (1987) Aspects tomodensitometriques des kystes synoviaux articulaires lombaires a developpement intrarachidien. J Radiol 68:519–526

Vorwerk D, Klose KC, Guenther RW, Loer F (1989) A new motor-driven percutaneous bone biopsy system: technical note. Cardiovasc Intervent Radiol 12:232–235

Weill A, Chiras J, Simon JM, Rose M, Sola-Martinez T, Enkaoua E (1996) Spinal metastases: indications for and results of percutaneous injection of acrylic surgical cement. Radiology 199:241–247

Yosipovitch G, Yosipovitch Z (1993) Acute calcific periarthritis of the hand and elbows in women. A study and review of the literature. J Rheumatol 20:1533–1538

Zeidman SM, Thompson K, Ducker TB (1995) Complications of cervical discography: analysis of 4400 diagnostic disc injections. Neurosurgery 37:414–417

8 Measurements and Related Examination Techniques in Orthopedic Radiology

N. Egund

CONTENTS

8.1
Introduction

Measurements in orthopedic radiology are easy to perform using a ruler and a goniometer, but the prerequisites for the measurements are commonly not available. Therefore it is the responsibility of the radiologist, in cooperation with his or her technicians, to ensure that examination techniques allow the measurements required and to recognize failures in related imaging techniques.

This chapter will focus on the principles underlying some common measurements in orthopedic radiology and the related examination techniques, which are of crucial importance for the measurements. Roentgen stereophotogrammetry represents the most accurate method of measurement in three dimensions (3-D), but it is invasive and available in

only a small number of hospitals, and will not be described in detail. Magnetic resonance imaging (MRI) and especially computed tomography (CT) represent tools for 3-D measurements not recognized in the literature and the general principles in respect of these measurements are described. Digital radiography may complicate the conventional measurements and software which substitutes film imaging is still undergoing development. Some measurements and examination techniques described have not been published specifically, but methods and results have appeared in the orthopedic literature and dissertations or are generally accepted. For a number of measurements reference should be made to specific textbooks: for scoliosis see Loenstein et al. (1995); for foot deformities see Ozonoff (1992); for bone dysplasias see Taybi and Lachman (1996), and for general skeletal measurements and their normal values see Keats (1990).

8.2
Measurement of Distances Using Conventional Radiography

The true length of an anatomical element can be measured directly on a single radiograph assuming that (a) the element is located parallel to the film plane and perpendicular to the central x-ray beam and (b) the magnification is known (Fig. 8.1). Such measurements, e.g., the length of extremities, may be accurately performed using an orthoradiographic technique (Fig. 8.1); for this purpose the points of interest, namely the centers of the joints, must be positioned equidistant to the film plane and exposures must be obtained precisely and separately at each point of interrest. These conditions are rarely fulfilled; for example, in the commonly used oblique lateral view of the proximal portion of the femur, magnification and positioning in relation to the focus and film plane are not controlled (Fig. 8.2). This oblique lateral view of the hip therefore cannot

N. Egund, MD, Professor, Røntgenafdeling R, Århus Kommunehospital, Nørrebrogade 44, DK-8000 Århus C, Denmark

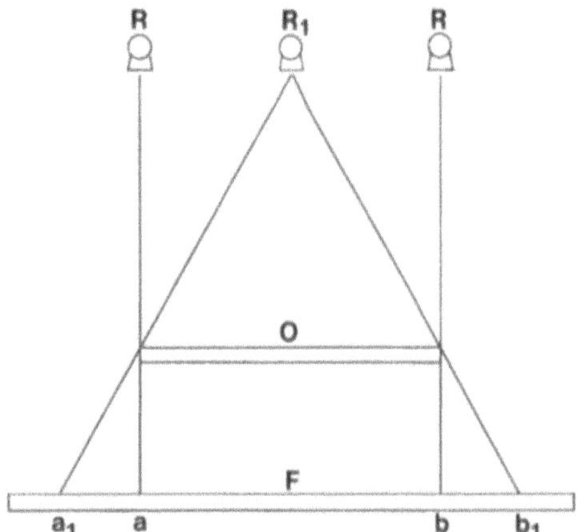

Fig. 8.1. Magnification of an object and measurement of distances on radiography. The length of the object O is measured using two separate exposures (R-a and R-b) of the same film. Both exposures are perpendicular to O and to the film (orthoradiographic technique). The length of O is equal to the distance that the x-ray table was moved between the exposures R-a and R-b. At a single exposure from R_1 a magnification of the object will occur, a_1-b_1. If the distance RO, between the tube R and the object O, and the distance RF, between the tube R and the film plane F, are known, the magnification, M, can be calculated: $M = RO/RF$, and thus the length of O on the film is a_1-b_1 = $a - b/M$

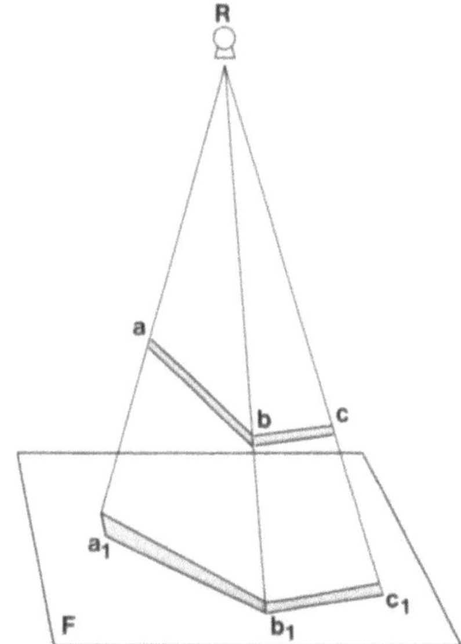

Fig. 8.2. Magnification of an object with an oblique position in relation to the film plane. The arms a-b and b-c of the object are not parallel to the film plane (F) and the magnification of their length from b_1-a_1 and from b_1-c_1 will vary, as well as the magnification of the diameter of the arms. The situation may represent the oblique lateral view of the hip. If in addition the film plane is not perpendicular to the x-rays, the situation is further complicated

be used for assessment of the necessary size of a prosthetic component prior to arthroplasty, but such assessment *is* possible by means of a true lateral view of the upper portion of the femoral bone with the patient in a lateral recumbent position. Templates of prosthetic components are available at defined magnifications, commonly 25%. Radiographs therefore must be obtained at exactly this degree of magnification in both planes, as may be ensured by use of a radiodense marker, e.g., a ring with a known diameter placed at the level of the bone to be measured. In the anteroposterior (AP) projection of the proximal femur antecurvation of the femoral diaphysis induces a difference in magnification of more than 5% between the lesser trochanter and the bone 10–15 cm more distally.

8.3
Measurement of Distances Using Digital Radiography

Computed radiography provides measuring facilities on the workstation with the original cassette size inherent in the system and also indicated on small hard copies.

In digital fluororadiography no landmarks are available for distance measurements. Therefore each image must contain a radiodense reference at the level of the anatomical structures to be measured, e.g., a circle of known diameter which is made of radiodense material. The reference has to be renewed in the measuring program of the workstation for each image. Software is available for image composition with fluorography and digital radiography, for measurements of long bones and scoliosis which may be performed on hard copies or directly at the workstation.

8.4
Measurement of Angles

Radiographic measurements of angles are independent of magnification. A prerequisite for direct measurement of angles is that both arms of the angle be positioned in the same plane, perpendicular to the central beam and parallel to the film plane (EDHOLM 1966). This condition is rarely fulfilled in the assessment of displaced fractures of long bones, and in such cases the correct angle is measured by a three-dimensional biplane technique which requires (a)

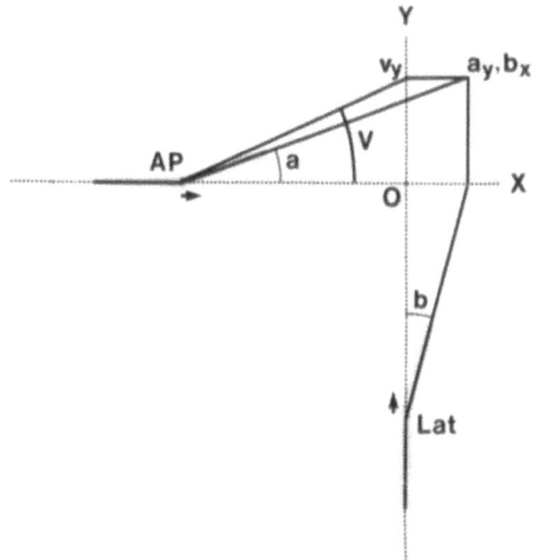

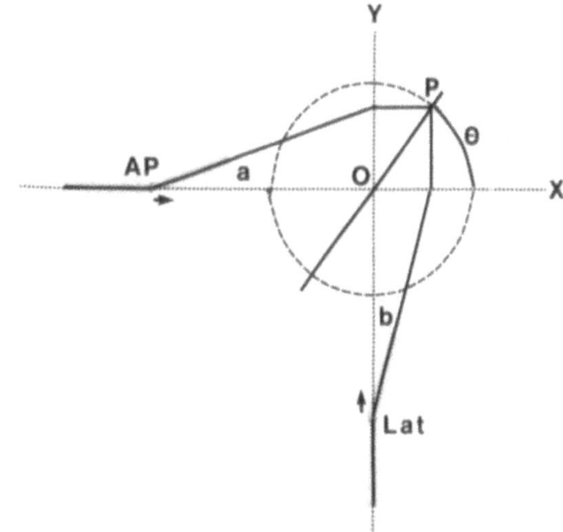

Fig. 8.3. Graphical construction of the total displacement with the biplanar radiographic technique. Two perpendicular lines, x and y, are drawn and at the same at arbitrary distances (*arrows*) from their intersection (O), the angles a and b are constructed, e.g., representing the displacement of a fracture of long bones in the anteroposterior (AP) and the lateral (Lat) view repectively. At the intersection between the arm of the angle b and line x, a line is drawn parallel to y and at its intersection with the arm of b (a_y, b_x) a new line parallel to x is continued to the intersection with y (V_y). The angle V represents the total displacement. At calculation, $V° = (a^2 + b^2)^{1/2}$

Fig. 8.4. Graphical construction and calculation of the direction of displacement of, for example, a fracture obtained from two perpendicular projections. The arms of the two angles $a°$ and $b°$ respectively are constructed on two perpendicular lines x and y, as in Fig. 8.4, and at their intersections lines are drawn parallel to x and y respectively to join each other at P, representing the site of the 3-D displacement in a 360° circle (e.g., the axial view of a leg). In the projection along OP no displacement is seen and in the projection perpendicular to OP the maximal displacement can be measured. The angle θ can be calculated by $\tan\theta = \tan b/\tan a$. At different directions of displacement of the arms, i.e., $-X,Y$, $-X,-Y$ and $X,-Y$, 90°, 180° and 270° respectively has to be added

knowledge of the direction of one of the arms of the angle and (b) that right angles are formed (1) between the directions of projections, (2) between the direction of projections and the known arm of the angle and (3) between the direction of projections and their respective films. From two perpendicular views, AP and lateral, a displacement of $a°$ in the AP view and $b°$ in the lateral view may be recorded. The total displacement $V°$ may be determined by a graphical construction (Fig. 8.3) or calculated by $\tan V° = (\tan^2 a + \tan^2 b)^{1/2}$ (Bogdanov) or simply $V° = (a^2 + b^2)^{1/2}$. The direction of displacement is evaluated by a graphical construction or calculated (Fig. 8.4).

8.5
Roentgen Stereophotogrammetry

Stereophotogrammetry refers to measurements in pictures that make it possible to reconstruct a 3-D object from two-dimensional (2-D) images. For roentgen stereophotogrammetry (RSA), two x-ray tubes are used, exposing the same object with simultaneous exposures (Fig. 8.5). Exposures on film of

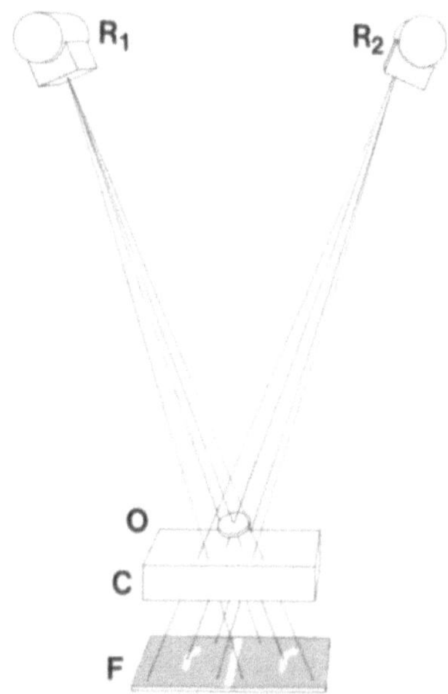

Fig. 8.5. Roentgen stereophotogrammetry. Determination of the position of the focus of the anode (R_1 and R_2). The images of the object (O) and the known position of a control point on the calibration cage (C) can be extrapolated towards the focus. A number of such lines intersect at the focus. F, Film plane

calibration cages with defined positions of metal markers at the top and bottom are used to identify the position of the two foci by computer calculation of the relative positions on the film(s) of each of the two layers of metal markers (SELVIG 1989). Identification of the position of each x-ray anode relative to the markers of the calibration cage allows definition of a 3-D space or x,y,z-coordinate system in which distances, angles, and volumes of rigid bodies can be measured. Identification of normal skeletal landmarks has low precision and therefore tantalum bullets of 0.5–2.0 mm in diameter are inserted in the objects to be analyzed making the method invasive. Two bullets, e.g., in a growing bone on each side of the physis, can allow determination of their 3-D distance and, at repeated examination, their change in distance with a precision of 2 µm. Rigid bodies are determined by not less than three bullets separated by distances and planes. At repeated examination, movements between two rigid bodies can be measured in terms of distance and rotation with a precision of 0.2 mm and 0.4° respectively (RYD 1986). RSA has many applications in developmental orthopedic surgery, the most important being the assessment of the mechanical stability of new components and their fixation to bone in joint replacement, and the method has become the gold standard for the accreditation of products in an increasing number of countries.

8.6
Measurements
Using Computed Tomography

Computed tomography is a useful tool for the measurement of distances, angles, and volumes but software programs for the measurements are not available from the product manufacturers although all the necessary information is available and used in 3-D surface- and volume-rendering reconstructions. Most 3-D measurements by means of CT in orthopedic radiology include only a few sections of relevant anatomical landmarks at low mAs, representing much lower radiation absorption than with conventional radiography, especially when the mAs is kept below 120 at 140 kV.

The *digital information* obtained by contiguous or helicon sectioning represents a rigid body of volume elements (voxels) in a defined 3-D coordinate system comparable to that of RSA (Fig. 8.6). In the sections on the screen each picture element (pixel) is defined by x,y coordinates obtained from

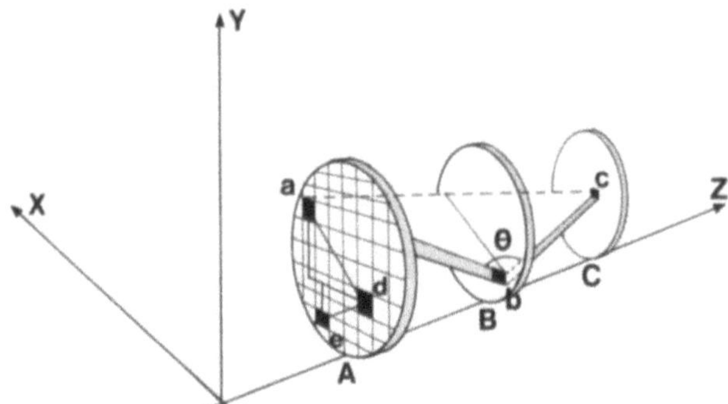

Fig. 8.6. Calculation of angles and distances on and between three separate sections A, B, and C along the examination table represent the z-axis in a 3-D coordinate system with known distances in millimeters between the sections. In each section anatomical landmarks (a, b, c, d, and e) are defined by one pixel with x,y coordinates obtained from the ROI. The distance between two pixels a and d (or d and e) are calculated by their respective x,y coordinates: $D = ((x_1 - x_2)^2 + (y_1 - y_2)^2)^{1/2}$. The distance L_1 (a-b) and L_2 (b-c) between two anatomical landmarks in different sections and at different x,y coordinates in each section is calculated by $L = ((x_1 - x_2)^2 + (y_1 - y_2)^2 + (z_1 - z_2)^2)^{1/2}$. Measurement of an angle (θ) between two lines, a-b-c, joining three pixels in different sections requires calculation of the length of the the two legs, L_1 and L_2 of the angle and the use of vector mathematics: $θ = (x_1 - a_1) (y_1 - a_1) + (x_2 - a_2)(y_2 - a_2) + (x_3 + a_3)(y_3 - a_3)/(L_1 * L_2)$. The coordinates of a, b, and c define a plane and in relation to that and an axis, e.g., a-c, the position of landmarks or pixels can be described in terms of distances and angles

the region of interest (ROI). The 3-D information is obtained when the table position, representing the remaining z-axis, is added (EGUND and PALMER 1984).

Distances. On the workstation and on the screen software for measurement of distances is available when setting the ROI between two pixels within anatomical landmarks. However, the window level is of crucial importance for the measurement. Before placing the ROI on an anatomical structure, the window level is adjusted to the Hounsfield value of that specific structure, so that measurement between a bony and a soft tissue structure needs two separate Hounsfield measurements and window settings. The distance D between two pixels is calculated by their respective x,y coordinates, and the distances between two anatomical landmarks in different sections and at different x,y coordinates are obtained by geometric calculation (Fig. 8.6).

Angles. Angles between lines joining three points in one x,y section are measured directly by means of the CT software program using ordinary trigonometric mathematics. Measurement of an angle (θ) between two lines joining three points in different sections (Fig. 8.6) requires calculation of the length of the two arms of the angle and the use of vector mathematics. A number of such formulas are available and easily inserted in calculators or PC worksheets. Unfortunately they are not available on CT workstations.

Volumes. Volumes can be calculated by a number of *x,y,z* coordinates at the surface of the body in different sections using more advanced but commercially available PC programs.

3-D measurements of angles and distances in relation to axis and planes. With the availability of *x,y,z* coordinates of any landmark within the scanning area, mathematical calculations in 3-D have no limitations and only a few sections may be necessary for advanced and accurate measurements, though the accuracy is not comparable to that of RSA. Guided by the scout view on CT two sections and, within these, three anatomical landmarks can define an axis and a plane in 3-D (EGUND and PALMER 1984). In terms of distances and angles, any anatomical structure and its position can be determined in relation to these (Fig. 8.6).

8.7
Measurements Using Ultrasonography

The development of high-frequency linear array real-time transducers has dramatically increased the capabilities of ultrasonography in the evaluation of the musculoskeletal system. This group of devices shares common characteristics of excellent near-field resolution, electronic focusing, and high transducer frequency. Measurements by means of ultrasonography are the most common in diagnostic radiology, the size of a lesion or a joint effusion or the thickness of a tendon being reported by the two transverse diameters in millimeters or centimeters. Distances are measured with high accuracy and with a theoretical axial resolution of 0.2 mm at 10 MHz and 0.4 at 5 MHz and a lateral (azimuthal) resolution of about 0.5 mm at optimal focusing. There are, however, a number of precautions that need to be kept in mind (EGUND and WINGSTRAND 1989). First, the direction of scanning has to be perpendicular to the surfaces between which the distance is measured. Secondly, the distance in relation to the scanning direction must be measured between the outer and the opposing inner surface of the structure or lesion (Fig. 8.7). Thirdly, the thickness of the echoes from a surface, e.g., a joint capsule, varies with transducer frequency and gain settings and should not be interpreted as the thickness of a fibrous capsule and the synovia.

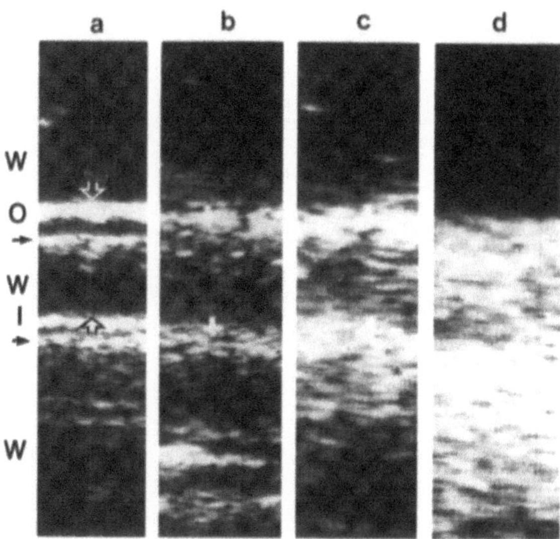

Fig. 8.7 a–d. Longitudinal ultrasonography of two cylinders inserted into each other and placed in water. The transducers used were 10 (**a**), 7.5 (**b**), 5.0 (**c**), and 3.5 (**d**) MHz. The echoes obtained represent the external surfaces of the outer (*O*) and the inner (*I*) cylinder. *W*, water. The further echogenic zones (*arrows*) between *O* and *I* and *I* and *W*, separated by hypoechoic zones, represent artifacts. There is a gradual broadening of the echoes of the outer surfaces from the use of 10-MHz to 3.5-MHz transducers. Distances are measured between the two outer surfaces (*open arrows*)

Ultrasonography of a surface may introduce a double-line pattern, consisting of an outer and an inner echogenic line separated by a relatively hypoechoic line, commonly interpreted as three separate anatomical structure layers. The double-line pattern represents an artifact and only the outer line can be used for measurements (NOLSØE et al. 1990).

8.8
The Lower Extremities

The most common measurements in daily routine concern conditions of the lower extremities including the hip, the knee, and the ankle. The examination technique employed for the measurements is decisive for the results.

8.8.1
Radiographic Examination Technique
for the Knee Joint in Standing

Routine standing knee examination in the adult is performed in total weight-bearing on the knee of

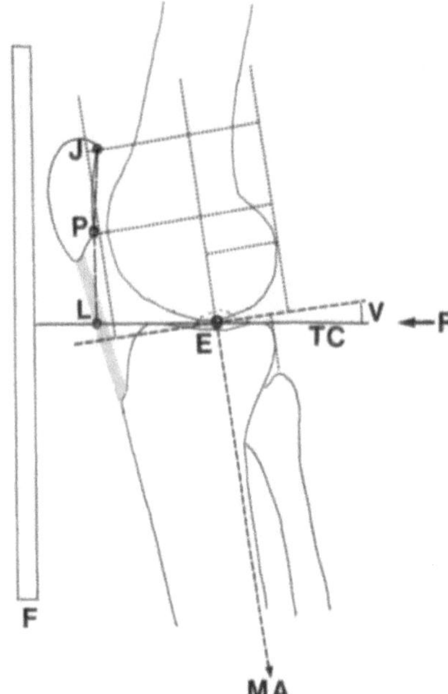

Fig. 8.8. Standing lateral view of the knee and related examination technique for the AP view. The mechanical axis of the lower leg (*MA*) connects the center of the tibial eminence (*E*) and the head of the talus. The dorsal tilt (*V*) of the tibial condyles is the angle between the line joining the proximal aspect of the joint surfaces of the tibial condyles (*TC*) and the line perpenducular to *MA*. To assess femorotibial joint spaces optimally, the PA/AP view of the knee is obtained with *TC* joining the central beam (*R*) and perpendicular to the film plane. Patellar height is measured by *PL/PJ*, *PL* and *PJ* being the height of the distal rim of the patellar joint surface above *TC* and the length of the patellar joint surface respectively. The normal value is 0.8 (SD 0.14). The *short broken lines* represent a model for analysis of the knee

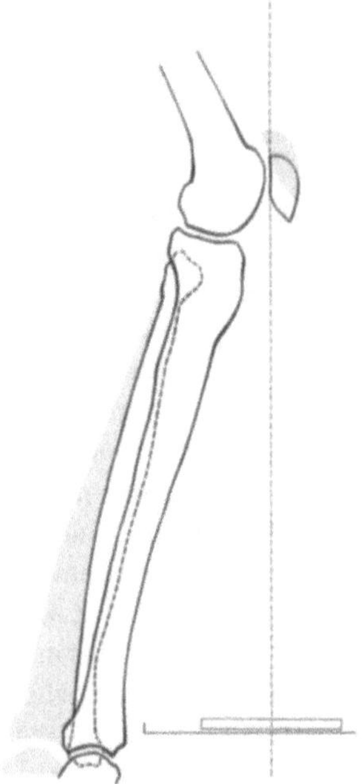

Fig. 8.9. The device for the radiographic examination of the patellofemoral joint. The support for the patella can be adjusted to the length along the two parallel rails. The angle between the vertical beam and the lower leg is 15°. The support for the angle is movable in the anterior/posterior direction, allowing adjustment for high and low riding patellas

interest. The frontal view is obtained in the PA position, supporting the knee against the casette holder and maintaining the lower leg in 5–10° of inclination and the knee in 20–25° of flexion (Fig. 8.8). The lateral view is obtained in weiht-bearing with the use of a support allowing 10–15° of inclination of the lower leg and 25–30° of knee flexion. Also the axial view of the patella is obtained in weight-bearing (Fig. 8.9) using a device by which the lower leg is positioned at 15° of inclination and the knee in varying degrees of flexion, most commonly 30–40° (Egund 1986).

These three routine images of the knee allow a number of measurements:

1. *Joint space.* The lateral femorotibial joint space (minimum 5 mm) is always larger than the medial femorotibial joint space (minimum 4 mm). Any right/left difference of the joint space is an accurate indicator of cartilage reduction. Meniscectomy, however, may result in a slight, but visible joint space reduction without reduction of cartilage thickness. The patellar joint space varies, but 4 mm should be considered the minimum variation, and any medio/lateral and right/left differences are indicative of pathology.

2. *Tibial condylar plane, lateral view (LTC).* In the lateral view the articular aspect of the tibial condyles has a posterior and distal slope of 10° (5–15°) this being the rationale for examination of the joint space in the AP/PA view (Fig. 8.8).

3. *Sagittal instability.* The lowest points of the distal aspect of the femoral condyles in the lateral view are normally, and at any degree of knee flexion, at the center of the tibial eminence (Fig. 8.10). At extension these points are located anterior to the center of the tibial eminence. This normal anatomy is independent of the examination technique used (supine or weight-bearing). In weight-bearing only (or on stress radiographs), a sagittal displacement may be registered and measured, most commonly representing an anterior displacement of the tibial eminence in anterior

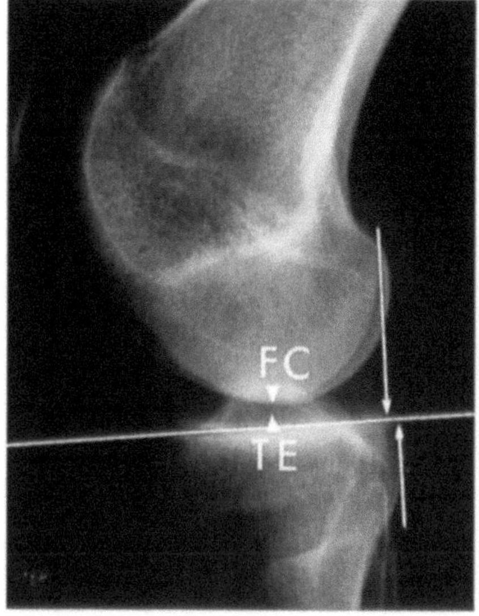
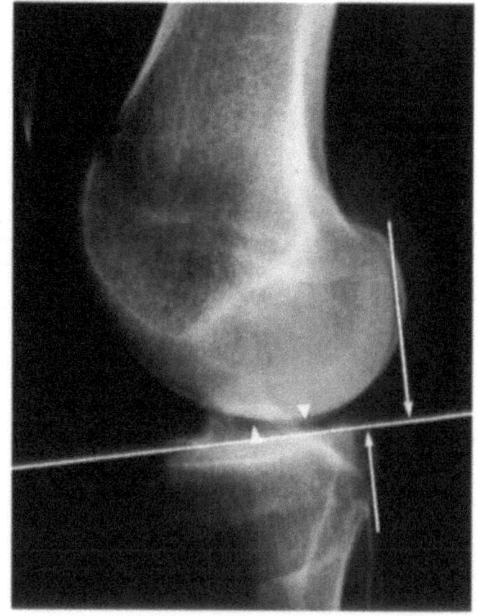

Fig. 8.10 a,b. Measurement of sagittal displacement of the knee joint. Lateral standing radiographs of a knee with anterior cruciate ligament deficiency. Without weightbearing (a) the site of the lowest point of the femoral condyles (*FC*) is at the center of the tibial eminence (*TE*), representing normal anatomy. In weightbearing (b) a dispalcement of 12 mm occurs between FC and TE. Measurements using the dorsal aspect of the femoral and tibial condyles (*arrows*) are less reliable in different degrees of knee flexion due to large normal variations

cruciate ligament deficiency (EGUND and FRIDÉN 1988). Any sagittal displacement exceeding 2 mm should be regarded as an indicator of ligamentous insufficiency.

4. *Patellar height.* The vertical position of the patella is defined as the site of the patella in relation to the anterior aspect of the femoral condyles or groove. The patellar ligament is under tension in weight-bearing and knee flexion only and patellar height should be measured in this position. Most commonly the Insall/Salvati index is used for assessment of patellar height, but this index is not reliable in the pre- and postoperative assessment of patellar surgery (EGUND et al. 1988).

Recommended is the method suggested by BLACKBURNE and PEEL (1977), which takes into account the patellar position above the plane of the femoral and tibial condyles and is especially reproducible if measurements are performed in relation to the condylar plane perpendicular to the long axis of the tibia (Fig. 8.8).

5. *Patellar shape and the femoral condylar groove angle.* Angles of the patellofemoral joint space and the femoral condylar angles at different degrees of flexion are measured according to BRATTSTRÖM (1964). Medial and lateral displacement can be measured from weight-bearing examinations only.

8.8.2
The Mechanical Axes of the Lower Extremities and Related Angles

Measurements of angles of the knee joint in the anteroposterior view include the hip and the ankle and the examination technique is of crucial importance for the accuracy and reproducibility of the measurements. A cassette with a 100–120 cm film or a cassette holder for three cassettes with film or CR-plates are used. The patient is positioned with the lower leg in 10° of inclination and the knee in 20° of flexion in complete weightbearing. A true lateral view of the femoral condyles is obtained at fluoroscospy and the AP view of the knee, hip, and ankle is obtained at a single perpendicular exposure with a film/focus distance of 2.0–2.5 m. Imaging obtained in extension may negate the prerequisites for the measurements. The mechanical axis of the femur and lower leg can be obtained separately using a knee/hip and knee/ankle view (Fig. 8.11).

8.8.2.1
The Mechanical Axis of the Lower Extremity

The mechanical axis of the lower extremity is represented by the line joining the center of the femoral

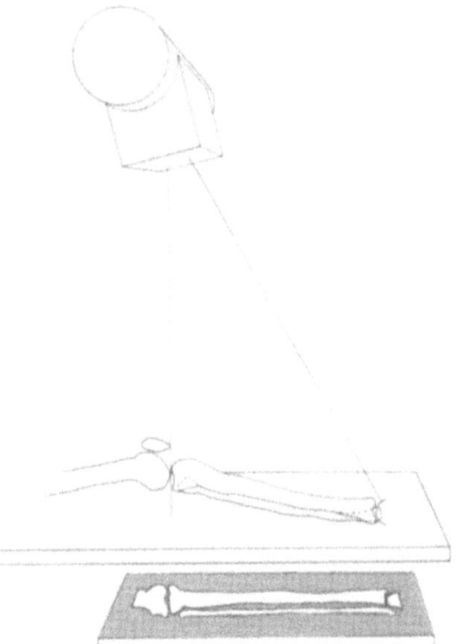

Fig. 8.11. Projections for separate measurements of angles of the knee and ankle and the knee and hip. The central beam is centered at the joint space of the knee placed in 15–20° of flexion. When the tube is tilted to cover the knee and lower leg (or hip), the knee is still projected perpendicular to the beams, being tangential to the joint surface of the tibial condyles with an inclination of the lower leg of 5–10°

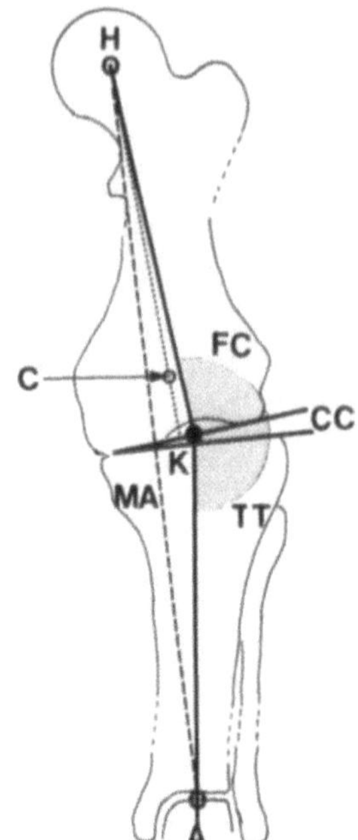

Fig. 8.12. Axes and angles of the lower extremity. *H*, *K* and *A* denote the center of the femoral head, the tibial eminence, and the head of the talus (ankle) respectively. *MA* is the mechanical weight-bearing axis. The hip-knee-ankle angle (HKA) is the lateral angle between *H*, *K*, and *A*. *FC* is the lateral angle between *HK* and the femoral condylar plane and *TT* is the lateral angle between *KT* and the tibial condylar plane. The *CC* angle represents the difference in width between the medial and the lateral joint space. HKA° = FC° + CC° + TT°. In medial arthrosis of the knee joint, the angles FC and TT remain unchanged if no bone attrition is present. Increase in the CC angle is the result of reduction of the medial joint space (1 mm = 1°) and increase in the lateral joint space = lateral instability. *C* is the center (midpoint) of the femoral condyles between their medial and lateral aspects. The distance along the femoral condylar plane between its intersection with the axis HC and K represents the femorotibial translantion

head and the center of the trochlea of the talus (Fig. 8.12).

8.8.2.2
The Hip-Knee-Ankle Angle

The hip-knee-ankle (HKA) angle is the lateral angle between the lines joining the center of the femoral head, the center of the tibial eminence, and the center of the head of the talus (Fig. 8.12; TJÖRNSTRAND et al. 1981). Angles larger than 180° are termed varus and angles less than 180° are termed valgus. With the examination technique used, normal values for females are 180° (SD 3°) and for males 182° (SD 4°).

8.8.2.3
The Femoral and Tibial Condylar Angles in the AP or PA View

The HKA angle is composed by three angles, the femoral and tibial condylar angles, FC and TC respectively, and the joint space angle (CC) in between FC and TC (Fig. 8.12). The equation HKA = FC + CC = TC is important in pre- and postoperative assessment of knee arthroplasties. Measurement of the

angle between the short axes through the femoral and tibial diaphyses, as obtained from a separate knee examination, has low accuracy and is not to be recommended, the angle being 6–7° of valgus in normal patients.

8.8.2.4
Translation Between the Femoral and Tibial Condyles in the AP/PA View

Instability in the coronal plane of the weight-bearing knee is common in osteoarthrosis and must be considered in the preoperative assessment. A line joining the center of the femoral head and the center of the femoral condyles passes medial to the center of the tibial eminence at a distance of 3 mm (Fig. 8.12).

8.8.2.5
Angles and Distances in the Assessment of Medial Osteoarthrosis of the Knee

Medial and lateral femorotibial arthrosis is associated with varus and valgus displacement, respectively, visualized at radiography with measurements of the HKA angle. Since the distance between the lower aspects of the medial and lateral femoral condyles is 4.5–5.5 cm, each millimeter of joint space reduction in one compartment will result in a 1° increase in the CC and and thus in the HKA angle. Lateral instability may further increase the CC angle, as may the medial translation of the femoral condyles (Fig. 8.12). Lateral instability and translation can be measured also by examination in the supine position with the knee in valgus stress.

8.8.2.6
Pre- and Postoperative Assessment of Knee Arthroplasty

When planning total knee replacement the radiologist has only two angles to consider, the FC and the TT angle, since the intended postoperative result is HKA = 180° = FC° + TT° (the CC angle = 0). For the orthopedic surgeon the situation is more difficult when attempting, during surgery, to combine the FC and TT angles to 180° by eye or using guide instruments. The use of guide instruments for the femoral component may require measurements of the angle between the femoral mechanical axis (HC) and the diaphyseal axis, taking into account variation in curvature of the diaphysis.

8.8.2.7
High Tibial Osteotomy

In high tibial osteotomy the mechanical axis of the lower extremity is transferred from passing through the medial compartment to the lateral compartment by a reduction of the TT angle, most commonly with the intention of achieving an HKT angle of 178–176° (2–4° of valgus). Three surgical methods are available for this purpose, namely dome, lateral closing wedge and medial opening osteotomy (PALEY et al. 1994) (Fig. 8.13). The surgical correction of the TT angle itself can be done with accuracy, but achieve-

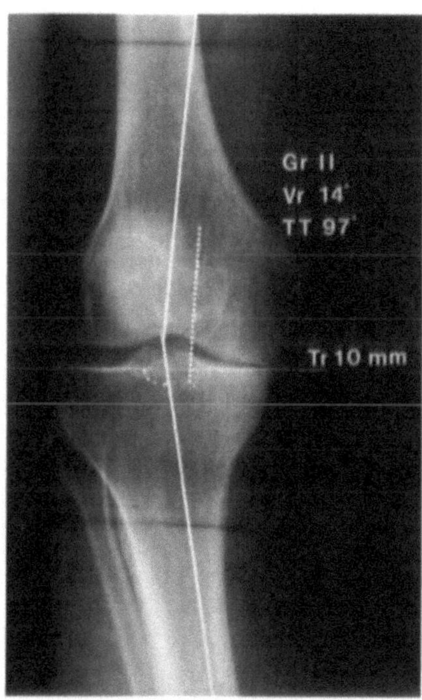

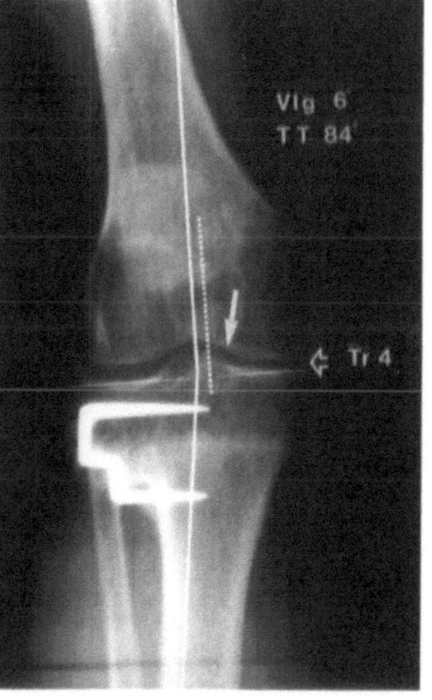

Fig. 8.13. Medial arthrosis stage II of the knee joint (a) with a varus displacement of 14° (HKA$_{pre}$ = 194°) and (b) the same knee in 6° of valgus (HKA$_{post}$ = 174°) after lateral closing high tibial osteotomy. HKA$_{pre}$ – HKA$_{post}$ = 20°, but TT$_{pre}$ – TT$_{post}$ = 13° of surgical correction. The remaining 7° of correction of HKA = 4 mm = 4° of lateral instability + 3 mm = 3° of reappearance of the medial joint space (*open arrow*) + 6 mm = 3° of reduced translation (distance between *broken line* and MA of femur) –3° due to lateralization of the tibial eminence. It appears that the line of MA of the tibia has turned from medial to lateral in relation to the tibial diaphysis. Reappearance of the medial joint is the result of reduced translation and unaffected cartilage centrally in the medial joint space (*arrow*)

ment of the intended change in the HKA angle requires carefully prepared radiographic examinations and measurements and calculations before surgery. After surgery lateral instability and translation are nullified and thus the starting point is to calculate the preoperative HKT_{pre} = FC + CC + TT – lateral instability – translation. Each of the aforementioned methods of osteotomy results in a lateralization of the point of measurement K in the center of the tibial eminence in relation to the diaphyseal axis of the tibia (Fig. 8.12), increasing the HKA_{pre} by $L° = 1–2°$ with lateral closing wedge and medial opening osteotomy and by up to 5° with dome osteotomy (measured with compasses placed at the center of rotation). The osteotomy is commonly associated with a reappearance of the medial joint space (Fig. 8.13); this is the result of reduced translation and gives rise to overcorrection, which can be controlled in medial opening osteotomy only. Thus the correction of the TT angle is $TT_{pre} – TT_{post}$ = $HKA_{pre} – HKA_{post}(176°) + L° (2°)$ – (eventual opening of medial joint space, 1 mm = 1°).

8.8.3
Rotational Angles of the Lower Extremities

Numerous radiological methods are available for assessment of normal and abnormal rotational positions around the vertical mechanical axis of the lower extremities, but few have proved of clinical significance. All measurements of in- or outward rotation of the femoral neck and the foot are related to the plane of the dorsal aspects of the femoral condyles (Fig. 8.14). A femoral diaphyseal or trochanteric fracture often results in a rotational displacement, and when the femoral neck anteversion in relation to the acetabulum is maintained at gait an outward rotation of the distal fragment will result in outward toeing of the foot and vice versa. The most common source of altered anteversion of the femoral neck is total hip replacement (HERMANN and EGUND 1997), but the significance of displacement in relation to function and mechanical loosening of prosthetic components has not yet been established.

8.8.3.1
The Anteversion Angle of the Femoral Neck

The CT technique proposed by MURPHY et al. (1987) has become the recommended method for measurement of the anteversion angle. Three CT sections at

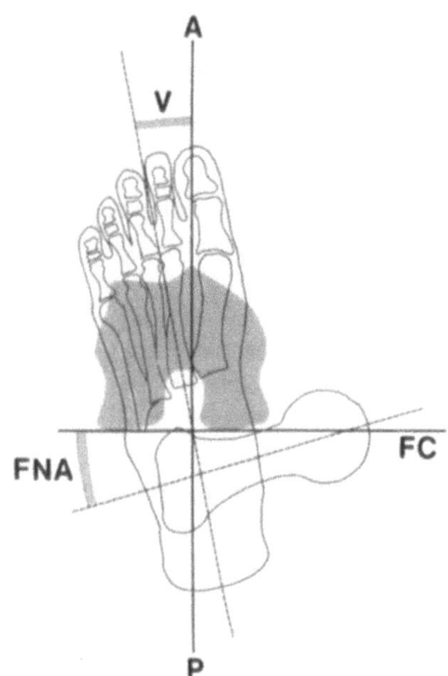

Fig. 8.14. Correlation between the position of the foot and the anteversion angle of the femoral neck (*FNA*) in relation to the dorsal aspect of the femoral condyles (*FC*) in 15–20° of knee flexion. The line through the middle of the heel and the interstitium between the second and the third toe is a functional indicator of tibial torsion (*V*) in relation to a true lateral view of the femoral condyles (*FC*) obtained at fluoroscopy. *AP* (anteroposterior) is perpendicular to *FC*

less than 120 mAs are used, one through the center of the femoral head, one through the lesser trochanter, and one through the femoral condyles. The three images are superimposed and the anteversion angle measured directly (Fig. 8.15). Prerequisites of the method are, however, that the centers of the sections through the lesser trochanter and the femoral condyles are located parallel to the examination table in both planes (identical x and y coordinates) (HERMAN and EGUND 1997). These prerequisites are never fulfilled during clinical measurements and therefore some simple mathematical adjustments have to be made using the 3-D possibilities of the CT information. With these adjustments the accuracy of the CT method is within ±1.5°, but without them it is ±8.8° (range –35° to +16°).

8.8.3.2
Measurement of the Anteversion Angle Using Conventional Radiography

Most routine radiographic examinations of the adult hip include AP and lateral oblique views and this is specifically true for postoperative examinations of

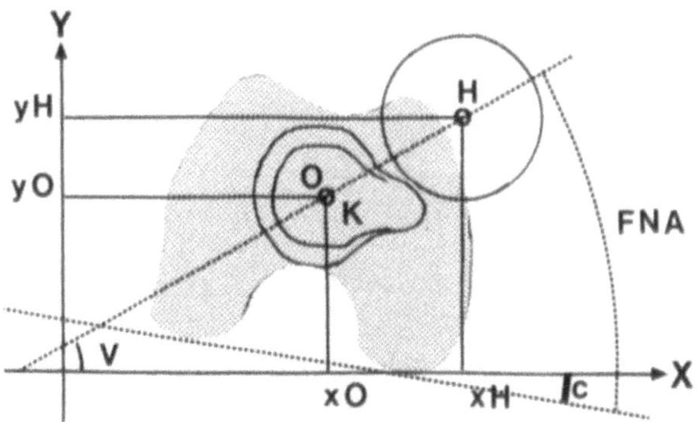

Fig. 8.15. The three CT sections necessary for the measurement of femoral neck anteversion (*FNA*), one through the center of the femoral head (*H*), one at the level of the lesser trochanter, and one through the femoral condyles. If the condition is fulfilled that the centers of the knee (*K*) and lesser trochanter (*O*) are located equidistant to horizontal and vertical, at identical *x* and *y* coordinates the three images are superimposed and the FNA between *OH* and the condylar plane is measured directly (*a*) and adjusted for knee rotation (*b*). Using the *x* and *y* coordinates of *H* and *O*, the angle can be calculated: $a = INV\ TAN\ (yH - yO)/(xH - xO)$, and similarly with knee rotation (*c*). FNA = $a + b$ at inward rotation and FNA = $a - b$ at outward rotation of the knee

fractures and total hip replacements. With one additional lateral radiograph of the femoral condyles of the knee, obtained with a horizontal beam direction and without alteration in the position of the extremity during the examination, an accurate measurement (SD and range) of the anteversion angle can be made. The measurement includes the same reference points employed in the CT measurement (Fig. 8.16).

8.8.3.3
In- and Outward Rotation of the Lower Leg

Measurement of inward and outward rotation of the lower leg by means of CT has been suggested, but the assessment has to be related to the clinical situation at gait using a standing weight-bearing position at 10–20° of flexion of the knee. At fluoroscopy a true lateral view of the dorsal aspect of the femoral condyles is maintained with the weight-bearing foot placed on a free rotatable plate. A line joining the middle of the heel and the interstitium of the first

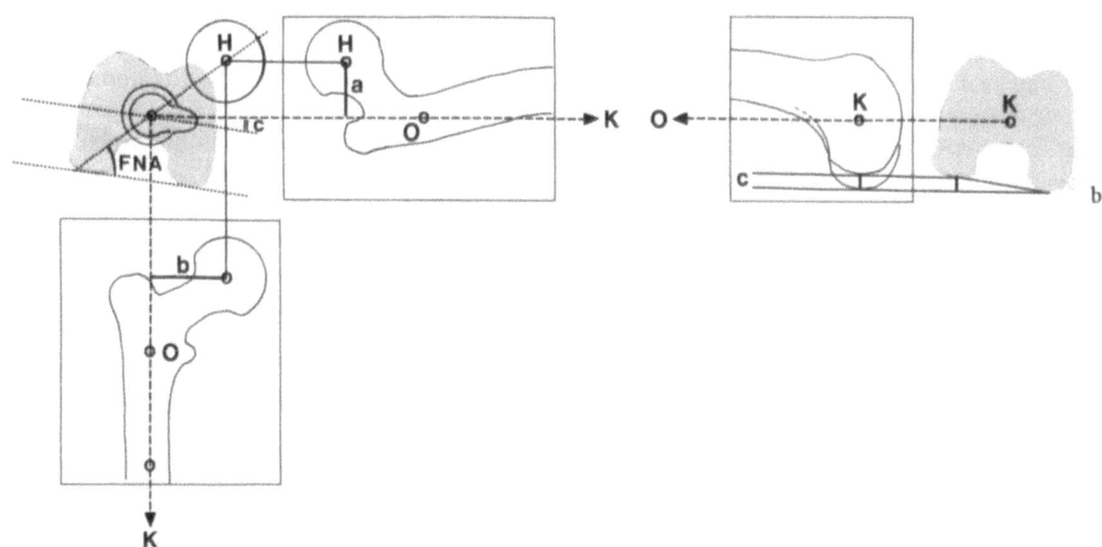

Fig. 8.16. Measurement of femoral neck anteversion (FNA) by means of a routine radiographic examination, demonstrating the similarity to measurement using CT. The radiographic examination consists of AP and 45° lateral oblique (a) and horizontal lateral views of the knee in internal rotation (b), the two films for the lateral views being placed equidistant to the examination table. All three exposures are obtained with the leg in the same position. The center of the femoral head (*H*) and the midpoint (*O*) of the femur at the lesser trochanter

are marked on the films in each view, as is the point (*K*) between the ventral and dorsal femoral condyles. The lateral views are placed parallel at a distance of 35 cm between *O* and *K*, through which the *broken line* is drawn. The diaphyseal axis (AP axis) is also created on the AP view. The distances *a* and *b* from the *O-K* axis in both planes to the femoral head are measured and $V = INV\ TAN\ (a/b)$. The distance (*c*) in mm between the femoral condyles = $c°$. The anteversion angle FNA° = $V° + c°$

and second toes indicates the functional torsion of the lower leg and ankle in relation to the sagittal plane perpendicular to the plane of the femoral condyles (Fig. 8.14). The measurement is highly reproducible. The assessment of in- and outward toeing can be performed without a plate with free rotation, bearing in mind that the normal knee in the weight-bearing position and at a moderate rotational stress allows outward/inward knee rotation of ±15°. In the presence of joint laxity these figures may be doubled and probably represent the main etiology of traumatic and recurrent patellar luxation.

8.8.3.4
Total Assessment of Rotational Deformities of the Lower Extremity

In total assessment of rotational deformities of the lower extremity the first step is a gait analysis in which the walking position of the foot (marks of the plantar surface of the foot on the floor) is registered and measured in degrees in relation to the walking direction. Secondly tibial torsion is measured by fluoroscopy and thirdly the anteversion angle of the femoral is measured by CT. In children inward toeing at gait is commonly a functional and transient phenomenon with normal tibial torsion and the anteversion angle within normal limits or at its upper level. Inward toeing in adults is usually associated with an abnormally high anteversion angle but normal tibial torsion. Outward toeing in adults is in most cases associated with increased outward tibial torsion with a normal anteversion angle. In the snapping hip with intoeing at gait, extremes of high femoral neck anteversion with normal tibial torsion can be recorded.

8.8.3.5
Direct Measurements of Rotational Positions

Rotational displacement around longitudinal axes (e.g., supracondylar fracture of the humerus) can be measured directly if two points of any anatomical surface are defined in each fragment (Fig. 8.17). Similarly, the rotational position of a humeral head prosthesis is measured at fluoroscopy with the elbow held in 90° of flexion. At a certain position of outward rotation of the humerus a tangential lateral view of the prosthetic head is obtained. In this position the angle of outward rotation of the humerus indicated by the forearm in relation to the direction

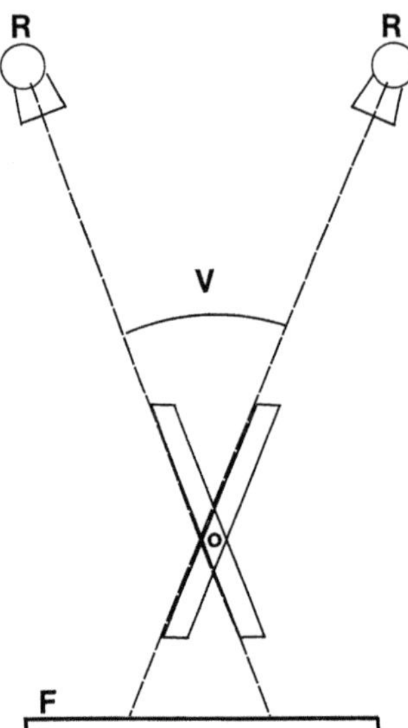

Fig. 8.17. Measurement of rotational displacement between fragments with defined surfaces. Rotational displacement is the angle (*V*) between the directions of the beams when they are tangential to each of the fragments. *F*, Film plane

of fluoroscopy represents the retroversion angle (40–50°).

8.9
Wrist and Hand Measurements

A variety of measurements are important when evaluating the injured or diseased wrist and hand. Because normal variations and abnormalities may be bilateral, when an abnormality is detected, comparison to the contralateral uninjured wrist can be of value. Measurements must in addition be assessed in relation to the overall clinical and radiographic evaluation of the patient. A prerequisite for assessing alignment and angles in the wrist and between the carpal bones is a full understanding of the importance of correct lateral and PA wrist radiographic projections.

8.9.1
Palmar Tilt of the Distal Radius

In the lateral view of the wrist a line joining the most distal points of the dorsal and ventral rims of the distal articular surface of the radius has a palmar tilt of 11° (2–22°) (Fig. 8.18). Reduction of the angle may indicate current or previous fracture of the distal radius. *Ulnar variance* is used to describe the relative

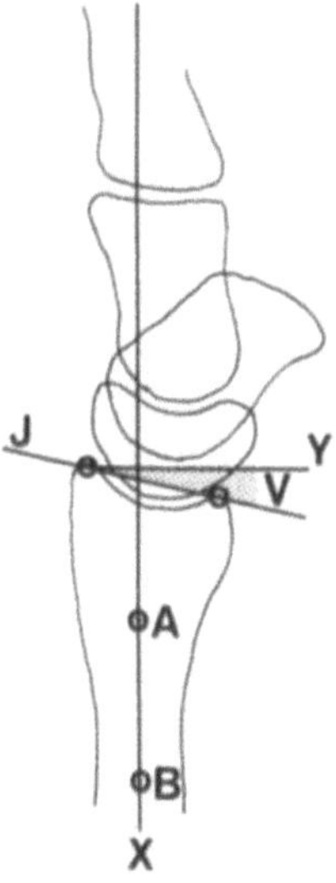

Fig. 8.18. Palmar tilt is determined by the line *J*, joining the most distal points of the dorsal and ventral rims of the distal articular surface of the radius. The degree of palmar tilt (*V*) is derived by the intersection of the line of palmar tilt (*J*) and a line perpendicular (*Y*) to the long axis of the radius (*X*), determined by a line through the center of its medullary space at 2 and 5 cm (*AB*) proximal to the radiocarpal joint

positions of the distal articular surfaces of the radius and ulna. A positive ulnar variance occurs when the distal cortical surface of the ulna projects more distally than the adjacent distal radial articular surface, and negative ulnar variance is present if it projects more proximally (-0.5 ± 1.5 mm).

8.9.2
Carpal Angles on Lateral Radiographs of the Wrist

The most precise measurements of the carpal bone angles are obtained by the use of the four axes suggested by Larsen in BARATZ and LARSEN (1996) (Fig. 8.19). Right/left differences of more than 5° can be considered significant. Palmar or dorsal tilt of the lunate axis in relation to the axes of the radius and scaphoid should be considered in any assessment of the lateral wrist radiographs.

8.9.3
Ulnar Translocation of the Wrist

Ulnar carpal translocation is a complication of rheumatoid arthritis and other synovial arthritides, but is also seen in rare cases of posttraumatic carpal instability. Measurements of translocation are most sim-

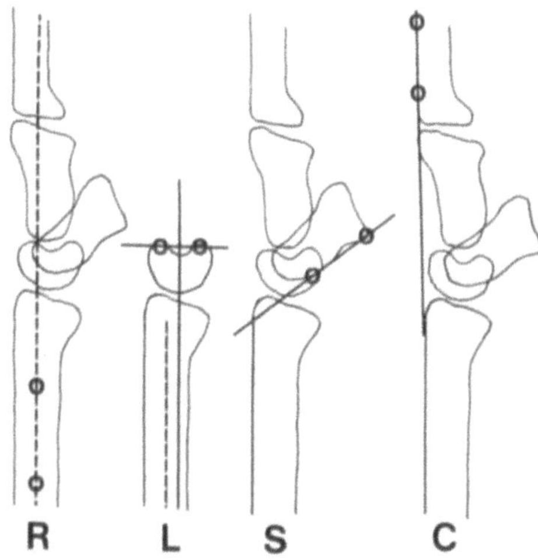

Fig. 8.19. The four axes providing the least observer variability for assessment of carpal alignment as recommended by Larsen et al. *R*, The long axis of the radius. Lunate (*L*) tilt is determined by the angle between the line joining the distal dorsal and ventral rims of the lunate and *R*. The scaphoid (*S*) tilt is measured using the ventral aspect of the bone in relation to *R*, and the capitate tilt (*C*) is measured using the dorsal aspect of the third metacarpal in relation to *R*

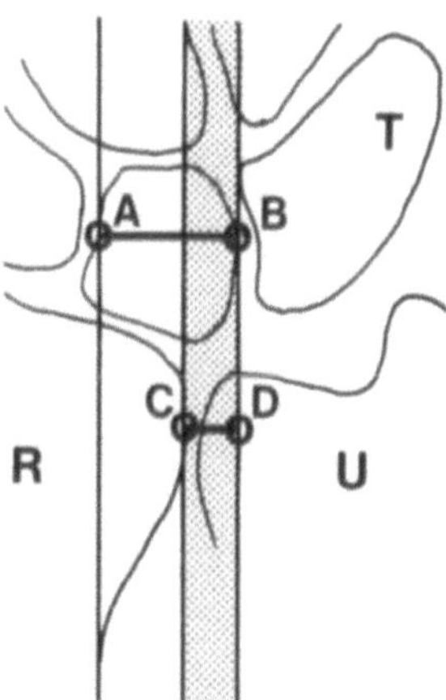

Fig. 8.20. Assessment of ulnar translocation of the carpus. The method of Gilula and Weeks indicates that ulnar translocation of the carpus is present if more than 50% of the lunate overhang is ulnar to the lunate fossa. The lunate overhang (*CD*) is divided by the lunate width (*AB*) to obtain a quantitative measurement (ratio). The semiquantitative measurement involves visual inspection of the radiograph and determination of whether the lunate is translated more than 50% ulnarly. *R*, Radius; *U*, ulna; *T*, triquetrum

ply performed as suggested by Gilula and Weeks (1978) (Fig. 8.20).

References

Baratz ME, Larsen CF (1996) Wrist and hand measurements and classification schemes. In: Gilula LA, Yin Y (eds) Imaging of the wrist and hand. Saunders, Philadelphia, pp 225–259

Blackburne JS, Peel TE (1977) A new method for measuring patellar height. J Bone Joint Surg [Br] 59:241–242

Brattström H (1964) Shape of the intercondylar groove normally and in recurrent dislocation of patella. Acta Orthop Scand Suppl 68

Edholm P (1966) Anatomic angles determined from two radiographic projections. Acta Radiol Suppl 259

Egund N (1986) The axial view of the patello-femoral joint. Description of a new radiographic method for routine use. Acta Radiol (Diagn) 27:101–104

Egund N, Fridén T (1988) Lesion of the anterior cruciate ligament and sagittal disalignment of the knee in weight-bearing. Acta Radiol 29:559–563

Egund N, Palmer J (1984) Femoral anatomy described in cylindrical coordinates using computed tomography. Acta Radiol (Diagn) 25:209–215

Egund N, Wingstrand H (1989) Pitfalls in ultrasonography in hip joint synovitis in the child. Acta Radiol 30:375–379

Egund N, Lundin A, Wallengren NO (1988) The vertical position of the patella. A new radiographic method for routine use. Acta Radiol 29:555–558

Egund N, Fridén T, Hjarbæk J, Lindstrand A, Stockerup R (1993) Radiographic assessment of sagittal knee laxity in weight bearing. A study on anterior cruciate-deficient knees. Skeletal Radiol 22:177–181

Gilula LA, Weeks PM (1978) Post-traumatic ligamentous instabilities of the wrist. Radiology 129:641–651

Hermann K, Egund N (1997) CT-measurement of femoral neck anteversion. The influence of femur positioning. Acta Radiol (in press)

Keats TE (1990) Atlas of roentgenographic measurement, 6th edn. Year Book Medical Publishers, Chicago

Loenstein JE, Bradford DS, Winter RB, Ogilvie JW (eds) (1995) Textbook of scoliosis and other spinal deformities. Saunders, Philadelphia

Murphy SB, Sheldon RS, Kijewski PK, Wilkinson RH, Griscom NT (1987) Femoral anteversion. J Bone Joint Surg [Am] 69:1169–1171

Nolsøe CP, Engel U, Karstrup S, Torp-Pedersen S, Garre K, Holm HH (1990) The aortic wall: an in vitro study of the double-line pattern in high-resolution US. Radiology 175:387–390

Ozonoff MB (1992) Pediatric orthopedic radiology. Saunders, Philadelphia

Paley D, Maar DC, Herzenberg JE (1994) New concepts in high tibial osteotomy for medial compartment osteoarthritis. Orthop Clin North Am 25:483–497

Ryd L (1986) Micromotion in knee arthroplasty. Acta Orthop Scand Suppl 220

Selvik G (1989) Roentgen stereophotogrammetry. A method for the study of the kinematics of the skeletal system. Acta Orthop Scand Suppl 232

Taybi H, Lachman RS (1996) Radiology of syndromes, metabolic disorders, and skeletal dysplasias. Mosby, St. Louis

Tjörnstrand B, Selvik G, Egund N, Lindstrand A (1981) Roentgen stereophotogrammetry in high tibial osteotomy for gonarthrosis. Arch Orthop Trauma Surg 99:73–81

9 Bone Densitometry

C. van Kuijk and H.K. Genant

CONTENTS

9.1
Introduction

Bone densitometry is the general term for the techniques that are used to quantify the amount or material properties of bone. This can be to assess either the actual density of bone (in g/cm^3) or the total amount in grams (total body bone mass), as well as other parameters, such as bone area density (in g/cm^2) or speed of sound (in m/s). Over the last 60 years the methodology has matured into a highly specialized field within skeletal radiology. The techniques currently used are diverse, ranging from

C. van Kuijk, MD PhD, Assistant Adjunct Professor of Radiology, Director of Radiographic Laboratory, Osteoporosis and Arthritis Research Group, Department of Radiology, M-392, University of California San Francisco, San Francisco, CA 94143-0628, USA
H.K. Genant, MD, Professor of Radiology, Medicine, Epidemiology and Orthopedic Surgery, Chief of the Musculoskeletal Section, Executive Director of the Osteoporosis and Arthritis Research Group, Department of Radiology, M-392, University of California at San Francisco, San Francisco, CA 94143-0628, USA

simple to complex in nature and from relatively inexpensive to rather expensive. All technical modalities in radiology are represented, from conventional radiology to computed tomography, ultrasound, and even magnetic resonance imaging.

The principal clinical uses of bone densitometry are in the diagnosis of primary and secondary osteoporosis and in the monitoring of treatment of these disease entities. Recently, however, new applications have been emerging. The measurement of periprosthetic bone loss is one of them. Furthermore, advanced imaging techniques and image processing methods are used to assess bone architecture: both high-resolution computed tomography and high-resolution magnetic resonance imaging are used for this purpose. However, such applications are still research tools and are beyond the scope of this chapter. The reader is referred to recent review literature discussing this matter (Genant et al. 1996).

9.2
Techniques

In the following sections the different techniques used in bone densitometry are discussed.

9.2.1
Conventional Radiography

Cortical thinning and increased intracortical porosity in tubular bones are well-known radiological features of diminished bone mass. Visually apparent loss of specific trabecular structures in the femoral neck and vertebral bodies is another well-recognized feature. However, it has been estimated that at least 30% of the skeletal calcium is lost before osteopenia has become detectable on conventional radiographs (Adran 1951; Finsen and Anda 1988; Kawashima and Uhthoff 1991; Mayo-Smith and Rosenthal 1991).

Semiquantitative grading methods have been developed to assess and evaluate these visually apparent features; for example, the Singh index (SINGH et al. 1970) is used to grade the appearance of the trabecular structure in the femoral neck. Although these "visual" methods certainly can give some information, the inter- and intraobserver reproducibility is disappointing. Consequently, more objective and precise techniques for quantifying the amount or physical properties of bone have been developed.

9.2.2
Radiogrammetry

One of the first papers to describe the use of this technique was that by BARNETT and NORDIN published in 1960. They measured the cortical thickness at different anatomical sites. Since then a large number of papers have been published on the use of bone dimension measurements in the assessment of osteoporosis (AGUADO et al. 1996; BLOOM et al. 1983; KALLA et al. 1989; MEEMA 1991; MEEMA and MEINDOK 1992; RICO et al. 1995). Several dimensions can be measured such as the total bone width, the cortical thickness, the ratio of cortical width to total bone width, and the cortical area. These measurements are usually performed on radiographs depicting tubular bones, such as the metacarpal bones and the radius. Usually the bone dimensions are measured with rulers and calipers. Recently, however, computer-aided techniques have been developed using image processing and analysis tools to perform these measurements in a (semi-)automated fashion.

A recent addition to this field is the measurement of the hip-axis length on standard radiographs of the hip as well as on images acquired by bone densitometers. The hip-axis length seems to be a prognostic factor for future hip fractures, independent of bone density at the hip (FAULKNER et al. 1993; GLÜER et al. 1994).

9.2.3
Radiographic Absorptiometry

In radiographic absorptiometry a standardized radiograph of the hand is made along with an aluminum reference wedge. In general with this technique, the density of the phalanges or metacarpals is determined and compared with that of the wedge using either an optical densitometer or a digital densito-

metric technique after digitizing the radiograph (COSMAN et al. 1991; TROUERBACH et al. 1985; STRID and KALEBO 1988). The results are given in aluminum-equivalent values. No distinction is made between the cortical and trabecular compartments of bone. Currently, several methods are used. One of them is a method (Osteogram) developed by CompuMed (Calif., USA). With this technique two anteroposterior (AP) radiographs of the hand are obtained at different kVp settings. The radiographs are analyzed at a central laboratory. Results from the two views are compared and if found to be in agreement (less than 3% difference), the results are averaged. If the difference between the measurements is more than 3%, the films are rejected and repeats are requested by the central laboratory. The short-term precision error is reported to be 1.5% (coefficient of variation) in vivo and about 1% in vitro (RAVN et al. 1996; YANG et al. 1994).

Other systems are those provided by NIM, Verona, Italy (Osteoradiometer; metacarpal bone and radius measurements); by Teijin, Tokyo, Japan (Bonalyzer; radius measurements), and by Chugai, Tokyo, Japan (metacarpal measurements) (ADAMI et al. 1996; SUK SEO et al. 1994). All these methods are based on the same principles and have reported precision errors of about 2% (YATES et al. 1995).

A slightly different technique was developed by TROUERBACH et al. (1985) at the Erasmus University Rotterdam, The Netherlands. In addition to the anteroposterior view of the hand an additional lateral view of the index finger is acquired on the same screen using a dedicated cassette. A linear aluminum wedge is used as reference (Fig. 9.1). By combining measurements on the same anatomical level in the middle phalanx using both views a real density value can be calculated and in addition provides for a sophisticated soft tissue correction.

9.2.4
Single-Photon Absorptiometry and Single X-ray Absorptiometry

In single-photon absorptiometry (SPA) a highly collimated photon beam from a radionuclide source is used to measure the photon attenuation of the measurement site (usually the radius or the os calcis), which is converted to bone mineral content in grams or area bone mineral density in grams/cm^2 using a known standard (CAMERON and SORENSON 1963). SPA scanning requires a constant tissue path length which is achieved by scanning the object of interest

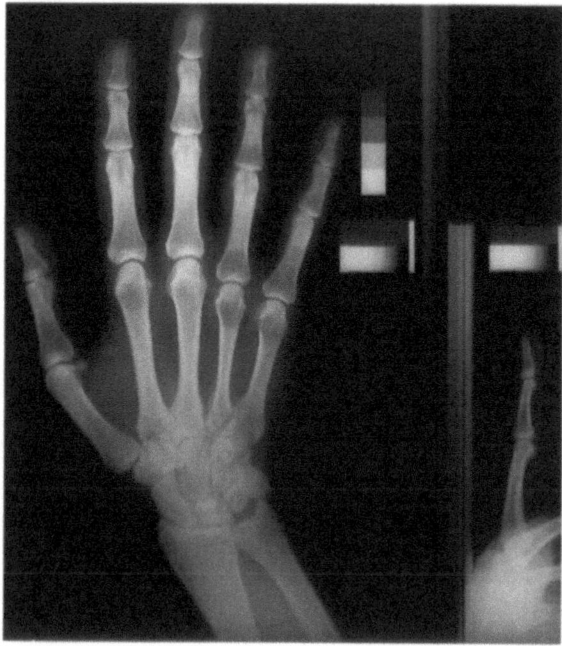

Fig. 9.1. Example of a hand x-ray made for radiographic absorptiometry with the aluminum wedge in place. For this method an additional lateral view of the index finger is acquired on the same film. (Copyright: Department of Experimental Radiology, Erasmus University Rotterdam, The Netherlands; courtesy of A.W. Zwamborn)

within a uniform soft tissue environment of known dimensions, usually a water bath. SPA measurements are therefore confined to the appendicular skeleton. As in radiographic absorptiometry, SPA cannot measure the cortical and trabecular compartment of bone separately. Recently, SPA has been superseded by single x-ray absorptiometry (SXA) (BJARNASON et al. 1995; GLÜER et al. 1992; KELLY et al. 1994). In SXA the nuclear photon source has been replaced by a stable x-ray tube. SXA has improved reproducibility and spatial resolution and reduced examination times.

9.2.5
Dual-Photon Absorptiometry and Dual X-ray Absorptiometry

Dual-photon absorptiometry (DPA) has been used for bone mass measurements in the central skeleton (femur, spine) or total body bone mineral content and fat content assessment. A radionuclide source emitting photons at two effective energies is used. The photon attenuation of the measurement site at the two energy levels is measured. Dual-energy scanning eliminates the need for a constant path length (KRØLNER and PORS NIELSEN 1980; PEPPLER and MAZESS 1981). Bone mass estimates are given as bone mineral content in grams or as bone mineral density in grams/cm^2. DPA cannot differentiate between cortical and trabecular bone.

Dual x-ray absorptiometry (DXA) is the modern, upgraded version of DPA (KELLY et al. 1988). The radionuclide source has been replaced by a stable x-ray tube. A dual-energy spectrum is generated by rapid switching of the tube voltage supply or by K-edge filtering. Examination times are reduced and the reproducibility of the measurements has improved compared with DPA (GLÜER et al. 1990; SLOSMAN et al. 1990). DXA technology has gained widespread acceptance and distribution. Examples of DXA scans are shown in Fig. 9.2. The in vivo precision is approximately 1%–2% (coefficient of variation) (LILLEY et al. 1991; ORWOLL and OVIATT 1991). DXA and DPA measurements of the spine in the AP projection are influenced by (intervertebral) osteoarthrosis, which falsely increases the measured bone mineral content (DRINKA et al. 1992), a major disadvantage when elderly patients are evaluated. Lateral DXA scanning of the spine has been developed as a potential solution to this problem (MAZESS et al. 1991). However, superposition of the ribs and of the iliac crest limits the measurement to one or two vertebral bodies when scans are made in the lateral decubitus position. When systems are used that have a rotating C-arm, lateral scanning is possible in the supine position, resulting in less scanning problems and better reproducibility when compared with lateral decubitus scanning (RUPICH et al. 1992; JERGAS et al. 1995a).

Recently, new software has made it possible to evaluate bone mass (g) and density (g/cm^2) at the forearm and the calcaneus (HAGIWARA et al. 1994; YAMADA et al. 1994) on regular DXA equipment. In addition, dedicated equipment for peripheral DXA has been developed.

9.2.6
Quantitative Computed Tomography

Quantitative computed tomography (QCT) is the only method that can estimate bone density separately in the trabecular and cortical bone compartments and the only method to give a true density (in g/cm^3) estimate. Usually, the vertebral body is the site of measurement (GENANT et al. 1982). A

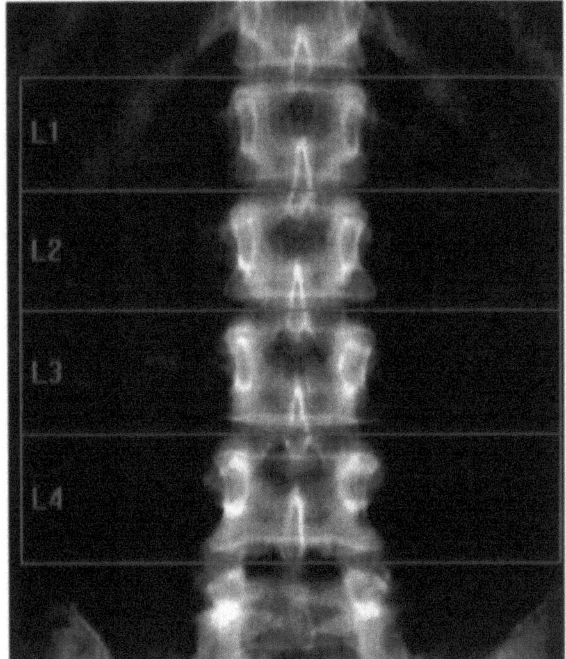

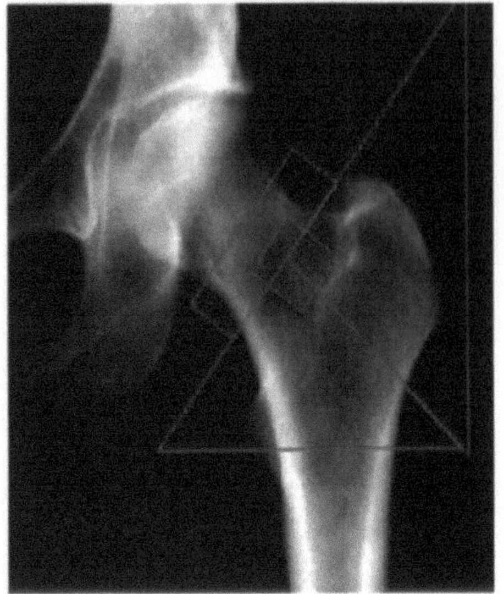

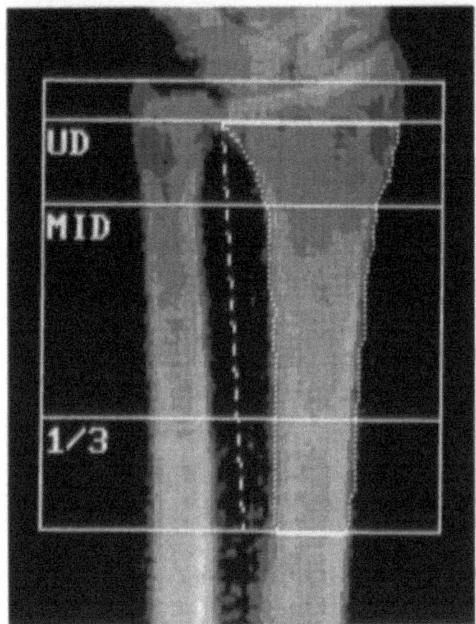

Fig. 9.2. Typical DXA examinations of **a** the spine (AP projection), **b** the femur, and **c** the forearm

reference standard is placed under the lumbar spine of the patient and scanned simultaneously. On a lateral scout view, a slice selection is made at the midvertebral levels of three to four consecutive vertebral bodies. The average attenuation value of the object of interest is measured in the image and compared with the attenuation values of the calibration standard (CANN and GENANT 1980; KALENDER et al. 1987). An example of a spinal QCT examination is shown in Fig. 9.3.

Single-energy QCT is the technique most widely used and recommended, although the intravertebral fat falsely lowers the measured bone mineral density (MAZESS 1983), which is also true for DXA measurements (KUIPER et al. 1996).

Dual-energy QCT is used to improve the accuracy of the bone density assessment as it potentially can correct for the fat error. Both preprocessing and postprocessing dual-energy QCT techniques have been developed. Dual-energy QCT, however, has increased radiation dose compared with single-energy QCT. Furthermore, technical difficulties, such as beam hardening and related correction algorithms, have to date limited postprocessing dual-

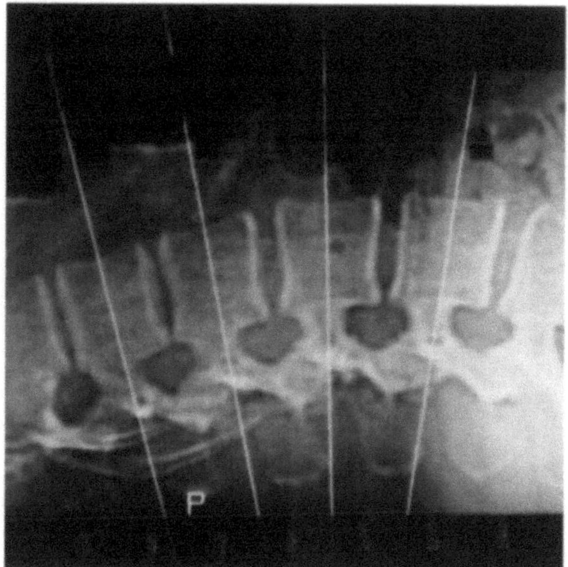

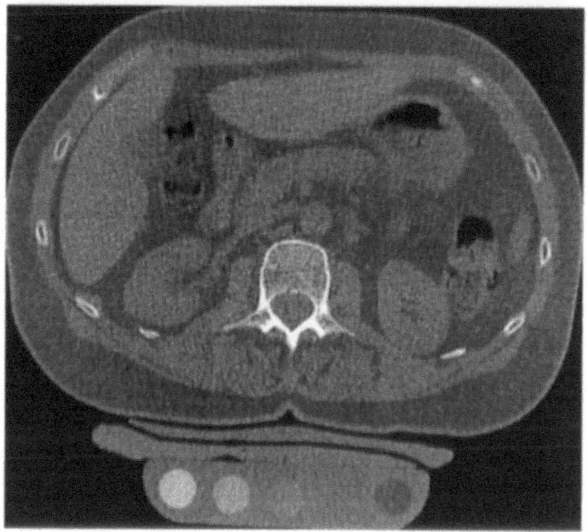

Fig. 9.3 a,b. Typical QCT examination of the spine. a Lateral scout view (also called scanogram) of the lumbar spine. This scout view is used to plan midvertebral slices typically through three or four consecutive vertebral bodies. b Typical axial slice through a vertebral body. Note the calibration material underneath the patient that is used to convert the measurements within the region of interest (usually an ellipse or pac-man shaped region within the trabecular part of the vertebral body) from Hounsfield units to bone-equivalent values (usually g/cm³ calcium hydroxyapatite)

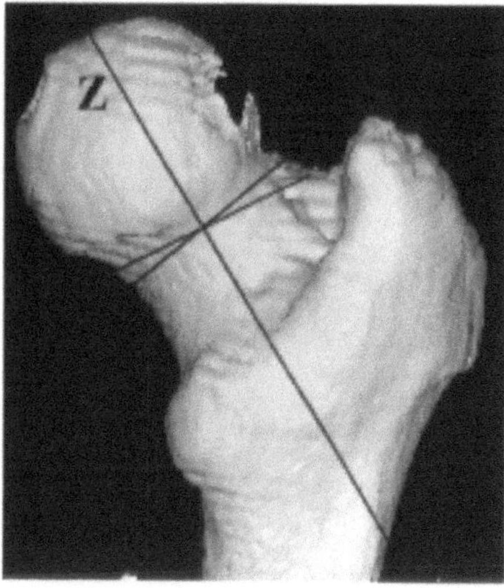

Fig. 9.4. Three-dimensional reconstruction of a femur from volumetric CT data. Density and geometrical parameters can be derived. (Courtesy of T. Lang, Department of Radiology, University of California San Francisco)

1986). Newer CT systems capable of spiral CT scanning allow a volumetric acquisition of imaging data from which a highly accurate three-dimensional reconstruction of vertebral bodies or femora can be made (LANG et al. 1996). In combination with advanced image analysis tools this allows for sophisticated density measurements in several regions of interest as well as for measurements of geometrical dimensions of the object of interest (Fig. 9.4). The clinical applicability of the volumetric QCT technique, however, is still under investigation.

Special-purpose CT systems have been developed for peripheral QCT (pQCT) of the radius and tibia (MÜLLER et al. 1989; RÜEGSEGGER et al. 1976; SCHNEIDER and BORNER 1991). The first generation of these pQCT systems used nuclear sources, but newer systems use x-ray sources. The reported precision of these methods is 1%–3%. An example of a pQCT examination is shown in Fig. 9.5.

energy QCT to research purposes (VAN KUIJK et al. 1990). Single-energy QCT, however, is widely used. The precision of QCT measurements is in the range of 2%.

Although primarily used for bone mass measurements in the spine, femoral QCT has been reported (ESSES et al. 1989; KUIPER et al. 1996; SARTORIS et al.

9.2.7
Quantitative Ultrasound

More recently, ultrasound velocity and attenuation measurements have been promoted for noninvasive measurement of bone quantity and structure (HEANEY et al. 1989; LANGTON et al. 1984; ORGEE et al. 1996; VENTURA et al. 1996). Measurements are confined to the appendicular skeleton and made at

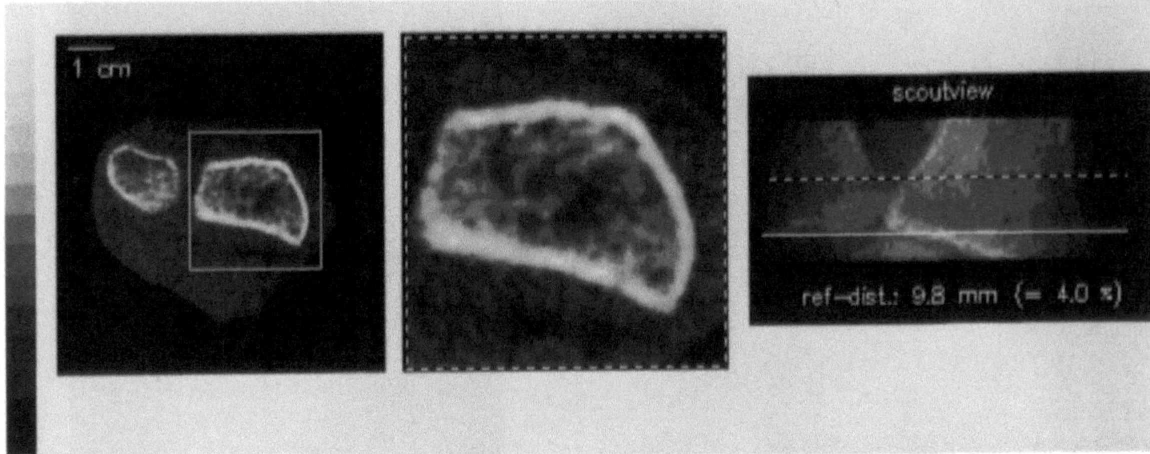

Fig. 9.5. Example of pQCT of the forearm. An axial slice of the radius is made at a level selected on the basis of the scout view

the calcaneus, tibia, patella, and phalangeal bones. Parameters measured are ultrasound transmission velocity and broadband ultrasound attenuation. These parameters are generally postulated to be determined by both bone density and bone architecture. Accordingly, this technique could provide additional information beyond just the amount of bone, as is measured with the other techniques. A clear advantage of quantitative ultrasound is its absence of radiation exposure. The precision of this technique is reported to be 1%–4%.

9.2.8
Quantitative Magnetic Resonance Imaging

As research tools, (high-resolution) magnetic resonance imaging (MRI) and MR microscopy are used to quantify the amount of bone as well as to study trabecular architecture. In quantitative MR the T2* relaxation time is measured. The differences in magnetic tissue parameters between the trabecular bone structures and the bone marrow content produce a distortion of the local magnetic field influencing the relaxation times. T2* is theoretically related to the density of the trabecular network and its geometry. This relation has been shown in different in vitro studies, proving the possible value of quantitative MRI in the future (JERGAS et al. 1995b; ROSENTHAL et al. 1990). The precision has been reported to be 4%–10%. This relatively large precision error, as well as the costs and availability of MR magnets, prohibits the use of this technique in routine clinical practice.

9.2.9
Summary

A summary of several features of the above-discussed techniques is given in Table 9.1. In this table, precision refers to reproducibility of the measurement technique in vivo and is given as coefficient of variation (%). Dose refers to the effective dose equivalent given in µSv. For comparison, the annual natural dose is approximately 3000 µSv (NCRP 1987).

9.3
Clinical Use

Obviously, a vast arsenal of bone densitometry tools is available to satisfy the needs of the clinician. However, confronted with this vast arsenal of methods one has to decide which to use under different circumstances; furthermore, the significance of the data obtained must be established.

9.3.1
Applications

There is now a general consensus that bone mass measurements are indeed very useful. Recently a consensus document was published summarizing some of the important issues in bone densitometry (MILLER et al. 1996). Twenty-two experts from Europe and the United States drafted this document. They stated that low bone mass in the asymptomatic

Table 9.1. Overview of different techniques

Technique	Anatomical sites of interest	Precision in vivo (%)	Accuracy error (%)	Estimated effective dose equivalent (μSv)
Radiogrammetry	Metacarpals, radius	1–3	?	<1
Radiographic absorptiometry	Phalanges	1–3	5	<1
Single-photon absorptiometry	Radius, calcaneus	1–3	4–6	<1
Single x-ray absorptiometry	Radius, calcaneus	1–2	4–6	<1
Dual-photon absorptiometry	Spine, femur, total body	2–5	3–10	3–5
Dual x-ray absorptiometry	Spine, femur, radius, total body	1–2	3–10	1–3
Quantitative computed tomography	Spine, femur	2–4	5–15 (single-energy)	50–100
Peripheral quantitative computed tomography	Radius, tibia	1–2	2–8	1–2
Quantitative ultrasound	Calcaneus, tibia, patella, phalanges	1–4	?	–
Quantitative magnetic resonance imaging	Calcaneus, spine	4–10	?	–

patient predicts fracture risk, just as high cholesterol or high blood pressure predicts the risk of heart disease or stroke. They concluded that there are several clinical situations in which an assessment of bone mass and fracture risk affects therapeutic decisions, including estrogen deficiency, vertebral deformities, radiographic osteopenia, asymptomatic primary hyperparathyroidism and long-term corticosteroid therapy. They also concluded that the appropriate technique and skeletal site to be measured should be chosen on the basis of the patient's circumstances. In summary, the following clinical applications are recommended (GENANT et al. 1989; JOHNSTON et al. 1991):

1. *Establishing a diagnosis of low bone mass and assessing its severity.* The absolute level of bone mass has been proven to be predictive for future fracture risk. Therefore, bone densitometry is recommended in those patients suspected for low bone mass caused either by aging (primary osteoporosis) or by other factors (secondary osteoporosis, as in Cushing's syndrome, hypogonadism, corticosteroid-induced osteoporosis, etc.).

2. *Assessment of perimenopausal women for initiation of therapy to prevent osteoporosis.* As bone loss is accelerated in women after the menopause and women achieve a lower peak bone mass compared with men, women are especially at risk for osteoporosis. Estrogen replacement therapy has been proven to preserve bone mass. It could be claimed that medication should be restricted to women with an increased fracture risk in order to maximize cost-effective use and to enhance compliance of preventive therapy. At present, bone densitometry provides the best prediction of fracture risk.

3. *Monitoring the efficacy of treatment or disease course.* If applied properly, modern bone densitometry techniques have a sufficient longitudinal reproducibility (1%–2%) to justify follow-up measurements in an individual to monitor disease course or therapeutic efficacy.

9.3.2
Interpretation

There remains the problem of how to interpret the data provided by bone densitometry tools. Usually the results are given in BMC (bone mineral content), BMD (bone mineral density), SOS (speed of sound), BUA (broadband ultrasound attenuation), etc., with a range of units used and their respective Z- and T-scores. As bone density decreases in aging, and differences in bone density exist between sexes and races, bone density measurements should be compared with those of age-, sex-, and race-matched controls. Therefore, a normative database is mandatory for interpreting the level of bone mineral content. Usually, the estimated bone density is given as a Z-score. The Z-score gives the patient's results as the deviation from the mean of age-matched controls divided by the standard deviation of this mean, which is an indication of the biological variability. In addition to the Z-score, the bone density of a patient is compared with the peak bone mass of young normal adults. Then, the estimation is given as the T-score. The T-score, like the Z-score, gives the patient's results as a deviation from the mean of young normal adults divided by the standard deviation of the mean. The T-score is predictive for fracture risk.

For most clinicians, however, these data are confusing. Recently a document was generated by the World Health Organization. This document provides working definitions for the use of bone densitometry (WHO 1994):

- *Normal* = a value for BMD/BMC not more than one standard deviation below the average value of young adults
- *Low bone mass or osteopenia* = a value for BMD/BMC more than one standard deviation below the young adult average, but not more than 2.5 standard deviations below
- *Osteoporosis* = a value for BMD/BMC more than 2.5 standard deviations below the young adult average
- *Severe osteoporosis* = a value for BMC/BMD more than 2.5 standard deviations below the young adult average value and the presence of one or more fragility fractures

9.4
Comparison of Different Techniques

As the clinician has a wide choice of techniques for bone densitometry, the question arises as to which technique to use. Several researchers have tried to determine the distinct values of these techniques. When the techniques are compared in terms of their discriminative power between normal healthy patients and (spinal) osteoporotics or between mild and severe osteoporotics, QCT has been reported to be the best technique, followed by DPA/DXA and SPA/SXA, although all techniques show a considerable overlap between normals and osteoporotics (GRAMPP et al. 1997; HEUCK et al. 1989; LAFFERTY and ROWLAND 1996; OTT et al. 1988; REINBOLD et al. 1986; VAN BERKUM et al. 1989; YU et al. 1995). However, some investigators have objected to this

method of validating these techniques and have argued that their real value lies in their predictive power in respect of future fractures rather than their ability to identify existing fractures (WASNICH 1990). As such, the techniques should be validated in large population-based longitudinal studies, and data are now available from such studies. These studies generally show that the ranking of tests based on prospective data parallels the results of cross-sectional studies. Radiographic absorptiometry, SPA, DPA/DXA, QCT, and quantitative ultrasound all seem to have some predictive power regarding future osteoporotic fractures (CUMMINGS et al. 1993; HUI et al. 1988; SEELEY et al. 1991; YATES et al. 1995).

The correlation between the different techniques is modest (typically $r = 0.6$–0.7). This precludes prediction of bone mass at one site by bone mass measurement at another site. This is due both to technical differences between the techniques and to differences in measurement sites, which have a different composition (ratio of cortical to trabecular bone) and differ in metabolic activity. The various techniques are therefore complementary rather than competitive. For clinical trials it is advised that bone mass be measured at a minimum of two skeletal sites. Priority should be given to measurement of sites of biological relevance (e.g., the spine for vertebral osteoporosis) (KANIS et al. 1991).

Precision (or reproducibility) is important when discussing these techniques, as is the rate of bone change expected in the skeletal part under investigation. If there is an expected change of bone mass of 1% a year, it takes 5.6 years to detect a significant bone change with a technique having a 2% reproducibility. If the expected change is 2% it takes 2.8 years with the same technique. However, a technique with 1% reproducibility will detect this 2% change in 1.4 years.

Usually changes (both loss and gain) have been found to be higher in trabecular compartments, which have much more bone surface for metabolic activity than cortical compartments of bone. Therefore the vertebral bodies in the spine with their large trabecular compartments are often chosen as the sites to be measured. In day-to-day clinical practice, however, the choice of technique will depend on the availability of techniques and specialists.

References

Adami S, Zamberlan N, Gatti D, Zanfisi C, Braga V, Broggini M, Rossini M (1996) Computed radiographic absorptiometry and morphometry in the assessment of postmenopausal bone loss. Osteoporosis Int 6:8–13

Adran GM (1951) Bone destruction not demonstrable by radiography. Br J Radiol 24:107

Aguado F, Revilla M, Hernandez ER, Villa LF, Rico H (1996) Behavior of bone mass measurements. Dual-energy X-ray absorptiometry total body bone mineral content, ultrasound bone velocity, and computed metacarpal radiogrammetry, with age, gonadal status, and weight in healthy women. Invest Radiol 31:218–222

Barnett E, Nordin BEC (1960) The radiological diagnosis of osteoporosis: a new approach. Clin Radiol 11:166–174

Bjarnason K, Nilas L, Hssager C, Christiansnen C (1995) Dual energy X-ray absorptiometry of the spine – decubitus lateral versus anteroposterior projection in osteoporotic women: comparison to single energy X-ray absorptiometry of the fore-arm. Bone 16:255–260

Bloom RA, Pogrund H, Libson E (1983) Radiogrammetry of the metacarpal: a critical reappraisal. Skeletal Radiol 10:5–9

Cameron JR, Sorenson J (1963) Measurement of bone mineral in vivo: an improved method. Science 142:230–232

Cann CE, Genant HK (1980) Precise measurement of vertebral mineral content using computed tomography. J Comput Assist Tomogr 4:493–500

Cosman F, Herrington BS, Himmelstein S, Lindsay R (1991) Radiographic absorptiometry: a simple method for determination of bone mass. Osteoporosis Int 2:34–38

Cummings SR, Black DM, Nevitt MC, et al. (1993) Bone density at various sites for prediction of hip fractures: the study of osteoporotic fractures. Lancet 341:72–75

Drinka PJ, DeSmet AA, Bauwens SF, Rogot A (1992) The effect of overlying calcification on lumbar bone densitometry. Calcif Tissue Int 50:507–510

Esses SI, Lotz JC, Hayes WC (1989) Biomechanical properties of the proximal femur determined in vitro by single-energy quantitative computed tomography. J Bone Miner Res 4:715–722

Faulkner KG, Cummings SR, Black D, Palermo L, Glüer CC, Genant HK (1993) Simple measurement of femoral geometry predicts hip fracture: the study of osteoporotic fractures. J Bone Miner Res 8:1211–1217

Finsen V, Anda S (1988) Accuracy of visually estimated bone mineralization in routine radiographs of the lower extremity. Skeletal Radiol 17:270–275

Genant HK, Cann CE, Ettinger B (1982) Quantitative computed tomography of vertebral spongiosa: a sensitive method for detecting early bone loss after oophorectomy. Ann Intern Med 97:699–705

Genant HK, Block JE, Steiger P, Glüer CC, Ettinger B, Harris S (1989) Appropriate use of bone densitometry. Radiology 170:817–822

Genant HK, Engelke K, Furst T, et al. (1996) Noninvasive assessment of bone mineral and structure: state of the art. J Bone Miner Res 11:707–730

Glüer CC, Steiger P, Selvidge R, Elliesen-Kliefoth K, Hayashi C, Genant HK (1990) Comparative assessment of dualphoton absorptiometry and dual-energy radiography. Radiology 174:223–228

Glüer CC, Vahlensieck M, Faulkner KG, Engelke K, Black D, Genant HK (1992) Site-matched calcaneal measurements of broad-band ultrasound attenuation and single X-ray absorptiometry: do they measure different skeletal properties? J Bone Miner Res 7:1071–1079

Glüer CC, Cummings SR, Pressman A, et al. (1994) Prediction of hip fractures from pelvic radiographs: the study of osteoporotic fractures. J Bone Miner Res 9:671–677

Grampp S, Genant HK, Mathur A, et al. (1997) Comparisons of non-invasive bone mineral measurements in assessing age-related loss, fracture discrimination, and diagnostic classification. J Bone Miner Res 12:697–711

Hagiwara S, Engelke K, Yang S-O, Dhillon MS, Guglielmi G, Nelson DS, Genant HK (1994) Dual x-ray absorptiometry forearm software: accuracy and intermachine relationship. J Bone Miner Res 9:1425–1427

Heaney RP, Avioli LV, Chesnut CH, Lappe J, Recker RR, Brandenburger GH (1989) Osteoporotic bone fragility: detection by ultrasound transmission velocity. JAMA 261:2986–2990

Heuck AF, Block J, Glüer CC, Steiger P, Genant HK (1989) Mild versus definite osteoporosis: comparison of bone densitometry techniques using different statistical models. J Bone Miner Res 4:891–900

Horsman A, Simpson M (1973) The measurement of sequential changes in cortical bone geometry. Br J Radiol 48:471–476

Hui SL, Slemenda CW, Johnston CC (1988) Age and bone mass as predictors of fracture in a prospective study. J Clin Invest 81:1804–1809

Jergas M, Breitenseher M, Glüer CC, et al. (1995a) Which vertebrae should be assessed using lateral dual-energy x-ray absorptiometry of the lumbar spine? Osteoporosis Int 5:196–204

Jergas M, Majumdar S, Keyak JH, et al. (1995b) Relationships between Young's modulus of elasticity, ash density and magnetic resonance imaging. MRI derived effective transverse relaxation time T2* in human tibial specimens. J Comput Assist Tomogr 19:472–479

Johnston CC Jr, Slemenda CW, Melton LJ III (1991) Clinical use of bone densitometry. N Engl J Med 324:1105–1109

Kalender WA, Klotz E, Suess C (1987) Vertebral bone mineral analysis: an integrated approach with CT. Radiology 164:419–423

Kalla AA, Meyers OL, Parkyn ND, Kotze TJvW (1989) Osteoporosis screening – radiogrammetry revisited. Br J Rheumatol 28:511–517

Kanis JA, Geusens P, Christiansen C (on behalf of the working party of the foundation) (1991) Guidelines for clinical trials in osteoporosis. A position paper of the European Foundation for Osteoporosis and Bone Disease. Osteoporosis Int 1:182–188

Kawashima T, Uhthoff HK (1991) A pattern of bone loss of the proximal femur: a radiologic, densitometric and histomorphometric study. J Orthop Res 9:634–640

Kelly T, Slovick D, Schoenfield D, Neer R (1988) Quantitative digital radiography versus dual photon absorptiometry of the lumbar spine. J Clin Endocrinol Metab 67:839–844

Kelly TL, Crane G, Barab DT (1994) Single x-ray absorptiometry of the forearm: precision, correlation, and reference data. Calcif Tissue Int 54:212–218

Krølner B, Pors Nielsen S (1980) Measurement of bone mineral content (BMC) of the lumbar spine. 1. Theory and application of a new two-dimensional dual-photon attenuation method. Scand J Clin Lab Invest 40:653–663

Kuiper JW, Van Kuijk C, Grashuis JL, Ederveen AHG, Schütte HE (1996) Accuracy and the influence of marrow fat on quantitative CT and dual-energy X-ray absorptiometry of the femoral neck in vitro. Osteoporosis Int 6:25–30

Lafferty FW, Rowland DY (1996) Correlations of dual-energy x-ray absorptiometry, quantitative computed tomography, and single photon absorptiometry with spinal and nonspinal fractures. Osteoporosis Int 6:407–415

Lang T, Heitz M, Keyak J, Genant HK (1996) A 3D anatomic coordinate system for hip QCT. Osteoporosis Int 6(S1):203

Langton CM, Palmer SB, Porter RW (1984) The measurement of broadband attenuation in cancellous bone. Eng Med 13:89–91

Lilley J, Walters BG, Heath DA, Drolc Z (1991) In vivo and in vitro precision of bone density measured by dual-energy x-ray absorption. Osteoporosis Int 1:141–146

Mayo-Smith W, Rosenthal DI (1991) Radiographic appearance of osteopenia. Radiol Clin North Am 29:37–47

Mazess RB (1983) Errors in measuring trabecular bone by computed tomography due to marrow and bone composition. Calcif Tissue Int 35:148–152

Mazess RB, Gifford CA, Bisek JP, Barden HS, Hanson JA (1991) DEXA measurement of spine density in the lateral projection. I. Methodology. Calcif Tissue Int 49:235–239

Meema HE (1991) Improved fracture threshold in postmenopausal osteoporosis by radiogrametric measurements: its usefulness in selection for preventive therapy. J Bone Miner Res 6:9–14

Meema HE, Meindok H (1992) Advantages of peripheral radiogrametry over dual-photon absorptiometry of the spine in the assessment of prevalence of osteoporotic vertebral fractures in women. J Bone Miner Res 7:897–903

Miller PD, Bonnick SL, Rosen CJ (1996) Consensus of an international panel on the clinical utility of bone mass measurements in the detection of low bone mass in the adult population. Calcif Tissue Int 58:207–214

Müller A, Rüegsegger E, Rüegsegger P (1989) Peripheral QCT. A low risk procedure to identify women predisposed to osteoporosis. Phys Med Biol 34:741–749

National Council on Radiation Protection and Measurements (1987) Ionizing radiation exposure of the population of the United States. NCRP report no. 93, Bethesda, MD

Orgee JM, Foster H, McCloskey EV, Khan S, Coombes G, Kanis JA (1996) A precise method for the assessment of tibial ultrasound velocity. Osteoporosis Int 6:1–7

Orwoll ES, Oviatt SK (1991) Longitudinal precision of dual-energy x-ray absorptiometry in a multicenter study. J Bone Miner Res 6:191–197

Ott SM, Kilcoyne RF, Chesnut CH (1988) Comparisons among methods of measuring bone mass and relationship to severity of vertebral fractures in osteoporosis. J Clin Endocrinol Metab 66:501–507

Peppler WW, Mazess RB (1981) Total body bone mineral and lean body mass by dual photon absorptiometry. I. Theory and measurement procedure. Calcif Tissue Int 33:353–359

Ravn P, Overgaard K, Huang C, Ross PD, Green D, McClung M, for the EPIC study group (1996) Comparison of bone densitometry of the phalanges, distal forearm and axial skeleton in early postmenopausal women participating in the EPIC study. Osteoporosis Int 6:308–313

Reinbold WD, Genant HK, Reiser UJ, Harris ST, Ettinger B (1986) Bone mineral content in early premenopausal and postmenopausal osteoporotic women: comparison of measurement methods. Radiology 160:469–478

Rico H, Revilla M, Hernandez ER, Villa LF, Alvarez de Buergo M (1995) Total and regional bone mineral content and fracture rate in postmenopausal osteoporosis treated with salmon calcitonin: a prospective study. Calcif Tissue Int 56:181–185

Rosenthal H, Thulborn KR, Rosenthal DI, Rosen BR (1990) Magnetic susceptibility effects of trabecular bone on magnetic resonance bone marrow imaging. Invest Radiol 25:173–178

Rüegsegger P, Elsasser U, Anliker M, Gnehm H, Kind H, Prader A (1976) Quantification of bone mineralization using computed tomography. Radiology 121:93–97

Rupich RC, Griffin MG, Pacifici R, Avioli LV, Susman N (1992) Lateral dual-energy radiography: artifact error from rib and pelvic bone. J Bone Miner Res 7:97–101

Sartoris DJ, Andre M, Resnick C, Resnick D (1986) Trabecular bone density in the proximal femur: quantitative CT assessment. Radiology 160:707–712

Schneider P, Borner W (1991) Peripheral quantitative computed tomography for bone mineral measurement with a new special purpose QCT-scanner. Fortschr Rontgenstr 153:292–299

Seeley DG, Browner WS, Nevitt MC, Genant HK, Scott JC, Cummings SR (1991) Which fractures are associated with low appendicular bone mass in elderly women? Ann Intern Med 115:837–842

Singh M, Nagrath AR, Maini PS (1970) Change in trabecular pattern of the upper end of the femur as an index of osteoporosis. J Bone Joint Surg [Am] 52:457–467

Slosman DO, Rizzoli R, Buchs B, Piana F, Donath A, Bonjour JP (1990) Comparative study of the performances of x-ray and gadolinium 153 densitometers at the level of the spine, femoral neck and femoral shaft. Eur J Nucl Med 17:3–9

Strid KG, Kalebo P (1988) Bone mass determination from microradiographs by computer assisted videodensitometry. I. Methodology. Acta Radiol 29:465–472

Suk Seo G, Shraki M, Aoki C, et al. (1994) Assessment of bone density in the distal radius with computer assisted X-ray densitometry (CXD). Bone Miner 27:173–182

Trouerbach WTH, Hoornstra K, Birkenhäger JC, Zwamborn AW (1985) Roentgendensitometric study of the phalanx. Diagn Imaging Clin Med 54:64–77

Van Berkum FNR, Birkenhäger JC, Van Veen LCP, et al. (1989) Noninvasive axial and peripheral assessment of bone mineral content: a comparison between osteoporotic women and normal subjects. J Bone Miner Res 4:679–685

Van Kuijk C, Grashuis JL, Steenbeek JCM, Schütte HE, Trouerbach WTH (1990) Evaluation of postprocessing dual-energy methods in quantitative computed tomography. 2. Practical aspects. Invest Radiol 25:882–889

Ventura V, Mauloni M, Mura M, Paltrinieri F, de Aloysio D (1996) Ultrasound velocity changes of the proximal phalanxes of the hand in pre-, peri- and posymenopausal women. Osteoporosis International 6:368–375

Wasnich RD (1990) Does current bone mass predict future fractures? In: Christiansen C, Overgaard K (eds) Osteoporosis 1990. Osteopress, Kopenhagen, pp 442–445

WHO (1994) Assessment of osteoporotic fracture risk and its role in screening for postmenopausal women. WHO Technical Reports Series, Geneva

Yamada M, Ito M, Hayashi K, Ohki M, Nakamura T (1994) Dual-energy x-ray absorptiometry of the calcaneus: comparison with other techniques to assess bone density and value in predicting risk of spine fractures. Am J Roentgenol 163:1435–1440

Yang SO, Hagiwara S, Engelke K, et al. (1994) Radiographic absorptiometry for bone mineral measurement of the phalanges: precision and accuracy study. Radiology 192:857–859

Yates AJ, Ross PD, Lydick E, Epstein RS (1995) Radiographic absorptiometry in the diagnosis of osteoporosis. Am J Med 98(2A):41S–47S

Yu W, Glüer CC, Grampp S, et al. (1995) Spinal bone mineral assessment in postmenopausal women: a comparison between dual X-ray absorptiometry and quantitative computed tomography. Osteoporosis Int 5:433–439

Practical Clinical Problems

10 The Shoulder

J. Hodler

CONTENTS

10.1
Introduction

The glenohumeral joint is a complex, inherently unstable articulation. A number of structures are required for adequate function and for maintenance of stability; these include the glenoid and the humeral head, the rotator cuff, the tendon of the long head of the biceps, the labrum, the joint capsule, and the glenohumeral ligaments.

The following sections discuss various groups of abnormalities specific to the shoulder. Each major section includes a definition of the disease in question, discussion of its pathogenesis, description of the morphological findings and their appearance on relevant imaging methods, and a suggestion for an imaging strategy.

J. Hodler, MD, Radiology, Balgrist Clinic, University of Zurich, Forchstrasse 340, CH-8008 Zurich, Switzerland

10.2
Shoulder Impingement Syndrome, Rotator Cuff Tears

10.2.1
Definitions and Pathogenesis

Shoulder impingement syndrome relates to abnormalities of the rotator cuff (mainly the supraspinatus) and the subacromial bursa caused by impingement of these structures underneath the coracoacromial arch in forward flexion of the arm. Impingement occurs at the anteroinferior part of the acromion, the coracoacromial ligament, and occasionally the acromioclavicular joint and the coracoid (Gerber et al. 1985). Three acromial types have been identified (type I, flat; type II, curved; and type III, hooked) (Bigliani et al. 1991). Types I and II are far more common than type III (22.8%, 68.5%, and 8.6%, respectively) (Getz et al. 1996). Types II and III are more commonly associated with shoulder impingement syndrome than is type I.

The classifical description of the various phases of shoulder impingement syndrome by Neer (1993) has summarized three stages of shoulder impingement syndrome. Stage I is mainly found in patients under the age of 25 years after excessive overhead motion related to work or sports. It is characterized by tendon edema and hemorrhage. These findings are reversible. The clinical differential diagnosis includes glenohumeral subluxation and abnormalities of the acromioclavicular joint. Stage II is more chronic and mainly occurs between 25 and 40 years. The bursa commonly is thickened and fibrotic. The tendon degenerates. Pain usually is worse during exercise. The clinical differential diagnosis includes frozen shoulder and hydroxyapatite deposition disease. In stage III, which mainly occurs above the age of 40, rotator cuff tears are present. The clinical differential diagnosis includes cervical radicular abnormalities and neoplasm. Beside impingement, other etiologies of rotator cuff tears have been discussed, including intrinsic rotator cuff abnormalities

(OGATA and UHTHOFF 1990), degeneration (STILES and OTTE 1993), and injury to an already abnormal tendon (NORWOOD et al. 1989). For small tears (less than 1 cm), pain is the most common symptom. With increasing tear size, pain becomes less predominant, and pseudoparalysis becomes the main symptom (GSCHWEND et al. 1988).

Os acromiale has been recognized to be associated with the shoulder impingement syndrome. PARK et al. (1994) examined ten patients between 35 and 68 years old with os acromiale using magnetic resonance (MR) imaging. The form of the os acromiale was variable due to the variability of the ossification centers of the tip of the acromion. In nine of the ten patients there were osteophytes at the margins of the acromial gap. Supraspinatus tendon degeneration was present in four, and a tear in six.

Posterosuperior impingement occurs commonly in athletes in the throwing disciplines and has to be differentiated from instability, which may produce very similar symptoms. Posterosuperior impingement has been described by WALCH et al. (1992). It is caused by impingement of the supraspinatus tendon underneath the posterosuperior glenoid during throwing. These patients have partial tears of the rotator cuff, and degeneration of the posterosuperior glenoid and labrum.

Lesions of the biceps tendon are more common in large than in small rotator cuff tears. They are present in 8% of rotator cuff tears smaller than 1 cm, in 25% of tears smaller than 2 cm, in 33% larger than 2 cm, and in 58% of advanced rotator cuff disease with osteoarthritis (PATTE et al. 1981).

The incidence of rotator cuff tears associated with glenohumeral dislocation increases with patient age. In a series of 31 patients with rotator cuff tears and shoulder dislocation, all patients were older than 40 years (mean: 57.5 years) (NEVIASER et al. 1988).

10.2.2
Imaging Findings

10.2.2.1
Standard Radiographs

Imaging in patients with suspected rotator cuff tears starts with standard radiographs. In intact rotator cuffs, the distance between the superior border of the humeral head and the acromion is approximately 9–10 mm on anteroposterior radiographs obtained in slight external rotation. If this distance is smaller than 6 mm in the middle-aged population, a rotator cuff tear should be considered (PETERSSON and REDLUND-JOHNELL 1984) (Fig. 10.1). Standard radiographs can be sufficient for the diagnosis of large tears without additional imaging when the humeral head is cranially migrated (KILCOYNE et al. 1989; RESNICK and NIWAYAMA 1995). They also demonstrate subtler signs of shoulder impingement syndrome, such as increased bone density in the region of the greater tubercle, calcification of the coracoacromial ligament (OGATA and UHTHOFF 1990), and obliteration of the peribursal fat plane (MITCHELL et al. 1988). They are important for the differential diagnosis by demonstrating calcific tendinitis, fractures, or osteoarthritis of the acromioclavicular and the glenohumeral joints. Standard radiographs also demonstrate abnormalities associated with shoulder impingement syndrome, such as a curved or hook-like acromion.

If obtained in a standardized fashion (WALCH et al. 1992), standard radiographs can be used for quantitative surgical planning and follow-up of subacromial decompression. The classification of the shape of the acromial arch varies depending on the exact radiographic technique (PEH et al. 1995). Therefore, the use of fluoroscopy may be required in order to obtain adequate radiographs (Fig. 10.2).

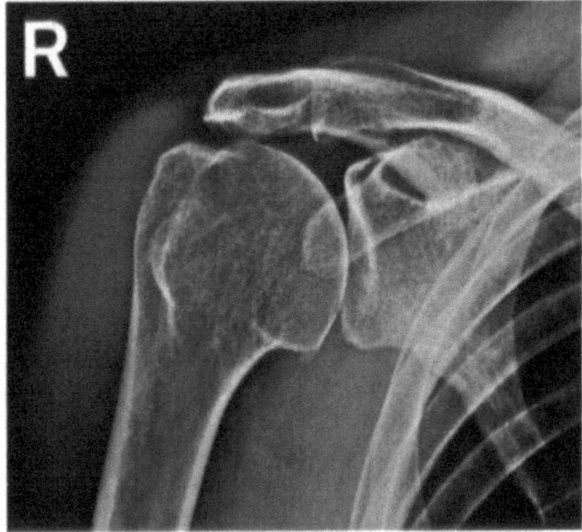

Fig. 10.1. Standard radiograph showing a rotator cuff tear. An osteophyte is present at the acromial edge; there is subchondral sclerosis of the greater tubercle, and decreased distance between humeral head and acromion

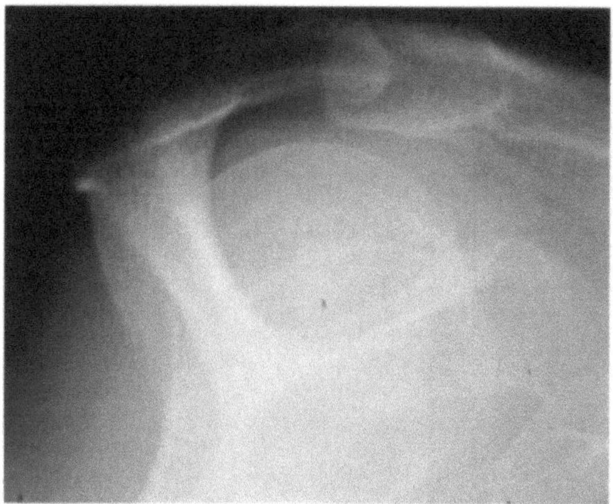

Fig. 10.2. Standard ("Neer's") radiograph. Type II acromion

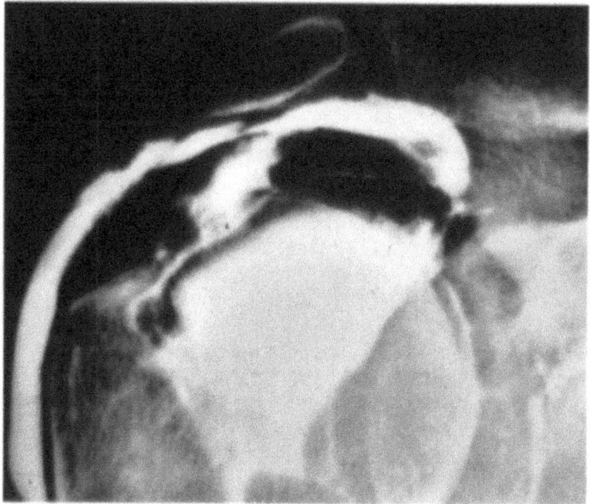

Fig. 10.3. Arthrogram in a patient with a small tear of the supraspinatus

10.2.2.2
Standard Arthrography (Fig. 10.3)

For assessment of the rotator cuff, standard athrography has been widely replaced by ultrasonography, MR imaging, or MR arthrography. Arthrograms have the advantage of a high sensitivity and specificity for full-thickness tears of the rotator cuff (MINK et al. 1985). However, the information on anatomic details of the rotator cuff obtained with standard arthrograms is limited. In large tears, contrast escapes early into the subacromial/subdeltoid bursa and obscures the rotator cuff tear.

10.2.2.3
Ultrasonography

The use of ultrasonography for the assessment of the rotator cuff has stirred some controversy. Although the value of ultrasonography has been debated especially in the U.S. literature (BURK et al. 1989; BRANDT et al. 1989; HALL 1989), it is a useful imaging tool for assessment of the rotator cuff in experienced hands. Ultrasonography has significant advantages such as low cost and high patient acceptance (HODLER et al. 1991; MIDDLETON 1993; WIENER and SEITZ 1993; FARIN and JAROMA 1995). Several problems may explain the problems with ultrasonography of the rotator cuff:

1. Part of the rotator cuff (especially the important supraspinatus) is hidden underneath the acromioclavicular arch. Even if this is corrected

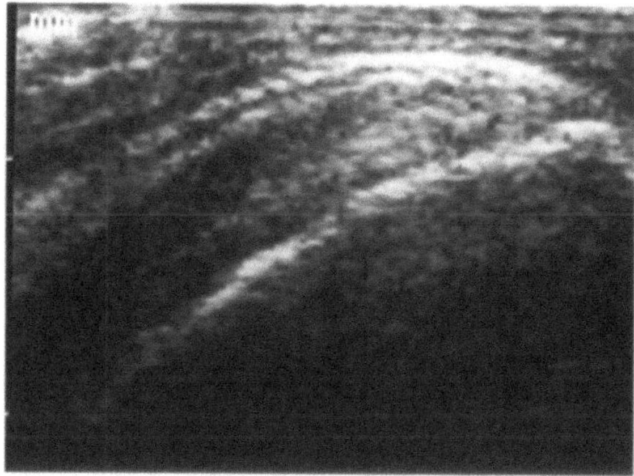

Fig. 10.4. Ultrasonography of a normal supraspinatus tendon (longitudinal image). Increased echogenicity of the distal portion is explained by the position of the transducer (perpendicular to the tendon fibers)

by placing the arm in internal rotation and retroversion, the entire rotator cuff is not accessible for ultrasonography.

2. Interindividual variability in the appearance of the rotator cuff is substantial. Irregular echogenicity can be found in tendon degeneration, in calcifications, in partial and small full-thickness tears, and in asymptomatic volunteers (Fig. 10.4). Moreover, the thickness of the rotator cuff is variable.

3. In the presence of a rotator cuff tear, bursal hypertrophy, tendon fragments, and blood may fill in the defect and mimic intact substance (although echogenicity is not normal in such cases). Therefore, tears can be difficult to detect

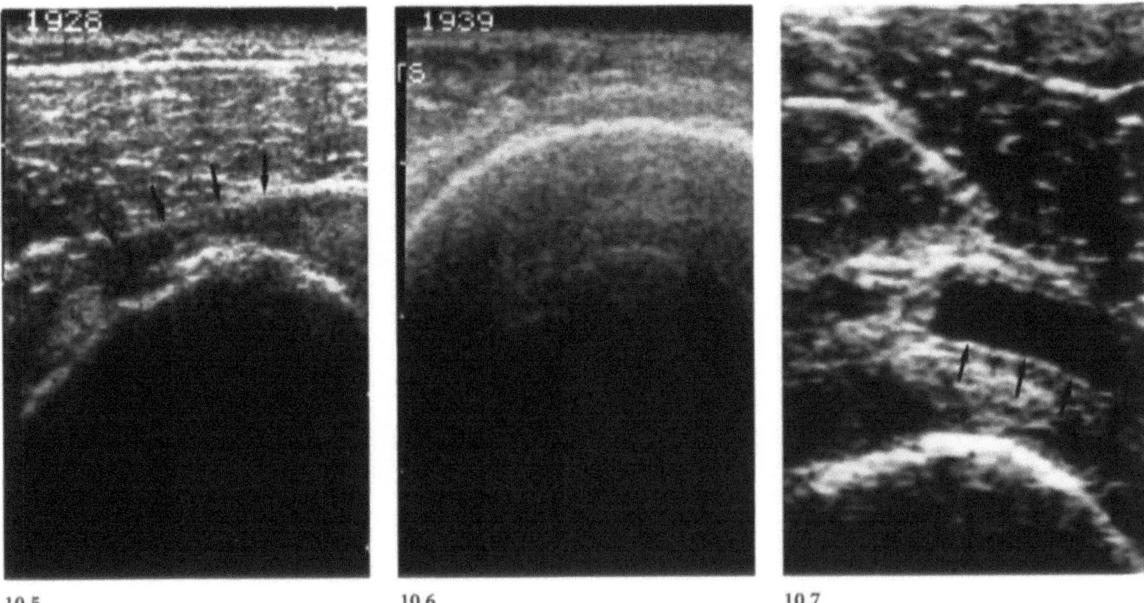

10.5   10.6   10.7

Fig. 10.5. Ultrasonography of a tear of the supraspinatus (transverse image). Missing tendon substance has been replaced by hypoechoic joint fluid (*arrows*)
Fig. 10.6. Ultrasonography of large rotator cuff tear (transverse image). Deltoid muscle is directly abutting the humeral head
Fig. 10.7. Ultrasonography in bursitis of the subacromial/subdeltoid bursa. A hypoechoic fluid collection is present between the supraspinatus and the deltoid (*arrows*)

or to exclude, especially for less experienced sonographers.

FARIN and JAROMA (1995) obtained excellent results in acute traumatic tears of the rotator cuff by using a limited number of well-defined criteria: They accepted the presence of a hypoechoic defect (Fig. 10.5), focal thinning, and complete nonvisualization of the rotator cuff (Fig. 10.6) as signs of a tear.

Ultrasonography is suitable for demonstration of bursitis of the subacromial/subdeltoid bursa (Fig. 10.7), hypertrophy of the joint capsule of the acromioclavicular joint, and fluid collections within the tendon sheath of the long biceps tendon. Fluid within the subacromial/subdeltoid bursa should prompt the attention of the sonographer to the rotator cuff, which is frequently torn in this situation (HOLLISTER et al. 1995). However, such fluid is not a specific sign of rotator cuff lesions (NEEDELL et al. 1996). Based on MR images it is present in 20% of asymptomatic volunteers (NEUMANN et al. 1992). It becomes more common in elderly patients (10% below 40 years, 48% above 40 years). Because the biceps tendon sheath represents a recess of the glenohumeral joint space, fluid can be present not only in biceps abnormalities but also in joint derangements. Biceps tendon sheath fluid may be present as a normal variant (KAPLAN et al. 1992) or in local abnormalities, such as tenovaginitis and biceps tendon tear.

Ultrasonography performed for assessment of the rotator cuff after injury occasionally detects fractures of the greater tuberosity that are overlooked on standard radiographs (PATTEN et al. 1992).

10.2.2.4
CT and CT Arthrography

Standard computed tomography (CT) has a limited role in the assessment of the rotator cuff, although tears of the supraspinatus may be correctly diagnosed, as demonstrated by DIHLMANN and BANDICK (1987) (Fig. 10.8). CT has been employed for the assessment of fatty infiltration associated with rotator cuff atrophy. This is clinically relevant, as atrophy is associated with reduced strength of the rotator cuff (GOUTALLIER et al. 1994). Moreover, the outcome of rotator cuff repair depends on the extent of atrophy and fatty infiltration. CT has been employed for demonstration of the rare subcoracoidal variant of the shoulder impingement syndrome (GERBER et al. 1985).

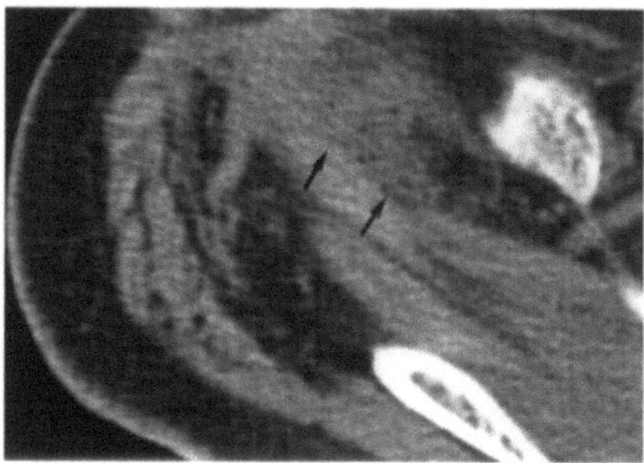

Fig. 10.8. CT of a torn supraspinatus. The anterior part of the tendon is missing (*arrows*)

Based on sagittal and coronal reformations, CT arthrography may contribute to the diagnosis of rotator cuff. Occasionally, direct sagittal CT arthrography of the rotator cuff has been employed successfully, with the patient sitting beside the gantry and the shoulder placed within the gantry (BELTRAN et al. 1986; BLUM et al. 1993).

10.2.2.5
MR Imaging

For decisions with regard to surgery (such as arthroscopy versus various open surgical techniques) it may be important to know the extent of the damage to the rotator cuff and to demonstrate additional abnormalities (such as tears of the biceps tendon, atrophy of the rotator cuff, osteoarthritis, and labral lesions). MR imaging allows for demonstration of many of these findings in a single step.

A large number of papers have been written about the value of MR imaging of the rotator cuff, citing variable results (STILES and OTTE 1993). With increasing experience and better equipment, however, the recent results have become more and more promising. When evaluating MR images of the rotator cuff it is important to recognize certain pitfalls which may mimic disease. KAPLAN et al. (1992) have demonstrated that increased signal within the supraspinatus tendon approximately 1 cm from the insertion is a normal finding (30 of 30 volunteers) on T1-weighted spin-echo and gradient-echo images. Similar results were published by LIOU et al. in 1993 (prevalence of increased signal: 95%). Several groups have investigated the etiology of this in-

creased signal. One possible explanation provided by VAHLENSIECK et al. (1993) is the presence of two different parts of the supraspinatus, an anterior fusiform and a posterior straplike part. A partial volume artifact and fat between these two parts could be eliminated as a reason for the increased signal by applying various imaging planes and fat suppression. These authors concluded that the difference in tendon orientation or differences in tissue relaxation times might be responsible. Another explanation was advocated by TIMINS et al. (1995). They attributed the signal to a magic angle effect based on the dependence of the signal on the position of the tendon. Increased signal within the rotator cuff is also present when the shoulder is examined in internal rotation of the arm, probably due to partial volume effects caused by soft tissue interposed between the supraspinatus and infraspinatus (DAVIS et al. 1991).

Obliteration of the subacromial fat plane has been considered abnormal. However, partial obliteration is also common in asymptomatic volunteers (KAPLAN et al. 1992: 57%; LIOU et al. 1993: 95%). Other indirect signs of rotator cuff abnormalities are not reliable (see discussion of fluid within the subacromial bursa and within the biceps tendon in Sect. 10.2.2.3). Intra-articular fluid is commonly present in normal volunteers. In an investigation including 20 shoulders in 12 asymptomatic volunteers, 14 had fluid collections (that were estimated to be less than 2 ml) (RECHT et al. 1994).

Commonly, three major types of rotator cuff abnormalities are identified on MR images: tendon degeneration, partial tears, and full-thickness tears (HODLER et al. 1992a). Tendon degeneration or tendinopathy is characterized by increased signal visible on T1-weighted or proton-density images. The signal commonly is not well demarcated (Fig. 10.9). On T2-weighted images, the signal is no longer visible. The term "tendinitis" is not correct in a strict sense to describe such findings, because histologically, degeneration is found, but not inflammation (KJELLIN et al. 1991).

In a partial tear, the signal as visible on T1-weighted and proton-density images is better demarcated and persists at least partially on T2-weighted images. The signal behavior is identical in full-thickness tears (Fig. 10.10). However, the abnormality extends throughout the entire substance. In large tears, the defect becomes directly visible on all imaging sequences. Apparently, employing fat suppression in combination with T2-weighted images can increase sensitivity for both full-thickness and

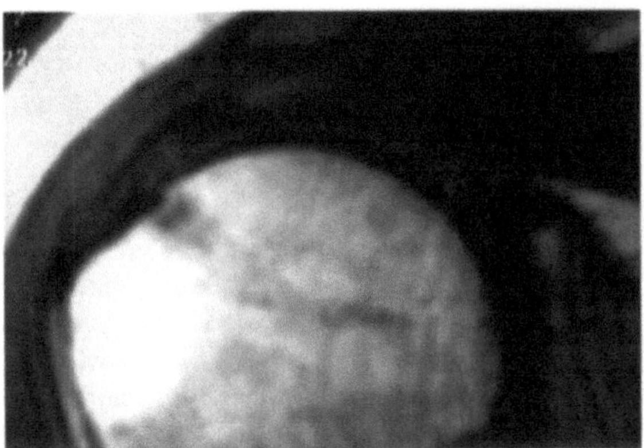

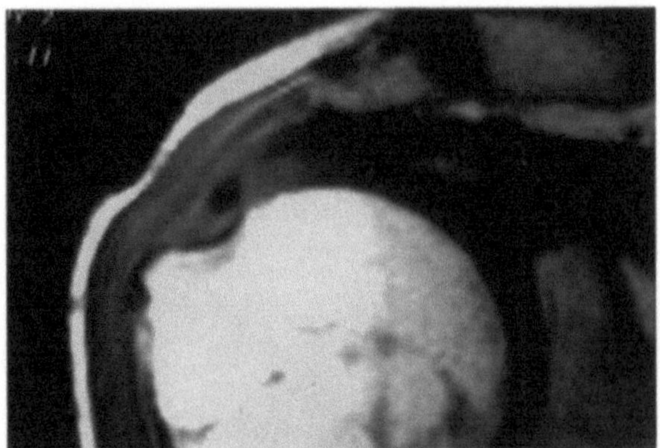

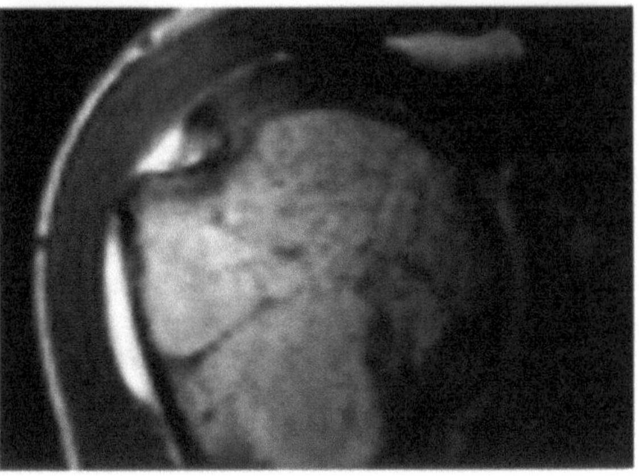

Fig. 10.9. Coronal proton-density weighted MR image of tendinopathy of the supraspinatus. There is increased signal in the distal part of the tendon, which is more extensive than would be expected in normal volunteers. Signal was no longer visible on the corresponding T2-weighted image (not shown)

Fig. 10.10. Coronal proton-density (a) and T2-weighted (b) turbo spin-echo images in a patient with a small tear of the rotator cuff. There are two zones of increased signal in a. The signal behavior in b, indicates that the distal part corresponds to full-thickness tear. Proximal signal abnormality represents tendinopathy

partial tears of the rotator cuff [according to REINUS et al. (1995), sensitivity is improved from 80% to 100% in full-thickness tears and from 15% to 35% in partial tears, with a slight decrease in specificity]. Use of fat-suppressed T2-weighted spin-echo imaging has also been supported by QUINN et al. (1995), who reported a sensitivity of 84% and a specificity of 97% with this method in a group of 20 complete and 11 partial tears. According to ROBERTSON et al. (1995), the MR diagnosis of full-thickness tears is far more reliable than a diagnosis of a partial tear or tendinopathy (sensitivities for detection of tendinopathy 13%–50%, for partial tears 19%–57%, and for full-thickness tears 81%–100%, for four different observers). The corresponding specificities varied between 73% and 89% for tendinopathy, between 85% and 93% for partial tears, and between 89% and 98% for full-thickness tears. Inter- and intraobserver variability was good for full-thickness tears and poor for the other diagnoses. In another recent paper, SINGSON et al. (1996) found a far better performance even for partial tears, with a sensitivity of 92% for fat-suppressed and 67% for standard T2-weighted spin-echo images.

Degeneration of the acromioclavicular joint is commonly visible on MR images obtained for abnormalities of the rotator cuff. It demonstrates capsular hypertrophy and degeneration of the disk within the joint. Joint fluid that appears hyperintense on T2-weighted images is common in osteoarthritis (SCHWEITZER et al. 1994). Because both the joint capsule and osteophytes are hypointense on MR images, these two structures cannot easily be differ-

entiated. The diagnosis of an osteophyte should preferably be made on standard radiographs. This is also true for the acromial edge: the origin of the deltoid muscle is tendinous and therefore appears hypointense and can mimic acromial osteophytes (KAPLAN et al. 1992).

The shape of the acromial arch can be assessed on sagittal MR images. However, the appearance of this shape varies significantly depending on the image used for the evaluation (PEH et al. 1995). Among 41 normal subjects no change in the acromial shape was visible from medial to lateral images in only nine. In 18/41 the acromial arch changed from type I to type II, in 2/41 from type I to type III, and in 11/41 from type II to type III, and in 1/41 from type II to type I. This variability probably explains the inferior interobserver variability of MR imaging compared with standard radiographs. HAYGOOD et al. (1994) found kappa values of 0.43 for standard radiographs and 0.23 for MR imaging).

10.2.2.6
MR Arthrography

Magnetic resonance arthrography can improve visualization of abnormalities of the rotator cuff (HODLER et al. 1992a), mainly with regard to the differentiation between tendinopathy, partial tears, and full-thickness tears. The best contrast is obtained with diluted (1:100–1:250 or 2–5 mmol/l) gadopentetate or other gadolinium-containing contrast media (Fig. 10.11). However, intra-articular use of gadopentetate has not been approved by the American Food and Drug Administration and corresponding European agencies. In most countries, the local ethics committee has to be consulted and informed consent has to be obtained from the patients prior to MR arthrography. This problem can be avoided by using intravenously injected gadopentetate, which leads to enhancement of the joint space (VAHLENSIECK et al. 1996). This method can result in excellent images, especially in combination with fat-suppressed sequences. Exercise improves intra-articular enhancement, which is adequate within 10–20 min. Disadvantages include limited distension of the joint and false-positive results caused by extra-articular enhancement, e.g., enhancement of the subacromial/subdeltoid bursa can mimic contrast leakage caused by a rotator cuff defect. Another possibility is the use of saline as a contrast medium (TIRMAN et al. 1993; ZANETTI and HODLER 1997). In our experience, however, contrast is not consistently of high quality with this method.

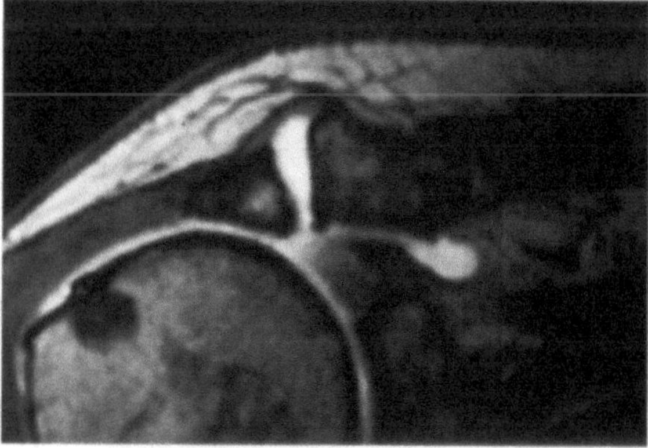

Fig. 10.11. Coronal MR arthrogram (intra-articular gadopentetate) in a patient with a large rotator cuff tear. The humeral head touches the acromial edge. A so-called geyser sign is present, with contrast leakage into a degenerated acromioclavicular joint

10.2.3
Imaging Strategy

In shoulder impingement syndrome or suspected rotator cuff tears the use of imaging depends on a working hypothesis and the possible therapeutic options. Standard radiographs are required for differential diagnosis and can demonstrate indirect signs of shoulder impingement syndrome. Standard MR imaging or MR arthrography can be used for a morphologically precise diagnosis and for preoperative planning. Ultrasonography is useful as a screening tool before additional, more expensive imaging methods are employed, especially when the clinical diagnosis is not clear.

10.3
Abnormalities of the Biceps Tendon

10.3.1
Definitions and Pathogenesis

The tendon of the long head of the biceps originates from the supraglenoid tubercle, close to the origin of the superior glenohumeral ligament and the superior labrum. Histologically, the relationship between the biceps tendon and these two other structures is close. There are a number of variants for this insertion. Not uncommonly, there is a deep recess at the base of biceps-labrum anchor which has to be differentiated from the true detachment [superior labrum anterior and posterior (SLAP)] lesion. This lesion occurs after a fall on the outstretched arm (SNYDER et al. 1990). Early lesions are characterized by simple fraying of the biceps anchor (grade I). In grade II lesions, complete detachment of this anchor is present. Grades III and IV are bucket-handle tears of the labrum alone (grade III) or of both the labrum and the biceps tendon (grade IV). The prevalence of the various grades in the original series was 11%, 41%, 33%, and 15%, respectively. Clinical symptoms include pain or "popping" in overhead motion. A true SLAP lesion has to be differentiated from the normal sublabral recess of the superior labrum (SMITH et al. 1996) and labral changes occurring with ageing (DE PALMA 1983). True lesions commonly are not exactly at the base of the labrum and/or are more irregular than a normal recess.

During the intra-articular course the biceps tendon is exposed to impingement. There may be fraying, swelling, and shredding of tendon fibers. Such changes start in the fourth decade and are present in

78% in the eighth and ninth decades. Biceps tendon tears are far more common in rotator cuff tears than in a control population of the same age group.

The biceps tendon is kept within the bicipital groove by a transverse ligament. When this complex apparatus is damaged, such as in lesions of the subscapularis, the biceps tendon subluxes medially (most commonly underneath the subscapularis tendon). The tendon may degenerate in this location. The form of the intertubercular groove may also be of importance. A narrow groove (narrowed by osteophytes or on a developmental basis) leads to increased tendon wear. A flat groove may increase the tendency for subluxation.

10.3.2
Imaging Findings

10.3.2.1
Standard Radiographs

In suspected abnormalities of the biceps tendon, standard radiographs are required for demonstration of the indirect signs of rotator cuff tears and other abnormalities. However, they do not contribute significantly to most of the diagnoses described above. Sulcus views for demonstration of the form of the intertubercular groove are now rarely used.

10.3.2.2
Ultrasonography

Ultrasonography is helpful in the demonstration of complete tears of the biceps tendon, demonstrates fluid (a nonspecific finding) of the biceps tendon sheath (Fig. 10.12), and may permit diagnosis of tendon degeneration (Middleton et al. 1985; Ptasznik and Hennessy 1995). It can demonstrate the form of the intertubercular groove. Ultrasonography is also employed for dynamic demonstration of tendon subluxation with the transducer placed transversely on the superior border of the intertubercular groove during external rotation of the arm (Farin et al. 1995). Figure 10.13 demonstrates a dislocated tendon.

10.3.2.3
CT Arthrography

By demonstrating contrast entering the base of the cranial labrum, CT arthrography is useful for the evaluation of a SLAP lesion (Hunter et al. 1992). However, MR imaging is probably superior in this regard because it allows direct coronal imaging. CT arthrography demonstrates subluxation or dislocation of the biceps tendon (Fig. 10.14) and the associated damage of the subscapularis-ligamentous

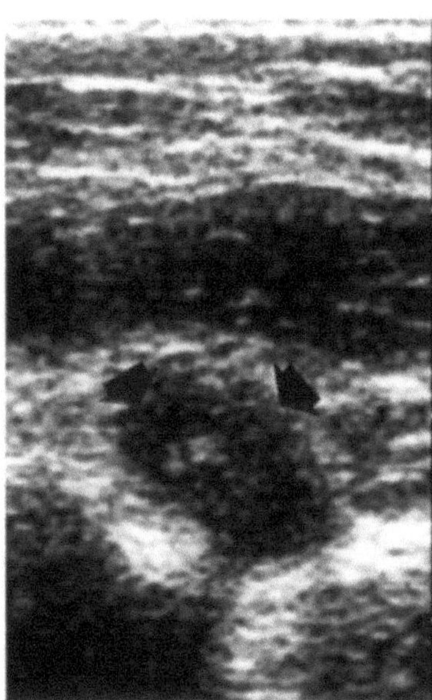

Fig. 10.12. Transverse ultrasonography of fluid within the biceps tendon sheath (*arrows*)

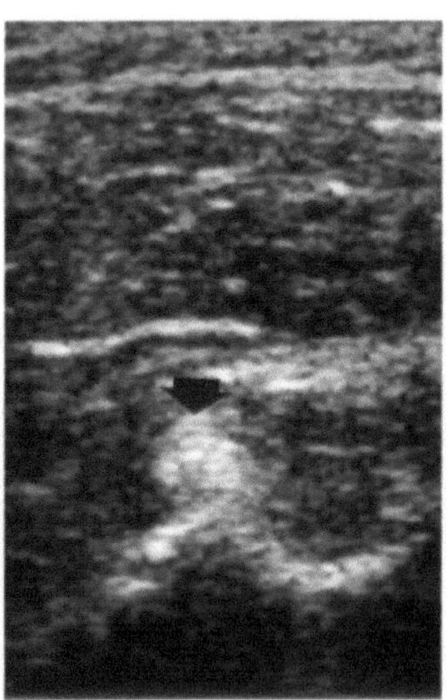

Fig. 10.13. Transverse ultrasonography of medially dislocated biceps tendon (*arrow*). x, Bicipital groove

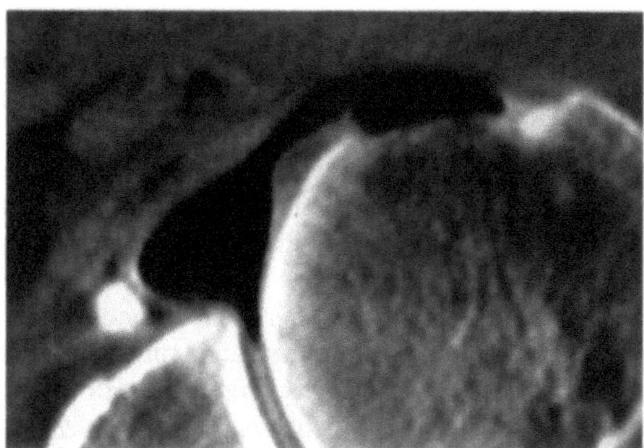

Fig. 10.14. CT arthrography in a patient with a medially dislocated biceps tendon (*arrow*)

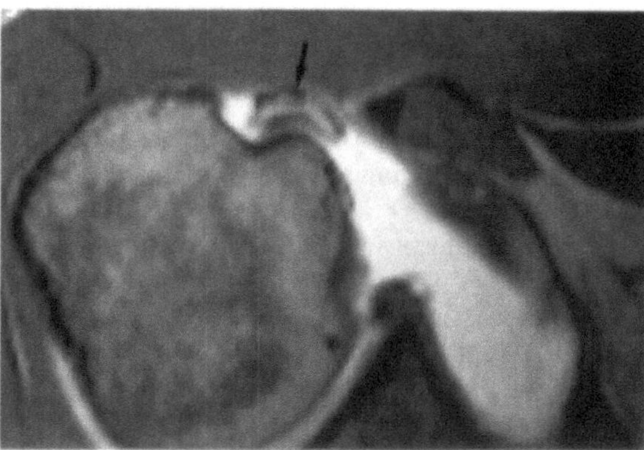

Fig. 10.16. MR arthrography of a medially subluxed and degenerated biceps tendon (*arrow*) and missing (torn) subscapularis tendon

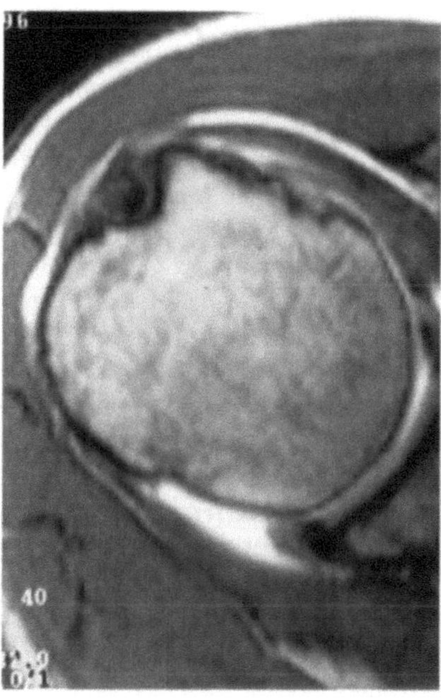

Fig. 10.15. MR arthrography demonstrating degeneration of the biceps tendon within the bicipital groove (increased signal)

complex with contrast leaking from the biceps tendon sheath into or underneath the subscapularis.

10.3.2.4
MR Imaging and MR Arthrography
(Figs. 10.15, 10.16)

Magnetic resonance imaging can be used with and without intra-articular contrast for the assessment of biceps tendon abnormalities, although MR

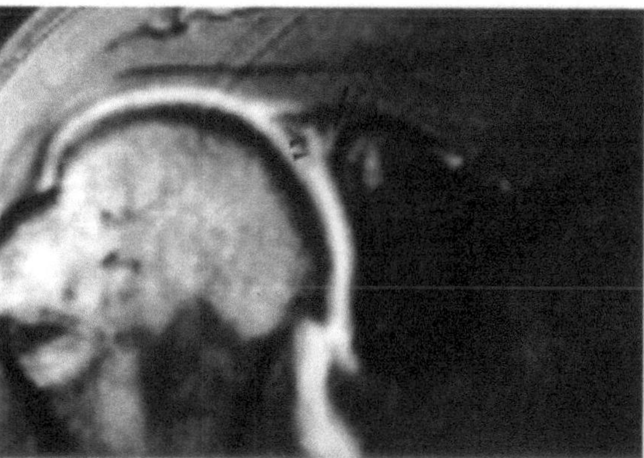

Fig. 10.17. MR arthrography of a SLAP lesion. Beside a normal recess (*arrow*) there is contrast medium entering the substance of the labrum (*arrowheads*)

arthrography may allow a more precise diagnosis, especially for the insertion into the glenoid (Fig. 10.17) and for the intra-articular course of the tendon.

The diagnosis of SLAP lesions has been discussed in Sect. 10.3.1. Tendon degeneration appears as increased signal within the tendon preferably on T1-weighted and proton-density images (TUCKMAN 1994). MR imaging is well suited for assessment of tears of the subscapularis tendon, which commonly are associated with dislocation of the biceps tendon (CERVILLA et al. 1991; PATTEN 1994).

10.3.3
Imaging Strategy

As in other abnormalities, standard radiographs represent the basis of imaging of the biceps tendon, mainly to diagnose or to exclude additional findings. In suspected tears and subluxation of the biceps tendon, ultrasonography is adequate. In suspected biceps tendon dislocation and associated injuries, CT arthrography, MR imaging, and MR arthrography are diagnostic. SLAP lesions are best diagnosed on MR arthrography or alternatively CT arthrography.

10.4
Instability, Glenohumeral Dislocation

10.4.1
Definitions and Pathogenesis

There are several ways to classify glenohumeral joint instability and dislocations (ROCKWOOD and WIRTH 1996): beside classification of the direction (anterior, posterior, or superior, with further differentiation of the anterior form into subcoracoid, subglenoid, subclavicular, or intrathoracic), assessment can be based on the degree of stability (dislocation versus subluxation), chronology (congenital, acute, chronic, locked, or recurrent), the force (traumatic versus atraumatic), and patient contribution (voluntary versus involuntary).

Based on the direction, anterior instability is the most common form. Among this type, the subcoracoid form is most frequently found. In the rare subglenoid form (luxatio erecta) the humeral head is dislocated underneath the inferior rim of the glenoid and the humerus assumes a vertical position. In the subclavicular form, the humeral head is displaced medial to the coracoid. The even rarer intrathoracic luxation is characterized by humeral head dislocation into the chest. These rare types of dislocation are associated with severe trauma and commonly demonstrate associated injuries of the greater tubercle or rotator cuff tears.

Posterior dislocation can be found in patients suffering from an epileptic attack or in alcoholics. Superior luxation is rare and is characterized by a tear of the rotator cuff and fractures of the superior structures such as the acromion. Multidirection instability indicates a generalized laxity of joint structures.

With regard to the degree of instability, dislocation (complete separation of the articular surfaces without immediate spontaneous relocation) has to be differentiated from subluxation (symptomatic translation of the humeral head, commonly transient). Hill-Sachs lesions have been found in 40% of patients with anterior subluxation (ROWE and ZARINS 1981).

Chronic dislocations are not rare (NEVIASER 1980). They are commonly found in patients older than 45–50 years. Complaints include loss of motion and pain. Many patients do not recall trauma, although a traumatic genesis is most probable. Chronic dislocation can be anterior or posterior. The posterior form of chronic dislocation is associated with epilepsy or alcoholism.

Another possibility to differentiate instability is to assess whether it is voluntary or involuntary. The first situation is especially common in multidirectional instability (ROCKWOOD and WIRTH 1996). In voluntary dislocation there are relatively frequently associated psychiatric problems.

In shoulder subluxation, the anteroinferior labrum is normally damaged. The labrum may be detached or completely missing. The humeral head has a posterolateral defect (a Hill-Sachs lesion) of variable depth and width. Other findings include chondral erosion of the rim of the anterior glenoid and attenuation of the inferior glenohumeral ligament (McGLYNN and CASPARI 1984). After recurrent dislocation, similar morphologic findings are present: There may be abnormalities of the anterior glenoid rim (rim fracture, rounding of the rim, or damage to the anterior glenoid cartilage), detachment with or without dislocation or wear of the anterior labrum, damage to the glenohumeral ligaments (missing or thinning), calcification of the anterior capsule, and a Hill-Sachs lesion (HILL and SACHS 1940). Hill-Sachs lesions are impression fractures of the posterosuperior humeral head caused by compression against the anteroinferior rim of the glenoid during dislocation. These lesions can be flat with slight deformity and discoloration of the articular cartilage. In more severe cases, the lesion can become trough-like. A "reversed Hill-Sachs lesion" may be present after posterior dislocation of the humeral head. More subtle signs of posterior subluxation and dislocation include reactive bone changes of the posterior glenoid and posterior capsular calcifications (FRONEK et al. 1989).

The labrum of the glenohumeral joint is composed of primitive mesenchymal tissue with only few chondrocytes in the fetal age. During the first years of life the few chondrocytes within the mesenchymal tissue modulate into fibrocartilage. In childhood and adults the labrum consists of fibrocartilage which is separate from the capsule (PRODROMOS et al. 1990). The labrum of the adult is highly variable. The superior labrum is histologically close to the long head of the biceps (COOPER et al. 1992). Anteriorly, the labrum has a close relationship to the glenohumeral ligaments and the joint capsule. The relationship with the glenoid is variable. The labrum may be completely attached to the glenoid, it may be meniscuslike, or it may be completely detached, especially at the level of the middle glenohumeral ligament (the sublabral hole) (TUITE and ORWIN 1996). The Buford complex is a rare variant of the anterosuperior labrum. In this abnormality, the anterosuperior labrum is missing. Instead, there is a cordlike middle glenohumeral ligament (TIRMAN et al. 1996). On cross-sectional images, the Buford complex can mimic an avulsion of the anterior labrum. The inferior and also the posterior labrum are far more constant than the superior and anterior counterpart (COOPER et al. 1992). They usually are firmly attached to the glenoid and demonstrate far fewer variations than anteriorly.

Beside the attachment to the bony glenoid, additional variants are commonly present. The size of the labrum is highly variable between individuals. Beside developmental factors, other etiologies influence the size of the labrum. On the one hand the labrum can demonstrate a decrease in size and fraying starting in the fifth decade, mainly inferiorly; on the other hand, superiorly and anteriorly there is a tendency toward increased labral size in the elderly due to synovial hypertrophy (DE PALMA 1983). Labral detachment, especially cranially, also increases in frequency with age; it is presumably caused by traction forces originating from the biceps tendon (DE PALMA 1983). Considering all these variants and more or less normal symptoms accompanying normal ageing, the correct diagnosis of an abnormality based on cross-sectional imaging can be difficult. Other factors, such as history and associated morphological findings (capsular lesion, Hill-Sachs lesion, glenoid rim fracture) have to be considered for a correct diagnosis. Reliable signs of abnormalities include complete detachment with and without dislocation of the labrum or a completely missing labrum.

The discussion of the capsule and the glenohumeral ligaments bears some resemblance to that about the labrum. These are quite variable structures, especially anteriorly. The insertion of the anterior capsule at the anterior glenoid was classified by ROTHMAN et al. (1975), who found three types of anterior capsular insertion: type 1 at or near the anterior labrum, type II approximately 1 cm from the labrum, and type III far medial (more than approximately 1 cm). The tentative cutoff of 1 cm does not appear in the original reference, however. Based on MR investigations, types I and II are common in asymptomatic volunteers. However, type III is uncommon (NEUMANN et al. 1991) in this subgroup. Posteriorly, variability is far smaller, and type I insertions are very common. In abnormal (unstable) shoulders, the capsule is widened and its insertion is commonly medial and perpendicular to the glenoid neck.

There are three glenohumeral ligaments: the superior, the middle, and the inferior. The superior one is the smallest glenohumeral ligament. It originates from the superior labrum and sends fibers towards the lateral base of the coracoid. It inserts at the humeral neck on the medial ridge of the intertubercular groove (DETRISAC and JOHNSON 1986). The middle glenohumeral ligament originates from the anterior superior labrum and glenoid and courses to the medial surface of the lesser tuberosity. The inferior glenohumeral ligament is the most important structure with regard to shoulder stability. It originates from the middle or inferior third of the anterior labrum and glenoid (COOPER et al. 1992) and courses inferiorly towards the inferior humeral head. Its superior band portion is quite constantly visible as an individualized ligament while the remainder of the ligament rather acts as a reinforcement of the capsule. In abnormal (unstable) joints, these ligaments can be torn, thinned, or completely missing.

Loose bodies are common after trauma and may originate from the articular surfaces, the labrum, or a free fragment of rotator cuff, tendon, or ligament (JOHNSON 1993). They are commonly multiple and may reside in the joint space (especially in the subscapularis recess) or be buried within the capsule.

10.4.2
Imaging Findings

10.4.2.1
Standard Radiographs

Standard radiographs represent the basis of imaging in unstable or dislocated shoulders. Beside the standard anteroposterior (Figs. 10.18–10.20) and Neer (Fig. 10.19) radiographs, additional radiographs may be required for assessment of the associated bony lesions. A number of projections have been described which are tangential to the posterosuperior head of the humerus and are able to demonstrate any Hill-Sachs lesion. For exact assessment of the glenoid rim, including fractures or rounding, as well as calcifications of the capsular insertion, routinely employed radiographs may be inadequate. One possibility to demonstrate such lesions has been shown by BERNAGEAU (1991). Using standardized radiographs, morphologic characteristics affecting shoulder stability can be measured, such as humeral retrotorsion (right and left: normal values 26.9° and 21.2°; recurrent anterior dislocation: 22.4° and 16.0°).

The values for glenoid retroversion are 8.0° and 7.1° for normal and 13.2° and 8.9° for abnormal shoulders (CYPRIEN et al. 1983).

10.4.2.2
Ultrasonography

Ultrasonography plays a limited role in the assessment of instability. It has been used to demonstrate the Hill-Sachs lesion (Fig. 10.21) (JEROSCH and MARQUARDT 1990). The access to the labrum is limited. JEROSCH et al. (1990) has suggested that ultrasonography may be used for quantification of instability by placing the transducer dorsally over the joint space and provoking subluxation of the humeral head by use of the examiner's free hand.

10.4.2.3
Computed Tomography

Standard CT demonstrates bone abnormalities associated with shoulder instability and dislocation. Beside the demonstration of the exact extent of a Hill-Sachs lesion, details of the glenoid [fractures (Fig. 10.22), calcification of the capsular insertion] are visible. This is important for surgical planning. CT may also be used for assessment of humeral head retroversion in instability.

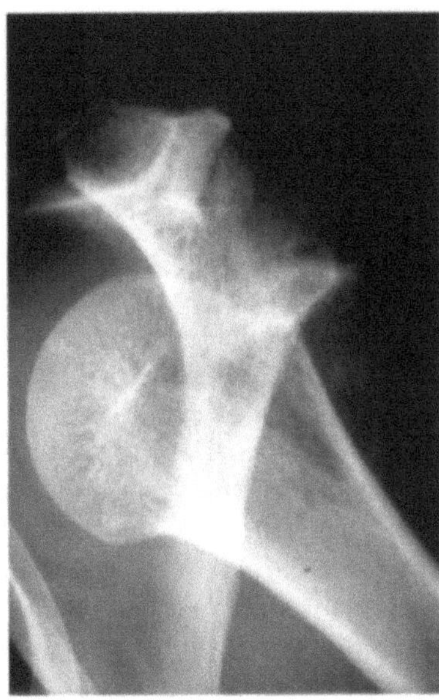

Fig. 10.18. Standard radiograph of anterior dislocation of the humeral head

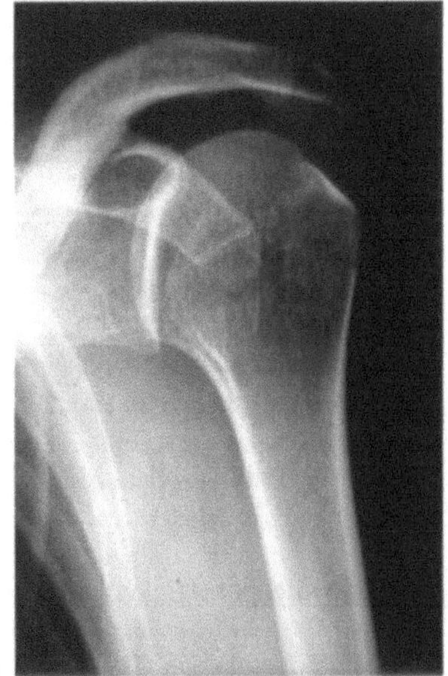
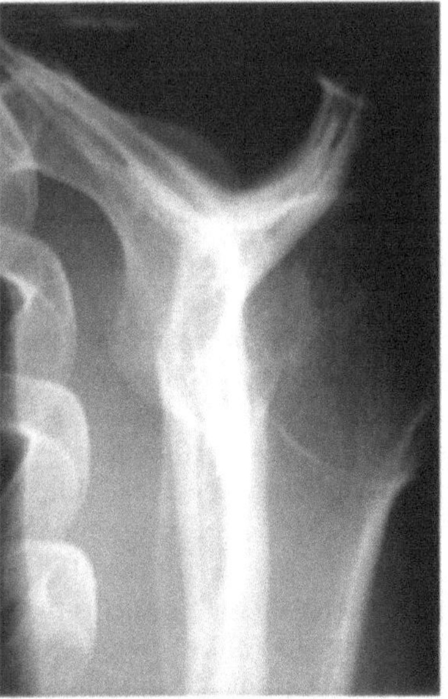

Fig. 10.19. Standard anteroposterior (**a**) and Neer (**b**) radiographs of posterior dislocation of the humeral head. In **a** overlapping of humeral head and glenoid rim is characteristic for the diagnosis

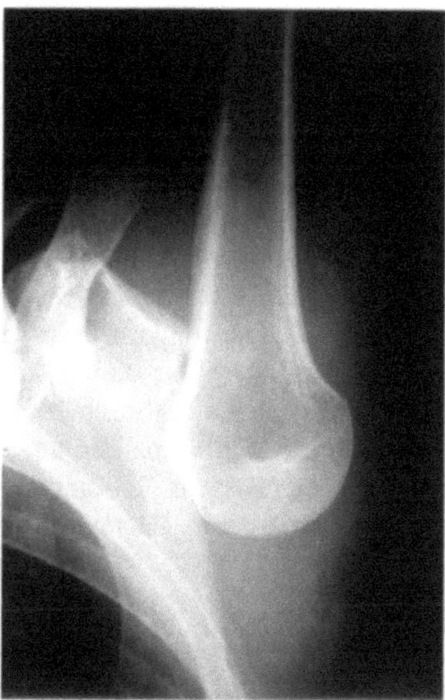

Fig. 10.20. Standard radiograph of luxatio erecta

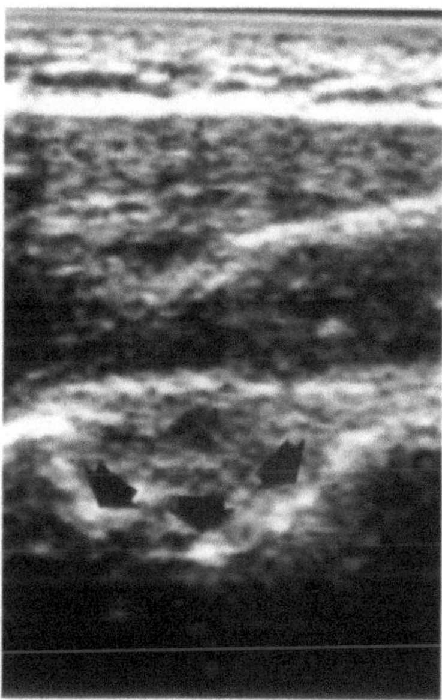

Fig. 10.21. Ultrasonography of Hill-Sachs lesion (*arrows*)

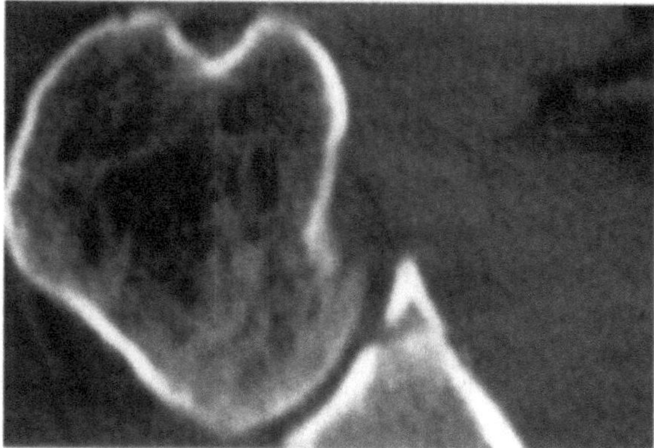

Fig. 10.22. CT in fracture of the anterior glenoid rim

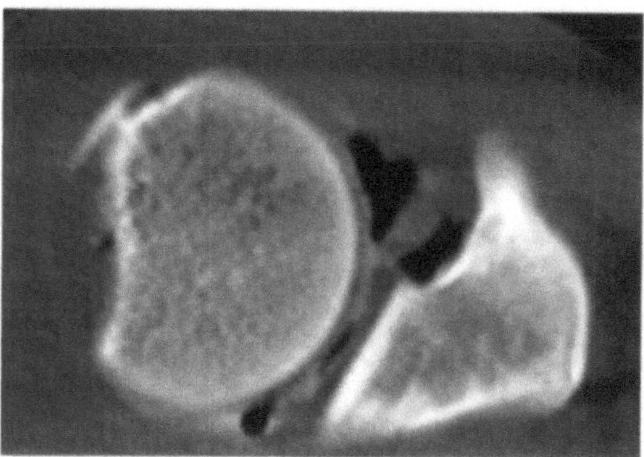

Fig. 10.23. CT arthrogram of a large but relatively flat Hill-Sachs lesion

10.4.2.4
CT Arthrography (Figs. 10.23–10.25)

For a long time CT arthrography has represented the basis of imaging in shoulder instability and dislocation (Kreitner et al. 1990). It still retains part of its role, although standard MR imaging and MR arthrography are now employed for these indica-

tions. The main advantages of CT arthrography in comparison to MR imaging are its wider availability and the better depiction of fractures and calcifications. CT arthrography can be obtained in various rotations of the arm in order to improve the diagnostic value. According to Pennes et al. (1989), most diagnoses are visible in internal rotation (Fig. 10.26). However, a modest increase (9%) in diagnostic yield can be expected when CT scans are also obtained in external rotation; for example, occasionally lesions of the anterior labrum or lesions of the posterior capsulolabral complex may be diagnosed.

10.4.2.5
MR Imaging

Standard MR imaging has been advocated for assessment of morphological abnormalities associated with instability and dislocation. MR imaging appears to be quite reliable with regard to abnormalities of the labrum (LEGAN et al. 1991), especially if suitable sequences are used, such as a combination of gradient-echo and spin-echo sequences (GUSMER et al. 1996). One problem consists in the differentiation of a tear (which appears hyperintense on commonly used MR sequences) at the base of the labrum from articular cartilage undercutting the labrum (KAPLAN et al. 1992) or a transitional zone of fibrocartilage (LOREDO et al. 1995). Assessment of the capsule by standard MR imaging is questionable, unless there is a significant amount of joint effusion, such as bleeding in recent trauma (KREITNER et al.

1992). Employing MR imaging in different rotations of the arm does not significantly improve the value of MR imaging in labral tears (TUITE et al. 1995).

A pitfall in the diagnosis of labral tears was described by KAPLAN et al. in 1992: a longitudinal tear of the anterior labrum can be mimicked by the middle glenohumeral ligament, which is very close to the anterior labrum. Such a finding was present in 11 of 30 shoulders of asymptomatic volunteers.

Shortly after trauma, MR imaging can demonstrate bone marrow edema and/or bleeding in the typical position of the Hill-Sachs lesion as a zone of decreased signal intensity on T1-weighted images and increased signal on T2-weighted images

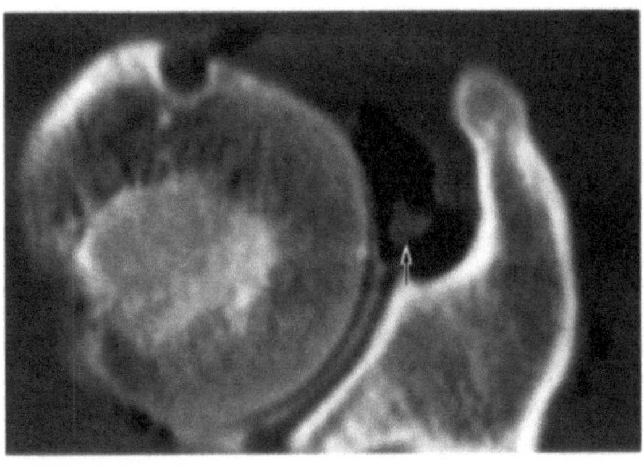

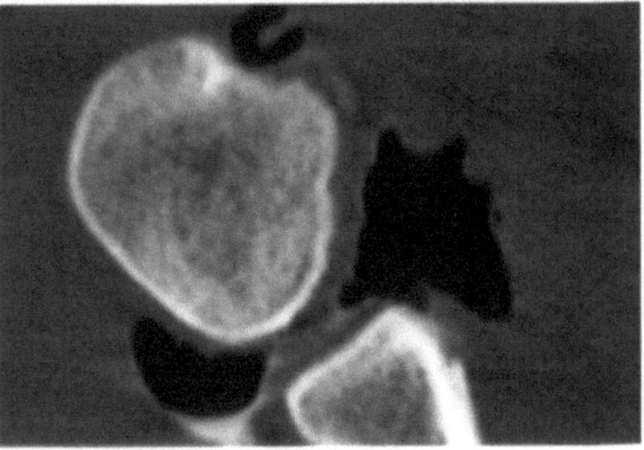

Fig. 10.24. CT arthrogram showing a detached anterosuperior labrum (*arrow*). The dorsolateral defect of the humeral head represents inferior extension of a Hill-Sachs lesion. At the coracoid level, differentiation of a smaller Hill-Sachs lesion and normal anatomy can be difficult

Fig. 10.25. CT arthrogram of glenohumeral instability. The anteroinferior labrum is missing completely. Articular cartilage is rounded at the anteroinferior glenoid. The anterior capsule is wide and irregular

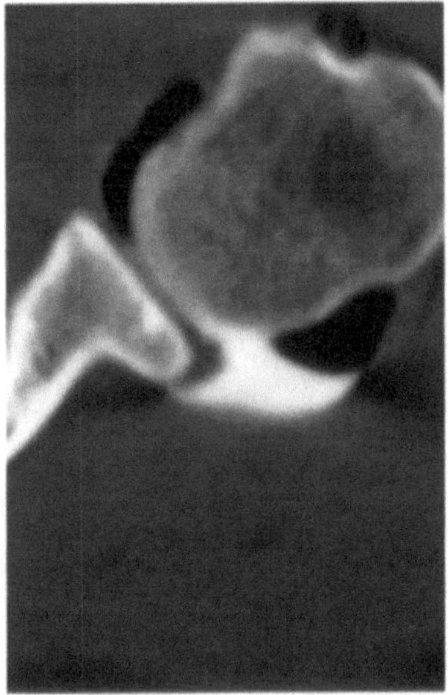

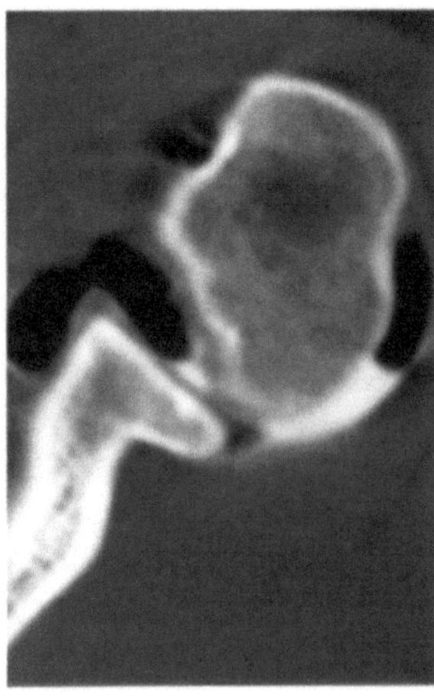

a

b

Fig. 10.26. CT arthrogram in neutral (**a**) and internal (**b**) rotation. The anterior labrum is only demarcated in internal rotation at this level

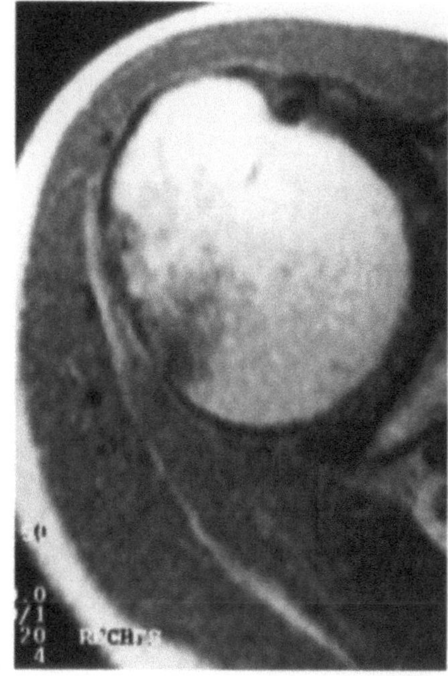

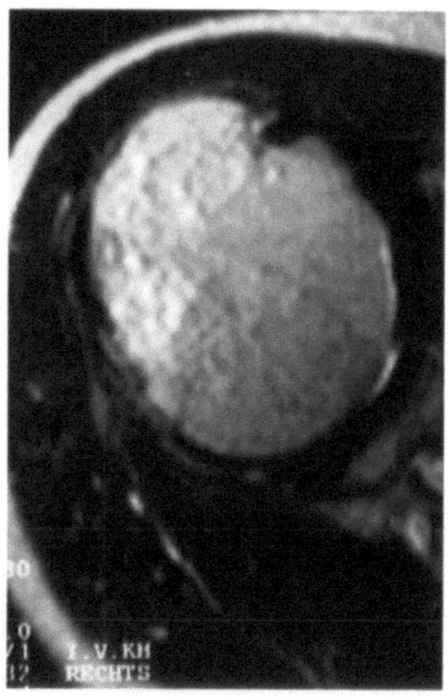

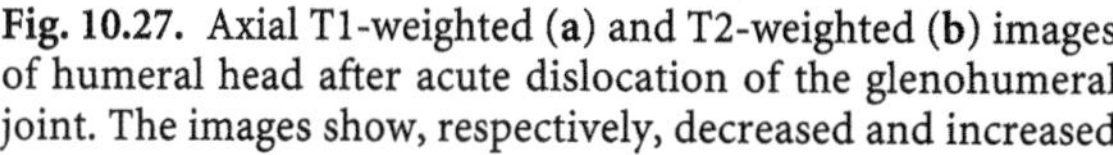

Fig. 10.27. Axial T1-weighted (a) and T2-weighted (b) images of humeral head after acute dislocation of the glenohumeral joint. The images show, respectively, decreased and increased signal intensity of the laterodorsal humeral head, corresponding to bone bruise in the expected location of a Hill-Sachs lesion

(Fig. 10.27). Later on, the Hill-Sachs lesion will appear as a deformity of the humeral head with normal underlying bone marrow. Hill-Sachs lesions are more cranial than the anatomic groove between the humeral head and neck, which can mimic a lesion (RICHARDS et al. 1994). Moreover, they are usually more lateral than the groove, although this difference is not statistically significant. The depth and width of the two structures is not different.

Magnetic resonance imaging is not suitable for detection of small fragments at the glenoid rim.

10.4.2.6
MR Arthrography (Figs. 10.28–10.33)

Magnetic resonance arthrography is not routinely employed for the reasons discussed above. However, MR arthrography has supporters, including this author, who believe that standard MR imaging is not adequate for the assessment of glenohumeral instability, especially with regard to labral and capsular lesions. Compared with CT arthrography, MR arthrography is superior in detection of certain injuries associated with glenohumeral dislocation, such as rotator cuff tears.

The diagnostic value of various abnormalities as visible on MR arthrograms was evaluated by PALMER and CASLOWITZ in 1995. Labral abnormalities had a sensitivity and specificity of 92% for anterior instability. The type of capsular insertion was not related to instability in this investigation ($P > 0.8$). Whereas inferior ligamentous problems were strongly associated with instability (<0.0001), noninferior labral-ligamentous abnormalities were associated with stable shoulders ($P = 0.01$).

10.4.3
Imaging Strategy

In instability and dislocation of the glenohumeral joint, standard radiographs represent the basis of imaging for direct demonstration of dislocation and of associated findings, such as a Hill-Sachs lesion, a fracture of the glenoid rim, or a fracture of the major tuberosity. Due to the complex anatomy, more specialized views are required in these patients, demonstrating the anterior rim of the glenoid or the posterolateral surface of the humeral head.

After standard radiographs, CT arthrography is probably the most recognized and accepted imaging

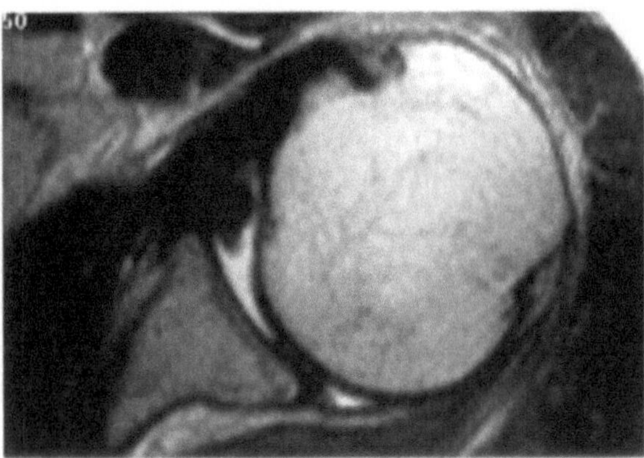

Fig. 10.28. MR arthrogram of a meniscus-like (nonpathological) anterior labrum

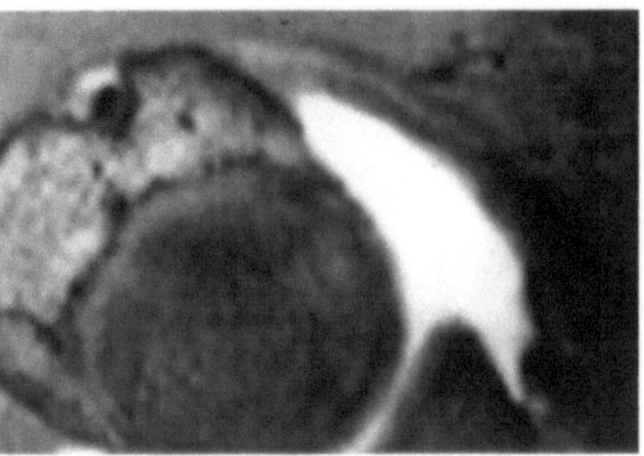

Fig. 10.31. MR arthrogram in recurrent instability with missing anterior labrum and irregularity of thinned anterior labrum

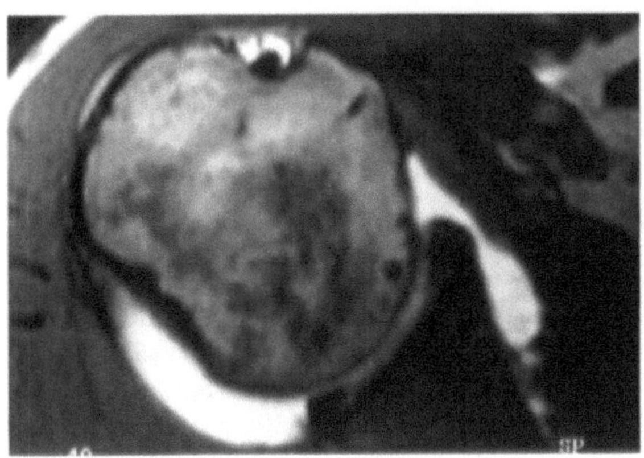

Fig. 10.29. MR arthrogram of a slightly rounded anterior labrum caused by slight synovial hypertrophy

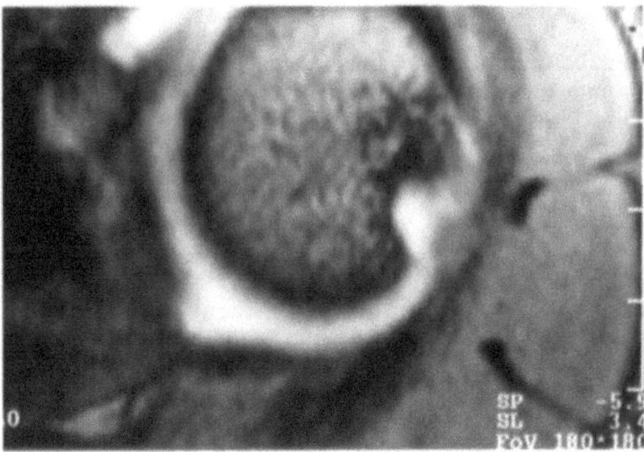

Fig. 10.32. MR arthrogram. Circumscribed Hill-Sachs lesion in the typical position

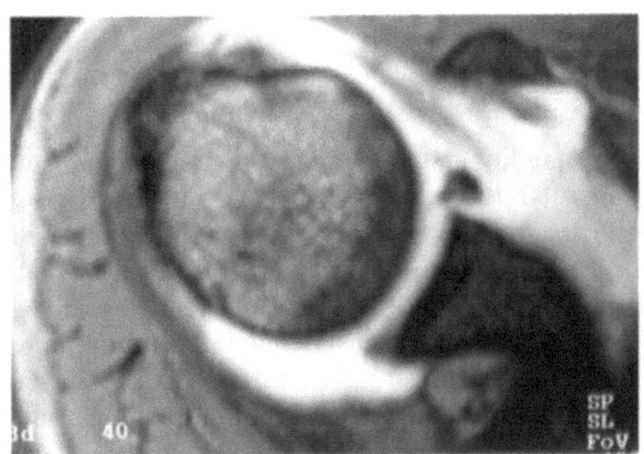

Fig. 10.30. MR arthrogram of a torn anterior labrum. The labrum is completely detached and rotated medially

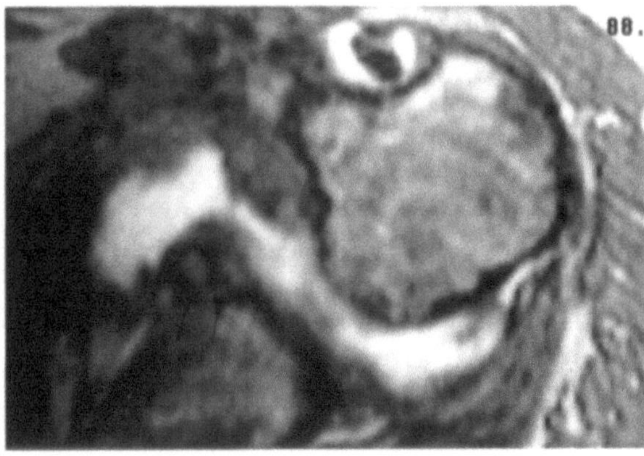

Fig. 10.33. MR arthrogram. Fracture of the anterior glenoid is not as easily visible on this gradient-echo image as on CT scans (*arrows*)

method for the assessment of both more subtle bone abnormalities and soft tissue injuries, such as capsular and labral abnormalities. MR arthrography can replace CT arthrography especially in clinically unclear situations with a broad differential diagnosis and when associated injury to the rotator cuff is suspected. Ultrasonography is indicated in the specific situation of proved shoulder dislocation requiring surgical intervention with an unclear situation with regard to the rotator cuff. Standard CT may be indicated for surgical planning in suspected glenoid rim fractures or other mainly bony abnormalities, such as locked dislocation of the shoulder.

10.5
Frozen Shoulder

Frozen shoulder occurs after trauma or idiopathically. Predisposing factors include hemiplegia, cerebral hemorrhage, diabetes mellitus, hyperthyroidism, and cervical disk disease (RESNICK 1995b). The joint capsule is thickened and retracted. The coracohumeral ligament has been found to represent an important factor in frozen shoulder. This ligament originates from the lateral surface of the base of the coracoid (NEER et al. 1991) and has a variable insertion into the rotator interval, the supraspinatus tendon, or the subscapularis tendon. It restricts external rotation. When it is sectioned in

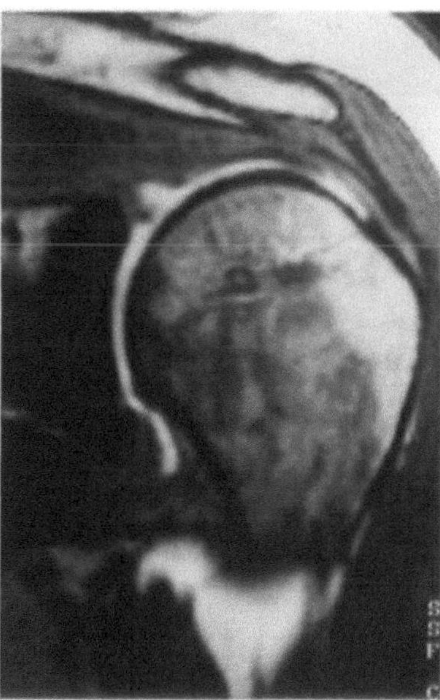

Fig. 10.34. MR arthrogram in frozen shoulder. Unusually narrow axillary recess

cadaveric specimens, external rotation increases by 32° (mean value).

Radiographs are mainly useful for excluding other disease. They may demonstrate osteopenia caused by inactivity. Moreover, the signs of previous trauma may be visible, such as a healed fracture. On arthrograms, CT arthrograms, and MR arthrograms, the axillaris recess is typically narrowed in patients with frozen shoulder (Fig. 10.34). The glenohumeral ligaments appear thickened on CT and MR arthrograms. MR images demonstrate increased capsular thickness (transverse diameter measured at the axillary recess: 5.2 mm versus 2.9 mm in frozen shoulder versus normal subjects), but are not able to demonstrate abnormalities of the coracohumeral ligament (EMIG et al. 1995).

10.6
Calcific Tendinitis

10.6.1
Definition and Pathogenesis

Calcific tendinitis is caused by calcium hydroxyapatite crystal deposition in a majority of cases. Uncommonly, calcium pyrophosphate dihydrate deposition (RESNICK et al. 1977) has been found about the shoulder joint. The hydroxyapatite crystal deposits initially are thin, cloud-like, and poorly defined and later become denser, better demarcated, and more sharply delineated (RESNICK 1995a). Patients are commonly between 40 and 70 years old. Bilateral disease is present in approximately half of the patients when they are followed up for several years. Approximately one-third of patients with calcifications have clinical symptoms. Beside the supraspinatus (approximately half of the locations in the shoulder), the other structures of the rotator cuff, the short and long heads of the biceps and the teres major can demonstrate calcifications due to hydroxyapatite deposition (RESNICK 1995a).

The calcifications can remain silent for years. In painful phases, calcifications are commonly less well demarcated than in the silent phases. When crystals are extruded into the subacromial bursa, clinical signs of bursitis appear. If the deposits are extruded into the bursa, calcifications in the distribution of the bursa can be visible.

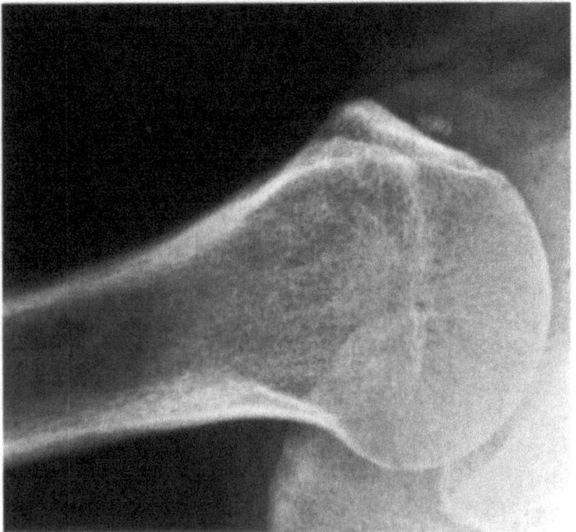

Fig. 10.35. Axial standard radiograph: a small calcification is present in the distal part of the supraspinatus tendon

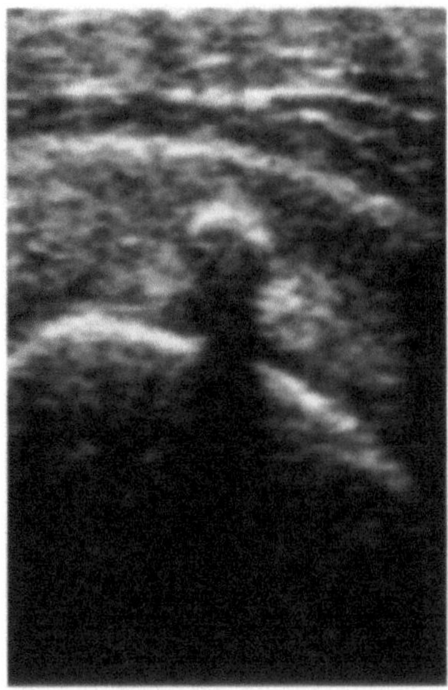

Fig. 10.36. Ultrasonography of calcific tendinitis. Typical calcification within the supraspinatus tendon with dorsal shadowing

10.6.2
Imaging Findings

10.6.2.1
Standard Radiographs (Fig. 10.35)

An anteroposterior view of the shoulder obtained in internal and external rotation of the arm is adequate for detection and follow-up of hydroxyapatite crystal deposition disease in most cases. Additional views may be required for better demonstration of calcifications in the subscapularis and infraspinatus tendons.

10.6.2.2
Standard Arthrography

Standard arthrography is not commonly indicated in hydroxyapatite deposition disease because rotator cuff tears are not typically associated with this disease and because the iodine-containing contrast medium interferes with the assessment of any calcifications present about the shoulder joint.

10.6.2.3
Ultrasonography (Fig. 10.36)

Ultrasonography may be more sensitive than standard radiographs with regard to tendon calcifica-

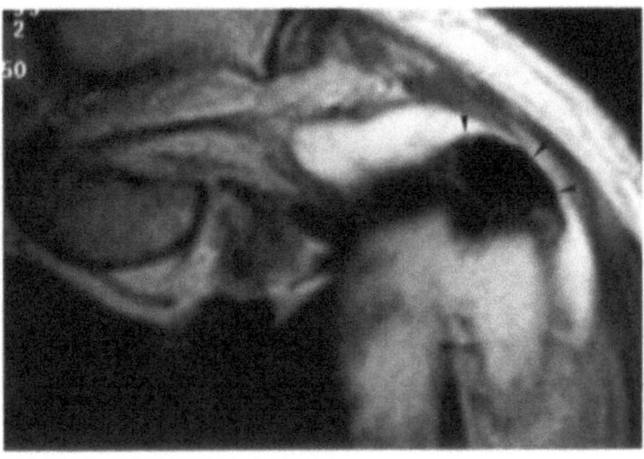

Fig. 10.37. MR arthrogram in calcific tendinitis; coronal oblique image showing the anterior border of the supraspinatus. A large hypointense deposit is present within the tendon (*arrowheads*)

tions. A major advantage of ultrasonography is certainly its better access to structures commonly obscured by the humeral head or other bones on standard radiographs (such as the infraspinatus and subscapularis tendons).

10.6.2.4
MR Imaging (Fig. 10.37)

Magnetic resonance imaging is not helpful in the evaluation of hydroxyapatite deposition disease.

Large deposits appear as hypointense zones. They are better recognized when they are near the surface of the rotator cuff because they are delineated by the fat accompanying the subacromial/subdeltoid bursa. Small calcifications can be missed when they are located within the substance of the tendon, which normally is hypointense. Commonly, however, there is a zone of hyperintensity on T1-weighted or proton-density images accompanying the deposits, presumably due to tendon degeneration. In the acute phase of hydroxyapatite deposition there may be fluid within the subacromial/subdeltoid bursa and/or within the glenohumeral joint which is easily recognized on T2-weighted images.

10.6.3
Imaging Strategy

Standard radiographs are usually adequate for the diagnosis and follow-up of tendon calcification. Ultrasonography can be used alternatively.

References

Beltran J, Gray LA, Bools JC, Zuelzer W, Weis LD, Unverferth LJ (1986) Rotator cuff lesions of the shoulder: evaluation by direct sagittal CT arthrography. Radiology 160:161–165

Bernageau J (1991) L'imagerie de l'épaule en 1991. Cahiers d'enseignement de la SOFCOT 40:111–115

Bigliani LU, Ticher JB, Flatlow WL, Soslowsky LJ, Mow VC (1991) The relationship of acromial architecture to rotator cuff disease. Clin Sports Med 10:823–828

Blum A, Boyer B, Regent D, Simon JM, Claudon M, Mole D (1993) Direct coronal view of the shoulder with arthrographic CT. Radiology 188:677–681

Brandt TD, Cardone BW, Grant TH, Post M, Weiss CA (1989) Rotator cuff sonography: a reassessment. Radiology 173:323–327

Burk DL, Karasick D, Kurtz AB, et al. (1989) Rotator cuff tears: prospective comparison of MR imaging with arthrography, sonography, and surgery. AJR 153:87–92

Butters KB (1996) Fractures and dislocations of the scapula. In: Rockwood DA, Green DP, Heckman JD, Bucholz RW (eds) Fractures in adults, 4th edition. Lippincott-Raven, Philadelphia, CD ROM

Cervilla V, Schweitzer ME, Ho C, Motta A, Kerr R, Resnick D (1991) Medial dislocation of the biceps brachii tendon: appearance at MR imaging. Radiology 180:523–526

Cooper DE, Arnoczky SP, O'Brien SJ, Warren RF, DiCarlo E, Allen AA (1992) Anatomy, histology, and vascularity of the glenoid labrum. J Bone Joint Surg [Am] 74:46–52

Cyprien JM, Vasey HM, Burdet A, Bonvin JC, Kritsikis N, Vuagnat P (1983) Humeral retrotorsion and glenohumeral relationship in the normal shoulder and in recurrent anterior dislocation (scapulometry). Clin Orthop 175:8–17

Davis SJ, Teresi LM, Bradley WG, Ressler JA, Eto RT (1991) Effect of arm rotation on MR imaging of the rotator cuff. Radiology 181:265–268

De Palma AF (1983) Surgery of the shoulder, 3d edn. Lippincott, Philadelphia

Detrisac DA, Johnson LL (1986) Arthroscopic shoulder anatomy. Pathologic and surgical implications. SLACK Incorporated, Thorofare, pp 37–84

Dihlmann W, Bandick J (1987) Computertomographie (CT) der Schulterweichteile. Fortschr Roentgenstr 147:147–151

Emig EW, Schweitzer ME, Karasick D, Lubowith J (1995) Adhesive capsulitis of the shoulder: MR diagnosis. AJR 164:1457–1459

Farin PU, Jaroma H (1995) Acute traumatic tears of the rotator cuff: value of sonography. Radiology 197:269–273

Farin PU, Jaroma H, Harju A, Soimakallio S (1995) Medial displacement of the biceps brachii tendon: evaluation with dynamic sonography during maximal external shoulder rotation. Radiology 195:845–848

Fronek J, Warren RF, Bowen M (1989) Posterior subluxation of the glenohumeral joint. J Bone Joint Surg [Am] 71:205–216

Gerber CF, Terrier F, Ganz R (1985) The role of the coracoid process in the chronic impingement syndrome. J Bone Joint Surg [Br] 67:703–708

Getz JD, Recht MP, Piraino DW, Schils JP, Latimer BM, Jellam LM, Obuchoswki NA (1996) Acromial morphology: relation to sex, age, symmetry, and subacromial enthesophytes. Radiology 199:737–742

Goutallier D, Postel J-M, Bernageau J, Lavau L, Voisin M-C (1994) Fatty muscle degeneration in cuff ruptures. Clin Orthop 304:78–83

Gschwend N, Ivoševic-Radovanovic D, Patte D (1988) Arch Orthop Trauma Surg 107:7–15

Gusmer PB, Potter HG, Schatz JA, Wickiewicz TL, Altchek DW, O'Brien SJ, Warren RF (1996) Labral injuries: accuracy of detection with unenhanced MR imaging of the shoulder. Radiology 200:519–524

Hall FM (1989) Sonography of the shoulder. Radiology 173:310

Hawkins RH, Dunlop R (1995) Nonoperative treatment of rotator cuff tears. Clin Orthop 321:178–188

Haygood TM, Langlotz CP, Kneeland JB, Iannotti JP, Williams GR, Dalinka MK (1994) Categorization of acromial shape: interobserver variability with MR imaging and conventional radiography. AJR 162:1377–1382

Hill HA, Sachs MD (1940) The grooved defect of the humeral head: a frequently unrecognized complication of dislocations of the shoulder joint. Radiology 35:690–700

Hodler J, Terrier B, von Schulthess GK, Fuchs WA (1991) MRI and sonography of the shoulder. Clin Radiol 43:323–327

Hodler J, Kursunoglu-Brahme S, Snyder SJ, et al. (1992a) Rotator cuff disease: assessment with MR arthrography versus standard MR imaging in 36 patients with arthroscopic confirmation. Radiology 182:431–436

Hodler J, Kursunoglu-Brahme S, Flannigan B, Snyder S, Karzel R, Resnick D (1992b) Injuries of the superior portion of the glenoid labrum involving the insertion of the biceps tendon: MR imaging findings in nine cases. AJR 159:565–568

Hollister MS, Mack LA, Patten RM, Winter TC, Matsen FA, Veith RR (1995) Association of sonographically detected subacromial/subdeltoid bursal effusion and intraarticular fluid with rotator cuff tear. AJR 165:605–608

Hunter JC, Blatz DJ, Escobedo EM (1992) SLAP lesions of the glenoid labrum: CT arthrographic and athroscopic correlation. Radiology 184:513–518

Jerosch J, Marquardt M (1990) Die Wertigkeit der sonographischen Diagnostik zur Darstellung von Hill-Sachs-Läsionen. Z Orthop 128:507–511

Jerosch J, Marquardt M, Winkelmann W (1990) Der Stellenwert der Sonographie in der Beurteilung von Instabilitäten des glenohumeralen Gelenks. Z Orthop 128:41–45

Johnson LL (1993) Diagnostic and surgical arthroscopy of the shoulder. Mosby, St. Louis, pp 231–275

Kaplan PA, Bryans KC, Davick JP, Otte M, Stinson WW, Dussault RG (1992) MR imaging of the normal shoulder: variants and pitfalls. Radiology 184:519–524

Kernwein GA (1965) Roentgenographic diagnosis of shoulder dysfunction. JAMA 194:1081–1085

Kilcoyne RF, Reddy PK, Lyons F, Rockwood CA (1989) Optimal plain film imaging of the shoulder impingement syndrome. AJR 153:795–797

Kjellin I, Ho CP, Cervilla V, et al. (1991) Alterations in the supraspinatus tendon at MR imaging: correlation with histopathologic findings in cadavers. Radiology 181:837–841

Kreitner K-F, Lehmann M, Zapf S, Wenda K, Schild HH (1990) Möglichkeiten der CT-Arthrographie in der Diagnostik von Schulterläsionen. Fortschr Roentgenstr 153:510–515

Kreitner K-F, Grebe P, Runkel M, Oberbillig C, Just M (1992) Stellenwert der MR-Tomographie bei akuten Schulterluxationen. Fortschr Roentgenstr 157:229–234

Legan JM, Burkhard TK, Goff WB, et al. (1991) Tears of the glenoid labrum: MR imaging of 88 arthroscopically confirmed cases. Radiology 179:241–246

Liou JTS, Wilson AJ, Totty WG, Brown JJ (1993) The normal shoulder: common variations that simulate pathologic conditions at MR imaging. Radiology 186:435–441

Loredo R, Longo C, Salonen D, et al. (1995) Glenoid labrum: MR imaging with histologic correlation. Radiology 196:33–41

Massengill AD, Seeger LL, Yao L, Gentili A, Shnier RC, Shapiro MS, Gold RH (1994) Labrocapsular ligamentous complex of the shoulder: normal anatomy, anatomic variation, and pitfalls of MR imaging and MR arthrography. Radiographics 14:1211–1223

McCarty DJ, Halverson PB, Carrera GF, Bruwer BJ, Kozin F (1983) "Milwaukee shoulder" – association of microspheroids containing hydroxyapatite crystals, active collagenase, and neutral protease with rotator cuff defects. I. Clinical aspects. Arthritis Rheum 24:464–473

McGlynn FJ, Caspari RB (1984) Arthroscopic findings in the subluxating shoulder. Clin Orthop 183:173–178

Middleton WD (1993) Sonographic detection and quantification of rotator cuff tears. AJR 160:109–110

Middleton WD, Reinus WR, Totty WG, Melson GL, Murphy WA (1985) US of the biceps tendon apparatus. Radiology 157:211–215

Mink JH, Harris E, Rappaport M (1985) Rotator cuff tears: evaluation using double-contrast shoulder arthrography. Radiology 157:621–623

Mitchell MJ, Causey G, Berthoty DP, Sartoris DJ, Resnick D (1988) Peribursal fat plane of the shoulder: anatomic study and clinical experience. Radiology 168:699–704

Needell SD, Zlatkin MB, Sher JS, Murphy BJ, Uribe JW (1996) MR imaging of the rotator cuff: peritendinous and bone abnormalities in an asymptomatic population. AJR 166:863–867

Neer CS (1993) Impingement lesions. Clin Orthop 173:70–77

Neer CS II, Craig EV, Fukuda H (1983) Cuff-tear arthropathy. J Bone Joint Surg [Am] 65:1232–1244

Neer CS, Satterlee CC, Dalsey RM, Flatow EL (1991) The anatomy and potential effects of contracture of the coracohumeral ligament. Clin Orthop 280:182–185

Neumann CH, Petersen SA, Jahnke AH (1991) MR imaging of the labral-capsular complex: normal variations. AJR 157:1015–1021

Neumann CH, Holt RG, Steinbach LS, Jahnke AH, Petersen SA (1992) MR imaging of the shoulder: appearance of the supraspinatus tendon in symptomatic volunteers. AJR 158:1281–1287

Neviaser RJ, Neviaser TJ, Neviaser JS (1988) Concurrent rupture of the rotator cuff and anterior dislocation of the shoulder in the older patient. J Bone Joint Surg [Am] 70:1308–1311

Neviaser TJ (1980) Old unreduced dislocations of the shoulder. Orthop Clin North Am 11:287–294

Norwood LA, Varrack R, Jacobson KE (1989) Clinical presentation of complete tears of the rotator cuff. J Bone Joint Surg [Am] 71:499–505

Obermann WR (1996) Optimizing joint-imaging: (CT)-arthrography. Eur Radiol 6:275–283

Ogata S, Uhthoff HK (1990) Acromial enthesopathy and rotator cuff tear. A radiologic and histologic postmortem investigation of the coracoacromial arch. Clin Orthop 254:39–48

Palmer WE, Caslowitz PL (1995) Anterior shoulder instability: diagnostic criteria determined from prospective analysis of 121 MR arthrograms. Radiology 197:819–825

Park JG, Lee JK, Phelps CT (1994) Os acromiale associated with rotator cuff impingement: MR imaging of the shoulder. Radiology 193:255–257

Patte D, Goutallier D, Debeyre J (1981) Ruptur der Rotatorenmanschette. Orthopäde 10:206–215

Patten RM (1994) Tears of the anterior portion of the rotator cuff (the subscapularis tendon): MR imaging findings. AJR 162:351–354

Patten RM, Mack LA, Wang KY, Lingel J (1992) Nondisplaced fractures of the greater tuberosity of the humerus: sonographic detection. Radiology 182:201–204

Peh WCG, Farmer THR, Totty WG (1995) Acromial shape: assessment with MR imaging. Radiology 195:501–505

Pennes DR, Jonsson K, Braunstein E, Blasier R, Wojtys E (1989) Computed arthrotomography of the shoulder: comparison of examinations made with internal and external rotation of the humerus. AJR 153:1017–1019

Petersson CJ, Redlund-Johnell I (1984) The subacromial space in normal shoulder radiographs. Acta Orthop Scand 55:57–58

Prodromos CC, Ferry JA, Schiller AL, Zarins B (1990) Histological studies of the glenoid labrum from fetal life to old age. J Bone Joint Surg [Am] 72:1344–1348

Ptasznik R, Hennessy O (1995) Abnormalities of the biceps tendon of the shoulder: sonographic findings. AJR 164:409–414

Quinn SF, Sheley RC, Demlow TA, Szumowski J (1995) Rotator cuff tendon tears: evaluation with fat-suppressed MR imaging with arthroscopic correlation in 100 patients. Radiology 195:497–501

Recht MP, Kramer J, Petersilge CA, et al. (1994) Distribution of normal and abnormal fluid collections in the glenohumeral joint: implications for MR arthrography. J Magn Reson Imaging 4:173–177

Reinus WR, Shady KL, Mirowitz SA, Totty WG (1995) MR diagnosis of rotator cuff tears of the shoulder: value of using T2-weighted fat-saturated images. AJR 164:1451–1455

Resnick D (1995a) Calcium hydroxyapatite crystal deposition disease. In: Resnick D, Niwayama G (eds) Diagnosis of bone and joint disorders, 3rd edn. (CD ROM). Saunders, Philadelphia

Resnick D (1995b) Internal derangements of joints. In: Resnick D, Niwayama G (eds) Diagnosis of bone and joint disorders, 3rd edn. (CD ROM). Saunders, Philadelphia

Resnick D, Niwayama G (1995) Degenerative disease of extraspinal locations. In: Resnick D, Niwayama G (eds) Diagnosis of bone and joint disorders, 3rd edn. (CD ROM). Saunders, Philadelphia

Resnick D, Niwayama G, Goergen TG, Utsinger PD, Shapiro RF, Haselwood DH, Wiesner KB (1977) Clinical, radiographic and pathologic abnormalities in calcium pyrophosphate dihydrate deposition disease (CPPD): pseudogout. Radiology 122:1–15

Richards RD, Sartoris DJ, Pathria MN, Resnick D (1994) Hill-Sachs lesion and normal humeral groove: MR imaging features allowing their differentiation. Radiology 190:665–668

Robertson PL, Schweitzer ME, Mitchell DG, Schlesinger F, Epstein RE, Frieman BG, Fenlin JM (1995) Rotator cuff disorders: interobserver and intraobserver variation in diagnosis with MR imaging. Radiology 194:831–835

Rockwood CA, Wirth MA (1996) Subluxations and dislocations about the glenohumeral joint. In: Rockwood DA, Green DP, Heckman JD, Bucholz RW (eds) Fractures in adults, 4th edn. Lippincott-Raven, Philadelphia, CD ROM

Rockwood CA, Williams GR, Young DC (1996) Injuries to the acromioclavicular joint. In: Rockwood DA, Green DP, Heckman JD, Bucholz RW (eds) Fractures in adults, 4th edn. Lippincott-Raven, Philadelphia, CD ROM

Rothman RJ, Marvel JP, Heppenstall RB (1975) Anatomic considerations in the glenohumeral joint. Orthop Clin North Am 6:341–352

Rowe CR, Zarins B (1981) Recurrent transient subluxation of the shoulder. J Bone Joint Surg [Am] 63:863–872

Schweitzer ME, Magbalon MJ, Frieman BG, Ehrlich S, Epstein RE (1994) Acromioclavicular joint fluid: determination of clinical significance with MR imaging. Radiology 192:205–207

Singson RD, Hoang T, Dan S, Friedman M (1996) MR evaluation of rotator cuff pathology using T2-weighted fast spin-echo technique with and without fat suppression. AJR 166:1061–1065

Smith DK, Chopp TM, Aufdemorte TB, Witkowski EG, Jones RC (1996) Sublabral recess of the superior glenoid labrum: study of cadavers with conventional nonenhanced MR imaging, MR arthrography, anatomic dissection, and limited histologic examination. Radiology 201:251–256

Snyder SJ, Karzel RP, Del Pizzo W, Ferkel RD, Friedman MJ (1990) SLAP lesions of the shoulder. Arthroscopy 6:274–279

Stiles RG, Otte MT (1993) Imaging of the shoulder. Radiology 188:603–613

Timins ME, Erickson SJ, Estkowski LD, Carrera GF, Komorowski RA (1995) Increased signal in the normal supraspinatus tendon on MR imaging: diagnostic pitfall caused by the magic-angle effect. AJR 164:109–114

Tirman PFJ, Stauffer AE, Crues JV, et al. (1993) Saline magnetic resonance arthrography in the evaluation of glenohumeral instability. Arthroscopy 9:550–559

Tirman PFJ, Feller JF, Palmer WE, Carroll KW, Steinbach LS, Cox I (1996) The Buford complex – a variation of normal shoulder anatomy: MR arthrographic imaging features. AJR 166:869–873

Tuckman GA (1994) Abnormalities of the long head of the biceps tendon of the shoulder: MR imaging findings. AJR 163:1183–1188

Tuite MJ, Orwin JF (1996) Anterosuperior labral variants of the shoulder: appearance on gradient-recalled-echo and fast spin-echo MR images. Radiology 199:537–540

Tuite MJ, De Smet AA, Norris MA, Orwin JF (1995) MR diagnosis of labral tears of the shoulder: value of T2*-weighted gradient-recalled echo images made in external rotation. AJR 164:941–944

Vahlensieck M, Pollack M, Lang P, Grampp S, Genant HK (1993) Two segments of the surpaspinous muscle: cause of high signal intensity at MR imaging? Radiology 186:449–454

Vahlensieck M, Peterfy CG, Wischer T, et al. (1996) Indirect MR arthrography: optimization and clinical applications. Radiology 200:249–254

Walch G, Boileau P, Noel E, Donell ST (1992) Impingement of the deep surface of the supraspinatus tendon on the posterosuperior glenoid rim: an arthroscopic study. J Shoulder Elbow Surg 1:238–245

Wiener SN, Seitz WH (1993) Sonography of the shoulder in patients with tears of the rotator cuff: accuracy and value for selecting surgical options. AJR 160:103–107

Zanetti M, Hodler J (1997) Contrast media in MR arthrography of the glenohumeral joint: intraarticular gadopentetate versus saline: preliminary results. Eur Radiol 7:498–502

11 The Hand and Wrist

W.R. Obermann and E.R. Tjin A Ton

CONTENTS

11.1
Introduction

The hand and wrist can be divided into three anatomical regions: the wrist, the midhand, and the fingers. The wrist and midhand are the link between the forearm and the fingers. The midhand allows little motion, whereas the wrist allows a lot of motion.

The wrist allows flexion-extension motion, deviation motion, and circumduction motion. The distal radioulnar joint provides pro- and supination motion, whereas wrist motion can take place in every pro- and supination motion or position, allowing infinite positions of the hand and fingers for optimal function.

W.R. Obermann, MD, PhD, Department of Radiology, University Hospital of Leiden, Albinusdreef 2, 2333 ZA Leiden, The Netherlands
E.R. Tjin A Ton, MD, Department of Radiology, University Hospital of Leiden, Albinusdreef 2, 2333 ZA Leiden, The Netherlands

The hand and wrist consist of tissues such as bone and joints, nerves, vessels, muscle, tendons, and ligaments, which react to disorders in the same manner as in the rest of the body. Furthermore, imaging of these disorders can be done in the same manner as in the other parts of the musculoskeletal system. Nevertheless, some disorders are more or less unique to the hand and wrist or have special features in this location. This chapter will focus on the imaging features of carpal instability and other disorders especially affecting the hand and wrist.

11.2
Imaging Modalities
and Algorithmic Approach

The main imaging modality for hand and wrist pathology is still the plain film, combined with special views like the scaphoid series (Fig. 11.1). In addition to plain films, fluoroscopy plays an important role in the case of joint instability. Arthrography and magnetic resonance imaging (MRI) are also important for joint disorders.

Tomography and computed tomography are important for analyzing fine bony changes and (post)traumatic changes. Bone scintigraphy can be helpful in the search for subtle bone pathology, e.g., osteoid osteoma.

Magnetic resonance imaging provides unique information about soft tissues, e.g., in patients with bone and soft tissue tumors, carpal tunnel syndrome, or tendinous or ligamentous disorders. It also permits assessment of bone vitality. Ultrasound is another modality which is useful in the imaging of tendons, ganglia, neurinoma, and vascular disorders, though angiography is an essential technique in respect of the last-mentioned. Figure 11.1 shows an algorithm based on clinical signs.

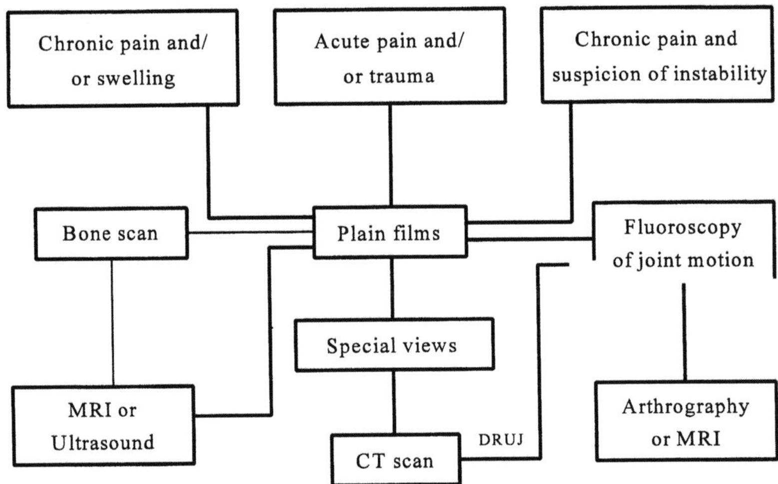

Fig. 11.1. Algorithm

11.3
Carpal Instability

The region between the forearm and midhand, the wrist, consists of a complex of eight differently shaped bones that have a complicated three-dimensional relationship with one another. They are held together by extrinsic and intrinsic ligaments and project differently, depending on the position of the hand.

11.3.1
Standard Views

Evaluation of carpal instability requires properly performed standard views. The posteroanterior (PA) and lateral radiographs of the wrist should be performed in neutral deviation and flexion. However, the natural resting position of the hand is in slight ulnar deviation, and dorsal flexion, and this should be corrected.

With the ulnar side 10–20° elevated, the joint space between the scaphoid and the lunate is better profiled and thus can be evaluated (MONEIM 1981) (Fig. 11.2). The normal carpal bone relationship results in parallel bone surfaces, so-called parallelism, when the joint space is tangentially viewed (GILULA 1979). The joint spaces must be about equal in width.

The carpal region is divided into a proximal carpal row consisting of scaphoid, lunate, triquetrum, and pisiform, and a distal carpal row consisting of trapezium, trapezoid, capitate, and hamate. In the frontal view, the proximal and distal borders of the proximal carpal row and the proximal border of the distal carpal row form smooth arcs (arc I–III) (GILULA 1979) (Fig. 11.2b).

The lateral view should be obtained with the joint spaces viewed tangentially between the lunate and capitate and between the lunate and radius (Fig. 11.3). In this manner the position of the ulnar head with respect to the radius can also be observed. The dorsal side of the hand should be in line with the forearm (OBERMANN 1994). In this projection, which clearly shows the scaphoid, lunate, and capitate as well as their axes, abnormal dorsal or volar flexion of the proximal carpal bones (in cases of carpal instability) can be judged. Because the capitate and metacarpal III are in line with the radius, measurement of the axis of the capitate is not necessary.

By deviating the hand to the ulnar or radial side, the proximal carpal bones will rotate in the dorsal or volar direction, respectively. This mimics a carpal instability pattern if one is not aware of the phenomenon; therefore, it is mandatory for the hand to be in a strictly neutral position.

One also has to keep in mind that after reduction of a Colles fracture the hand is always kept in ulnar deviation while in plaster. Consequently, a dorsal rotation of the lunate will result and should not be judged as abnormal (DISI).

The shape and relationship of the carpal bones will differ between individuals. On standard views it is possible to recognize individual wrists; there is also a striking left-right similarity. An exception may be the length of the ulna and the shape of the ulnar head in the same individual.

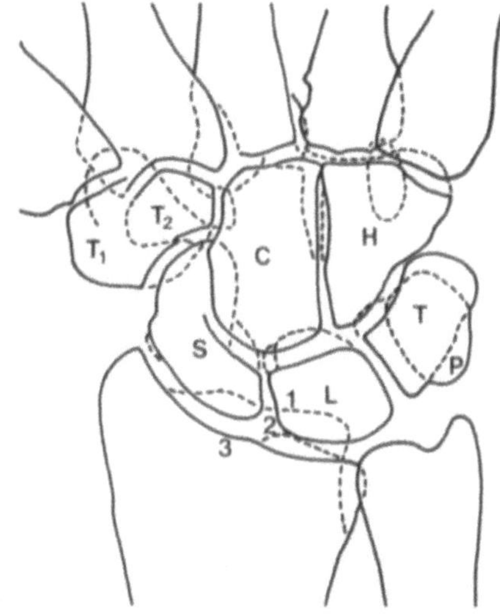

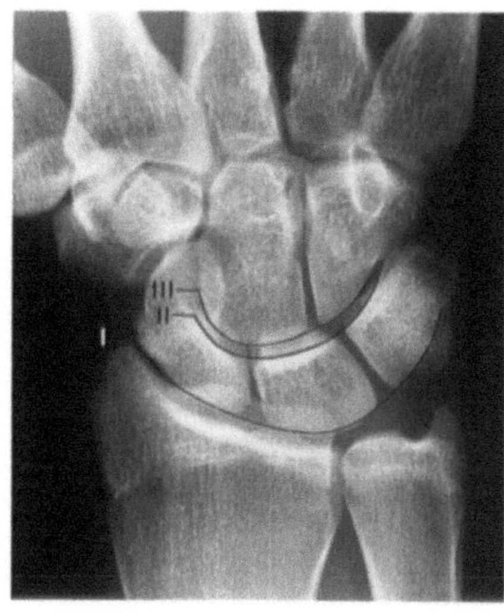

Fig. 11.2. **a** PA radiograph and **b** schematic representation. Projection of bones with the "modified" PA view in which the ulnar side is elevated about 10° for better visualization of the scapholunate joint space. Also drawn are the three carpal arcs according to GILULA (1979). *C*, Capitate; *H*, hamate; *L*, lunate; *P*, pisiform; *S*, scaphoid; *T1*, trapezium; *T2*, trapezoid; *T*, triquetrum. The margin of the dorsal (*1*) and volar (*2*) lip of the distal radius is visible and can be distinguished from the deepest curvature (*3*) of the distal articulating surface of the radius. [Reprinted from OBERMANN WR (1994) Radiology of carpal instability: a practical approach, 1st edn, with kind permission from Elsevier Science – NL, Sara Burgerhartstraat 25, 1055 KV Amsterdam, The Netherlands]

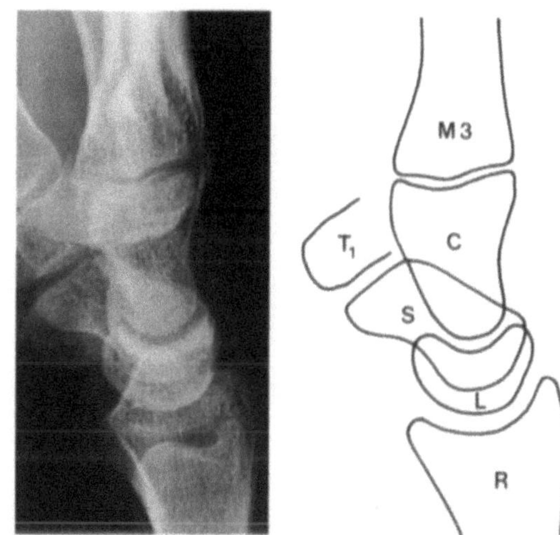

Fig. 11.3. **a** Lateral radiograph and **b** schematic representation. **a** The third metacarpal and the capitate are in line with the radius; between the capitate and the radius is the intercalated lunate. **b** Schematic representation of these carpal bones; they are clearly distinguishable and their axes are therefore easy to determine. *M3*, Metacarpal III; *R*, radius. For other abbreviations, see Fig. 11.2. [Reprinted from OBERMANN WR (1994) Radiology of carpal instability: a practical approach, 1st edn, with kind permission from Elsevier Science – NL, Sara Burgerhartstraat 25, 1055 KV Amsterdam, The Netherlands]

11.3.2
Definitions

Traumatic instability of the wrist is defined as "a carpal injury in which loss of normal alignment of the carpal bones develops early or late" (LINSCHEID et al. 1972; DOBYNS et al. 1975), either at rest [static instability or carpal instability dissociated (CID)] or under the influence of axial loading or certain wrist motions [dynamic instability or carpal instability nondissociated (CIND)] (COONEY et al. 1990). LINSCHEID et al. (1972) recognized four major groups: dorsiflexion instability or dorsiflexed intercalated segment instability (DISI), volar flexion instability or volarflexed intercalated segment instability (VISI), ulnar translocation, and dorsal subluxation.

The proximal carpal row has no muscle attachments and is therefore called the intercalated segment (between the forearm and the distal carpal row). The lunate is the bone whose position determines the volar flexed or dorsiflexed malposition. In CID one of the (intra-articular) interosseous ligaments of the proximal carpal row is completely disrupted, which gives rise to a dissociation between these bones, e.g., in scapholunate dissociation the lunate rotates to dorsal and the scaphoid to volar (Fig. 11.4).

In CIND one or more radiocarpal and/or midcarpal (capsular) ligaments are disrupted or

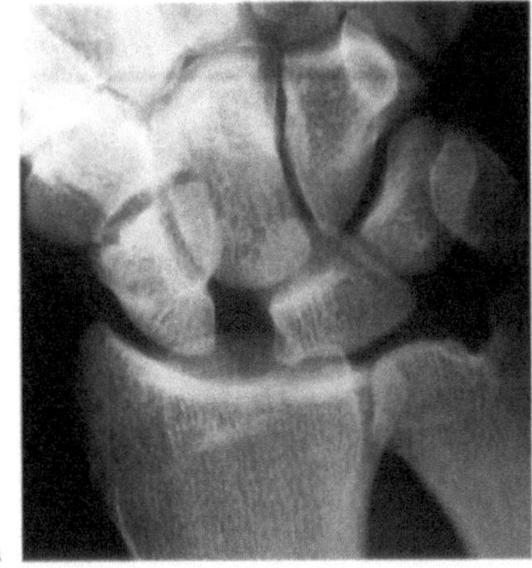 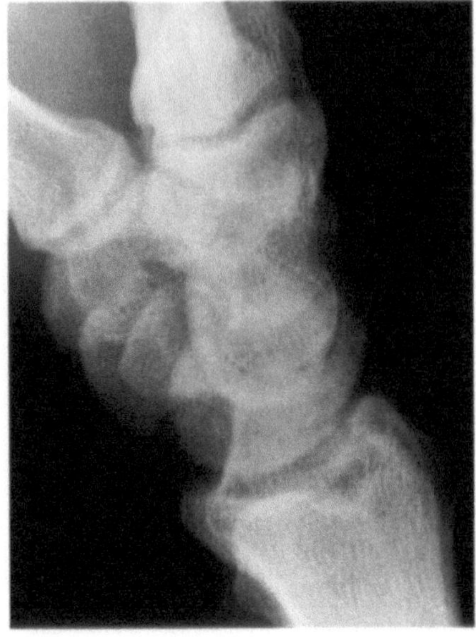

Fig. 11.4 a,b. Scapholunate dissociation as a late result of a trauma. **a** Wide scapholunate joint space. The scaphoid is foreshortened because of increased volar flexion. The lunate shows an elongated trapezium configuration which indicates too much dorsiflexion (DISI). **b** Lateral view: same abnormal rotational position of the scaphoid and lunate. DISI con-figuration (compare with Fig. 11.6). [Reprinted from OBERMANN WR (1994) Radiology of carpal instability: a practical approach, 1st edn, with kind permission from Elsevier Science – NL, Sara Burgerhartstraat 25, 1055 KV Amsterdam, The Netherlands]

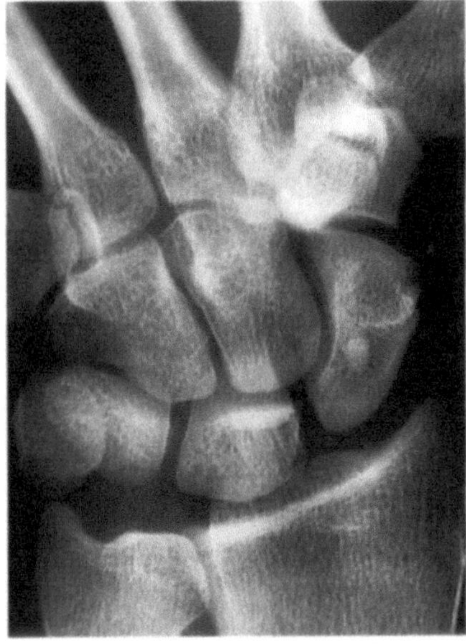 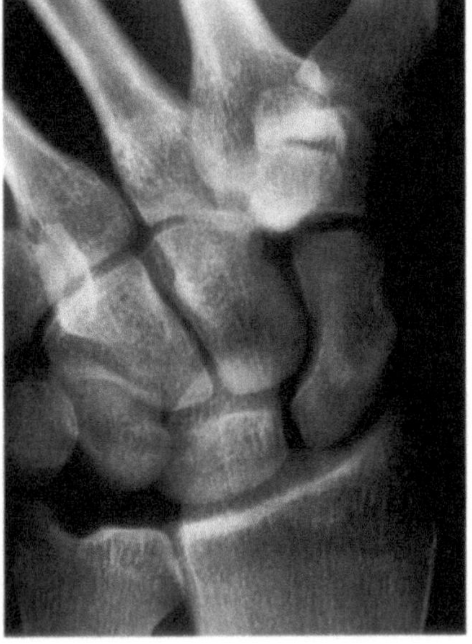

Fig. 11.5 a,b. Posttraumatic wrist with a snap. Dynamic (midcarpal) instability. Radiological diagnosis may only be made by fluoroscopy. **a** In voluntary ulnar deviation the proximal carpal row enters a volar flexed position due to muscle tension (foreshortened scaphoid and an elongated triangular configuration of the lunate). Normally the proximal carpal row should be in a dorsiflexed position in ulnar deviation. There is diastasis between the scaphoid and radius. **b** In ulnar deviation motion, after a loud snap the proximal row sud-denly rotates from an abnormally volar flexed position to the normal dorsiflexed position of ulnar deviation of the hand. This snap is associated with pain and with a dorsal shift of the wrist. Note the change of projection of the scaphoid and lunate. [Reprinted from OBERMANN WR (1996) Wrist injuries: pitfalls in conventional imaging. Eur J Radiol 22:11–21, with kind permission of Elsevier Science – NL, Sara Burgerhartstraat 25, 1055 KV Amsterdam, The Netherlands]

stretched. In CID or static instability the instability is always detectable on standard radiographs provided they are acquired in a proper way. In CIND or dynamic instability there is sometimes an association with a VISI deformity, but often the plain films are normal and the diagnosis can only be made under fluoroscopy (Fig. 11.5).

Dynamic instability usually comprises midcarpal instability because there is often an instability crossing the midcarpal joint, whereby the whole proximal row moves in an abnormal way with respect to the distal carpal row (and also to the forearm) (Fig. 11.5). This abnormal motion has many variations, but the main common feature is that the proximal row as a whole moves abnormally. In dynamic instability there is often also a snap (or clunk) in the wrist. Very rarely there is a combination of dynamic instability with CID in which the plain films are normal and the dissociation is only recognizable on fluoroscopy (Fig. 11.6).

On well-positioned standard radiographs of the wrist a DISI instability pattern is defined by a dorsal angulation of the lunate of more than 15° (LINSCHEID et al. 1972; SEBALD et al. 1974). A DISI pattern usually indicates a scapholunate dissociation in which the interosseous ligament between the scaphoid and the lunate and also the radioscapholunate ligament are completely disrupted, and the scaphoid is detached from the volar radiocapitate ligament (TALEISNIK 1980). As a result the joint space between the scaphoid and lunate widens (Fig. 11.4). The scaphoid rotates to volar by its attachment to the trapezium and trapezoid and the lunate rotates to dorsal by its attachment to the triquetrum. As a result on the lateral view the angle between the scaphoid and lunate becomes greater than 70° (LINSCHEID et al. 1972).

Another cause of a DISI pattern is a pseudarthrosis of the scaphoid in which the lunate (and triquetrum) together with the proximal pole of the scaphoid are loose from the distal part of the scaphoid and hence tend to dorsiflex. In longstanding cases resorption of the fracture-bordering bone at the volar part can take place, giving rise to

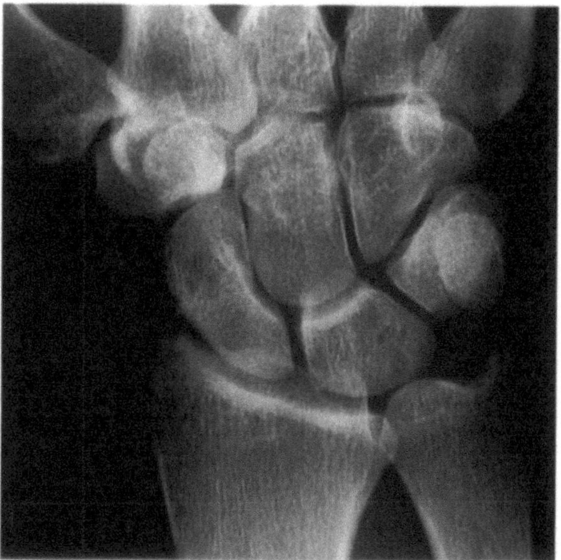

a

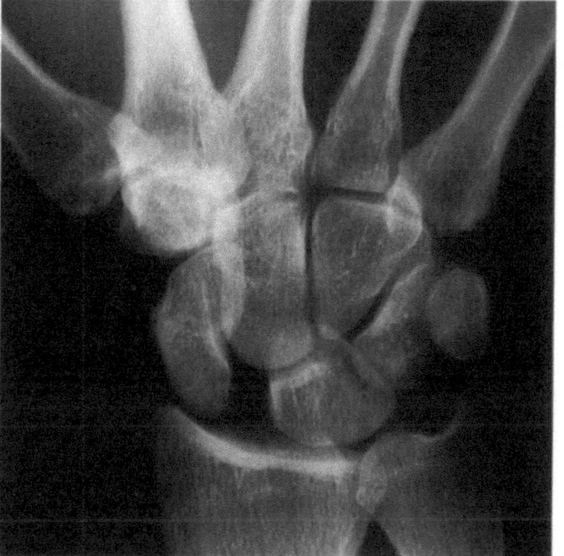

b

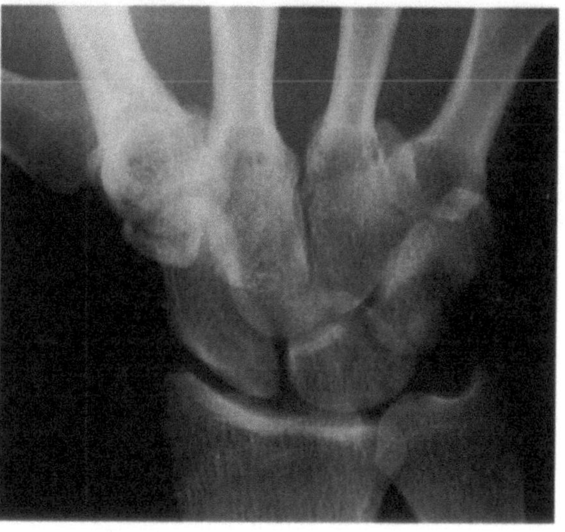

c

Fig. 11.6. a Normal position and projection of carpal bones. **b** By muscle contraction and the motion from slight volar flexion toward dorsiflexion the joint space between the scaphoid and lunate widens. **c** In the motion further to dorsiflexion the scaphoid jumps back to the lunate with a loud snap. This represents dynamic scapholunate dissociation

angulation of the scaphoid and malposition of the proximal part with humpback deformity and resultant restricted dorsal flexion (TALEISNIK 1985) (Fig. 11.7). These malrotation positions of scaphoid nonunion are best analysed by CT scan (Fig. 11.8)

and should be corrected by surgical therapy. Another cause of a DISI instability pattern can be a malunion of a Colles fracture.

A VISI pattern is defined by a volar angulation of the lunate of more than 20° on the standard neutral

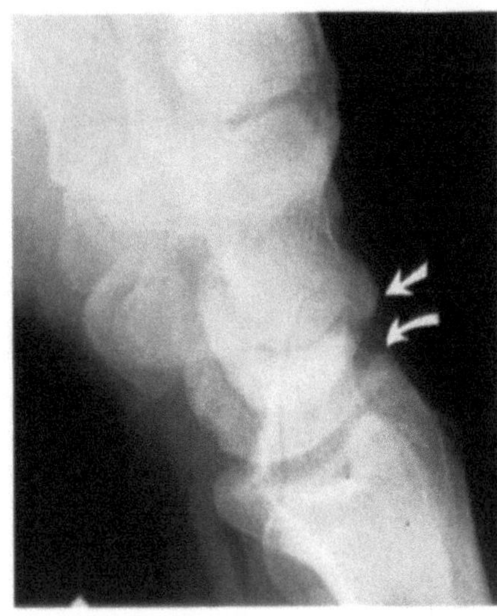

Fig. 11.7. a Lateral radiograph and b schematic representation. Pseudarthrosis of the scaphoid with humpback deformity (*arrow*) caused by malrotation position of the two parts of the scaphoid. The humpback abuts the dorsal lip of the radius (*curved arrow*). This malposition should be redressed before fixation of the scaphoid nonunion. [Reprinted from OBERMANN WR (1994) Radiology of carpal instability: a practical approach, 1st edn, with kind permission from Elsevier Science – NL, Sara Burgerhartstraat 25, 1055 KV Amsterdam, The Netherlands]

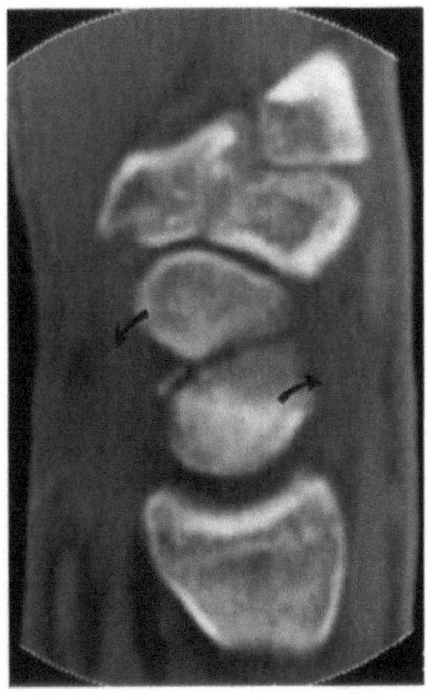

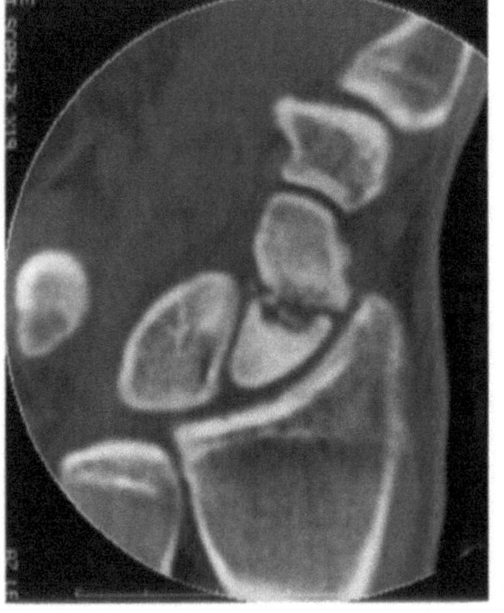

Fig. 11.8. CT scan scaphoid nonunion in the lateral (a) and PA (b) views. Note the malrotation of the proximal and distal parts (*curved arrows*) and the slight dislocation at the scaphoradial joint on the PA view

lateral view (LINSCHEID et al. 1972; SENNWALD 1987). A VISI pattern is often associated with dynamic or midcarpal instability.

11.3.3
Fluoroscopy

In cases of carpal instability in which the diagnosis is not clear on the plain films and in all cases with the clinical sign of a snap or clunk in the wrist, fluoroscopy of wrist motion is mandatory (if necessary also with video registration). Under fluoroscopy abnormal motion of one or more bones can be observed. Radiographs of the end position of these motions can be taken for documentation (Figs. 11.5, 11.6).

Fluoroscopy of wrist motion can best be performed with a universal fluoroscopy unit or a C-arc with the patient prone, the arm along the head, and the investigator sitting at the end of the table.

11.3.4
Arthrography and MRI

In patients in whom the plain films and fluoroscopy of wrist motion cannot explain certain (pain) complaints an arthrogram of one or more compartments of the wrist may be obtained with the patient in the same position as during fluoroscopy. Mixing contrast medium with an anesthetic provides a second diagnostic test. When pain disappears quickly after injection, there is most likely a relationship between the pain and structures in or closely around the joint.

An arthrogram can show (scapholunate and lunotriquetral) interosseous ligament ruptures (Fig. 11.9), disc ruptures, ganglia (Fig. 11.10), adhesive capsulitis, synovial proliferation, free bodies (Fig. 11.11), etc. The aspect of the interosseous ligament ruptures (or sometimes perforations) and disc ruptures should be judged; this is achieved by stressing the ligament by deviation or distraction (Fig. 11.9). The ruptures can be small (perforation) or complete or intermediate.

Except for the pain testing, MRI can provide approximately the same information. Also stressing of the ligaments during MRI improves visualization of the ruptures (TJIN A TON et al. 1995) (Fig. 11.12). The articular disc often has a slitlike perforation on the radial side, but it is unclear whether this is of clinical significance.

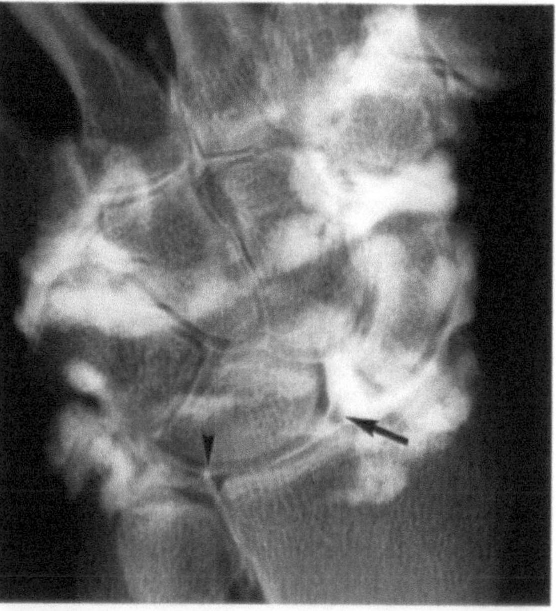
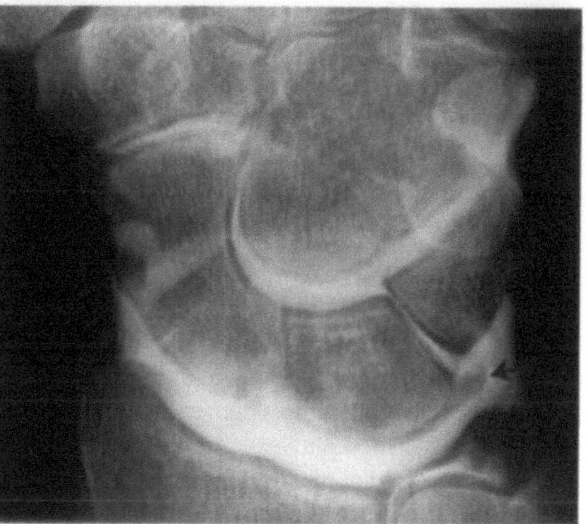

Fig. 11.9 a,b. Wrist arthrograms. a Scapholunate dissociation with separation of the scaphoid from the lunate and a remnant of the interosseous ligament near the lunate (*arrow*). Note also an articular disc perforation (*arrowhead*). b The completely ruptured lunotriquetral interosseous ligament is well seen on this distraction view, with a remnant attached to the lunate (*arrow*)

Magnetic resonance imaging of the wrist requires a good high-resolution image because of the small structures and the complex anatomy. A high field strength magnet (1–1.5 T) and a dedicated wrist coil are necessary to produce thin slices (3 mm or less) with a small field of view (8 cm or less). The coronal plane is best for the evaluation of interosseous ligaments and the triangular fibrocartilage, typically using a proton density and T2-weighted spin-echo (SE), turbo spin-echo (TSE), or gradient-echo se-

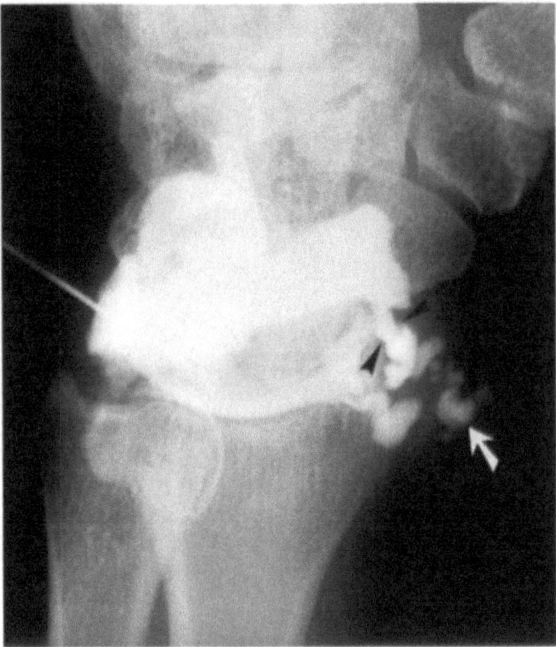

Fig. 11.10. Arthrogram of the radiocarpal joint. A residual ganglion is present at the volar side of the wrist. The ganglion (*arrow*) originates from the radiocarpal joint, in particular from the sulcus interligamentum (*arrowheads*). [Reprinted from OBERMANN WR (1994) Radiology of carpal instability: a practical approach, 1st edn, with kind permission from Elsevier Science – NL, Sara Burgerhartstraat 25, 1055 KV Amsterdam, The Netherlands]

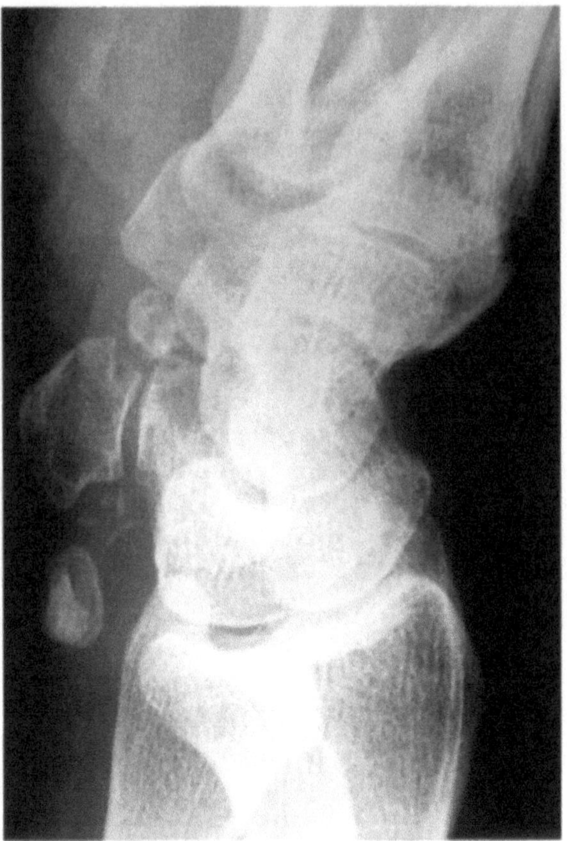

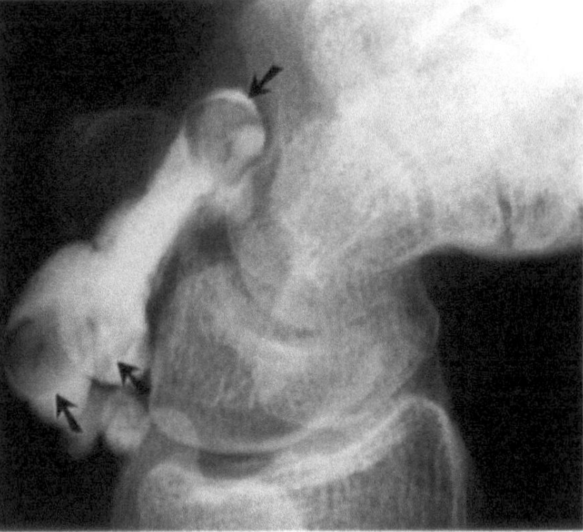

Fig. 11.11 a,b. Plain oblique film and arthrogram of the pisotriquetral joint in a patient with locking symptoms of the wrist. Free bodies are present in the pisotriquetral joint (*arrows*)

quence (GRE). SE and TSE sequences in general are less sensitive but more specific in demonstrating abnormalities. GRE provides a higher spatial resolution and is more sensitive but less specific for abnormalities. Increased signal in GRE sequences can be due to tendinitis, tendinosis, (partial) tear, or an artifact. The carpal tunnel and tendons are best evaluated with an axial T2-weighted TSE sequence with or without fat suppression.

11.3.5
Treatment

Partial tears of the interosseous ligament can be treated by arthroscopic debridement (RUCH and POEHLING 1996) while complete tears can require partial wrist arthrodesis (WATSON and RYU 1986) or soft tissue repair and/or reconstruction (TALEISNIK 1985). Treatment for midcarpal instability can comprise a partial wrist arthrodesis between one of the bones of the distal carpal row and one of the bones of the proximal carpal row, providing bridging of the midcarpal joint in order to achieve stability

(Fig. 11.13). In cases of midcarpal instability and lunotriquetral interosseous ligament rupture a more extensive arthrodesis should be performed (four-quarter arthrodesis: hamate-capitate-lunate-triquetrum).

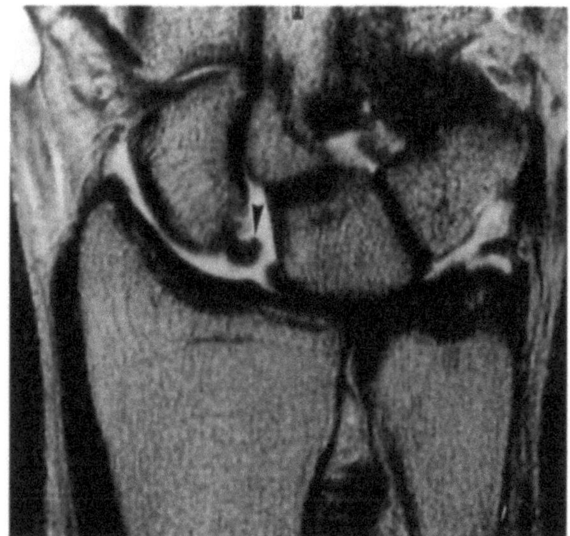

a

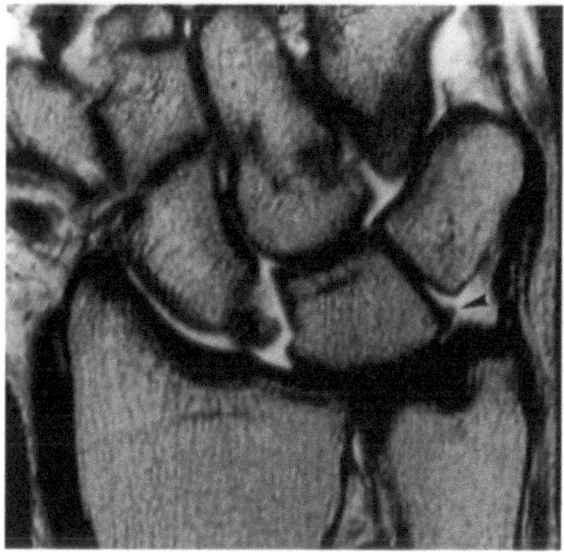

b

Fig. 11.12 a,b. MRI of the wrist with the hand in a stress device. Coronal T2-weighted TSE images. **a** Neutral position. There is clear delineation of the broken scapholunate interosseous ligament still attached to the scaphoid (*arrowhead*). **b** Same wrist in radial deviation. Now a lunotriquetral interosseous ligament rupture surrounded by fluid can also be evaluated. A remnant of ligament is attached to the lunate (*arrowhead*). [Reprinted from TJIN A TON et al. (1995) Interosseous ligaments: device for applying stress in wrist MR imaging. Radiology 196:863–864. RSNA Publications]

Even in cases of carpal coalition instability can develop (Fig. 11.14). Carpal coalitions can be complete (Fig. 11.14) or incomplete (RESNIK et al. 1986), and many variations exist (Fig. 11.15). Incomplete carpal coalitions can be painful (painful synostosis), requiring an arthrodesis at that site.

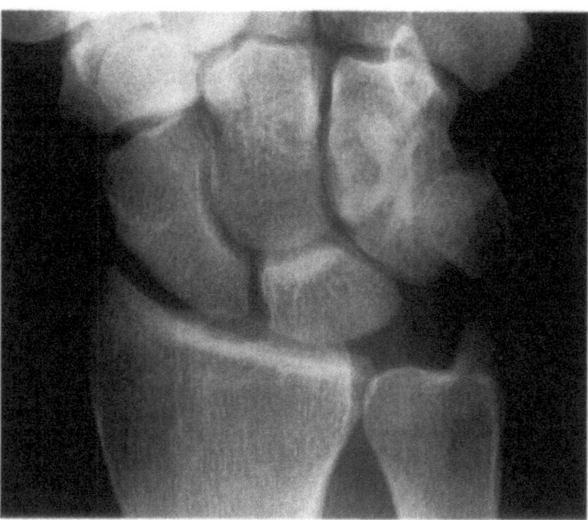

Fig. 11.13. PA view. Triquetrohamate arthrodesis as treatment of midcarpal instability. Note the bony bridge between the triquetrum and hamate

11.3.6
Late Sequelae

A late sequela of CID is the so-called SLAC wrist (scapholunate advanced collapse) (WATSON and RYU 1984) (Fig. 11.16); this is caused by malpositioning of the scaphoid in its elliptical articulation with the radius, giving rise to osteoarthritis, and by imbalance of the midcarpal articulation due to separation of the scaphoid from the lunate. The lunate-radial articulation is spared because this is a spheroidal articulation in which a malrotational position of the lunate does not provoke osteoarthritis.

The same collapse of the wrist can occur in cases of long-standing scaphoid pseudoarthrosis, so-called SNAC (scaphoid nonunion advanced collapse). A SLAC pattern can also occur in the presence of coalitions with scapholunate dissociation (Fig. 11.17) and is frequently seen in calcium pyrophosphate dihydrate crystal deposition disease (CHEN et al. 1990).

11.4
Scaphoid Nonunion

The projection of the scaphoid on a PA view is foreshortened by the normal volar flexed position in neutral deviation. As a consequence, a scaphoid fracture or nonunion can be missed (Fig. 11.18). In ulnar deviation the scaphoid rotates dorsally and elongates to fill the space between the radius and trapezium. In

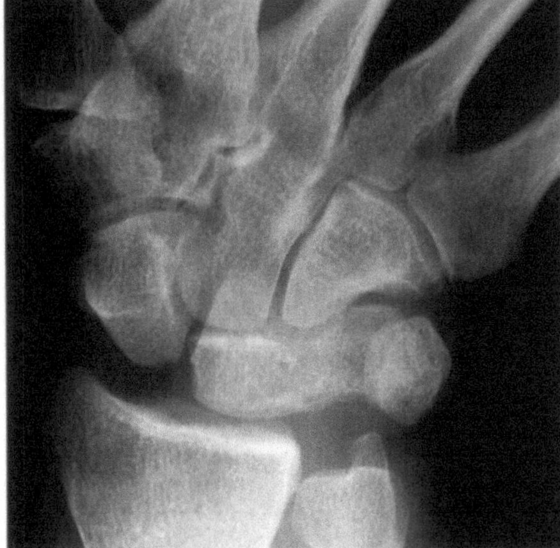

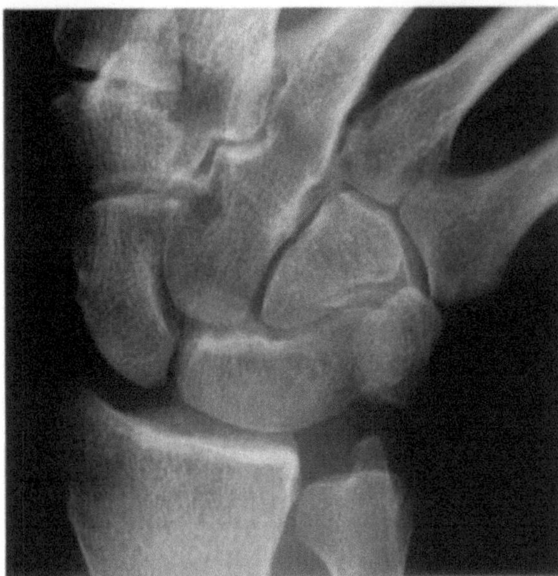

Fig. 11.14 a,b. Midcarpal (dynamic) instability in a wrist with carpal coalitions (lunate–triquetrum, capitate–metacarpal 3, and trapezoid–metacarpal 2). **a** In ulnar deviation the proximal carpal row enters a volar flexed position due to muscle tension. There is diastasis between the scaphoid and radius. **b** In ulnar deviation motion, there is a loud snap as the proximal row suddenly rotates to the normal dorsiflexed position of ulnar deviation of the hand (cf. Fig. 11.5)

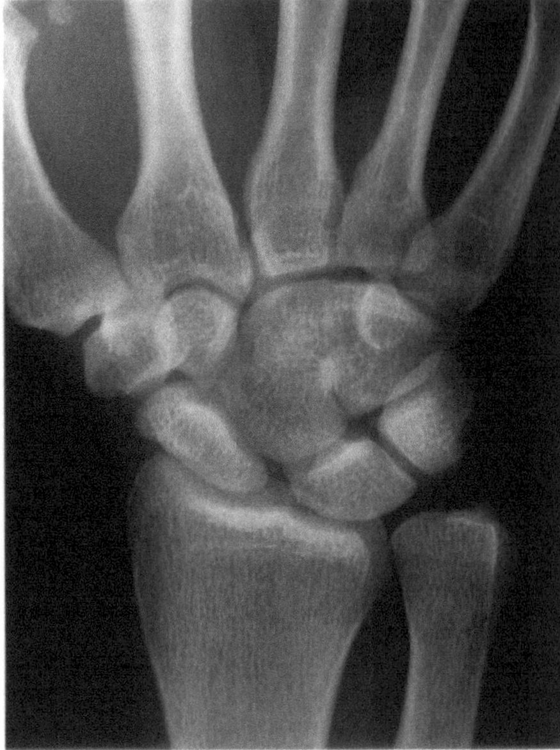

Fig. 11.15. Example of carpal coalition, between the hamate and triquetrum; the scaphoid and lunate are somewhat underdeveloped. The scapholunate joint space is somewhat widened, as is sometimes also seen in other coalitions (METZ et al. 1993)

Table 11.1. Recommended scaphoid series to diagnose fractures and nonunions

Hand in ulnar deviation
1. PA view
2. PA view with radial side elevated 30°
3. PA view with radial side elevated 60°
4. PA view with the thumb under the fingers (children's fist)
5. AP view

Hand in neutral position
6. Lateral view

this position the scaphoid is seen in its whole length and so a nonunion of the proximal pole can become evident (Fig. 11.18).

By obtaining radiographs at different angles with the hand in ulnar deviation the scaphoid can be better analyzed (Table 11.1). CT is best suited for visualization of malposition of the scaphoid nonunion if surgery is to be performed (BIONDETTI et al. 1987; PENNES et al. 1989; BAIN et al. 1995) (Fig. 11.8). Healing of treated scaphoid nonunions also can be assessed by CT. The principle in positioning the scaphoid in the CT scanner is to ensure that the scaphoid with its long axis is in the plane of the scan cuts and at the same time to avoid scanning of the whole forearm and hand, which would adversely affect the image (Fig. 11.8).

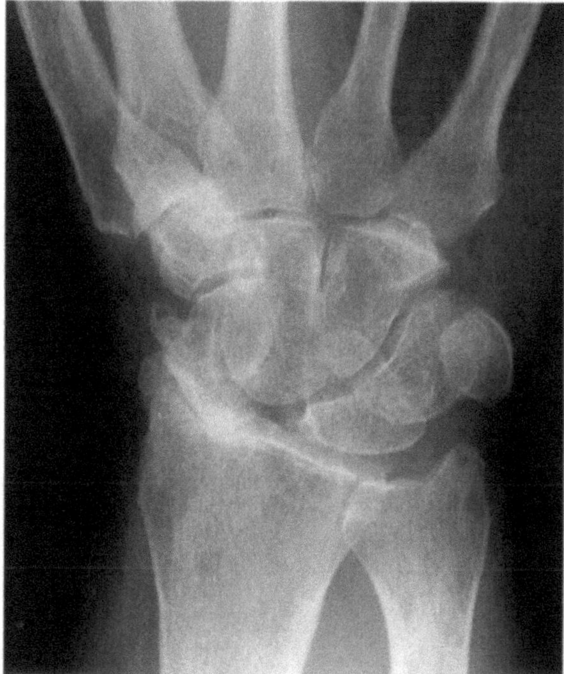

Fig. 11.16. SLAC wrist. Late sequela of scapholunate dissociation with radioscaphoid osteoarthritis and collapse of the carpus, in which the capitate approaches the radius

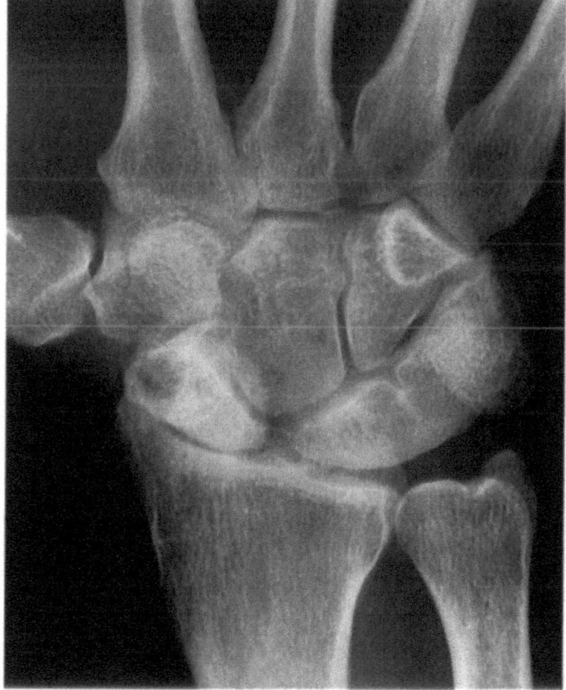

Fig. 11.17. Carpal coalition (lunotriquetrum) with scapholunate dissociation and secondary osteoarthritis between the radius and scaphoid: SLAC pattern

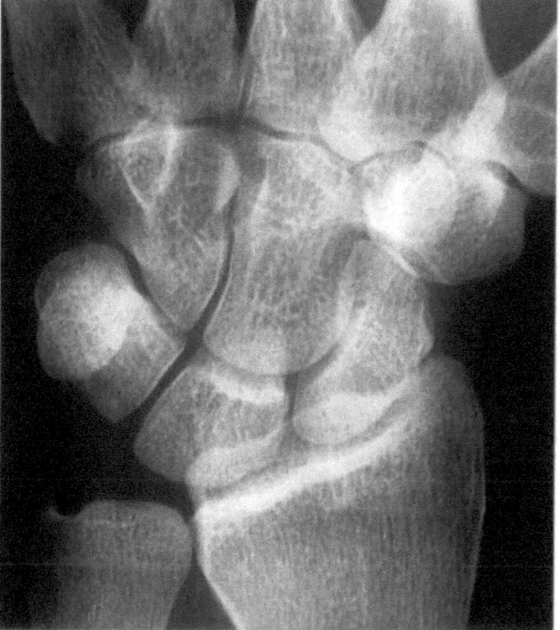

a

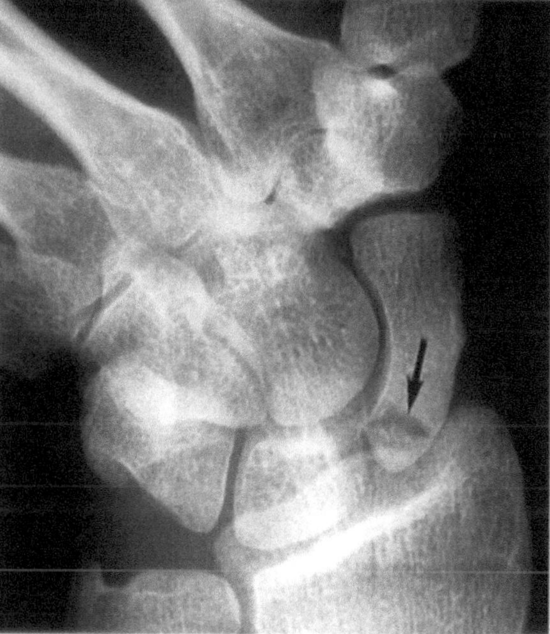

b

Fig. 11.18 a,b. Patient with wrist trauma 12 weeks previously. No abnormality was found on PA and lateral views. **a** Repeated PA view shows no abnormality. The scaphoid is in a normal volar flexed position and therefore foreshortened. **b** In ulnar deviation the old scaphoid fracture becomes obvious (*arrow*). The scaphoid rotates dorsally in ulnar deviation and as a result elongates, better showing a fracture. [Reprinted from OBERMANN WR (1996) Wrist injuries: pitfalls in conventional imaging. Eur J Radiol 22:11–21, with kind permission of Elsevier Science – NL, Sara Burgerhartstraat 25, 1055 KV Amsterdam, The Netherlands]

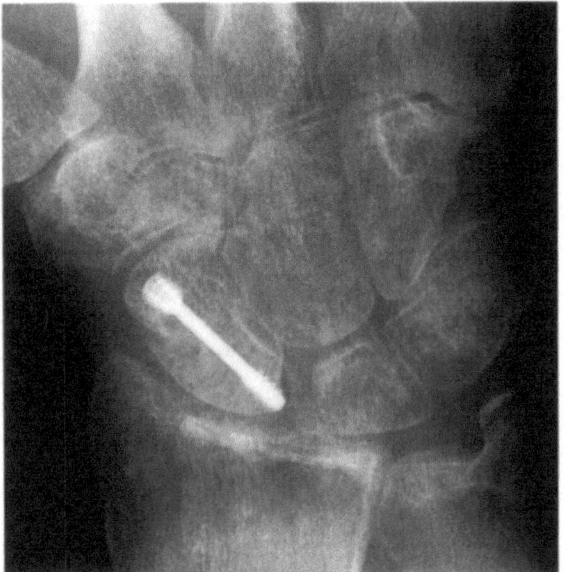

Fig. 11.19. Herbert screw fixation of a scaphoid fracture 6 weeks after immobilization causing severe disuse osteopenia

Treatment of a scaphoid nonunion can be by screw fixation (Fig. 11.19) or bone grafting (Matti-Russe) (Fig. 11.20).

11.5
Distal Radioulnar Joint

Distal radioulnar joint dislocations, subluxation, and locking features are best analyzed by CT scanning without (MINO et al. 1983; WECHSLER et al. 1987) (Fig. 11.21) or with a stress device (PIRELA-CRUZ et al. 1991).

11.6
The Carpal Boss

Another malformation in the carpometacarpal region is a bony extrusion on both sides of the CMC III joint at the dorsal side, sometimes with an ossicle in between, the so-called styloid bone (KOOTSTRA et al. 1974; KAULESAR SUKUL et al. 1986) (Fig. 11.22). This malformation can cause pain complaints either by itself or through tendinitis provoked by rolling over the extensor tendons (KAULESAR SUKUL et al. 1986). Treatment is by removal of the bony extrusions and the styloid bone.

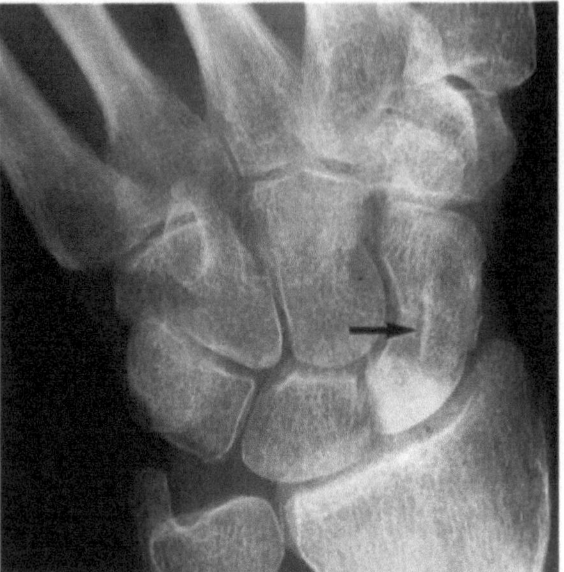

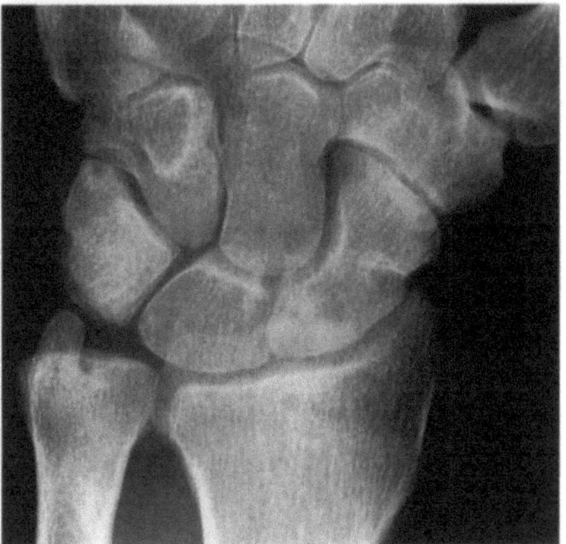

Fig. 11.20 a,b. Matti-Russe inlay grafting treatment of a pseudarthrosis of the scaphoid. **a** Appearance 9 months after treatment. Note the residual inlay graft (*arrow*). There is complete consolidation and the proximal pole is still osteosclerotic. **b** Wrist joint 22 years after surgery

11.7
Soft Tissues

11.7.1
Ganglia

Ganglia can be an extrusion of the joint capsule or tendon sheath, intra-articular, or even intraosseous. In cases of residual ganglia surgeons like to know the

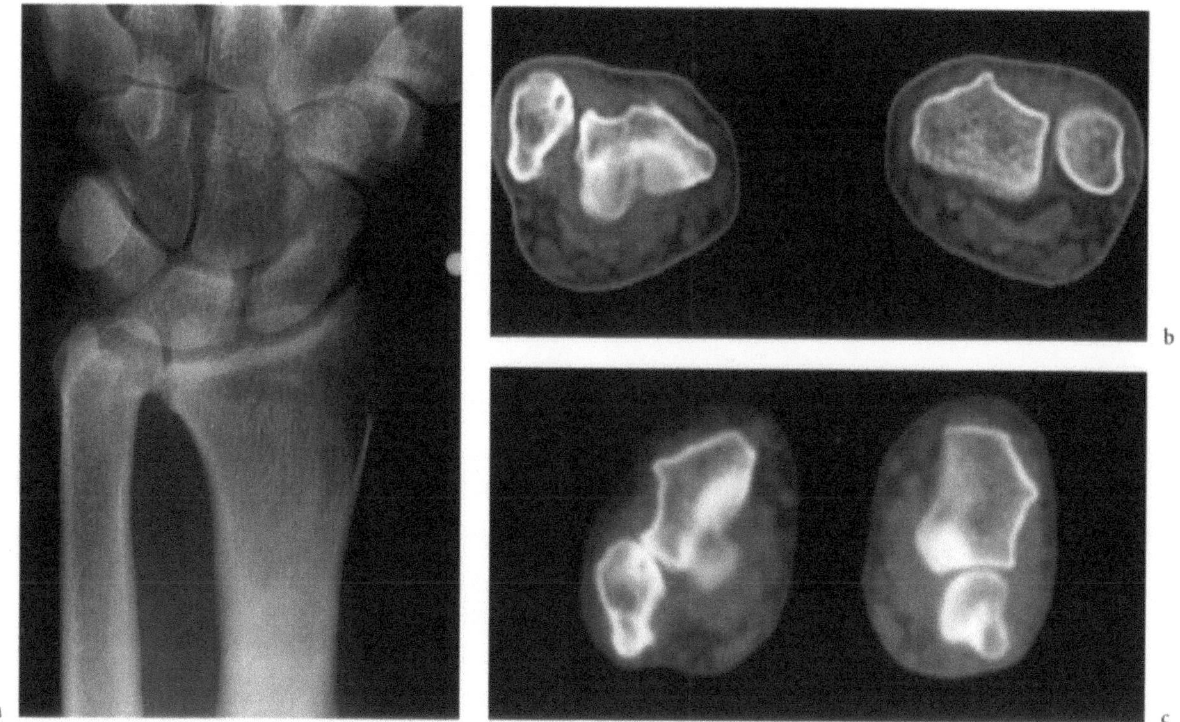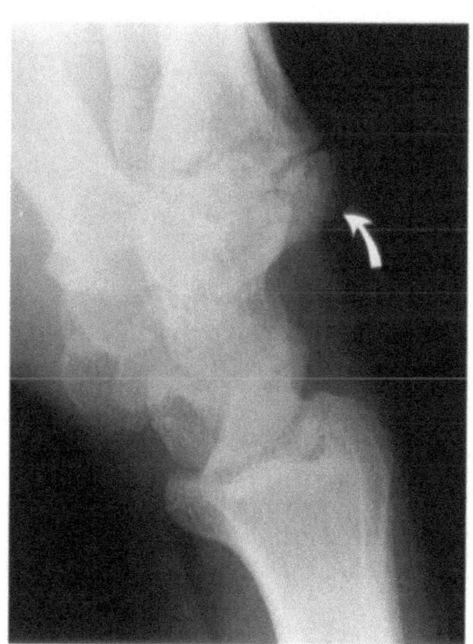

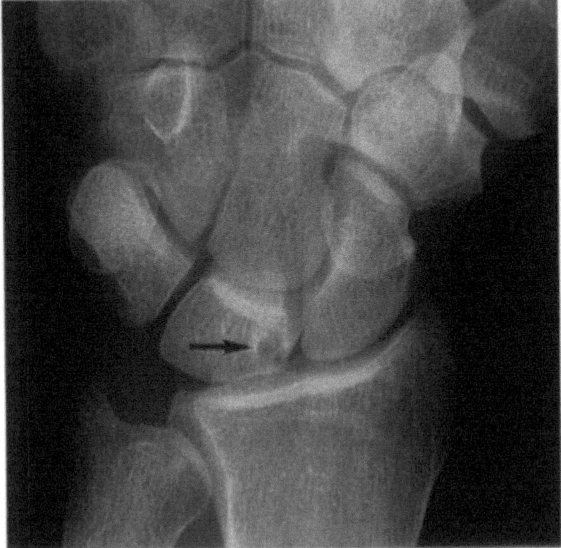

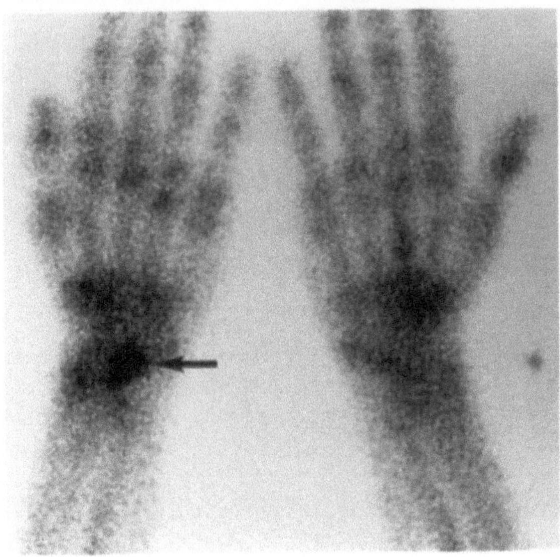

Fig. 11.23 a,b. Symptomatic intraosseous cyst or ganglion. a Plain film with cyst in the lunate (*arrow*). b Bone scan of both hands; imaging from the volar side. There is an obvious hot spot caused by cyst activity (*arrow*)

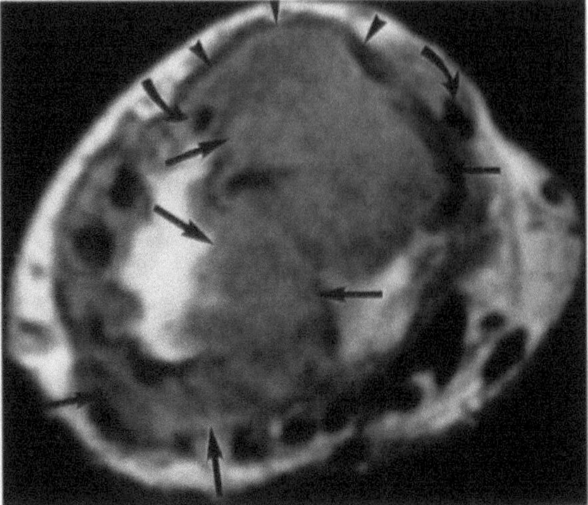

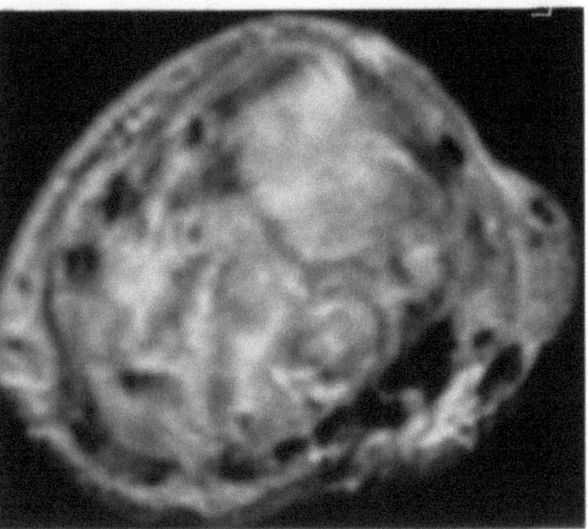

Fig. 11.24 a,b. Transverse MR images at the level of the radiocarpal joint in rheumatoid arthritis. a T1-weighted SE. b T1-weighted SE after administration of gadolinium-DTPA. A huge pannus is present around the joint and tendons (*arrows*). There is enhancement of the pannus after gadolinium administration as a sign of activity. Ruptured (missed) tendons are present in the fourth and fifth extensor tendon compartments (*arrowheads*). At the border of the pannus on the left is the extensor pollicis longus (*left, curved arrow*) and on the right, the extensor carpi ulnaris (*right, curved arrow*)

tendinitis calcaria of the shoulder, wrist soft tissue calcification can be treated by puncture and irrigation (Fig. 11.27).

11.7.3
Carpal Tunnel Syndrome

In carpal tunnel syndrome (CTS) MRI is diagnostic (and also shows the cause) (Fig. 11.28), but one should be aware of the anatomical variations and the variations depending on the position of the hand (flexion–extension) (MIDDLETON et al. 1987; ZEISS et al. 1989; MESGARZADEH et al. 1989a,b). Diffuse swelling of the median nerve at the entrance of the carpal tunnel, flattening of the median nerve at the level of the hamate, volar bowing of the flexor retinaculum, and increased signal in the median nerve on T2-weighted images have been described as changes pathognomonic for CTS. Despite its diagnostic efficacy, use of MRI is in general restricted to cases

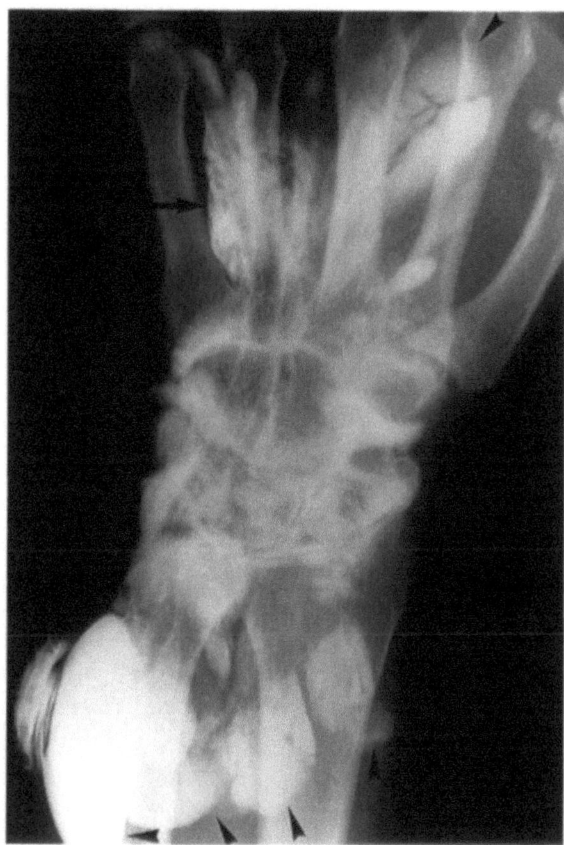

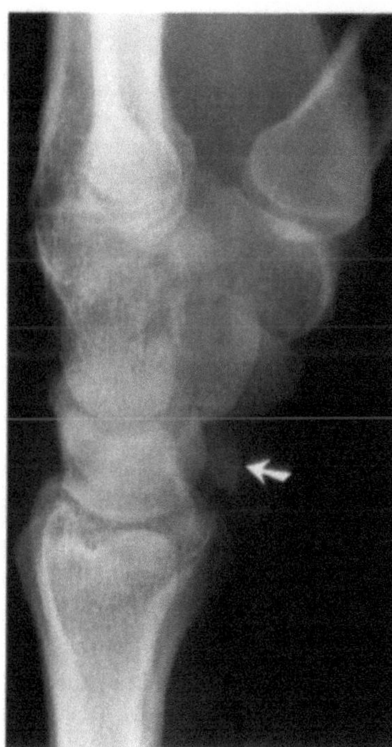

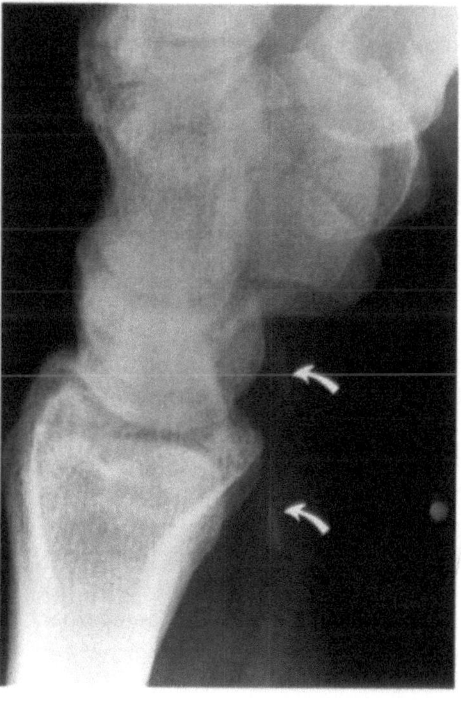

Fig. 11.25. Tendon sheath synovitis. Tenogram extensor tendons. Synovial proliferations (*arrow*) are present in combination with excessive synovial fluid production, which resulted in huge sacciform dilatation of the tendon sheaths (*arrowheads*)

Fig. 11.26 a,b. Pain in the wrist after trauma. **a** Calcification at the volar aspect of the carpal bones (*arrow*). **b** The lateral view 5 days later showing the calcification migrating into and spreading along the flexor tendon sheath of the third finger (the calcium moved when moving the third finger) (*curved arrows*). Diagnosis: acute tendinitis. [Reprinted from OBERMANN WR (1996) Wrist injuries: pitfalls in conventional imaging. Eur J Radiol 22:11–21, with kind permission of Elsevier Science – NL, Sara Burgerhartstraat 25, 1055 KV Amsterdam, The Netherlands]

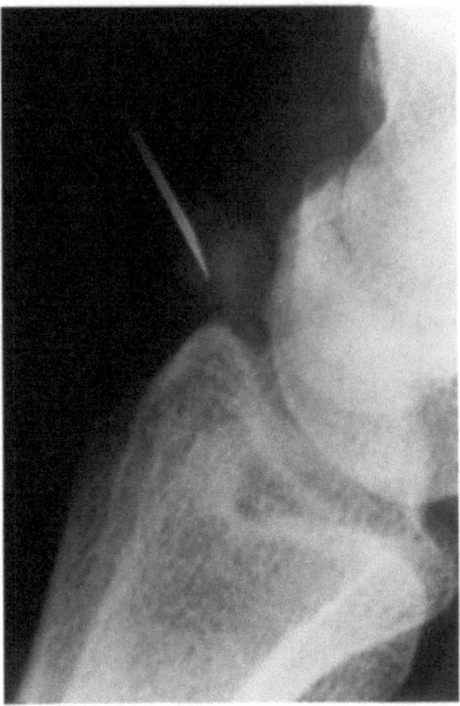

Fig. 11.27. Soft tissue calcification of the dorsal side of the wrist. Treatment was performed with needle puncture and irrigation ("barbotage"). On this image the needle is in the calcification

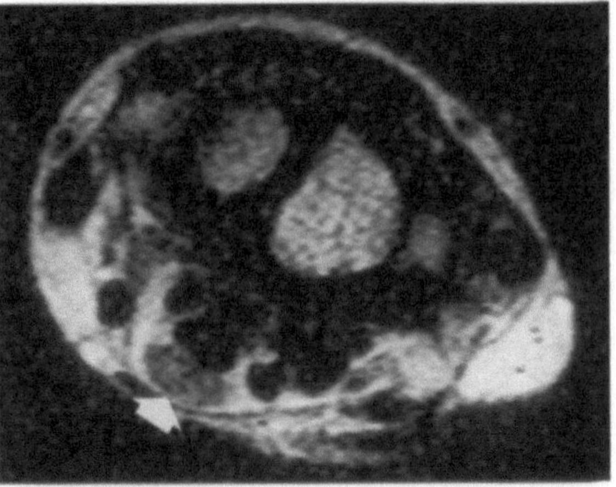

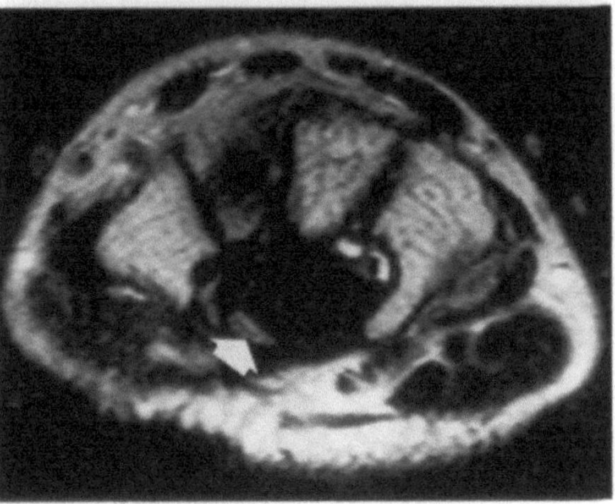

Fig. 11.28 a,b. Carpal tunnel syndrome. Transverse T2-weighted TSE images. **a** At the proximal side of the carpal canal. The thickened median nerve (*arrow*) at this level is characteristic. **b** At the distal side of the canal the median nerve is flattened (*arrow*)

in which electromyography is inconclusive, a space-occupying lesion is suspected, or symptoms persist after surgery.

11.7.4
Ligaments

As well as the interosseous ligaments in the wrist (TJIN A TON et al. 1995; see above) the articular disc (GOLIMBU et al. 1989; TOTTERMAN and MILLER 1995) and the capsular ligaments can be judged (ZLATKIN et al. 1989; SMITH 1993a,b; TOTTERMAN et al. 1993) by MRI, although for the capsular ligaments the findings are difficult to categorize in the nonacute stage.

In the MCP I joint MRI nicely differentiates between nondisplaced and displaced tears of the ulnar collateral ligaments in the acute gamekeeper thumb (SPAETH et al. 1993), which is important for treatment planning.

11.8
Avascular Necrosis

In order to assess the vitality of parts of a bone in avascular necrosis and the revascularization after fixation and/or bone grafting, e.g., in scaphoid nonunion and Kienböck's disease, MRI with gadolinium enhancement can be performed (Fig. 11.29).

The lunate bone and proximal pole of the scaphoid (after a scaphoid fracture) are the common sites of avascular necrosis (AVN). In early AVN bone marrow is hypointense on T1-weighted images, while the signal intensity on T2-weighted images is variable. High signal on T2-weighted images in general indicates viable tissue, which indicates a

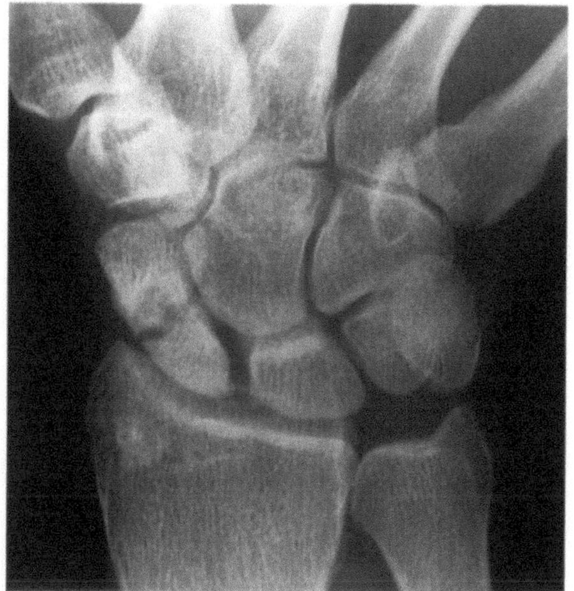

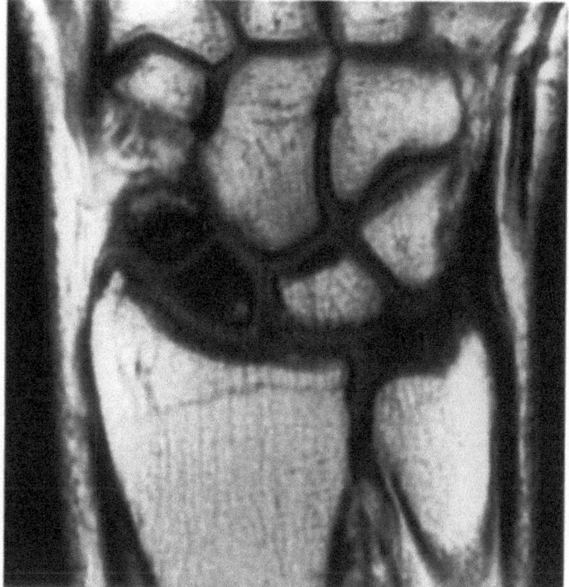

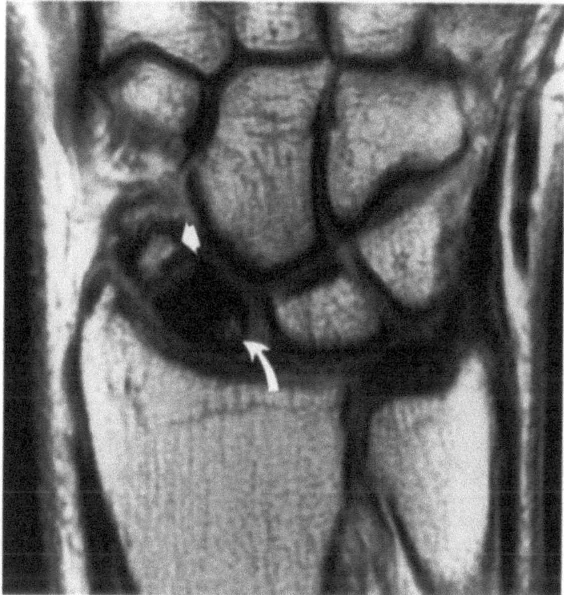

Fig. 11.29 a–c. Pseudarthrosis of the scaphoid. a Plain film. b Coronal T1-weighted SE image. c Coronal T1-weighted SE image after gadolinium-DTPA administration. There is enhancement of the distal pole (*arrow*) and to a lesser degree of part of the proximal pole (*curved arrow*), indicating viable tissue

better prognosis for healing. T1-weighted images after administration of Gd-DTPA might better delineate areas of retained vascularity or revascularization.

11.9
Bone Tumors

Bone tumors are rare in the hand and wrist region. Of the malignant bone tumors 1.5% are in the hand and wrist region (especially in the phalanges and metacarpal bones). Of the benign bone tumors a higher percentage (14%) occur in the hand and wrist region, especially in the phalanges. Of the benign tumors of the carpal bones about 35% are osteoid osteomas or osteoblastomas (MULDER et al. 1993). Plain films and MRI are mandatory for characterization and assessing tumor extension.

Sometimes one encounters a rare painful entity in which radiographs appear normal. A positive bone scan and subsequently thin-slice CT will reveal an *osteoid osteoma* (Fig. 11.30). The treatment of choice

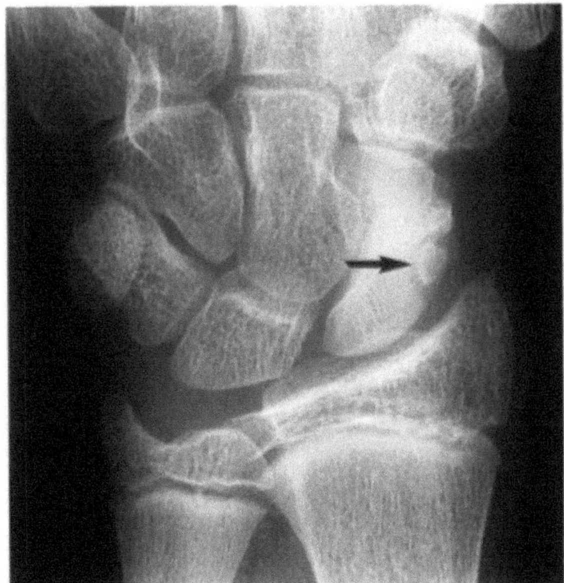

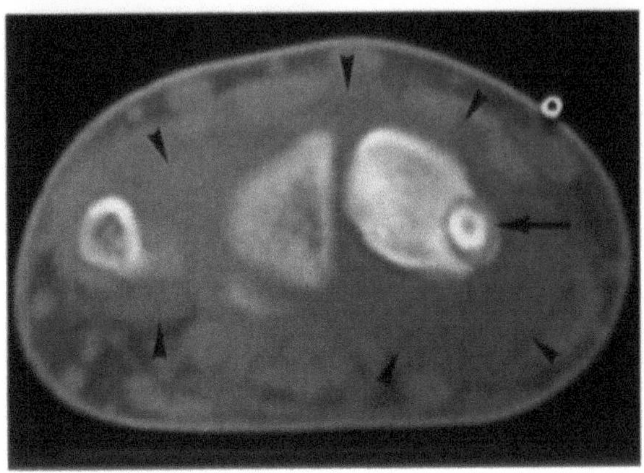

Fig. 11.30 a,b. Osteoid osteoma of the scaphoid (*arrow*). **a** Plain film with marked reactive sclerosis of the scaphoid. **b** CT scan with leadball marking on the skin for percutaneous thermocoagulation. A huge amount of synovial fluid is present (*arrowheads*)

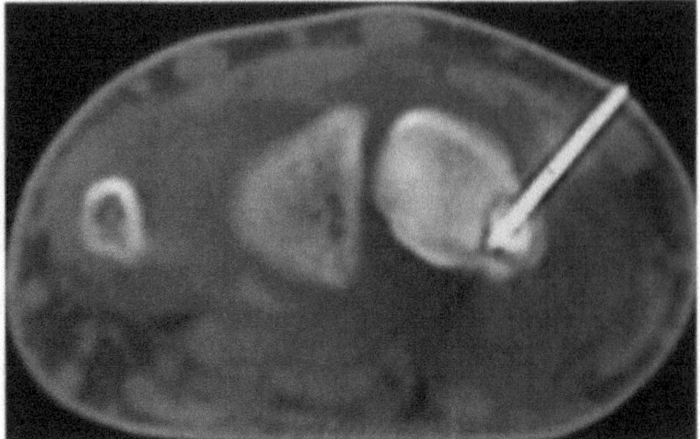

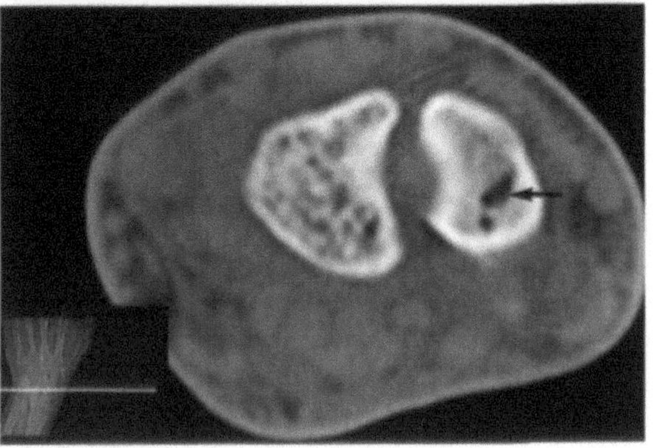

Fig. 11.31 a,b. Same patient as Fig. 11.30. **a** Thermocoagulation needle and probe in osteoid osteoma. **b** Two years after treatment extensive repair with only some residual cancellous bone loss is seen (*arrow*)

in our opinion is thermocoagulation (Rosenthal et al. 1992; De Berg et al. 1995) in which, under CT guidance, a fine drill is entered in the nidus and a probe is inserted through the hollow drill into the centre of the osteoid osteoma. Depending on the size of the probe, maximally a sphere of 1 cm in diameter can be destroyed by 4 min of heating at 90°C (Fig. 11.31).

References

Bain GI, Bennett JD, Richards RS, Slethaug GP, Roth JH (1995) Longitudinal computed tomography of the scaphoid: a new technique. Skeletal Radiol 24:271–273

Binkovitz LA, Berquist TH, McLeod RA (1990) Masses of the hand and wrist: detection and characterization with MR imaging. AJR 154:323–326

Biondetti PR, Vannier MW, Gilula LA, Knapp R (1987) Wrist: coronal and transaxial CT scanning. Radiology 163:149–151

Cardinal E, Buckwalter KA, Braunstein EM, Mih AD (1994) Occult dorsal carpal ganglion: comparison of US and MR imaging. Radiology 193:259–262

Chen C, Chandnani VP, Kang HS, Resnick D, Sartoris DJ, Haller J (1990) Scapholunate advanced collapse: a common wrist abnormality in calcium pyrophosphate dihydrate crystal deposition disease. Radiology 177:459–461

Cooney WP, Dobyns JH, Linscheid RL (1990) Arthroscopy of the wrist: anatomy and classification of carpal instability. Arthroscopy 6:133–140

De Berg JC, Pattynama PMT, Obermann WR, Bode PJ, Vielvoye GJ, Taminiau AHM (1995) Percutaneous computed-tomography-guided thermocoagulation for osteoid osteomas. Lancet 346:350–351

Dobyns JH, Linscheid RL, Chao EYS, Weber ER, Swanson GE (1975) Traumatic instability of the wrist. AAOS Instr Course Lect 11:182–199

Gilula LA (1979) Carpal injuries: analytic approach and case exercises. AJR 133:503–517

Golimbu CN, Firooznia H, Melone CP Jr, Rafii M, Weinreb J, Leber C (1989) Tears of the triangular fibrocartilage of the wrist: MR imaging. Radiology 173:731–733

Hall FM (1995) Intraosseous ganglia of the wrist. Letters to the editor. Radiology 196:546

Kaulesar Sukul DMKS, Steinberg PJ, Lichtveld PLM (1986) The carpal boss. Neth J Surg 383:90–92

Kootstra G, Huffstadt AJC, Kauer JMG (1974) The styloid bone, a clinical and embryological study. The Hand 6:185–189

Linscheid RL, Dobyns JH, Beabout JW, Bryan RS (1972) Traumatic instability of the wrist. J Bone Joint Surg [Am] 54:1612–1632

Magee TH, Rowedder AM, Degnan GG (1995) Intraosseus ganglia of the wrist. Radiology 195:517–520

Mesgarzadeh M, Schneck CD, Bonakdarpour A (1989a) Carpal tunnel: MR imaging. I. Normal anatomy. Radiology 171:743–748

Mesgarzadeh M, Schneck CD, Bonakdarpour A, Mitra A, Conaway D (1989b) Carpal tunnel: MR imaging. II. Carpal tunnel syndrome. Radiology 171:749–754

Metz VM, Schimmerl SM, Gilula LA, Viegas SF, Saffar P (1993) Wide scapholunate joint space in lunotriquetral coalition: a normal variant? Radiology 188:557–559

Middleton WD, Kneeland JB, Kellman GM, et al. (1987) MR imaging of the carpal tunnel: normal anatomy and preliminary findings in the carpal tunnel syndrome. AJR 148:307–316

Mino DE, Palmer AK, Levinsohn EM (1983) The role of radiography and computerized tomography in the diagnosis of subluxation and dislocation of the distal radioulnar joint. J Hand Surg [Br] 8:23

Moneim MS (1981) The tangential posteroanterior radiograph to demonstrate scapholunate dissociation. J Bone Joint Surg [Am] 63:1324–1326

Mulder JD, Schütte HE, Kroon HM, Taconis WK (1993) Radiologic atlas of bone tumors, 2nd edn. Elsevier, Amsterdam, pp 9–27

Obermann WR (1994) Radiology of carpal instability: a practical approach, 1st edn. Elsevier, Amsterdam

Obermann WR (1996) Wrist injuries: pitfalls in conventional imaging. Eur J Radiol 22:11–21

Pennes DR, Jonsson K, Buckwalter KA (1989) Direct coronal CT of the scaphoid bone. Radiology 171:870–871

Pirela-Cruz MA, Goll SR, Klug M, Windler D (1991) Stress computed tomography analysis of the distal radioulnar joint: a diagnostic tool for determining translational motion. J Hand Surg [Am] 16:75–82

Resnik CS, Grizzard JD, Simmons BP, Yaghmai I (1986) Incomplete carpal coalition. AJR 147:301–304

Rosenthal DI, Alexander A, Rosenberg AE, Springfield D (1992) Ablation of osteoid osteomas with a percutaneously placed electrode: a new procedure. Radiology 183:29–33

Ruch DS, Poehling GG (1996) Arthroscopic management of partial scapholunate and lunotriquetral injuries of the wrist. J Hand Surg [Am] 21:412–417

Sebald JR, Dobyns JH, Linscheid RL (1974) The natural history of collapse deformities of the wrist. Clin Orthop 104:140–148

Sennwald G (1987) The wrist, 1st edn. Springer, Berlin Heidelberg New York, pp 47–63

Smith DK (1993a) Dorsal carpal ligaments of the wrist: normal appearance on multiplanar reconstructions of three-dimensional Fourier transform MR imaging. AJR 161:119–125

Smith DK (1993b) Volar carpal ligaments of the wrist: normal appearance on multiplanar reconstructions of three-dimensional Fourier transform MR imaging. AJR 161:353–357

Spaeth HJ, Abrams RA, Bock GW, et al. (1993) Gamekeeper thumb: differentiation of nondisplaced and displaced tears of the ulnar collateral ligament with MR imaging (work in progress). Radiology 188:553–556

Taleisnik J (1980) Posttraumatic carpal instability. Clin Orthop 149:73–82

Taleisnik J (1985) The wrist, 1st edn. Churchill Livingstone, New York, chapter 6

Tjin A Ton ER, Pattynama PMT, Bloem JL, Obermann WR (1995) Interosseous ligaments: device for applying stress in wrist MR imaging. Radiology 196:863–864

Totterman SMS, Miller RJ (1995) Triangular fibrocartilage complex: normal appearance on coronal three-dimensional gradient-recalled-echo MR images. Radiology 195:521–527

Totterman SMS, Miller R, Wasserman B, Blebea JS, Rubens DJ (1993) Intrinsic and extrinsic carpal ligaments: evaluation by three-dimensional Fourier transform MR imaging. AJR 160:117–123

Watson HK, Ryu J (1984) Degenerative disorders of the carpus. Orthop Clin North Am 15:337–353

Watson HK, Ryu J (1986) Limited triscaphoid intercarpal arthrodesis for rotatory subluxation of the scaphoid. J Bone Joint Surg [Am] 68:345–349

Wechsler RJ, Wehbe MA, Rifkin MD, Edeiken J, Branch HM (1987) Computed tomography diagnosis of distal radioulnar subluxation. Skeletal Radiol 16:1–5

Zeiss J, Skie M, Ebraheim N, Jackson WT (1989) Anatomic relations between the median nerve and flexor tendons in the carpal tunnel: MR evaluation in normal volunteers. AJR 153:533–536

Zlatkin MB, Chao PC, Osterman AL, Schnall MD, Dalinka MK, Kressel HY (1989) Chronic wrist pain: evaluation with high-resolution MR imaging. Radiology 173:723–729

12 The Hip

E.G. McNally

CONTENTS

12.1
Introduction

In this chapter we will explore the more common hip problems at various ages from neonate to old age. Although this division is arbitrary, it does allow a description of the most common hip disorders and provides a focused approach to the investigation of an individual patient.

12.2
Anatomy

The anatomy of the hip joint is not particularly complex. The acetabulum is normally anteverted approximately 17° in the adult. This represents a gradual progression from 0° at birth. The acetabulum is deepened by a fibrous structure, termed the labrum, which is deficient inferiorly. The gap is filled by the inter transverse ligament, which gives origin to the ligamentum teres, the latter inserts into the femoral head at the fovea. The head itself is well contained within the acetabulum with little lateral uncovering. Femoral anteversion averages 13°, but is larger by approximately 6° in patients with osteoarthritis (Reikeras et al. 1983). The entire joint is invested by synovium and covered by a fibrous capsule. Anteriorly, the joint capsule extends to the inter trochanteric line, but only two-thirds of this distance on the posterior aspect. The capsule is reinforced by three ligaments: the iliofemoral, the ischiofemoral and the pubofemoral.

12.3
Techniques

All current radiological techniques can be applied to the hip. The relative importance of each in the diagnosis of pathological conditions will be discussed in the text.

12.3.1
Plain Radiography

Standard projections include AP and lateral images. The latter can be obtained either as a shoot through

E.G. McNally, MD, Consultant Musculoskeletal Radiologist, Nuffield Orthopaedic Centre, Headington, Oxford OX3 7LD, UK

following elevation of the contralateral limb or as a frogleg view, which is particularly useful in children. Gonadal protection should not be used on the first film except in males where it can be certain that lead protection does not obscure bone detail. Lead protection should always be used on subsequent films if symptoms do not change.

The acetabulum comprises two columns and two walls. The anterior column is defined by the iliopectineal line and the posterior by the ilio-ischial line. The latter forms part of the teardrop. The anterior and posterior wall lines can be identified on good quality radiographs, with the posterior wall line the more lateral of the two. Oblique radiographs (Judet views) obtained at 45° demonstrate these features to better effect, and are particularly important following trauma where fractures of the acetabulum are suspected. The illiac oblique shows a flattened ilium and depicts the posterior column and anterior wall. The obturator oblique shows the obturator foramen, the anterior column and the posterior wall.

Plain x-rays of the femoral neck demonstrate two major trabecular bundles, separated by an apparent radiolucent triangle called Ward's triangle. The area of thick cortex along the inferomedial margin of the femoral neck is called the calcar.

A number of common variants of normal anatomy are recognised. Bone islands are well-defined areas of bone sclerosis that predominantly occur in the femoral neck. They are well demarcated from adjacent normal bone and show no change on successive radiographs. On magnetic resonance imaging (MRI) they are seen as low signal lesions on all sequences. Cartilage rests are also well demarcated on plain radiographs, but are lytic rather than sclerotic. Synovial pits occur on the surface of the femoral neck and represent ingrowths of the synovium. Well-defined lucencies are characteristic. They have a typical target appearance on MRI, containing both synovium and fat. They are degenerative in aetiology.

12.3.2
Arthrography

Hip arthrography can be carried out by a number of approaches. Direct anterior puncture over the femoral neck is straightforward following careful palpation of the femoral vessels. Anterior osteophyte formation can, however, obstruct the needle and bony contact can be erroneously interpreted as having entered the joint. Failure to screen laterally will

result in an extra-articular injection. It is also more difficult to manipulate the needle within the joint via this anterior approach. An oblique anterior approach with the skin puncture site overlying the mid point of the intertrochanteric line and the needle tip directed towards the femoral head avoids some of these problems. An oblique anterior approach also allows the needle tip to be redirected towards the medial aspect of the neck – head junction, which is a useful location to aspirate synovial fluid. A direct lateral approach, over the greater trochanter, while screening in the AP plane is also useful. The procedure can be extended to include synovial biopsy by passing a Tru-cut needle parallel to the arthrogram needle. An approach along the lateral border of the femoral neck near the neck – head junction is favoured.

Following cannulation of the joint, local anaesthetic can be instilled. This is useful to confirm the hip as a source of the patient's symptoms. A dose of 2.5 ml of 0.5% bupivicaine is recommended. Gadolinium-DTPA can also be injected for MR arthrography. A dose of 0.2 ml is all that is required.

12.3.3
Ultrasound

The technique of hip ultrasound will be described in more detail in subsequent sections. As with most musculoskeletal applications, linear array probes of maximum frequency should be used. In most cases, a minimum of 7.5 MHz is recommended, though in obese adult patients 5 MHz will often be required to view the joint.

12.3.4
Magnetic Resonance Imaging

Magnetic resonance imaging offers several advantages over computed tomography (CT), particularly with regard to its multiplanar capability. Each examination will be tailored to the individual patient, but coronal T1-weighted and T2-weighted images followed by axial fat-suppressed images provide a useful screen. Coronal T2* images are necessary to examine the acetabular labrum. The conspicuity of this structure can be improved using intra-articular gadolinium. Image resolution will be considerably improved if attention is directed to the symptomatic hip only using dedicated surface coils. The exception is avascular necrosis where images of both hips should be obtained with coronal T1-weighted images, to detect asymptomatic contralateral disease.

12.4
Hip Disorders in the Neonate

12.4.1
Developmental Dysplasia of the Hip

Hip abnormalities are common at birth in the United Kingdom and are present in 15–20/1000 live births. Only a small proportion of these will show persistent dislocation (1–2/1000), with a similar number showing signs of dysplasia (standing Medical Advisory Committee, DHSS 1986). The world-wide incidence is not known, but there are well-recognised ethnic differences, with a higher incidence in several American Indian tribes and in regions of Japan and Scandinavia. Conversely, it is almost unknown in Africa. Developmental dysplasia of the hip (DDH) is more common in breech presentations, female infants and first-born infants. A family history is common, implying a hereditary component, and there is a higher incidence in infants with congenital foot and neck disorders. The left hip is more commonly affected than the right, possibly reflecting intra-uterine positioning.

The neonatal hip is assessed using ultrasound with plain radiography having little, if any, role. Although a variety of measurements are described in relationship to the plain film, they are all dependent on standard radiographic positioning, which is difficult in infants. Observer errors of up to 6% occur in the assessment of the acetabular index, a measurement which is considered one of the most useful. Competent ultrasound examination should detect all abnormalities that are clinically apparent while clinical examination alone will miss all infants with stable dysplasia.

A coronal image is most commonly employed to assess the neonatal hip using ultrasound. High-resolution probes (>7.5 MHz) accurately depict the bony, fibrous, fibrocartilage and hyaline cartilage landmarks (Fig. 12.1).A superficial layer of articular cartilage is separated from the bony acetabulum by a band of hyaline cartilage. Augmenting acetabular depth is a peripheral rim of fibrocartilage called the cartilaginous labrum. As hyaline and articular cartilage contain 80% water they appear of low reflectivity, as opposed to the reflective fibrocartilage of the labrum. The inferior part of the labrum is deficient, completed by the transverse acetabular

ligament. The ligamentum teres arises from this ligament and insets into the fovea centralis of the femoral head. This intra-articular structure is readily seen in neonates with high-resolution ultrasound as a broad reflective band within the joint. The tri-radiate cartilage is identified as a low reflective structure between the reflective leading echoes of the ilium, ischium and pubis. The cartilaginous femoral head is spherical. The entire joint is contained within the capsular-ligamentous complex, which is reflective.

Assessment of acetabular depth is limited by progressive ossification of the femoral head. The ossification centre normally appears between the 2nd and 8th month, typically earlier in girls. The earliest change on ultrasound is due to a confluence of tiny vessels within the head and is seen as an area of increased reflectivity. This is followed by calcification and finally ossification, which progresses to the periphery of the femoral head. The appearance and development of the ossification centre are often de-

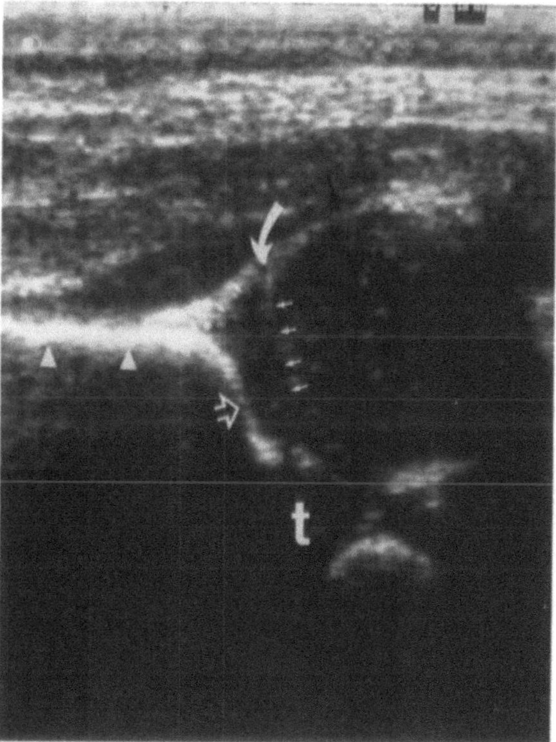

Fig. 12.1. Normal hip ultrasound. The image is obtained in the coronal plane. Note the reflective surface of the ilium (*arrowheads*), the bony roof line (*open arrow*), the unossified roof cartilage (*small white arrows*) and reflective labrum (*curved arrow*) blending with joint capsule (*black arrow*). The tri-radiate cartilage (*t*) separates the bony acetabular roof from the ischium

layed in DDH, although some asymmetry in size and position also occurs in the normal population.

12.4.1.1
Technique

Complete assessment involves a static examination of joint morphology, followed by a dynamic assessment of the relationship between the femoral head and the acetabulum during stress. A variety of positions have been described. For screening a large number of infants, the author uses a single coronal view obtained with the infant in the lateral decubitus position with the hip and knee slightly flexed. In this position the normal anatomical structures are easily identified, and the relationship of the femoral head to the acetabulum reliably assessed. The femoral head has been likened to a ball resting on the spoon of the acetabulum, with the handle of the spoon represented by the reflective lateral border of the ilium. The transducer position should be adjusted so that the ilium forms a straight line when viewed horizontally. This position must be achieved for accurate assessment of acetabular depth. If the plane of scanning is too far anterior or posterior, acetabular depth may be underestimated.

An alternative approach, popular in the United States (HARCKE and GRISSOM 1990), is to evaluate the hip in a variety of additional positions, beginning with a transverse view with the hip in neutral, moving to transverse, hip flexed, in both abduction and adduction and ending with a coronal view with the hip in flexion. On the axial projection, approximately equal portions of the femoral head lie anterior and posterior to the tri-radiate cartilage; however, displacement of 2–3 mm in an otherwise normal hip is acceptable. Using multiple planes may give a better assessment of stability during movement, and allow abnormalities found in one position to be confirmed in another. The examination is prolonged, however, which is important if a screening program is being considered. A prolonged examination may also result in the child becoming agitated, thus obscuring subluxation on the stress test due to increased muscle tension. SCHULER et al. (1990) found that the addition of transverse scanning as advocated by Harcke did not provide any additional diagnostic information.

Graf recommends a coronal scan plane only. He describes four morphological types (I–IV) based on the relationship of the bony roof line and cartilage roof line to a standard baseline. The baseline is represented by a line drawn along the horizontal lateral iliac margin. The bony roof line is drawn to include the angle between the ilium and acetabular roof and the superolateral margin of the tri-radiate cartilage, and the cartilage roof line is drawn to include the angle formed by the ilium and bony acetabular roof and the tip of the labrum. The angle subtended by the baseline and the bony roof line is designated the *alpha* or *bony angle*, while the angle between the baseline and the cartilage roof line is called the *beta* or *cartilage angle*. The normal hip has an alpha angle of greater than 60° and a beta angle of less than 77°. A small alpha angle indicates a shallow bony acetabulum. Graf's classification is outlined in Table 12.1.

Other authors advocate a subjective assessment of femoral head coverage by the bony and cartilaginous acetabulum. For a given femoral head diameter, the percentage of femoral head coverage will be roughly proportional to the bony angle. The proportion of femoral head that lies above and below the baseline can be assessed from the static picture, often without the need for actual measurements. This assessment correlates well with the absolute measurements of the acetabular angle (MORIN et al. 1985) on

Table 12.1. Classification of acetabular dysplasia according to Graf

Type	Subtype	Bony angle	Cartilage angle	Notes
I	a	Normal (>60°)	Normal (<77°)	Sharp promentary
	b	Normal (>60°)	Normal (<77°)	Rounded promentary
II	a	Decreased (>50°)	Normal (<77°)	Age under 3 months
	b	Decreased (>50°)	Normal (<77°)	Age over 3 months
	c	Deficient (>43°)	Normal (<77°)	
	d	Deficient (>43°)	Increased (<77°)	
III	a	Poor (<43°)	Increased (<77°)	Normal roof cartilage
	b	Poor (<43°)	Increased (<77°)	Reflective roof cartilage
IV		Poor (<43°)	Increased (<77°)	Dislocated

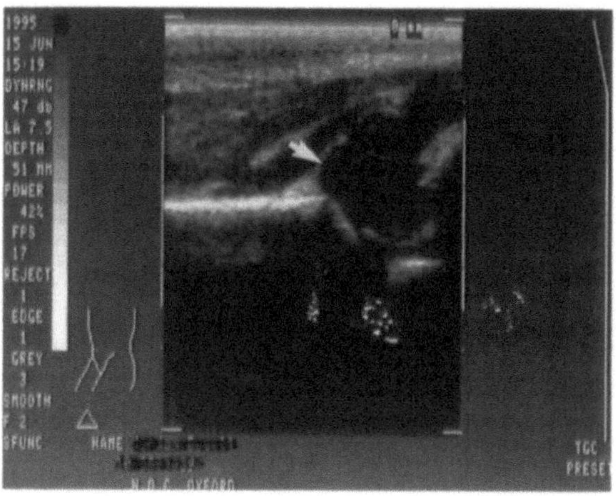

Fig. 12.2. Dysplastic acetabulum with reduced femoral head coverage. The proportion of femoral head that underlies the bony roof is approximately 40%. Note the slight upward subluxation of the unossified roof cartilage (*arrow*) with preservation of its ultrasound signal

plain films. Morin et al. determined that the radiographic acetabular angle was always normal in hips where femoral head coverage exceeded 58% and always abnormal when coverage was less than 33% (Fig. 12.2). The acetabular angle itself, however, is position dependent and can be altered by tilting the pelvis.

Percentage femoral head coverage and the bony roof angle do not correspond exactly. It is possible to have a shallow bony roof angle and normal femoral head coverage. This occurs in the presence of a large unossified cartilage anlage.

Examination of the acetabulum in the presence of a dislocated head can be difficult, as the bony femoral metaphysis casts an acoustic shadow over the acetabulum. The dislocated femoral head is identified lateral, posterior and superior to the acetabulum. The joint capsule may also be thickened and the acetabular cartilage may be displaced superiorly and stretched, with increased reflectivity. Gentle abduction in these infants usually reduces the hip, unless reduction is prevented by an inverted labrum, hypertrophy of the ligamentum teres, an intra-articular fat pad, an hour-glass deformity or the transverse acetabular ligament.

12.4.1.2
Assessment of Stability

Assessing acetabular depth is only part of the examination and is followed by an assessment of femoral head stability during stress. A variety of stress manoeuvres have been described. A simple method is to perform a modification of the Barlow test. To examine the left hip the infant is placed in the right lateral decubitus position. The tip of the probe is gripped between the thumb and index finger of the right hand. The palm and remaining fingers are placed against the buttock and lower back to provide posterior support. The hip and knee are flexed, and the knee rests in the palm of the left hand with the middle finger extended along the inner left thigh. This is the same position that is used to acquire a static coronal image. When a true coronal image through the centre of the tri-radiate cartilage has been obtained, gentle posterior pressure with the left palm on the knee, followed by abduction pressure on the inside of the left thigh with the examiner's left middle finger is applied. Abnormal posterior and lateral motion of the femoral head is easily detected. The pressure required is minimal and should not upset the infant; indeed, a false-negative examination may result if the child becomes distressed, as muscular contraction can stabilise an otherwise subluxable femoral head. It is also important to refrain from applying counter-pressure with the probe, as this too will prevent subluxation.

Smaller degrees of movement are more readily appreciated when successive pressure/relaxation cycles are applied. In many cases increased echogenicity will be noted within the joint during the stress manoeuvre; this is normal and represents the appearance of multiple microbubbles of nitrogen, the ultrasound equivalent of the vacuum phenomenon.

12.4.1.3
Ultrasound in the Management of DDH

Clinical screening (Ortolani and Barlow) alone will only detect 50% of cases of DDH; an equal number will present as late "missed" dislocation. Other clinical signs, including restriction of abduction, poor movement of the affected limb and asymmetric skin creases, are all non-specific. The tendency to nurse infants on one side will result in decreased abduction on that side, matched by increased abduction of the other limb.

All clinically unstable hips are detected by competent ultrasound. The hip that is both dysplastic and both clinically and sonographically unstable is usually treated. A controlled trial comparing treatment with non-treatment in this group has yet to be performed.

Dysplastic but stable hips will only be detected by ultrasound. Ultrasound finds many more "abnormalities" in clinically normal babies and it has been assumed that, within this group, are the hips that, had they not been detected by ultrasound, would have progressed to late dislocation. This assumption has yet to be firmly proven. Ultrasound will also detect femoral head instability. The question as to whether dysplasia or instability, if either, is the more important remains unresolved, despite a vast literature.

Morphologically "abnormal" hips are more prevalent at birth, and are seen in about 10% of infants. This falls to about 5% over 2 weeks. Minimal subluxation may also be noted under normal circumstances, particularly in children less than 30 days old. KELLER et al. (1988) measured the degree of subluxation in 40 normal neonates during the first 2 days of life and noted a mean subluxation of 3.2 mm (range 1–6 mm). The left hip was more likely to sublux than the right; therefore asymmetry of subluxation cannot be used to predict late instability. Subluxation beyond the neonatal period is uncommon but to what degree it is important is unclear. The author regards less than 2 mm as normal, but this is entirely arbitrary. More than this requires ultrasound follow-up, during which the majority of cases are seen to resolve. Findings in femoral head subluxation also include a thickened stretched joint capsule and flattening of the posterior bony acetabulum (BOAL and SCHWENKTER 1985).

The higher prevalence of both abnormal morphology and ultrasound instability at birth suggests that if screening is to be performed, then it is best done when the infant is at least 4 weeks of age. After this age, physiological subluxation is uncommon and the number of immature hips will be reduced. Abnormalities present at this stage are often treated as many orthopaedic surgeons feel that the earlier that treatment is instituted, the better the outcome. Better results may, however, simply reflect the inclusion of large numbers of babies who would have resolved spontaneously without any form of treatment.

The natural history of ultrasound abnormalities in clinically stable hips has been assessed in a number of small studies with varying results. GARDINER and DUNCAN (1992) found 15 morphologically abnormal hips (Graf IIc or worse) in 158 clinically normal high-risk babies. None were treated and all were normal at 6 months. ROSENDAHL et al. (1992) did not treat 50% of a group of babies with abnormal morphology and, again, all developed normally. CASTELEIN and co-workers (1992), however, evaluated a group of 101 babies that were clinically normal but had abnormal morphology on ultrasound. None were treated and four were described as "having dysplasia" at 6 months. The severity of dysplasia at birth did not correlate with outcome. In Gardiners group, there were also three morphologically normal hips that demonstrated instability on ultrasound (GARDINER and DUNCAN 1992). On follow-up, one of these became morphologically abnormal and was treated.

Advocates of neonatal hip screening claim that if all babies are examined close to birth, late DDH is eradicated. In some centres, however, the percentage of screened babies that are treated is high, thus increasing the risk of iatrogenic avascular necrosis. Controlled trials are lacking, and until they are available controversy regarding whom to treat and whom to continue to observe with ultrasound will remain.

Conservative treatment implies some form of harness. Several different types are in use, including the Pavlik, the Von Rosen, the modified Dennis Brown and the Frejka Pillow. The last-mentioned has been associated with a particularly high rate of avascular necrosis. It is generally recommended that treatment with a harness should be considered to have failed if the hip has not reduced within 3 weeks of treatment. A high risk of failure is present in hips that do not reduce at first assessment and in bilateral dislocation in infants over 7 weeks of age at the onset of treatment. This group often fail with other methods of closed reduction.

Ultrasound can be used to assess reduction and stability within the harness, and can be used to confirm normality at the end of the treatment period. Harnessing appears to be effective in the management of the subluxed, but not the dislocated hip, which requires more effective splinting. Regular monitoring of stability using ultrasound is recommended, particularly in the older age groups and in those with higher grades of dislocation. The incidence of avascular necrosis depends on the degree of dislocation at diagnosis, ranging from 1% for minor to approximately 15% for severe dislocation.

Ultrasound can also be used to assess reduction following open or closed reduction and spica casting.

A window is cut medial or lateral to the affected hip. Windows are immediately repaired following the ultrasound examination. To allay concerns over cast instability, caused by cutting a window, CT or, preferably, MRI, is now used to assess reduction following casting.

12.4.1.4
Screening for Acetabular Dysplasia

Clinical examination alone at birth is insufficient to detect all cases of acetabular dysplasia or instability (CLARKE et al. 1989) An ultrasound examination detects very many more abnormalities, but, as discussed above, only a relatively small proportion of these will require treatment (CLARKE et al. 1989). Treatment is generally with some form of harness, for which there is a small but definite association with avascular necrosis. On the other hand, ultrasound can reduce unnecessary treatment in other areas. Berman showed that the demonstration of a normal acetabulum on ultrasound meant that 14 of 17 infants with clinical instability were not treated and developed normally (BERMAN and KLENERMAN 1986).

The cost of screening all infants is high. In Oxford 19 babies are born every day. Repeat scan rates have been variously reported at 10%–40%, depending on experience, and in a proportion of these cases, several follow-up scans may prove necessary. Despite this, the cost may be less than the cost of treating missed DDH.

One approach that has been suggested is to recommend ultrasound on a subpopulation of babies with a known risk factor for DDH. Risk factors include: a first-degree family history of DDH, breech presentation at birth, an abnormality (click or clunk) found on physical examination, or an associated congenital anomaly (spinal, foot anomaly or torticollis). While this has been the policy that has been adopted in many centres, there is evidence that limiting ultrasound to infants with these risk factors fails to detect all cases of DDH. Approximately 10% of all infants will be examined if a "high-risk" policy is applied. In Clarke's group of 448 infants, this resulted in the treatment of 17 infants (3.7 per 1000) of whom five were clinically normal. To justify a high-risk screening policy, these five patients should represent the patients who would have presented with late DDH had they not been detected by ultrasound. Unfortunately there were as many late cases of DDH

during the year of Clarke's study as in previous years. This single study suggests that, not only does high-risk screening fail to reduce late DDH, but as many as one-third of infants treated in such a program are treated unnecessarily.

12.4.2
Proximal Focal Femoral Deficiency

Proximal focal femoral deficiency (PFFD) is one of the proximal femoral dysgeneses, where the femur is short and where there is a pseudarthrosis between the proximal cartilage anlage and the shortened femur. The condition is classified according to whether a femoral head is present or not, and whether there is associated acetabular dysplasia. Plain films demonstrate the unossified femoral shaft. A tapered bone end suggests there is no communication between shaft and head, and that ultimately a pseudarthrosis will develop. Ultrasound and MRI also have a role in staging this uncommon condition and in determining whether the femoral head is reduced or dislocated.

12.4.3
Coxa Vara

Coxa vara is a deformity of the proximal femur manifested as a decrease in the femoral neck to shaft angle. It can result from a variety of disorders, particularly those associated with bone softening. The condition may be the result of trauma but there is also an idiopathic variety. The usual presentation is a painless limp. A neck shaft angle of less than 120° on plain films is diagnostic. Congenital coxa vara is also associated with a typical lucent band vertically oriented across the femoral neck and, for this reason, the condition is often considered a variant of proximal femoral dysgenesis. Acetabular dysplasia may also be associated.

12.5
Hip Disorders in Young Children

Painful hip is one of the commonest causes of non-traumatic, acute paediatric presentations in orthopaedic practice and ultrasound plays a pivotal role in its assessment. In the majority of cases the underlying cause is benign, usually transient synovitis, a disorder that is still incompletely understood but with self-limiting and short-lived symptoms. The majority of cases will settle within 5 days without specific treatment.

12.5.1
Investigation of the Irritable Hip

Plain films are of no value in the assessment of effusion. Although various lines have been described on plain films, they are non-specific and unreliable. Ultrasound should be the primary investigation. In the author's centre, local anaesthetic cream is applied to the skin anterior to the affected hip as soon as the child presents to hospital. The optimal time for ultrasound examination is 1.5–2 h later. The hip is examined with the child supine using an anterior approach. A high-frequency linear array transducer is optimal and the probe is aligned along the femoral neck by rotating it slightly clockwise. Easily identified landmarks include the femoral capital epiphysis, the physis itself and the femoral neck (Fig. 12.3). Echo-poor cartilage is identified overlying the femoral head. More superficially still, a reflective band representing the anterior capsule is identified and can be traced inferiorly to its insertion on the femoral neck. Anterior displacement of the capsule occurs in the presence of an effusion. The degree of displacement is measured by the maximum distance between the anterior capsule and the

femoral neck. The thickness of articular cartilage overlying the femoral head can also be measured at this time.

Opinions vary as to what constitutes a normal anterior joint space. ADAM et al. (1986) suggest 2 mm as the upper limit of normal. ALEXANDER et al. (1989) define less than 2.3 mm as normal, but note that greater distances may be normal if symmetrical. In the author's view, asymmetry is the most useful sign and a greater than 2 mm difference between sides should be regarded as abnormal. The size of an effusion does not determine its nature, though the largest effusions are most often seen in association with benign transient synovitis.

There are several pitfalls for the unwary. The iliopsoas muscle, which overlies the anterior capsule, is relatively echo-poor, and can masquerade as an effusion. A true effusion, however, has a convex border rather than being flat as is the iliopsoas. The undistended capsule is concave upwards. Compression of the anterior synovial space occurs with the hip in external rotation, and the space may also decompress if the femoral head is subluxed as a consequence of the effusion. The subluxed hip can be identified by an increased distance between the acetabular rim and the physis compared with the normal side; however, the operator must be alert to the possibility to detect this pitfall.

If an effusion is present, aspiration using a 21 gauge needle is carried out. An anterior approach is recommended. The probe is placed vertically over the most distended part of the capsule. As the needle is to be inserted at right angles to the skin, the probe must be held vertically over the point of maximum distension (Fig. 12.4). The mid-points of the ends and sides of the probe are then marked on the skin, and the probe removed. The skin is punctured at the centre of the four marked points and the needle advanced slowly in a single motion. Gentle traction is applied on the syringe plunger as soon as the skin is breached. Scanning during needle insertion and advancement is not necessary. Aspiration should be as complete as possible, as this may result in more rapid resolution of symptoms and a shorter hospital stay. There is good correlation between the size of the effusion (BICKERSTAFF et al. 1990), pain intensity and restriction of movement, with intra-articular pressure.

Occasionally a bleeding disorder can present for the first time as hip haemarthrosis. Aspiration in these circumstances does not appear to be associated with any increased risk of complications.

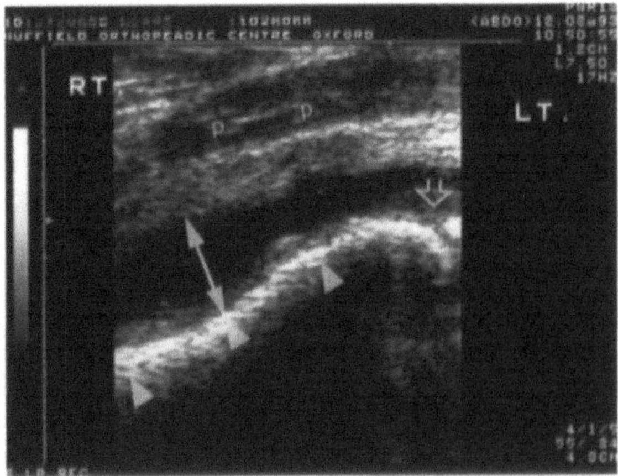

Fig. 12.3. Hip effusion. The ultrasound probe is oriented along the femoral neck (*arrowheads*). Easily identifiable landmarks include the physis (*open arrow*) and psoas tendon (*p*). There is wide separation between the femoral neck and the anterior capsule due to an effusion (*double-headed arrow*)

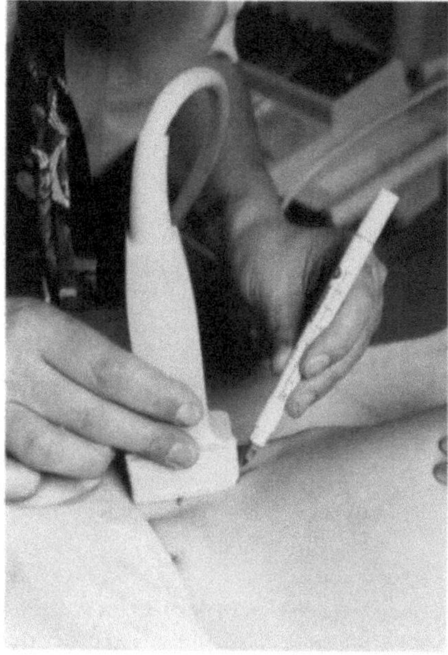
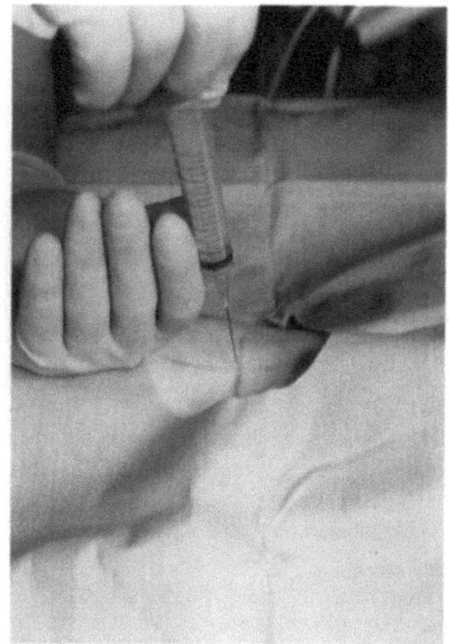

Fig. 12.4 a,b. Technique for aspiration hip effusions using ultrasound. a The probe is held vertical to the skin over the point of maximal capsular distension. The centre points of the probe are marked along its four sides. Lines joining opposite points insect at the puncture site. b The needle is inserted vertical to the skin surface and the puncture point

12.5.2
Transient Synovitis

The diagnosis of transient synovitis is only suggested when other conditions are excluded, and signs and symptoms subside without complications. Children present with a 1- to 7-day history of pain and limitation of movement. The incidence is approximately 0.2% and up to 25% will have recurrent episodes (ALEXANDER et al. 1989). A seasonal variation has been reported, with more cases occurring when respiratory tract infection is common, though this has not been the experience in Oxford. Approximately 75% have an effusion, and the condition is more common in boys, with a ratio to girls of 2.5:1. There is no side predisposition. Resolution occurs in 75% within 2 weeks. Persisting effusion is suggestive of another cause, and Perthes disease should then be considered.

12.5.3
Septic Arthritis

Because of the devastating consequences of bacterial infection within the joint, early diagnosis is mandatory. Typically the effusion is reflective (ZIEGER et al.

1987) with synovial thickening. Thickening of the anterior capsule is seen in about 50% (DORR et al. 1988) but this sign is also common in transient synovitis (MIRALLES et al. 1989). While the majority of effusions in septic arthritis are reflective, some are echo-poor (SHIV et al. 1990). As there are no absolute differences between the clinical presentation or sonographic appearance of a septic effusion and transient synovitis, aspiration is recommended in all cases. Indeed, if septic arthritis is strongly suspected clinically, aspiration should be performed if no effusion is demonstrated, though this is unusual. Particulate matter in the aspirate in the absence of blood is suggestive of sepsis.

12.5.4
Perthes Disease

Perthes disease is an osteonecrosis of the femoral capital epiphysis which occurs in the first decade. The mean age of patients with this disease is 7 years, and it carries a male to female ratio of 5:1. The incidence is approximately 1:10000 in the United Kingdom. Risk factors include low birth weight, lower socio-economic group and low birth order. Approximately 20% of cases are bilateral, but symmetrical

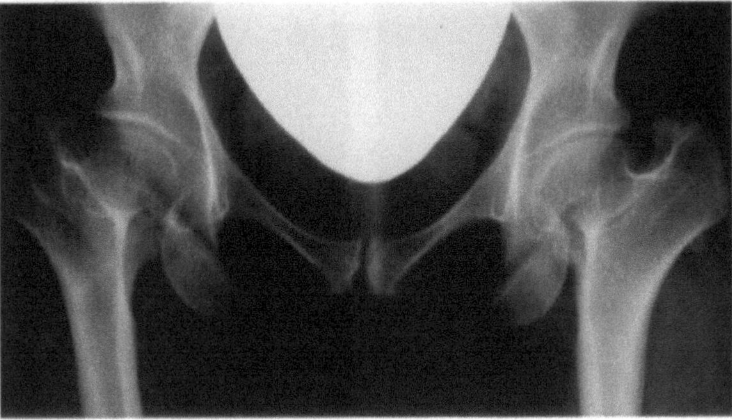

Fig. 12.5. Epiphyseal dysplasia

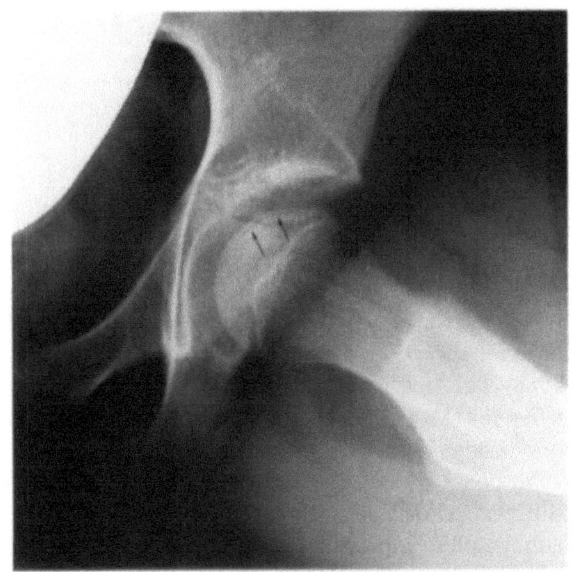

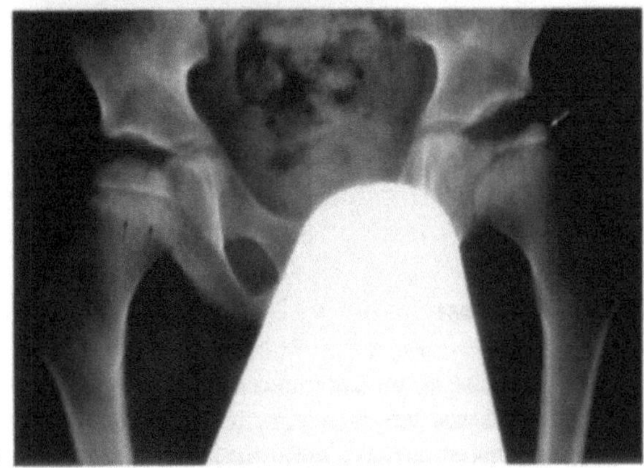

Fig. 12.6. a The early frogleg lateral view shows a subchondral fissure (*arrows*). b Eight months later there is more severe femoral head flattening, subchondral sclerosis and lateral migration of the ossified epiphysis on the left (*white arrow*). Earlier disease is present on the right with epiphyseal sclerosis and subtle flattening. Note also the early cystic change in the metaphysis (*black arrows*)

disease is rare. Symmetrical disease should suggest an alternative condition such as Meyer dysplasia, hypothyroidism, epiphyseal dysplasia (Fig. 12.5) or another cause of avascular necrosis.

The characteristic radiographic appearances are sclerosis and fragmentation within the ossification centre of the femoral capital epiphysis. An effusion is present in approximately half of patients; however, this is more reliably detected on ultrasound than on plain films. Displacement of the pericapsular fat pad on plain radiography has been described as a sign of effusion; however, this is highly positional dependent and non-specific. Widening of the medial joint space has also been described and has been attributed variously to synovial thickening, effusion, true cartilage hypertrophy and apparent cartilage hypertrophy due to lateralisation of the femoral capital epiphysis.

Signs within the ossification centre itself that may precede fragmentation include a subchondral fissure (Fig. 12.6), which may contain gas, and decreased height of the epiphysis. Later signs include fragmentation and lateral migration of the epiphysis, seen as laterally positioned calcification.

A horizontal growth plate has been described. Changes within the metaphysis include metaphyseal "cysts" due to unossified cartilage and a short wide femoral neck.

The ultrasound findings in Perthes disease include effusion, thickening of femoral head articular cartilage and fragmentation of the epiphysis and increased femoral anteversion (Fig. 12.7). An effusion is more common in the early stages of the disease (WIRTH et al. 1992).

The prognosis of Perthes disease is good, with up to 60% of cases resolving with conservative treatment. Prognostic factors have been suggested

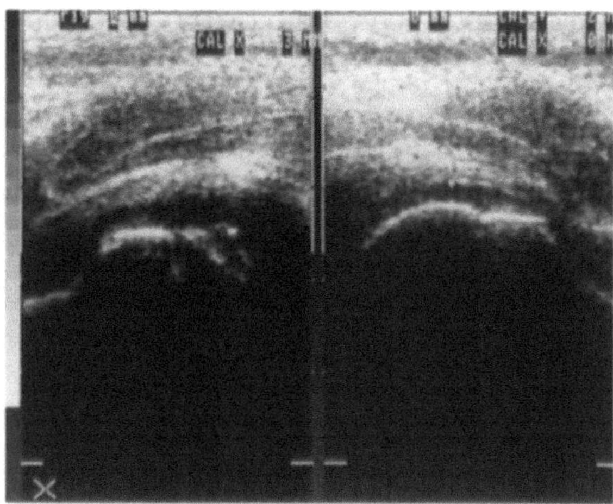

Fig. 12.7. Perthes disease. Note the effusion, femoral head fragmentation and cartilage thickening of the right hip (*left of figure*) compared to be normal left side. Greater than 3 mm discrepancy between cartilage thickness is significant. Care must be taken to ensure that only femoral head cartilage is included in the measurement. For this reason, the measurement should be obtained on a portion of the head that is uncovered by acetabulum, as indicated by the *asterisks*

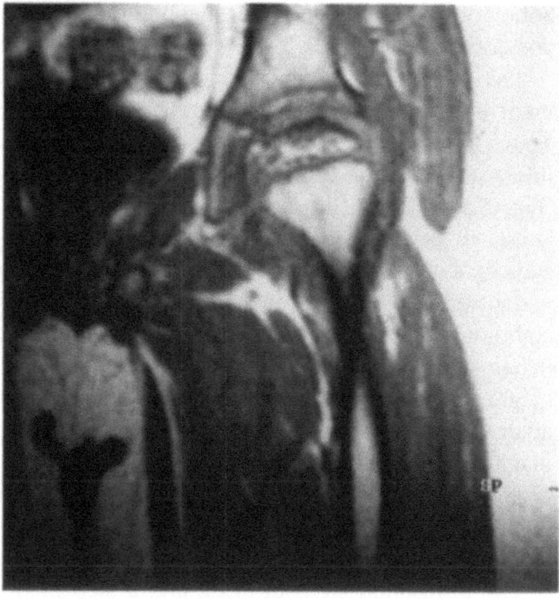

Fig. 12.8. Perthes disease. MRI demonstrates fragmentation of the femoral capital epiphysis and shows the relative preservation of cartilage

(MURPHY and MARSH 1978), and if more than two of the radiological signs outlined above are present the head is considered at risk. CATTERALL (1980) graded the severity of epiphyseal involvement on the basis of its extent, the presence of sequestra, the crescent sign and collapse, and the degree of metaphyseal involvement. Four types are defined, with the poorest results present in types III and IV. A typical type III hip shows a subchondral fracture line extending into the posterior half of the epiphysis, a clear margin between involved and uninvolved epiphyseal segments and diffuse metaphyseal changes. A type IV hip shows more diffuse epiphyseal involvement and sclerotic demarcation between involved and uninvolved segments. The role of MRI has yet to be fully established in this condition (Fig. 12.8). Long-term effects include coxa magna or other femoral head deformity.

12.5.5
Meyer Dysplasia

Meyer dysplasia is a condition of unknown aetiology in which both femoral heads show sclerosis and fragmentation. Unlike Perthes disease, Meyer dysplasia is symmetrical. The hips invariably become normal with no sequelae. The condition is not associated with an effusion on ultrasound.

12.6
Hip Disorders in Older Children

12.6.1
Slipped Capital Femoral Epiphysis

Children with slipped capital femoral epiphysis (SCFE) are older than those with transient synovitis. The mean age from several series is 11 years, whereas the mean age in transient synovitis and Perthes disease is 6.7 years. The condition is more common in boys, with a ratio to girls of approximately 3:1. Regional variations are observed throughout the world and amongst different races. At presentation, the left hip is involved more frequently than the right. Bilateral involvement is present in between one-third and one-half of cases. A seasonal variation has been reported. Children with SCFE are often above average weight for age and skeletal age is frequently less than expected for chronological age.

Pain is the commonest presenting symptom. The affected limb is often held in external rotation. Laboratory investigations are usually normal, though the occasional patient will have an underlying metabolic disorder.

The epiphysis is most frequently displaced posteriorly and medially. This is termed a varus slip. With chronicity, remodelling may occur, with new bone formation beneath the stripped periosteum posteriorly.

The plain film findings enable a diagnosis to be made in the majority of cases. On the AP film, a line drawn along the lateral margin of the femoral neck intersects a portion of the epiphysis in normal hips (Fig. 12.9). When a slip has occurred, this line passes lateral to be femoral head. Under normal circumstances, a small portion of the medial margin of the metaphysis overlaps the inferior margin of the acetabulum. This is not seen with slipped epiphysis. Other signs on the AP film include slight loss of height of the femoral capital epiphysis and an irregular physis. A frogleg lateral view confirms the diagnosis (Fig. 12.10). Occasionally, a child of appropriate age presents with hip pain and the only plain film finding is slight metaphyseal lucency. This has been termed pre-slip.

The ultrasound signs of SCFE are: a step in the anterior physeal outline and diminished distance between the anterior acetabular rim and the femoral metaphysis compared with the contralateral side. An effusion is seen in about half the patients, and is more likely when the onset is acute. The AP film fails to show displacement in 14% of cases; a frogleg view is necessary to detect these. Ultrasound provides an accurate measurement of the physeal step and the degree of metaphyseal shortening in the acute slip, without the need for ionising radiation. In chronic SCFE measurements of the physeal step are unreliable due to metaphyseal remodelling.

The natural history of the condition is not well understood. It has been suggested that without treatment approximately 30% will develop early osteoarthritis. Whilst this is more likely to occur with major slips, occasionally patients with relativity minor ones develop complications.

Slips can be graded according to the fraction of epiphysis (usually divided into thirds) that remains in contact with the metaphysis. Treatment is directed at preventing further slip and is surgical. Both intra- and extra-articular procedures have been used; however, the former are associated with an additional risk of avascular necrosis, and are therefore probably not appropriate for more modest slips.

Two types of pins are generally employed for extra-articular fixation, depending on residual growth potential. If further growth is needed, smooth rather than threaded pins are used as the latter tend to promote epiphysiodesis. Postoperative imaging is necessary to confirm that the tips of the pins lie within the epiphysis and that the joint surface has not been compromised. Ideally, the pins should reach the inferior portion of the epiphysis, where the bone is most dense. Follow-up radiographs are scrutinised to exclude complications such as avascular necrosis, chondrolysis and osteoarthritis.

12.6.2
Juvenile Chronic Arthritis

Juvenile chronic arthritis or inflammatory arthropathy in the presence of an unfused physis can lead

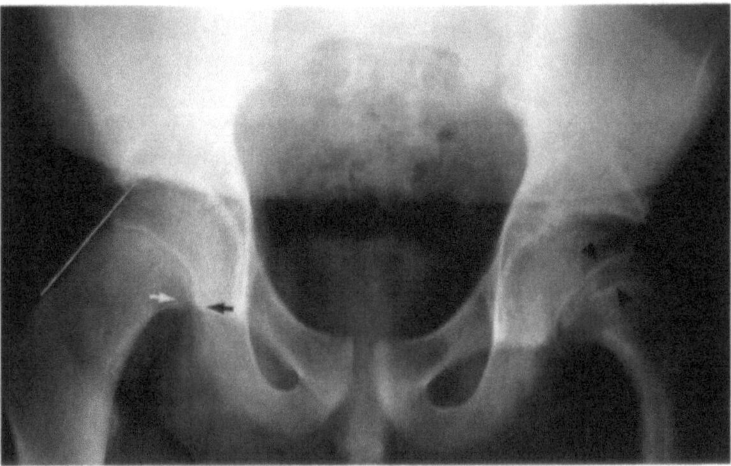

Fig. 12.9. Bilateral slipped epiphysis. Note that the line drawn along the lateral border of the left femoral neck passes lateral to the epiphysis. This is abnormal and confirms a slipped epiphysis. Note in addition how the medial margin of the metaphysis (*white arrow*) no longer overlaps the ischium (*black arrow*) and the widening of the growth plate (*arrowheads*)

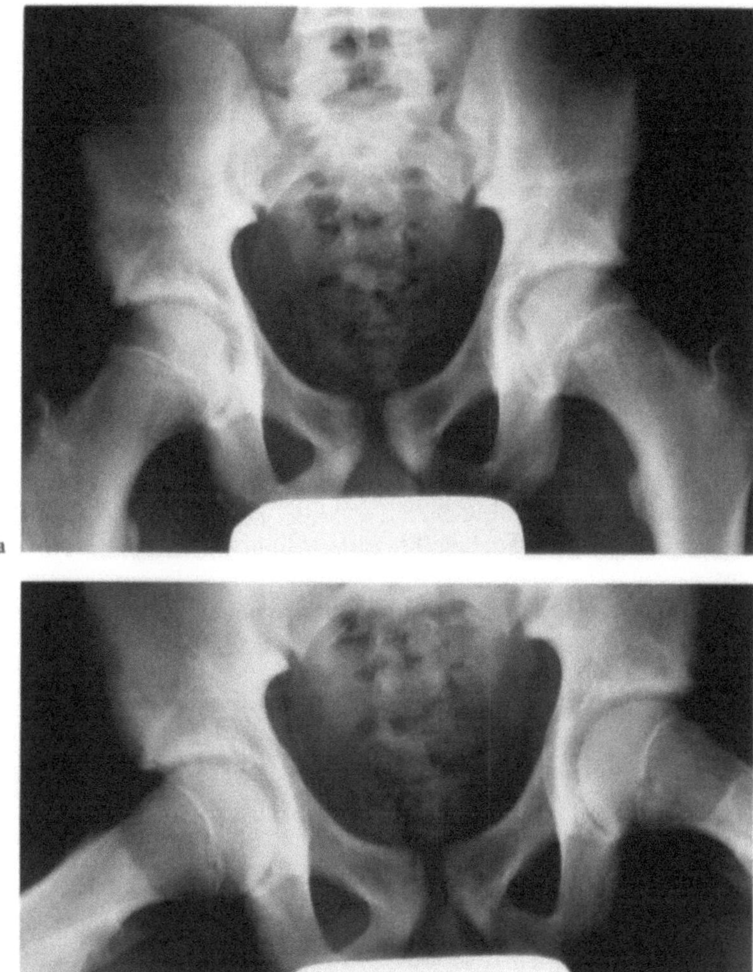

Fig. 12.10. a Slipped epiphysis on the left. Compare with the normal right side. **b** Frogleg lateral view confirms the diagnosis

to hyperaemia-induced epiphyseal enlargement. Ultimately premature fusion occurs, leading to growth arrest. The hip is involved with other large joints most commonly in the pauciarticular and juvenile sacroiliac variants. A characteristic sign is elongation of the lesser trochanter due to psoas traction. Cartilage loss is late, protrusio can occur and ankylosis is described in the juvenile rheumatoid variant.

12.6.3
Chondrolysis

Chondrolysis is a poorly understood condition characterised by periarticular osteoporosis and cartilage loss. Some cases follow trauma including slipped epiphysis, while no cause is apparent in others. The condition is more common in females, with a ratio of 5:1, and amongst people of African origin. Patients present in their early teens with hip pain, limp and restricted movement (HUGHES 1985). Treatment is conservative; however, growth disturbances and bony ankylosis may occur.

12.6.4
Imaging Protocol in Children with a Painful Hip

The best protocol for imaging children with hip pain is one that detects all important abnormalities yet minimises exposure to ionising radiation. Ultrasound is excellent at detecting effusions and is at present the most appropriate first investigation. Children with effusions are aspirated and the fluid is examined urgently with a gram stain and for its microscopic appearances. It is then cultured in suitable media. If symptoms settle promptly, and do not recur, then further imaging is rarely necessary. A rare occurrence is the co-existence of a sterile effusion secondary to underlying osteomyelitis. This has been proposed as a reason to obtain an x-ray on all children with irritable hip, though most will be normal. Even those with osteomyelitis may have normal plain films in the early stages.

All children over the age of 8 years require plain films to exclude SCFE; however, these can be limited to a frogleg lateral. Whilst an ultrasound examination can detect an acute epiphyseal slip, chronic slips are more difficult. With time the effusion disappears and metaphyseal moulding can occur, obscuring the usual ultrasound signs.

Children with irritable hip and without effusions on ultrasound, and those whose symptoms fail to settle promptly, should undergo radiography to exclude Perthes disease or a more sinister condition.

The role of MRI remains to be established, but where available it could even replace ultrasound as the first-line investigation. In the majority of children over the age of 5 years, the examination can be carried out without any form of sedation. Further compliance can be obtained by a parent accompanying the child into the bore of the magnet. In our centre, sedation has not proved necessary for the vast majority of children who have been scanned. Effusions can be detected and osteomyelitis can be excluded with confidence. Using MRI it may also be possible to differentiate effusions due to transient synovitis or other benign causes from those due to septic arthritis. Firm data on this is lacking, but if confirmed this approach would have a significant impact on the number of children undergoing aspiration. A T1-weighted image orientated along the femoral neck provides an elegant view of the displaced femoral head in SCFE and the angle that it forms with the femoral neck. New bone formed at the site of periosteal stripping can also be assessed. These findings may help with the surgical management.

Magnetic resonance imaging has the additional advantage of not using ionising radiation, but the disadvantage of expense and limited availability. With software improvements and more rapid and therefore cheaper examination times, MRI is likely to assume its correct role as the first investigation in children with irritable hip.

12.7
Hip Disorders in Young Adults

12.7.1
Osteoid Osteoma

Osteoid osteoma is a benign bone-forming tumour occurring most commonly between the ages of 7 and 25 years. The femur is the commonest bone to be affected, making hip pain a common presenting symptom. Pain is typically worse at night and dramatic relief with salicylates is described. Typically the lesion appears on plain radiographs as a sclerotic area (Fig. 12.11) within which a lucent nidus may be identified. The nidus is generally better depicted on CT (Fig. 12.12) or MRI. When osteoid osteoma occurs within the hip joint, synovitis, joint effusion, limitation of motion and growth disturbances occur (GIUSTRA and FREIBERGER 1970).

Symptoms of intra-articular lesions may also be atypical and the usually characteristic night pain may be absent in a high proportion of cases (GOLDBERG and JACOBS 1975). An intra-articular location is also associated with less sclerotic response

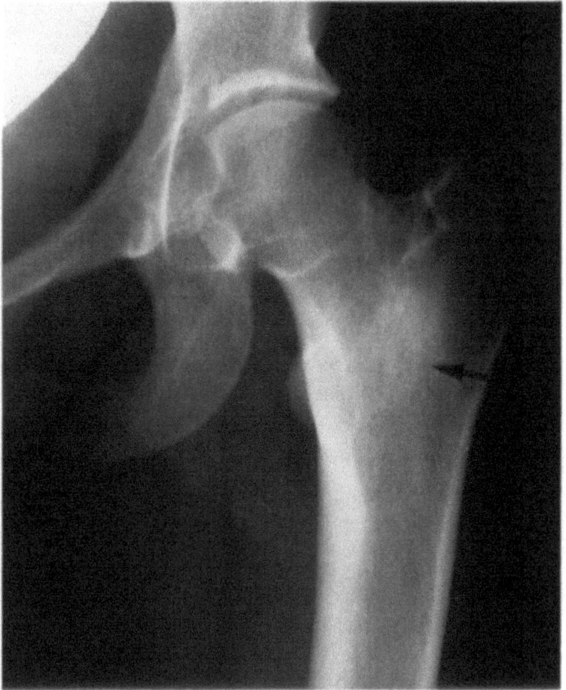

Fig. 12.11. Osteoid osteoma. Note the poorly defined area of sclerosis within the intertrochanteric region of the right hip (*arrow*). Bone sclerosis obscures the nidus

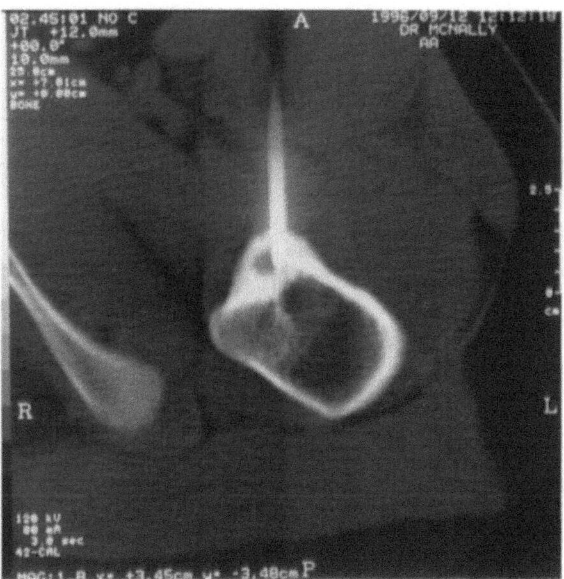

Fig. 12.12. Axial CT section through the proximal femur lateral to the base of the lesser tuberosity. Note the marked bony sclerosis with a central lucent nidus typical of osteoid osteoma. The margin of the nidus has been cannulated prior to radiofrequency ablation

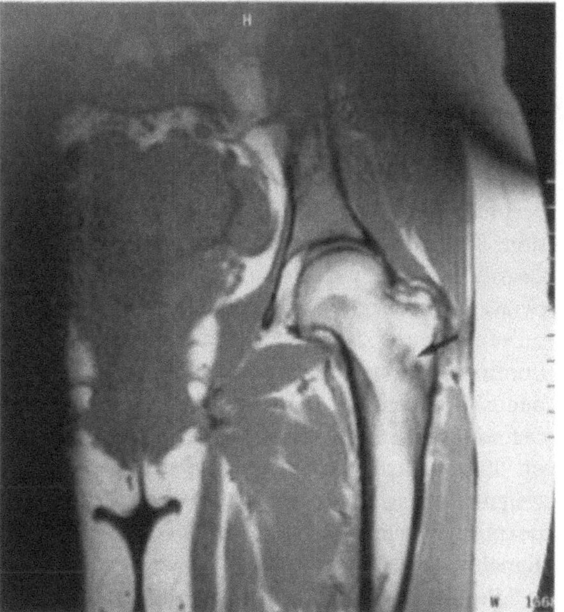

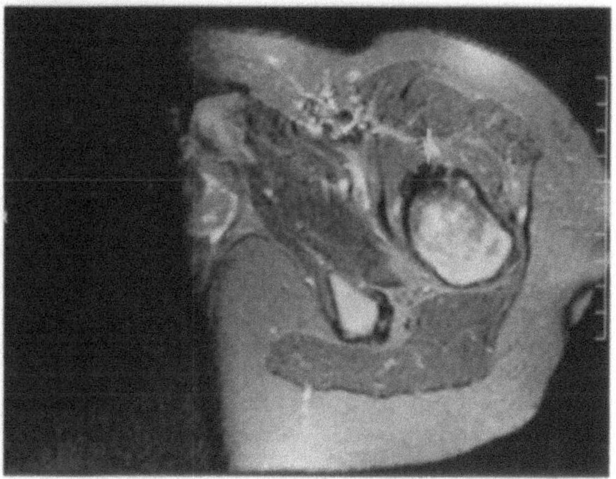

Fig. 12.13. a Coronal T1-weighted image showing medullary oedema and low-signal nidus (*black arrow*). b Axial STIR image through the proximal femur, showing marked medullary oedema and cortical sclerosis (*white arrow*)

and periosteal reaction than extra-articular lesions. For these reasons, a negative plain radiograph is not conclusive. Osteoid osteomas are typically extremely hot scintigraphically, producing a target appearance with the densest uptake in the highly vascular central nidus. The vascular character of the lesion also makes it highly conspicuous on MRI (Fig. 12.13) where, in inexperienced hands, the extensive surrounding oedema has been mistaken for a more aggressive tumour. Both MRI and CT (Muscolo et al. 1995) are useful in the identification of the nidus. The author favours MRI in patients in whom the clinical history suggests possible osteoid osteoma as it is a better screening examination than CT. Where a lesion has been identified by another imaging technique, for example plain films or scintigraphy, thin-section CT is preferred to confirm the diagnosis.

Treatment of lesions around the hip is best achieved percutaneously, to limit the amount of bone resected and avoid secondary fracture. A small nidus can be excised using standard bone biopsy techniques. Larger lesions can be treated by a combination of biopsy and ablation, using either radiofrequency or thermoablation.

12.7.2
Osteochondromatosis

Occurring in the 20- to 50-year age group, synovial osteochondromatosis is a synovial metaplasia that results in the production of multiple cartilaginous and osseous loose bodies. Along with the knee and the elbow, the hip is one of the most commonly involved joints. Pain is a typical presenting symptom.

Careful scrutiny of plain radiographs may reveal multiple intra-articular loose bodies (Fig. 12.14). These can be subtle (Figs. 12.15, 12.16) and a negative plain film does not exclude the diagnosis. Arthrography will demonstrate both the loose bodies and the associated synovitis. Increased uptake on scintigraphy has been described but the appearances are non-specific. On MRI, three distinct patterns are seen which are probably dependent on the relative preponderance of synovitis and loose body formation. The most common pattern in a series of 21 cases described by KRAMER et al. (1993) was a synovial mass associated with foci of signal void on all pulse sequences. Ten percent of cases are synovitis dominant lesions appearing as a lobulated homogeneous intra-articular mass with signal characteristics that are isointense to slightly hyperintense to muscle on T1-weighted images and hyperintense on T2-weighted images. More extensive calcification appears as areas of low signal, and ossification as low signal masses with central fat (KRAMER et al. 1993).

Like other synovial processes in the hip joint, osteochondromatosis can produce large erosions of the femoral neck, which in extreme cases produce an "apple core" appearance. Similar lesions are seen in

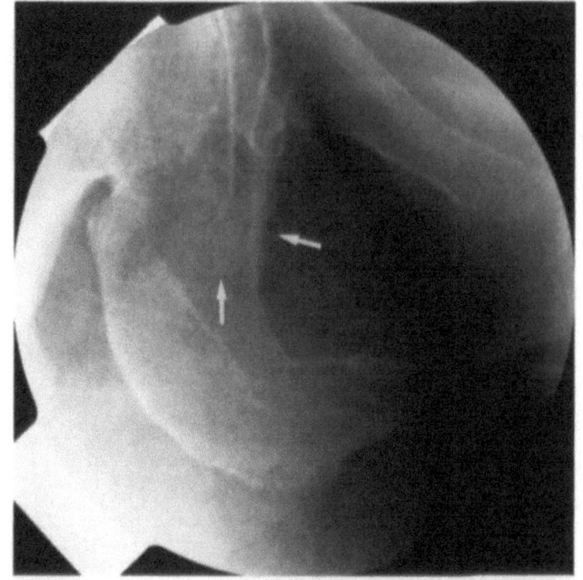

a

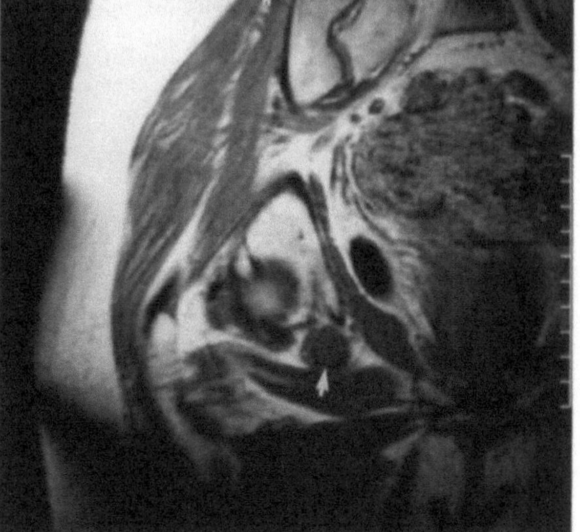

b

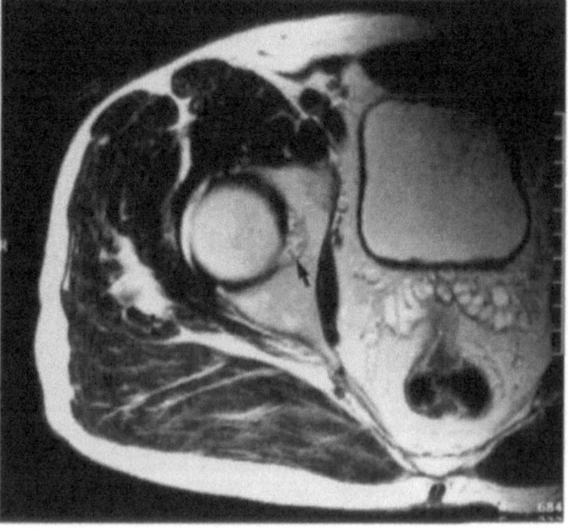

c

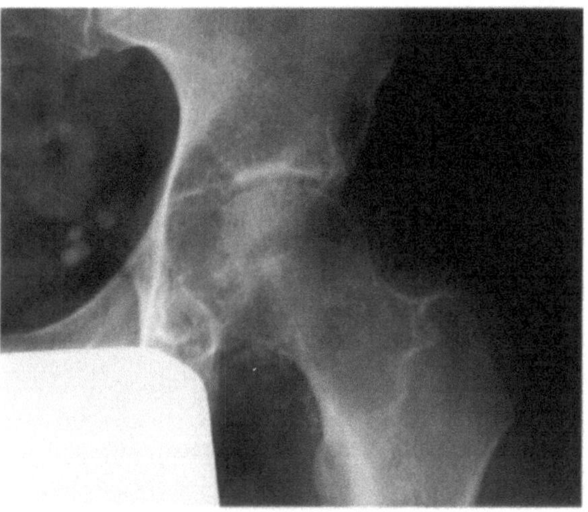

Fig. 12.14. Synovial osteochondromatosis. Note the small calcific densities and the poor definition of the femoral head inferomedially. The latter feature can be an early finding preceding the appearance of loose bodies

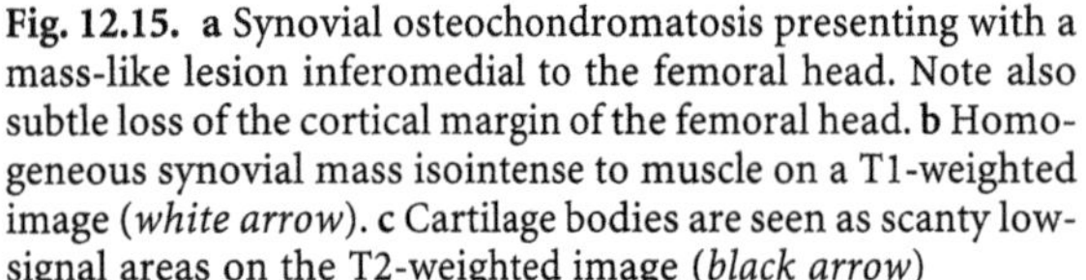

Fig. 12.15. a Synovial osteochondromatosis presenting with a mass-like lesion inferomedial to the femoral head. Note also subtle loss of the cortical margin of the femoral head. **b** Homogeneous synovial mass isointense to muscle on a T1-weighted image (*white arrow*). **c** Cartilage bodies are seen as scanty low-signal areas on the T2-weighted image (*black arrow*)

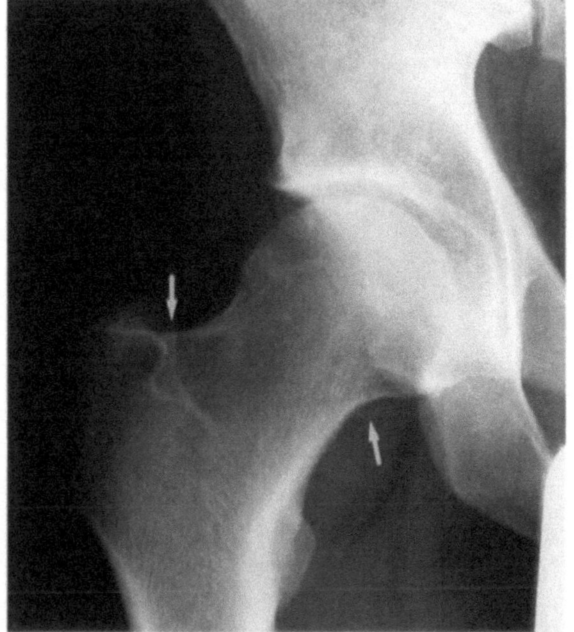

a

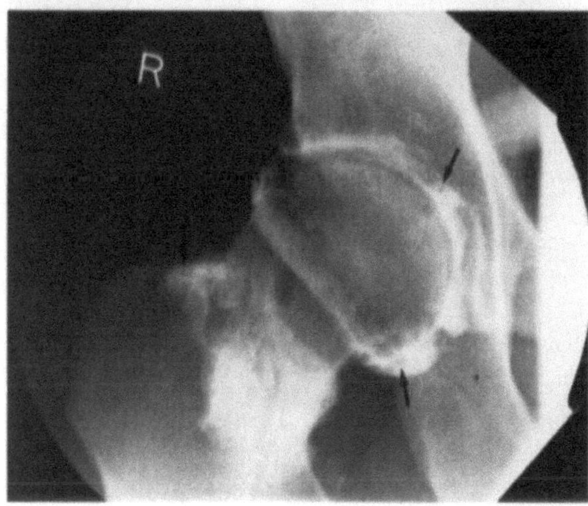

b

Fig. 12.16 a,b. Synovial osteochondromatosis. a Calcific densities along the femoral neck both superiorly and inferiorly (*white arrows*). b Arthrography demonstrates multiple filling defects (*black arrows*), consistent with calcific bodies

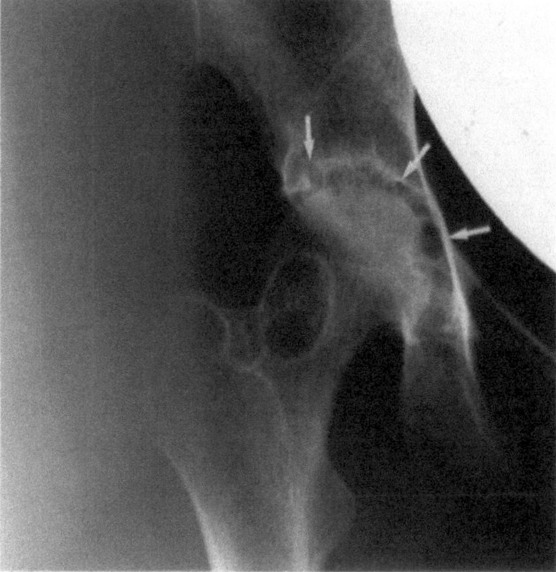

Fig. 12.17. Pigmented villonodular synovitis. Note the well-defined erosion occupying much of the femoral neck (apple core lesion). Note also the multiple small cysts within the acetabulum (*white arrows*), confirming the arthropathic nature of the process

pigmented villonodular synovitis (Fig. 12.17), rheumatoid arthritis and amyloidosis (GOLDBERG et al. 1983).

12.7.3
Pigmented Villonodular Synovitis

Pigmented villonodular synovitis (PVNS) is a benign synovial proliferation that derives its name from a characteristic brown-yellow appearance on gross specimens that results from haemosiderin deposition. The age of onset of this condition is usually between 30 and 50 years, with pain and limitation of motion being the most common presenting features. Occasionally a palpable mass may be felt in the groin (COTTEN et al. 1995). Plain radiographs may demonstrate subarticular lytic areas that are often large and more commonly located on the femoral side of the joint (Fig. 12.17). Joint space loss can occur late in the disease (FLIPO et al. 1994). Arthrography shows the synovial mass, which can involve the joint in a diffuse or focal manner (GOLDMAN and DiCARLO 1988). MRI is a less invasive and more specific means of confirming the diagnosis. The characteristic appearance is of synovial deposits that have low signal and appear black on T2-weighted sequences (Fig. 12.18) secondary to the paramagnetic effect of haemosiderin. Occasionally haemosiderin-poor lesions will not demonstrate this phenomenon. A low-signal mass on T2 weighting needs to be differentiated from a fibrous or amyloid mass, which should be considered in renal transplant patients. Haemosiderin deposition will also occur following haemarthrosis, particularly if repeated as a result of synovial haemangioma or haemophilic arthropathy.

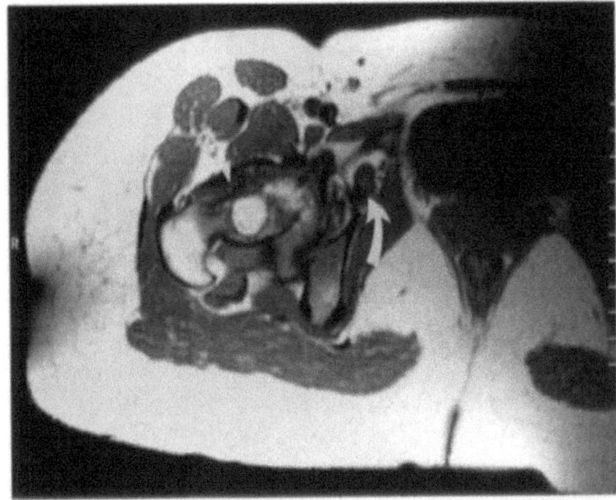

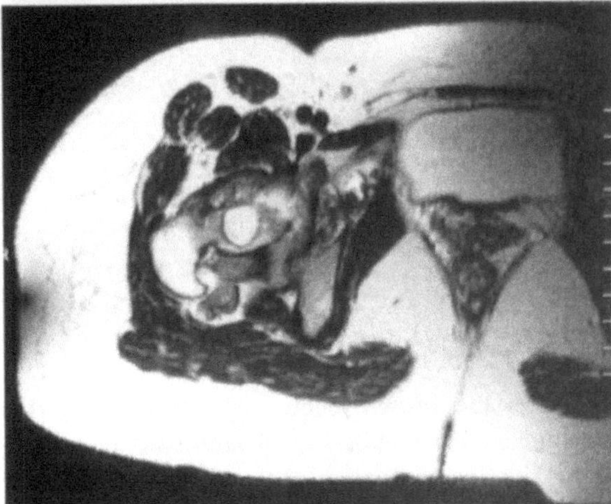

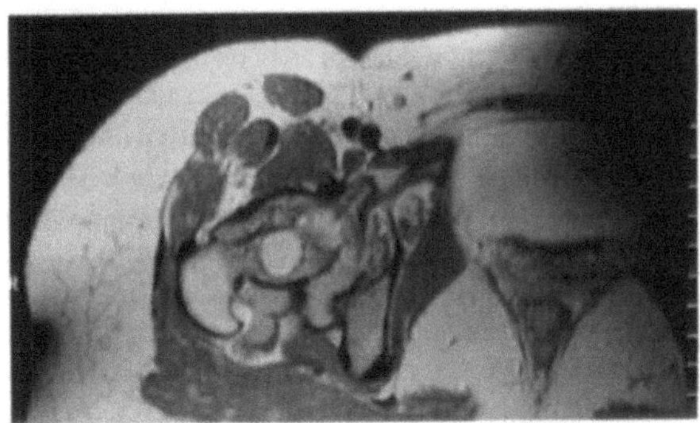

Fig. 12.18 a–c. Pigmented villonodular synovitis. a T1-weighted image showing a large fat-containing lesion within the femoral neck and associated synovial thickening (*white arrow*). Note extension of mass through the obturator foramen (*curved arrow*). b On the T2-weighted image the synovial thickening remains of low signal. This occurs due to the paramagnetic effect of haemosiderin deposited within the synovial tumour. c Following gadolinium, the enhancing synovial mass becomes more conspicuous

12.7.4
Avascular Necrosis

The unique blood supply to the adult femoral head renders it prone to avascular necrosis. This may follow a variety of stimuli including intracapsular fracture of the femoral neck, drugs (steroids, chemotherapy, immunosuppressant therapy), barotrauma and haematological disorders that increase blood viscosity (sickle cell disease, polycythaemia rubra vera and cryoglobulinaemia). The earliest changes occur in the anterosuperior part of the femoral head (Fig. 12.19) and progress in a superior and posterior direction to involve the entire weight-bearing portion. An early plain film clue is preservation of normal bone density in the face of surrounding disuse osteoporosis occurring as a result of pain. This is particularly common in avascular necrosis following fractures of the femoral neck. Bone sclerosis (Fig. 12.20) is followed by subcortical fracture, seen as lucent lines deep to the subchondral cortex. Late stages include femoral head collapse and secondary osteoarthritis. A hallmark of the process is relative preservation of the joint space and absence of changes within the acetabulum until secondary osteoarthritis supervenes.

In the early stages of the process, skeletal scintigraphy may demonstrate a focal cold spot. Once bone reparation around the lesion begins, increased intensity is seen. The characteristic MRI appearances are of a geographical area of abnormal signal surrounded by a double line on T2-weighted images, the outer line of low signal and the inner line of increased signal representing granulation tissue. The lesions themselves can show a variety of signal

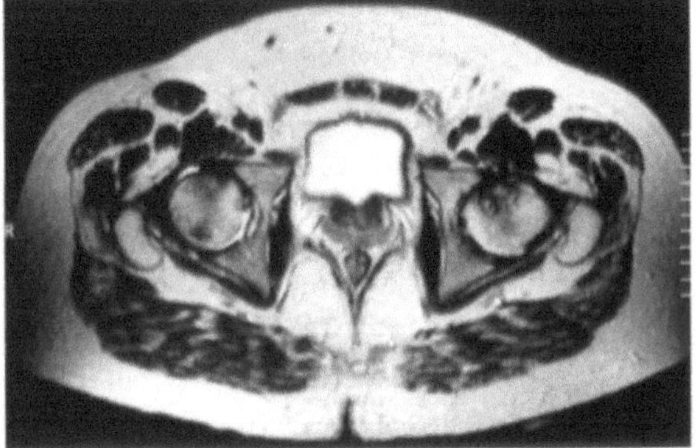

Fig. 12.19. Avascular necrosis of both hips. Axial sections demonstrate the earliest changes to occur in the anterosuperior part of the femoral head

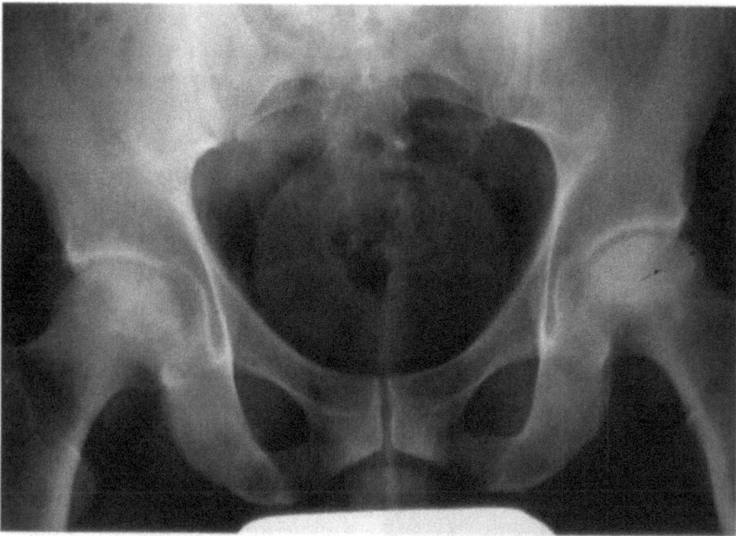

Fig. 12.20. Avascular necrosis of the left hip with sclerosis of the femoral head and a lytic area superolaterally (*black arrows*). Note also early sclerosis on the contralateral side, confirming early avascular necrosis

patterns depending on their age. In the early stages (MRI type A) the lesion is composed largely of fat and therefore returns high signal on both T1 and T2 weighting. With increasing age, there is a gradual transition between a more fluid type signal (intermediate on T1 and high on T2; type B) through to a fibrotic appearance (low on both T1 and T2; type C). This MRI-based classification does not appear to be useful in clinical practice, however, and as yet has not been shown to be important either in defining the choice of treatment or in determining prognosis. Of the several classifications used, the best known is the FICAT classification, which is based on plain radiographic changes; however, even with this classification large intra- and inter-observer differences are common (KAY et al. 1994). Others have defined various measurements that assist in determining whether a femoral head is more or less likely to collapse. These include the angle of the femoral head involved by avascular necrosis, the proportion of the weight-bearing head that is involved and the proportion of the total femoral head volume that is avascular (SUGANO et al. 1994). Involvement of more than 33% of the weight-bearing surface or 43% of the total femoral head volume carries a poor prognosis for massive collapse; however, large studies have not been carried out to test these indices prospectively. More recently STEINBERG et al. (1995) have proposed a grading process which is also based on a combination of radiological change and extent of involvement. Normal imaging is rated grade 0. An abnormal bone scan or MRI is grade 1, if the plain film is normal. Plain film changes confined to sclerosis and cysts only is grade 2. Subchondral collapse is grade 3, or 4 if there is femoral head flattening. Secondary osteoarthritis is grade 5, or 6 if the acetabulum is involved. Each of these ratings is further subdivided into mild, moderate and severe, depending on whether less than 15% or more than 30% of the head is involved. Others have found percentage involvement of the weight-bearing surface to be a poor predictor and suggest that more diffuse high signal throughout the head on fat-suppressed imaging (as opposed to high signal limited to the periphery of the avascular area as in the double line sign) might indicate a good prognosis (GILBERT 1997).

12.7.5
Transient Osteoporosis

Transient osteoporosis refers to one of a spectrum of poorly understood disorders characterised by pain, stiffness and acute osteoporosis. Osteoporotic changes on plain films occur later than increased uptake on scintigraphy and diffuse bone marrow oedema on MRI. The earliest plain film finding is the loss of the subchondral cortex beneath the femoral head. Fat-suppressed MRI is the most sensitive technique, showing diffuse, poorly defined high signal in the femoral head, neck and upper part of the femoral

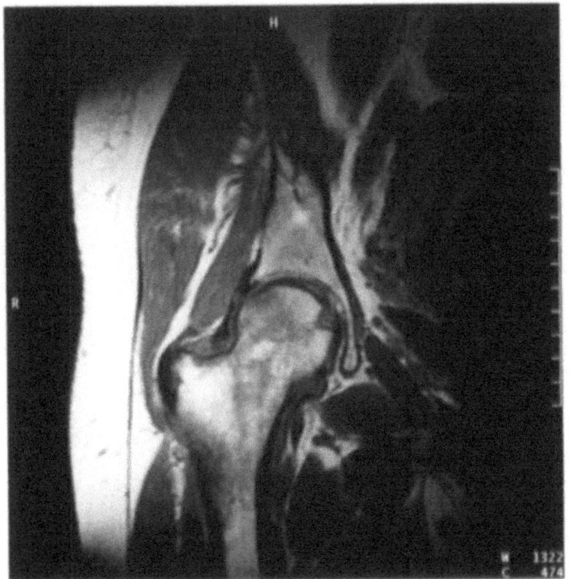

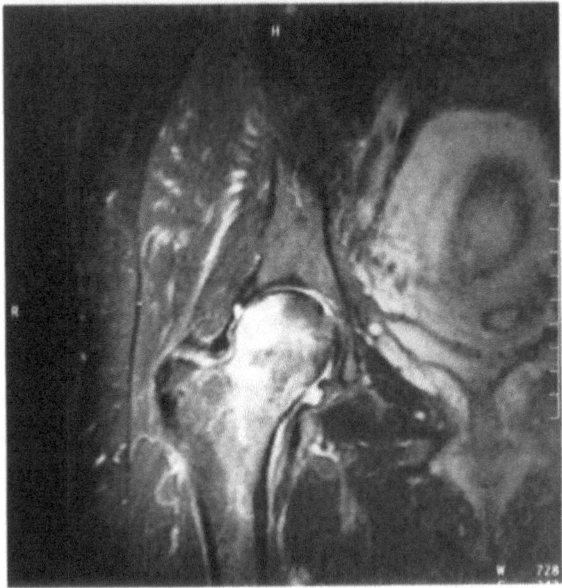

Fig. 12.21 a,b. Transient osteoporosis of pregnancy with more unusual involvement of the right hip. **a** T1-weighted image showing poorly defined trabecular oedema in the femoral head and neck. There is a small effusion, seen as slight distension of the inferior recess. Unlike in avascular necrosis, the abnormal area is not well demarcated and does not show the typical double line sign on T2-weighted imaging. Note the flame-like extension into the metaphysis, and the gravid uterus. **b** STIR imaging shows the diffuse oedema more clearly. In this particular case the acetabulum is not involved; acetabular oedema can also occur but is usually not as prominent as oedema within the femoral head

shaft, where, possibly as a result of the orientation of the primary trabecular lines, it is seen as a flame-like extension into the metaphysis. The MRI picture is entirely different from the well-defined geographical defect that is seen in avascular necrosis, where the degree of surrounding oedema is more subtle, if present at all.

Transient osteoporosis of the hip occurs in two distinct clinical settings: middle-aged men and women in the third trimester of pregnancy (Fig. 12.21), where the left hip is almost invariably affected. The aetiology is unknown; aborted ischaemia has been postulated as a cause but this is unproved. The condition is self-limiting but care must be taken to reduce weight bearing and prevent fracture. If it occurs in pregnancy, it most commonly resolves following parturition. In other cases, it may resolve in one joint to appear in another, in which case the condition is termed regional migratory osteoporosis. The MRI appearances are also easily distinguished from tumour, and biopsy, which may precipitate fracture, can be avoided.

12.7.6
The Clicky Hip

A clicking hip is a common presentation to the orthopaedic clinic. In some cases this is only felt by the patient; in others a loud click is present which is both audible and palpable.

The common causes of the clicky hip in adolescence are snapping iliotibial band, iliopsoas tendon snapping and labral tears. In many cases no anatomical cause for clicky hips is determined. The clinical features of all three aforementioned forms are a palpable or audible click and, sometimes, associated pain. The pain associated with a labral tear tends to last hours to days, as opposed to minutes to hours for other causes.

In many cases, no imaging is required to diagnose a snapping iliotibial band or psoas tendon and the diagnosis is a clinical one. Dynamic ultrasound can demonstrate tendon movement in some individuals and may show an associated iliopsoas bursa, which may also be seen on axial MRI sections.

Labral tears are best visualised on MRI arthrography (Fig. 12.22). Plain MRI can be considered the first line of investigation as it is non-invasive. If positive, it obviates the need for an invasive procedure; however, most studies suggest that less than 50% will be positive and in some studies this figure is as low as 30%.

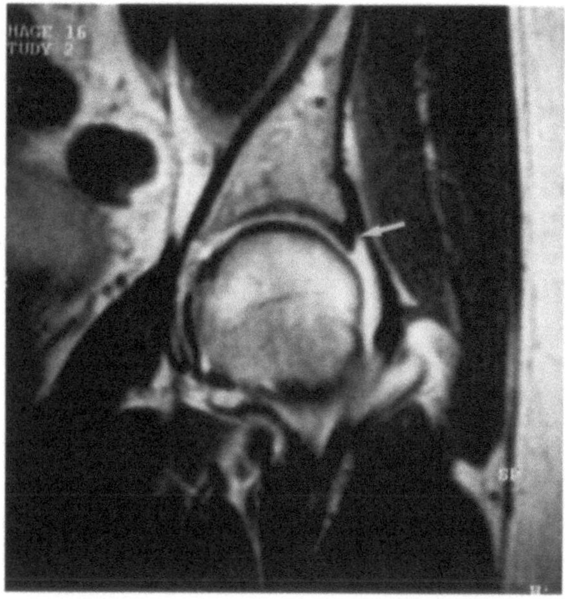

Fig. 12.22. T1-weighted image following gadolinium arthrogram, showing the normal labrum (*arrow*)

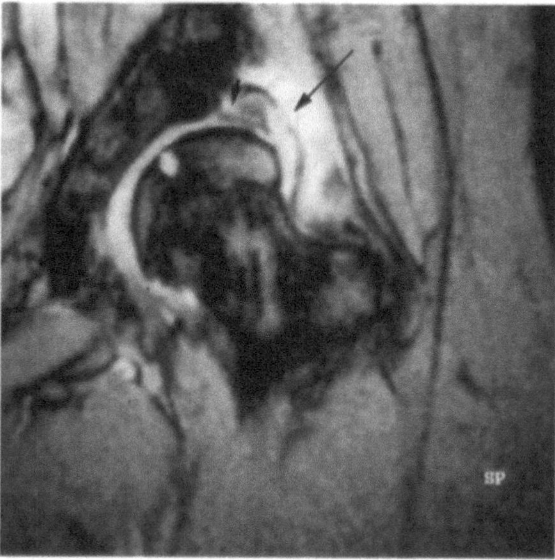

Fig. 12.23. Coronal gradient echo T2* image showing labral tear; acetabular dysplasia with secondary osteoarthritis. Note fluid within the labrum (*short black arrow*) and disruption of the iliofemoral ligament (*long arrow*)

Gadolinium MR arthrography can be combined with standard arthrography to improved diagnostic yield. Conventional arthrography can be carried out initially. A small dose of gadolinium is introduced into the joint at the same time. The recommended dose is 0.25 ml mixed with 10–15 ml of conventional contrast medium. The arthrogram is completed with the instillation of local anaesthetic into the joint. Symptom alleviation following this manoeuvre confirms the hip joint as the pain source, which can be useful when symptoms are atypical. MR images should include a coronal T1-weighted sequence oriented perpendicular to the acetabulum. Sagittal and axial images have also been suggested, though there is no evidence that they convey any additional diagnostic information. The high contrast between fibrocartilage and hyaline cartilage also makes T2* gradient-echo imaging very useful, particularly in association with fat suppression. Tears are seen as high signal fissures within the labrum (Fig. 12.23), extending to the surface. Occasionally intra articular-gadolinium is absorbed into the mid substance of the labrum. The significance of this finding is unclear but it probably represents labral degeneration. The majority of patients with labral injuries have concurrent developmental dysplasia of the hip. Pain can occur as a result of the labral tear itself, or from associated labral cysts or early osteoarthritis.

12.7.7
Trochanteric Bursitis

Trochanteric bursitis is usually a clinical diagnosis with pain and tenderness located inferior and slightly posterior to the greater trochanter. Plain films may show some calcification; this is better appreciated when films are obtained with the hip in various degrees of rotation. Ultrasound can also be diagnostic. Symptoms from enlargement of the iliopectineal bursa are more anterior and a mass may be palpated. Pain is increased on hip extension. Large lesions may irritate the femoral nerve. Sciatic nerve irritation may be a result of ischiogluteal bursitis.

12.8
Hip Disorders in Adults

12.8.1
Rheumatoid Arthritis

The hip is involved late in the rheumatoid process and therefore poses less of a diagnostic dilemma than disease of the small joints of the hand or feet. Rheumatoid arthritis (RA) can be differentiated from the very much more common osteoarthritis

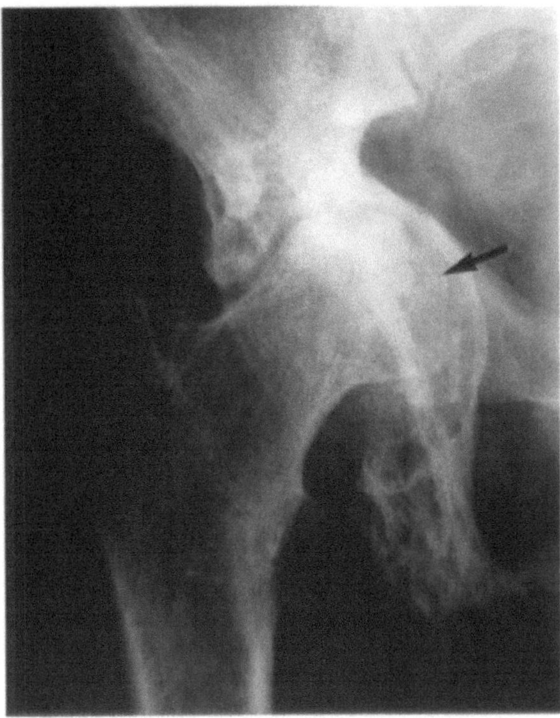

Fig. 12.24. Symmetrical loss of articular cartilage, periarticular osteoporosis and protrusio characterise rheumatoid involvement of the hip. The paucity of osteophyte formation is an additional clue. Note the large erosion in the region of the fovea (*black arrow*)

porosis and protrusio (Fig. 12.25) are characteristic findings, though the latter can also occur with OA, with other inflammatory arthropathies, with metabolic conditions that soften bone, including osteoporosis and Paget's disease, and as an idiopathic phenomenon (when it is referred to as Otto's pelvis). Protrusio is present when the medial border of the acetabulum lies medial to the medial margin of the teardrop by at least 3 mm in males and 6 mm in females.

12.8.2
Seronegative Spondyloarthropathy

Hip involvement is more frequent and occurs at an earlier stage of disease than in RA. It can be present in approximately 25% of patients at presentation. Abnormalities are usually bilateral and symmetrical. One of the earliest features is a characteristic superolateral osteophyte (Fig. 12.26). Progression of osteophytosis results in a typical appearance that has been likened to the ruffles of Tudor dress. Involvement is bilateral in more than 90% of cases and is usually symmetrical. The combination of symmetrical joint space loss and osteophytosis helps to separate a seronegative arthropathy from RA, where osteophytes are uncommon, and OA, where joint space loss is not usually symmetrical. Ultimately the joint may undergo bony fusion. The hip is less frequently involved in psoriasis, Reiter's syndrome and the enteropathic spondyloarthritides than with ankylosing spondylitis.

(OA) by the symmetrical involvement, both within each joint and between the two sides. Cartilage loss involves the entire surface (Fig. 12.24) and does not dominate on the weight-bearing surfaces. Osteo-

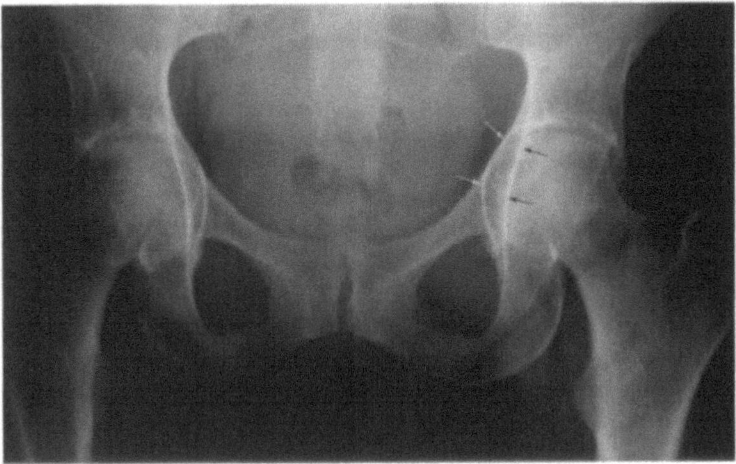

Fig. 12.25. Bilateral protrusio acetabuli in RA. Note how the medial margin of the acetabulum (*white arrows*) projects medial to the ilio-ischial line (*black arrows*) by more than 6 mm; 3 mm is sufficient to make the diagnosis in males

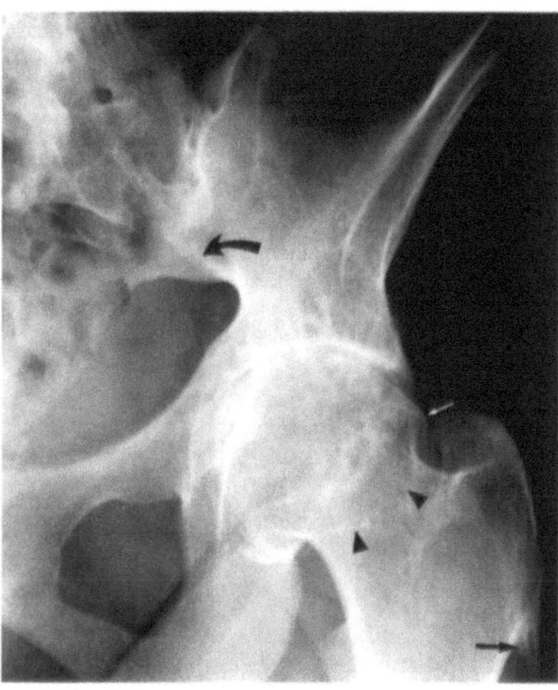

Fig. 12.26. Ankylosing spondylitis. Note the typical lateral osteophyte (*white arrow*). Extension around the margin of the femoral head is seen as a band of increased sclerosis, the so-called Tudor ruff sign (*arrowheads*). Symmetrical joint space loss and the large enthesophyte from the greater trochanter (*black arrow*) are also features. Note also the sacroiliac fusion (*curved arrow*)

12.8.3
Osteoarthritis

The hallmarks of OA are asymmetrical joint space loss, osteophyte formation, subchondral bone sclerosis and the formation of periarticular cysts. The latter can be a dominant feature even in the early stages of the disease. A large cyst in the anterolateral acetabulum is a common pattern observed on plain films and the term "Egger's cyst" has been applied. The patient may present with groin pain and relatively little reduction in the range of motion. Occasionally these cysts can rupture through the lateral wall and involve the soft tissues, when they resemble para-labral cysts (Fig. 12.27).

Several distinct patterns of OA of the hip are observed. Superolateral osteophytes may fill in the normal concavity between the femoral head and the neck, resulting in the pistol-grip deformity. With increasing severity, the femoral head may migrate in a superior (Fig. 12.28), medial or axial (Fig. 12.29) direction. Superior migration is the most common of

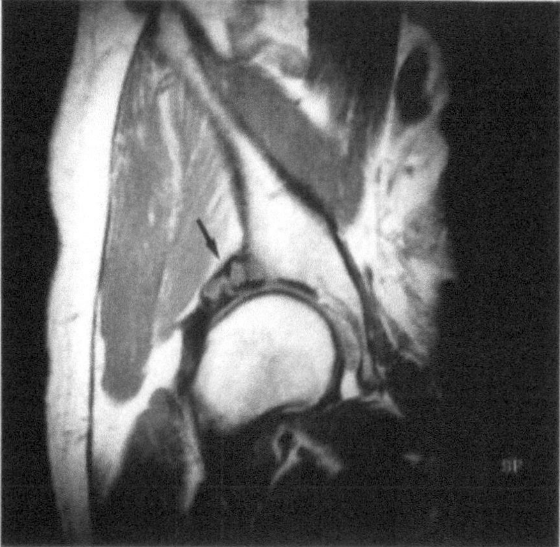

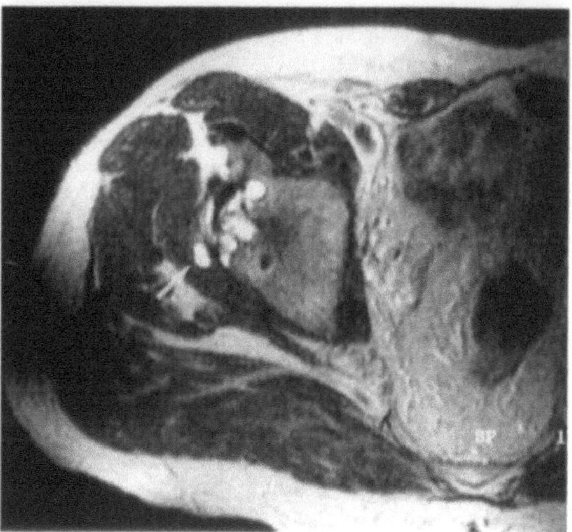

Fig. 12.27. **a** Coronal T1-weighted image showing a large acetabular cyst breaching the lateral cortex. Note the pistol grip deformity of early OA. **b** Axial T2-weighted image depicting rupture of the cyst through the lateral cortex and capsule and into the gluteus minimus, resembling a para-labral cyst

these patterns and can be further subdivided into superomedial and superolateral. The former is more common in men and is bilateral. The latter is more common in women and is frequently unilateral. Superolateral migration is also the typical pattern when there is associated acetabular dysplasia. On transaxial CT sections, superolateral migration has been shown to be associated with anterior migration whereas posterior migration is the typical pattern associated with medial migration (HAYWARD et al.

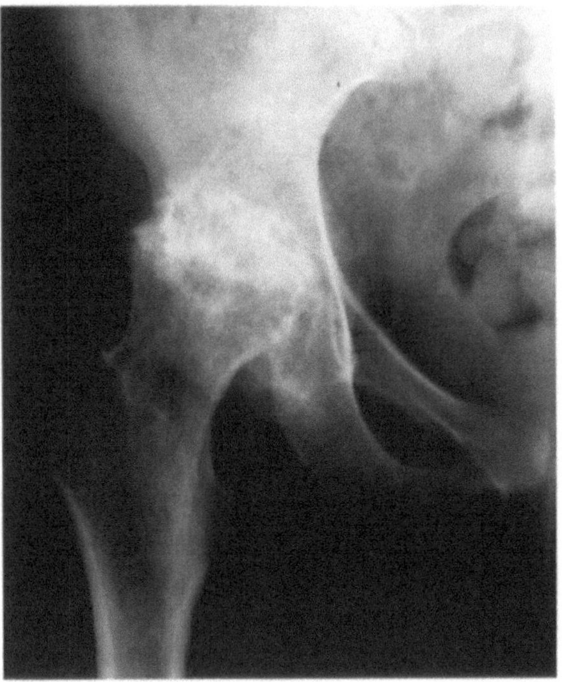

Fig. 12.28. Severe OA with superior and lateral femoral head migration. Large subarticular cysts have formed and there is marked flattening and sclerosis of the femoral head. Note the increased distance between the head and tear drop (*arrows*)

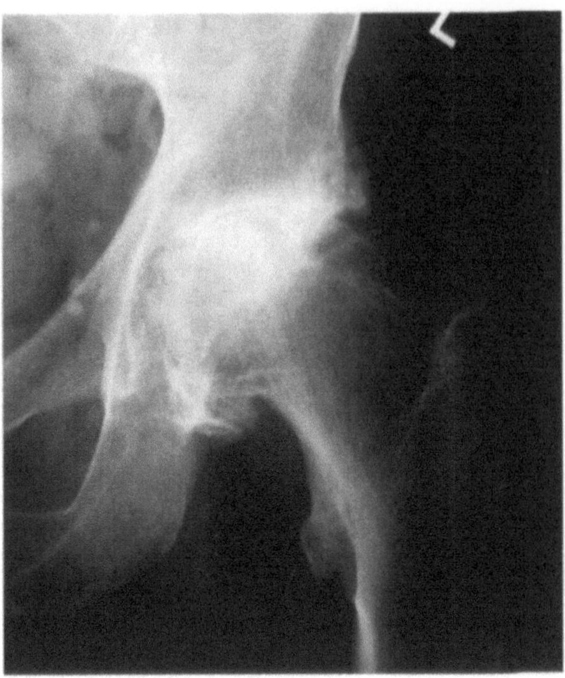

Fig. 12.29. Osteoarthritis. Compare with Fig. 12.28. Medial migration and axial joint space loss are present in this case

1988). True medial migration is also more common in women.

Rapidly destructive OA of the hip has been previously attributed to use of certain non-steroidal anti-inflammatory agents. Though this hypothesis is now disproved, the aetiology of this aggressive variant is still unknown. Infection, avascular necrosis and co-existent crystal arthropathy all need to be considered in the differential diagnosis pre-operatively. The typical patient is an elderly female with a short duration of symptoms despite rapid progression of radiological changes (Fig. 12.30) (ROSENBERG et al. 1992). Osteophytosis tends to be a minor component, probably reflecting the aggressive nature of the disorder with relatively little time for new bone formation. Fragmentation and debris occur in a manner similar to neuropathic osteoarthrosis. Occasionally the joint space appears widened, mimicking avascular necrosis. The degree of acetabular involvement helps to differentiate these two conditions. The excised femoral heads show typical appearances of OA (ROSENBERG et al. 1992).

12.8.4
Hip Arthroplasty

While the symptoms of degenerative OA can be controlled temporarily and its progress occasionally delayed by the judicious use of osteotomy, the ultimate outcome for many severely degenerate hips is arthroplasty. Hip arthroplasty must rank as one of the most successful operations in the surgical armamentarium. The indication for surgery is based almost entirely on symptoms. Although a variety of radiographic classifications of OA exist, there are no features that correlate with clinical symptoms and occasionally even the most severe radiographic changes can be associated with minimal symptoms. Hip arthroplasty is discussed in more detail in Chap. 18.

Acknowledgements. I am particularly grateful to Alison Davies and Mary Morgan for their invaluable assistance in the preparation of this chapter.

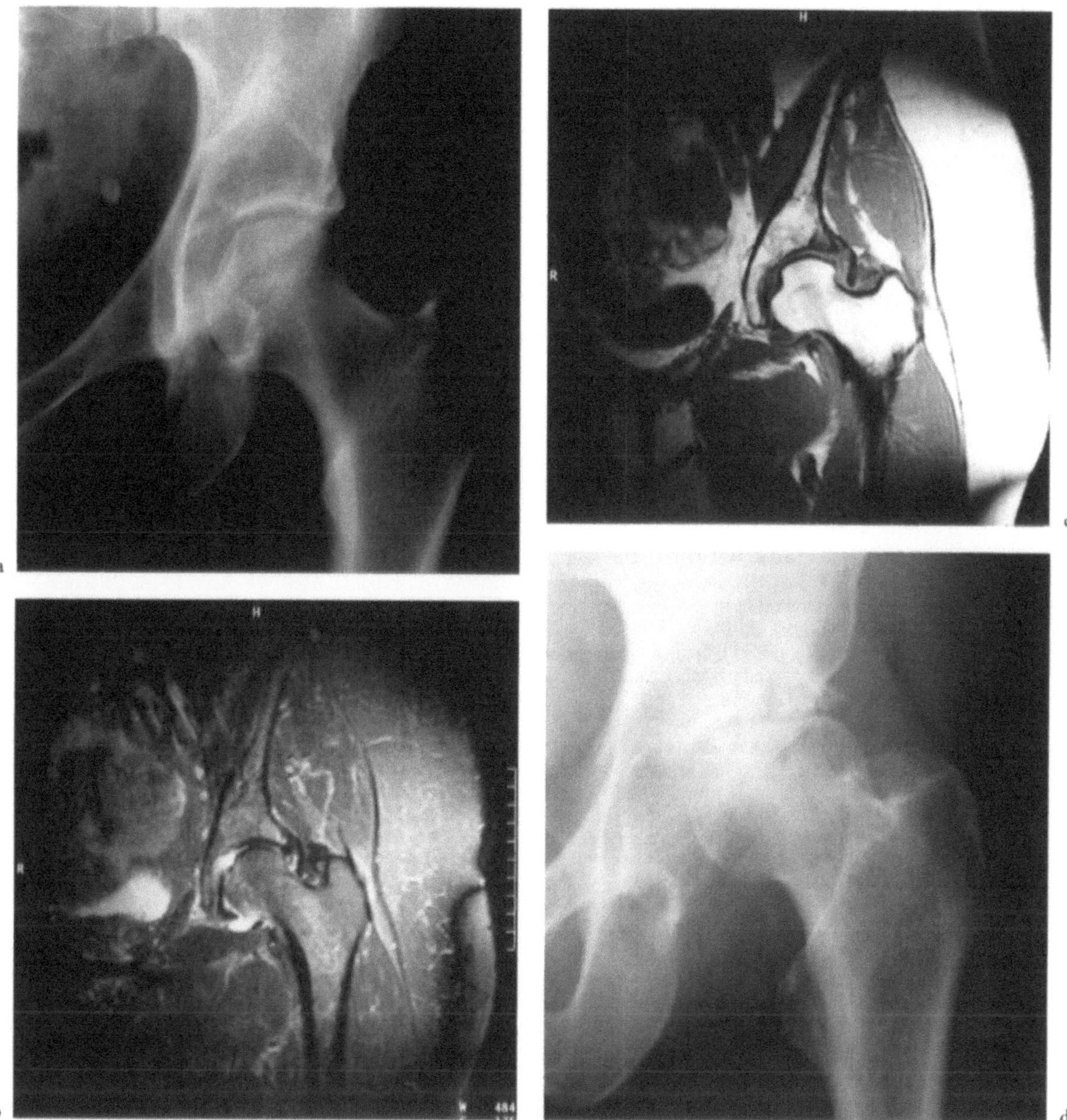

Fig. 12.30 a–d. This 60-year-old female underwent MRI for unexplained left hip pain. **a** The initial plain film is essentially normal. **b** T1-weighted coronal MRI showing small effusion and synovial thickening particularly along the superior femoral neck and in the inferior recess. **c** STIR imaging demonstrates low-signal synovial thickening. Diagnostic possibilities include PVNS, amyloid or fibrosis. Synovial biopsy following arthrography revealed cellular inflammatory exudate only. **d** There was rapid progression of osteoarthritis over the next 18 months, culminating in total hip replacement. Microscopy of the excised head at demonstrated OA with a large inflammatory component. There was no evidence of infection. Diagnosis: rapidly progressive OA

References

Adam R, Hendry GMA, Moss J, et al. (1986) Arthrosonography of the irritable hip in childhood: a review of 1 year's experience. Br J Radiol 59:205–208

Alexander JE, Seibert JJ, Glasier CM, et al. (1989) High-resolution hip ultrasound in the limping child. J Clin Ultrasound 17:19–24

Berman L, Klenerman L (1986) Ultrasound screening for hip abnormalities: preliminary findings in 1001 neonates. Br Med J 293:719–722

Bickerstaff DR, Neal LM, Booth AJ, Brennan PO, Bell MJ (1990) Ultrasound examination of the irritable hip. J Bone Jt Surg [Br] 72:549–553

Boal DKB, Schwenkter EP (1985) The infant hip:assessment with real-time US. Radiology 157:667–672

Castelein RM, Sauter AJM, De Vlieger M, Van Linge B (1992) Natural history of ultrasound hip abnormalities in clinically normal newborns. J Pediatr Orthop 12:423–427

Catterall A (1980) Natural history, classification, and x-ray signs in Legg-Calvé-Perhes' disease. Acta Orthop Belg 46:346–351

Clarke NMP, Clegg J, AL Chalabi AN (1989) Ultrasound screening of hips at risk for CDH. Failure to reduce the incidence of late cases. J Bone Jt Surg [Br] 71:9–12

Cotten A, Flipo RM, Chastanet P, Desvigne Noulet MC, Duquesnoy B, Delcambre B (1995) Pigmented villonodular synovitis of the hip:review of radiographic features in 58 patients. Skeletal Radiol 24:1–6

Dorr U, Zieger M, Hauke H (1988) Ultrasonography of the painful hip. Prospective studies in 204 patients. Pediatr Radiol 19:36–40

Flipo RM, Desvigne Noulet MC, Cotten A, et al. (1994) Pigment villonodular synovitis of the hip. Results of a French National Survey (with 58 cases). Rev Rhum Engl Ed 61:77–87

Gardiner HM, Duncan AW (1992) Radiological assessment of the effects of splinting on early hip development: results from a randomised controlled trial of abduction splinting vs sonographic surveillance. Pediatr Radiol 22:159–162

Gilbert F (1997) Prognostic factors in femoral head AVN. Radiology 97, Birmingham, UK

Giustra PE, Freiberger RH (1970) Severe growth disturbance with osteoid osteoma. A report of two cases involving the femoral neck. Radiology 96:285–288

Goldberg RP, Weissman BN, Naimark A, Braunstein E (1983) Femoral neck erosions: sign of hip joint synovial disease. AJR Am J Roentgenol 141:107–111

Goldberg VM, Jacobs B (1975) Osteoid osteoma of the hip in children. Clin Orthop 106:41–47

Goldman AB, DiCarlo EF (1988) Pigmented villonodular synovitis. Diagnosis and differential diagnosis. Radiol Clin North Am 26:1327–1347

Harcke HT, Grissom LE (1990) Performing dynamic sonography of the infant hip. Am J Roentgenol 155:837–844

Hayward I, Bjorkengren AG, Pathria MN, Zlatkin MB, Sartoris DJ, Resnick D (1988) Patterns of femoral head migration in osteoarthritis of the hip: a reappraisal with CT and pathologic correlation. Radiology 166:857–860

Hughes A (1985) Idiopathic chondrolysis of the hip: a case report and review of the literature. Ann Rheum Dis 49:268–272

Kay RM, Lieberman JR, Dorey FJ, Seeger LL (1994) Inter- and intraobserver variation in staging patients with proven avascular necrosis of the hip. Clin Orthop Relat Res 307:124–129

Keller MS, Weltin GG, Rattner Z, Taylor KJW, Rosenfield NS (1988) "Normal instability of the hip in the neonate: US standards." Radiology 169:733–736

Kramer J, Recht M, Deely DM, et al. (1993) MR appearance of idiopathic synovial osteochondromatosis. J Comput Assisted Tomogr 17:772–776

Miralles M, Gonzalez G, Pulpeiro JR, et al. (1989) Sonography of the painful hip in children: 500 consecutive cases. Am J Roentgenol 152:579–582

Morin C, Harcke HT, MacEwen GD (1985) The infant hip: real-time US assessment of acetabular development. Radiology 157:673–677

Murphy R, Marsh H (1978) Incidence and natural history of head at this factor in Perthes disease. Clin Orthop 132:102

Muscolo DL, Velan O, Acero GP, Ayerza MA, Calabrese ME, Araujo ES (1995) Osteoid osteoma of the hip: percutaneous resection guided by computed tomography. Clin Orthop Relat Res 310:170–175

Reikeras O, Bjerkreim I, Sortland O (1983) Fluoroscopy in measurement of femoral neck anteversion. Acta Radiol 24:81–83

Rosenberg ZS, Shankman S, Steiner GC, Kastenbaum DK, Norman A, Lazansky MG (1992) Rapid destructive osteoarthritis: clinical, radiographic, and pathologic features. Radiology 182:213–216

Rosendahl K, Markestad T, Lic RT (1992) Ultrasound in the early diagnosis of congenital dislocation of the hip: the significance of hip stability versus acetabular morphology. Pediatr Radiol 22:430–433

Schuler P, Feltes E, Kienapfel H, Griss P (1990) Ultrasound examination for the early determination of dysplasia and congenital dislocation of neonatal hips. Clin Orthop Relat Res 258:18–26

Shiv VK, Jain AK, Taneja K, Bhargava SK (1990) Sonography of hip joint in infective arthritis. J Can Assoc Radiol 41:76–78

Steinberg ME, Hayken GD, Steinberg DS (1995) A quantitative system for staging avascular necrosis. J Bone Jt Surg [Br] 77:34–41

Sugano N, Ohzono K, Masuhara K, Takaoka K, Ono K (1994) Prognostication of osteonecrosis of the femoral head in patients with systemic lupus erythematosus by magnetic resonance imaging. Clin Orthop Relat Res 305:190–199

Wirth T, LeQuesne GW, Paterson DC (1992) Ultrasonography in Legg-Calve-Perthes disease. Pediatr Radiol 22:498–504

Zieger MM, Dorr U, Schulz RD (1987) Ultrasonography of hip joint effusions. Skeletal Radiol 16:607–611

13 The Knee

C. Masciocchi and M.V. Maffey

CONTENTS

13.1 Introduction

The knee is a complex articulation particularly vulnerable to direct traumas and to forces provoking torsions and exaggerations of common movements; its stability, furthermore, depends on the sturdiness of the surrounding muscles and ligaments. The increase in sporting activity and in workplace and car accidents has resulted in an increase in lesions involving the knee.

The study of knee pathologies (SMILLIE 1980), whether traumatic or degenerative, requires an integrated course of evaluation which includes an accurate clinical examination as a first step. A traumatic event involving the knee often brings about pluristructural involvement and a complex clinical picture due to the anatomical and biomechanical complexity of this area. In these cases it is necessary to perform a global evaluation of the articulation of both the intra- and extra-articular structures in order to decide upon the appropriate therapeutic approach.

A complete radiographic examination consists of comparative anterior-posterior projections of the intercondylar notch. Oblique projections may also be employed.

In the case of acute or chronic knee pathologies, the radiographic examination is often the first instrumental investigation to be performed, with the aim of excluding or diagnosing the presence of fractures. Apart from serious traumatic lesions, such as multifragmentary fractures involving the femur, the tibia, or the fibula, there exists a series of moderately serious fractures that need to be identified by means of radiography, including tibial plate fractures, torn tibial spine fractures or fractures involving the entire intercondylar eminence, and Segond fractures, which are pathognomonic of anterior cruciate ligament rupture. Radiography also permits identification of patellar osteochondral fractures, fractures involving tearing of the tibial insertion of the posterior cruciate ligament (which is well identified in the correct lateral projection), and fractures of the peroneal head (which may reveal a lesion of the posterior capsuloligamentous lateral complex).

It is important to pay attention to the importance of indirect radiological signs in the traumatic knee. Often, in fact, moderately serious fractures may not be perceptible on radiographs, but only presumed on the basis of certain indirect signs, such as the presence of increased density in soft tissue areas and the dislocation of fat pads owing to effusions or hemorrhagic extravasation.

In patients with chronic pathologies, radiographs provide a large quantity of information concerning the history of the condition such as the presence of peri- or intra-articular calcifications and the reduction of one or both femorotibial compartments due to the presence of arthrotic damage; in such cases,

C. MASCIOCCHI, MD, Professor, Università degli Studi dell'Aquila, Facoltà di Medicina e Chirurgia, Cattedra di Radiologia, Ospedale Collemaggio, I-67100 L'Aquila, Italy
M.V. MAFFEY, MD, Università degli Studi dell'Aquila, Facoltà di Medicina e Chirurgia, Cattedra di Radiologia, Ospedale Collemaggio, I-67100 L'Aquila, Italy

it is necessary to complete the investigation with an orthostatic examination. Calcifications in the meniscal area need to be distinguished from the multiple calcifications which are typical of synovial chondromatosis. Intra-articular mobile bodies also have to be carefully searched for and distinguished from the sesamoid or supernumerary bones. Peri- or intra-articular calcifications can indicate the presence of a central pivot or medial collateral ligament injury.

Finally, the radiographic examination can both exclude and reveal the presence of a heteroplastic process; hence it is of prime importance in the study of the anatomical and mechanical axis of the knee in the event of surgery, allowing evaluation of the type and extent of correction that is required.

The chief limitation of the radiographic examination is its inability to directly visualize the menisci, ligaments, and cartilages, which are the structures most frequently involved in knee trauma.

13.2
The Menisci

The meniscal structures, together with the ligamentous tendinous formations and the articular heads, play an important role in the articular biomechanics of the knee. In acute meniscal pathologies, caused by an asynchronism between the flexion-extension and rotation movements of the knee, the symptom/sign that characterizes the clinical picture is the articular block involving the articular rim, accompanied by pain and evident effusion. In chronic pathologies, the clinical picture is dominated by pain which cannot always be localized by the patient and is sometimes attributed to the contralateral compartment. The meniscal syndrome is completed by the presence of the so-called mechanical symptomatology, i.e., particular articular blocks, jerks perceived as clicks (which may be more or less painful), and instability which is generally typical of a ligamentous relaxation.

Similar symptoms occur when lesions of the posterior horn of the medial meniscus destabilize the articulation. It is furthermore important to point out that young, adult, and elderly patients generally have different types of lesions, given that in the young meniscal lesions are often due to sports injuries while in the elderly they are typically of a chronic, degenerative nature.

Although the diagnosis of meniscal syndrome may be achieved by accurate clinical investigation and anamnesis, the introduction of computed tomography (CT) and subsequently magnetic resonance imaging (MRI) has markedly improved reliability in the determination of meniscal pathologies (MASCIOCCHI et al. 1993a) and in the global evaluation of the knee joint (PASSARIELLO et al. 1983a). The reliability demonstrated by CT in the study of normal and pathological anatomy in the knee joint (PASSARIELLO et al. 1983) depends on the proper application of a simple but accurate examination technique which entails the use of high resolution programs with thin layer density and, for the study of menisci, the possible use of partial slice overlapping. Study of the knee with MRI can be considered complete when imaging is performed in each of the scan planes. Even though MRI in the longitudinal plane permits easier and more immediate imaging, the slice thickness of 2–3 mm in the axial plane is of fundamental importance in the evaluation of some strategic and critical areas of this articulation, such as the intercondylar notch and the femoropatellar joint. The sagittal plane is utilized for the study of the meniscal horns and of the interactions between the medial meniscus and posteromedial shell; the coronal plane, by contrast allows for perfect evaluation of the meniscal body and is indispensable in the identification of possible loose meniscal fragments.

These imaging methods, therefore, permit correct documentation of both the medial and the lateral menisci (which are easily distinguishable on the basis of their dimensions and morphology), and of their anatomical interactions with the capsular structures. In particular, the body of the medial meniscus shows direct contact with the deep fibers of the medial collateral ligament, while the posterior horn is attached to the posterior medial shell; on the other hand, the lateral meniscus is attached to the tendon of the popliteal muscle and to the meniscofemoral ligament at the level of the posterior horn (Fig. 13.1). According to pathogenetic criteria, meniscal pathologies are divided into the following types: traumatic, degenerative, cystic, and congenital. However, in numerous cases various conditions coexist.

13.2.1
Traumatic Pathology

Lesions of a traumatic origin more frequently involve the medial meniscus, which, due to its intimate connections with the capsular structures, is less free to swing. The main cause of medial meniscus lesions is the pinching of the posterior horn which

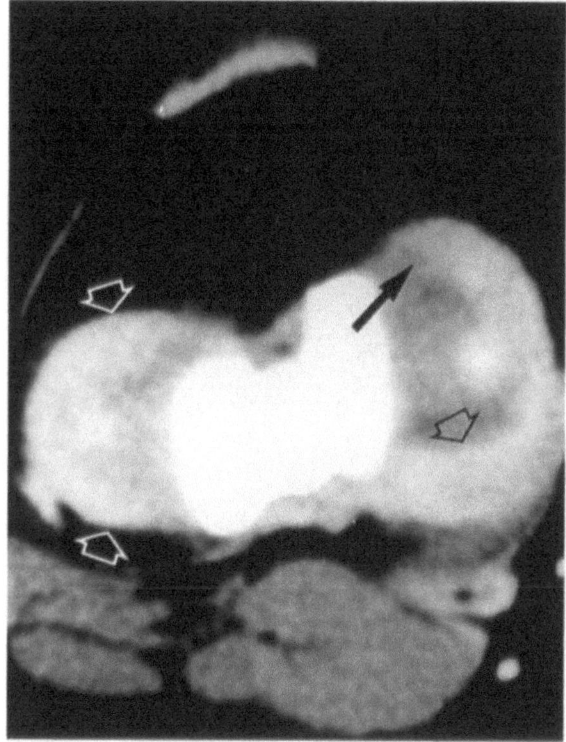

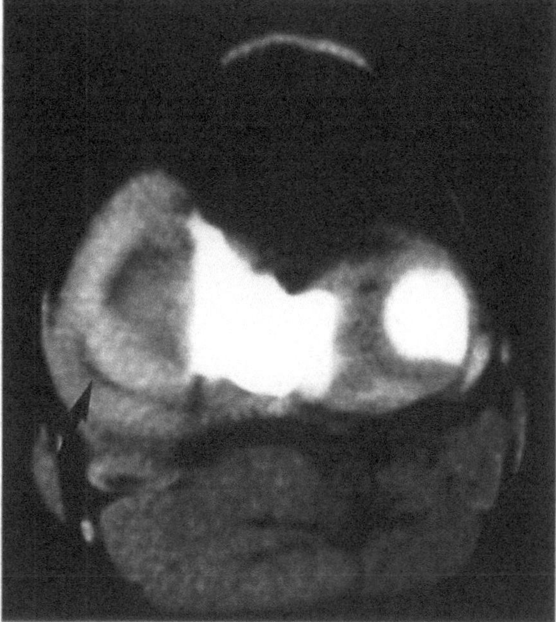

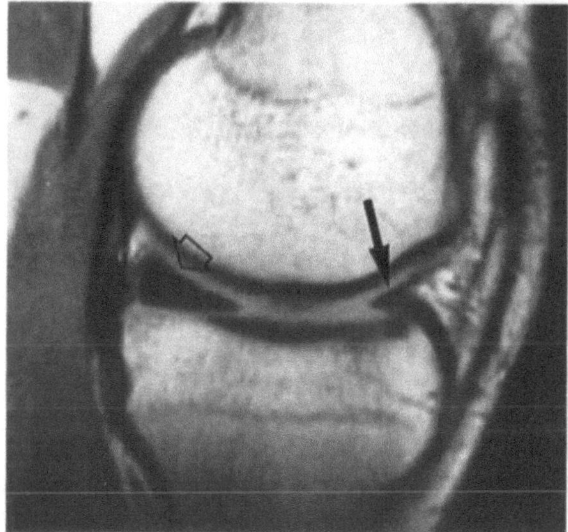

Fig. 13.1 a,b. Normal anatomy. a CT; b MRI. Posterior horn
of the medial meniscus (*open black arrow*), anterior horn
the medial meniscus (*solid black arrow*), and posterior and
anterior horns of the lateral meniscus (*open white arrows*)

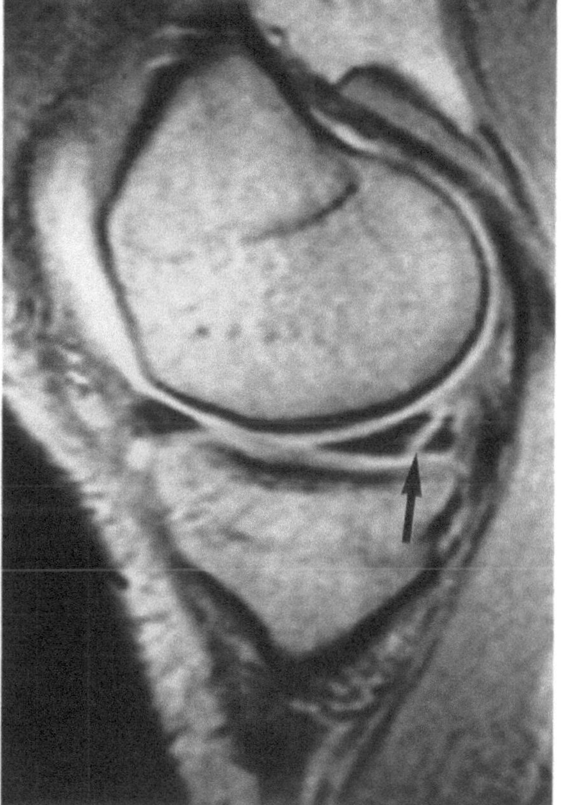

Fig. 13.2 a,b. Lesion of the posterior horn of the medial
meniscus (*arrow*). a CT; b MRI

results from squeezing between the articular heads.
It must be remembered at this point that the medial
meniscus may be involved by disinsertion of the capsular element at the level of its posterior body and
horn.

The lateral meniscus undergoes the most microtrauma because of its higher mobility, its adaptation
to rotation, and its sliding movements. It is also
more prone to degenerative involution owing to the
scarce vascularization resulting from the limited
interaction with the capsular element.

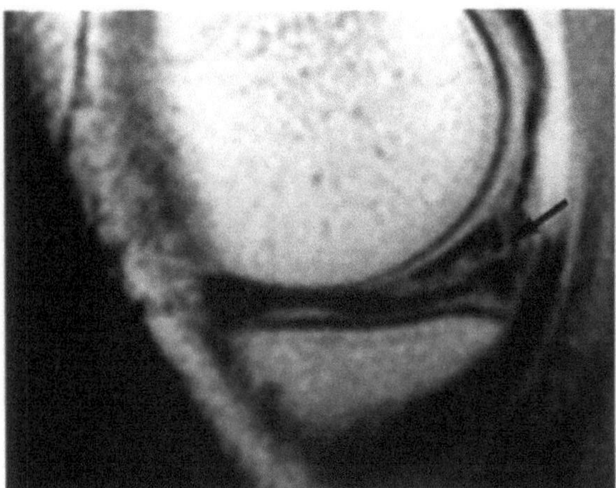

a

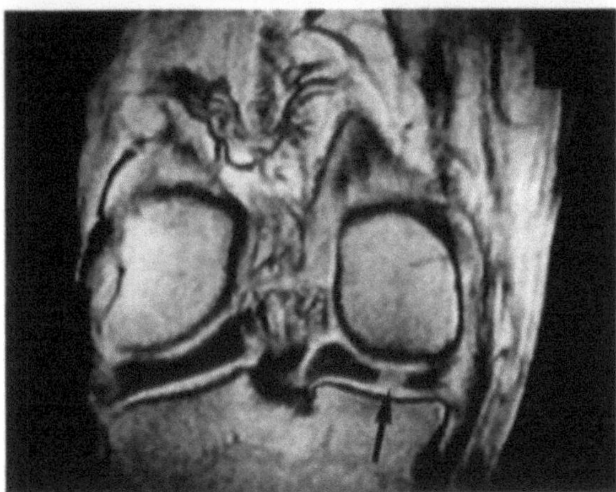

b

Fig. 13.3 a,b. MRI shows a complete lesion of the posterior horn of the medial meniscus (*arrow*). a Sagittal view; b coronal view

On CT, traumatic lesions are characterized by longitudinal hypodense striae with margins that are quite clean-cut or shaded, according to the presence of granulation tissue or fluid, if the direction of the lesion is vertically oriented. If, on the other hand, the direction is oblique, the area of hypodense striae will shift into a laterolateral or anterior-lateral orientation in the following scan planes. With MRI (Fig. 13.2), these lesions can be easily identified as areas of increased signal intensity (SI) which spread to at least one of the articular surfaces. Lesions can be considered complete (Fig. 13.3) when they involve both meniscal surfaces, and partial when only one surface is involved. In the latter case the inferior meniscal surface is usually affected, and since the arthroscopic evaluation may reveal a normal meniscus, palpation of the tibial meniscal surface is required (CRUES et al. 1987).

If a longitudinal lesion is complete, the meniscus remains anchored only to the insertions of the meniscal horns, creating the conditions for a diastasis between the central and peripheral portions and a possible dislocation called the "bucket handle" tear. In this case, a voluminous flap forms, dislocating into the intercondylar notch, and a typical "knot-like" aspect is observed in the seat of the flap reflexion. This type of lesion is, without doubt, more frequent in the medial meniscus (Fig. 13.4), although it is sometimes found in the lateral meniscus. As already mentioned, capsular disinsertions can occur in the medial area, where traumatic damage can be limited to the capsular structure, with more or less evident involvement, on one side, of the medial collateral ligament and the posterior oblique ligament (POL or Hughston ligament). Together these structures make up the so-called posterior-interior angle point, and in traumatic conditions there will be a serosanguinous infarction of the posteromedial capsular shell. This condition may be identified with MRI using high-contrast sequences sich as spin-echo (SE) and gradient-echo (GE) T2-weighted sequences.

Capsular lesions and those involving the posterior wall can result in a reparative evolution as a result of the vascularization features of these structures. Lesions of the more mobile lateral meniscus will display a wider variation in course, from radial (more frequent between the anterior horn and body) to horizontal (Fig. 13.5). Complex lesions, characterized by the presence of numerous fracture rims, generally involve both the medial and the lateral meniscus, and the formation of multiple flaps may occur. These flaps may show a peduncular connection with the meniscal structure and arrange themselves variously in the articular space where they are free to move, resulting in block or pseudoblock.

13.2.2
Degenerative Pathology

Meniscal degenerative changes include all fibrocartilaginous tissue alterations, with or without the presence of a macroscopic lesion. They are all associated with old age and/or functional overload and are therefore typically found in patients older than 45–50 years or in those who subject the knee to particular stress through either sports or professional activities. Although not infrequently the changes are asymptomatic, they often cause pain, swelling, failure, and restrictions in articulation, and the diagno-

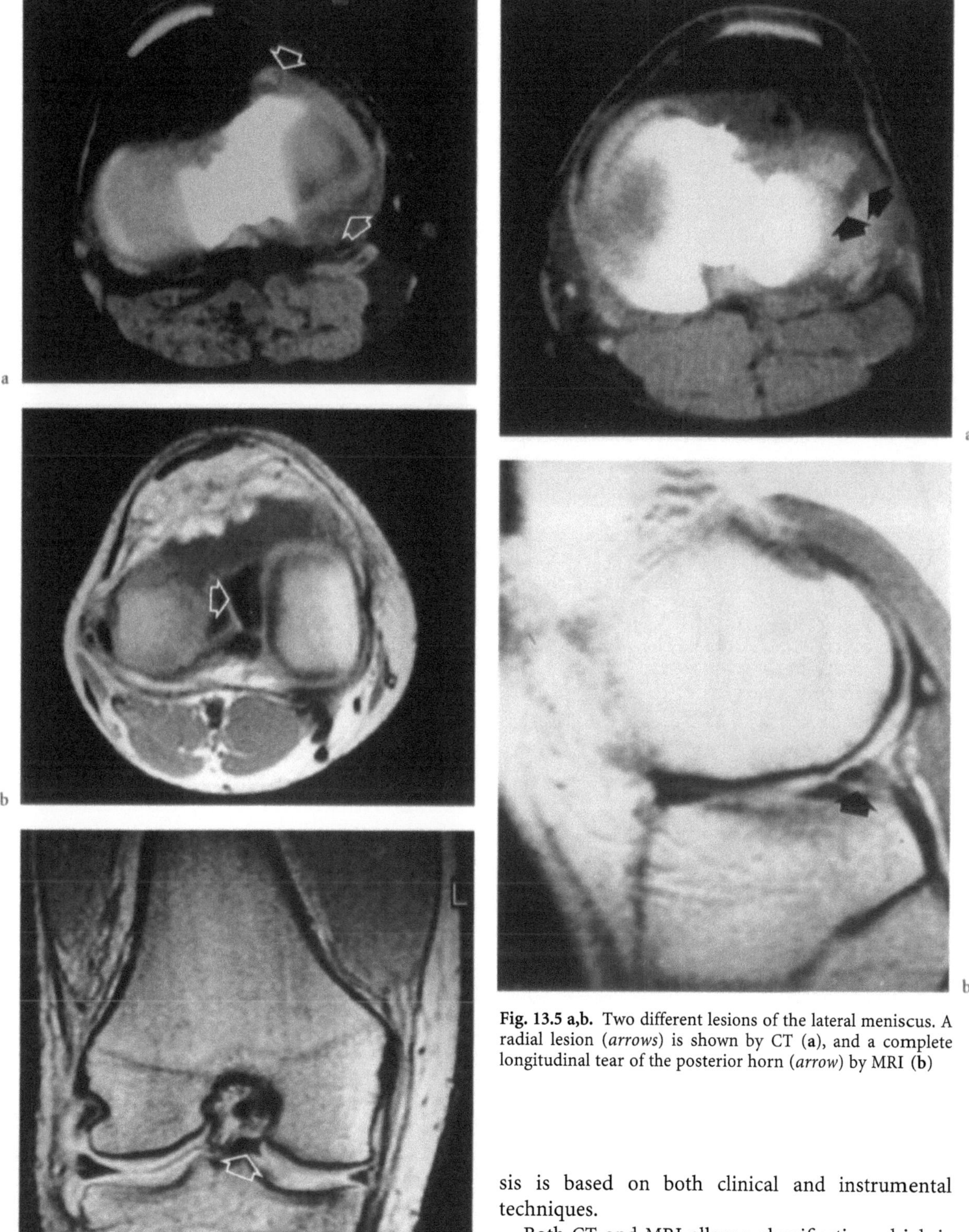

Fig. 13.4 a-c. Bucket-handle lesion of the medial meniscus (*arrow*). **a** CT; **b,c** MRI: axial (**b**) and coronal (**c**) views

Fig. 13.5 a,b. Two different lesions of the lateral meniscus. A radial lesion (*arrows*) is shown by CT (**a**), and a complete longitudinal tear of the posterior horn (*arrow*) by MRI (**b**)

sis is based on both clinical and instrumental techniques.

Both CT and MRI allow a classification which is similar to the arthroscopic classification, in which lesions are categorized as ranging from simple degenerative changes to those associated with other lesions (e.g., chondral). As a preliminary consideration, it needs to be borne in mind that both in

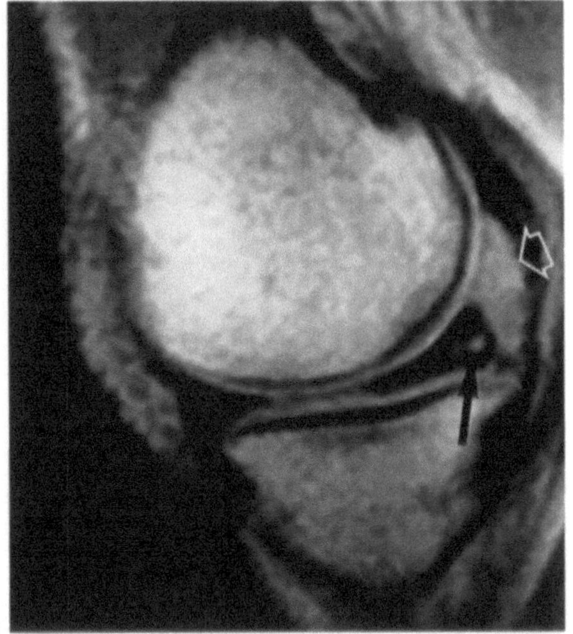

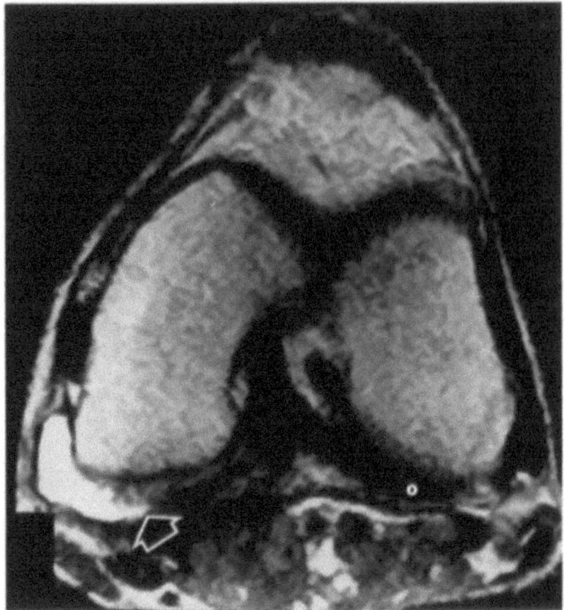

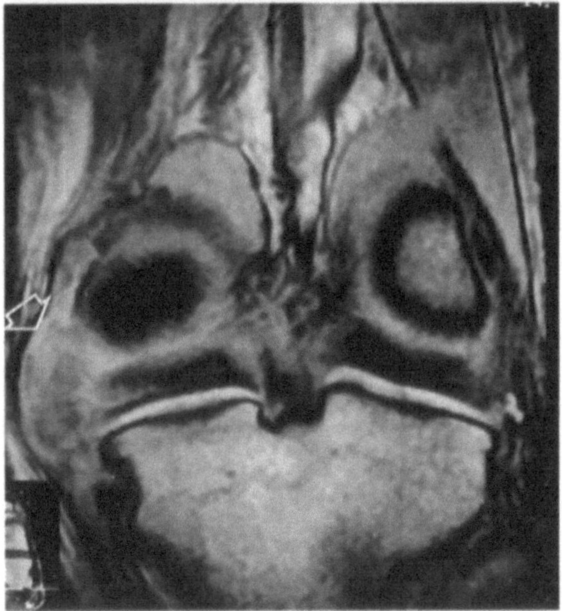

Fig. 13.6 a–c. Lesion of the posterior horn of the medial meniscus (*black arrow*) associated with cystic degeneration (*open white arrows*)

elderly patients and in young subjects who practice sports, it is possible to identify degenerative changes in the central portion of the meniscus which do not extend superficially. In these areas vascularization is scarce and they are considered paraphysiological because neither clinical signs nor complete meniscal lesions are evident. By contrast, irregularity of the meniscal morphology represents a different condition, usually at the level of the posterior horn, in which both the density on CT and the signal intensity on MRI are irregular. This condition may be consid-

ered an advanced degenerative process in which multiple meniscal flaps and cleavages are present. These pictures are often associated with extensive osteochondral alterations which need to be examined carefully.

Particular attention also needs to be given to meniscal cysts. These are more frequent at the lateral meniscus which, as mentioned above, displays limited vascularization and greater mobility, and therefore often undergoes repeated microtrauma which will result in colliquative processes, mucoid

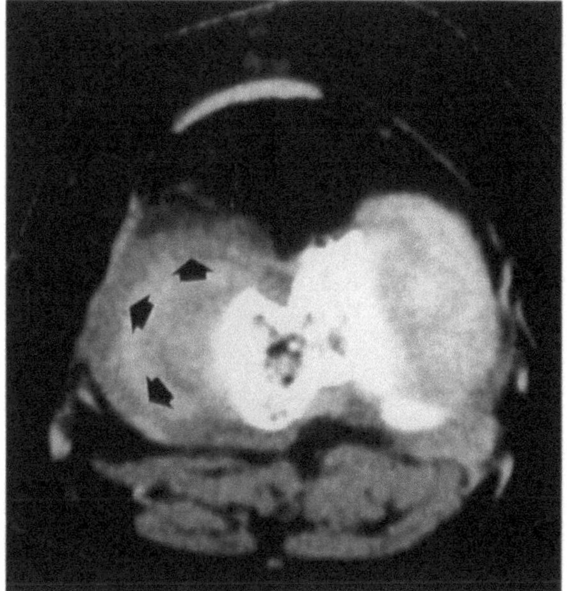

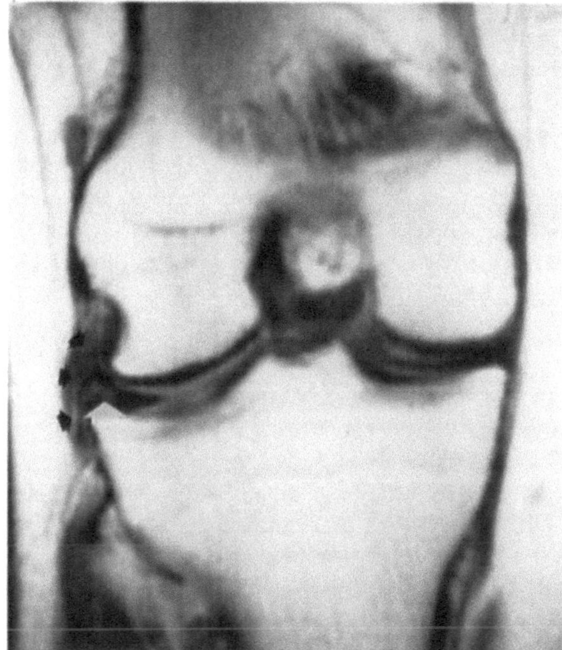

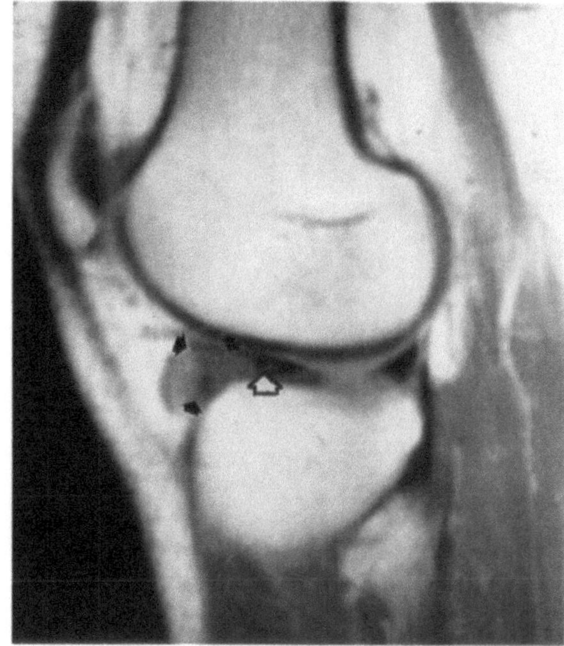

Fig. 13.7. **a** CT shows degenerative change of the lateral meniscus (*arrows*); on sagittal (**b**) and coronal (**c**) MRI scans the associated lesion is well evident (*arrows*)

change, and, as a consequence, real cystic formations. At the level of the medial meniscus such cystic degeneration is rare; if present, it can be identified at the level of the posterior horn (Fig. 13.6).

On CT, cysts appear as nonhomogeneous hypodense areas due to myxoid degeneration and their multilocular nature. On MRI, SE T1-weighted and GE T2-weighted sequences in the sagittal plane allow direct imaging of the cyst. The presence of an area of increased signal intensity confirms its degenerative nature (Fig. 13.7).

13.2.3
Congenital Pathology

Congenital meniscal disorders usually involve the lateral meniscus, which can be affected by dysplastic conditions, such as the discoid lateral meniscus syndrome. In this syndrome the total and subtotal discoid forms represent the classical configuration, although anterior and posterior megahorn may also be encountered. The clinical pattern is dominated by jerky conditions which can cause a real articular

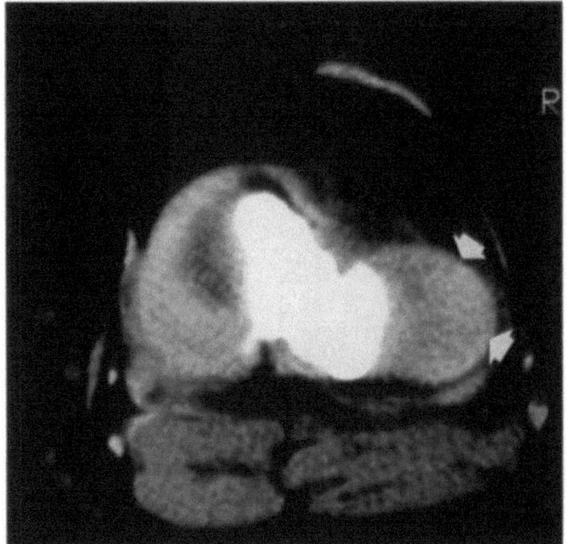

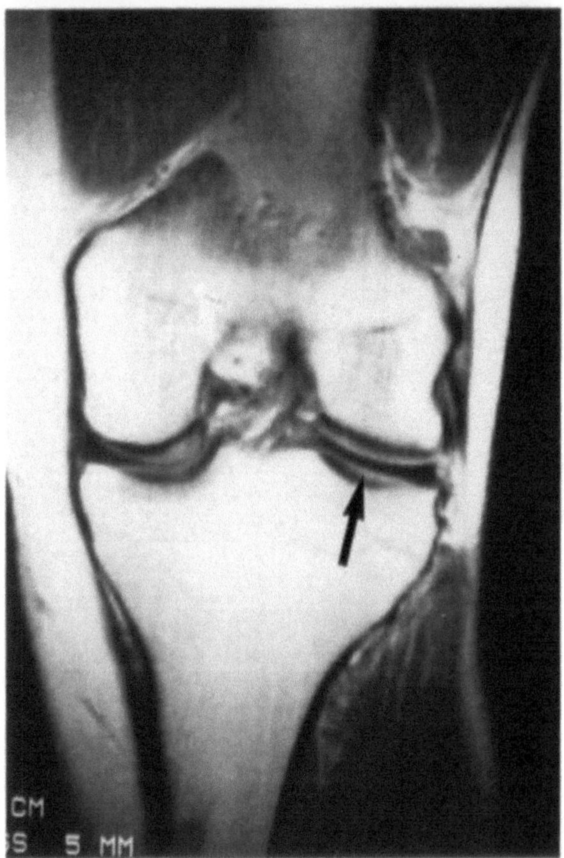

Fig. 13.8. Complete discoid dysplasia of the lateral meniscus on CT (**a**, *white arrows*) and MRI (**b**, *black arrow*)

block and, in degenerative conditions, lateral articular pain. Both CT and MRI may demonstrate this condition (DICKAUT and DeLEE 1982). With CT the meniscal outline is easily identifiable, so that it is possible to establish both the presence and the type of dysplasia (Fig. 13.8). However, MRI provides more information, since coronal scans will reveal the presence of a lamellar structure, completely separating the femoral from the tibial articular space. Furthermore, it is possible to demonstrate degenerative changes which are the result of a modification of the articular mechanism due to the presence of a fibrocartilaginous disc which hinders normal rotation and sliding movements and to a weakness of the meniscal structure caused by the poor nutrition.

13.3
The Ligaments

The knee utilizes three fundamental elements for articular stabilization. Two of these are important for passive stabilization, namely the articular congruence and the meniscal and ligamentous structures (PASSARIELLO et al. 1986), while the third is involved in active stabilization and consists of the tendinomuscular structures. All these biomechanical elements, interacting synergistically, contribute to the articular stability of the knee. The central pivot, represented by the anterior cruciate ligament (ACL) and the posterior cruciate ligament (PCL), plays a fundamental role in terms of passive stabilization potentiality.

13.3.1
Central Pivot

The central pivot is so called because it is the joint which allows for flexion-extension and internal-external rotation movements of the knee. The ACL is made up of the anteromedial and posterolateral bands (ARNOCZKY and RUSSEL 1988). These bands are inserted at the medial wall of the lateral femoral condyle proximally, while they reach the anterior spine of the tibia distally. In complex movement phases, the two main bands behave in a different biomechanical manner; in fact in 90° flexion, while the anteromedial band is in tension, the posterolateral band undergoes complete relaxation. The PCL has two fundamental elements (medial and lateral) and is reinforced by an accessory band, called the meniscofemoral ligament, which originates from the posterior horn of the external meniscus and reaches the internal femoral condyle, either posterior (Wrisberg ligament) or anterior (Humphrey ligament) to the posterior cruciate ligament itself. In biomechanical terms, the accessory band plays a different role than the principal bands since it intervenes above all in the stabilization of the external meniscus.

An important anatomical aspect is the vascularization of the central pivot, which depends on branches of the median geniculate artery. These branches penetrate between the two ligaments, creating an anastomosis. While the PCL is furnished by four branches which supply it homogeneously, the anterior ligament only receives a main branch at its medium level and therefore it is often affected by ischemic events.

The pathogenesis of a lesion involving the ACL involves the following four common traumatic events: external valgus rotation, internal varus rotation, hyperextension, and sudden contraction of the quadriceps (BESSETTE and HUNTER 1990). The traumatic event firstly results in a stretching of the ligamentous structure and only secondarily in a rupture involving the entire ligament or just a part of it. While the elastic stretching allows for anatomical recovery, a complete lesion will lead to ischemic and atrophic events owing to the above-mentioned vascularization problems.

While clinical examination achieves a high level of accuracy in the diagnosis of lesions involving the ACL, both in the acute and in the chronic phase, it is less accurate in respect of lesions of the PCL. It is therefore necessary to resort to imaging methods for the evaluation of ligamentous damage and associated lesions, whether meniscal or osteochondral. In both anatomical and pathological situations, CT and MRI allow good identification of ligamentous structures in general, and the ACL in particular. Both modalities accordingly achieve a diagnostic accuracy ranging between 94% and 96%.

With CT investigation the ligamentous structures of the central pivot (PAVLOV et al. 1979) are clearly visible over their entire course and show a homogeneous aspect, with density values of around 55–70HU for the ACL and 85–100HU for the PCL (REISER et al. 1981). The ACL will appear in the medial-proximal tract (Fig. 13.9), displaying the typical horseshoe shape created by the twisting of the bands, while in the distal tract the insertions will appear to be well separated. The PCL, on the other hand, appears egg-shaped in its distal course, while it is triangular in the medial-proximal tract.

On MRI (Fig. 13.10), the central pivot formations can and must be studied in both the sagittal and the axial plane. This allows complete documentation of these structures, showing their anatomical

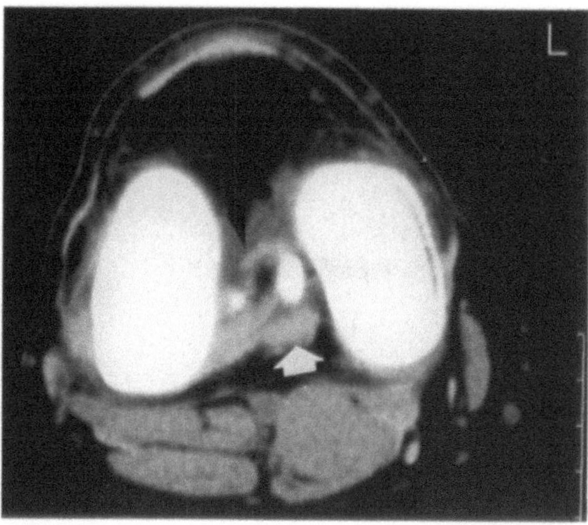

Fig. 13.9. Normal anatomy of the cruciate ligaments on CT: anterior (*black arrow*) and posterior (*white arrow*) cruciate ligaments

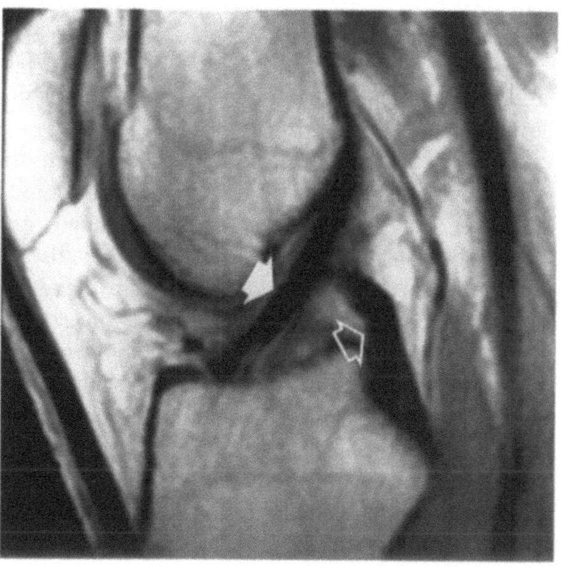

Fig. 13.10. MRI of the anterior (*white arrow*) and posterior (*open white arrow*) cruciate ligaments

bands (with the same characteristics as on CT) and the morphological alterations caused by traumatic events and related to the absence of partial volumes.

T1-weighted spin-echo sequences are usually used because they provide good visualization of the ligaments, which have a low and homogeneous signal intensity. The utilization of high contrast sequences sometimes completes the pathological picture. The aspect will differ according to (a) the extension and location (partial and total) of the lesions and (b) the time interval between the trauma

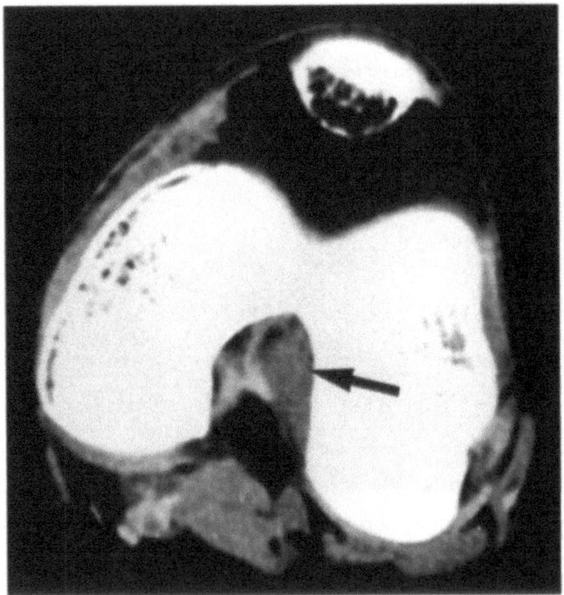

Fig. 13.11. CT showing an acute lesion of the ACL (*arrow*)

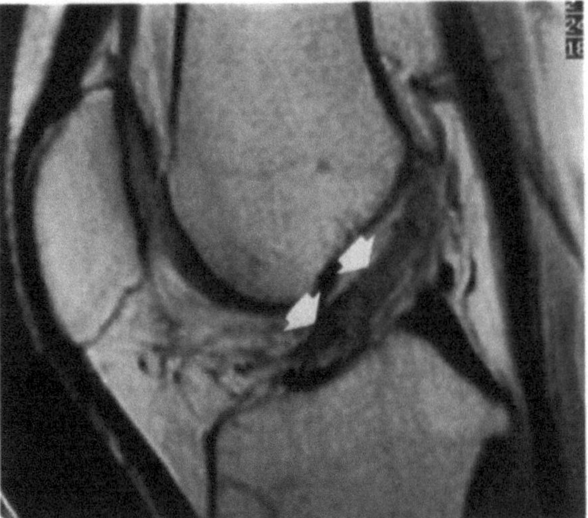

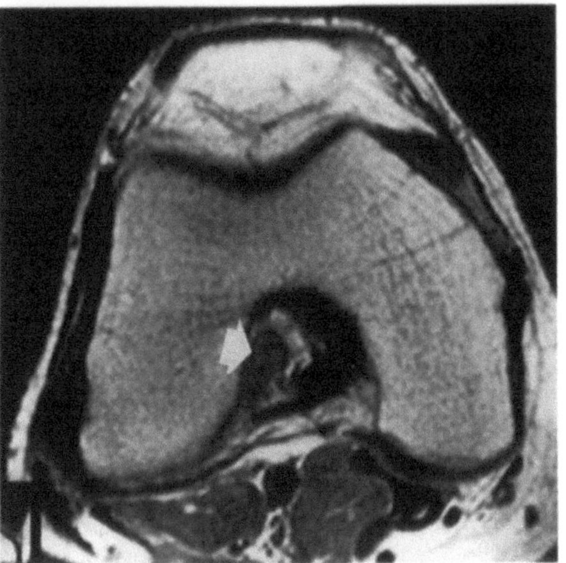

Fig. 13.12 a,b. MRI in a patient with an acute traumatic lesion of the ACL (*arrows*): sagittal (**a**) and axial (**b**) views

and the examination (acute, subacute, and chronic lesions may be distinguished).

On CT, a complete lesion of the ACL in the acute phase (Fig. 13.11) appears nonhomogeneous, enlarged, and hypodense, due to edema and hemorrhage, and is characterized by the absence of ligamentous structures on some scan planes, in accordance with the extension of the lesion.

If the lesion occurs in the intrasynovial proximal insertion of the ligament, the swelling will cause a severe hypodensity of the ligament itself, which will appear thin at this level; it is also possible for there to be an enlargement of the distal portion due to the dislocation of the ligamentous structures.

In the case of a mild traumatic event with only a slight edematous reaction, as in partial lesions, it is sometimes possible to distinguish a residual portion of the ligament.

In the acute phase, the axial MRI sequences will reveal an enlarged and hyperintense structure with irregular and nonhomogeneous signal intensity (SI) due to the synovial reaction, which contrasts with the normal SI of the PCL (Fig. 13.12).

On T2-weighted SE images of a partial lesion (Fig. 9.13) it is possible to see a residual portion of the ligament, which will appear with low SI and for this reason is clearly distinguishable from the edema and hemorrhage surrounding of the damaged portion. On the other hand, in the case of a complete ligamentous lesion it will be impossible to see any residual portion; furthermore the entire structure will show a

high SI using these sequences. Subsequently, with the regression of edema and hemorrhage, T1-weighted sequences may very well demonstrate the lesion and, if present, the residual portion of the ligament, though the latter is again better evidenced on the T2-weighted sequences. In fact, the possibility of demonstrating this residual portion is of vital importance both in medical jurisprudence and in those cases where clinical signs are not helpful. However, MRI, like other imaging techniques, cannot define the biomechanical residual functionality of a damaged ligament, even in the presence of a residual portion, especially when the traumatic event has caused stretching of this structure without any rupture. In these cases, the ligament will be evident in

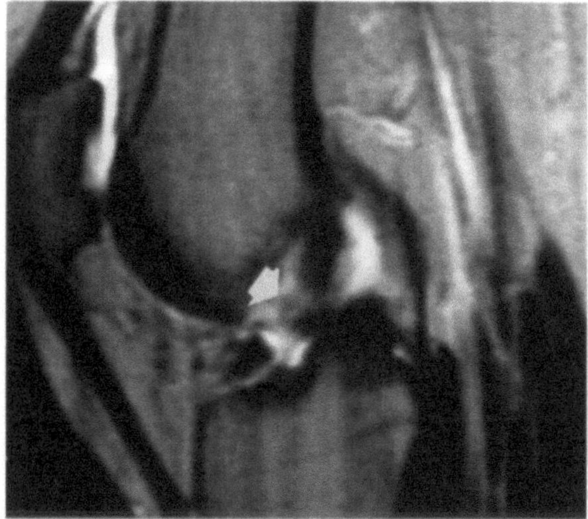

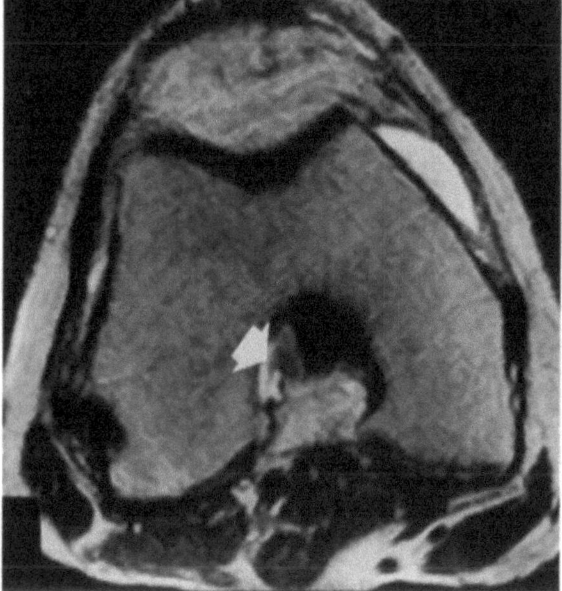

Fig. 13.13. Acute partial tear of the ACL. The residual portion of the ligament (*arrow*) is demonstrated on both the sagittal (**a**) and the axial (**b**) T2-weighted scans

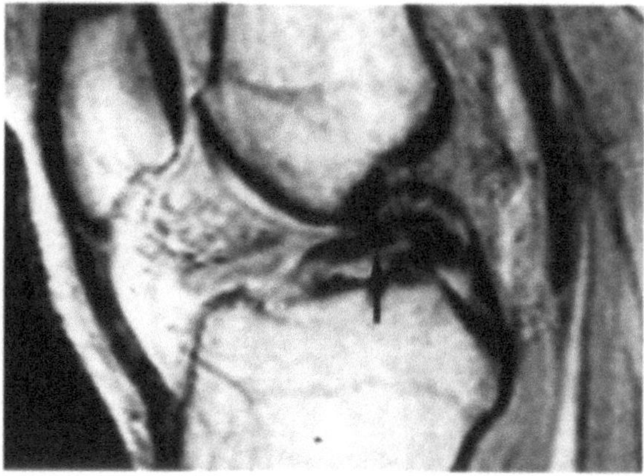

Fig. 13.14. Complete detachment of the ACL, which appears deflexed on the intercondylar eminence (*arrow*)

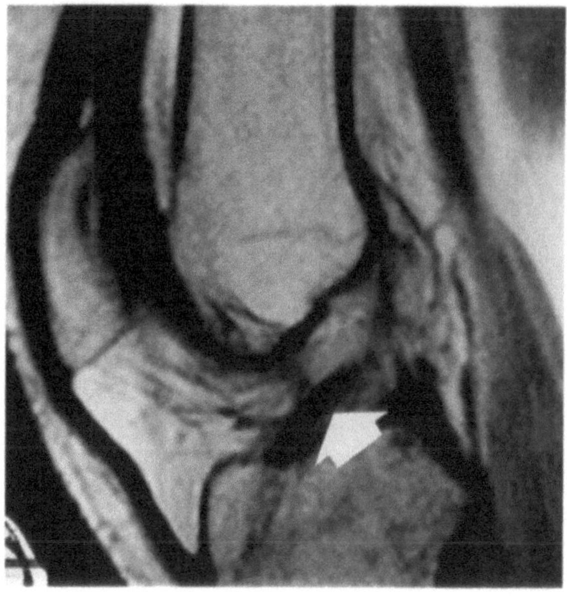

Fig. 13.15. Residual distal portion (*arrow*) of the ACL with atrophic degeneration

both the axial and the sagittal plane, but it will show an altered morphology, such as a deflexion, and also an altered SI.

Absence of the ACL can be well visualized both by CT and by MRI in the axial plane, the presence of the PCL resulting in the characteristic appearance known as a "blind notch."

Occasionally, the distal stump inserts into the synovial membrane of the PCL where it scars, thereby producing the so-called suckling appearance. This is observed on MRI scans in the sagittal plane, where the residual portion appears at the level of the tibia (Figs. 13.14, 13.15).

Lesions of the PCL (Loos et al. 1981) are less frequent and more often partial than complete. They are caused by a variety of traumatic events, ranging from sudden and violent hyperextension of the knee to accidents in which other articular structures are also involved (Masciocchi et al. 1993b; Mink et al. 1988). In the case of incomplete acute lesions, on CT the ligament appears as a hypodense, nonhomogeneous, and enlarged structure, while in cases of complete rupture it is interrupted (Fig. 13.16). The subsequent scarring produced by a partial lesion will result in a nonhomogeneous and enlarged pattern of the PCL, whereas after a com-

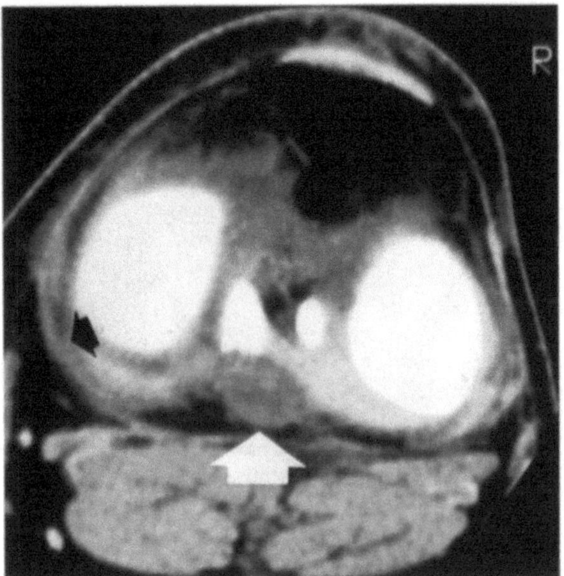

Fig. 13.16. Acute tear of the PCL (*white arrow*) associated with a lesion of the postero-oblique ligament (*black arrow*)

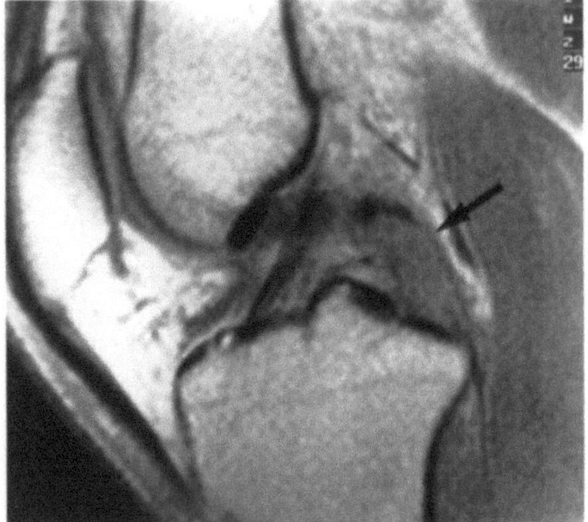

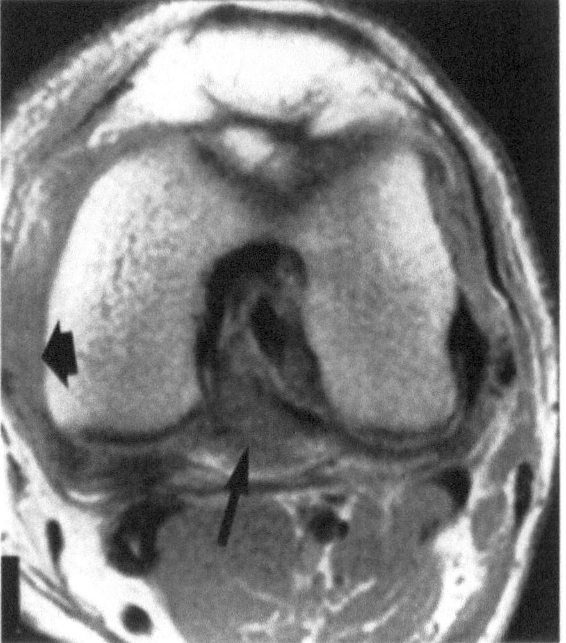

Fig. 13.17. Sagittal (**a**) and axial (**b**) MRI scans in a patient with an acute lesion of the PCL (*thin arrow*) associated with high-grade tear of the medial collateral ligament (*thick arrow*)

plete interruption the ligament will display a hypodense appearance in the stabilized phase.

On MRI, T1-weighted SE sequences, especially in the sagittal plane, seem best able to permit evaluation of the PCL, which normally demonstrates an arched course and a homogeneous and low SI. Following an acute traumatic event, the PCL will appear enlarged and nonhomogeneous in the sagittal plane, with a high SI due to serous-hemorrhagic phenomena, while the axial plane will allow exact evaluation of the damage (Fig. 13.17).

As mentioned above, due to its vascularization the PCL can show a recovery of the trophism and of the SI, sometimes associated with increased thickness of the ligament brought about by the tissue scarring.

13.3.2
Medial and Lateral Ligamentous Compartments

The medial ligamentous compartment is structurally complex and is composed of the posteromedial capsuloligamentous system, formed by the medial collateral ligament, the posteromedial capsule with the oblique posterior ligament, and the semimembranous muscle with its five insertions. The medial collateral ligament opposes external valgus rotation movement with a flexed knee, especially at

the level of its deep fibers, which are the first to be damaged.

In the acute phase, the trauma may cause serosanguinous imbibition until a partial or a complete lesion occurs. In this case, a focal or extended hypodensity, associated with an increased volume of the ligament, will appear on CT, while on MRI a nonhomogeneous pattern is seen due to the presence of hyperintense areas (DEUTSCH and MINK 1988).

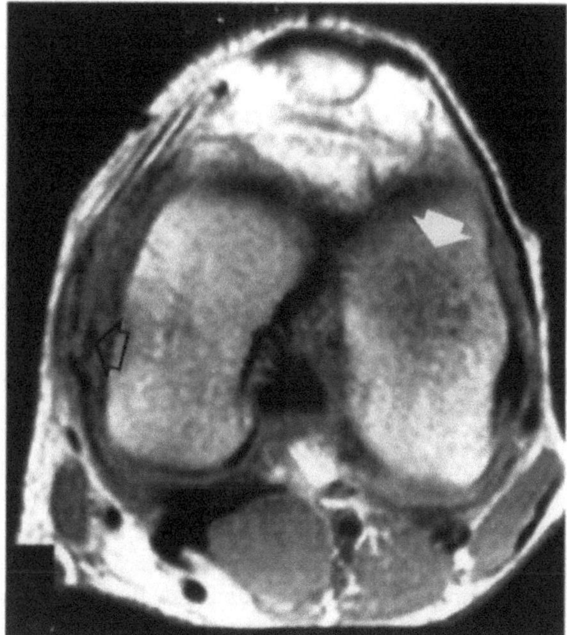

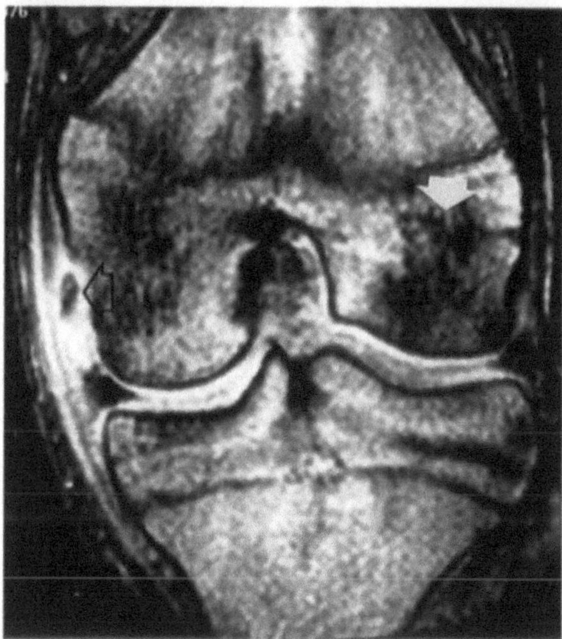

Fig. 13.18 a,b. High-grade tear of the medial collateral ligament with an osteochondral avulsion (*open black arrows*) associated with bone bruise of the lateral femoral condyle (*white arrows*)

If the traumatic event is severe, the oblique posterior ligament and the posteromedial capsule are affected, appearing enlarged and fringed. They are sometimes interrupted, with high SI areas more evident on the axial T2-weighted images (Fig. 13.18). In the stabilized chronic phase, by contrast, the liga-

ment will appear thickened, with a low SI on both T1- and T2-weighted images due to scar tissue. On CT, this condition is characterized by the presence of a thickened ligamentous structure with mild hyperdensity, sometimes associated with the presence of small calcifications.

The lateral ligamentous compartment consists of the lateral collateral ligament, the femoral biceps, the popliteal muscle, and the arcuate complex. From an anatomopathological point of view, lesions of the lateral and of the medial collateral ligament are identical; they are localized in the distal portion of the ligament, especially at the level of the peroneal head, such that both CT and MRI will show the alterations just described. The popliteal tendon participates in the external complex of the knee and, along its intra-articular passage, may be involved by degenerative changes which are finely visualized on MRI examination.

Another interesting pathological condition is the ileotibial band friction syndrome (IBFS) (MURPHY et al. 1992), which often affects runners. This impingement is located at the level of the ileotibial tract, upon the lateral femoral condyle, and affects not only the tendon but also its bursa and the underlying periosteal surface. The condition is due to a biomechanical overload, in particular during the flexion-extension of the knee, or to direct contusive trauma; both these conditions may provoke an inflammatory reaction and subsequently a degenerative alteration of the tendinous structure, associated with a synovial reaction (Fig. 13.19). Usually the diagnosis is based on clinical signs, but sometimes, when the pain is not well localized, it may be necessary to differentiate this condition from others affecting the lateral compartment of the knee.

Both CT and MRI may depict the friction of the ileotibial band, while the x-ray evaluation is often useless. CT will show a hypodense band localized between the lateral femoral condyle and the ileotibial band, which is also thickened. Both T1- and T2-weighted MR images in the axial and coronal planes demonstrate well the involutional and degenerative changes of the tendinous structure, while the corresponding capsulosynovial portion appears to be enlarged and fringed in the case of fibrotic and hyperplastic processes. Consequently both CT and MRI are useful in achieving correct diagnosis of these anatomopathological conditions, allowing, for example, identification of the tendinous and synovial inflammatory reaction which characterizes this syndrome.

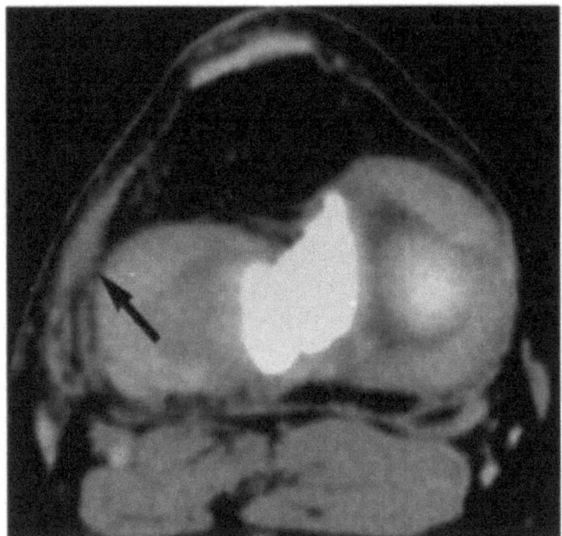

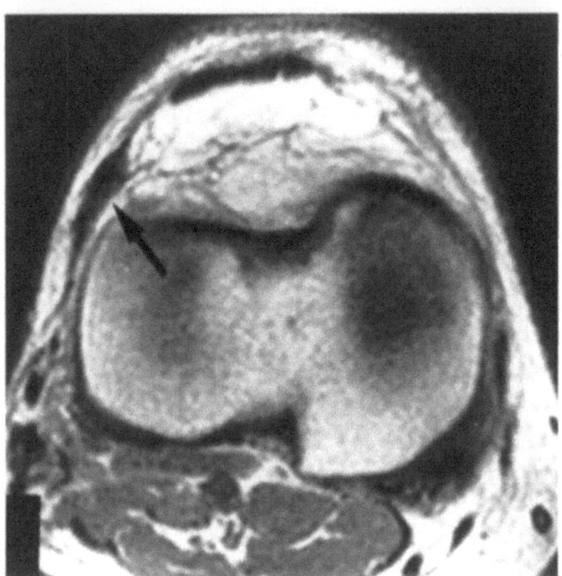

Fig. 13.19. IBFS (*arrows*) shown on CT (**a**) and MRI (**b**)

13.4
Synovial Disorders

The synovial membrane of the knee is particularly large and covers the deep face of the articular capsule, inserting into the femur, patella, and tibia. Alterations of this membrane play a fundamental role in painful syndromes involving the knee and may have a primary or secondary etiology, as in the case of trauma (Deutsch and Mink 1988).

13.4.1
Plicae

Synovial plicae can persist at the end of embryogenetic development. Such plicae may be divided into suprapatellar, mediopatellar, infrapatellar (or ligament of Hoffa's body), or lateral. Under physiological conditions they do not cause any biomechanical problems as they adapt perfectly to movements. However, they can become pathogenic and symptomatic, usually in response to frequent rubbing in flexion-extension. Degenerative, reactive, posttraumatic and inflammatory phenomena give rise to edema and widening in these structures. With time, involutional changes occur, leading to the formation of fibrotic scar tissue and the creation of truly pathogenic plicae.

This condition can be diagnosed by accurate clinical examination and instrumental investigations such as CT and MRI. CT never fails to identify the four types of plicae and their possible associations. Under normal conditions a plica appears as a thin laminar formation which is regular and moderately hyperdense in relation to the surrounding adipose tissue. In pathological conditions, both direct and indirect morphological signs can be identified. The direct signs consist of enlargement and hypodensity which, with time, turn into fringing and hyperdensity owing to fibrotic and fibrocalcific evolution; the latter condition cannot be assessed on MRI. The most important indirect indication is chondritis produced by the plica (Fig. 13.20).

On MRI under normal conditions the synovial plicae appear as lamellar structures with a low SI on both T1- and T2-weighted sequences (Fig. 13.21). When reactive processes are present, be they of an inflammatory or a chronic hyperplastic type, these structures become larger and have a lower SI. A typical nodular appearance may then be seen.

Magnetic resonance imaging permits accurate evaluation of the chondropathy caused by pathogenic plicae, which is a secondary complication of the mechanical damage occurring on the articular surfaces. It is important to recognize the high diagnostic reliability of MRI with regard to this serious complication, as will be seen in Sect. 13.5.

13.4.2
Cysts

The pathogenesis and physiopathology of synovial cysts are complex. According to the anatomical

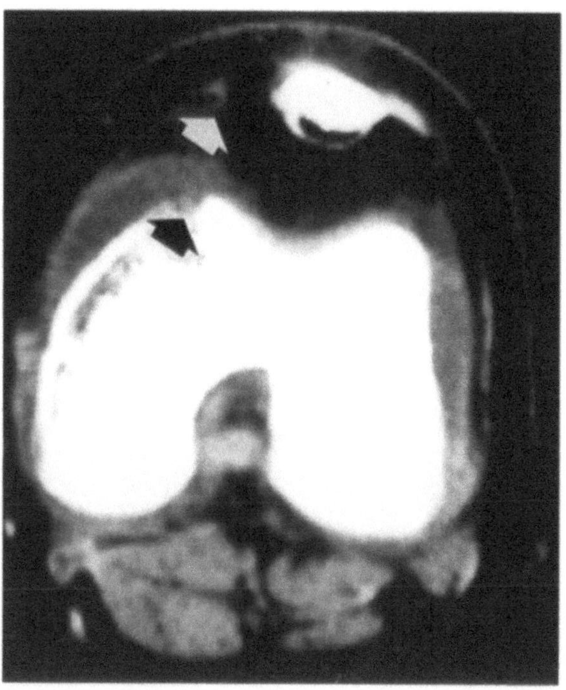

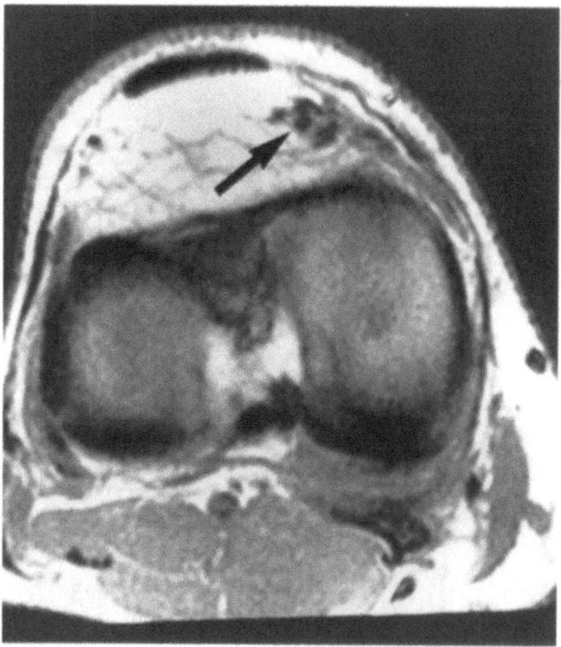

Fig. 13.20. Synovial plica syndrome: medial synovial plica (*white arrow*) inducing a high osteochondral erosion (*black arrow*)

Fig. 13.21. Hyperplastic and fibrotic medial synovial plica (*arrow*)

site, one can distinguish between synovial extra-articular cysts, which may or may not communicate with the articular cavity, and intra-articular cysts. Noncommunicating extra-articular cysts usually arise from synovial ectopic areas; through their compressive or dislocating effect they cause patho-logical conditions, with modification of the normal relationships between the various muscular ele-ments and adjacent nervous ones. The best known noncommunicating cyst is undoubtedly Baker's cyst, although this term encompasses a series of cystic formations, all located in the popliteal fossa. Baker's cyst occurs secondary to mechanical capsular loos-ening due to an articular degenerative process with an increase in the synovial liquid pressure, as in rheumatoid arthritis.

The most common cystic formation involving the popliteal fossa is cystic bursitis of the gastrocnemiosemimembranous bursa; here, too, there is an increase in endoarticular pressure because of an increase in synovial fluid.

The diagnosis of such cystic formations is often based on clinical examination and ultrasonography, but they can frequently be identified on CT, showing low density values ranging from 5 to 25 HU. To evaluate the site, the extension, the interactions with the surrounding structures and the characteristics of the wall, high contrast MRI sequences are helpful.

13.4.3
Synovitis

Synovitis may occur due to unknown inflammatory processes that are unrelated to inflammatory and metabolic diseases and are characterized by negative laboratory tests. Clinical signs are effusion, articular restriction, and pain, while anatomopathologically these conditions are characterized by the presence of small villous formations covering the synovial membrane. In particular, edema and hyperemia can occur, both in septic or post-traumatic forms and in algodystrophic forms. In the case of hemorrhage following vascular alterations, there will be iron precipitation in the form of hemosiderin: this is the so-called hemorrhagic synovitis.

On CT, the inflammatory and edematous com-ponents are associated with increased density and thickening of Hoffa's body due to hypertrophy of the synovial structures, forming fluffy fringes and presenting a frosty glass aspect.

According to the anatomopathological alter-ations, MRI reveals two different SI patterns: if the hyperemic conditions predominate, there will be a prevalence of liquid with a subsequent elevated SI on the T2-weighted sequences, whereas iron precipita-tion will result in a low SI on the same sequences due to the paramagnetic effect. The areas of low SI are irregularly distributed and they rarely conglomerate in nodular structures.

The presence of homogeneous and regular villonodular formations is typical for pigmented villonodular synovitis (PVNS), which resembles giant cell synovial tumor insofar as giant cells, probably of dystrophic origin, are present. CT diagnosis is based on the identification of the nodular formation and on its specific density, with values ranging from 55 to 75 HU in accordance with the iron content; in addition widespread microcalcifications are sometimes present. The MRI appearance of this pathology is nonspecific because the signal characteristics alone do not distinguish between chronic synovitis evolving into scar nodules and PVNS, due to the fact that both are characterized by low SI on T1- and T2-weighted sequences (Fig. 13.22).

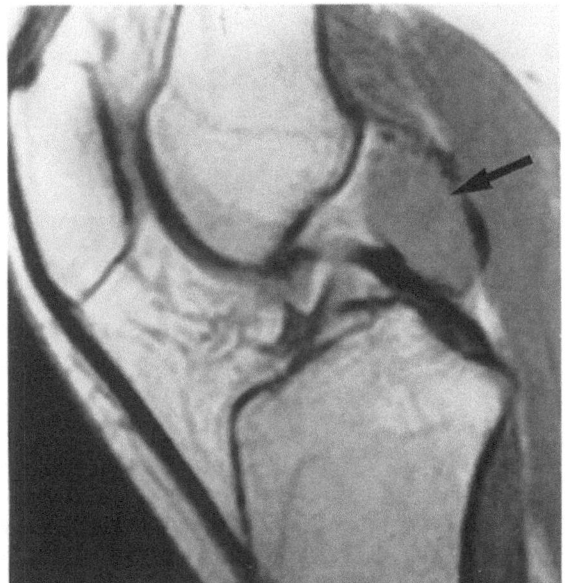

a

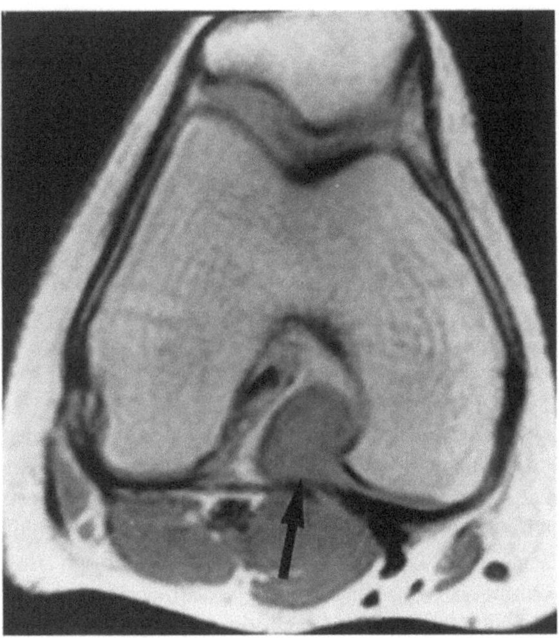

c

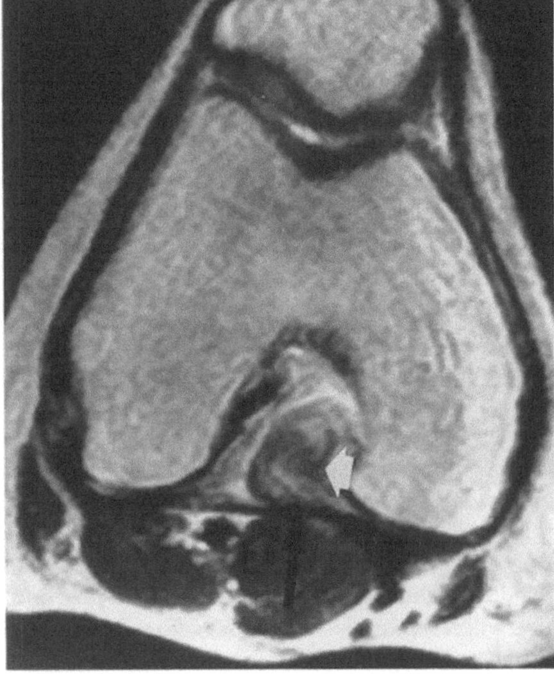

b

Fig. 13.22. PVNS. A synovial cyst (*arrow*) localized at the posterior capsular recess is well demonstrated on sagittal (**a**) and axial (**b**) MRI scans. On axial T2-weighted images (**c**) a hemorrhagic nodular formation (*white arrow*) is shown

Malignant synovial tumors are infrequent in the knee, and they appear as irregular masses infiltrating the capsule and spreading into extra-articular sites; on this basis it is possible to distinguish them from other forms of synovitis. At the level of the knee, the most frequent pseudotumoral synovial condition is so-called synovial chondromatosis, which is characterized by a synovial chondroid metaplastic alteration. The consequent formation of hyaline cartilage in the synovial membrane produces endoarticular chondral bodies, more frequently located at the level of the synovial sheath of the central pivot, at the level of the tendon sheaths, or in the synovial plicae.

Computed tomography easily reveals the early phases of these conditions owing to the typical high density; with calcium precipitation, a high and homogeneous density will occur, particularly at the level of the anterior space, where these formations are larger. The presence and the dimensions of calcification limit the MRI diagnosis of the pseudotumoral forms. In particular, the hypertrophic synovial membrane shows a low SI on T1-weighted images and a homogeneous and high SI on T2-weighted images if edema and hyperemia predominate, while the presence of ossifications will be characterized by low SI on both T1- and T2-weighted images.

13.5
Osteochondral Pathology

Osteochondral pathologies result from both traumatic and degenerative causes; sometimes, indeed, traumatic and degenerative factors coexist and result in anatomopathological situations which completely modify the articular biomechanical arrangement of the knee. Because of the possibility of exploring areas which are difficult to evaluate with other modalities, MRI demonstrates very high sensitivity in recognizing alterations affecting these structures.

13.5.1
Traumatic Lesions

Following trauma involving the bone, some types of fractures (DAFFNER 1978) present diagnostic problems due to the difficulty in identifying them with instrumental examinations, such as radiography and CT, and for this reason they are usually called "occult" (MINK and DEUTSCH 1989); medullary bone impact fractures are usually thus defined. Occult

fractures are actually osteochondral and subchondral post-traumatic lesions caused by bruising of the articular cartilage, of the subchondral bone, and of the medulla. Given the clinical difficulty in precisely defining the painful area during the acute phase, MRI is the method of choice in the study of such alterations. It has the advantages of allowing imaging in the longitudinal plane and of providing high contrast resolution. On MRI occult fractures have a typical pattern characterized by hemorrhage and edema. The normally high SI on T1-weighted sequences, which derives from the adipose components, is replaced by a low SI owing to water content (KAPLAN et al. 1992). It is the reactive inflammatory event, therefore, rather than the trabecular or osseous lesion, which is visible on MRI. It is important to recognize that in the case of occult fractures only the medullary bone is affected without involvement of the cortical bone, which will appear hypointense (Fig. 13.23).

On MRI (DIPAOLA et al. 1991), the identification of the osteochondral fragment in the acute phase is limited by its dimensions and by the amount of the chondral and osseous components; furthermore, if the loose body is small and mostly cartilaginous, visualization is difficult because it will have the same SI as the surrounding fluid. A certain amount of osseous component is therefore necessary to identify the fragment, and in this case it will appear with a low SI. It is important to remember that T2-weighted GE sequences are preferable to T2-weighted SE sequences (Fig. 13.24) for the identification of osteochondral fragments owing to the better contrast that they offer.

While small fragments are gradually corroded by the synovial fluid, larger ones have a tendency to reach strategic articular recesses, such as the suprapatellar and popliteal recesses, and because of the space occupation they can cause rubbing and blocking conditions. CT images, obtained only in the axial plane, are of little value in the study of chondral surfaces but can identify and locate detached and dislocated fragments, even when the chondral component is larger than the osseous one.

13.5.2
Degenerative Conditions

Chondral pathologies (HAYES and CONWAY 1992), which are often painful and difficult to evaluate clinically, entail four phases of degenerative damage. The first phase is termed "chondromalacia" and is

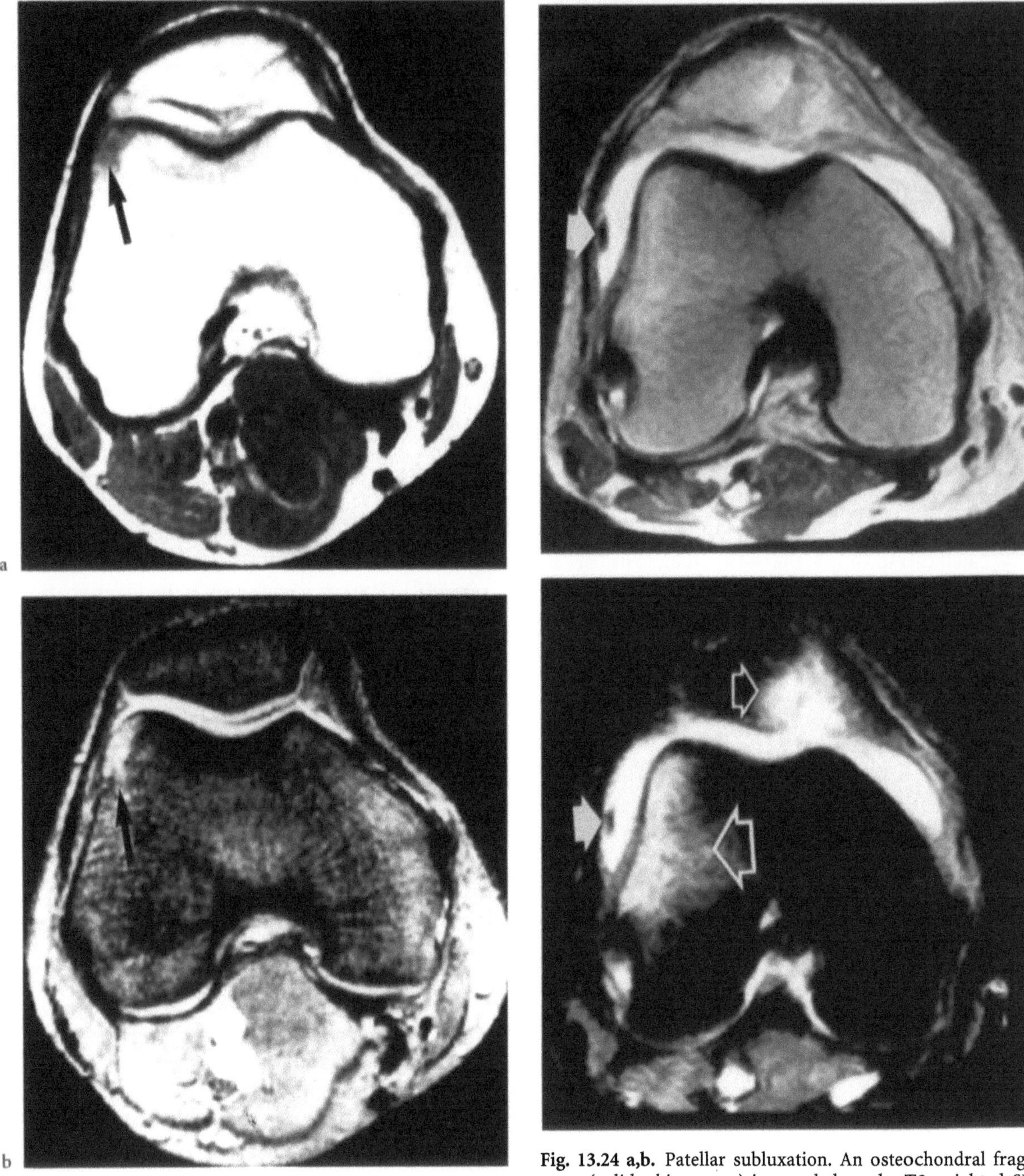

Fig. 13.23 a,b. Occult fracture of the lateral femoral condyle (*arrow*). a T1-weighted SE and T2-weighted GE MRI scans

Fig. 13.24 a,b. Patellar subluxation. An osteochondral fragment (*solid white arrow*) is revealed on the T2-weighted SE image (**a**). Fat suppression technique (**b**) depicts posttraumatic bone bruise (*open white arrows*)

characterized by the absence of a real lesion of the cartilage; however, the increased fluid content causes the cartilage to become saturated, and if it is arthroscopically palpated it feels softer than is normal. This phase cannot be assessed on MRI because the soaking is insufficient to significantly modify the signal intensity. If cartilaginous damage

is greater, the second or "fraying" phase takes place; this is clearly evident on MR images, which reveal an irregular cartilaginous surface with small fringes. In the third and fourth phases, termed "erosion" and "subchondral bone exposure," respectively, progressive loss of cartilaginous portions takes place, with the eventual formation of ulcers and craters which in turn expose the underlying bone (Fig. 13.25). These two phases are well documented on MRI because the cartilage is no longer visible and the subsequent reaction to this alteration is the exposure of the subchondral bone (GYLYS-MORIN et al. 1987). Involutional and fibrotic events, caused by the presence of microfractures and compressions in these chondral lesions, will give a homogeneous and regular SI.

This condition, although similar in signal behavior, must be distinguished from osteonecrosis, which is characterized by an initial bone alteration followed by a secondary involvement of the cartilage, which shows normal morphology and SI in the initial phase. At first, osteonecrosis appears as a nonhomogeneous lesion with hyperintense areas, surrounded by a peripheral rim of low SI on T1-weighted sequences that is caused by the greater resistance of the adipose cellular component to the ischemic event. In the following phases, when the process has become estabilished and the necrosis has involved all the cellular components, reparative and involutional events will result in an area of low SI on both T1- and T2-weighted sequences, while the peripheral rim will appear with a high SI on T2-weighted images due to revascularization (late phase).

Sometimes CT of the knee is carried out as a first investigation in cases of osteonecrosis when a sudden pain, resembling a meniscal-like syndrome, occurs. Due to the possibility of utilizing windows suitable for the study of the bone, CT documents the presence of a fragment site and of lamellar bone crowding, surrounded by an osteosclerotic rim, at the level of the medullary bone.

Comparing radiography, CT, and MRI, early and exact diagnosis of degenerative osteochondral conditions is possible only with MRI, which can clearly define the characteristics of the pathological area and any chondral damage associated with it.

13.5.3
Osteochondritis Dissecans

Although osteochondritis dissecans may be caused by osteonecrotic vascular alterations, it is characterized by a more frequent incidence at the level of femoral condyles and in young males. The lesion displays a particular morphology. Radiographically, it is possible to observe a well-defined osteochondral fragment located in a condylar "nidus" delimited by an osteoslerotic rim. CT permits delimitation of the subchondral area and the fragment; however, while the localization and dislocation of the fragment are clearly evident, CT provides poor visualization of the cartilaginous surface of the femoral condyle. MRI shows an altered and nonhomogeneous SI area surrounding the lesion, which appears semilunar, and a subchondral site, clearly evident on the longitudinal planes.

In cases of osteochondritis dissecans, it is of vital importance to ascertain whether the fragment is stable because instability means loss of bone

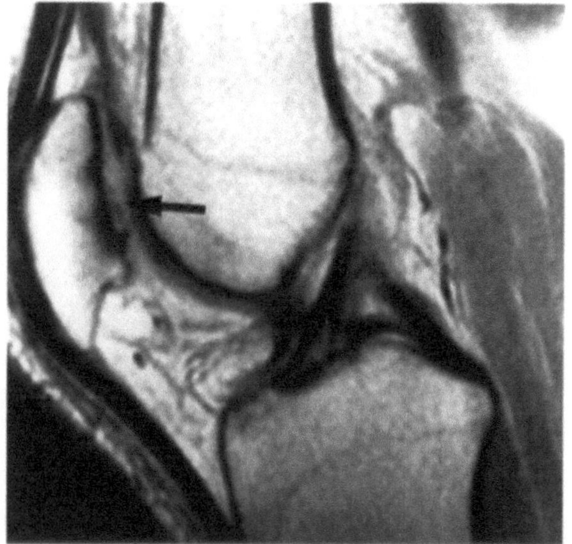

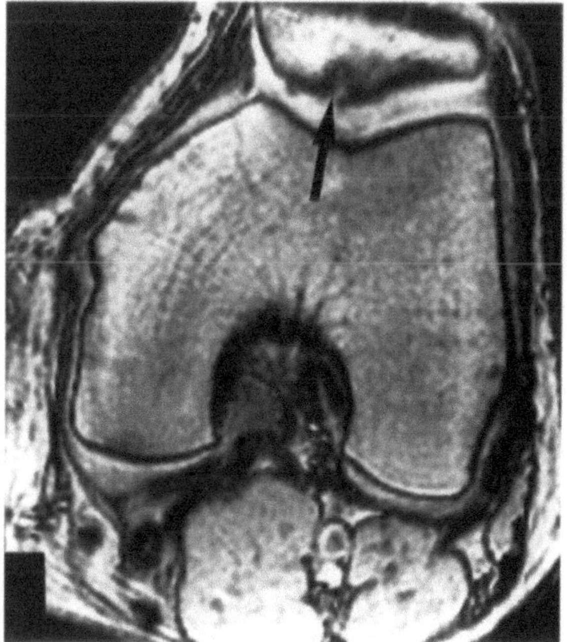

Fig. 13.25 a,b. Grade IV chondropathy with subchondral erosion (*arrow*)

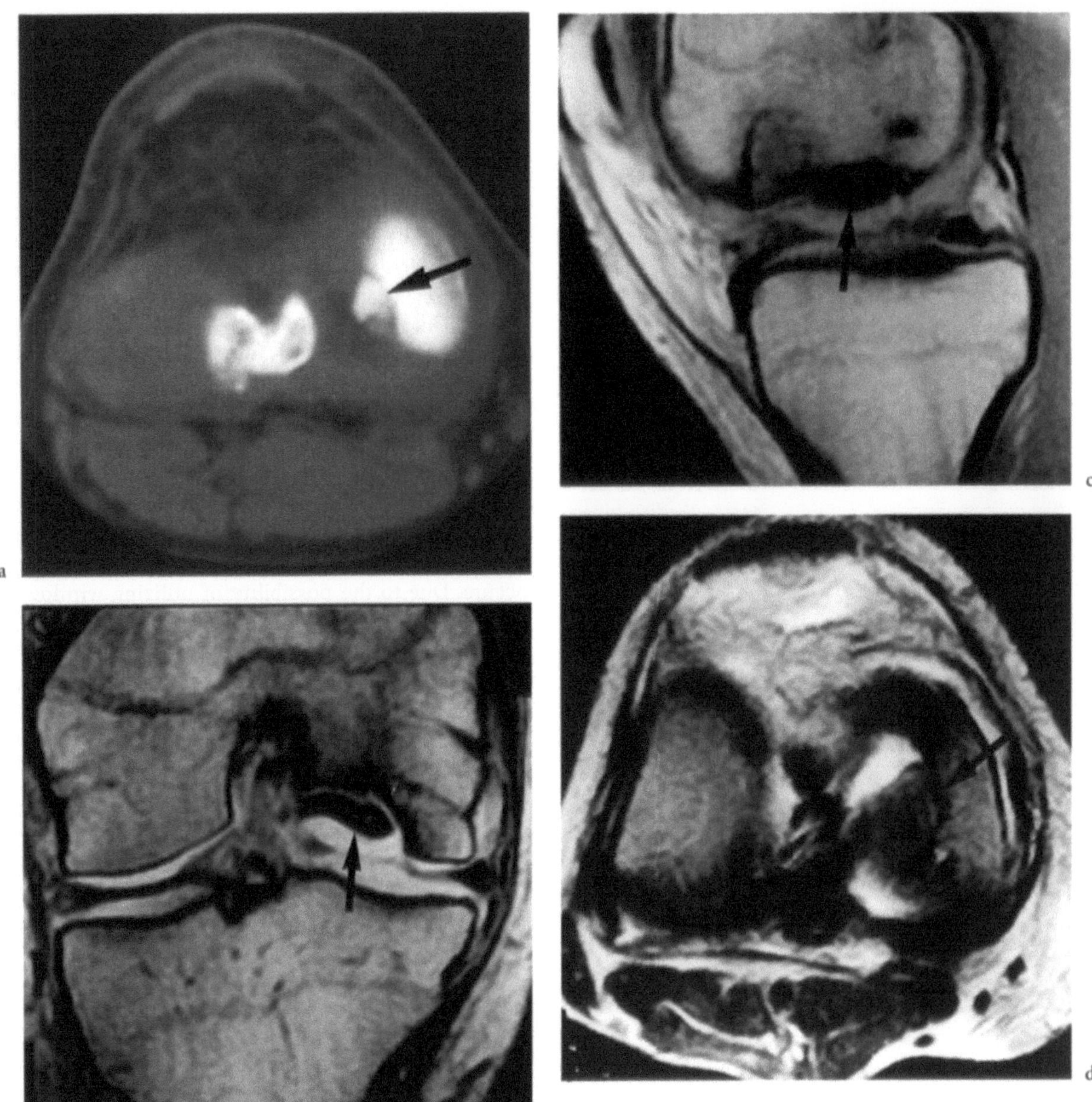

Fig. 13.26 a–d. Osteochondritis dissecans in the medial femoral condyle. An unstable osteochondral fragment (*arrow*) is identified on CT (**a**). MRI shows, with more diagnostic accuracy, the dimensions of the fragment (*arrow*) and its relationship with the condylar nidus (**b–d**)

connections and the potential formation of loose bodies, which could affect the therapeutic approach (Fig. 13.26).

13.6
Conclusions

The knee is certainly the most complex articulation in the human body and we have seen how it is necessary to use various imaging modalities in order to investigate it completely and correctly. Among these modalities, CT (Passariello et al. 1983b) and above all MRI (Mink et al. 1987) have brought about significant changes in the approach to the diagnosis of this joint (Masciocchi et al. 1988).

Arthroscopy (Quinn and Brown 1991) is now considered the "gold standard," but it has some disadvantages such as invasiveness and morbidity. Moreover it is an expensive and operator-dependent technique that does not allow the visualization of some areas of the joint. In fact, it is impossible to detect lesions of the inferior surface of the medial meniscus or incomplete horizontal lesions associated with cystic degenerative changes. In the central pivot, this technique cannot evaluate distal lesions of the posterior cruciate ligament. In the case of pathological alterations involving the posterior recesses of the knee, such as PVNS (Fig. 13.22), and subchondral bone pathologies (TYRREL et al. 1988) without chondral involvement, such as osteonecrosis, post-traumatic ischemic degeneration, occult fractures, and osteochondritis dissecans, arthroscopy is not able to identify the lesion.

Magnetic resonance imaging is particularly helpful in the evaluation of traumatic (both acute and chronic) injuries and degenerative lesions, which are often difficult to identify with other imaging methods. This is of fundamental importance because accurate and early diagnosis of the various pathological conditions affecting the knee increases the likelihood of achieving its functional recovery.

References

Arnoczky SP, Russel RF (1988) Anatomy of the cruciate ligaments. In: Feagin JA (ed) The cruciate ligaments. Churchill Livingstone, New York.

Bessette GC, Hunter RE (1990) The anterior cruciate ligament. Orthopedics 13:551–562

Crues JV, Mink JH, Levy TL, et al. (1987) Meniscal tears of the knee: accuracy of MR imaging. Radiology 164:445–448

Daffner RH (1978) Stress fractures: current concepts. Skeletal Radiol 2:221–229

Deutsch AL, Mink JH (1988) MRI of musculoskeletal trauma. Radiol Clin North Am 27:983–1002

Dickaut SC, DeLee JC (1982) The discoid lateral meniscus syndrome. J Bone Joint Surg [Am] 64:1068–1073

Dipaola JD, Nelson DW, Colville MR (1991) Characterizing osteochondral lesions by MRI. Arthroscopy: J Arthroscopic Related Surg 7:101–104

Gylys-Morin VM, Hajek PC, Sartoris DJ, et al. (1987) Articular cartilage defects: detectability in cadaver knees with MR. AJR 148:1153–1157

Hayes CW, Conway WF (1992) Evaluation of articular cartilage: radiographic and cross-sectional imaging techniques. Radiographics 12:409–428

Kaplan PA, Walker CW, Kilcoyne RF, et al. (1992) Occult fracture patterns of the knee associated with anterior cruciate ligament tears: assessment with MR imaging. Radiology 183:835–838

Loos WC, Fox JM, Blazina ME, et al. (1981) Acute posterior cruciate ligament injuries. Am J Sports Med 9:86–92

Masciocchi C, de Paulis F, Fascetti E, et al. (1988) Raffronto TC-RM nello studio della patologia articolare del ginocchio. Radiol Med 75:4–11

Masciocchi C, Barile A, Fascetti E (1993a) Il ruolo della diagnostica per immagini nella patologia meniscale: il pensiero del radiologo. Artroscopia & Ginocchio 1:25–32

Masciocchi C, Barile A, Fascetti E (1993b) La diagnostica per immagini del legamento crociato anteriore. In: Puddu G, Cerullo G (eds) La patologia del legamento crociato anteriore: diagnosi e trattamento. Il Pensiero Scientifico, pp 53–60

Mink JH, Deutsch AL (1989) Occult osseous and cartilaginous injuries about the knee: MR assessment, detection and classification. Radiology 170:823–829

Mink JH, Reicher MA, Crues JV (1987) Magnetic resonance imaging of the knee. Raven Press, New York

Mink JH, Levy T, Crues JV III (1988) Tears of the anterior cruciate ligament and menisci of the knee: MR imaging evaluation. Radiology 167:769–774

Murphy BJ, Hechtman KS, Uribe JW, et al. (1992) Iliotibial band friction syndrome: MR imaging findings. Radiology 185:569–571

Passariello R, Trecco F, de Paulis F, et al. (1983a) Computed tomography of the knee joint: technique of study and normal anatomy. J Comput Assist Tomogr 7:1035–1042

Passariello R, Trecco F, de Paulis F, et al. (1983b) Computed tomography of the knee joint: clinical results. J Comput Assist Tomogr 7:1043–1049

Passariello R, Trecco F, de Paulis F, et al. (1986) CT demonstration of capsuloligamentous lesions of the knee joint. J Comput Assist Tomogr 10:450–456

Passariello R, Masciocchi C, Barile A (1992) CT and MRI of the knee. Acta Radiologica Portuguesa IV, 15:117–119

Pavlov H, Hirschy JC, Torg JS (1979) Computed tomography of the cruciate ligaments. Radiology 132:389–393

Quinn SF, Brown TF (1991) Meniscal tears diagnosed with MR imaging versus arthroscopy: how reliable a standard is arthroscopy? Radiology 181:843–847

Reiser M, Rupp N, Karpf PM, et al. (1981) Evaluation of the cruciate ligaments by CT. Eur J Radiol 1:9–15

Smillie IS (1980) Diseases of the knee joint. Churchill Livingstone, New York

Tyrrel RL, Gluckert K, Pathria M, et al. (1988) Fast three-dimensional MR imaging of the knee: comparison with arthroscopy. Radiology 166:865–872

14 The Ankle and Foot

H. Imhof, M. Breitenseher, S. Trattnig, F. Kainberger, and J. Haller

CONTENTS

14.1
Trauma

Ankle and foot traumata are the most common injuries encountered in a trauma department. While in previous times bony lesions were the most frequent

H. Imhof, MD, Professor, Osteologie und MR-Einrichtung der Medizinischen Fakultät, Universitätsklinik für Radiodiagnostik, Allgemeines Krankenhaus der Stadt Wien, Lazarettsgasse 14, A-1090 Wien, Austria, and L. Boltzmann Institut für rad.-phys. Tumordiagnosis
M. Breitenseher, MD, Docent, MR-Einrichtung der Medizinischen Fakultät, Universitätsklinik für Radiodiagnostik, Allgemeines Krankenhaus der Stadt Wien, Lazarettsgasse 14, A-1090 Wien, Austria
S. Trattnig, MD, Docent, MR-Einrichtung der Medizinischen Fakultät, Universitätsklinik für Radiodiagnostik, Allgemeines Krankenhaus der Stadt Wien, Lazarettsgasse 14, A-1090 Wien, Austria
F. Kainberger, MD, Docent, Osteologie, Universitätsklinik für Radiodiagnostik, Allgemeines Krankenhaus der Stadt Wien, Lazarettsgasse 14, A-1090 Wien, Austria
J. Haller, MD, Docent, Zentralröntgen, Hanuschkrankenhaus, A-1090 Wien, Austria

indications for imaging, nowadays soft tissue, ligamentous, cartilage, and bone marrow abnormalities occupy this position. This change has been brought about by the introduction of magnetic resonance imaging (MRI) and computed tomography (CT) in routine clinical work.

14.1.1
Ligamentous Injuries

In the foot and ankle the lateral collateral ligaments are the most commonly injured ligaments (anterior talofibular ligament > calcaneofibular ligament > posterior talofibular ligament), followed by the sinus tarsi ligaments. Injuries of the medial collateral ligament (deltoid ligament) or the distal syndesmosis (anterior, posterior, and inferior tibiofibular ligaments), especially without a fracture, are rare traumatic disorders. Other ligamentous lesions in the foot and ankle are uncommon.

14.1.1.1
Collateral Ankle Ligament

The diagnosis of lateral collateral ankle ligament trauma is based on patient history and clinical examination with inspection, palpation, and clinical stress tests. If the clinical stress test is negative, no further imaging is needed. Conventional radiographs in anteroposterior and lateral views should be obtained to exclude a fracture or a disruption of the tibiofibular syndesmosis (MARDER 1994). If the clinical stress test is positive, stress radiography might be performed (MARDER 1994).

Lateral stress radiography will be performed with inversion stress of the foot (MARDER 1994; WAGNER and DANN 1995; GEISSLER et al. 1996). The talar tilt is measured in both joints and the difference used to classify patients into three groups to indicate the severity of lateral ankle ligament injury: ≤5°, intact ligaments: short-term immobilization; 6–14°, single-ligament tear: casting; and $\geq 15°$, two to three torn ligaments: surgery. The talar tilt angles are obtained in both ankle joints to exclude idiopathic ligamentous laxity.

There is no consensus in the literature as to (a) the usefulness of stress radiography in acute ankle sprain, (b) the cut-off talar tilt angle beyond which a two-ligament rupture would be certain [estimates range from 5° (BUCK 1972; COX and HEWES 1979) to 9° (KELIKIAN and KELIKIAN 1985), 15° (MARDER 1994), and as much as 30° (DZIOB 1956)], and (c) which patients should undergo casting or surgery (BROSTROM et al. 1965; BROSTROM 1965). Various attempts have been made to enhance the usefulness of stress radiography by the use of supplementary techniques, without significant improvements; these techniques have included anterior drawer (sagittal) stress radiography, added to the inversion maneuvers (JOHANNSEN 1978), anesthesia to relieve pain-induced muscle splinting (OLSON 1969), and conventional arthrography or tenography for the detection of ligamentous injuries (OLSON 1969; ALA-KETOLA et al. 1984; SAUSER et al. 1983; SPIEGEL and STAPLER 1975). Furthermore the development of chronic instability in 10%–15% of cases is not avoided by the use of stress radiography.

The role of magnetic resonance imaging (MRI) in acute ligamentous injuries is unclear at present. Today MRI is not used in this indication, but in future MRI could be used similarly to its frequent application in the knee joint for the evaluation of acute ligamentous injuries, prior to surgery or arthroscopy. With controlled positioning of the foot and with defined sections, MRI can visualize with great certainty the lateral collateral ankle ligaments (SCHNECK et al. 1992a) as well as injuries of these structures (SCHNECK et al. 1992b; CHANDNANI et al. 1994; CARDONE et al. 1993; RIJKE et al. 1993; VERHAVEN et al. 1991).

The diagnosis of a *complete tear* is predicated upon the demonstration of discontinuity, ligamentous stumps, or local fluid collections (Figs. 14.1, 14.2) (MARDER 1994; BROSTROM 1965; SPIEGEL and STAPLER 1975; DELACEY and BRADBROOKE 1979; ERICKSON et al. 1991). A *partial tear* is diagnosed on the basis of irregular thickening of the ligament, with occasional circumscribed thinning out of the ligament, wavy contours, and an increased signal intensity within the ligament (Figs. 14.3, 14.4). An intact ligament appears as a structure of low signal intensity in all images, with uniform width and good delineation from the adjacent fatty tissue (SCHNECK

et al. 1992a,b; BELTRAN et al. 1986; MESGARZADEH et al. 1989; NOTO et al. 1989). Advantages of MRI include the potential to detect additional bony and tendinous lesions. Furthermore, the lateral stress

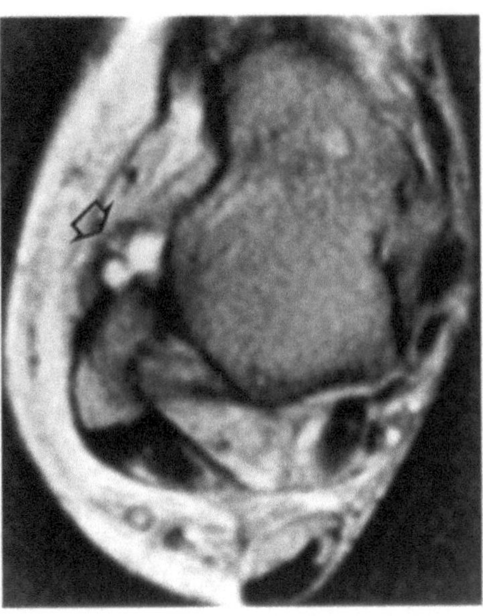

Fig. 14.1. Axial T2-weighted MR image with the foot in 10–20° dorsiflexion. A complete midsubstance tear of the anterior talofibular ligament is present (*arrow*). Frayed stumps and discontinuity are well visualized on the T2-weighted sequence by virtue of the high-intensity surrounding fluid

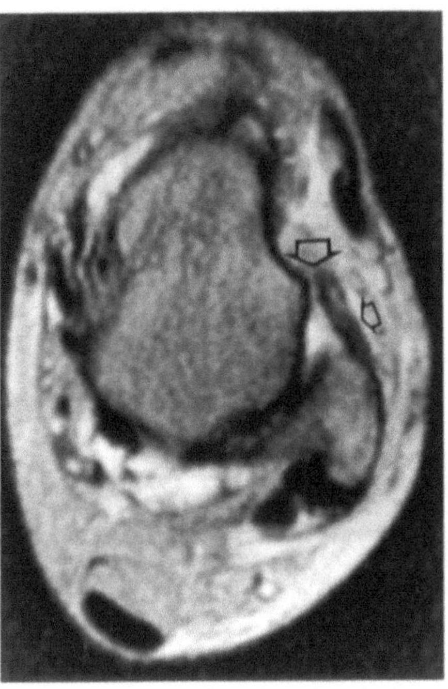

Fig. 14.2. Axial T2-weighted MR image with the foot in 10–20° dorsiflexion. A complete tear of the anterior talofibular ligament is present at the fibular attachment site (*large arrow*). A widened, high-intensity stump is seen (*small arrow*)

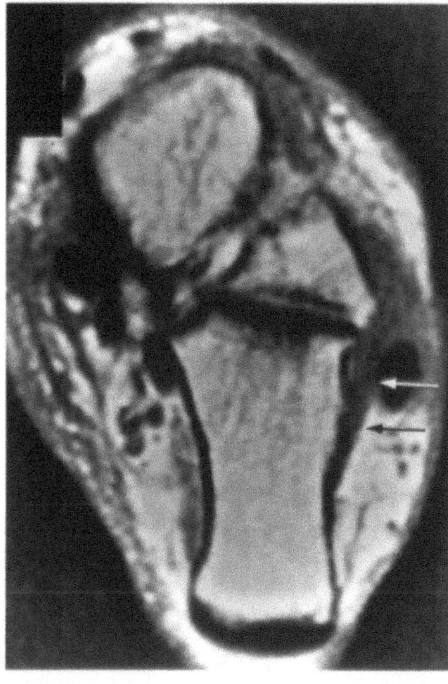

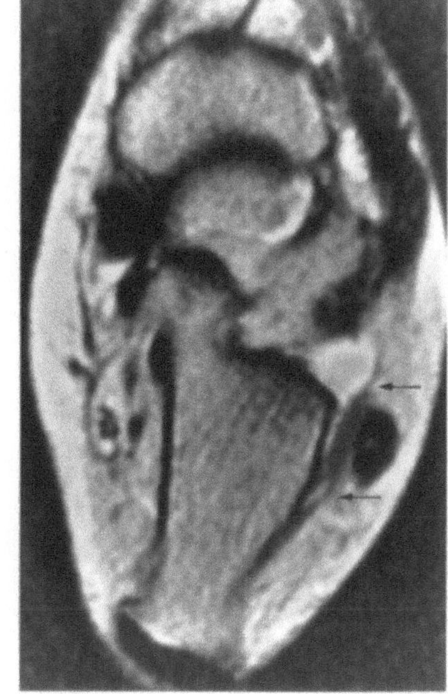

Fig. 14.3. Axial MR images (**a** T1-weighted; **b** T2-weighted) with the foot in 40–50° plantar flexion. The lesion was thought to be a partial tear with altered intensity pattern, ligament of uneven width and markedly thinned out at the rupture site

(*black arrows*) and under the peroneal tendons (*white arrow*). At surgery, the ligament was found to have ruptured completely

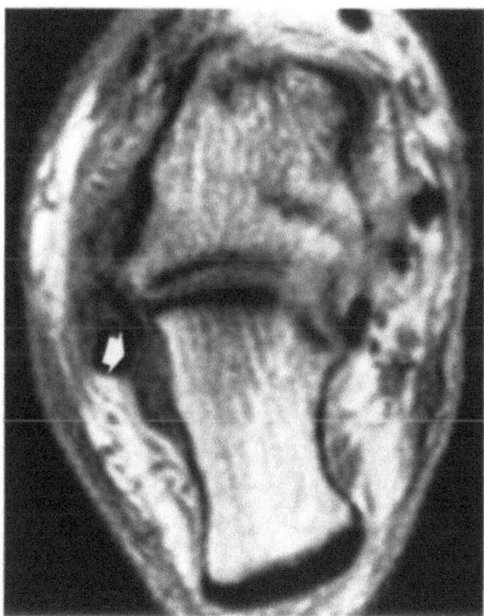

Fig. 14.4. Axial T1-weighted MR image with the foot in 40–50° plantar flexion. There is a complete midsubstance tear of the calcaneofibular ligament; one can see no MR signal, wavy ligament stumps, and complete discontinuity in the middle portion (*arrow*). Confirmed at surgery

position of the ankle joint is not necessary for MRI evaluation.

14.1.1.2
Sinus Tarsi Ligaments

The sinus tarsi is an anatomic space that is bounded by the talus and calcaneus and the talonavicular and posterior subtalar joints, and continues medially into the tarsal canal. The contents of the sinus tarsi include the inferior extensor retinaculum, anteriorly the cervical ligament, and centrally the talocalcaneal interosseous ligament (BELTRAN et al. 1990; KLEIN and SPREITZER 1993; KJAERSGAARD-ANDERSEN et al. 1988, 1989; SCHNECK et al. 1992a). The sinus tarsi ligaments constitute a unit with the lateral ankle ligaments, serving to stabilize the lateral aspect of the ankle and the hindfoot. Inversion of the heel without dorsiflexion or extension provokes rupture of the anterior talofibular and calcaneofibular ligaments first, followed by rupture of the talocalcaneal interosseous ligament (SCHNECK et al. 1992a).

Anteroposterior stress radiography, stress tomography, and arthrography of the subtalar joint have occasionally been used to diagnose subtalar

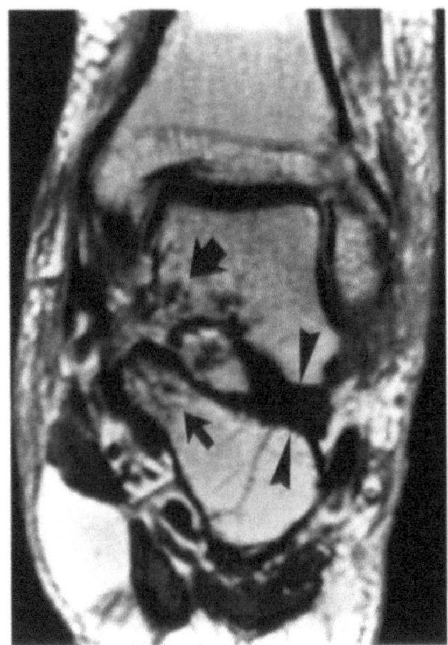
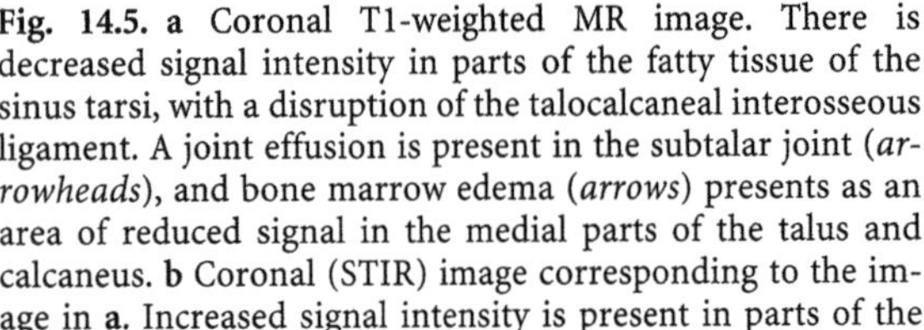
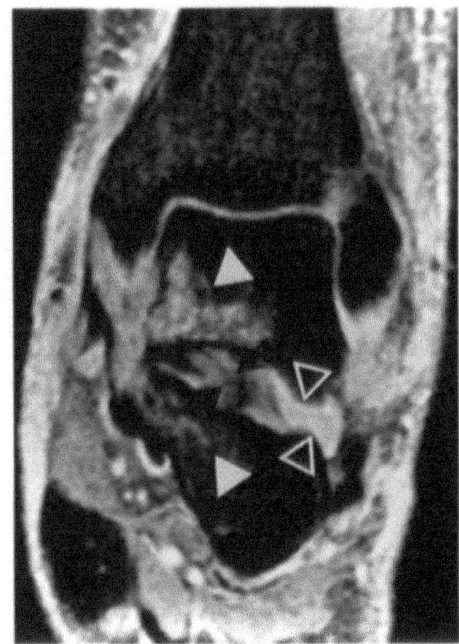

Fig. 14.5. a Coronal T1-weighted MR image. There is decreased signal intensity in parts of the fatty tissue of the sinus tarsi, with a disruption of the talocalcaneal interosseous ligament. A joint effusion is present in the subtalar joint (*arrowheads*), and bone marrow edema (*arrows*) presents as an area of reduced signal in the medial parts of the talus and calcaneus. b Coronal (STIR) image corresponding to the image in a. Increased signal intensity is present in parts of the fatty tissue of the sinus tarsi, with ligament thickening and signal increase from a disrupted talocalcaneal interosseous ligament (*black arrows*). The joint effusion in the subtalar joint presents high signal intensity (*white open arrowheads*), and areas of increased signal due to bone marrow edema are seen in the medial parts of the talus and calcaneus (*white arrowheads*). Diagnosis: acute ankle sprain injury

instability in sinus tarsi syndrome. These methods, however, are limited because stress radiography and stress tomography provide only functional and thus indirect information about the sinus tarsi, while arthrography is invasive and insensitive to the sinus tarsi syndrome (LAURIN et al. 1968; RUBIN and WITTEN 1960; BROSTROM et al. 1965; GOOSSENS et al. 1989).

Acute sinus tarsi changes are not evident clinically, since swelling and pain of the entire ankle are present with acute ankle sprain injury, but ruptures of the sinus tarsi ligaments can be evaluated by MRI (Fig. 14.5). After ankle injury, changes of the sinus tarsi may persist, becoming chronic (MEYER et al. 1988) and possibly resulting ultimately, after months or even years, in chronic sinus tarsi syndrome. Clinically, the sinus tarsi syndrome, first described by O'CONNER (1958), is characterized by lateral foot pain, focal pain over the tarsal sinus in response to palpation, and hind foot instability. This chronic disease is related to a history of inversion trauma in 70% of patients (KLEIN and SPREITZER 1993; KJAERSGAARD-ANDERSEN et al. 1989; MEYER et al. 1988) and 39% of patients with chronic lateral ankle ligament tears have been reported to show an abnormal sinus tarsi (KLEIN et al. 1993). MRI has been described as the imaging method of choice for the evaluation of abnormalities associated with chronic sinus tarsi syndrome (KLEIN and SPREITZER 1993).

14.1.2
Tendinous Injuries

In the foot and ankle the most common disorders under traumatic conditions are rupture of the Achilles tendon (due to a combination of degeneration and trauma), rupture of the posterior tibial muscle tendon (typically acute ruptures in young men with a sports history and chronic ruptures in elderly women), and ruptures of the peroneal tendons.

Achilles tendon injuries are less obvious, with up to 25% being diagnosed incorrectly. On physical examination, swelling often obscures the presence of a palpable tendinous defect (INGLIS et al. 1976). The Thompson test may also remain negative with a partial tendinous tear.

Ultrasound is the imaging modality of choice, since it provides the best resolution of the tendon itself. If ultrasound does not solve the diagnostic problem, MRI should be used: MRI provides high contrast in acute injuries and visualizes the surrounding soft tissues and other tissues.

Magnetic resonance imaging of the normal tendons demonstrates long, thin, hypointense structures owing to the extremely low water content, with high contrast between the dark tendon and the bright surrounding fat. The magic angle phenomenon of a tendon can simulate a tear, but differentiation is possible using a T2-weighted sequence or another position of the tendon. Tendon rupture is of three types: (a) partial tendon rupture combined with tendon hypertrophy and thickening, (b) partial tendon rupture combined with tendon attenuation, and (c) discontinuity with complete rupture.

A *partial tendon tear* presents as an intratendinous lesion with a higher signal intensity on T2-weighted images than on T1-weighted or proton density images. MRI signs of *complete tendon rupture* are large areas of signal increase on T2-weighted images that fill the gap in the tendon, retraction of the distal and proximal ends of the tendon, widening of the contour of the remaining tendon, and a "mop-end" appearance of the tendon edges. Edema and hemorrhage of the soft tissue or the tendon sheath accompany the tendon rupture. Longitudinal ruptures are found in the tibialis posterior tendon and in the peroneal tendons. *Subluxation* and *dislocation* will be found at the peroneal tendons. Dislocation of the peroneal tendons may be combined with tendonitis, partial tear, or bone injury.

14.1.3
Osseous Injuries

Foot and ankle fractures are common injuries which in most cases can be diagnosed by plain radiography. For this purpose an established terminology is used on a routine basis, as outlined below.

14.1.3.1
Fractures of the Malleoli and Fibula

Fractures of the malleoli and fibula are designated using the Weber classification:

- Weber A fractures comprise fractures of the lateral malleolus at or distal to the tibiofibular joint without injury to the tibiofibular complex. They can be combined with a medial malleolus fracture.
- Weber B fractures comprise fractures of the lateral malleolus at the level of the ankle joint with partial disruption of the tibiofibular complex (Fig. 14.6). They can be combined with a medial malleolus fracture.
- Weber C fractures comprise fibular fractures proximal to the ankle joint with tears of the tibiofibular complex; they are associated with pure ligamentous tears and fractures of the anterior (Tillaux-Chaput) or posterior tubercles (Volkmann) of the distal tibia. The medial

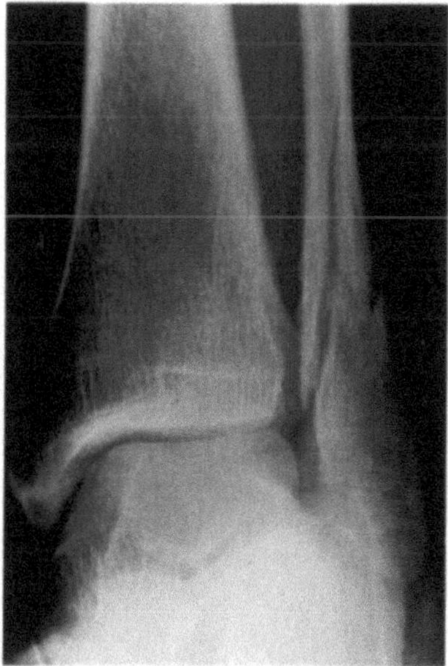
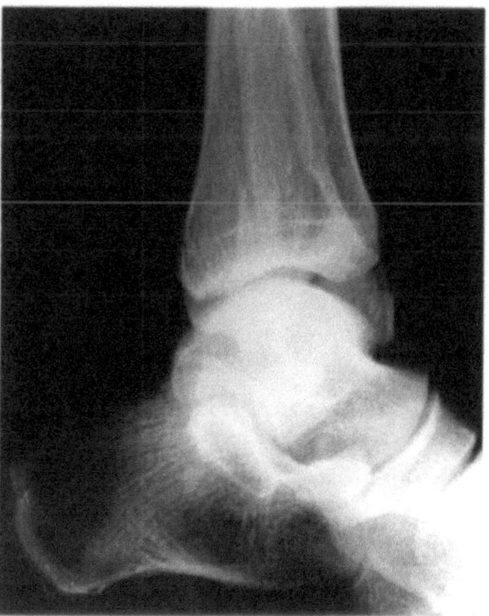

Fig. 14.6. Anteroposterior (**a**) and lateral (**b**) radiographs demonstrating a Weber B fracture with a fracture line of the lateral malleolus at the level of the ankle joint

malleolus is fractured or the deltoid ligament torn. Medial malleolar fracture combined with a fracture of the proximal third of the fibula is called Maisonneuve fracture.

14.1.3.2
Fractures of the Malleoli and Tibia

Fractures of the tibia may occur in conjunction with (a) fractures of the medial malleolus, (b) bimalleolar fractures (combined with fracture of the lateral malleolus), and (c) trimalleolar fractures (combined with fracture of the posterior tubercle).

14.1.3.3
Juvenile Fractures

The juvenile Tillaux fracture is a fracture of the lateral portion of the distal tibial epiphysis (Salter 3). The triplan fracture (vertical: epiphysis; horizontal: epiphyseal plate; oblique: metaphysis) involves, in addition to the Tillaux fracture, a posterior triangular metaphyseal fragment.

14.1.3.4
Fractures of the Calcaneus

Fractures of the calcaneus are classified as intra- or extra-articular. A decreased Böhler angle implies an intra-articular fracture.

14.1.3.5
Fractures of the Talus

Osteochondral fractures are seen at the dome of the talus, medially, laterally, or bilaterally.

14.1.3.6
Navicular Fractures

Navicular fractures are typically stress fractures.

14.1.3.7
Lisfranc Fracture-dislocation

Dorsal dislocation of the tarsometatarsal joints associated with avulsion fractures (lisfranc fracture-dislocation) is more likely to be due to diabetic Charcot (neuropathic) joints than to trauma.

14.1.3.8
Metatarsal Fractures

Transversal fracture of the base of the fifth metatarsal is also termed Jones' fracture or dancer fracture. Second or third metatarsal fracture is the most common stress fracture (march fracture).

14.1.3.9
Occult Fractures and Other Subtle Disorders

Other imaging modalities beyond plain radiography are required for the demonstration of occult fractures, stress fractures, osteochondritis dissecans, and posttraumatic osteonecrosis.

An *occult fracture* is defined by clinical suspicion of a fracture but negative initial radiographs. The diagnosis of fracture is delayed until weeks after trauma, when follow-up radiographs demonstrate the "initially occult" fracture because of resorption and better demarcation around the fracture line (YOUNG et al. 1988). Correct early diagnosis provides the benefit of early commencement of definitive treatment and decreases the rate of complications such as delayed union, nonunion, or avascular necrosis; furthermore it can potentially shorten the duration of hospital stay, reduce morbidity, and decrease costs (QUINN and MCCARTHY 1993; HARAMATI et al. 1994).

Radionuclide bone scans have been considered in the past to be the imaging technique of choice for the diagnosis of occult fractures (TIEL-VAN-BUUL et al. 1992), but they are unspecific and lack spatial resolution. In occult fractures MRI is exquisitely sensitive to marrow abnormalities and is therefore superior to CT, rendering even nondisplaced fractures obvious (DEUTSCH et al. 1989; QUINN and MCCARTHY 1993). Additionally MRI can show fracture lines, both cortical and trabecular, better than do plain films (LANG et al. 1992). These findings are present immediately after trauma. MRI has already demonstrated diagnostic utility in radiographically occult fractures such as fractures of the proximal femur (DEUTSCH et al. 1989; QUINN and MCCARTHY 1993; HARAMATI et al. 1994). The best diagnostic strategy in the management of clinically suspected fractures consists in initial radiography followed by MRI in patients with negative radiographs rather than repeated radiography, CT, or bone scans.

In addition to occult "complete" fractures that breach the cortex, several different types of subtle or radiographically occult fractures can be diagnosed definitively with MRI, including *osteochondral fractures* and *stress fractures* which are causes of bone marrow edema (YAO and LEE 1988; KAPLAN et al. 1992; MINK and DEUTSCH 1989). Late stages of *osteochondritis dissecans* (OD) can be evaluated by radiographs, but in the diagnosis of early OD MRI is the modality of choice. Since exact staging of OD is necessary to decide upon the appropriate therapeutic procedure, intra-articular MR-arthrography would appear the best modality.

Early *posttraumatic osteonecrosis* cannot be seen with radiographs but MRI is diagnostic in these early stages. For the evaluation of a subchondral fracture in OD, CT or radiographs are helpful in addition to MRI.

14.2
Infection of the Foot

Infections of the foot and ankle represent common problems. Older patients are frequently affected because of such risk factors as venous disease, soft tissue edema, and decreased lymphatic drainage. Early diagnosis is necessary to allow initiation of optimal therapy which will prevent the development of many complications associated with infections in this region. Infections of the foot should be separated into soft tissue and bone infection. The distinction between soft tissue infection and involvement of bone is critical in the management of patients presenting with suspected infections of the foot and ankle. While soft tissue infections are commonly managed by local wound care and limited antibiotic therapy, osteomyelitis is more refractory to treatment and needs prolonged intravenous antibiotic therapy and in many cases bone debridgement (EDMONS 1986; KAUFMAN et al. 1987; ROBSON and EDSTROM 1977).

14.2.1
Soft Tissue Infection

14.2.1.1
Pathogenesis

Pathophysiologically the first signs of soft tissue infection are hyperemia, edema, swelling, and acute infiltration with white blood cells; these are followed by necrosis, possible development of ulcers and sinus tracts, invasion of fibrovascular tissue, demarcation, and scarring. In many cases restitutio ad integrum is possible; in others severe fibrous scars remain.

Soft tissue infection may involve cutaneous, subcutaneous, muscular, fascial, tendinous, ligamentous, or bursal structures. The plantar region of the foot is most commonly affected, the causes including skin ulcerations from weightbearing and foreign bodies and, in diabetics, soft tissue necrosis over pressure points which provides the site of entry for different organisms (RESNICK 1995).

Soft tissue dissemination of infection can occur via the three plantar muscle compartments: medial, lateral, and intermediate. The intermediate compartment additionally provides a pathway for spread of infection involving the plantar aspect of the foot into the lower leg via the tendon for the flexor hallucis longus muscle. The posterior tibial tendon also allows spread of infection from the lower leg to the foot (RESNICK 1995).

Soft tissue infections can be subdivided into different morphologies: cellulitis, ulceration and sinus tracts, and abscesses and other localized fluid collections.

The initial contamination of skin and subcutaneous tissues can rapidly progress to infective osteitis, osteomyelitis, and septic arthritis.

14.2.1.2
Imaging of Soft Tissue Infection

14.2.1.2.1
CONVENTIONAL X-RAY FILMS
The most important signs on conventional radiography are soft tissue swelling and unsharp borders of affected soft tissue. In some cases gas may be detected within the inflamed tissue.

14.2.1.2.2
COMPUTED TOMOGRAPHY
Higher soft tissue contrast in comparison to conventional radiography allows visualization of cellulitis as soft tissue replacing subcutaneous fatty tissue, ulcers, and abscess formation. In particular, contrast enhancement of infected soft tissue is helpful in delineation.

14.2.1.2.3
ULTRASOUND

Infection of soft tissue poses a problem with ultrasound since findings are unspecific. However, liquid areas in abscess formation can be easily identified.

14.2.1.2.4
SCINTIGRAPHY

While radionuclide methods are sensitive in the detection of soft tissue infection, differentiation of drainable abscess from cellulitis is not possible (BELTRAN et al. 1988).

14.2.1.2.5
MAGNETIC RESONANCE IMAGING

Examination for suspected soft tissue infection versus osteomyelitis should be performed by using a high-resolution small field of view study. Evaluation should be limited to the forefoot, midfoot, or hindfoot or possibly even to one or two phalanges in order to obtain small fields of view for optimal resolution. A field of view should range from 8 to 14 cm and surface coils are mandatory.

T1-weighted spin-echo sequences, STIR sequences, and T2-weighted fast spin-echo sequences with frequency-selective fat saturation are used. After intravenous contrast administration a T1-weighted spin-echo sequence is applied in two planes, with additional frequency-selective fat saturation imaging in at least in one plane.

Cellulitis appears as a diffuse infiltrative pattern within the subcutaneous tissue of the plantar region and demonstrates a diffuse signal alteration replacing the normal high signal intensity subcutaneous fat on T1-weighted sequences. On long T2-weighted and STIR sequences areas of cellulitis demonstrate increased signal intensity consistent with edema. The involved soft tissues are typically thickened (Fig. 14.7). The most common etiology of cellulitis is arterial hypoperfusion or venous congestion. The hypoxic tissue loses local resistance and is predisposed to infection. However, the presence of edema within soft tissues is nonspecific. Soft tissue edema, which is the most common finding when imaging diabetic feet, cannot be differentiated from cellulitis by MRI. Soft tissue edema in the absence of infection may be due to uneven distribution of body weight secondary to peripheral neuropathy with stasis and fluid accumulation (YUH et al. 1989). The presence of distortion of soft tissues in addition to the increased signal intensity may be helpful in the distinction between cellulitis and noninflammatory edema (MASON et al. 1989).

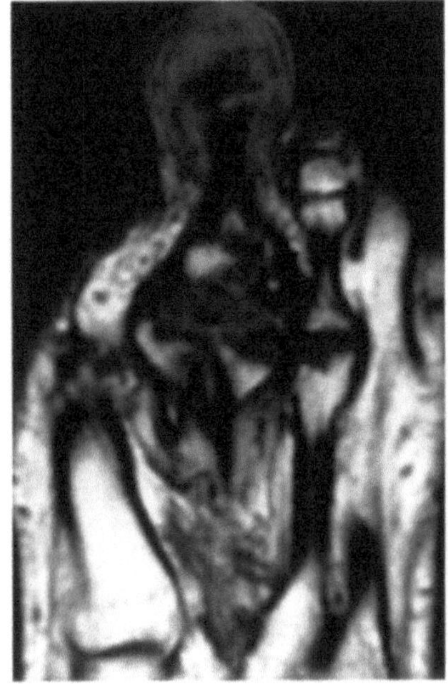
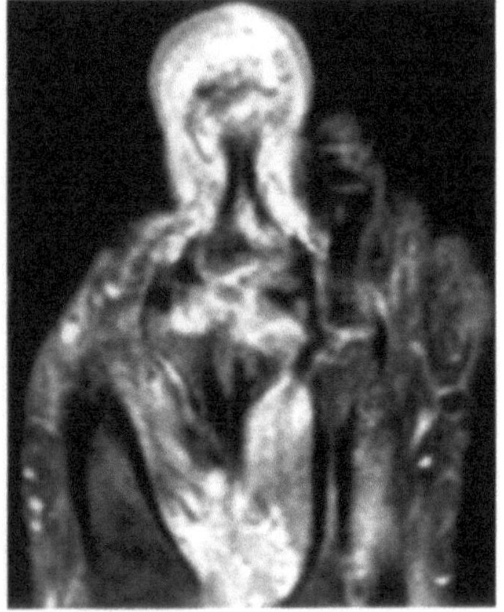

Fig. 14.7. a Coronal T1-weighted MR image showing partial hypointensity and destruction of the second phalanx including the metatarsophalangeal joint. There is marked swelling of the soft tissue surrounding the affected phalanx, with a hypointense signal alteration. b T1-weighted MR image obtained after i.v. contrast administration and frequency-selective fat saturation: Marked contrast enhancement of bony lesions is visible. Surrounding soft tissue also shows significant contrast enhancement representing cellulitis. Diagnosis: acute osteomyelitis of the second phalanx

The identification of ulcers or sinus tracts is useful to draw attention to adjacent areas of abscess or osteomyelitis. Gadolinium enhancement can be identified in the walls of sinus tracts associated with abscesses and osteomyelitis (DANGMAN et al. 1992).

Abscesses can be detected as well-defined regions of intense high signal intensity on T2-weighted or STIR images, indicating fluid (YUH et al. 1989). They can be distinguished from cellulitis by virtue of a smooth margin, homogeneous higher signal intensity, and typically convex borders in relation to the surrounding background edema. Gas content within abscesses is demonstrated on MRI by discrete areas of signal loss on all pulse sequences. Gadolinium enhancement is also helpful in distinguishing focal areas of cellulitis from abscess: areas of cellulitis enhance following the administration of gadolinium, whereas abscesses produce rim enhancement (DANGMAN et al. 1992).

14.2.2
Bone Infection

14.2.2.1
Pathogenesis

In the adult foot spread of infection from a contiguous source represents the most important mechanism of infection. Direct implantation of infectious material into the bone or joint, as with puncture wounds and penetrating injuries, is a further important mechanism of infection. In hematogenous spread of infection the direction of spread is opposite in that medullary bone is involved at an early stage with later extension towards the cortex and periosteum. Pathophysiologically early signs are hyperemia and edema; these signs are followed by osteoporosis, osteolysis, necrosis, possible sequestration, development of fibrovascular tissue, demarcation, restitutio ad integrum, or scar formation of bone representing sclerosis.

Different stages of osseous involvement can be distinguished. Infected periostitis comprises involvement of the periosteum bordering the bone. Subperiosteal accumulation of organisms can lead to infective osteitis and osteomyelitis (RESNICK 1995). Infective osteitis comprises infection of the bone cortex; this can represent an isolated stage but more commonly is associated with osteomyelitis.

14.2.2.2
Imaging of Acute Osteomyelitis

14.2.2.2.1
CONVENTIONAL X-RAY FILMS
Conventional radiography may reveal normal findings or different stages of osteoporosis, bone destruction with focal osteolysis, unsharp and irregular borders of lesions, and periosteal reactions. Moreover pronounced soft tissue swelling is characteristic. Increasing sclerosis is a sign of healing.

14.2.2.2.2
COMPUTED TOMOGRAPHY
CT plays no significant role in the evaluation of acute osteomyelitis. The findings are similar to those observed on conventional x-rays. Early healing (sclerosis) may be easily identified with CT.

14.2.2.2.3
SCINTIGRAPHY
Three-phase bone scans and immunoscintigraphy with technetium-99m labeled white blood cell antibodies provide high sensitivity which is equal to that of MRI; however, they lack specificity due to the inferior spatial resolution (LARTOS et al. 1991). The appearance may be pathologic within 7–10 days after the onset of infection. A sign of healing is reduced radioactivity uptake.

14.2.2.2.4
MAGNETIC RESONANCE IMAGING
MRI demonstrates not only changes in medullary bone in patients with osteomyelitis of the foot, but also changes within the periosteum and bony cortex, which is important since in the foot soft tissue infection commonly involves adjacent bone. Laminated periosteal reaction may be seen in some cases of osteomyelitis, this being recognized as concentric low signal intensity lines paralleling the outer cortical margins of the bone. Interposed between and beyond the periosteal changes and bony cortex, high signal intensity changes can be seen on T2-weighted and STIR images, representing pus or noninflammatory edema (GREENFIELD et al. 1991). On high-resolution images initial involvement of bone from an adjacent soft tissue infection can be appreciated. This stage of infective osteitis is best diagnosed by increased signal intensity on T2-weighted or STIR sequences, often in the form of a linear or bandlike pattern confined to the cortex of the involved bone. Extension of signal alteration to the medullary bone appears as a decreased signal

intensity within the intramedullary spaces on T1-weighted images and as increased signal intensity on T2-weighted or STIR images (YUH et al. 1989; TANG et al. 1988) (Fig. 14.8). In tarsal and metatarsal bones, in which fatty marrow predominates, T1-weighted sequences are generally sufficient for detection of osteomyelitis. T2-weighted or STIR sequences are more sensitive for detection of osteomyelitis in regions of hematopoietic marrow, particularly in children. STIR sequences have been reported to provide the greatest sensitivity for the detection of osteomyelitis in regions of both fatty and hematopoietic marrow (UNGER et al. 1988). Cortical bone may appear mostly normal on MRI in cases of acute hematogenous osteomyelitis, although periosteal elevation with subperiosteal pus may be demonstrated (ERDMAN et al. 1991). In cases of acute osteomyelitis extensive changes are present in surrounding muscle and fascial planes, with increased signal intensity on T2-weighted and STIR sequences. The soft tissue changes most likely reflect a combination of infectious and noninfectious edema. The signal changes associated with acute osteomyelitis are nonspecific and have to be distinguished from signal changes associated with recent or subacute trauma (UNGER et al. 1988). First signs of healing are a conversion into fatty marrow, less contrast enhancement and reduced soft tissue swelling (Table 14.1).

Gadolinium enhancement is extremely helpful. Enhancement is almost always indicative of infection (MORRISON et al. 1993), although all that is bright on T2-weighted or STIR images or all that is dark on T1-weighted images may not be infected. A variable amount of sympathetic edema is often seen with osteomyelitis. In addition, patients with septic arthritis may have adjacent reactive edema in bone which is not infected. Contrast administration is the best technique to differentiate inflammatory edema from abscesses, and may also help to differentiate between inflammation and tumor (BOHNDORF 1996).

14.2.2.3
Imaging of Chronic Osteomyelitis

14.2.2.3.1
CONVENTIONAL X-RAY FILMS
Findings can vary from osteolytic lesions with surrounding sclerosis, to mixed osteolytic-osteosclerotic lesions, to pure osteosclerosis. Compacta-spongiosa borders are unsharp. Sequestra may also be identified. The shape of bone may be irregular.

14.2.2.3.2
COMPUTED TOMOGRAPHY
In cases in which identification of sequestra and a sinus tract is not possible on conventional radiographs or MRI, e.g., in posttraumatic chronic osteomyelitis, CT is helpful.

14.2.2.3.3
SCINTIGRAPHY
For diagnosis of reactivation of chronic osteomyelitis scintigraphic techniques are equivalent to MRI; however, for therapeutic planning the essential determination of extension of lesion is best achieved by MRI.

14.2.2.3.4
MAGNETIC RESONANCE IMAGING
Remodeling of the cortex and the medullary cavity of long bones with chronic osteomyelitis is well demonstrated on MRI. The cortical changes appear as a low signal intensity expansion of the cortex (ERDMAN et al. 1991). The remodeled medullary cavity may demonstrate areas of fat signal intensity presumably related to areas of regenerated or healed marrow. T2-weighted and STIR sequences are best used for depiction of foci of active disease which will demonstrate high signal intensity contrasted against the lower signal intensity of the thickened surround-

Table 14.1. Healing signs in acute osteomyelitis

Conventional X-ray	CT	Scintigraphy	MRI
Sclerosis	Condensation	Decrease in activity	Decrease in abnormal signal on T1/T2-weighted images; reconversion of fatty bone marrow
Reduction of soft tissue swelling	Reduction of soft tissue swelling		

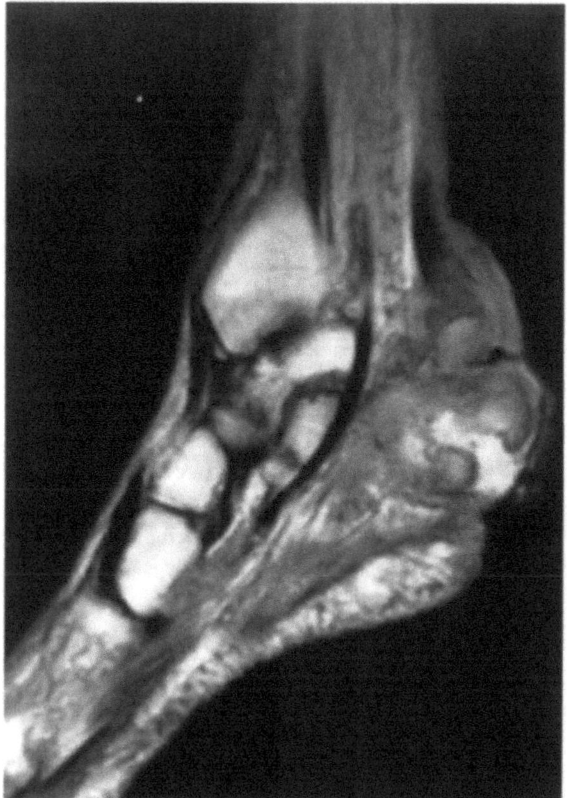

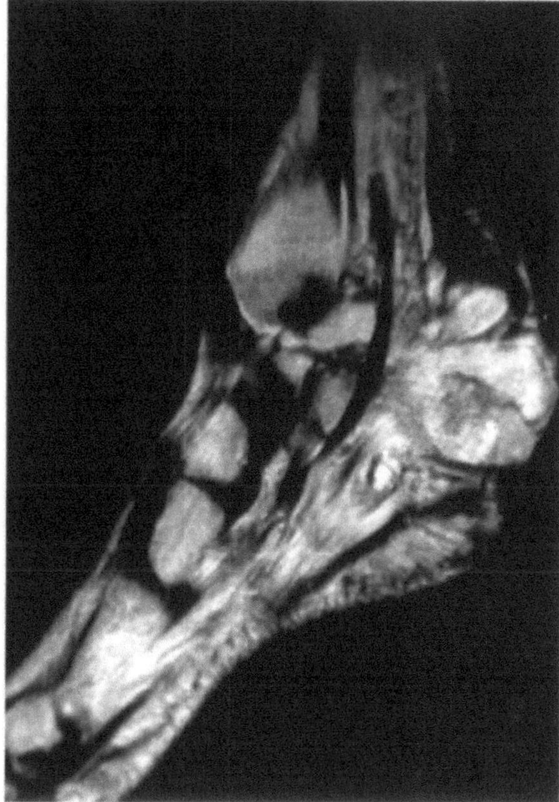

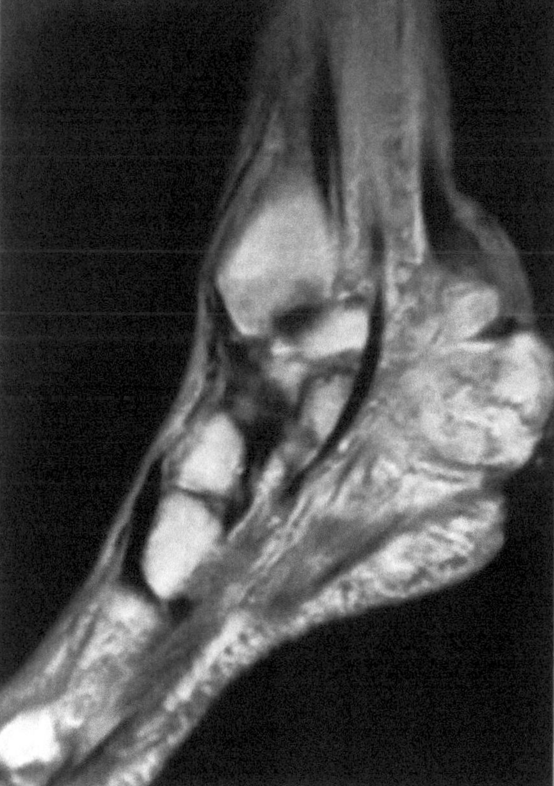

Fig. 14.8. a Sagittal T1-weighted MR image revealing a defect in the dorsal aspect of the calcaneus, with hypointense signal alteration of the remaining bone marrow. The adjacent soft tissue shows a large skin ulcer. A displaced bony fragment attached to the Achilles tendon is shown above the calcaneus. The Achilles tendon is thickened and retracted with associated soft tissue swelling. **b** Sagittal T2-weighted MR image. The calcaneus and displaced bony fragment demonstrate a marked hyperintense signal alteration representing bone marrow edema. **c** Sagittal T1-weighted MR image obtained following contrast administration. The calcaneus and the superiorly displaced bony fragment as well as skin ulcer reveal marked contrast enhancement. Diagnosis: osteomyelitis of the hindfoot with avulsion fracture of the Achilles tendon

ing bone (UNGER et al. 1988). These foci of high signal intensity may not necessarily represent frank pus, but rather regions of infected material. Sinus tracts are identified as linear areas of increased signal on T2-weighted sequences that extend from the bone to the skin surface. The site of the disruption of the cortex may be well demonstrated. In patients with previous surgery or extensive soft tissue deformity secondary to trauma the differentiation of a sinus tract from a retracted scar may be difficult if continuity with the site of bone infection cannot be clearly demonstrated (MASON et al. 1989). Sequestra are typically sharply marginated bone fragments located in the medullary aspect of tubular bones surrounded by granulation tissue. On MRI sequestra appear as areas of diminished signal intensity or areas of similar intensity to cortical bone within and contrasted against the high-intensity foci of infection on T2-weighted sequences (MORRISON et al. 1993). Foci of chronic osteomyelitis may demonstrate a rim sign consisting of a well-defined rim of low signal intensity surrounding the area of focal active disease on MRI scans (ERDMAN et al. 1991). Gadolinium may be helpful in defining the presence of intraosseous abscesses and sequestra (DANGMAN et al. 1992).

Findings in the healing phase of osteomyelitis on follow-up examinations employing different imaging techniques are listed in Table 14.1 (VORBECK et al. 1996).

14.3
Diabetic Foot

The diabetic foot is an entity which develops as the result of four influencing factors due to diabetes mellitus: neuropathy, angiopathy, mechanical overload, and bacterial superinfection. It results in abnormalities of the bones or of the soft tissues of the foot with the potential for extensive destruction.

14.3.1
Pathogenesis

It is generally agreed that of the aforementioned four parameters which influence the development of a diabetic foot, neuropathy is the most important. However, the mechanisms of neuropathic damage of the bones and joints of the foot are not understood in detail. It is assumed that microangiopathy of the

nutritive vessels of the supporting nerves with or without loss of axons leads to structural damage of the nerve and, probably more important, of the nerve sheath. Due to loss of vasoconstrictive neural impulses, active hyperemia occurs in circumscribed parts of the bones. Other sequelae of neuropathy are sensory loss with respect to the skin and the joints, motoric deficits, and disturbances of autonomic nerve function. Hyperemia eventually results in various patterns of active bone resorption. These neurologic abnormalities and the corresponding imaging signs should not be confused with those due to palsies or spasms of the skeletal muscles. Normalization of the insulin metabolism will generally improve the neuropathic changes of the bones within weeks.

Macroangiopathy of the arterial vessels may lead to chronic or acute ischemia, an important trigger mechanism in the development of diabetic foot. In recent studies it has been reported that normalization of arterial blood supply may improve abnormalities in about one-third of patients (JUNG 1996).

Mechanical overload, particularly on the tarsal and metatarsal bones, is the result of neurologic deficits due to improper muscle control and impaired vegetative control mechanisms in the joint capsules. The atactic movements of the foot are clinically referred to as "diabetic gait."

Bacterial superinfection, though in many cases the first and the most impressive clinical feature, has to be regarded as a final complication in the development of a diabetic foot. It results from painless skin ulceration due to sensory loss, lowered production of sweat, abnormal movements during gait, or hypoxia due to inadequate arterial blood supply. Osteomyelitis only rarely develops as a sequela of septicemia; rather, in more than 90% of

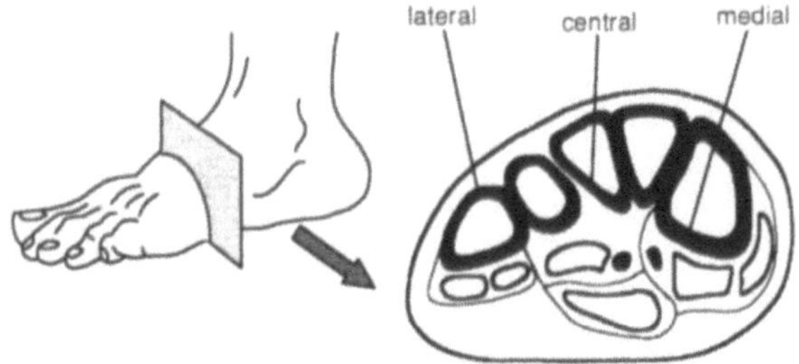

Fig. 14.9. Compartments of the foot. Spread of infection is strongly oriented towards major septa which generally confine inflammatory edema. Minor septa, in contrast, contain openings for spread of inflammation or tumors. (Modified after GOODWIN et al. 1995)

cases it occurs via continuous infection of the bone via cutaneous, soft tissue, and periosteal spread (BAMBERGER et al. 1987). Because of the compartmental orientation of the foot, spread of infection is generally guided by septa between the medial, intermediate, and lateral compartments (Fig. 14.9) (GOODWIN et al. 1995). It is a matter of fact that foot complications of diabetes are the most common cause of nontraumatic lower extremity amputation in the United States and in Western Europe. Foot complications in diabetic patients account for more hospital days than other aspects of their disease (VESTRING et al. 1995).

14.3.2
Imaging

14.3.2.1
Plain Film Radiography

Findings on plain films of the ankle and the foot mainly reflect the neuropathic changes of the bones, joints, and surrounding soft tissues. Descriptive imaging terms like "anarchic appearance" of bones and joints or "organized chaos" reflect the extensive destruction in cases of neurogenic osteopathy and arthropathy (DIHLMANN and BANDICK 1995). Because of sensorineurologic deficits, great discrepancies may exist between minor clinical symptoms and x-ray findings (Fig. 14.10). Partly depending on the patient's metabolic parameters, a more destructive and osteolytic pattern or a more sclerotic pattern of radiographic findings may be found. These two patterns are referred to as atrophic and hypertrophic forms of neurogenic osteoarthropathy (Table 14.2). According to SINHA et al. (1972) 46% of the lesions are located in the tarsal bones, 27% in the tarsometatarsal bones, and 27% in the

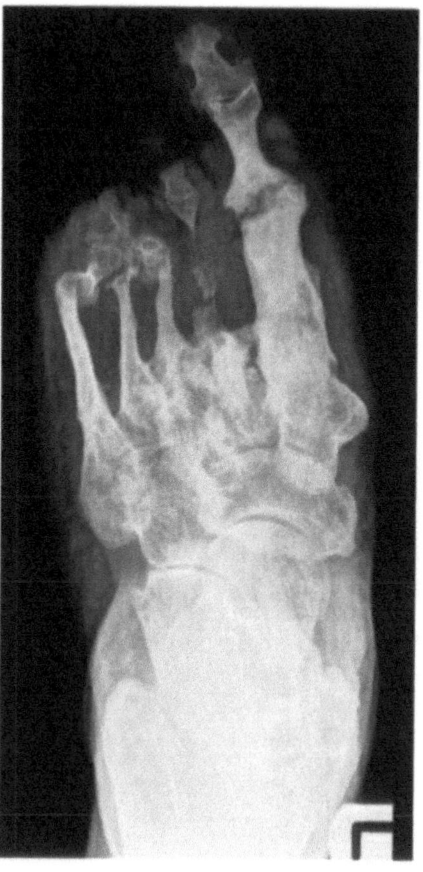

Fig. 14.10. Digital radiograph (dorsoventral) of the forefoot showing extensive destruction of metatarsal and phalangeal bones. Osseous bridges are present between the bases of the metatarsals. Heterotopic calcifications are seen in the area of the completely resorbed second metatarsal shaft, and a "sucked candy-stick" appearance of the third to fifth metatarsals can be observed. There is periosteal thickening of the first metatarsal with sclerosis of cancellous bone. An insufficiency fracture of the fourth metatarsal and erosions of MTP joint I are remnants of a former atrophic phase. Diagnosis: hypertrophic form of diabetic neuropathy

Table 14.2. Plain film findings in diabetic neurogenic osteoarthropathy

	Atrophic changes	Hypertrophic changes
Deviations, deformations	Subluxation, luxation	Ankylosis, osseous bridges between bones
Soft tissue abnormalities	Soft tissue swelling with or without emphysema	Heterotopic calcification or ossification
Joint damage	Juxta-articular osteoporosis, osteolysis, erosions	Sclerosis of the cancellous bone, "pencil and cup" deformities
Bone lesions	Insufficiency fractures, fragmentation	Bone sclerosis, pseudarthrosis with excessive callus formation, periosteal thickening, "sucked candy-stick" appearance of metatarsals

metatarsophalangeal bones. Therefore, concomitant imaging of the ankle and the foot is necessary in patients referred for evaluation of diabetic foot. The main goal of plain film radiography is to diagnose and document the extent of bone and joint destruction not detectable by clinical means alone. Diagnosis of neuropathic bone changes will assist in the earliest possible institution of proper management. Moreover the form of neurogenic osteopathy, atrophic or hypertrophic, should be defined to support therapeutic decisions.

The diagnostic value of plain film radiography in diagnosing bacterial osteomyelitis is reported to be low. In a meta-analysis, GOLD et al. (1995) reported a sensitivity of 28%–72% (with a single study reporting 93%) and a specificity of 50%–92%. These poor results are in part due to the fact that resorptive or sclerotic bone changes may be the result of neurogenic osteopathy, bone infection, or both. Air bubbles in the soft tissue of the foot or skin defects indicating ulcers may be associated with bacterial superinfection but serve neither to confirm nor to exclude osteomyelitis.

14.3.2.2
Scintigraphy

The sensitivity of bone scintigraphy with three-phase ^{99m}Tc-methylene diphosphonate bone scans in the diagnosis of osteomyelitis is reported to be 69%–100%. However, specificity is low because of the high rate of false-positive results due to osteoblastic reactions in neurogenic osteopathy (18%–79%) (GOLD et al. 1995). The value of indium-labelled leukocytes is a matter of controversy. This technique displays the highest sensitivity of all radionuclide studies for osteomyelitis, but as many as 31% of cases have been reported to be false-positives.

14.3.2.3
Magnetic Resonance Imaging

Both structural damage due to neurogenic osteopathy and osteomyelitis usually appear as hyperintense changes of the bone marrow on T2-weighted images. By contrast they are generally hypointense on T1-weighted images. The most sensitive means to detect abnormalities of the bone and soft tissues in diabetic feet is use of a T2-weighted fat-suppressed sequence (STIR) (Fig. 14.11). To differentiate infectious from noninfectious bone changes in the diabetic foot it is

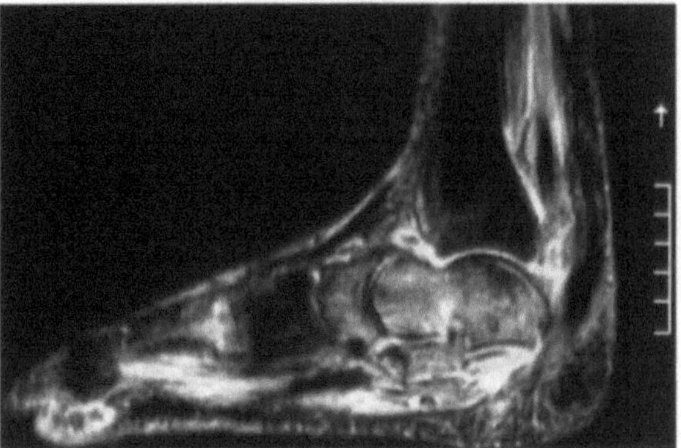

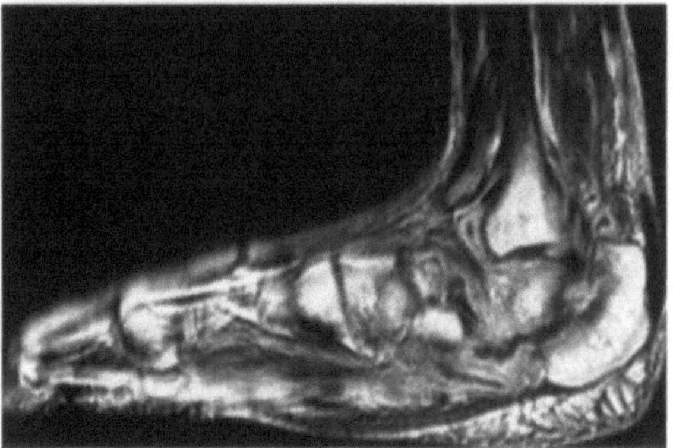

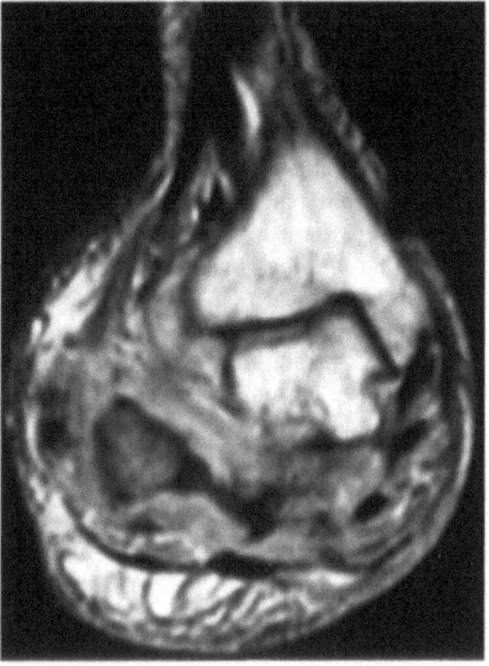

Fig. 14.11. a Sagittal (STIR) MR image demonstrating collapse of the middle and hindfoot with increased signal intensity within the talus and navicular bone. Soft tissues in the dorsal aspect of the plantar region also reveal hyperintense signal alteration. b Sagittal and c coronal T1-weighted images after i.v. contrast administration show marked contrast enhancement of bony and soft tissue lesions with abscess formations. Diagnosis: Diabetic neuroarthropathy complicated by osteomyelitis

necessary to analyze the anatomic orientation of the abnormalities. Spread of infection is regarded in most of the cases to be continuous and compartment oriented. Therefore, documentation of skin ulcers, soft tissue swelling with or without abscess formation, periosteal thickening and subperiosteal inflammation, and bone marrow edema strongly suggests osteomyelitis. Another hint supporting the diagnosis of osteomyelitis is the "ghost sign." When there is extensive disruption of the shape of the tarsal bones the shadow of these bones may be visible on STIR or gadolinium-enhanced images.

14.3.2.4
Angiography

Angiography for imaging of the arterial vessels of the lower leg is, in conjunction with interventional techniques like percutaneous transluminal angioplasty, suitable for improving ischemia of the bones and the soft tissues of the foot. Concerning magnetic resonance angiography (MRA), UNGER et al. (1995) reported in a preliminary study that this technique might be of value in demonstrating abnormalities in the major arterial and venous vessels of the foot; to date, however, MRA is not superior to conventional arteriograms in all cases.

14.3.2.5
Computed Tomography

Computed tomography (CT) was used prior to the routine application of MRI to assess the extent of bone and soft tissue infection. Although there has been no large-scale investigation comparing the usefulness of CT and MRI, the latter modality is gener-

ally believed to show the extent of infection more clearly.

14.3.3
Differential Diagnosis

Other rare forms of neurogenic osteoarthropathy have to be considered because they may mimic changes in the diabetic foot (Table 14.3). It is of practical value that calcifications of the vessel wall of the pedal arteries are strongly associated with diabetes mellitus and its complications. Only rarely do these calcifications occur together with other forms of Charcot's joint or pseudo-Charcot arthropathy.

Other causes of excessive joint damage have to be considered (Table 14.4). However, in clinical practice, the information about proven diabetes mellitus and the typical localization (Lisfranc joint, Chopart joint, ankle) strongly suggests the diagnosis of diabetic foot.

In summary, plain film radiography should be the initial imaging examination. If plain films are positive and osteomyelitis is clinically suspected, MRI should be performed. If plain films are normal and there is a strong clinical suspicion of osteomyelitis, scintigraphic studies should be performed, and in the event of a positive result these should be followed by MRI.

14.4
Neoplastic Disorders of the Foot

Primary and secondary bone tumors of the foot and tarsals are uncommon, accounting for 3% of all bone tumors. Most tumors involve the calcaneus, the metatarsals, and the talus although many

Table 14.3. Charcot's joint: differential diagnosis

	Disease entity	Typical location
Metabolic	Diabetes mellitus	Foot
	Hypercorticism	Hip, knee, shoulder
	Amyloid neuropathy	Ankle and foot
Inflammatory	Leprosy	Various
	Tabes dorsalis	Knee, hip, foot
Congenital	Syringomyelia	Shoulder, elbow
	Spinal dysraphism	Ankle and foot
Traumatic	Spine, peripheral nerves	Dependent on level of injury
Idiopathic		Various

Table 14.4. Differential diagnosis of excessive structural damage of joints

Neurogenic osteo-arthropathy	Diabetic foot Pseudo-Charcot joint Charcot joint: other forms
Degenerative joint disease (osteoarthritis)	Rapid destructive osteoarthritis
Crystal-induced arthropathies	Calcium pyrophosphate dihydrate deposition disease (CPPD) Gout
Infectious arthritis	Bacterial arthritis (various forms) Madura foot
Neoplastic	Malignant tumors with contact to joints (bone metastasis) Gorham's disease
Traumatic	Posttraumatic osteolysis

enchondromas and osteochondromas are located in the phalanges (KRICUN 1993; RESNICK et al. 1988). The highest incidence of all malignant bone tumors has been observed for chondrosarcoma and Ewing's sarcoma, while cartilaginous tumors as a group, and specifically osteochondromas, account for most benign bone tumors (KRICUN 1993; RESNICK et al. 1988).

Morton's neurofibroma, which is a tumorlike lesion occurring adjacent to the head of the third and fourth metatarsals, is by far the most common tumorous entity that affects the soft tissues of the foot (BERLIN 1980). Hemangiomas, lipomas, and fibromatoses are other tumors which may affect soft tissue compartments of the foot and can be diagnosed on the basis of their tumor-specific behavior with different imaging modalities (STOLLER et al. 1993). Malignant soft tissue tumors are extremely rare. The great number of tendon sheaths in the foot might be the reason for the relatively high incidence of synovial sarcomas (Fig. 14.12).

14.4.1
Imaging

Analysis of roentgenograms with respect to lesion location, pattern of destruction, classification of tumor matrix, periosteal new bone formation, and the tumor border may provide tumor-specific information on neoplasms of the musculoskeletal system (LODWICK 1965). Tumors located in short tubular bones are in principle similar to those discovered in long tubular bones. However, because of their size location plays no role in the evaluation of lesions of the short bones of the midfoot and the phalanges. By

contrast, location does play an important part in the diagnosis of tumors in the calcaneus (Figs. 14.13–14.16). The calcaneus develops from two centers of ossification, a main ossification center for the body and an apophysis adjacent to the posterior calcaneus. Thus, there is an epiphyseal equivalent, the apophysis, a metaphyseal equivalent, the region of the body of the calcaneus near the cartilage plate, and a diaphyseal equivalent corresponding to the body of the calcaneus (KRICUN 1993).

In contrast to their potential specificity, conventional radiographs are not extremely sensitive in the detection of small amounts of bone destruction, especially if the destructive focus is located in the cancellous bone (RESNICK et al. 1988). Ultrasound and CT are valuable tools for evaluation of tumor extension in the soft tissues. Ultrasound permits accurate assessment of the cartilage cap of exostoses. The detection rate and measurement accuracy of ultrasound are higher than those of CT and comparable to those of MRI (MALGHEM et al. 1992). Hematoma, synovial cyst, ganglion, and lipoma may be diagnosed by CT although malignant transformation cannot be ruled out. Tumor size may be overestimated due to surrounding edema (SARTORIS and RESNICK 1988). Nevertheless, CT is helpful in preoperative staging, guided biopsy of tumor tissue, monitoring of chemotherapeutic response, and postoperative observation.

Vascular supply to the extremity must be maintained to achieve a satisfactory surgical result. For this reason, preoperative angiography is important, as it can demonstrate vessel displacement, compromised status of distal run-off vessels, and the source and amount of neovascularity (MITTY 1993).

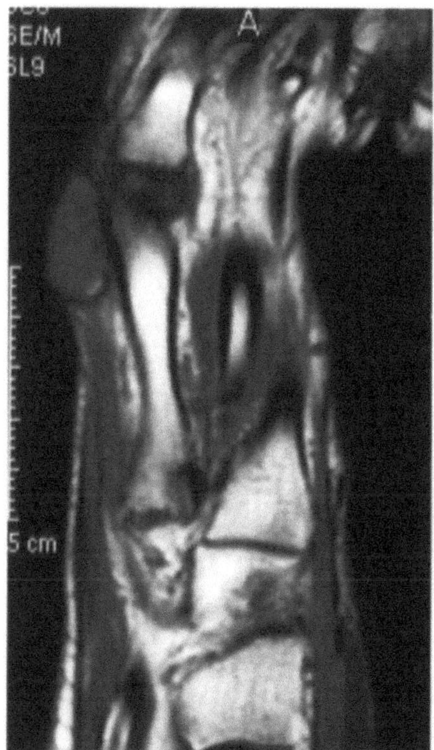
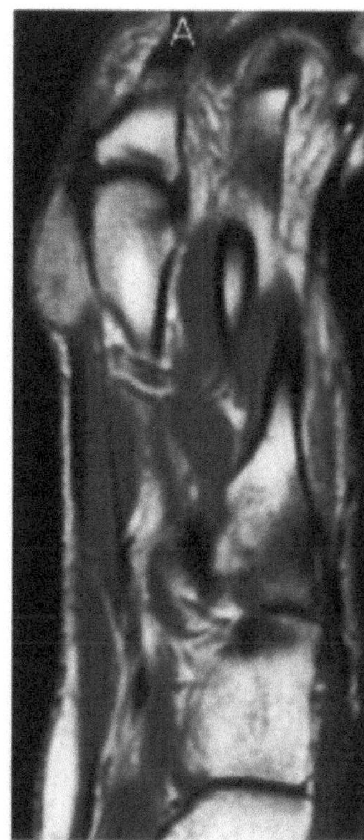
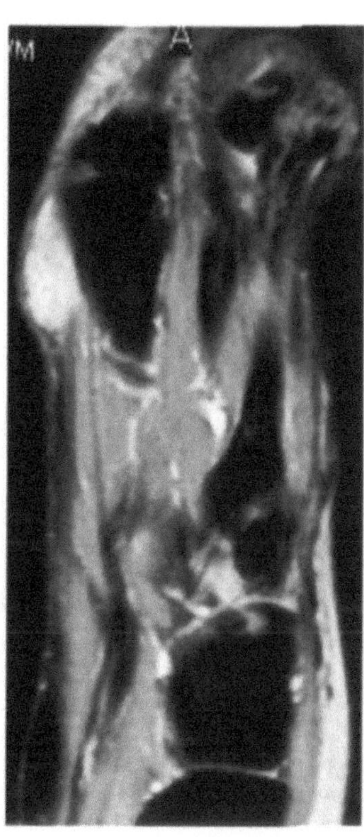

a

b, c

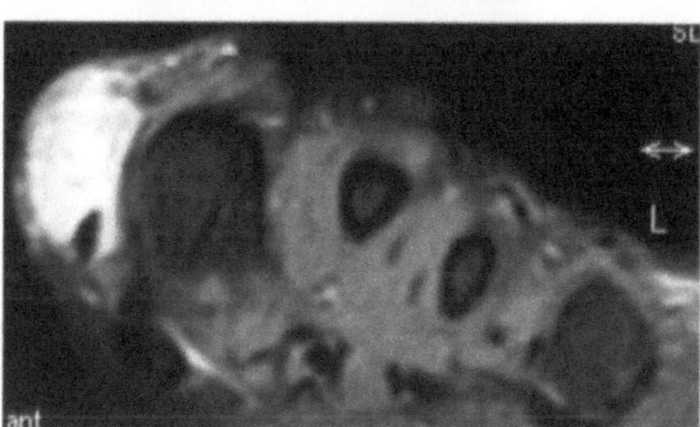

d

Fig. 14.12 a–d. Coronal and axial MR images of the midfoot demonstrating an ovoid tumor extending from the tendon sheet of the abductor hallucis longus. **a** T1-weighted image showing muscle isointense ovoid tumor tissue extending from the tendon sheet of the abductor hallucis longus. **b** T1-weighted image obtained after i.v. contrast administration. **c** STIR image revealing bright signal in the lesion. **d** Coronal T1-weighted image with fat saturation, showing hyperintensity within the tumor. Diagnosis: malignant synovial sarcoma

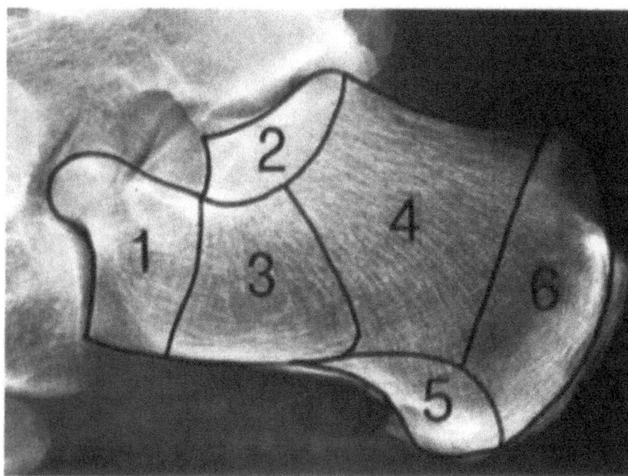

Fig. 14.13. Lateral view of the calcaneus with schematic drawing of the regions predominantly involved by different tumor entities. Region 1: chondroblastoma, giant cell tumor; region 2: osteoid osteoma, chondroblastoma; region 3: simple bone cyst, lipoma; region 4: metastasis, Ewing's sarcoma; region 5: chondromyxoid fibroma; region 6: aneurysmal bone cyst, chondroblastoma, giant cell tumor, metastasis

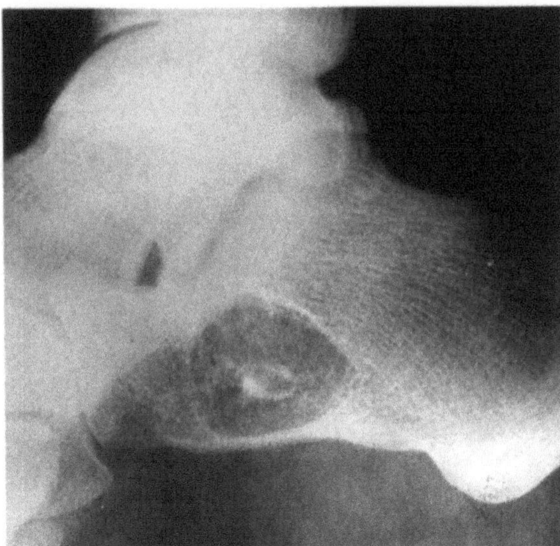

Fig. 14.14. Calcaneus (lateral view). An intraosseous lipoma is present in region 3. Typical central calcification in a radiolucent lesion surrounded by a thin sclerotic rim

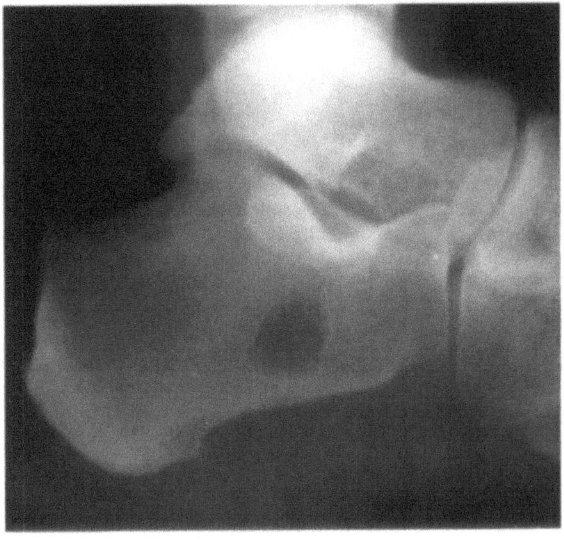

Fig. 14.15. Calcaneus (lateral view). A simple bone cyst is shown in region 3. The radiolucent lesion displays a thin sclerotic rim without matrix calcification

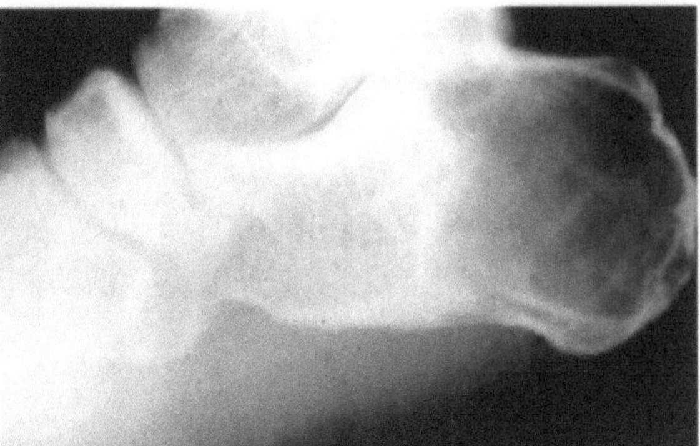

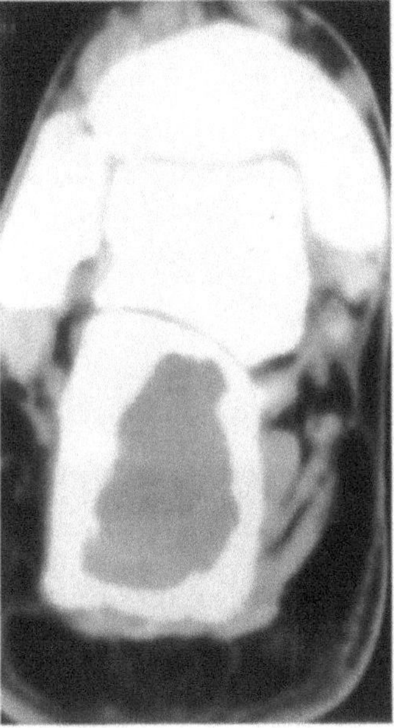

Fig. 14.16 a,b. Calcaneus. **a** Lateral view demonstrating a radiolucent lesion – an aneurysmal bone cyst – with a prominent sclerotic rim in region 6. **b** Coronal view of the hindfoot: fluid-fluid level in an aneurysmal bone cyst

Noninvasive diagnosis of infiltration of bone marrow in leukemia and lymphomas and early demonstration of tumor recurrence are a domain of MRI due to its superior contrast resolution. A majority of bone tumors are characterized by low signal intensity on T1-weighted and high signal intensity on T2-weighted spin-echo sequences (Fig. 14.17).

Differentiation of tumor tissue and surrounding perifocal edema is simply achieved using T2-weighted image sequences and administration of contrast media (ERLEMANN et al. 1990). Although no differences in signal intensity have been found for benign versus malignant bone tumors, a higher uptake of gadolinium-DTPA is usually seen with malignant transformation. Tumor-specific diagnosis with MRI is limited to certain entities like hemangiomas, lipomas, and bone cysts. Furthermore blood degradation products may lead to the correct diagnosis of a hemangioma of the synovial membrane or pigmented villonodular synovitis.

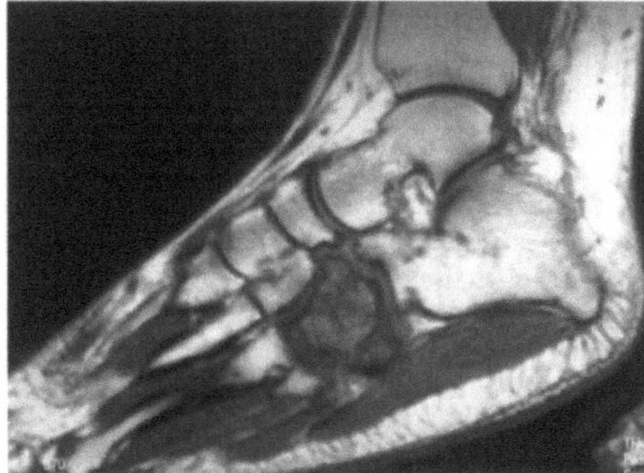

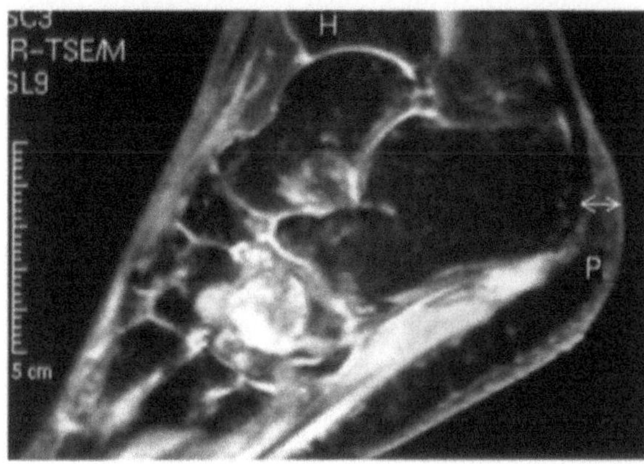

Fig. 14.17. a Sagittal T1-weighted MR image of the midfoot, revealing decreased signal intensity in the marrow space of the cuboid. b Sagittal STIR image. The neoplasm turns bright. Partial destruction of the cortex indicates invasion of adjacent joints. Diagnosis: malignant fibrous histiocytoma

14.4.2
Cartilaginous Tumors

Osteochondromas are composed of a covering cartilage cap and a bony attachment to the underlying host bone. A broad-based attachment is characteristic of a sessile osteochondroma. Conversely, if the lesion arises from a narrow pedunculated stalk it is more commonly called an exostosis. The different morphology is of prognostic significance as malignant transformation of an exostosis is extremely unusual, whereas the sessile osteochondroma is more likely to undergo malignant transformation. Prior to skeletal maturity the cap of the osteochondroma is completely cartilaginous. The thickness of the cap varies, but in the adult it typically ranges from 1 to 6 mm in thickness. In cases of chondrosarcomatous transformation, destruction of the cortex and inva-

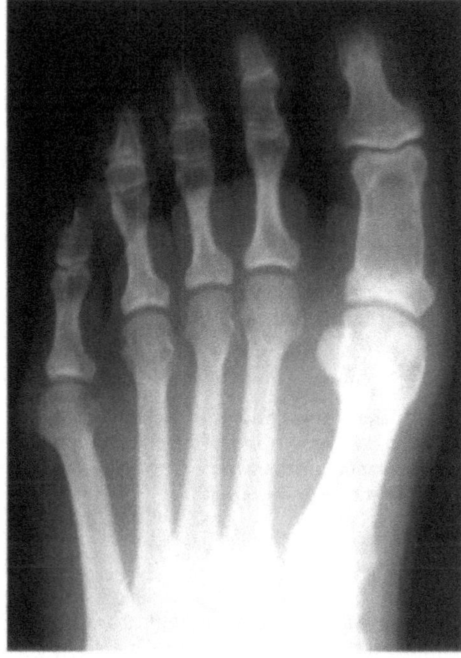

Fig. 14.18. Expansive well-marginated lesion in the proximal phalanx of the great toe without matrix calcifcation. Diagnosis: enchondroma

sion of adjacent soft tissue can be observed. Distinction between a sessile osteochondroma and periosteal osteosarcoma can be problematic on the basis of radiographs; MRI can resolve this dilemma by identifing marrow that is contiguous between the host bone and the lesion, thereby establishing the diagnosis of osteochondroma (GIUDICI et al. 1993).

Fifteen percent of all tumors of the skeleton of the foot are enchondromas (SCHAJOWICZ 1981) (Fig. 14.18). These tumors are predominantly located in the phalanges and metatarsals, presenting as geographic osteolytic metaphyseal lesions which may demonstrate typical popcorn matrix calcification. Rare complications in childhood are fractures and sarcomatous transformation in Ollier's disease (LIU et al. 1987).

Chondroblastomas involve the epiphysis, most frequently the talar neck (Fig. 14.19), and the dorsal aspect of the calcaneus adjacent to the subtalar joint. Recent studies explain this tumor location by incorporation of physeal hyaline cartilage, which may demonstrate minor calcification (KRICUN 1993).

Cortical destruction or at least endosteal resorption is almost always visible with chondrosarcoma, which may demonstrate typical ring or arclike calcification of matrix. Most chondrosarcomas are

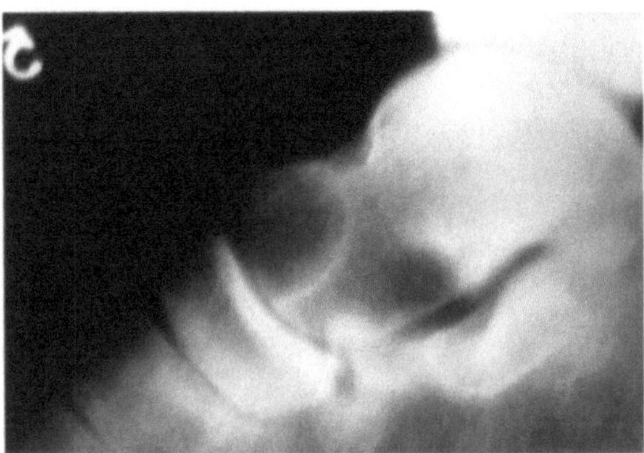

Fig. 14.19. Lateral conventional tomogram of the talus. A radiolucent lesion is seen in the subchondral region of the talar head, expanding into the neck. The lesion is well marginated. Diagnosis: chondroblastoma

located in the hindfoot and account for up to 3.5% of all primary tumors of the foot (SCHAJOWICZ 1981).

14.4.3
Osteogenic Tumors

The neck of the talus is the most common site of osteoid osteomas, although the subtalar joint and the calcaneus occasionally may be involved. Osteoid osteomas arising in the cortex usually appear as a radiolucent nidus surrounded by extensive sclerosis. Osteoid osteomas arising in cancellous bone are usually osteolytic without surrounding sclerosis, whereas subperiosteal osteoid osteomas may appear as a small regular osteolytic defect in an intact cortex (KRICUN 1993). The rare osteosarcoma of the foot most often develops in the mid- and hindfoot.

Radiographically osteosarcomas appear osteolytic, osteosclerotic, or as mixed lesions within an aggressive pattern of bone destruction. About 0.5%–1.5% of all osteosarcomas occur in the foot. Parosteal osteosarcoma is rare in the foot. These lesions appear similar to their counterparts in long bones, but they are much smaller. A cleavage plane may or may not be visible between the tumor mass and underlying bone (KRICUN 1993).

14.4.4
Myelogenic Tumors

Lipomas of the foot typically occur in the anterior third to mid portion of the calcaneus and may show

diagnostic matrix calcification on conventional roentgenograms (Fig. 14.14). The well-defined lesion with a thin, sharp sclerotic rim resembles a unicameral bone cyst (Fig. 14.15). Differential diagnosis is easy when using MRI.

Calcification is present in 80% of cases and may be caused by fatty infarction or mucinous degeneration (LAGER 1980). Simple cysts are particularly rare in the foot under the age of 17 years. Almost all reported simple bone cysts develop in the anterior third of the calcaneus. The triangular shaped lesions never show matrix calcification and can be differentiated against pseudocysts in the calcaneus by virtue of their distinct sclerotic margin (KRICUN 1993).

The ability of MRI to detect abnormal marrow cellularity with high sensitivity is due primarily to replacement of the normally bright marrow fat (short T1) by darker cellular infiltrate (longer T1). In the accelerated and blast phases of chronic myelocytic leukemia a marked diffuse decrease in marrow signal reflects replacement of marrow fat by leukemic infiltrate. The focal lesions in the marrow of patients with Hodgkin's disease differ significantly from the pattern seen in leukemia. Cases of severe anemia and reactive reconversion of adult yellow marrow to hematopoietic red marrow are probably not distinguishable from leukemias. T2-weighted series and clinical shift imaging may be helpful for differentiation of leukemic infiltration from myelofibrosis (OLSON et al. 1986). Conventional radiographs rarely demonstrate osteosclerotic or osteolytic lesions in the foot.

Ewing's sarcomas are evenly distributed between the tubular bones and the tarsals. Radiographically Ewing's sarcoma is an aggressive, poorly defined osteolytic or osteosclerotic lesion; it may be mistaken clinically and radiographically for osteomyelitis or avascular necrosis (Fig. 14.20). In the calcaneus, osteomyelitis usually develops posteriorly near the cartilage plate, whereas Ewing's sarcoma may develop anywhere in the calcaneus (KRICUN 1993).

Carcinomas of the colon, kidney, and lung are the tumors that most frequently metastasize to the foot, involving primarily the calcaneus and the metatarsals. The lesions are usually osteolytic with an aggressive pattern of bone destruction but may be well defined with sharp sclerotic margins. Periosteal reaction may be present. Purely osteosclerotic lesions can occur (KRICUN 1993).

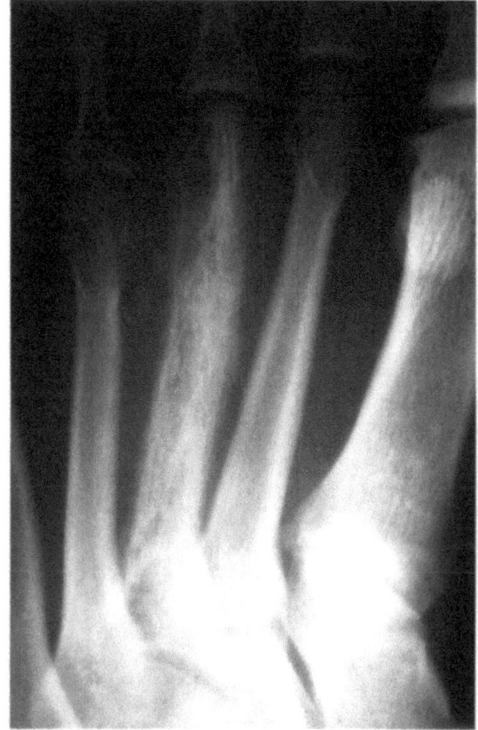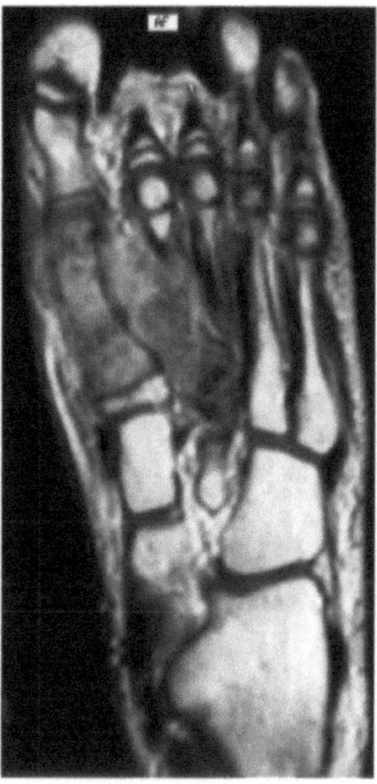

a

b

Fig. 14.20. a Conventional radiograph of the midfoot (oblique view), revealing permeative destruction of the third metatarsal. b T1-weighted anteroposterior MR image obtained after i.v. contrast administration. The medullary cavity of the first metatarsal is enhanced; destruction of the cortex is present, surrounded by soft tissue tumor. There is no infiltration of the proximal epiphysis. Diagnosis: Ewing's sarcoma

14.4.5
Aneurysmal Bone Cysts and Giant Cell Tumors

The tumorlike aneurysmal bone cyst most often involves the posterior aspect of the calcaneus. Fluid levels can be found in the majority of these osteolytic lesions which may complicate chondroblastomas or giant cell tumors (Fig. 14.16). Most aneurysmal bone cysts occur in the metatarsals or involve the posterior aspect of the calcaneus. The highly expansive osteolytic lesions may be poorly defined or display a sharp sclerotic margin.

Giant cell tumors account for about 5% of the primary tumors of the foot (KRICUN 1993). The tumor most often appears osteolytic and expansive with ill-defined or sharp nonsclerotic margins and no periosteal reaction. Giant cell tumors of the talus occur most frequently in the body.

14.4.6
Soft Tissue Tumors

Magnetic resonance imaging is the method of choice for evaluation of soft tissue tumors. The most common lesion of the soft tissues of the foot is Morton's neurofibroma. MRI can easily differentiate this degenerative fibrosing process from true neurinoma (STOLLER et al. 1993).

Only rarely may conventional radiographs identify large lipomas by virtue of the low tissue density, which is radiolucent relative to adjacent muscles.

Hemangiomas and other vascular tumors may demonstrate rounded phleboliths, thus providing a clue as to their vascular nature. The MRI appearance of hemangiomas is sufficiently characteristic to suggest a diagnosis, with signal intensity approximating that of skeletal muscle on T1-weighted and high signal intensity on T2-weighted pulse sequences (Fig. 14.21). Schwannomas and neurofibroma tend to be homogeneous on T1-weighted pulse sequences with a signal intensity somewhat greater than that of skeletal muscle (KRICUN 1993).

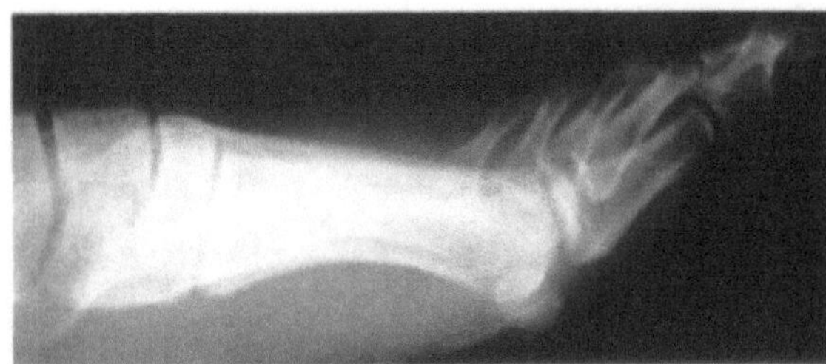

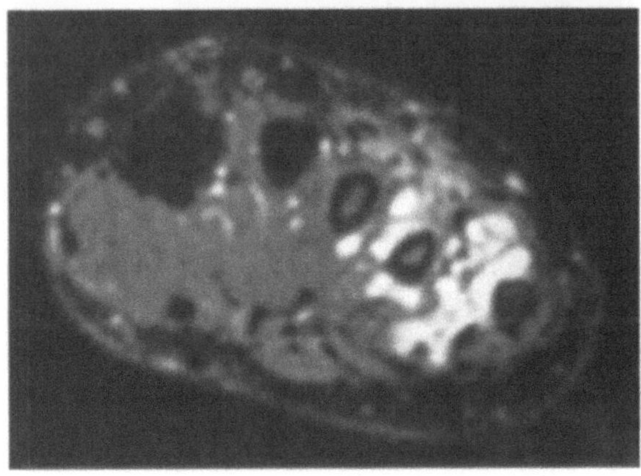

Fig. 14.21. a Lateral radiograph of the forefoot. A tiny calcification on the plantar aspect of the midfoot represents a phlebolite in a hemangioma. **b** Coronal T2-weighted MR image demonstrating a high signal intensity alteration of the soft tissue adjacent to the third and fourth metatarsals, representing methemoglobin. Diagnosis: soft tissue hemangioma

It should be noted that while images of benign fatty tumors, pigmented villonodular synovitis, hemangiomas, and hematomas can be quite char acteristic there are no reliable criteria to distinguish malignant from benign tumors or tumorlike lesions.

References

Ala-Ketola L, Keski-Nisula L, Haapannen A (1984) Ankle arthrography in acute injuries. Ann Clin Res 16:10–13

Bamberger DM, Daus GP, Gerding DN (1987) Osteomyelitis in the feet of diabetic patients: long-term results, prognostic factors, and the role of antimicrobial and surgical therapy. Am J Med 83:653–660

Beltran J, Noto AM, Mosure JC, Sharman OM, Weiss KL, Zuelzer WA (1986) Ankle: surface coil MR imaging at 1.5 T. Radiology 161:203–205

Beltran J, McGhee RB, Shaffer PB, et al. (1988) Experimental infections of the musculoskeletal system: evaluating with MR imaging and Tc-99m MDP and Ga-67 scintigraphy. Radiology 167:167–172

Beltran J, Munchow AM, Khabiri H, Magee DG, McGhee RB, Grossman SB (1990) Ligaments of the lateral aspect of the ankle and sinus tarsi: an MR imaging study. Radiology 177:455–458

Berlin SJ (1980) A review of 21 720 lesions of the foot. Pediatr Assoc 70:318–324

Bohndorf K (1996) Imaging of acute and chronic osteomyelitis. Radiologe 36:786–794

Brostrom L (1965) Sprained ankles. III. Clinical observations in recent ligament ruptures. Acta Chir Scand 130:560–569

Brostrom L, Liljedahl SO, Lindvall N (1965) Sprained ankles. II. Arthrographic diagnosis of recent ligament ruptures. Acta Chir Scand 129:485–499

Buck RL (1972) It's only a sprained ankle. Am Fam Pract 6:68–75

Cardone BW, Erickson SJ, Den Hartog BD, Carrera GF (1993) MRI of injury to the lateral collateral ligamentous complex of the ankle. J Comput Assist Tomogr 17:102–107

Chandnani VP, Harper MT, Ficke JR, Gagliardi JA, Rolling L, Christensen KP, Hansen MF (1994) Chronic ankle instability: evaluation with MR arthrography, MR imaging, and stress radiography. Radiology 192:189–194

Cox JS, Hewes TF (1979) Normal talar tilt angle. Clin Orthop 140:37–41

Dangman BC, Hoffer FA, Rand FF, et al. (1992) Osteomyelitis in children: gadolinium-enhanced MR imaging. Radiology 182:743

DeLacey G, Bradbrooke S (1979) Rationalizing requests for x-ray examination of acute ankle injuries. Br Med J I:1597–1598

Deutsch AL, Mink JH, Waxman AD (1989) Occult fractures of the proximal femur: MR imaging. Radiology 170:113–116

Dihlmann W, Bandick J (1995) Die Gelenksilhouette – Das Informationspotential der Röntgenstrahlen. Springer, Berlin Heidelberg New York

Dziob JM (1956) Ligamentous injuries about the ankle joint. Am J Surg 91:692–698

Edmons ME (1986) The diabetic foot: pathophysiology and treatment. Clin Endocrinol Metab 15:889

Erdman WA, Tamburro F, Jayson HT, et al. (1991) Osteomyelitis: characteristics and pitfalls of diagnosis with MR imaging. Radiology 180:533–539

Erickson SJ, Smith JW, Ruiz ME, et al. (1991) MR imaging of the lateral collateral ligament of the ankle. Am J Roentgenol 156:131–136

Erlemann R, Sciuk J, Bosse A, Ritter J, Kusnierz-Glaz CR, Peters PE, Whismann R (1990) Response of osteosarcoma and Ewing sarcoma to preoperative chemotherapy: assessment with dynamic and static MR imaging and skeletal scintigraphy. Radiology 175:791–796

Evans GA, Hardcastle P, Frenyo AD (1984) Acute rupture of the lateral ligament of the ankle – to suture or not suture? J Bone Joint Surg [Br] 66:209–212

Geissler WB, Tsao AK, Hughes JL (1996) Fractures and injuries of the ankle. In: Rockwood CA, Green DP, Bucholz RW, Heckmann JD (eds) Rockwood and Green's fractures in adults. Lippincott-Raven, Philadelphia, pp 2201–2266

Giudici MA, Moser RP, Kransdorf MJ (1993) Cartilaginous bone tumors. Radiol Clin North Am 31:237–258

Gold RH, Tong DJF, Crim JR, Seeger LL (1995) Imaging the diabetic foot. Skeletal Radiol 24:563–572

Goodwin DW, Salonen DC, Yu JS, Brossmann J, Trudell DF, Resnick D (1995) Plantar compartments of the foot: MR appearance in cadavers and diabetic patients. Radiology 196:623–630

Goossens M, De Stoop N, Claessens H, Van der Straeten C (1989) Posterior subtalar joint arthrography. A useful tool in the diagnosis of hindfoot disorders. Clin Orthop 249:248–455

Greenfield GB, Warren DL, Clark RA (1991) MR imaging of periosteal and cortical changes of bone. Radiographics 11:611–623

Haramati N, Staron RB, Barax C, Feldman F (1994) Magnetic resonance imaging of occult fractures of the proximal femur. Skeletal Radiol 23:19–22

Inglis AE, Scott TP, Sculco TP, Patterson AH (1976) Rupture of the tendon achillis. An objective assessment of surgical and nonsurgical treatment. J Bone Joint Surg [Am] 58:990–993

Johannsen A (1978) Radiological diagnosis of lateral ligament lesions of the ankle. A comparison between talar tilt and anterior drawer sign. Acta Orthop Scand 49:295–301

Jung V (1996) Salvage of the diabetic foot. Zentralbl Chir 121:387–393

Kaplan PA, Walker CW, Kilcoyne RF, Brown DE, Tusek D, Dussault RG (1992) Occult fracture patterns of the knee associated with anterior cruciate ligament tears: assessment with MR imaging. Radiology 183:835–838

Kaufman J, Breeding L, Rosenberg N (1987) Anatomic location of acute diabetic foot infection. Its influence on the outcome of treatment. Am Surg 53:109

Kelikian H, Kelikian AS (1985) Disorders of the ankle. Saunders, Philadelphia, pp 339–363

Kjaersgaard-Andersen P, Wethelund JO, Helmig P, Soballe K (1988) The stabilizing effect of the ligamentous structures in the sinus and canalis tarsi on movement in the hindfoot: an experimental study. Am J Sports Med 16:512–516

Kjaersgaard-Andersen P, Soballe K, Andersen K, Pilgaard S (1989) Sinus tarsi syndrome: presentation of seven cases and review of the literature. J Foot Surg 28:3–6

Klein MA, Spreitzer AM (1993) MR imaging of the tarsal sinus: normal anatomy, pathologic findings, and features of the sinus tarsi syndrome. Radiology 186:233–240

Kricun ME (1993) Imaging of bone tumors: tumors of the foot. Saunders, Philadelphia

Lager R (1980) Case report 128. Skeletal Radiology 5:257–269

Lang P, Genant HK, Jergesen HE, Murray WR (1992) Imaging of the hip joint: computed tomography versus magnetic resonance imaging. Clin Orthop 274:135–153

Lartos G, Brown ML, Sutton RT (1991) Diagnosis of osteomyelitis of the foot in diabetic patients: value of ^{111}In-leukocyte scintigraphy. AJR 157:527–531

Laurin CA, Quellet R, St. Jacques R (1968) Talar and subtalar tilt: an experimental investigation. Can J Surg 11:270–279

Liu J, Hudkins PG, Swee RG, Unni KK (1987) Bone sarcomas associated with Ollier's. Cancer 59:1376

Lodwick GS (1965) A systemic approach to the roentgen diagnosis of bone tumors. In: Tumors of bone and soft tissues. Papers, MD, Anderson Hospital. Chicago Year Book, Chicago

Malghem J, Van de Berg B, Noel H, Maldague MD (1992) Benign osteochondromas and exostotic chondrosarcomas evaluation of cartilage cap thickness by ultrasound. Skeletal Radiol 21:33–37

Marder R (1994) Current methods for the evaluation of ankle ligament injuries. J Bone Joint Surg [Am] 76:1103–1111

Mason MD, Zlatkin MB, Esterhai JL, et al. (1989) Chronic complicated osteomyelitis of the lower extremity: evaluation with MR imaging. Radiology 173:335–359

Mesgarzadeh M, Schneck CD, Bonakdarpour A (1989) Magnetic resonance imaging of the knee: correlation with normal anatomy. Radiographics 8:707–733

Meyer JM, Garcia J, Hoffmeyer P, Fritschy D (1988) The subtalar sprain: a roentgenographic study. Clin Orthop 226:169–173

Mink JH, Deutsch AL (1989) Occult cartilage and bone injuries of the knee: detection, classification, and assessment with MR imaging. Radiology 170:823–829

Mitty HM (1993) Musculoskeletal neoplasms. Role of angiography in diagnosis and interpretation. Semin Intervent Radiol 10:277–283

Morrison WB, Schweitzer ME, Bock GW, et al. (1993) Diagnosis of osteomyelitis: utility of fat-suppressed contrast-enhanced MR imaging. Radiology 189:251–257

Noto AM, Cheung Y, Rosenberg LS, Leeds NE (1989) MR imaging of the ankle: normal variants. Radiology 170:121–124

O'Conner D (1958) Sinus tarsi syndrome: a clinical entity. J Bone Joint Surg [Am] 40:720–726

Olson DO, Shields AF, Scheurich CJ, Porter BA, Moss AA (1986) Magnetic resonance imaging of the bone marrow in patients with leukemia, aplastic anemia and lymphoma. Invest Radiol 21:540–546

Olson RW (1969) Arthrography of the ankle: its use in evaluation of ankle sprains. Radiology 92:1439–1446

Quinn SF, McCarthy JL (1993) Prospective evaluation of patients with suspected hip fracture and indeterminate radiographs: use of T1-weighted MR images. Radiology 187:469–471

Resnick D (1995) Osteomyelitis, septic arthritis, and soft tissue infection: the mechanisms and situations. In: Diagnosis of bone and joint disorders. Saunders, Philadelphia, pp 2524–2619

Resnick D, Kyriakos M, Greenway GD (1988) Tumors and tumorlike lesions of bone: imaging and pathology of specific lesions. In: Resnick D, Niwayama G (eds) Diagnosis of bone and joint disorders. Saunders, Philadelphia, p 3616

Rijke AM, Goitz HT, McCue FC, Dee PM (1993) Magnetic resonance imaging of injury to the lateral ankle ligaments. Am J Sports Med 21:528–534

Robson MC, Edstrom LE (1977) The diabetic foot: an alternative approach to major amputation. Surg Clin North Am 57:1089–1099

Rubin G, Witten M (1960) The talar-tilt angle and the fibular collateral ligaments. A method for the determination of talar tilt. J Bone Joint Surg [Am] 42:311–326

Sartoris D, Resnick D (1988) Computed tomography of the lower extremity. Part III. Orthop Rev 17:20–24

Sauser DD, Nelson RC, Lavine MH, Wu CW (1983) Acute injuries of the lateral ligaments of the ankle: comparison of stress radiography and arthrography. Radiology 148:653–659

Schajowicz F (1981) Tumors and tumorlike lesions of bone and joints. Springer, Berlin Heidelberg New York

Schneck CD, Mesgarzadeh M, Bonakdarpour A, Ross GJ (1992a) MR imaging of the most commonly injured ankle ligaments. Part I. Normal anatomy. Radiology 184:499–506

Schneck CD, Mesgarzadeh M, Bonakdarpour A (1992b) MR imaging of the most commonly injured ankle ligaments. Part II. Ligament injuries. Radiology 184:507–512

Sinha S, Minichoodappa CS, Kozak GP (1972) Neuroarthropathy (Charcot joints) in diabetes mellitus (clinicial study of 101 cases). Medicine (Baltimore) 51:191–210

Spiegel PK, Stapler OS (1975) Arthrography of the ankle joint: problems and diagnosis of acute lateral ligament injuries. Radiology 114:587–590

Stoller DW, Steinkirchner TM, Porter BA (1993) Bone and soft-tissue tumors. Magnetic resonance imaging in orthopedics and sports medicine. Lippincott, Philadelphia, p 1031

Tang JS, Gold RH, Bassett LW, et al. (1988) Musculoskeletal infection of the extremities: evaluation with MR imaging. Radiology 166:205–209

Tiel-Van-Buul MM, Van Beek EJ, Van Dongen A, Van Royen EA (1992) The reliability of the 3-phase bone scan in suspected scaphoid fracture: an inter- and intraobserver variability analysis. Eur J Nucl Med 19:848–852

Unger EC, Moldofsky PJ, Gatenby RA, et al. (1988) Diagnosis of osteomyelitis by MR imaging. AJR 150:605–610

Unger EC, Schilling JD, Awad AN, et al. (1995) MR angiography of the foot and the ankle. J Magn Reson Imaging 5:1–5

Verhaven EF, Shahabpour M, Handelberg FW, Vaes PH, Opdecam PJ (1991) The accuracy of three-dimensional magnetic resonance imaging in the diagnosis of ruptures of the lateral ligaments of the ankle. Am J Sports Med 19:583–587

Vestring T, Fiedler R, Greitemann B, Scivk J, Peters PE (1995) The diabetic foot. Radiologe 35:447–455

Vorbeck F, Morscher M, Ba-Ssalamah A, Imhof H (1996) Infectious spondylitis in adults. Radiologe 36:795–804

Wagner M, Dann K (1995) Sprunggelenk. In: Rüter A, Trenz O, Wagner M (eds) Unfallchirurgie. Urban & Schwarzenberg, München, pp 851–880

Yao L, Lee JK (1988) Occult intraosseous fracture: detection with MR imaging. Radiology 167:749–751

Young MR, Lowry JH, McLeod NW, Crone RS (1988) Clinical carpal scaphoid injuries. BMJ 296:825–826

Yuh W, Corson J, Baraniewski H, et al. (1989) Osteomyelitis of the foot in diabetic patients: evaluation with plain film, ^{99m}Tc-MDP bone scintigraphy and MR imaging. AJR 152:795–800

15 The Spine

I.W. McCall

CONTENTS

15.1
Introduction

This chapter addresses the clinical presentation of conditions affecting the axial skeleton and its associated soft tissues but which may also affect the neurogenic structures that pass through the spine. The most appropriate methods of initial and subsequent investigation are discussed in relation to the nature of the clinical problem but there is an inevitable bias towards the use and appearances of magnetic resonance imaging (MRI), which has become the most important imaging modality for spinal pathology.

The most common presentation for clinical conditions affecting the spine is pain, which may be localised to a specific area of the spine or be more diffuse with referred pain to the appendicular skeleton or to other areas of the trunk. The pain may be acute in onset or develop gradually and may become continuous or intermittent. Occasionally, patients with spinal disorders may present with a gradual loss of function of a limb or abnormal sensation. More acute neurological function loss usually results from trauma but rapid onset of neurology may result from disc prolapse or tumour, or in association with a vascular insult. Deformity of the spine usually develops gradually, and at varying ages, with lateral or sagittal curvature of the spine or a combination thereof. The investigation of the spine may involve a number of different imaging techniques, all of which have their own value. MRI has significantly enhanced the imaging range, providing great detail of all the soft tissue components of the spine, including the cord and nerve roots. Spinal investigation, however, identifies not only pathological processes but also the normal features of ageing and these have to be differentiated from the changes that may be more significant.

15.2
Disc Degeneration

The most common clinical presentation of spinal disorders in adults is that of non-specific pain, due to degenerative changes in the main components of the motion segments of the spine, namely the intervertebral disc and facet joints. The process of degeneration is similar throughout the spine, although certain levels may be more susceptible owing to biomechanical features. This is particularly the case for the intervertebral discs in the cervical spine at the C5 to C7 level and in the lumbar spine at the L4/5 and L5/S1 levels. Disc degeneration is also a feature of the dorsal spine, in particular at the dorsolumbar junction, but seems less significant in terms of clinical symptoms than the mid-cervical and lower lumbar spine. Pain associated with disc degeneration tends to be chronic; it may refer to the shoulders or arms from the cervical spine and the buttocks and lower limbs from the lumbar spine and may be relatively diffuse in nature. The quality of the pain varies from a dull ache to a sharp stabbing pain and usually develops gradually, although acute onset may occur.

I.W. McCall, MD, Professor, Department of Radiology, Robert Jones and Agnes Hunt Orthopaedic & District Hospital NHS Trust, Oswestry, Shropshire SY10 7AG, UK

Occasionally, paraesthesia may be associated with the presence of the pain but wasting of muscles is not usually a feature of non-specific neck or back pain, unless due to disuse.

The outer fibres of the annulus of the disc are innervated and stretching or irritation of these nerve endings is a possible a source of pain (BOGDUK 1992). Clefts develop in the outer fibres of the annulus, particularly posteriorly, and these may be concentric or radial. Complete radial tears occur either posteriorly or posterolaterally. In the cervical spine these radial clefts in the discs are common after the age of 20 and link with the neurocentral joints. They are also found with increasing incidence with age in the lumbar spine. Granulation tissue may occur around complete tears in the posterior annulus and may lead to pain (YU et al. 1988). As the degenerative process proceeds, the nucleus loses water content and the disc height reduces with bulging of the annulus and the posterior longitudinal ligament, which may also be a source of pain.

The plain radiograph remains the most common initial investigation although the changes are usually non-specific in that they may be seen in patients with no pain as frequently as in those with pain (MAGORA and SCHWARTZ 1976). The features are those of disc space narrowing which may be associated with some bulging of the annulus, and which produces a reactive bone remodelling of the margins of the vertebra in the form of osteophytes. The disc space narrowing may be mild, moderate or severe and may be associated with some irregularity of the vertebral end plate and in some cases sclerosis in the vertebral bodies, in both the cervical and the lumbar spine, may be present (Fig. 15.1). In the dorsal spine, the degenerative process tends to occur anteriorly, in association with disc height loss, kyphosis and osteophyte formation (Fig. 15.1). Calcification may occasionally occur in the disc and is most common in the thoracic spine. However, it is not always identified on the plain films. In severe cases of disc degeneration, particularly at the lowest mobile segment of the lumbar spine, complete disc resorption may occur, with sclerosis of the adjacent vertebral bodies. In some degenerative discs, the presence of a vacuum sign may be seen, particularly on extension in both the cervical and the lumbar spine. Radiographic examination is not indicated in the first 6 weeks of symptoms as in the majority of patients the pain settles within 3 weeks. The exception is if the pain is severe, continuous and present throughout 24 hours of the day, when the possibility of more severe pathology must be considered.

A more detailed examination of degenerative discs may be achieved with MRI. In the early stages of disc degeneration, there is a loss of the high signal of the nucleus on the T2-weighted images. This is best demonstrated throughout the spine on the turbo spin-echo (TSE) T2-weighted images, which differentiate between the low signal of the annulus and the high signal of the nucleus whereas the gradient-echo images demonstrate increased signal throughout the annulus and nucleus and are less sensitive to early degenerative changes. The relative narrowness of the disc in the cervical spine makes interpretation of the early degenerative changes more difficult than in the lumbar spine. As degeneration progresses, the annulus will bulge, showing slight indentation of the anterior margin of the dural sac and CSF on the T2-weighted sagittal sequence on all slices in the sagittal plane (Fig. 15.2) or a generalised increase in the curved outline of the disc around the whole vertebra in the axial plane. The latter may be more difficult to appreciate than the former.

Tears in the outer annulus may produce a more acute clinical syndrome of back or neck pain and these can be identified on MRI. In the lumbar spine, they present as high-signal zones through the posterior part of the annulus on the T2-weighted TSE sequence and may produce focal bulges of the disc (Fig. 15.3). Annular tears may enhance following the injection of gadolinium and in the lumbar spine have been more closely related to the presence of back pain (APRILL and BOGDUK 1992). These high-intensity zones are separated from the nucleus by low-signal annulus and have been shown to be related to posterior radial tears, in association with concentric tears (APRILL and BOGDUK 1992). The pain reproduction on discography has been demonstrated to occur in 80% of patients with high-intensity zones (SCHELLHAS et al. 1996). In the cervical and thoracic spine, high-intensity zones are not commonly seen but occasionally may be demonstrated (Fig. 15.3). The posterior annular tears in the cervical spine may occur either centrally or radially but there is no clear relationship between the presence of a posterior annular tear and pain, and the demonstration of a tear is difficult due to the small size of the disc on MRI. Posterolateral radial tears in the cervical spine may result in the development of osteophyte formation around the neurocentral joints, which may be an important feature of chronic cervical disc disease.

Disc degeneration may also be associated with changes in the adjacent marrow. In the cervical and

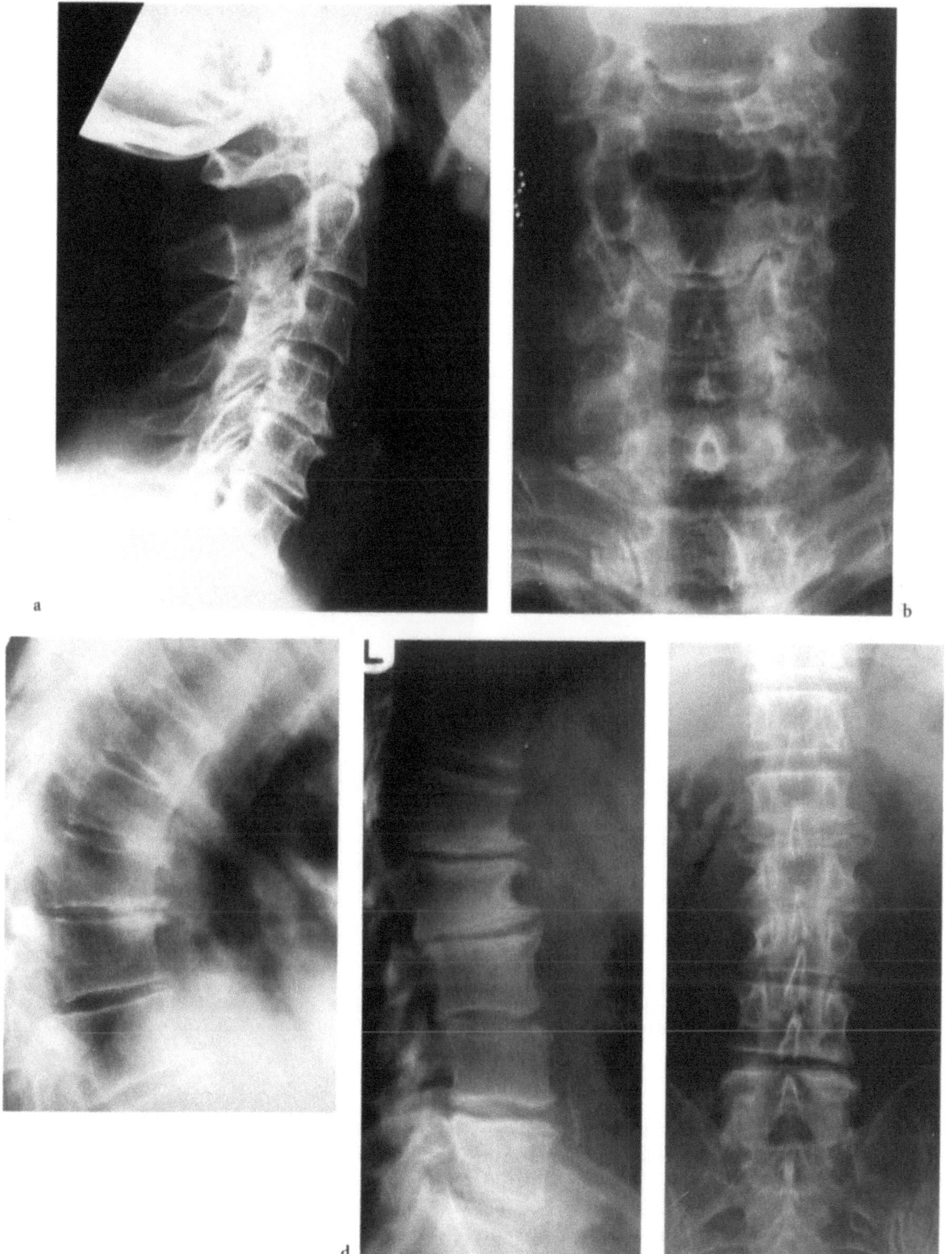

Fig. 15.1 a–e. Degenerative changes on plain films. a There is marked disc space narrowing at the C4/5 and C5/6 levels, with osteophyte formation and vertebral sclerosis. Joint space loss with sclerosis is present in the facet joints. b The AP view shows osteoarthritic changes in the neurocentral joints. c Lateral view of the thoracic spine shows narrowing of the disc spaces anteriorly with sclerosis of the end plates and anterior osteophyte formation. d There is narrowing of most lumbar disc spaces, with some sclerosis of the end plates and osteophyte formation (e). The facet joints show sclerotic changes on the AP view

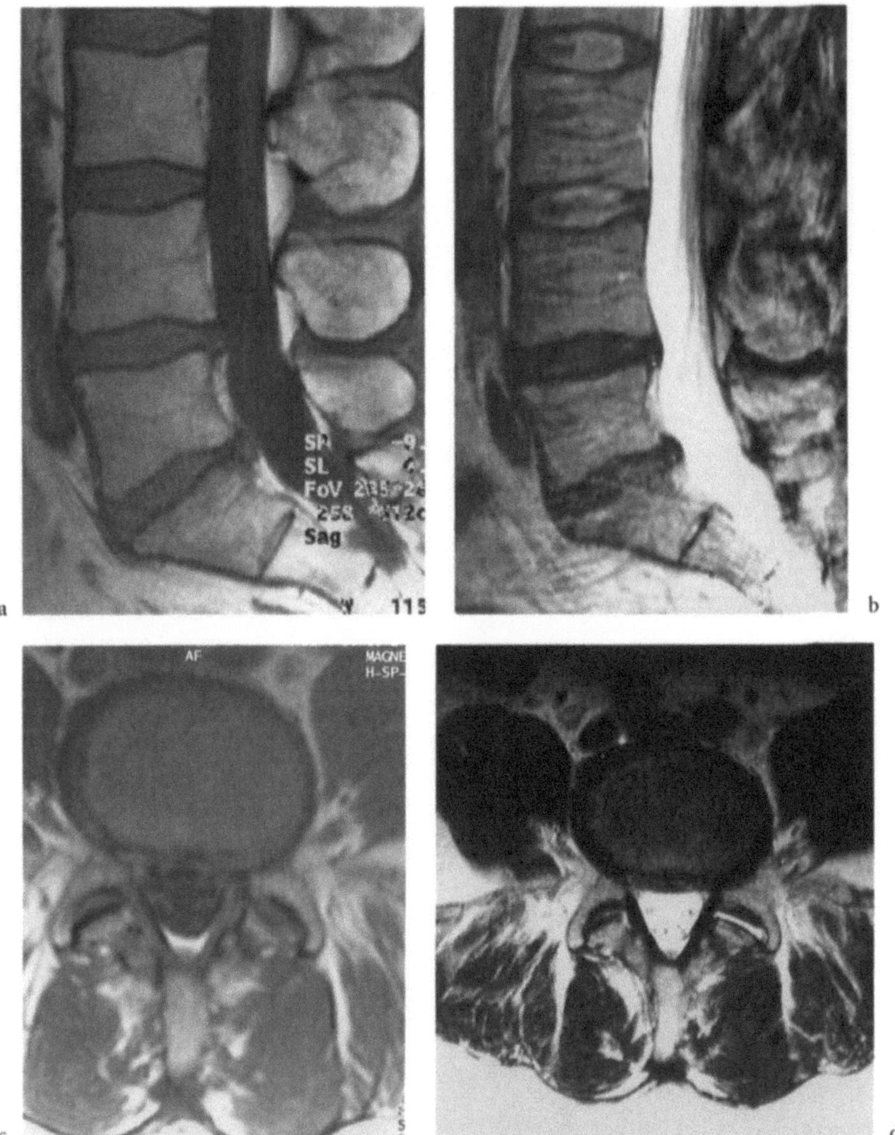

Fig. 15.2 a–d. Disc degeneration and annular bulging. **a** The T1-weighted sagittal image shows narrowing of the disc space, with bulging of the disc both posteriorly and anteriorly at the L4/5 level. **b** The T2-weighted sagittal image shows loss of signal from the nucleus of the disc. **c,d** The axial T1- (**c**) and T2-weighted (**d**) scans show an even convexity of the outline of the posterior annulus, with no evidence of a localised protrusion

thoracic spine, marrow changes are unusual but vertebral sclerosis may be seen on plain films. Vertebral end plate irregularity is not uncommon in the thoracic spine, varying from localised Schmorl's nodes to marked changes involving the whole end plate. In the lumbar spine, marrow changes appear more commonly and have been described as forming three types (MODIC et al. 1988). Type 1 represents an area of low signal on T1-weighted images and increased signal on T2-weighted images, particularly seen on fat suppression sequences (Fig. 15.4). These changes correlate with pathological evidence of fibrovascular infiltration of the marrow (TOYONE et al. 1994). Type 2 changes show increased signal on both T1- and T2-weighted images and represent increase in the fat within the marrow, often associated with thickening of the individual trabeculae (Figs. 15.3, 15.4). The third type is that of markedly thickened trabeculae and replacement of marrow by trabecular bone owing to chronic sclerotic changes. These changes are often associated with marked narrowing of the disc space. The significance of these changes is not

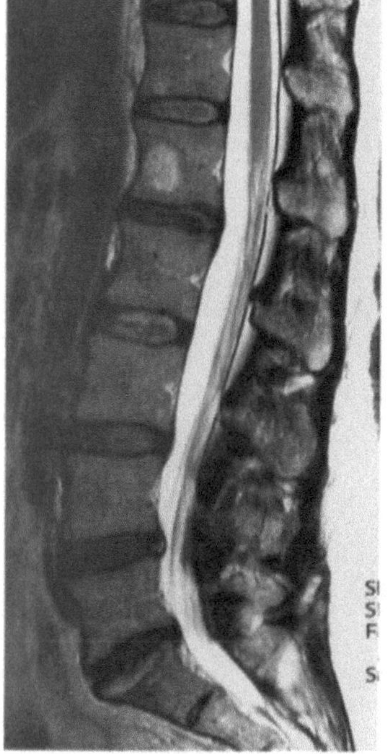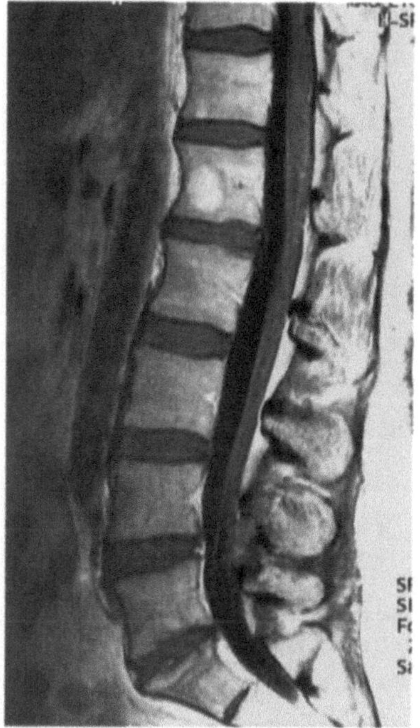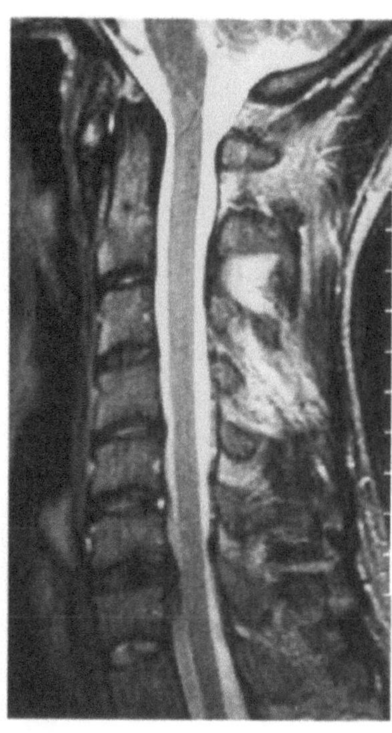

a,b   c

Fig. 15.3. a Sagittal T2-weighted TSE image of the lumbar spine shows a localised high-intensity zone in the posterior annulus at the L4/5 level. **b** There is enhancement on the T1-weighted post-gadolinium sagittal scan. High signal is seen in the end plate of L5/S1, associated with disc space narrowing on both the pre- and postenhanced T1 and T2 sequences. A focal area of high signal is seen in the L1 vertebral body, due to a haemangioma. **c** A focal high-intensity zone is seen at the C5/6 level on the T2-weighted TSE sequence. A disc herniation is seen at C6/7

fully understood but there is evidence to suggest that the type 1 changes are more commonly related to the presence of back pain and that the type 2 changes reflect more chronic disc degeneration and have less clinical significance (TOYONE et al. 1994; McCALL et al. 1997). Although the changes of MR lack specificity, some of these features may point to a level of the symptomatic source.

15.3
Discography

In the small number of patients who have intractable pain, in whom surgery may considered to be the only solution, the isolation of individual disc levels as pain sources may require investigation by means of discography. The discogram involves the injection of contrast into the nucleus, with the evaluation of pain occurring during the injection. The radiological appearances following contrast injection vary depending on the degree and type of annular disruption (Fig. 15.5). Tears in the annulus may be radial or concentric or a combination thereof. In the presence of degeneration the contrast extends throughout most of the disc and there may be associated disc bulging. In the lumbar spine the extent of concentric and radial tears is best seen on computed tomography (CT) following the disc injection and can be recorded using the Dallas grading system (SACHS et al. 1987). In the cervical spine the disc space is too narrow for CT to be of value. Following the injection of contrast and the stimulation of symptomatic pain, injection of a long-acting analgesic such as bivucaine will allow evaluation of the pain relief from the disc area although this feature is less sensitive as a diagnostic test than the pain production, as pain from stretching of the outer fibres of the annulus will not be relieved by the local anaesthetic injection. Discography, as an investigation, has advocates and antagonists (NACHEMSON 1989; North American Spine Society 1988). In the lumbar spine, there is a generally accepted view that it is helpful in the evaluation of the patient in whom surgery is being contemplated, and accuracy rates of 85% have been recorded for discography, with a high sensitivity but a relatively low specificity, based on the outcome of surgery (COLHOUN et al. 1988). The

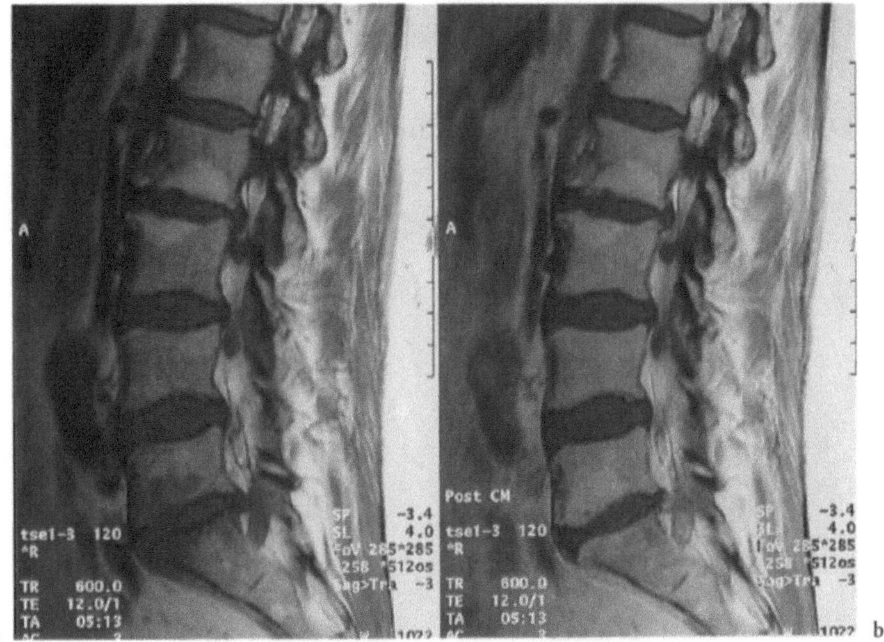

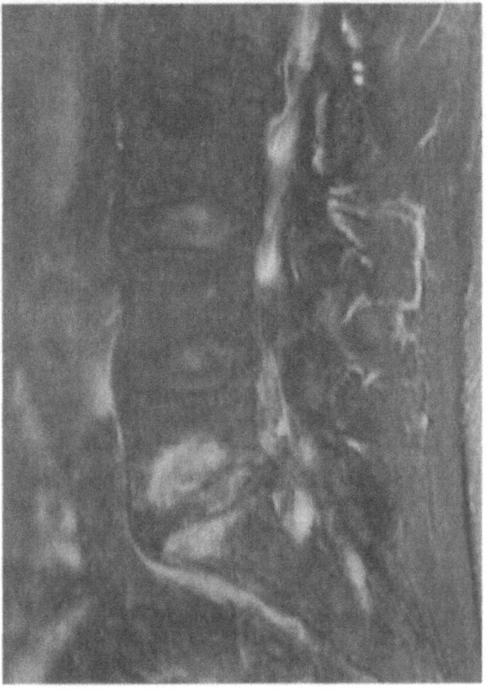

Fig. 15.4. a Low signal on unenhanced T1-weighted images is seen at L5/S1. High signal is seen at L2/3. After Gd-DTPA the L5/S1 signal increases but the L2/3 signal is unchanged. **b** Fat saturation sequence confirms the high signal at L5/S1 of the Modic type 1 change but at L2/3 the increased yellow marrow is suppressed, confirming Modic type 2 changes

value of cervical discography is more contentious although our own correlation with surgical success at 1 year has shown an 84% accuracy, which drops after 2 years to 70%, based on excellent or good results at the levels predicted by discography. Discog-raphy carries a potential complication of inducing infection. In the cervical spine, considerable care is required to avoid contamination of the disc and infection rates have been reported as high as 3%. Individual cases of quadriplegia following cervical

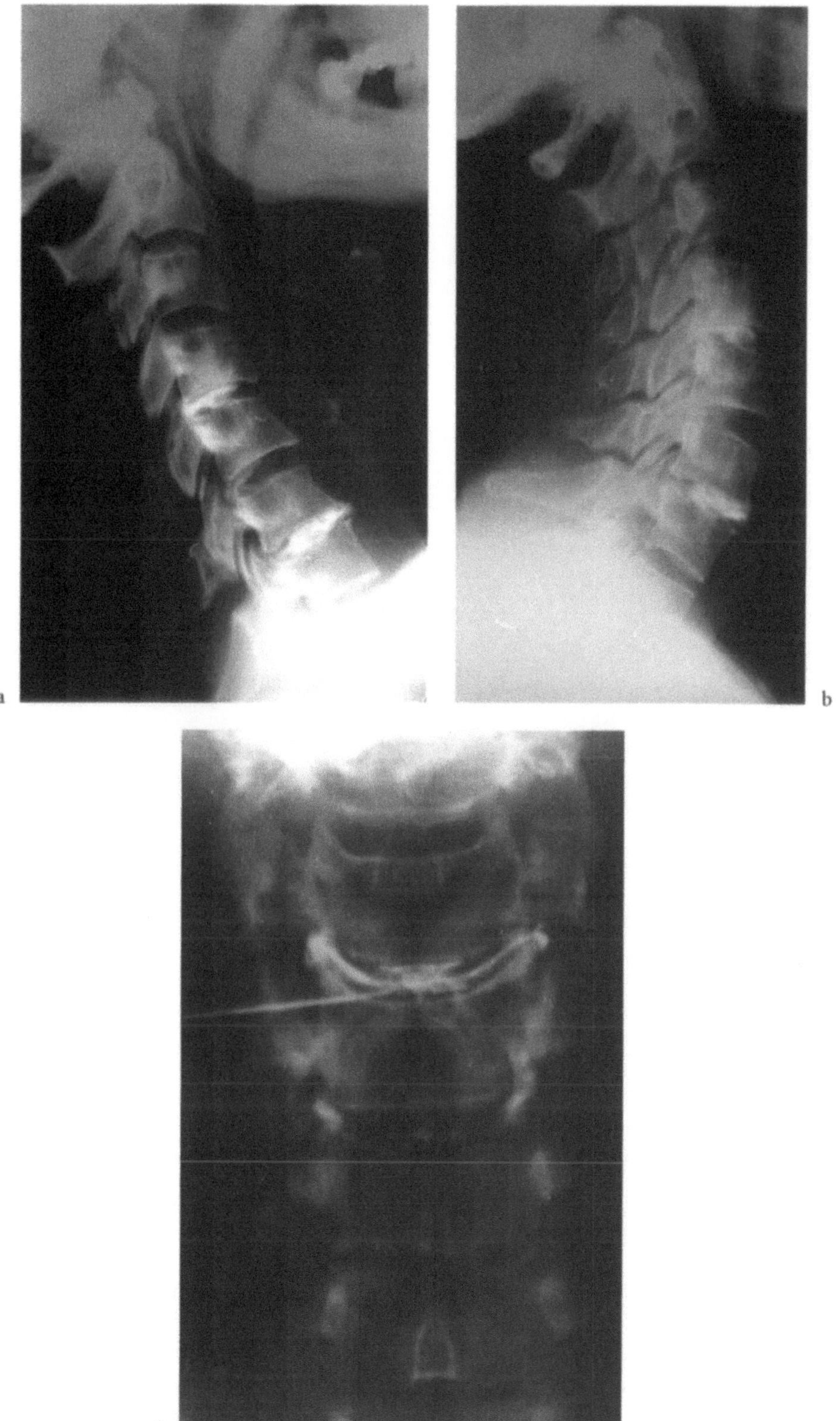

Fig. 15.5 a,b. Discogram: Three-level cervical discography has been performed with **a** flexion and **b** extension views. There is a contained posterior annular tear at C4/5, internal annular disruption at C5/6 and a bulky nucleus, with anterior annular disruption and osteophyte formation at C6/7. **c** The AP view shows contrast extending into the neurocentral joints. Lumbar CT discography shows a normal discogram (**d,e**), a posterior annular tear (**f,g**) and degenerate disc (**h,i**) with contrast throughout the disc

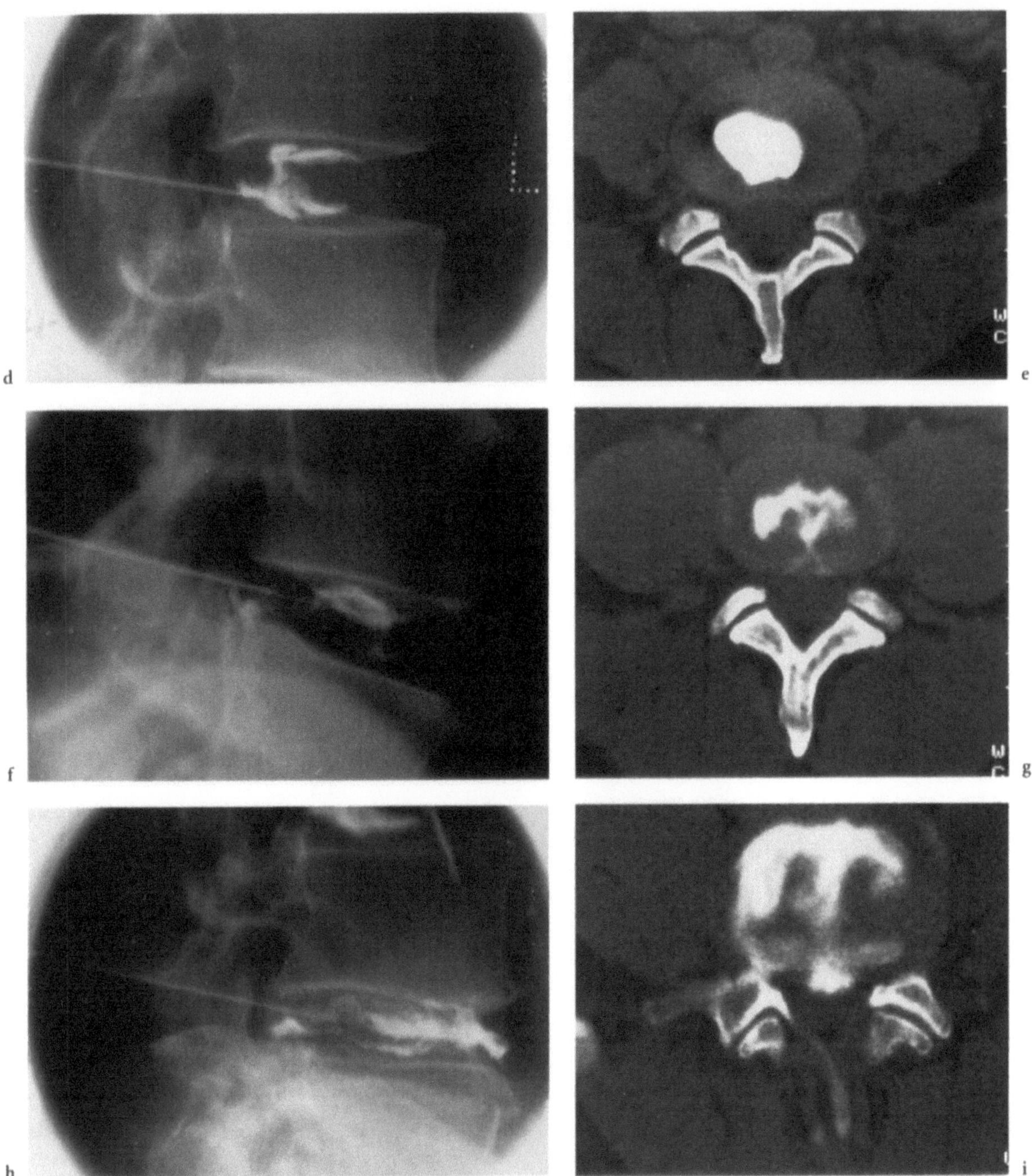

Fig. 15.5 d–i

discography have been reported, secondary to the inducement of infection. In the lumbar spine, the infection rate, with a double-needle technique and skilled operators in properly sterile surroundings, is lower, at approximately 0.4% (FRASER et al. 1989).

Injection of intravenous or intradiscal antibiotics just prior to the injection of contrast medium may reduce the infection rate to a very low level, although this does not completely exclude the development of infection.

15.4
Intervertebral Facet Joints

Chronic pain in the spine may also originate from the facet joints. Studies stimulating the facet joints in the cervical and lumbar spine, using hypertonic saline, have induced pain in the form of a dull ache or cramp which has been predominantly in the region of the stimulation or referred to the proximal aspect of the adjacent limb (DWYER et al. 1990; MCCALL et al. 1979). There is considerable overlap of pain from different stimulated levels owing to the multilevel innervation by medial branches of the dorsal rami (BOGDUK 1982; BOGDUK et al. 1982). The pain may be increased by movement where the facets are stressed such as extension, rotation or lateral bending but there is also evidence to indicate that pain is worse at rest and eased by motion (FAIRBANK et al. 1981). The facets are usually evaluated on the plain radiographs and, due to the varying orientation throughout the spine, may require different views at different levels for optimised demonstration. In the cervical spine, the lateral radiographs show the joint space, which is orientated horizontally (Fig. 15.1a), and the AP view will demonstrate lateral marginal osteophyte formation. Oblique views are rarely required unless the intervertebral foramen is to be evaluated. The thoracic facets are difficult to visualise as the ribs overly them on the lateral view and the orientation on the AP view is not optimal, while in the lumbar spine orientation varies with level and AP, lateral and oblique views may all be required. Osteoarthritis is demonstrated as joint space narrowing, sclerosis of the subchondral bone and the presence of osteophyte formation, but early joint space loss may be difficult to evaluate, due to the curved nature of the joint and the varied rotation of the joint and radiograph. The cartilage of the facet joint is best demonstrated by MRI in the sagittal plane in the cervical and thoracic spine and in the axial plane in the lumbar spine, where it is seen as intermediate signal on T1-weighted images but mild or localised cartilage loss is difficult to recognise. Osteophyte formation around the joint is more easily recognised on CT (Fig. 15.6) and differentiation between ligamentum flavum calcification and osteophytes can also be better achieved. Extensive marginal bone formation, which may rarely mimic bone-forming tumour, can be identified. There is no correlation between the demonstration of facet osteoarthritis on plain radiographs or CT and the presence of pain (SCHWARZER et al. 1995). The use of technetium-99m methylene diphosphonate (^{99m}Tc-

MDP) has been suggested to identify joints with more active arthritis, which may be symptomatic, and increased uptake in some joints, particularly using single-photon emission computed tomography (SPECT), has been demonstrated. While some authors have found a good correlation between successful facet blocks with local anaesthetic and positive isotope uptake, our own experience is that the sensitivity and specificity are low.

Magnetic resonance imaging using T2-weighted TSE sequences may demonstrate an effusion as a high signal in some joints and occasionally cystic dilatation of the capsule into the canal may be seen. These facet cysts may compress the neighbouring nerve root, causing sciatica, which can be evaluated with CT or MRI. CT is valuable to show the presence of calcification in the wall or within the cyst. The cysts may rupture spontaneously but occasionally surgical removal is required.

Precise identification of the symptoms of neck or back pain to the facet joints requires the use of focused local anaesthetic injections either directly into the facet joints or around the medial branch of the dorsal ramus supplying the joint. Facet joint injections or nerve blocks are usually performed using fluoroscopy but CT guidance may be utilised. A small spinal needle is inserted just above the junction of the transverse process and the lamina for nerve blocks, usually from a posterolateral approach, and local anaesthetic injected into the nerve. Intrafacet joint injections require a small quantity of non-ionic water-soluble contrast agent to confirm the intra-articular position and then 0.5 cc of long-acting local anaesthetic (bivucaine) is injected. The quantity of local anaesthetic should be kept low to avoid extensive diffusion or capsular rupture, which will reduce the anatomical specificity of the procedure. A positive result to the injection is judged as the relief of symptomatic pain. In the lumbar spine this is unlikely to occur in more than 20% of an unselected group of patients with back pain (MCCALL et al. 1990), although levels of immediate response of up to 67% have been reported (JACKSON et al. 1988). False-positive results have been reported using intramuscular injections of saline in 32% of patients undergoing facet injections (SCHWARZER et al. 1992), and a double-injection technique has been advocated with the initial successful lidocaine injection being followed on the return of pain by a confirmatory bivucaine injection. No correlation between CT findings of osteoarthritis and positive facet injections has been found (SCHWARZER et al. 1995). Blocks of the facets and the medial branch of the

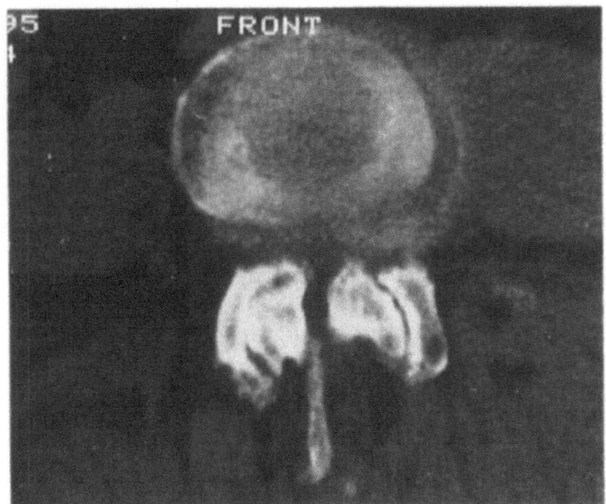

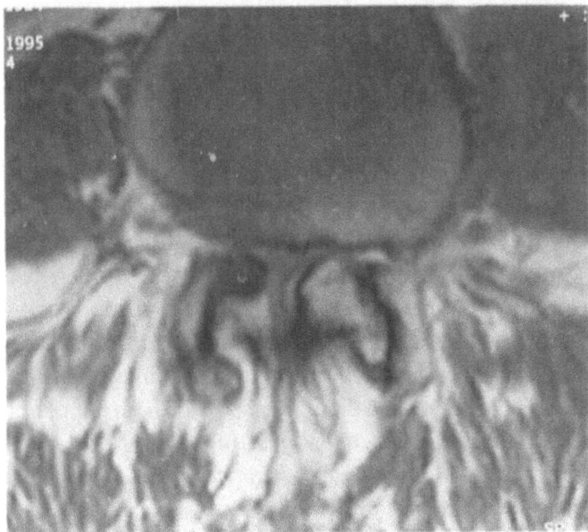

Fig. 15.6 a,b. Facet OA. **a** The CT scan shows marked joint space loss and irregularity of the articular surface, with sclerosis of the subchondral bone and osteophytes. **b** Similar features may be seen on the T1-weighted axial MR sequence but the subchondral sclerosis is less easily identified. Severe spinal stenosis is also present, due to the degenerative changes

posterior primary ramus have been found to be positive in 60% of patients with chronic symptoms following whiplash injury and are proposed as a diagnostic and in some cases a treatment process (LORD et al. 1996).

15.5
Adolescent Spinal Pain

Spinal pain in adolescents is uncommon and should be evaluated with care. Onset may be related to physical exercise, particularly if it is intense or involving hyperextension in the lumbar spine, presenting as low back pain. In this situation, the possibility of acute stress fracture of the pars interarticularis should be considered. Lateral radiographs of the lumbar spine usually demonstrate the defect in established cases but early stress fractures and a unilateral defect may be masked by the normal or sclerotic intact side. If doubt exists, a 45° oblique or AP view, angled 20° caudad, may demonstrate the defect but a CT scan, particularly using a reversed angle along the line of the pars, is the most effective method of demonstration (Fig. 15.7) and evidence of repair or established non-union, with sclerosis on either side of the defect, is also clearly shown (Fig. 15.7).

In the early stages of a stress-related defect, the plain films and CT may appear normal. In these circumstances, a ^{99m}Tc-MDP SPECT study should be performed as this will demonstrate increased activity in the region of the pars interarticularis (RABY and MATTHEWS 1993). The presence of increased activity also indicates potential for healing, provided stabilisation is achieved (Fig. 15.7). In established pars interarticularis defects clearly demonstrated on plain films, the SPECT scan is usually negative unless there are secondary degenerative changes in the facet joints.

The lateral radiograph performed with the patient standing will highlight any associated spondylolisthesis (LOWE et al. 1976) although the degree of forward displacement is usually mild up to 25° and does not progress significantly in adulthood. Severe displacement may occur in adolescents in the growth spurt but these patients usually have a degree of hypoplasia of the posterior elements, associated with the spondylolytic defect.

The most common form of spondylolytic defect develops in early childhood is not related to excessive stress and has not been found to be a significant cause of symptoms (FREDERICKSON et al. 1984). The defect itself was not considered to be a pain source in the past but recent histochemical studies have demonstrated nerve fibres within the fibrous tissue of the pseudarthrosis but not in the pseudo-synovial lining membrane (NORDSTROM et al. 1994). The established defects commonly connect with the adjacent facet joints but local anaesthetic injected within the joints rarely results in abolition of the low back pain (PARK et al. 1985).

Magnetic resonance imaging will demonstrate the defect in the pars on sagittal T1-weighted scans passing through the pars, which show the intermediate signal gap in continuity of the marrow of the pars. A reverse angle axial view of the pars or three-dimensional acquisition with thin slices through the

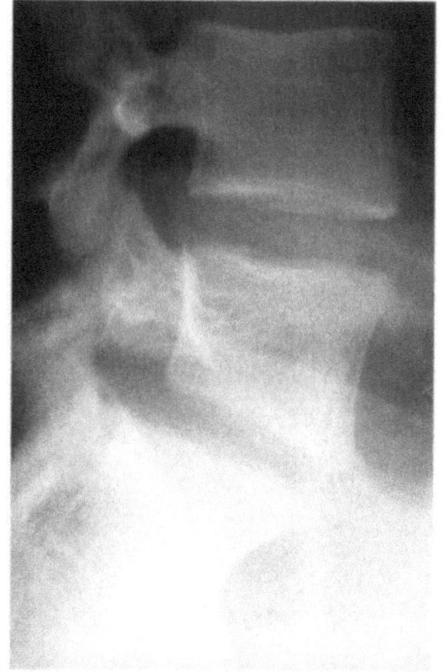

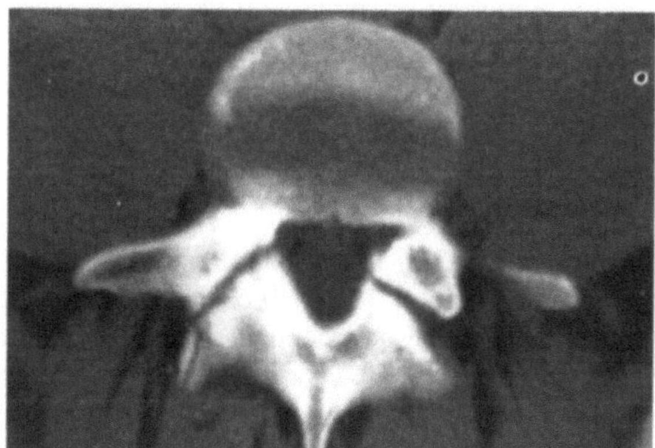

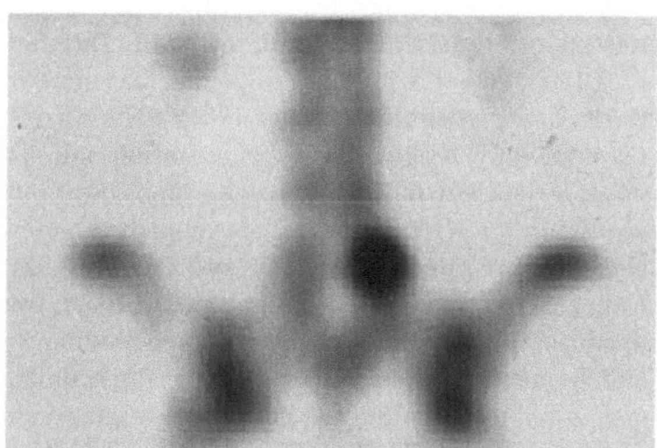

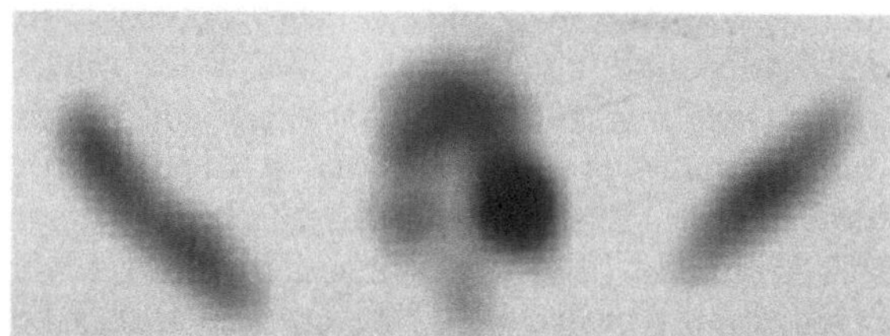

pars may increase the accuracy of detection. On the T2-weighted images, there may be increased signal in the pars defect, particularly on the gradient-echo sequences, owing to the presence of fibrous tissue (JOHNSON et al. 1988). The pars defect may still be difficult to see on MRI if it is thin or sclerotic, and the partial volume effect may include a degenerative spur from the articular process of the facet, which may simulate a defect. The main value of MRI lies in the demonstration of other causes of back pain in the presence of a defect, and especially in the clarification of the status of the discs above and below. Degenerative changes in the disc below the lysis are common after the age of 40 but the disc above may also undergo degenerative changes and in a series of patients with spondylolisthesis, associated with a spondylolysis, the L4/5 disc above the defect was found to be the source of the patient's symptoms on discography in 50% of cases (HENSON et al. 1987). MRI has demonstrated posterior annular tears and disc protrusions at these levels, which have been shown to be symptomatic on subsequent discography. MRI will also establish whether the adjacent discs are normal in order that repair of the lysis can be undertaken. Spondylolysis in the cervical spine also occurs but is not activity related, rather being developmental in aetiology. The relationship between the defects and neck pain is not clear.

Thoracic or upper lumbar pain may be due to Scheuermann's disease. The plain x-rays will usually provide the diagnosis, showing wedging of three or more thoracic vertebral bodies, associated with irregularity of the vertebral end plates, often with sclerosis and also disc space narrowing. The vertebral bodies show increase in AP width and a kyphosis may develop (Fig. 15.8). End plate irregularity associated with disc degeneration may occur without significant vertebral wedging and may be limited to one or two levels. Discography performed at levels with interosseous disc herniations through defective end plates in adolescents with persistent pain has identified them as pain sources (McCALL et al. 1985). The combination of lumbar disc degeneration and Scheuermann's disease has been shown to

Fig. 15.7 a–d. Spondylolysis. **a** The lateral radiograph of the lumbar spine shows a 10% spondylolisthesis, with a vertical lucency across the pars interarticularis of L5. **b** The reverse angle CT scan shows bilateral spondylolysis, with sclerosis in the bone on either side of the defect, especially on the right. **c,d** SPECT (**c** coronal and **d** axial) scans shows high uptake on the right at L5

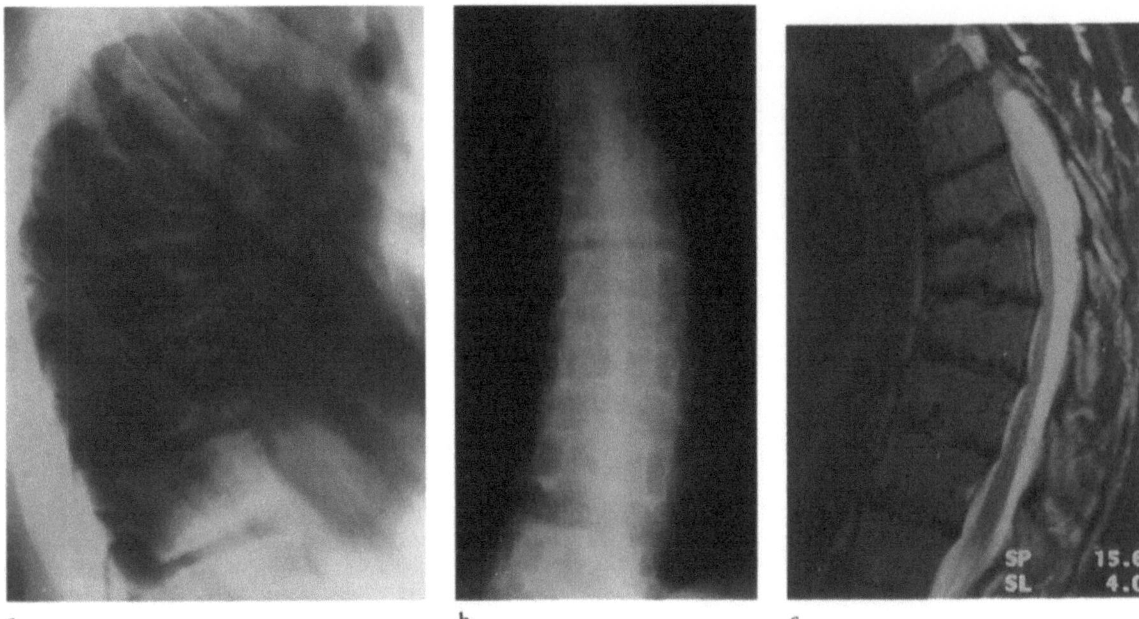

a   b   c

Fig. 15.8 a–c. Scheuermann's disease. **a** The lateral radiograph of the thoracic spine shows an increased kyphosis, with wedging of the vertebral bodies, irregularity of the end plates and disc narrowing. **b** The AP view shows a mild scoliosis. **c** Sagittal T2-weighted TSE MR sequence shows marked loss of signal in the discs and demonstrates the end plate changes clearly. There is no evidence of localised disc herniation or focal cord compression

occur in 9% of patients referred for MRI of the lumbar spine, with a relatively high percentage in the younger age group, suggesting an underlying structural weakness (Heitoff et al. 1994). ^{99m}Tc-MDP studies have little value but MRI may demonstrate an occasional complication of disc protrusion at the apex of the curve and can assess the effect of the thoracic curve on the cord.

Neck pain is uncommon in children and adolescents but patients may occasionally develop acute torticollis with pain. Plain radiographs confirm the curvature but rarely show any underlying features. Rotational subluxation at the atlanto-axial level may be demonstrated. If this is suspected on plain films, then CT is the investigation of choice for confirmation. Soft tissue swelling may also be seen in the prevertebral space if the aetiology is infection.

15.6
Acute Nerve Root Pain

Acute nerve root pain presents as pain in the back or neck and extends down the upper or lower limb, usually in the distribution of a specific nerve root, although the pain may not follow the complete anatomical distribution. The onset of pain may be acute or gradual and may be accompanied by paraesthesia and numbness in the distribution of the nerve. Weakness in the muscles served by the relevant nerve roots may be identified and occasionally this may be the presenting feature, with drop foot or wasting of the small muscles of the hand. The compression is due to a prolapse of the nuclear material through the posterior annular fibres. The nuclear material may be contained by the outer annular fibres or may extrude completely through the annular disruption. Occasionally, the extruded fragment becomes separate from the disc and migrates away from the disc, while rarely it may erode through into the dural sac. Compression of the nerve roots may be either central or posterolateral, giving either unilateral or bilateral symptomatology. Plain radiographs of the spine are of little value in the diagnosis. They may be normal or may show some narrowing of the disc space and occasionally evidence of muscle spasm, with a loss of lordosis or a torticollis in the cervical spine or a non-rotated scoliosis in the lumbar spine. The investigation of choice is now MRI (Fig. 15.9). The sagittal T1-weighted images will demonstrate a protrusion of the disc, which will be of

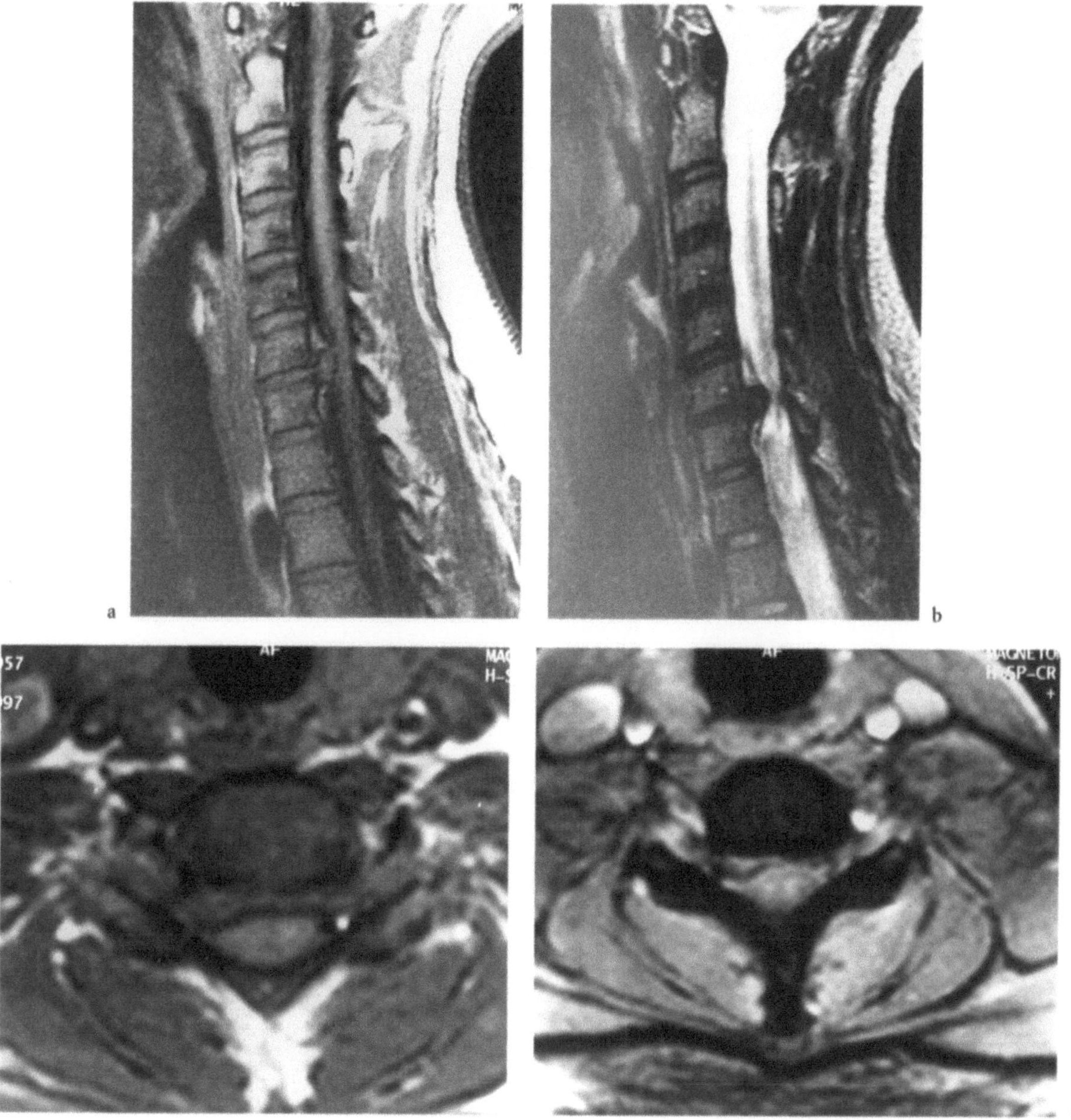

Fig. 15.9 a–k. Disc prolapse. **a–d** Cervical spine: The sagittal T1- (**a**) and T2-weighted (**b**) scans show a large disc herniation, which is elevating the posterior longitudinal ligament away from the vertebra. An axial scan shows the right sided disc herniation compressing the nerve roots (**c,d**). **e–g** Thoracic spine. Sagittal T1- (**e**) and T2-weighted (**f**) scans show a large disc herniation in the mid thoracic spine, with compression of the cord. The herniation contains very low signal material within it, which is shown on CT to be calcification (**g**). **h–j** Lumbar spine. The sagittal (**h**) and axial (**i**) T1-weighted sequence shows a disc herniation at the L5/S1 level, with a sequestrated fragment behind S1. The sagittal (**j**) and axial (**k**) post-gadolinium T1-weighted studies outline the fragment more clearly and show localised enhancement of the nerve root on the left compared to the right side

similar intermediate signal to the remainder of the intervertebral disc and which has an outer margin of continuous low signal due to the intact outer fibres of the annulus/posterior longitudinal ligament complex. If the outline is disrupted, this indicates an extruded disc. The low-intensity signal of the dural sac would be indented in all but the most far out disc prolapses.

On the T2-weighted images, the high signal of the nucleus will be seen to be extending through the posterior annulus and be situated within the prolapse. If the prolapse is recent, then high signal

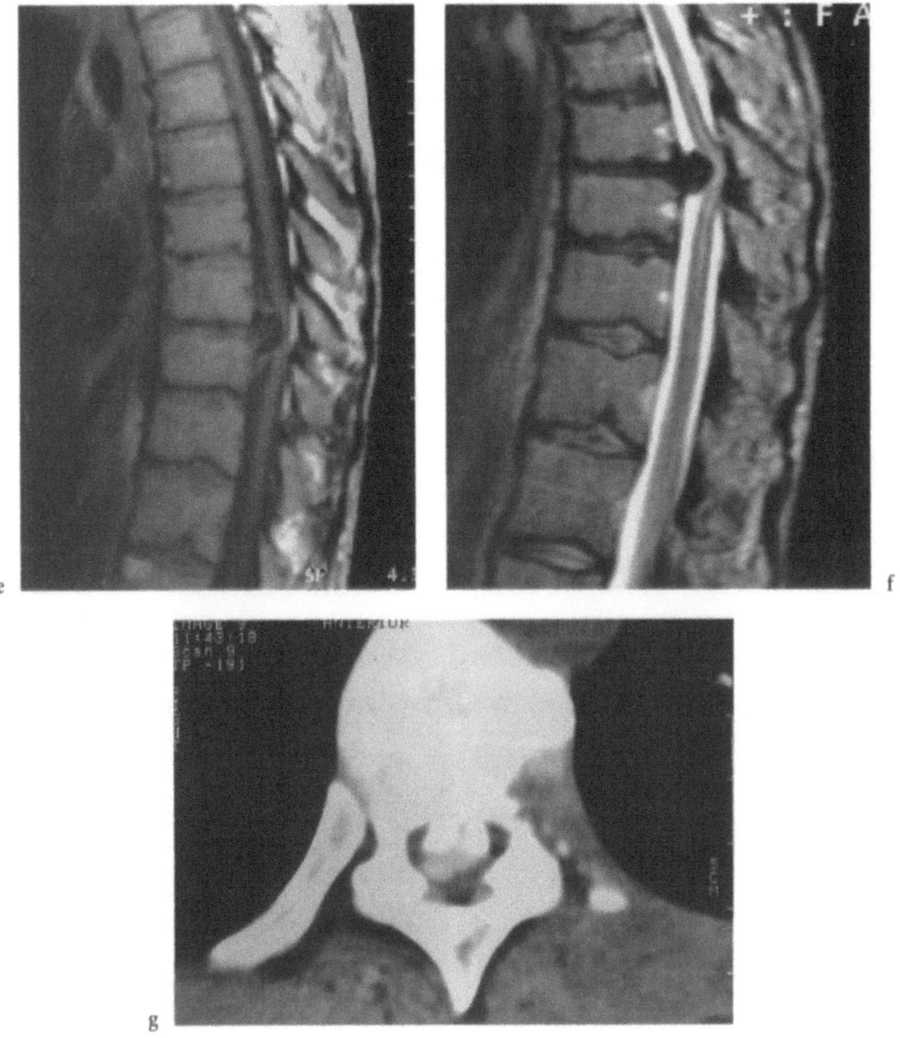

Fig. 15.9 e–g

will be preserved but over a period of time the water content of the disc prolapse, and thus the signal intensity, will reduce. However, accuracy of differentiation between a protrusion and extrusions by demonstration of the low-signal outer fibres has been reported to be low (SILVERMAN et al. 1995).

On the axial scans, the T1- and T2-weighted images will enable the nerve roots to be visualised and compression of the individual nerve roots to be identified. The T2-weighted TSE axial view is particularly valuable for demonstrating individual nerve roots in the lumbar spine. In the cervical spine, the gradient-echo axial T2 views are more appropriate. In the thoracic spine, cord compression may occur from the disc prolapse and cause mild indentation of the cord, while occasionally severe compression of the cord may occur (Fig. 15.9).

Sequestrated disc fragments may be seen as low- or intermediate-signal fragments situated in the entry zone of the nerve root canal. Gadolinium has been used in acute disc herniations, resulting in enhancement of the rim of the herniation; such enhancement was reported in 93% of cases in a recent series, but its persistence following improvement of symptoms calls its value into question (MODIC et al. 1995). In a percentage of cases of nerve root compression caused by disc prolapse, gadolinium-DTPA (Gd-DTPA) will result in focal enhancement of the nerve root which is being compressed (Fig. 15.9) and enhancement may extend proximally in the lumbar spine, even to the conus. There would appear to be a greater incidence of enhancement in cases of large extruded or sequestrated fragments of disc (TYRRELL et al. 1997). This

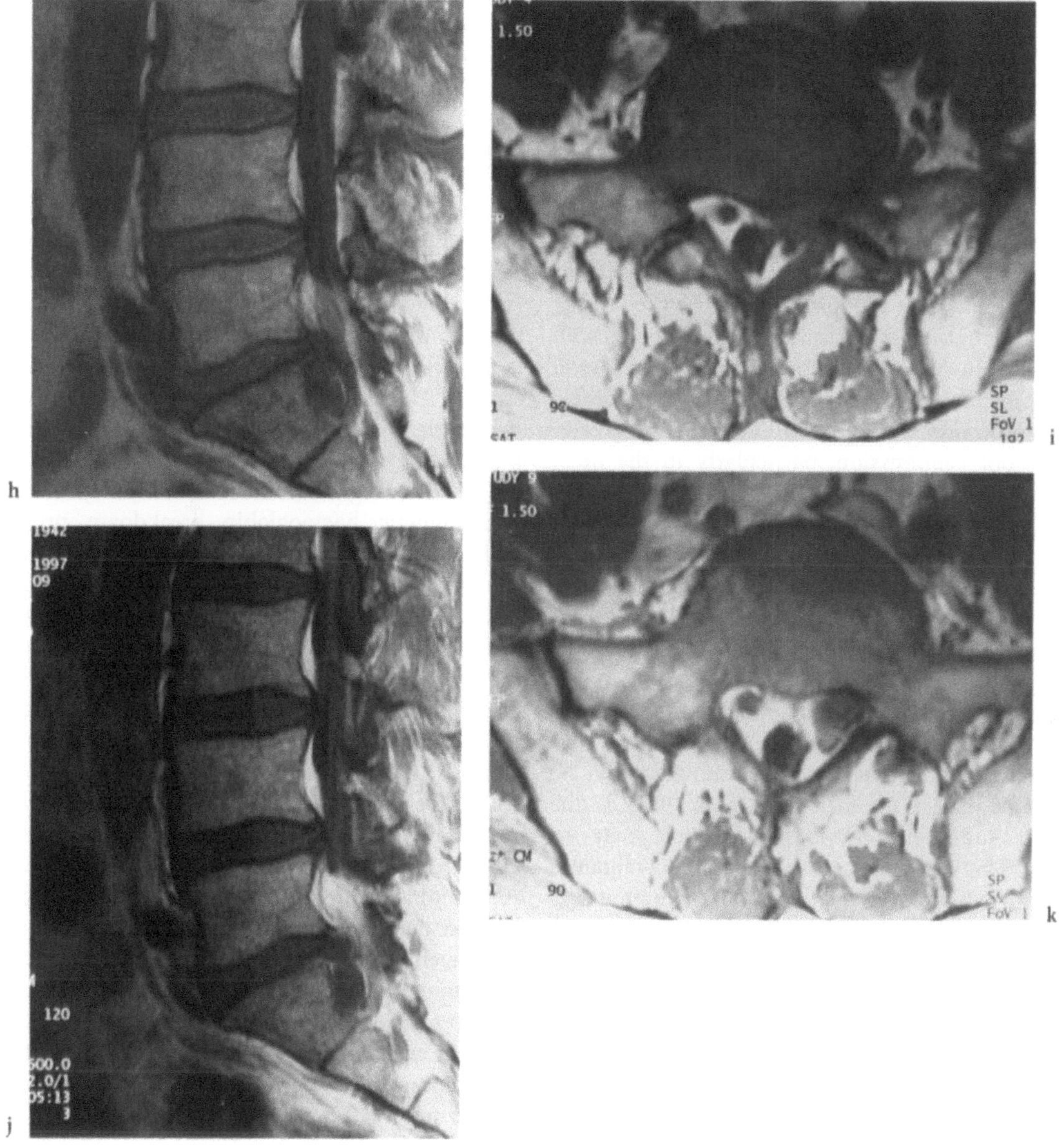

Fig. 15.9 h–k

may be related to the inflammatory chemical contents of the disc prolapse, including prostoglandins and nitrous oxide and interleukin-6 (KANG et al. 1996). However, it may also be related to the degree of compression (KOBAYASHI et al. 1993) and be time related, with more chronic compression showing less nerve root enhancement. It has also been shown, by means of saturation pulses across the spine, that some apparent nerve root enhancement up to the conus is due to gadolinium within the radicular veins (LANE et al. 1994). There is no evidence to suggest that gadolinium assists in the diagnosis of cervical or thoracic disc prolapses.

Computed tomography may also be used to demonstrate disc prolapse, although the differentiation between a protrusion and an extruded disc prolapse may be difficult. In the cervical spine, the differentiation between the dural sac and cord and the disc may be hindered by the relative absence of epidural fat. An accurate diagnosis of cervical protrusion can be assisted by using the highlighting mode concentrating on Hounsfield numbers between −20 and +20, which will outline the dural sac and enable the disc prolapse to be more clearly defined. The evaluation can also be improved by multiplanar reconstruction of a block of the cervical spine with highlighting.

Spiral CT enables this multiplanar reconstruction to be achieved more satisfactorily. Similar problems may occur in the thoracic spine and again the highlighting process may assist in diagnosis. In the lumbar spine, the presence of epidural fat enables the disc prolapse to be more clearly defined, and accuracy rates for the diagnosis by means of unenhanced CT scan vary from 70% to 90% (JACKSON et al. 1989). Water-soluble myelography is now rarely used for the examination of the disc prolapse in the lumbar spine although it may occasionally be combined with a CT scan. In the cervical spine, either a direct lateral C1/2 puncture or a run-up cervical water-soluble myelogram may be required to define the degree of nerve root compression, particularly in the presence of osteophyte formation associated with disc prolapse. However, a low-dose injection of contrast run up to the cervical spine, combined with CT, may be entirely adequate for the examination of the nerve roots and cord in the cervical spine. Special MR techniques, with paraxial views, may also be of value in demonstrating nerve root compression in the cervical spine. The use of T1-weighted 3D acquisition, associated with contrast enhancement, has been advocated and selected paraxial views may be valuable to assess the foramina (Ross 1995). In the lumbar spine, the myelographic effect may be achieved by using a heavily T2-weighted sequence with superimposition of slices and maximum intensity processing.

The natural history of disc prolapses is to gradually resolve on conservative treatment, with loss of signal from the nuclear material within the disc prolapse and retraction of the prolapse. Resorption of sequestrated fragments appears to occur faster than protrusions (MODIC et al. 1995), this being explained by the ability of epidural macrophages to encase and resorb the prolapsed nuclear material. Patients, however, may have significant acute pain and be unable to tolerate the slow process of conservative therapy. In these circumstances, in the lumbar spine, the option of intradiscal therapy or surgery may be considered.

Injection of chymopapain into the nucleus causes disruption of the protoglycans and subsequent loss of water in the disc. This leads to retraction of the disc prolapse and narrowing of the disc space. Following chymopapain injection there is reduction in the T2-weighted signal from the disc, and there may also be changes in the end plate which are similar to those described by MODIC, comprising low signal on T1 and increased signal on T2 (MASARYK et al. 1986).

Percutaneous discectomy may be undertaken, which may decrease the T2-weighted signal in the centre of the nucleus. There is no immediate change in the size of the disc prolapse and there is no clear relationship between the size of the disc prolapse and the initial relief of pain. Gradual resorption and retraction of disc prolapse will, however, occur over a longer period.

Finally, surgery may be undertaken, which in the cervical and thoracic spine is usually performed anteriorly, while in the lumbar spine microdiscectomy is undertaken posteriorly through the ligamentum flavum. The immediate postoperative appearances in the lumbar spine may demonstrate a persistent mass of intermediate or high signal on T1 and low or intermediate signal on T2, due to the dephasing by paramagnetic haemoglobin breakdown products within the haematoma at the operation site (DINA et al. 1995). At a later stage, scar tissue develops and becomes homogeneous with intermediate signal around the dural sac or in the lateral recesses, which may appear as a recurrent disc herniation. These will also be intermediate or low signal on T2. Some distortion of the dural sac may also occur and the nerve root will not be clearly visualised. The injection of Gd-DTPA enables the differentiation between recurrent or persistent disc prolapse and epidural fibrosis (Fig. 15.10). The latter enhances, rapidly outlining the unenhanced nerve root and dural sac and defining the disc margin, whereas disc herniations, although enhancing around the rim, will have a central portion of unenhanced tissue (Fig. 15.10). The use of fat suppression following injection of Gd-DTPA may enhance visualisation. Differentiation between disc material and fibrosis using Gd-DTPA has been reported to be successful in 90% of 44 patients at 50 operated sites (Ross et al. 1990).

Computed tomography may be required to investigate patients who are unsuitable for MRI. In the immediate postoperative period a clearly delineated shadow protruding into the canal and of similar attenuation to disc may be seen in nearly 50% of asymptomatic patients (MONTALDI et al. 1988). Heterogeneous material of lower attenuation than disc with blurred margins thought to represent haemorrhage may also be seen. In the later stages of maturity, scar tissue develops and appears as diffuse areas of tissue of lower attenuation than disc, which also enhances after intravenous contrast injection (CERVELLINI et al. 1988).

Distortion of the dural sac must also be distinguished from arachnoiditis as the latter may cause severe pain. Although arachnoiditis is more likely to

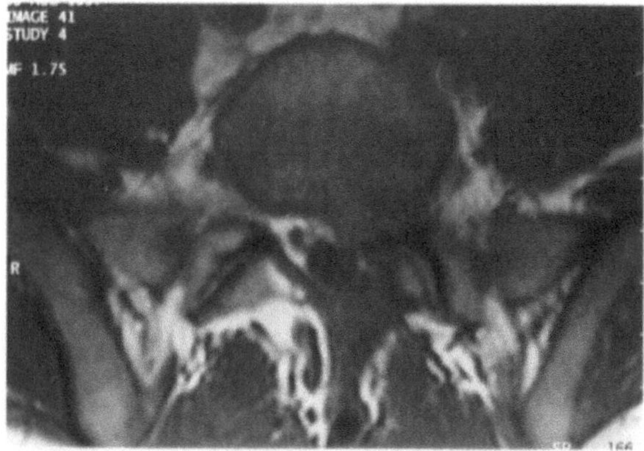

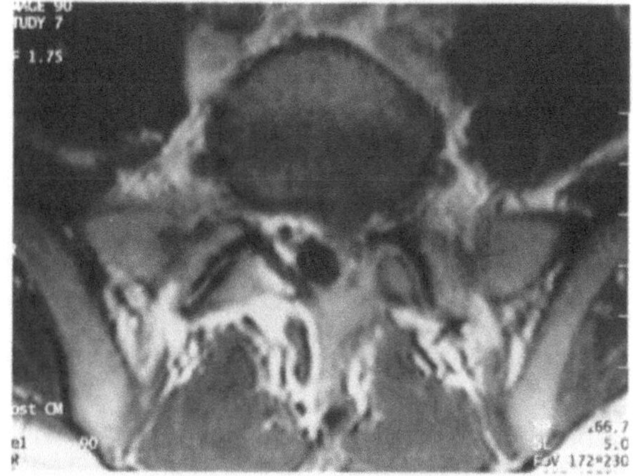

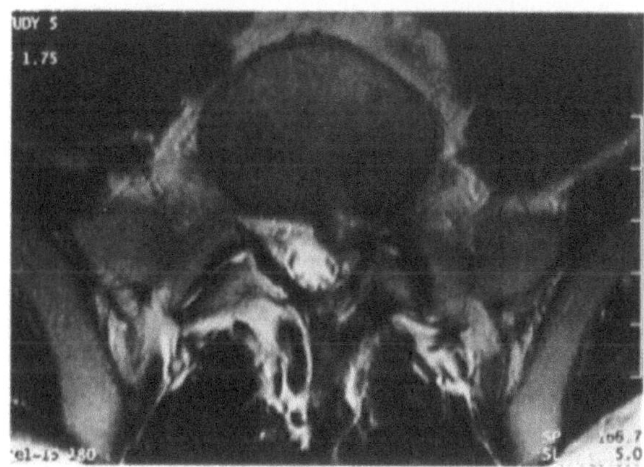

Fig. 15.10 a–c. Postoperative fibrosis. **a** The T1-weighted axial image shows a loss of fat signal around the left S1 nerve root and behind the dural sac. **b** Following the injection of Gd-DTPA the fibrosis posteriorly and around the nerve root enhances, while fragments of recurrent disc herniation, which is displacing the nerve root, remain of low signal. **c** The T2-weighted sequence shows high-signal nuclear material within the disc prolapse

result from the use of oil-based contrast agents, previous spinal surgery may also be a cause. MRI is the initial investigation of choice and T1-weighted images demonstrate clumping of the nerve roots as

a mass of intermediate signal within the low-signal CSF or as nerve roots adherent to the periphery of the dural sac producing the appearance of an empty dural sac. These features are also demonstrated on the sagittal and axial T2-weighted scan with high-signal CSF. Gd-DTPA enhancement of affected nerve roots has also been demonstrated in arachnoiditis but the degree is variable and does not aid diagnosis (JOHNSON and SZE 1990). Enhancement and clumping of nerve roots with increased thickness may also be seen in metastatic infiltration and lymphoma. If MRI is inappropriate, water-soluble myelography with CT will satisfactorily demonstrate the clumping or adhesion of nerve roots.

Postoperative assessment of fusion may also be required. In the cervical spine interbody fusion is usually performed and comparison of lateral flexion and extension radiographs will show evidence of movement. Lateral tomography will allow the assessment of continuity of bone across the vertebra and also union of the facets in a posterior fusion. If metal has been used for fixation, this may be the only means of assessment due to artefactual effect on CT or MRI. If metal is not present, T1- and T2-weighted MR sequences may also show marrow continuity across a fusion mass. In the lumbar spine anterior interbody fusions may be assessed in a similar manner to those in the cervical spine but posterior or posterolateral fusions may prove more difficult to evaluate. Fixation with wire, bars or pedicle screws are often present and the bone mass may be irregular in shape and thickness. Comparison with flexion-extension lateral radiographs or AP tomography is valuable. CT with multiplanar reconstruction is the method of choice if there is no metal present (LANG et al. 1988). The use of ^{99m}Tc-MDP scanning has shown a substantial number of false-negative studies when compared to direct visualisation although the use of SPECT has increased the sensitivity of this modality but not its utilisation.

15.7
Spinal Stenosis

The clinical presentation of spinal stenosis depends primarily on the type of compressive neurology. In the cervical and thoracic spine, central stenosis may result in myelopathy, which is usually of insidious onset. Initially, sensory changes including paraesthesia and a gradual weakness in the lower limbs occur, with evidence of long tract disruption such as clonus and up-going plantar reflexes. In central cervical stenosis, there is variable involvement of the arms. However, central stenosis in the cervical spine is usually accompanied by nerve root compression in the foramina, which results in paraesthesia, weakness and muscle wasting at the relevant level. Pain is also a feature in the lumbar spine. Both central and lateral compression cause lower motor neurone signs and symptoms, including pain often induced by exercise, which is relieved by resting and particularly by bending forward in the case of central stenosis. Weakness and paraesthesia are also features. Physical signs may be limited but areas of numbness and muscle weakness and atrophy occur. Physical signs may be enhanced by repeating the examination after exercise.

Plain radiographic examination will identify the AP diameter of the bony canal but does not provide significant information regarding the area of the canal and particularly the true capacity for the cord and nerve roots. Measurement should be made from the osteophyte formation on the posterior rim of the vertebral end plates to the line along the base of the spinous process in the cervical and lumbar spine. The range in the cervical spine varies from 20 mm at the C1 level to 11–15 mm at C5/6. In the lumbar spine, the range is 13–18 mm at L4. Direct measurements of the thoracic spine are difficult, due to the overlying ribs. Oblique views at 45° in the cervical spine will demonstrate narrowing of the foramina by osteophytes but the relationship between size and symptoms is poor owing to the additional impact of soft tissue. In the lumbar spine, the foramina are superimposed and estimates of foraminal size are inaccurate.

Evaluation of spinal stenosis is best achieved by computed imaging, either CT or MRI. Sagittal T2-weighted sequences on MRI will demonstrate the true AP diameter of the dural sac and will show the soft tissue encroachment, including disc bulging and ligamentum flavum buckling, which produce a pinching effect on the dural sac, as well as the bony narrowing. In the cervical and thoracic spine, where significant compression is present, a localised zone of high signal in the cord may also be demonstrated in some patients (Fig. 15.11), which is often more marked on the gradient-echo sequences than on TSE, although both sequences demonstrate these changes. Histological studies show that these changes indicate the presence of cystic degeneration of the cord with myelomalacia and are due to chronic compression (TAKAHASHI et al. 1989). These authors reported that the frequency of the presence of high-intensity changes was directly proportional to the

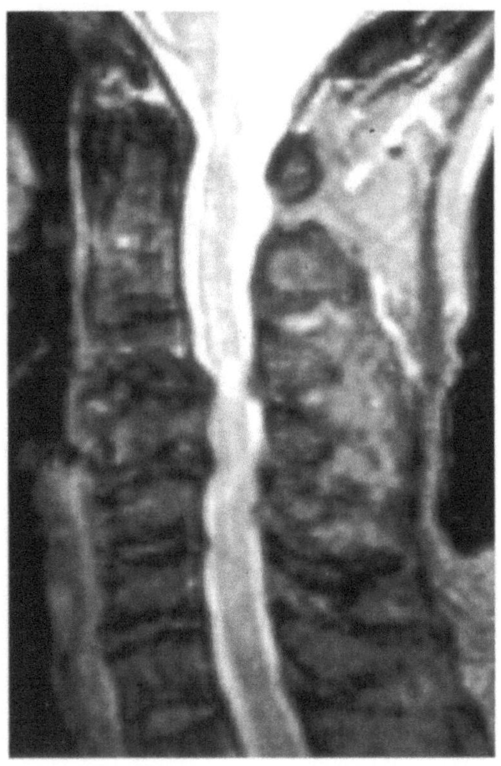

Fig. 15.11. Cervical cord compression. The sagittal T2-weighted scan shows marked narrowing of the cervical canal due to bony stenosis. The cord is compressed and shows a focal high-intensity area within it

severity of the clinical myelopathy and the degree of spinal canal compression on MRI (TAKAHASHI et al. 1989). However, other authors have not found a relationship with clinical severity (WADA et al. 1995). A significant correlation between an axial dural sac area of 50 mm^2 or less and cord compression and clinical myelopathy has also been reported (NAGATA et al. 1990). These authors compared the cord on MRI before and after decompressive surgery and although patients with residual cord atrophy did less well clinically compared to those with recovery of cord deformity, the difference was not statistically significant (NAGATA et al. 1990). TAKAHASHI et al. (1989) also demonstrated that those patients with a high signal focus, which reverted to normal after cord decompression, had reversal of long tract signs, whereas those patients in whom the high-signal zone persisted did not recover. This work requires further confirmation and at the present time the clinical importance of high signal in the cord remains unclear.

In the lumbar spine, the sagittal scans will demonstrate narrowing of the dural sac and compression of the cauda equina but the axial scans are the more important as they demonstrate the true capacity of the dural sac and the degree of nerve root compression. T2-weighted sequences show a loss of high-signal CSF around the nerve roots and a reduction in the area of the dural sac to less than 75 mm^2 is likely to be related to clinical symptoms (ULLRICH et al. 1980). Central compression is maximal at the disc level and is due to posterolateral dural indentation by the osteoarthritic facet joints, with thickening and invagination of the ligamentum flavum-facet-capsule complex and anterior indentation by disc bulging, with or without osteophyte formation (Fig. 15.12). If there is doubt about the nature of the compression, CT will differentiate between disc bulge or protrusion and osteophyte formation. The gold standard for degree of compression remains CT myelography, where the evaluation of contrast flows and the effect of lordosis on the dural sac can be evaluated. Following contrast injection into the dural sac, lateral views of the cervical or lumbar spine are taken with the spine in flexion and extension. These differentiate the compression effect of bulging discs and ligamentum flavum, which is relieved in flexion compared to the persistent effect of osteophyte or bony compression. CT following myelography will enable the capacity of the canal to be accurately evaluated and will demonstrate levels of total obstruction, which may be difficult to confirm on MRI.

15.8
Foraminal Stenosis

Compression of the nerve roots in the nerve root canal in the cervical spine is usually due to osteophyte formation on the neurocentral joints. Thickening of the ligamentum flavum may occur and this may also ossify, causing further stenotic effect. Forward displacement of one cervical vertebra on another will narrow the central canal and foramen. The mechanism is similar in the thoracic spine although the degenerative process is more commonly anterior. However, ossification of the ligamentum flavum or calcification in an old disc prolapse may cause chronic nerve compression. The entry zone of the lumbar nerve root canal may be narrowed by thickening of the undersurface of the lamina and ligamentum flavum. Facet osteoarthritis and osteophyte formation will narrow the mid zone of the canal, and osteophytes on the neurocentral joints and rim of the end plate will affect the mid and exit zones of the foramen. Loss of disc height, with over-riding of the facet, will produce cephalad/caudad compression, and spondylolisthesis, associated with facet degeneration, may cause significant increase in both central and nerve root canal stenosis (Fig. 15.12).

CT has proved to be a very accurate method for the assessment of stenosis in the nerve root canal. The examination should be undertaken with con-

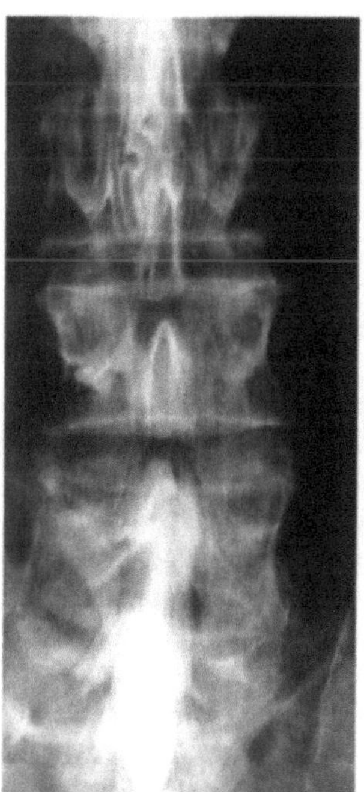

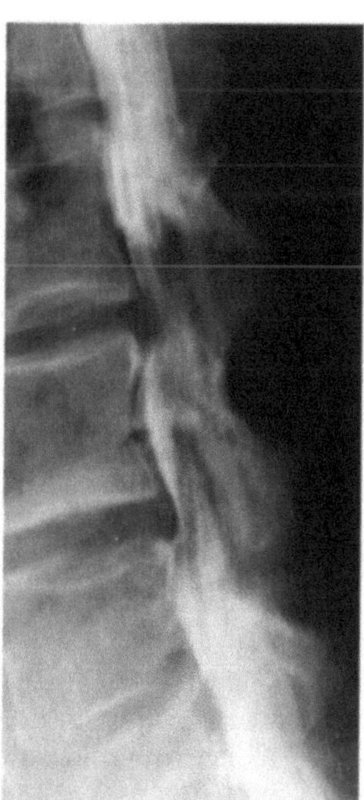

Fig. 15.12 a–f. Lumbar stenosis. The water-soluble myelogram shows narrowing of the contrast column from L3/4 to L4/5 on the AP (a) and lateral view (b). There are distended veins above the obstruction. c Post-myelogram CT shows marked crowding of the nerve roots within the compressed dural sac. d Sagittal T2-weighted TSE sequence shows a severe narrowing of the canal at L3/4 and L4/5, due to osteoarthritic changes in the facets, disc bulging and a degenerative spondylolisthesis. e The axial T2-weighted image shows no CSF around the nerve roots in the compressed dural sac. f The sagittal T1 SE sequence shows marked narrowing of the foramen due to the osteoarthritic changes in the facets. The high-signal fat around the nerve root at L3/4 is absent at L4/5

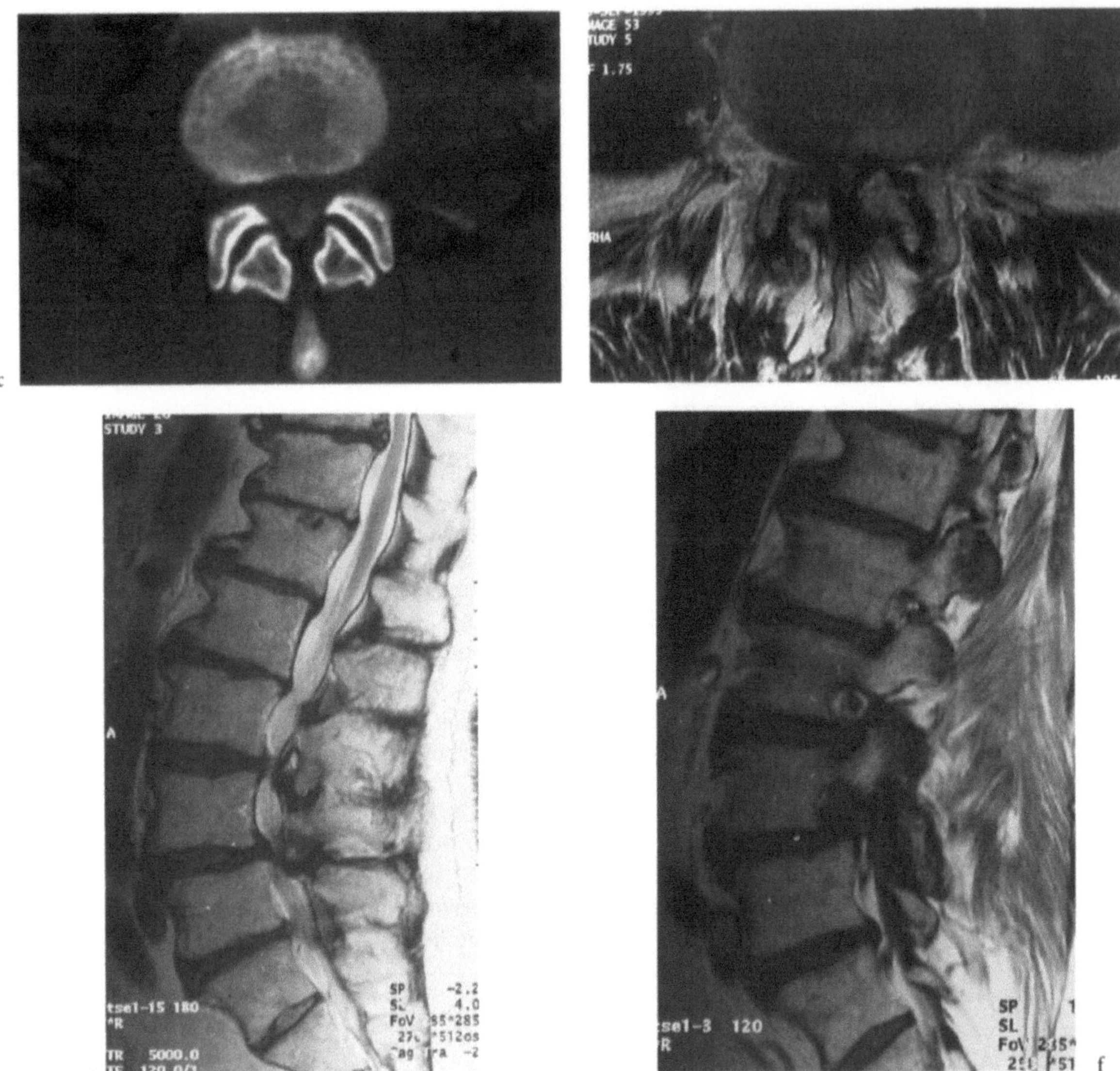

Fig. 15.12 c–f

tiguous slices in order that multiplanar reconstruction may be performed. The axial scans should be imaged on both bone and soft tissue window settings to assess the true bone dimensions of the foramen and the effect of total foraminal contents on the nerve roots. The nerve root should be followed out slice by slice and also viewed sagittally following reconstruction for up/down stenosis. The highlighting mode using the bone setting and −20 to +20 HU will be helpful in assessing the relationship of the nerve root to the canal. In the cervical spine, the relationship of nerve to foramen may be more difficult to assess owing to the relative absence of epidural fat, and CT myelography may be helpful in a full assess-

ment. A small dose of non-ionic water-soluble contrast may be run up after insertion via lumbar puncture and CT may be performed following a full water-soluble contrast examination of the spine. The multiplanar reconstruction in the cervical spine is best performed paraxially to assess the foramina.

Magnetic resonance imaging will also evaluate foraminal stenosis satisfactorily. In the cervical spine, either paraxial acquisitions or 3D acquisition with paraxial reconstruction will be required to accurately assess the foraminal capacity. The 3D sequence commonly used is a gradient-echo T2 but a T1 gradient echo with gadolinium enhancement has been proposed as a satisfactory solution (Ross 1995).

The resolution of the reconstruction will not be as good as the direct paraxial images but the slices will be 1 mm thick. In the thoracic and lumbar spine, the alignment of the foramina allows them to be satisfactorily assessed from the sagittal scans, and the T1-weighted images will show the nerve root as intermediate signal in the foramen outlined by fat. A loss of the fat and compression of the nerve root by the adjacent bone will indicate significant foraminal stenosis (Fig. 15.12). On the axial scans, the nerve roots are clearly demonstrated and compression by bone and soft tissue will be recognised.

15.9
Spinal Trauma

Acute spinal injury may result from many types of trauma and it is important where possible for the radiographic interpretation to appreciate the mechanism of injury. In some cases such as high-velocity motor vehicle accidents, the injury may be complex and therefore difficult to relate to the type of injury.

The initial investigation remains plain radiography, which should be undertaken without moving the patient. The standard AP and shoot-through lateral may be supplemented by supine obliques or a lateral in the swimmer's position to show the lower cervical region and to help evaluate the pedicles and lateral mass. CT is valuable to assess fractures of the facets and pedicles in the cervical spine and the lamina in the thoracic and lumbar spine. The scout view of the whole spine may also be useful to identify secondary levels of trauma. The role of MRI in acute trauma is still being evaluated. The main advantage is the demonstration of the cord and intrasegmental soft tissues. Cord haemorrhage can be identified depending on its temporal stage. Initially, when deoxyhaemoglobin is present, the cord may be isointense on T1 and show reduced signal on T2 owing to a dephasing local paramagnetic effect (KULKARNI et al. 1987). The cord is swollen and there may be a rim of hypointense signal due to oedema around the haemorrhage. Conversion of the haemoglobin to methaemoglobin results in an increase in signal in the cord on T1, with a reduced or isointense signal on T2. The demonstration of haemorrhage in the cord has been shown to be a poor prognostic feature, whereas oedema alone, which is iso- or hypointense on T1 and hyperintense on T2, may indicate potential for neurological recovery (FLANDERS et al. 1990) (Fig. 15.13). The demonstration of soft tissue compression of the cord may

also be of importance as decompressive surgery may be considered. Epidural haematoma can be differentiated from prolapsed disc material and the presence of a disc herniation is a contraindication to manipulation to reduce a dislocation.

Acute dislocation and fracture are usually easily diagnosed following adequate imaging evaluation. Fractures may be treated conservatively or by operative fixation and 3D CT reconstruction may be valuable in the latter case. Operative decompression or fixation has not been shown to significantly change the neurological outcome but more rapid mobilisation may be possible. Remodelling of the fracture site may also occur following conservative treatment, with gradual reduction in the compressive effect of fracture fragments. The healing of ligamentous injuries may be less successful and further deformity may result from a failure to recognise significant ligamentous injury. This is particularly evident in hyperflexion injuries in the cervical spine, where bone trauma may be absent or limited to a small anterior rim compression fracture of the vertebral body. Initial displacement does not cause dislocation and spontaneous reduction takes place; the resultant instability is masked by the protective effect of muscle spasm. Plain radiographs may initially show some widening of the prevertebral space due to oedema but the main feature will be widening of the interspinous distance (Fig. 15.14) and, when present, a small tear-drop fracture of the anterior vertebral end plate (WEBB et al. 1976). Flexion and extension films taken under carefully controlled conditions, following valium to reduce muscle spasm, will demonstrate increased movement and interspinous distance at the affected level. Magnetic resonance studies are invaluable to demonstrate loss of continuity of the posterior longitudinal ligament and ligamentum flavum in the more chronic state, best seen on T2-weighted images. In the more acute state, high signal may be present in the interspinous space (Fig. 15.14) and in the facet joint capsules, due to ligamentous injury, which is best demonstrated on STIR or fat-suppressed T2-weighted sequences.

The presence of ruptures of the posterior ligamentous complex is a clear indication for posterior fusion of the affected segment and if surgery is not undertaken, careful follow-up of patients with significant hyperflexion injuries is essential as kyphosis of the cervical spine may occur. Wedge fractures of the thoracolumbar region may also lead to persistent and progressive kyphosis, due to posterior ligament injury and loss of support from vertebral columns. The degree of initial wedging of the

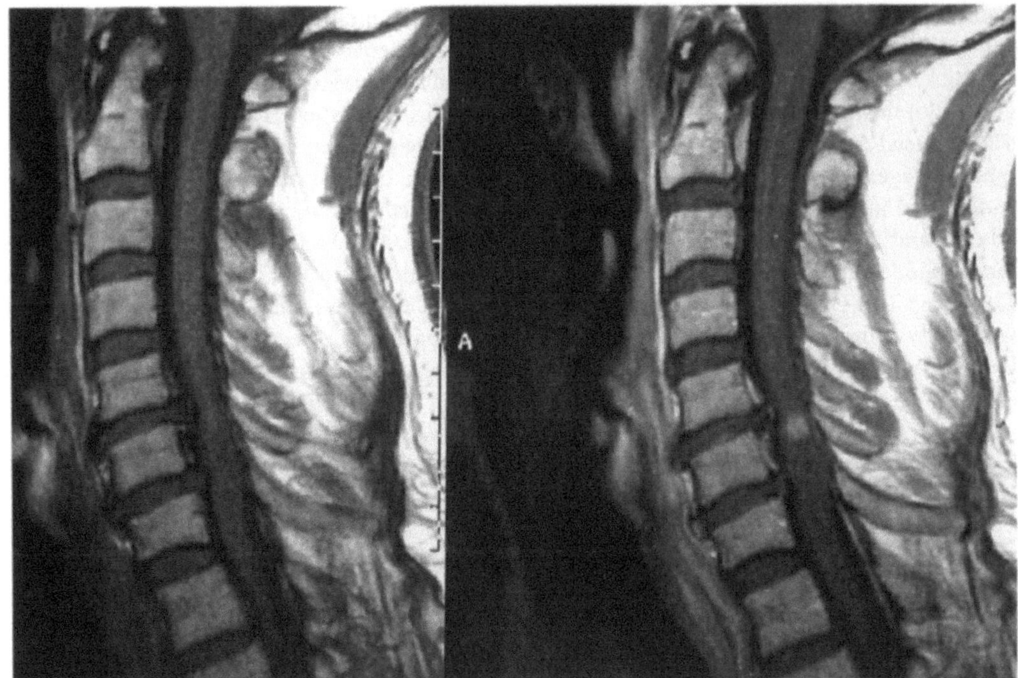

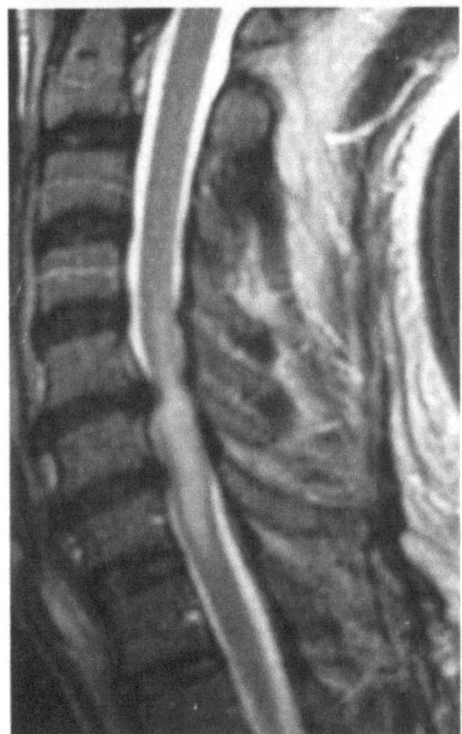

Fig. 15.13. Severe disc herniation in the cervical spine shows increased signal focally in he cord after gadolinium injection (**a**) but there is extensive increased signal on T2-weighted images due to cord oedema (**b**)

vertebral body may be related to the development of kyphosis.

Hyperextension injuries often related to posterior vehicle impaction may result in neck pain, which may develop a few hours after the injury and may be persistent. Plain films are often normal or demonstrate non-specific degenerative changes. In severe cases, some prevertebral soft tissue swelling may be

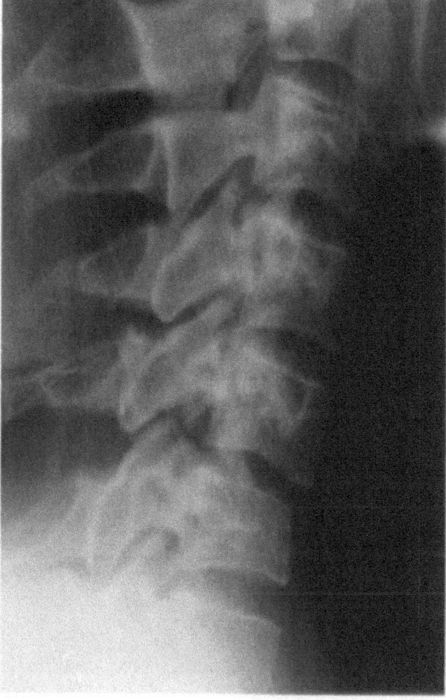

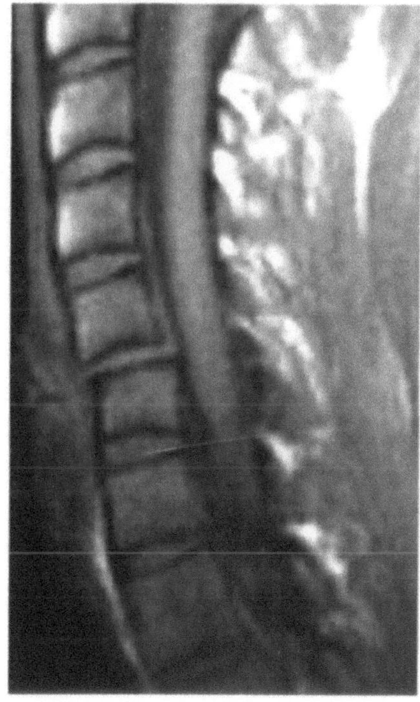

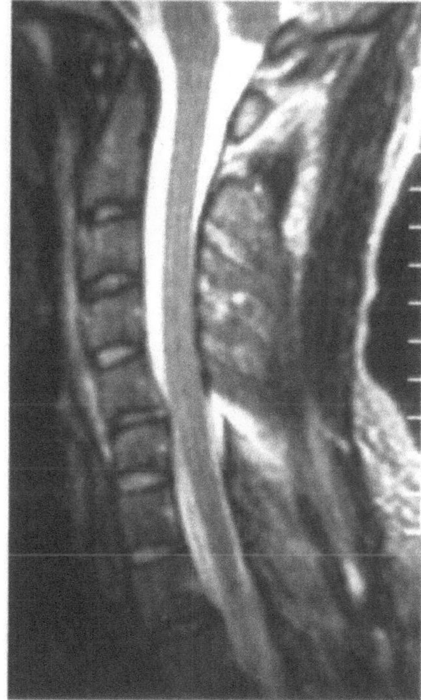

Fig. 15.14 a–c. Hyperflexion injury. **a** Lateral cervical radiograph showing widening of the interspinous space, with a mild degree of forward displacement and kyphotic angulation. **b** The T1-weighted image shows forward displacement of C5 on C6. There is a loss of continuity of the anterior and posterior longitudinal ligaments. **c** The T2-weighted sagittal image shows complete rupture of the ligamentum flavum and high signal in the interspinous ligament, indicating rupture. Increased signal is also seen in the prevertebral soft tissue

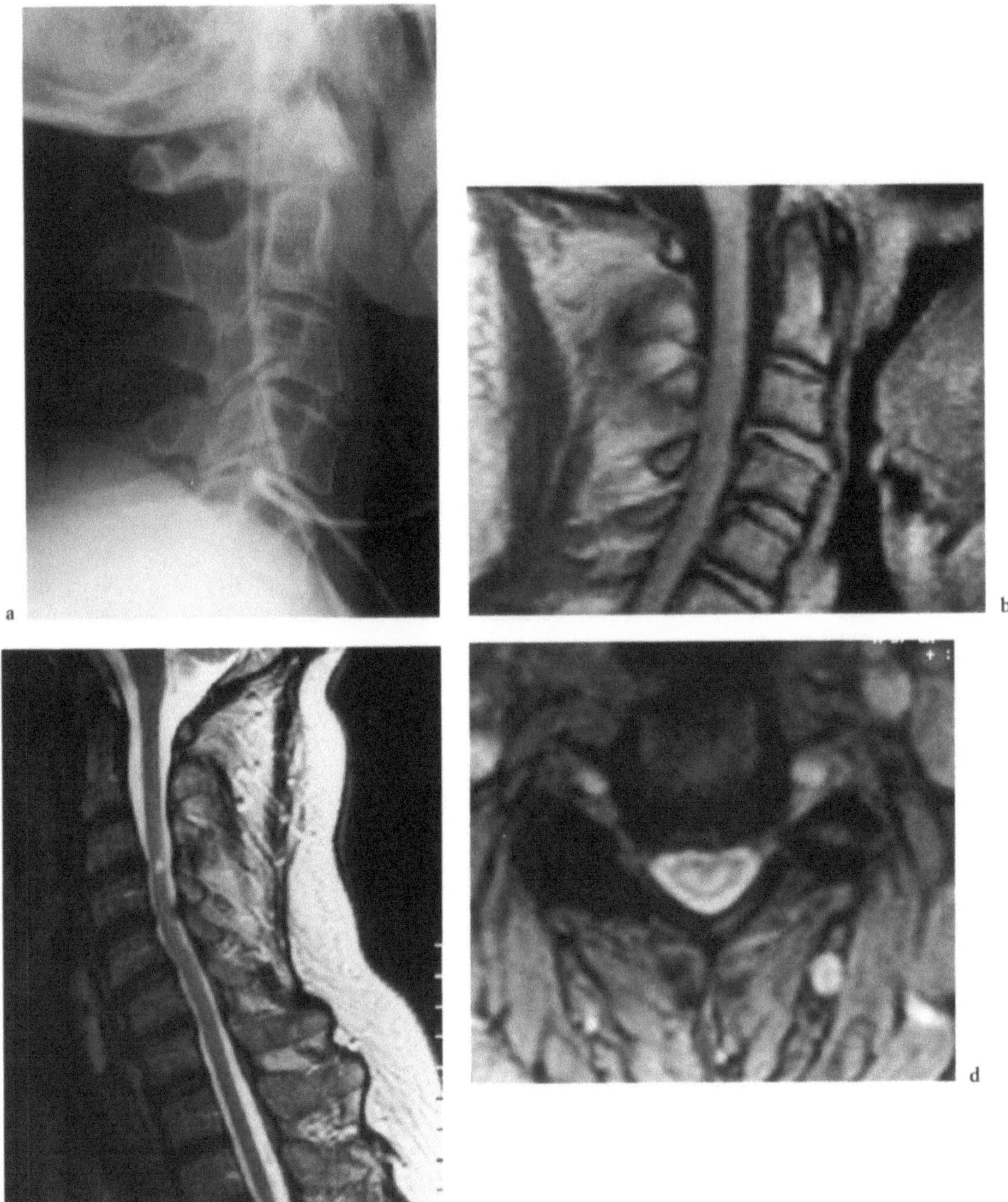

Fig. 15.15 a–d. Hyperextension injury. **a** Plain film showing widening of the anterior disc space at C3/4, with widening of the prevertebral soft tissue shadow. **b** T1-weighted MRI shows a tear in the anterior longitudinal ligament. Follow-up MRI shows disc height loss, with high signal in the cord, due to residual myelomalacia, seen on both the sagittal (**c**) and axial (**d**) views

present but this is rare (PENNIE and AGAMBAR 1991). Compression fractures of the posterior articular process may occur but are rarely visualised on plain films. Some limitation of movement in flexion and extension is possible, and a loss of normal cervical lordosis is seen but is not a specific sign of injury. MRI has been studied as a means of evaluating whiplash injuries. Occasionally, anterior longitudinal ligament tears may be seen (Fig. 15.15) and in patients with evidence of radiculopathy following

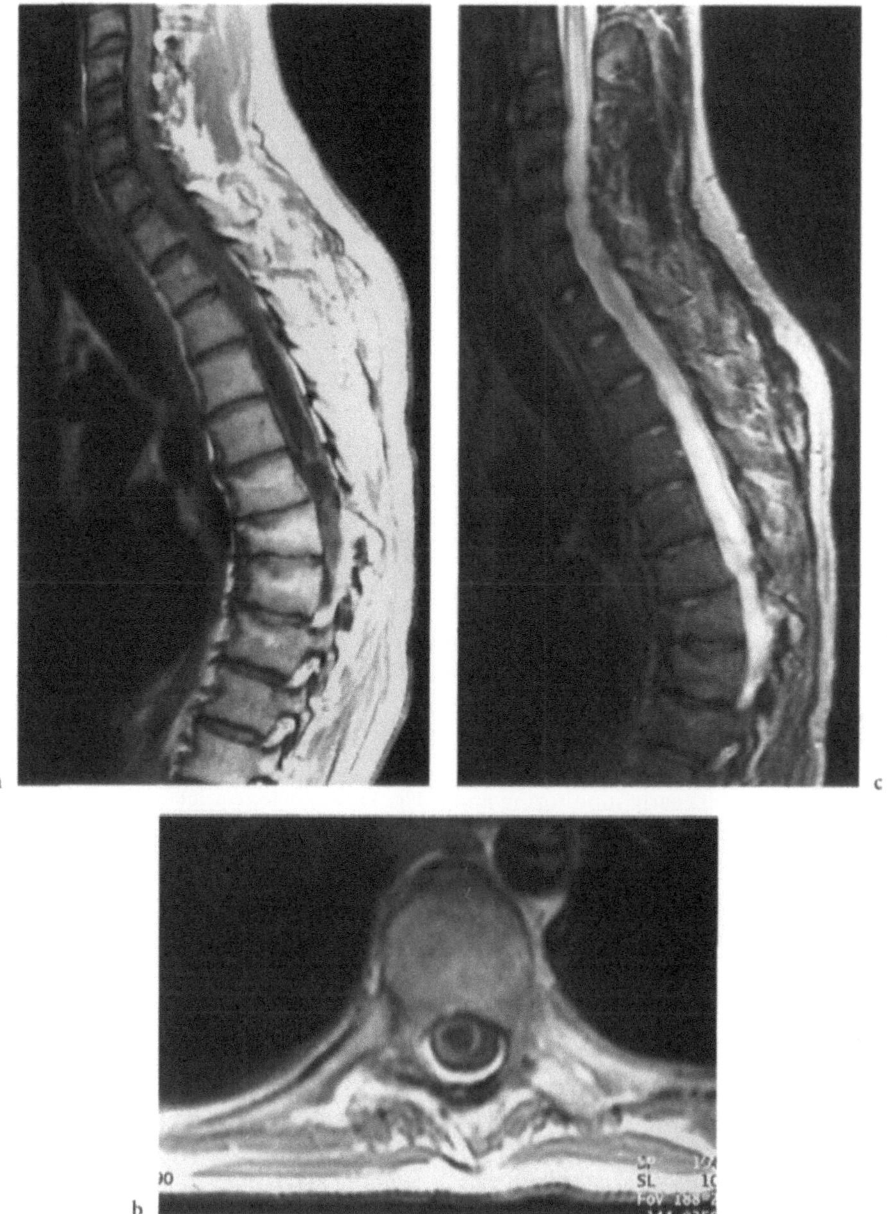

Fig. 15.16 a–c. Post-traumatic syrinx. Old traumatic wedging of T7, T8 and T9 is demonstrated. There is an extensive post-traumatic syrinx in the centre of the cord, which is seen as low signal on the T1-weighted sagittal (a) and axial (b) sequences and as high signal on T2 (c). The high signal extends into the cervical cord

injury, a high incidence of disc prolapse has been reported (JONSSON et al. 1994); in less severe injuries, however, MRI has not been found to be helpful in defining a clear injury. Nevertheless, MRI performed soon after the injury may serve as a baseline for evaluating future appearances in those patients whose symptoms persist, despite appropriate treatment.

Non-union of fractures may lead to persistent in-stability, especially in the odontoid. Careful evaluation of this region should be undertaken, either with CT combined with sagittal reconstruction of the axial slices, or by means of MRI, which will show clear interruption of the marrow signal of the odontoid and body of C2. Flexion and extension lateral views of the cervical spine may demonstrate the degree of instability.

Long-term follow-up of patients with spinal cord

injury may demonstrate deterioration of neurological level or function. Such deterioration may be due to syrinx formation and MRI will show a low-signal zone in the centre of an expanded cord on the T1-weighted images extending over many segments (Fig. 15.16). This may be best seen on the axial scans and will appear as increased signal on T2-weighted sequences and must be differentiated from myelomalacia and cystic degeneration of the cord by the site in the centre of the cord and the extent. Surgical decompression of the syrinx will usually alleviate the neurological deterioration.

15.10
Spinal Deformity

The most common clinical presentation of adolescent idiopathic scoliosis is the recognition of a rib hump or thoracic deformity by the patient or patient's family, often because clothes do not fit properly. This deformity is accentuated by forward bending. However, the active use of screening programmes in schools has enabled the early recognition of scoliosis and thus its early monitoring and treatment. It is important to recognise that adolescent idiopathic scoliosis is predominantly a painless condition and develops relatively slowly over a period of 2 or 3 years, accelerating during the growth spurt. Therefore, the presence of pain or a rapid development of the curvature must be investigated carefully as an underlying cause for the curvature may be present.

A curvature presenting at a younger age may also be idiopathic but is more likely to be due to congenital anomalies of the spine or may be secondary to neural pathology, which may be associated with neurological deficits or foot deformity.

The investigation of scoliosis relies initially on the plain AP radiograph of the spine. The examination is undertaken using a long cassette to include the whole spine, preferably with the patient standing and using varying intensifying screens within the cassette to improve uniformity of film density and contrast. A lateral view is also performed. Measurement of the angle of the curve is undertaken between the end plates of the vertebrae at either end of the curve – the Cobb angle. This is an underestimation of the maximum curve as rotation takes place during curve development. Screening of the patient into a position where there is a true AP of the vertebra at the apex of the curve, the Stagnara view, will allow an accurate measurement of curve angle (Fig. 15.17). A lateral

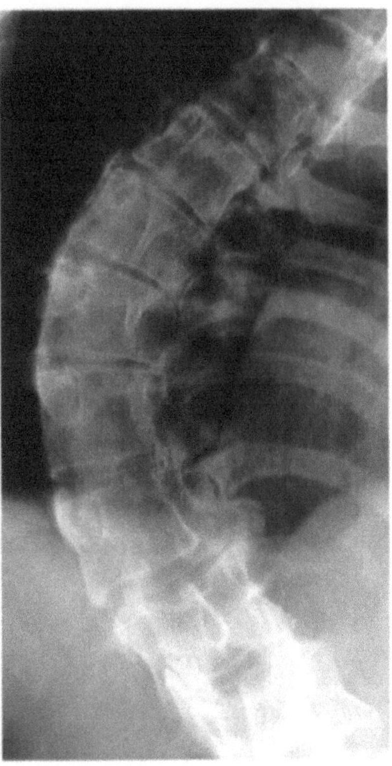

Fig. 15.17. Adolescent idiopathic scoliosis. The patient has been rotated so that the pedicles at the apex are almost a true AP. The so-called Stagnara view, which enables the true curvature to be measured, is greater than the apparent curve on the AP view. The wedging of the vertebral bodies into the concavity of the curve, with displacement of the nucleus of the disc to the convexity, is well demonstrated

view taken at the same time will show that the thoracic spine is in lordosis in idiopathic scoliosis. Side bending films or an AP view with the patient in traction from their own weight, or supine with traction, will provide an assessment of curve flexibility and help differentiate between the more structured curves and compensatory curves. Assessment of rotation and rib deformity may be undertaken using CT of the thorax but this is not routinely performed in most units. Monitoring of the progress of scoliosis is undertaken clinically and with further AP x-ray films. PA views may be undertaken, which reduce the radiation dose to the breast, but conversion factors for vertebral angle and magnification may be required.

If the initial radiograph demonstrates a congenital anomaly such as hemivertebra or evidence of total or partial vertebral fusion, linear tomography may be required to evaluate the extent of the anoma-

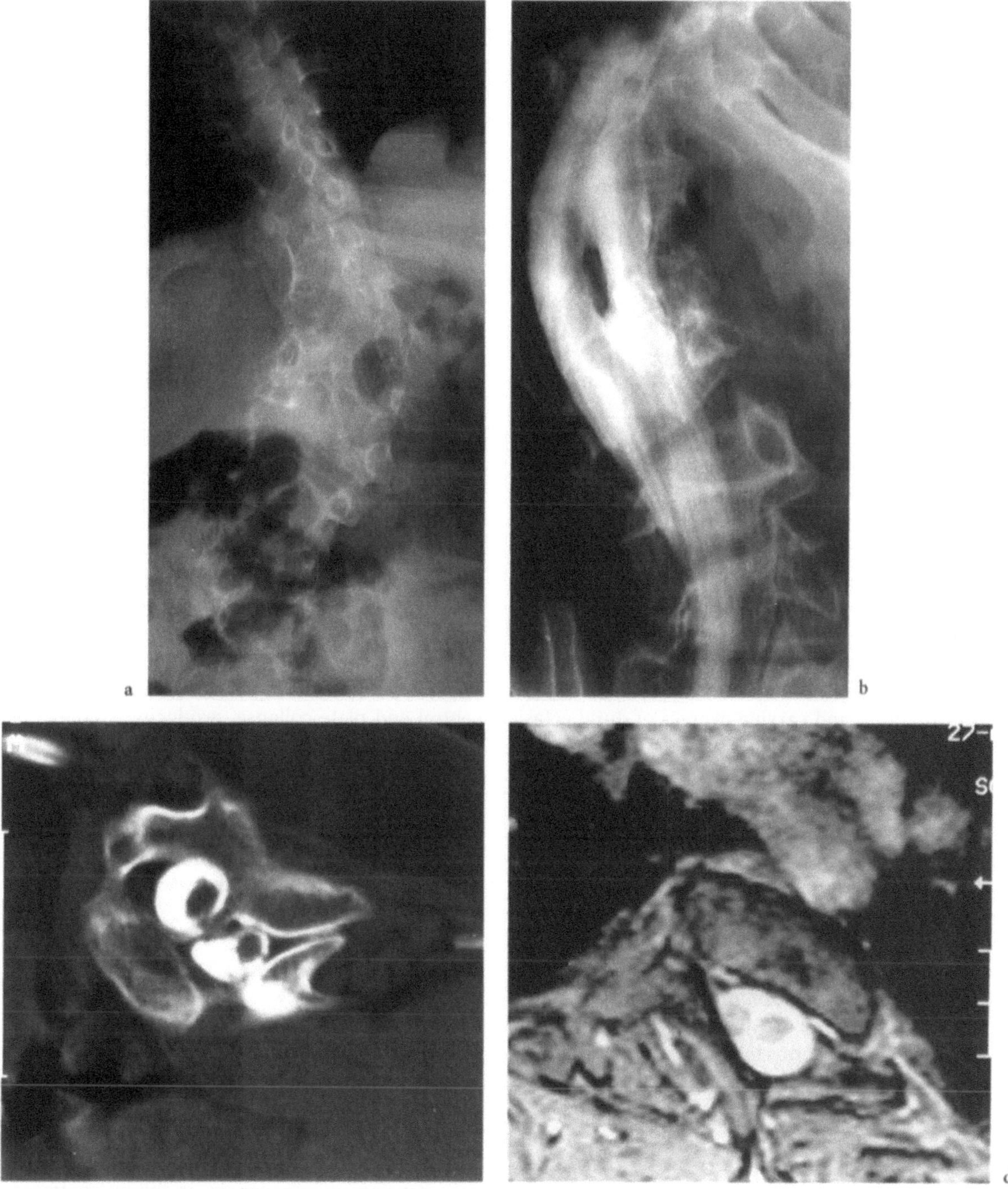

Fig. 15.18 a–d. Diastematomyelia. **a** Plain film showing multiple vertebral anomalies and a lumbar scoliosis. **b** Intradural water-soluble contrast shows the dural sac split by a bony spur. **c** This is best demonstrated on the CT scan, which shows the two halves of the cord in separate dural sacs, separated by a bony spur. **d** Axial T2-weighted gradient-echo sequence also demonstrates a split cord

lies. CT, with multiplanar reconstruction, may also be useful, particularly in the assessment of unsegmented bars in the posterior elements. The presence of pain, neurological anomalies or rapid development of curvature requires further investigation. The presence of a left-sided idiopathic scoliosis also has a higher incidence of an underlying causation and café au lait spots will suggest neurofibromatosis. MRI of the spine has become the investigation of choice if any complicating feature is suspected. A ^{99m}Tc-MDP isotope scan is also of value to identify a focal bone lesion such as an osteoid osteoma or osteoblastoma, which may be painful and may be resulting in a focal scoliosis. The MR examination should image the whole spine, including the craniocervical junction, with T1- and T2-weighted sequences. If the curve is severe, multiple angled views may have to be taken to show the whole cord and the coronal projection may be the most useful as it is most commonly in the cord plane. Axial views should be obtained where there is evidence of cord lesions, which may be either a tumour or diastematomyelia, with cord splitting and tethering (Fig. 15.18). The injection of Gd-DTPA may be required if a cord tumour is suspected, and the sacral region should be included to identify evidence of cord tethering by persistent filum terminale of a meningocele or of dural ectasia.

Studies of a series of patients with idiopathic scoliosis have produced a varied incidence of incidental abnormalities, but in a series of patients with atypical idiopathic scoliosis the most common lesions include a mild syrinx of the cord, Chiari I malformations and varying degrees of dural ectasia (Barnes et al. 1993). Spinal tumours are rare as an incidental finding without clinical features. The true association of mild syrinx and scoliosis is not clear as large-scale population MRI studies have not been performed but the presence of a scoliosis has been reported in 63% of patients with syringomyelia (McRae and Standen 1966). The presence of a diastematomyelia without evidence of a congenital anomaly of the vertebra is not recognised. All patients with congenital abnormalities and idiopathic scoliosis with any suspicious features should undergo MRI or CT myelography prior to surgery.

15.11
Kyphosis

Kyphosis of the dorsal spine may be due to failure of formation of the vertebral body which results from a disturbance in development at the fifth to sixth embryological week or in a failure to ossify in the later stages of development. Kyphosis may also develop from growth abnormalities of the end plate in Scheuermann's disease (Fig. 15.8) or as a result of vertebral collapse due to osteoporosis. Congenital dorsal hemivertebrae will be easily demonstrated on lateral plain x-ray films and should be evaluated

with MRI to assess cord compression. Kyphosis of the dorsal or dorsolumbar spine due to a dorsal hemivertebra poses a particular risk of cord injury and even relatively minor degrees of kyphosis can cause significant cord pressure due to its focal nature.

If MRI is not possible owing to the deformity or for other reasons, water-soluble myelography should be performed with linear tomography as CT may be difficult to interpret because of the deformity.

The radiological features of Scheuermann's disease have been described but quite marked kyphosis may develop, particularly at the time of the growth spurt (Fig. 15.8). Cord compression may occur owing to disc herniation at the kyphosis, although this is rare. MR scans will demonstrate the status of the discs in this condition and should be performed if there is a prominent kyphosis (Fig. 15.8).

Osteoporosis is the main cause for kyphosis in the elderly and apart from the curvature, may be asymptomatic. Pain does occur, however, and this may be associated with vertebral collapse. The plain radiograph will demonstrate a loss of bone density, with a sharp pencil-like outline to the vertebral body. In severe cases, the vertebral end plate will collapse; this may be in the form of a wedge collapse or due to central end plate collapse, producing a curved end plate.

Estimation of early bone density loss is difficult on plain films and bone densitometry using dual-energy x-ray absorptiometry or quantitative CT measurements against known standards is the most accurate method of measuring the degree of osteoporosis.

15.12
Spinal Infection

The most common clinical presentation of spinal infection is severe relatively localised pain. In a pyogenic infection, the onset may be acute with continued pain at night, accompanied by fever and an increased erythrocyte sedimentation rate. The pain may develop spontaneously or follow an invasive procedure on the spine, such as surgery or discography. Pain may also be more gradual in onset, particularly in tuberculosis, and occasionally the patient may present with evidence of neurological compromise and long track signs or evidence of cauda equina syndrome.

If the pain has been present for a few weeks the initial investigation of choice remains the plain radiography; this will demonstrate loss of sharpness of the vertebral end plates, which become irregular and

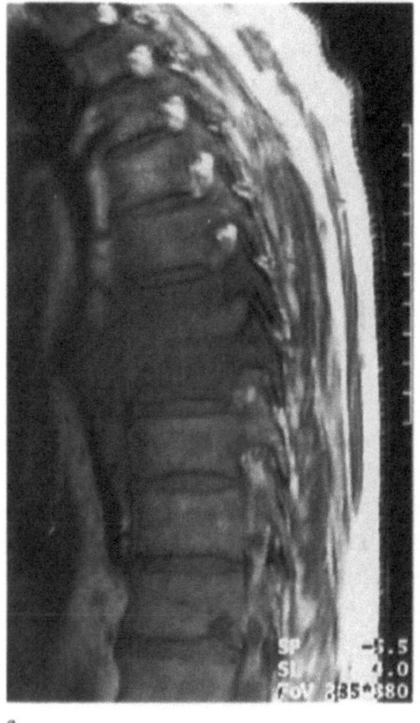 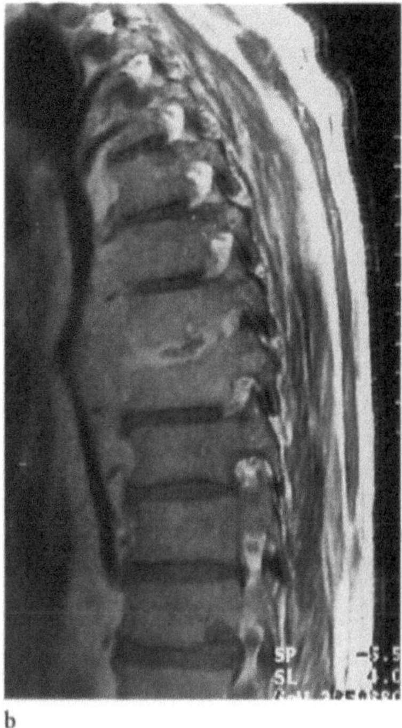 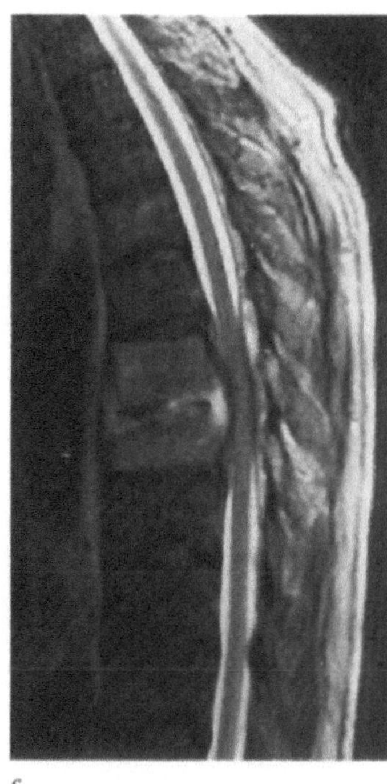

a   b   c

Fig. 15.19 a–c. Infection. **a** The T1-weighted sequence shows disc space loss, end plate destruction and low signal in the vertebral bodies and foramen. **b** Following Gd-DTPA administration, there is enhancement of the disc space, vertebra and foramen. **c** The midline T2-weighted TSE sequence shows increased signal in the disc space and an abscess under the posterior longitudinal ligament, which is indenting the dural sac and cord. The signal from the vertebra is also increased

eventually sclerotic as the infection progresses. The disc height will be reduced. The paravertebral soft tissues may be widened, as manifested by the prevertebral space in the cervical spine and the lateral paravertebral shadow in the thoracic spine. Loss of psoas outline is less predictable but widening may occur if a psoas abscess is present, particularly in tuberculosis. Pus may collect under the anterior longitudinal ligament, causing scalloping of the vertebral body, and may track over a series of vertebrae. In severe cases, extensive vertebral body and disc destruction results in collapse of the segment and a local kyphos.

In the early stages of infection, there may be no visible change on the plain radiographs. In the presence of persistent pain, particularly following an invasive procedure, MRI is the initial investigation of choice. In experimental studies, changes have been shown to precede the development of increased activity on ^{99m}Tc-MDP bone scanning (SZPRYT et al. 1988) although the use of SPECT may increase sensitivity and assist in localisation. In the early stages of infection, MRI will show decreased signal from the marrow fat of the vertebral end plate and adjacent vertebra and disc on T1-weighted images, due to inflammatory exudate and oedema. On T2-weighted images the signal will be increased owing to an increased proton density and prolonged T2 relaxation times (Fig. 15.19).

The use of the short tau inversion recovery sequence (STIR) will increase the intensity of the signal from the vertebra and disc, suppressing the signal from normal fat, which may assist in diagnosis in early cases and will differentiate them from the similar vertebral Modic 1 changes that occur in some cases of disc degeneration as the latter do not show increase in signal from the disc. Extension of infection with epidural abscess is well demonstrated by MRI, appearing as a well-defined soft tissue mass with tapered edges which is typically isointense with spinal cord and cauda equina on T1 and hyperintense on T2, and which may indent or compress the cord. The abscess may be localised within the epidural space and have an enhancing rim on T1-weighted sequences following the injection of Gd-DTPA (Fig. 15.19), though a more generalised diffuse homogeneous or heterogeneous enhancement may also occur. Differentiation of infection from vertebral metastatic disease, which produces similar signal changes in the vertebral body, is usually easy owing to the absence of disc involvement, but occasional cases of disc involvement in metastatic disease have been reported.

Spinal tuberculosis may produce substantial paravertebral abscess formation and fragmentation of bone within it is a particular feature, thought to be due to the absence of requisite bone-resorbing enzymes. Fragments of bone within the abscess are best appreciated on CT. Atypical cases may also only involve the vertebral body in the initial stages and destruction of the posterior element may occur, with or without vertebral body involvement. This will be well shown by both MRI and CT. The most valuable isotope study is the whole-body bone scan, which may highlight other levels of infection in the spine or, in rare cases of multifocal chronic infection, lesions in the appendicular skeleton as well. Indium-111 or ^{99m}Tc-hexamethylprophylene amine oxime labelled white cell scanning provides a more specific diagnostic test for an acute spinal abscess but there is relatively low accumulation in more chronic infective cases and this technique has therefore not been of great value. The role of Tc-labelled human immunoglobulin has not been fully evaluated but it appears to be of less value in the spine than in the appendicular skeleton owing to the accumulation of the isotope in overlying abdominal organs.

Attempts should be made to identify the infecting organism in all cases of spinal infection. If the blood culture is negative, a vertebral biopsy should be performed, if possible prior to antibiotic treatment; histology and culture of the tissue or pus should be undertaken. The success of culture varies between 50% (STOKER and KISSIN 1985) and 90% (FYFE et al. 1983) using a 2-mm needle.

15.13
Inflammatory Diseases

Patients with inflammatory disorders of the spine often present with characteristic backache and stiffness in the morning, which improves with movement. Low back pain that is initially dull over the region of the sacroiliac joints may be difficult to discriminate from mechanical pain of disc degeneration but the gradual easing of pain with activity is uncharacteristic of mechanical pain, which usually increases as the day progresses. Neck stiffness may also be an early feature and as the disorder progresses, reduction in the range of movement and thoracic cage expansion gradually occur. Pain in the peripheral tendon and ligament attachments may be present simultaneously or prior to the spinal changes and confirms the diagnosis.

The earliest radiological manifestation of inflammatory spondylitis is usually in the sacroiliac joints and the initial investigation remains the plain radiograph, which is preferably performed prone, with AP and lateral views of the lower thoracic and lumbar spine. The earliest feature of sacro-iliitis is loss of sharpness of the subchondral line of the synovial part of the sacroiliac joint, which may be associated with increased subchondral bone density. The joint space may initially appear widened but at a later stage in the disease process, narrowing occurs. If the plain films are suspicious but equivocal, a CT scan of the joints may show irregularity of the joint surface more clearly. Alternatively, ^{99m}Tc-MDP scanning using SPECT provides a method of directly assessing sacroiliac activity. The normal range of activity is wide but quantitative analysis may prove useful. The role of MRI in the early diagnosis of sacroiliitis is not as yet fully clarified. Irregularity of cartilage thickness in the synovial portion of the joint may be seen and high signal in the subchondral bone on STIR or fat-suppressed T2-weighted sequences indicates the presence of inflammatory change. Dynamic assessment following injection of Gd-DTPA has shown rapid increases in signal on fast gradient-echo T1-weighted sequences in patients with active sacroiliitis, while normal controls have a relatively mild increase in signal. A high sensitivity has been claimed by these studies (BOLLOW et al. 1995) but others have found a high false-positive rate (WITTRAM and WHITEHOUSE 1995).

Inflammatory changes in the remainder of the spine may develop simultaneously or at a later stage and may affect all ligaments and synovial joints. Early erosions of the anterior longitudinal ligament enthesis are best recognised on the lateral spinal radiographs and result in squaring of the anterior vertebral border, associated with focal sclerosis of the end plates (Fig. 15.20). Fat-suppressed TSE T2-weighted MR images or STIR sequences may show increased signal at the enthesis in these cases but ^{99m}Tc-MDP scans are often negative. Increased uptake may be seen in the costovertebral and costotransverse joints before radiographic changes are evident, however (Fig. 15.20). In more advanced cases, erosion of the vertebral end plates may be more extensive and these are best demonstrated on

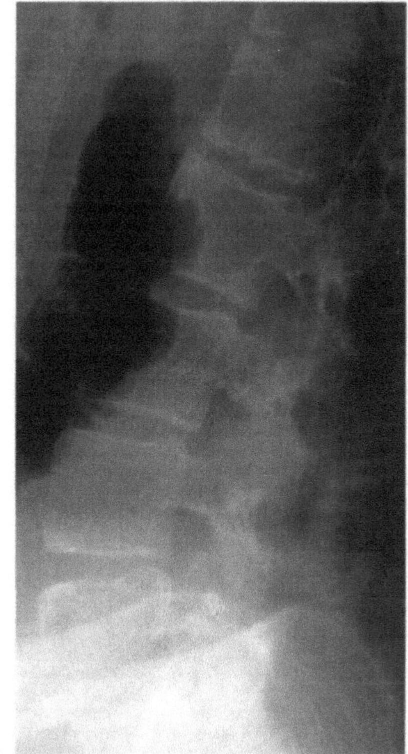

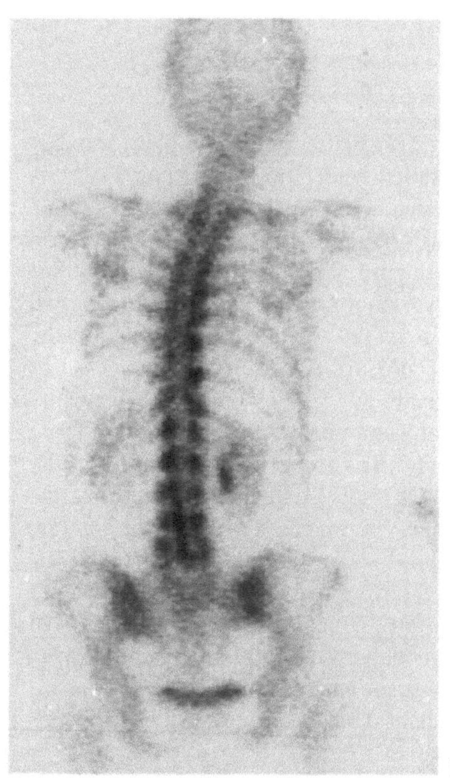

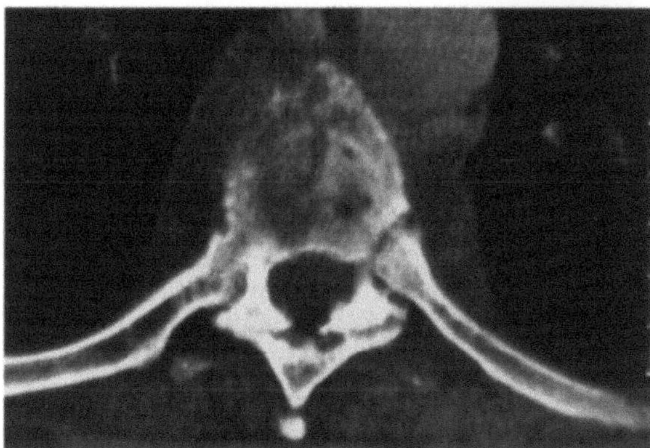

Fig. 15.20 a–e. Ankylosing spondylitis. **a** The plain film shows squaring of the vertebral bodies due to resorption at the anterior rim. A sclerotic focus is seen at the rim of L4/5 and L1/2, the "shiny" corner sign. **b** ^{99m}Tc-MDP spinal scan shows increased isotope uptake in the facet joints in the lumbar spine and the costovertebral joints in the thoracic spine. **c** CT shows extensive erosions of the costovertebral and facet joints. **d,e** Anderson lesion: There is low signal on T1-weighted sequences (**d**) in the vertebral body, with end plate irregularity, and high signal on the T2-weighted TSE sequence (**e**) which mimics infection. The destruction is due to a pseudarthrosis in the posterior ankylosed spine at this level

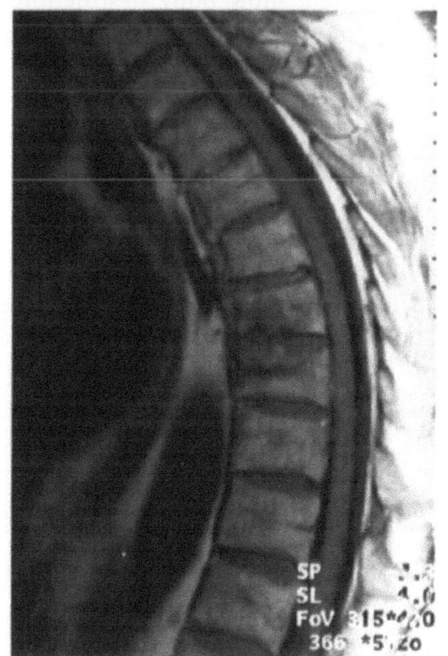

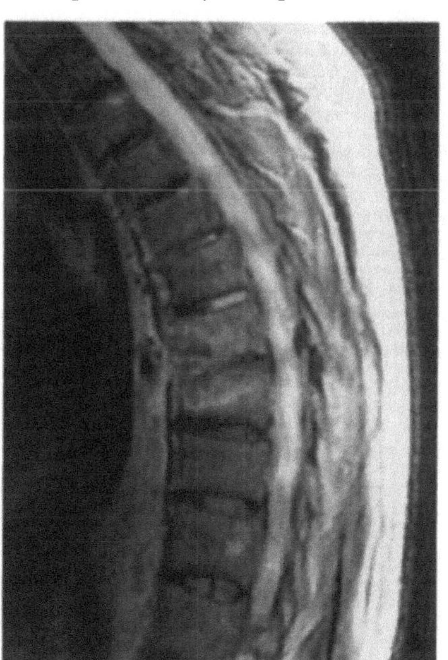

MRI. Extensive destruction of the end plate and disc space, the so-called Anderson lesion (Fig. 15.20), is usually associated with a pseudarthrosis or single residual mobile segment and the disc and vertebral destruction with granulation tissue is due to inflammation and hypermobility. ^{99m}Tc-MDP studies will demonstrate increased uptake at the site of the pseudarthrosis and are useful in established cases with renewed pain.

Neurological complications of longstanding ankylosing spondylitis are all well demonstrated by MRI, although CT may also be of value. In the cauda equina syndrome, T2-weighted images demonstrate an enlarged dural sac with multiple dorsal diverticula, lamina thinning and nerve roots adhering to the posterior dura (Tullous et al. 1990). The fused cervical spine is vulnerable to minor trauma owing to the osteoporosis and fracture may cause cord oedema or haemorrhage, both of which are well demonstrated on MRI.

15.14
Spinal Neoplasm

Plain films have been the main method of initial investigation of primary and metastatic tumours in the spine. In the young, tumours are rare but the appearances may be diagnostic.

Eosinophilic granuloma is classically seen as a flattened vertebral body, so-called vertebra plana, but a bubbly lytic expansile lesion of both the vertebral bodies and posterior elements without significant collapse may occur. The intervening disc space is preserved. Healing results in partial reconstitution of vertebral height which may be associated with some sclerosis and trabecular coarsening. The degree of restitution of vertebral height depends on the age at onset.

Aneurysmal bone cysts of the spine are typically osteolytic and expansile and often involve the posterior elements, resulting in a loss of pedicular outline, but they may also extend into the vertebral body. In addition, involvement of adjacent posterior elements and vertebral bodies sometimes occurs. CT scanning or MRI will demonstrate the extent of the tumour and may also demonstrate fluid levels within the mass.

Patients presenting with pain which is particularly pronounced at night and which responds to aspirin may have an osteoid osteoma, seen as a small sclerotic focus in the posterior elements. Scoliosis may also be a presenting feature with associated localised pain. CT will usually be required to demonstrate the osteoid osteoma, which on CT scans appears as a small lucent focus containing mineralisation with surrounding bony sclerosis. If a clear lesion is not immediately visible, a ^{99m}Tc-MDP bone scan will identify a localised high-uptake focus.

Osteoblastomas may also present with pain, which is generally mild and may be accompanied by muscle spasm and scoliosis. The posterior elements are most commonly involved by a well-defined expansile osteolytic lesion that is extensively calcified or ossified and the features are best evaluated on CT. The tumours may become large unless totally removed and can result in nerve root or cord compression which is most appropriately evaluated by MRI.

Haemangiomas are usually asymptomatic and are demonstrated in the vertebral body as thickened sclerotic vertical striations which may extend into the pedicles and lamina. They may be small and round and are often only seen on MRI; on T1-weighted images they have increased signal with a mottled appearance due to interspersed thickened trabeculae and T2-weighted sequences also demonstrate increased signal intensity (Fig. 15.3). Pain may result from vertebral collapse and rarely expansion results in cord compression.

Primary malignant tumours of the spine are rare but include osteosarcoma, chondrosarcoma, Ewing's sarcoma and chordoma. Chordoma arises from remnants of the primitive notocord and typically presents in middle age with local pain. Chordomas arise most commonly in the sacrum or clivus and plain radiographs show bony destruction with areas of amorphous calcification. CT demonstrates the calcification, often with a paravertebral and epidural soft tissue mass. MRI will show the soft tissue extent of the tumour, which is iso- or hypointense on T1 and hyperintense on T2 with low-signal septa (Sze et al. 1988). Ewing's sarcoma primarily affects children and young adults but spinal involvement is uncommon. Plain radiographs usually show vertebral destruction but occasionally osteosclerosis is observed and soft tissue paravertebral mass is seen on CT and MRI. The marrow involvement appears with low signal on T1-weighted images and increased signal on T2-weighted images. Osteosarcomas rarely arise in the spine but plain radiographs then typically show a destructive mass with periosteal and tumour new bone. The margins of the tumour and the extent of bone formation are best shown by CT, while MRI shows the epidural involvement and neural compression.

15.15
Spinal Metastasis

Although MRI is ideal for imaging vertebral body structure, bone scintigraphy remains the sole technique to provide a sensitive survey of the whole skeleton and conventional radiographs provide an inexpensive and rapid assessment of any symptomatic site. Increased uptake of isotope is not uncommonly seen in severe degenerative changes in the spine, particularly in the facet joints or in vertebral sclerosis, secondary to severe disc degeneration. Differentiation of uptake from metastases may be difficult but extension of activity beyond the lateral vertebral margin, particularly if it is bilateral, is likely to represent a metastatic deposit. Correlation with plain radiographs is essential. The spine is the most frequently investigated region because it contains a high proportion of haematopoietic marrow in adults.

On MRI, focal or diffuse marrow involvement generally modifies the balance between fat and non-fat marrow components, which is more easily detected on T1-weighted images (Fig. 15.21). Fat-suppressed sequences also facilitate the detection of vertebral lesions, which, particularly on STIR sequences, appear as high signal against the low signal intensity background (Fig. 15.21) (MEIROWITZ et al. 1994). The lesions may be focal, with a clear margin, but active lesions may have a rim of increased activity on T2-weighted sequences owing to local infiltration (SCHWEITZER et al. 1993). Such conditions include multiple myeloma, high-grade non-Hodgkin's lymphoma and Hodgkin's disease (Fig. 15.22) and a single metastasis. These conditions may also present with a diffuse, homogeneous decrease in signal intensity compared to muscle and may make the intervertebral disc appear bright on T1-weighted images (CASTILLO et al. 1990). Differentiating diffuse infiltration from normal or hypercellular marrow may be difficult and the demonstration of high signal on fat-suppressed sequences (Fig. 15.22) or following contrast enhancement helps to confirm diffuse involvement. Interpretation from normal heterogeneous haemopoietic marrow distribution can remain difficult and interobserver reproducibility may be relatively low (STABLER et al. 1996). Biopsy will therefore be required for accurate diagnosis.

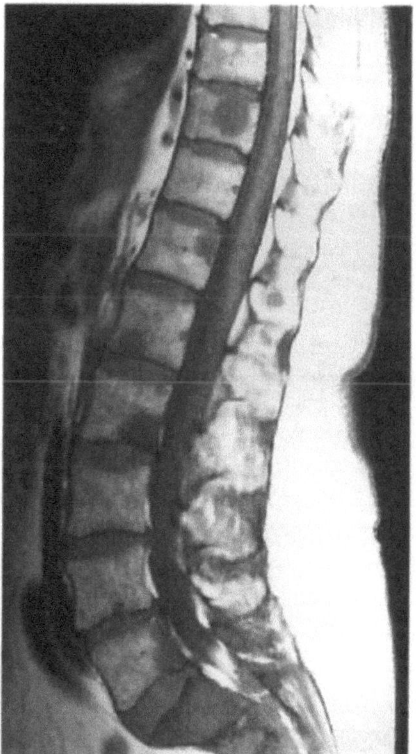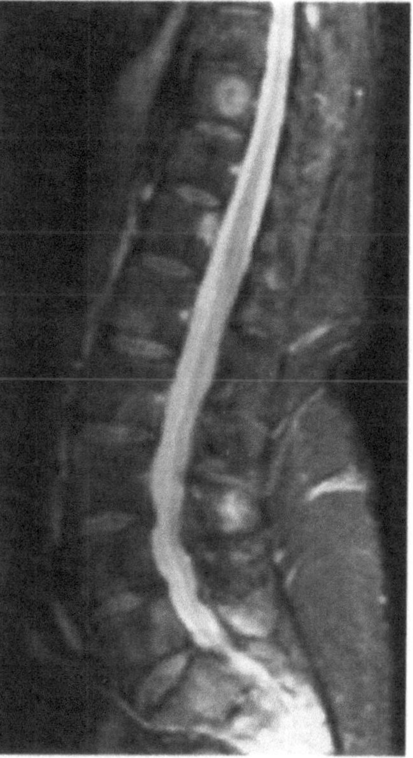

Fig. 15.21 a,b. Metastasis. **a** Multiple areas of low signal are demonstrated within the vertebral bodies on the T1-weighted sequence. **b** The STIR sequence shows the same areas as high signal surrounded by the suppressed vertebral yellow marrow

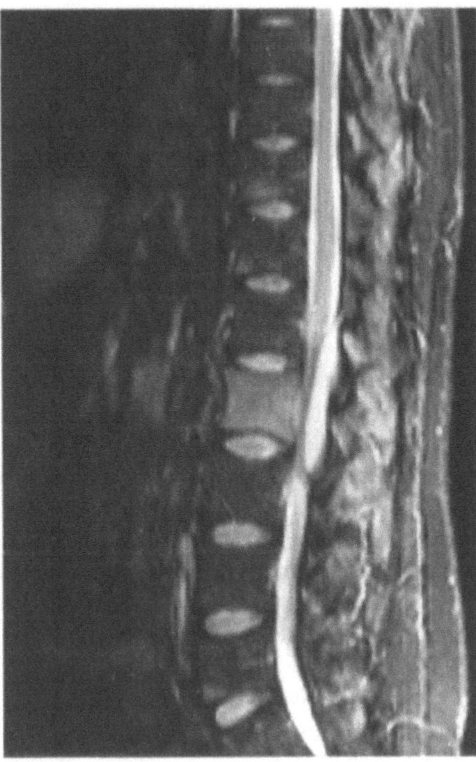

Fig. 15.22. Lymphoma. The STIR sagittal sequence shows increased signal throughout the L2 vertebral body. A high signal epidural mass is seen posteriorly, which was confirmed as lymphomatous tissue

Lesion conspicuity on MRI is decreased by the presence of red marrow in children and in adults with treatment-induced red marrow hyperplasia. Focal lesions of multiple myeloma and angioma may have a relatively high signal on T1 and be less conspicuous. Fat-saturated T2-weighted TSE or STIR sequences will demonstrate the lesions (MEIROWITZ et al. 1994) but differentiation between red marrow hyperplasia and marrow infiltration can occasionally be difficult, if not impossible (TANAKA et al. 1996).

The place of spinal MRI in routine workup of asymptomatic patients with neoplastic disorders is not yet clear and is likely to depend on treatment strategies. The specificity of MRI is also still not certain, with autopsy comparisons suggesting a degree of overestimation due to the presence of reactive marrow changes (PETREN-MALLMIN et al. 1992).

Isotope bone scan with correlative radiography is the present imaging method in the assessment of response to therapy but MRI may prove to be more valuable. The indication for MRI of the spine in prognosis and assessment of response to therapy remains to be fully evaluated. In multiple myeloma,

asymptomatic patients with stage 1 disease and marrow abnormalities on MRI have a shorter time lag before onset of aggressive disease than those with normal MRI studies. MRI does not provide additional information in the evaluation of response to treatment (MOULOPOULOS et al. 1995). Symptomatic patients with non-Hodgkin's lymphoma can have epidural lymphomatous involvement alone or in association with vertebral or paraspinal involvement, which may be the presenting feature or arise during the disease course (Fig. 15.22). MRI demonstrates the compression of the dural sac and may show intradural lesions and enhancement of the neoplastic tissue and nerve roots. Early detection of vertebral lesions causing spinal cord compression is vital as radiotherapy or corporectomy may avoid irretrievable neurological damage. MRI is the investigation of choice as it will demonstrate the state of the cord and also the extent of the soft tissue or bony mass. Computed tomography may also provide useful information, particularly about bone destruction, but the inferior soft tissue contrast and the inability to demonstrate the cord without intrathecal contrast make it a secondary choice modality.

Differentiation of osteoporotic vertebral fracture from fractures caused by multiple myeloma is difficult on conventional radiographs. MRI may allow that distinction but difficulties remain and the shape of the posterior vertebral cortex may be of value as it is more likely to be concave in osteoporosis and convex in malignant vertebral compression fractures. The shape of marrow changes on MRI in osteoporosis is more likely to be horizontal beneath the fractured vertebral end plate and to be delineated from normal surrounding bone marrow by a convex margin, whereas malignant marrow replacement is more extensive. Linear bands of isotope uptake, which are restricted to the extent of the vertebral margin, are seen in acute vertebral fractures in osteoporosis. Increased signal in vertebral marrow may also be seen on T2-weighted fat-suppressed sequences initially following radiation therapy, which is followed by an increase in signal intensity on T1-weighted images. The areas of signal change will correspond to the irradiated portals. Cord changes may occasionally occur.

References

Aprill C, Bogduk N (1992) High intensity zone: a diagnostic sign of painful lumbar disc on magnetic resonance imaging. Br J Radiol 65:361–369

Barnes PD, Brody JD, Jaramillo D, et al. (1993) Atypical idiopathic scoliosis MR imaging evaluation. Radiology 186:247–253

Bogduk N (1982) The clinical anatomy of the cervical dorsal rami. Spine 7:319–330

Bogduk N (1992) The sources of low back pain in the lumbar spine and back pain. In: Jayson MIV (ed) The lumbar spine and back pain, 4th edn. Churchill Livingstone, Edinburgh, p 83

Bogduk N, Wilson AS, Tynan W (1982) The human lumbar dorsal rami. J Anat 134:383–397

Bollow M, Braun J, Hamm B, et al. (1995) Early sacroiliitis in patients with spondyloarthropathy: evaluation with dynamic gadolinium enhanced MR imaging. Radiology 194:529–536

Castillo M, Malko JA, Hoffman JC Jr. (1990) The bright intervertebral disc: an indirect sign abnormal spinal bone marrow on T1 weighted MR images. Am J Neuradiol 11:23–26

Cervellini P, Curri D, Volpin L, et al. (1988) Computed tomography for epidural fibrosis after discectomy: a comparison between symptomatic and asymptomatic patients. Neurosurgery 23:710–713

Colhoun E, McCall IW, Williams W, et al. (1988) Provocative discography as a guide to planning operations on the spine. J Bone Joint Surg [Br] 70:267–271

Dina TS, Boden SD, Davis DO (1995) Lumbar spine after surgery for herniated disk: imaging findings in the early postoperative period. Am J Roentgeol 164:665–671

Dwyer A, Aprill C, Bogduk N (1990) Cervical zygapophyseal joint pain patterns. 1: A study in normal volunteers. Spine 15:453–457

Fairbank JCT, Park WM, McCall IW, O'Brien JP (1981) Apophyseal injection of local anaesthetic as a diagnostic aid in primary low back pain syndromes. Spine 6:598–605

Flanders AE, Schaefer DM, Doan HT (1990) Acute cervical spine trauma: correlation of MR imaging findings with degree of neurological deficit. Radiology 177:25–33

Fraser RD, Osti OL, Vernon-Roberts B (1989) Iatrogenic discitis: the role of intravenous antibiotics in prevention and treatment: an experimental study. Spine 14:1025–1031

Frederickson BE, Baker D, McHolick WJ, et al. (1984) The natural history of spondylolysis and spondylolisthesis. J Bone Joint Surg [Am] 66:699–707

Fyfe IS, Henry AP, Mulholland RC (1983) Closed vertebral biopsy. J Bone Joint Surg [Br] 65:140–143

Heitoff KB, Gundry CR, Burton CV, et al. (1994) Juvenile discogenic disease. Spine 14:335–340

Henson J, McCall IW, O'Brien JP (1987) Disc damage above a spondylolisthesis. Br J Radiol 60:69–72

Jackson RP, Jacobs RR, Montesano P (1988) Facet joint injections in low back pain: a prospective statistical study. Spine 13:966–971

Jackson RP, Cain JE, Jacobs RR, et al. (1988) The neuroradiographic diagnosis of lumbar herniated nucleus pulposus. A comparison of computed tomography (CT), myelography, CT myelography and magnetic resonance imaging. Spine 14:1362–1367

Johnson CE, Sze G (1990) Benign lumbar arachnoiditis: MR imaging with gadopentate dimeglumine. Am J Roentgenol 155:873–880

Johnson DW, Farnum GN, Latchaw RF, et al. (1988) MR imaging of the pars interarticularis. Am J Neuroradiol 9:1215–1220

Jonsson H Jr, Cesarini K, Sahlstedt B, Rauschning W (1994) Findings and outcome in whiplash-type neck distortions. Spine 19:2733–2743

Kang JD, Georgescu HI, MiIntyre-Larkin L, et al. (1996) Herniated lumbar intervertebral discs spontaneously produce matrix metalloproteinases, nitrous oxide, interleukin-6, and prostoglandin E2. Spine 21:271–277

Kobayashi S, Yoshizawa H, Hachiya Y, Ukai T, Morita T (1993) Vasogenic oedema induced by compression injury to the spinal nerve root. Spine 18:1410–1424

Kulkarni MV, McArdle CB, Kopanicky D, et al. (1987) Acute spinal cord injury: MR imaging at 1.5T. Radiology 164:837–843

Lane JI, Koeller KK, Atkinson JLD (1994) Enhanced lumbar nerve roots in the spine without prior surgery, radiculitis or radicular vein. Am J Neuroradiol 15:1317–1325

Lang P, Genant HK, Chafetz N, et al. (1988) Three dimensional computed tomography and multiplanar reformations in the assessment of pseudarthrosis in posterior lumbar fusion patients. Spine 13:69–75

Lord SM, Barnsley L, Wallis BJ, Bogduk N (1996) Chronic cervical zygapophyseal joint pain after whiplash. A placebo-controlled prevalence study. Spine 21:1737–1745

Lowe RW, Hayes TD, Kaye J, et al. (1976) Standing roentgenograms in spondylolisthesis. Clin Orthop 117:80–85

Magora A, Schwartz A (1976) Relation between low back pain syndrome and X-ray findings. Scand J Rehabil Med 8:115–125

Masaryk TJ, Boumphrey F, Modic MT, et al. (1986) Effects of chemonucleolysis demonstrated by MR imaging. J Comput Assist Tomogr 10:917–923

McCall IW, Park WM, O'Brien JP (1979) Induced pain referral from posterior lumbar aliments in normal subjects. Spine 4:441–446

McCall IW, Park WM, O'Brien JP, et al. (1985) Acute traumatic interosseous disc herniation. Spine 10:134–137

McCall IW, Colhoun E, Pullicino VC (1990) The facet joints in chronic low back pain. Poster presentation, International Society for Study of Lumbar Spine, Boston, USA

McCall IW, Cassar-Pullicino VN, Tyrrell PNM (1997) MR vertebral end plate changes and back pain. Proceedings of The International Society for the Study of the Lumbar Spine, Singapore

McRae DL, Standen J (1966) Roentgenologic findings in syringomyelia and hydromyelia. Am J Roentgenol 98:695–703

Meirowitz SA, Apicella P, Reinus WR, Hammerman AM (1994) Imaging of bone marrow lesions: relative conspicuousness on T1 weighted fat suppressed T2 weighted and STIR images. Am J Roentgenol 162:215–221

Modic MT, Steinberg PM, Ross JS, et al. (1988) Degenerative disk disease: assessment of changes in the vertebral body marrow with MR imaging. Radiology 166:193–199

Modic MT, Ross JS, Obuchowski NA, Browning KH, Cianflocca AJ, Mazanec DJ (1995) Contrast enhanced MR imaging in acute radiculopathy: a pilot study of the natural history. Radiology 195:429–435

Montaldi S, Frankhouser M, Schnyder P, et al. (1988) Computed tomography of the post-operative intervertebral disc and lumbar spinal canal. Neurosurgery 22:1014–1022

Moulopoulos LA, Dimopoulos MA, Smith JL, Weber DM, Delasalle KB, Libstritz HI, Alexanian R (1995) Prognostic significance of magnetic resonance imaging in patients with asymptomatic multiple myeloma. J Clin Oncol 13:251–256

Nachemson A (1989) Editorial comment: Lumbar discography – where are we today? Spine 12:555–557

Nagata K, Kiyonaga K, Ohashi T, Sagara M, Miyazaki S, Inoue A (1990) Clinical value of magnetic resonance imaging for cervical myelopathy. Spine 15:1088–1096

Nordstrom D, Santavirta S, Seitsalo S. et al. (1994) Symptomatic lumbar spondylolysis: neuroimmunologic studies. Spine 19:2752–2758

North American Spine Society Executive committee (1988) Position statement on discography. Spine 13:1343

Park WM, McCall IW, Benson D, et al. (1985) Spondyloarthrography: the demonstration of spondylolysis by apophyseal joint arthrography. Clin Radiol 36:427–430

Pennie BH, Agambar LJ (1990) Whiplash injuries. A trial of early management. J Bone Joint Surg [Br] 72:277–279

Petrin-Mallmin M, Nordstrom B, Andreasson I, Nyman R, Jonssen M (1992) MR imaging with histopathological correlation in vertebral metastases of breast cancer. Acta Radiol 33:213–220

Raby N, Mathews S (1993) Symptomatic spondylolysis: correlation of CT and SPECT with clinical outcome. Clin Radiol 48:97–99

Ross JS (1995) Three-dimensional magnetic resonance techniques for evaluating the cervical spine. Spine 20:1099–1102

Ross JS, Masaryk TJ, Modic MT (1990) MRI of the post operative spine: further assessment. Am J Neuroradiol 11:771–776

Sachs B, Vanharanta H, Spivey MA, et al. (1987) Dallas discogram description: a new classification of CT/discography in low back disorders. Spine 12:287–294

Schellhas KP, Pollei SR, Gundry CR et al. (1996) Lumbar disc high intensity zone: correlation of magnetic resonance imaging and discography. Spine 21:79–86

Schwarzer AC, Inang S, Laurent R, et al. (1992) The role of the zygapophyseal joint in chronic low back pain. Aust N Z J Med 22:185

Schwarzer AC, Wang S, O'Driscoll D, et al. (1995) The ability of computed tomography to identify a painful zygapophyseal joint in patients with chronic low back pain. Spine 20:907–912

Schweitzer ME, Levine C, Mitchell DG, Gannon FH, Gomella LC (1993) Bull's eyes and halo's useful MR discriminators of osseous metastases. Radiology 188:249–252

Silvermann CS, Lenchik L, Shimkin PM, et al. (1995) The value of MR in differentiating subligamentous from supraligamentous lumbar disk herniations. Am J Neuroradiol 16:571–579

Stabler A, Baur A, Bartl R, Munker R, Lamerz R, Reiser MF (1996) Contrast enhancement and quantitative signal analysis in MR imaging of multiple myeloma: assessment of focal and diffuse growth patterns in marrow correlated with biopsy and survival rates. Am J Roentgenol 167:1029–1036

Stoker DJ, Kissin CK (1985) Percutaneous vertebral biopsy: a review of 135 cases. Clin Radiol 36:569–577

Sze G, Vichan LS, Brant-Zawadzki M, et al. (1988) Chordomas: MR imaging. Radiology 166:187–191

Szpryt EP, Hardy JG, Hinton CE, et al. (1988) A comparison between magnetic resonance imaging and scintigraphic bone imaging in the diagnosis of disc space infection in an animal model. Spine 13:1043–1049

Takahashi M, Yamashita Y, Sakamoto Y, Kojima R (1989) Chronic cervical cord compression: clinical significance of increased signal intensity on MR images. Radiology 173:219–224

Tanaka O, Ichikawa T, Kobayashi Y, Matsuura K, Nagai J, Takagi S (1996) MR relaxation times in diffuse bone marrow disorders: evaluation of their clinical usefulness in differentiation between leukemia and anemia. Nippon Acta Radiol 56:539–545

Toyone T, Takahashi K, Kitahara M, et al. (1994) Vertebral bone marrow changes in degenerative lumbar disc disease: an MRI study of 74 patients with low back pain. J Bone J Surg [Br] 76:757–764

Tullous MW, Skerhut HEI, Storey JL, et al. (1990) Cauda equina syndrome of longstanding ankylosing spondylitis: case report and review of the literature. J Neurosurg 73:441–447

Tyrrell PM, Cassar-Pullicino VN, McCall IW (1997) The incidence and significance of gadolinium enhancement of symptomatic nerve roots in MRI of the lumbar spine. Eur Radiol (to be published)

Ullrich CG, Binet EF, Sanecki MG, et al. (1980) Quantitative assessment of the lumbar spinal canal by computed tomography. Radiology 134:137–143

Wada E, Ohmura M, Yonenobu K (1995) Intramedullary changes of the spinal cord in cervical spondylotic myelopathy. Spine 20:2226–2232

Webb JK, Broughton RBK, McSweeney T, et al. (1976) Hidden flexion injury of the cervical spine. J Bone J Surg [Br] 58:322–327

Wittram C, Whitehouse GH (1995) Normal variation in magnetic resonance imaging appearances of the sacroiliac joints: pitfalls in diagnosis of sacroiliitis. Clin Radiol 50:371–376

Yu S, Sether LA, Ho PSP, et al. (1988) Tears in the annulus fibrosus: correlation between MR and pathologic findings in cadavers. Am J Neuroradiol 9:367–370

16 Polyarthritis

I. WATT

CONTENTS

16.1
Introduction

The purpose of this chapter is to describe the place of clinical radiology in the diagnosis, assessment and management of polyarthritis. Whilst the commoner diseases will be mentioned, each will not be described in detail; for that reference to a standard textbook is recommended (SUTTON 1992; RESNICK 1997; DIEPPE and KLIPPEL 1993). Section 16.2 will concentrate on how to make a differential diagnosis of a polyarthropathy, and Sect. 16.3 addresses the role of further imaging techniques in the detection and assessment of polyarthropathy.

I. WATT, FRCP, FRCR, Consultant Clinical Radiologist, Department of Clinical Radiology, Bristol, Royal Infirmary, Bristol, BS2 8HW, UK

16.2
A Basic Approach to the Differential Diagnosis of Polyarthritis

Any joint must be seen as a whole organ, comprising capsule, synovium, cartilage, bone, enthesis and joint fluid; each component does not exist in isolation. At least three approaches to plain film diagnosis are possible:

1. *The "Aunt Minnie" method* ("I have seen this pattern before and the diagnosis turned out to be. . . .") works well, but requires care. Simply recognising a pattern stops thought about what is going on, and the difference between one arthropathy and another may not be clear-cut.

2. *The "target joint" approach.* For example, rheumatoid disease and osteoarthritis rarely involve the adult ankle joint, whereas haemophilia and haemochromatosis do. However, just because a particular joint rarely gets a given disease, it does not mean that it can be excluded, or vice versa.

3. *Consideration of what part of the joint is involved primarily*; this approach is strongly recommended. According to the site of the abnormality as seen on a plain film, three basic categories of disease may be distinguished:

- *Synovial diseases*, where the synovium directly contacts bone adjacent to hyaline cartilage (the "bare" area) (MARTEL et al. 1980). For example, this is where rheumatoid pannus erodes bone.
- *Cartilage diseases*: the articular surface is covered by hyaline cartilage and all those conditions whose primary effect is on cartilage and subchondral bone, including osteoarthritis (OA), septic arthritis and relapsing polychondritis, fall into, this group.
- *Enthesis diseases*: occur primarily at an enthesis (defined as those sites where capsule, ligament or tendon is inserted into bone). Classical examples include ankylosing spondylitis, where enthesis erosion is the hallmark, and Forestier's disease (or DISH, diffuse idiopathic skeletal

hyperostosis) (RESNICK et al. 1975), where bone proliferation occurs without erosion.

At this point a few words of warning are appropriate. It is much easier to make a diagnosis early in disease evolution, when most radiological signs are at a similar stage. In advanced disease it is best to select which of the three sites is predominantly affected. In end-stage disease it may not be possible to be didactic; for example, little difference may be seen between burnt-out rheumatoid disease and psoriasis. It may be thought of this way: some joints end up being atrophic, others hypertrophic (see Sect. 16.2.2.4). Conversely, one should avoid trying to make a diagnosis too early in disease evolution. Only about 30% of patients with active rheumatoid disease show erosion of a joint at 1 year, so the chances of showing erosion in an individual patient at 3 months is slight (see below). Also, word pictures to describe diagnostic features should be avoided. The "cup and pencil" deformity of psoriasis is not totally diagnostic. Finally, having one arthropathy does not protect against another. Hence an elderly patient with OA may develop rheumatoid disease.

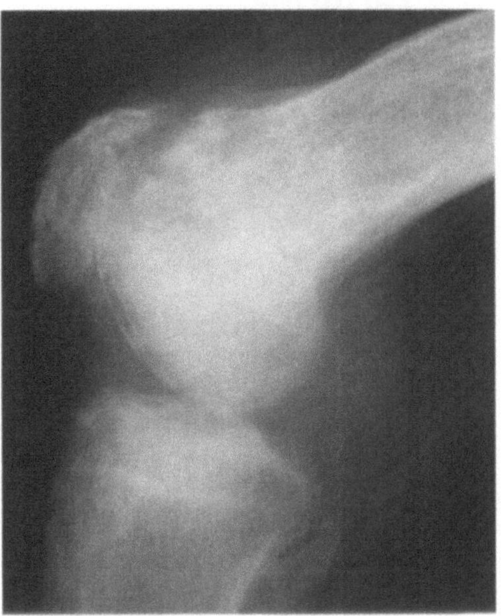

Fig. 16.1. Infectious arthritis. A lateral view of the knee demonstrates considerable synovial thickening, the outline of which is very ill defined. Also obvious acute bone loss is shown with cortical destruction, consistent with an infected joint

16.2.1
Synovial Diseases

Synovial disease is accompanied by soft tissue swelling of the joint and thus a good-quality radiograph is needed to see the soft tissue planes. The cause of the swelling may be thickened synovium, a joint effusion or both. If the swelling is ill-defined, an inflammatory cause should be considered (Fig. 16.1). To fully distinguish between synovial thickening and a joint effusion, intra-articular contrast medium, ultrasound or an MRI scan is required.

16.2.1.1
Opaque Synovium

Soft tissue swelling that is unusually dense or radiopaque suggests a high atomic number substance has been deposited. The possibilities are:

1. Iron from repeated intra-articular haemorrhage, as in haemophilia, Christmas disease or synovial haemangioma (Fig. 16.2).
2. Calcium. If the calcification has a structure, especially rings of calcification, synovial chondromatosis should be considered. Coarse, amorphous calcification suggests the deposition

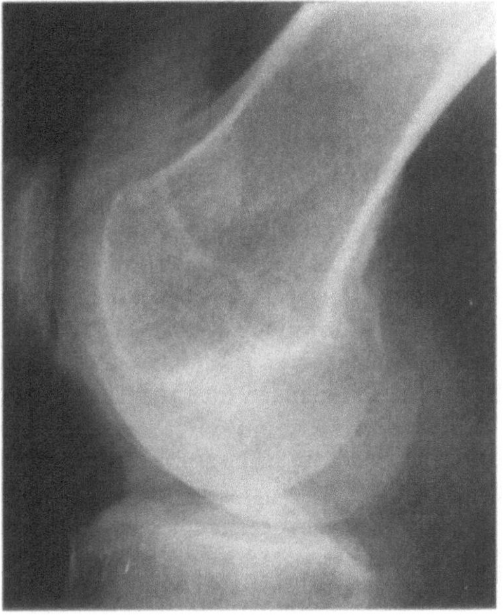

Fig. 16.2. Opaque synovium. A lateral view of the knee demonstrates radio-opacity in the suprapatellar pouch in a patient with haemophilic arthropathy

of calcium pyrophosphate dihydrate crystals (CPPD).

3. Iatrogenic causes, including previous intra-articular contrast medium, lead debris (from bullet fragments for example) and disintegrating

barium-impregnated cement or metal from prosthetic joint failure.

16.2.1.2
Symmetrical or Asymmetrical Soft Tissue Swelling

Symmetrical soft tissue swelling occurs in many causes of synovitis and is thus not specific. The more indistinct the outline, the more actively inflammatory is the cause. However, asymmetrical soft tissue swelling is a feature of the depositional states and occurs with other synovial "mass" lesions. Causes include rheumatoid nodules, gouty tophi, xanthomata and amyloid deposition (Fig. 16.3). If only one or two joints are involved, more localised synovial mass lesions such as synovial chondromatosis and pigmented villonodular synovitis (PVNS) should be considered.

16.2.1.3
Is Bone Erosion Present at the Bare Areas?

Radiologically an erosion is diagnosed on an x-ray when the articular cortex on one or either side of a joint cannot be seen. Thus the "white line" of the cortex is disrupted with trabeculae appearing to be uncovered, producing a "hair on end" or "paintbrush" appearance (Fig. 16.4). This sign should be sought initially where the synovium directly contacts cortex without overlying hyaline cartilage, the "bare" area. Thus, synovial erosion may seem to occur well away from the main articular surfaces, dependent upon local anatomy.

Erosions take time to develop. In infective arthritis they occur rapidly, but in rheumatoid disease only about 30% of patients will develop erosions in the first year of their disease, and 60% by their second anniversary (BROOK and CORBETT 1977). Erosions should not be confused with normal anatomical markings or intense focal osteopenia where the "white line" of the cortex remains intact.

16.2.1.4
Well-Defined or Ill-Defined Erosions?

A major difference in significance is determined by the margin of the erosions. When well defined, by a line of cortical bone or sclerosis, an inert lesion is likely, either slow growing or healed. Ill definition indicates that the disease process is still active.

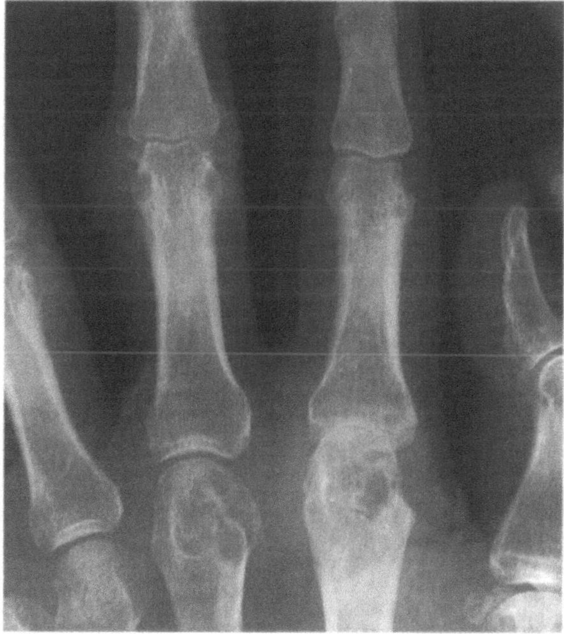

Fig. 16.3. Asymmetric soft tissue swelling. In this case a mass lesion is demonstrated adjacent to the proximal interphalangeal joint of the middle finger. Note the well-defined erosions of the metacarpal heads. This patient has rheumatoid arthritis with nodules involving the hands. Gout would be a reasonable differential diagnosis

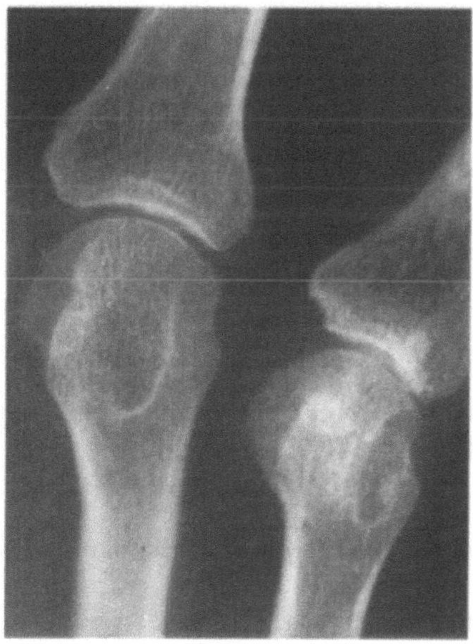

Fig. 16.4. Rheumatoid arthritis, early erosions. Localised views of the middle, ring and little finger metacarpophalangeal joints demonstrate evidence of cortical loss at the bare area of the ring finger metacarpal head particularly

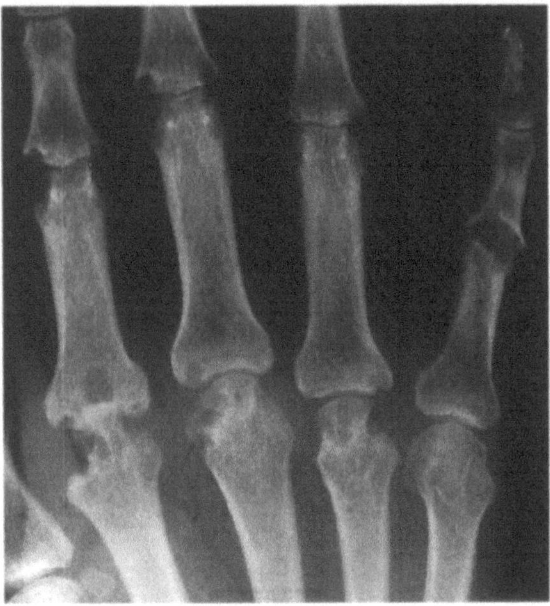

Fig. 16.5. Well-defined erosions. In this patient with inactive rheumatoid arthritis well-defined synovial erosions are shown at several joints without significant soft tissue swelling, implying inactive disease

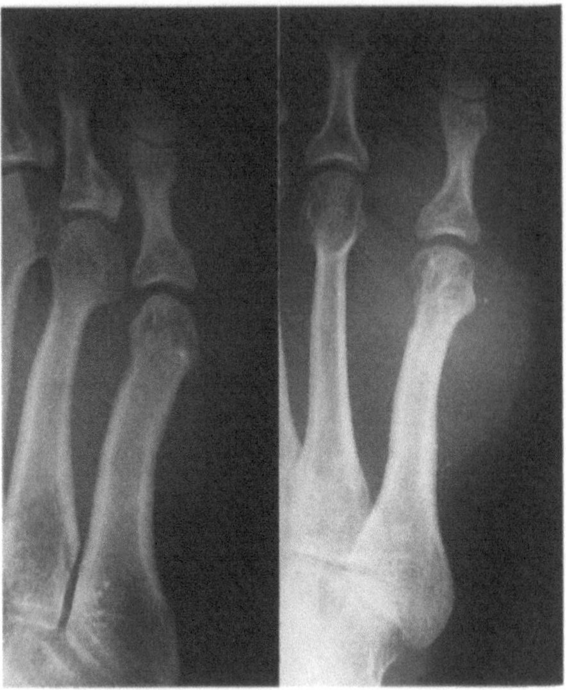

Fig. 16.6. Pigmented villonodular synovitis. A synovial mass lesion involves the little toe metatarsophalangeal joint. Note the slight radio-opacity of the soft tissue swelling on the AP view, well-defined marginal erosions on both views and an apparently wide joint space width

When dealing with well-defined erosions, consideration should be given to:

1. Old, inactive rheumatoid disease (Fig. 16.5)
2. Deposits in the synovium, including gouty tophi, xanthomatous deposits (either focal or generalised, as in multicentric reticulohistiocytosis), "pressure defects" (grossly thickened capsule in lupus erythematosus) and local synovial disease such as chondromatosis or PVNS (Fig. 16.6)

In practice the only difficult pair to separate is gouty arthritis and "robust" rheumatoid arthritis, in which bone density is often preserved, the erosions are relatively scanty and focal synovial masses may suggest tophi.

16.2.1.5
Ill-Defined Erosions
With or Without New Bone Formation?

The presence of new bone formation in association with active, ill-defined erosions will permit further differential diagnosis. Erosions in adult rheumatoid disease are not accompanied by new bone formation whereas they are in other seronegative arthritides, in particular psoriasis and Reiter's syndrome (Fig. 16.7). The latter group are very similar radiologi-

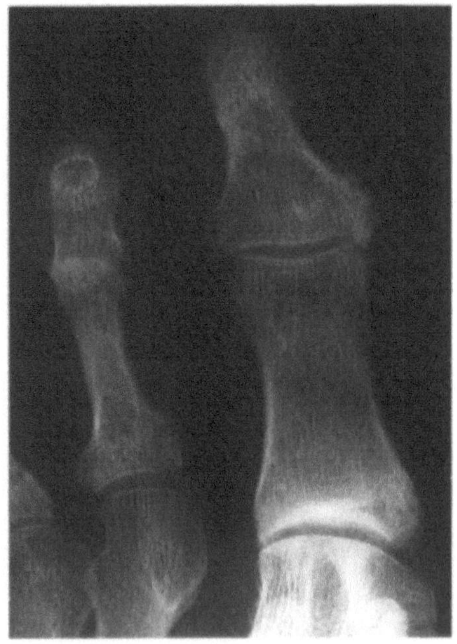

Fig. 16.7. Psoriatic arthropathy exhibiting proliferative bone erosion

cally. However, hand disease in a middle-aged woman suggests psoriasis, whereas foot disease in a younger man is more likely to be Reiter's syndrome. Another useful finding in such cases is medullary sclerosis, sometimes known in the digits as "ivory phalanges".

To summarise, rheumatoid disease is a non-bone-forming, atrophic arthritis, whereas psoriasis is the opposite, proliferative, bone forming and sclerotic.

16.2.2
Cartilage Diseases

Cartilage cannot be assessed fully on plain film. Indirect "cartilage thickness" or "joint-space width" is gauged by reference to the distance between two bony articular cortices, which may not represent true cartilage thickness. Modifications of technique have been employed, for example, weight-bearing, slightly flexed films in the case of the knee; however, the only other joint that may need to be assessed weight-bearing is the hip (CONROZIER et al. 1997). True cartilage thickness may be measured if contrast medium is introduced into the joint, or non-invasively by magnetic resonance imaging (MRI).

How may cartilage respond to disease? Basically it may thicken, thin or become calcified.

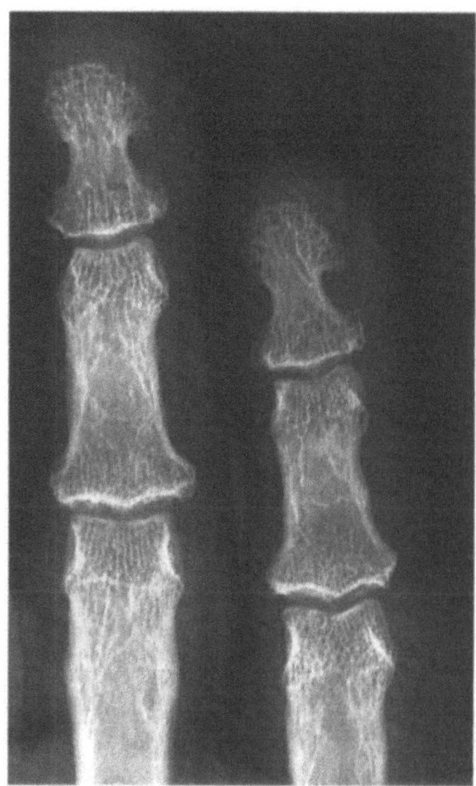

Fig. 16.8. Increased joint space thickness. Here, in a patient with acromegaly, the joint space of the interphalangeal joints of the ring and little fingers is disproportionately wide. Overgrowth of bone has also occurred with marked tufting of the terminal phalanges

16.2.2.1
Is Cartilage Thickness Increased?

An apparently wide joint-space width is said to occur transiently with early cartilage degeneration and small joint effusions. However, this is not a reliable radiological plain film sign, although it may be shown by ultrasound or MRI. Generally thick cartilage occurs in acromegaly [when it may the earliest diagnostic sign (Fig. 16.8)], cretinism and hypothyroidism. In some erosive arthropathies, despite marked erosion, joint space width may appear relatively wide as in gouty arthritis, PVNS and multicentric reticulohistiocytosis.

16.2.2.2
Is Cartilage Thin?

Thin cartilage is a very non-specific sign and occurs in many disorders which may be thought to "poison" cartilage. Causes include infectious arthritis (especially if concentric thinning occurs quickly with articular cortical loss; see Fig. 16.1). Other non-infective causes are inflammatory arthritis (e.g. juvenile chronic arthritis or rheumatoid disease), cartilage lysis associated with a metal hemiarthroplasty and diseases of unknown aetiology such as relapsing polychondritis (Fig. 16.9) (BOOTH et al. 1989). However, the most usual cause of cartilage disease is OA and its variants (see Sect. 16.2.2.4).

16.2.2.3
Is Hyaline Cartilage Calcified?

Opaque calcified cartilage (chondrocalcinosis) is usually due to either CPPD or basic calcium phosphate (BCP or calcium hydroxyapatite) deposition. In most cases it is CPPD. CPPD deposition is an age-related phenomenon, not necessarily associated with symptoms or arthritis. It may be associated with episodes of crystal shedding ("pseudogout"), many features of which clinically resemble septic arthritis. In this case crystals are released as the result of shedding of the superficial layers of hyaline cartilage itself (Fig. 16.10). Typically pseudogout occurs in older patients who have been stressed (such as following hospitalisation or after diuretic therapy). Chondrocalcinosis is associated with other metabolic disorders, especially hyperparathyroidism, haemochromatosis or abnormalities of magnesium metabolism. Again, it is not necessarily associated with either symptoms or a progressive destructive arthropathy. Chondrocalcinosis occurs in those joints which also have fibrocartilages in them, typically the knee and wrist. The reason for this is unknown.

Calcification due to BCP is very rare and is associated with collagen vascular diseases such as scleroderma.

16.2.2.4
What Is Happening in Subchondral Bone?

It is insufficient to diagnose a cartilage disease. Changes in subchondral bone can give further differential diagnostic information. Usually hyaline cartilage disease is part of OA in which the associated bone changes show a spectrum from a hypertrophic response, to the skeleton which appears unable to respond to a joint insult (WATT 1994). Hypertrophic bone formation is seen in conjunction with CPPD deposition (Fig. 16.11) (DIEPPE et al. 1982). The atrophic joint may be associated with BCP deposition (Fig. 16.12) (DIEPPE et al. 1984). Table 16.1 summarises the main features of the three crystals associated with the bulk of disease in which crystalline material is implicated.

The exact role of crystals in the pathogenesis of arthritis is debatable. Originally it was suggested that a distinctive form of hypertrophic OA occurred in some patients suffering from the pseudogout

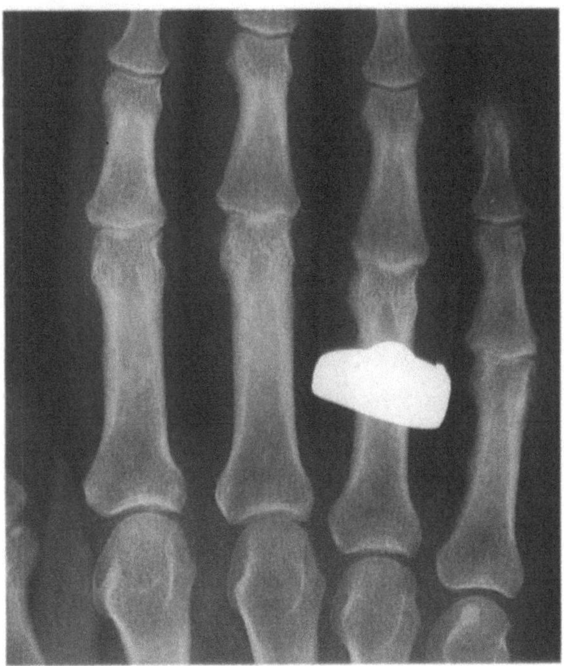

Fig. 16.9. Relapsing polychondritis. In this example, all of the proximal interphalangeal joints show complete loss of hyaline cartilage thickness, but without any other evidence of arthropathy

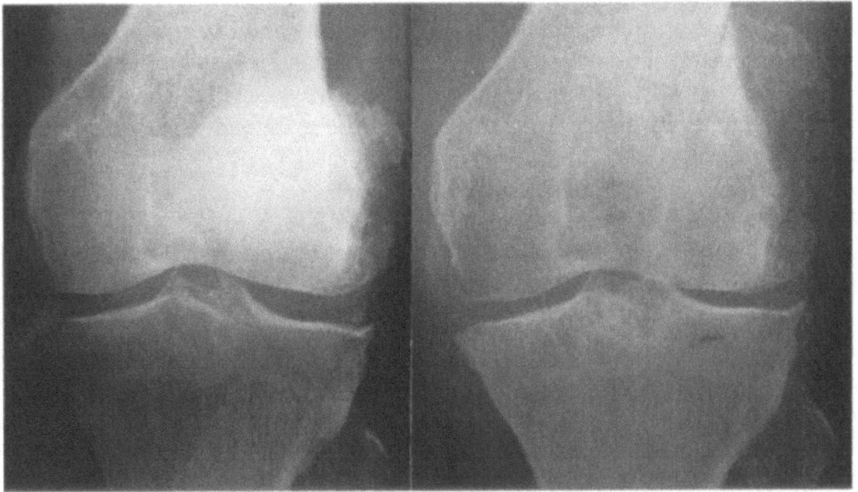

Fig. 16.10. Calcium pyrophosphate dihydrate crystal shedding. This middle-aged man had well-marked chondrocalcinosis (*left*), but following several episodes of acute knee pain the crystals are no longer present (*right*). Note also, however, that hyaline cartilage thickness has been reduced due to associated shedding of superficial cartilage layers

syndrome (MARTEL et al. 1970). Later it was suggested that this was a distinctive arthropathy, and named pyrophosphate arthropathy (RESNICK et al. 1977). Similarly BCP was noted in some patients with a destructive, atrophic OA of the shoulder (HALVERSON et al. 1984). Initially such arthritis was thought to be due to a cocktail of proteolytic enzymes, but later this was shown not to be the case (CAMPION et al. 1988). What then is the interplay between crystals and arthritis?

1. Crystals may cause arthritis; however, evidence suggests that this is frequently not the case.

2. Crystals may form as the result of joint degeneration. Old trauma or operative meniscectomy may provoke crystal deposition (DOHERTY et al. 1982). It is tempting to link OA and CPPD deposition, but not all patients with OA have CPPD and vice versa.

3. Crystals and joint damage result in a vicious circle, one provoking the other.

4. Crystals and arthritis are both caused by other factors. Currently this is the most favoured explanation. Hence hypertrophic OA may be associated with CPPD crystal deposition, but this is not always the case. Similarly, atrophic OA is frequently associated with the deposition of BCP in joint fluid. Thus the spectrum of OA appearances may reflect more generalised bone-regulating factors than just local joint damage. Crystals are thought to be a disease marker, not the cause or the effect. Only a single exception to this general rule occurs and that is

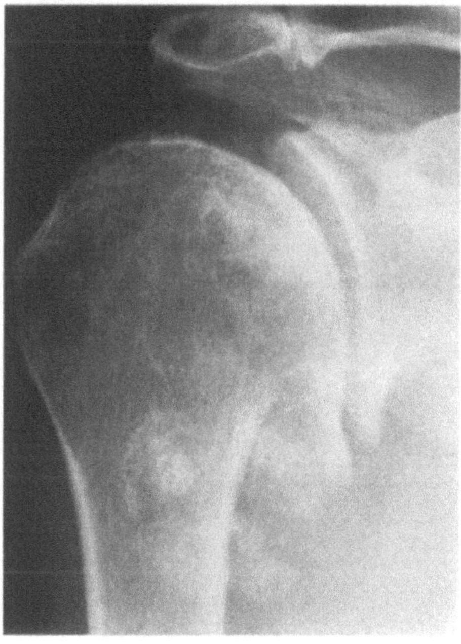

Fig. 16.11. Calcium pyrophosphate dihydrate-associated osteoarthritis. In this example hypertrophic osteoarthritis of the shoulder is demonstrated with pronounced osteophytosis and multiple separate osteochondral bodies. The shoulder is rarely primarily involved with osteoarthritis, another feature of this disease association

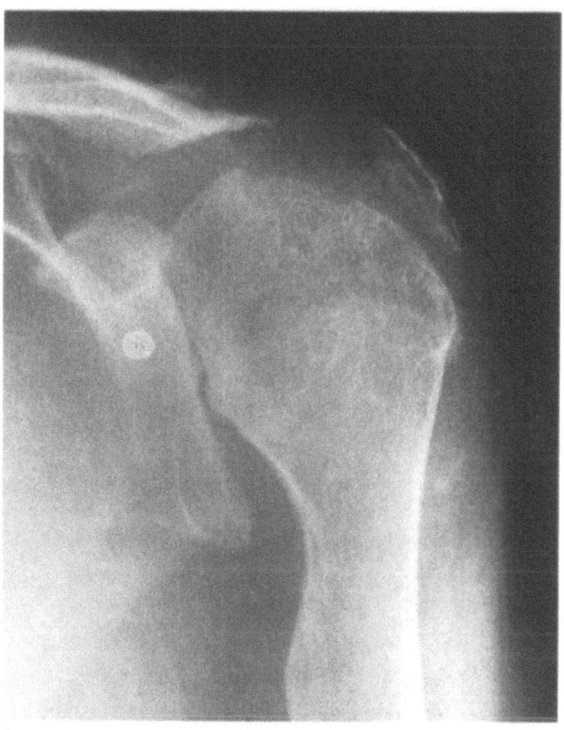

Fig. 16.12. Atrophic destructive osteoarthritis of the shoulder. Note the considerable attrition of bone in the absence of secondary bone response in this elderly lady

Table 16.1. Major features of the crystal-related arthritides

Crystal	Site	Distribution	Acute disease	Chronic form
Sodium biurate (BU)	Articular and periarticular	Peripheral (hands and feet)	Acute gout (feet)	Tophaceous deposits
Calcium pyrophosphate (CPPD)	Mainly articular	Intermediate (knees and wrists)	Pseudogout	Hypertrophic OA
Basic calcium phosphate (BCP)	Mainly periarticular	Central (shoulders and hips)	Acute periarthritis	Atrophic OA

diabetic osteoarthropathy. Here, the hypertrophic form of Charçot joint seems to combine bone formation and destruction.

5. A number of conditions exist in which multiple subchondral "cysts" occur. Apart from pyrophosphate-associated OA, marked hyperparathyroidism and haemochromatosis exhibit multiple cysts. Iron and calcium chemistry are closely related. Hence, the incidence of CPPD deposition is higher in these conditions. Features that distinguish haemochromatosis from CPPD deposition in the presence of multiple small subchondral "cysts" (AXFORD et al. 1991) are the absence of associated features of hypertrophic OA and involvement of joints that do not usually suffer idiopathic OA, such as the ankle.

Another subset of OA that sometimes causes confusion is erosive OA (EOA). This purely radiological subset of OA has no specific clinical or laboratory markers (COBBY et al. 1990) and is confined largely to the interphalangeal joints of the hands. Distinction between EOA and psoriatic arthropathy may cause difficulty. However, distinction is straightfor-

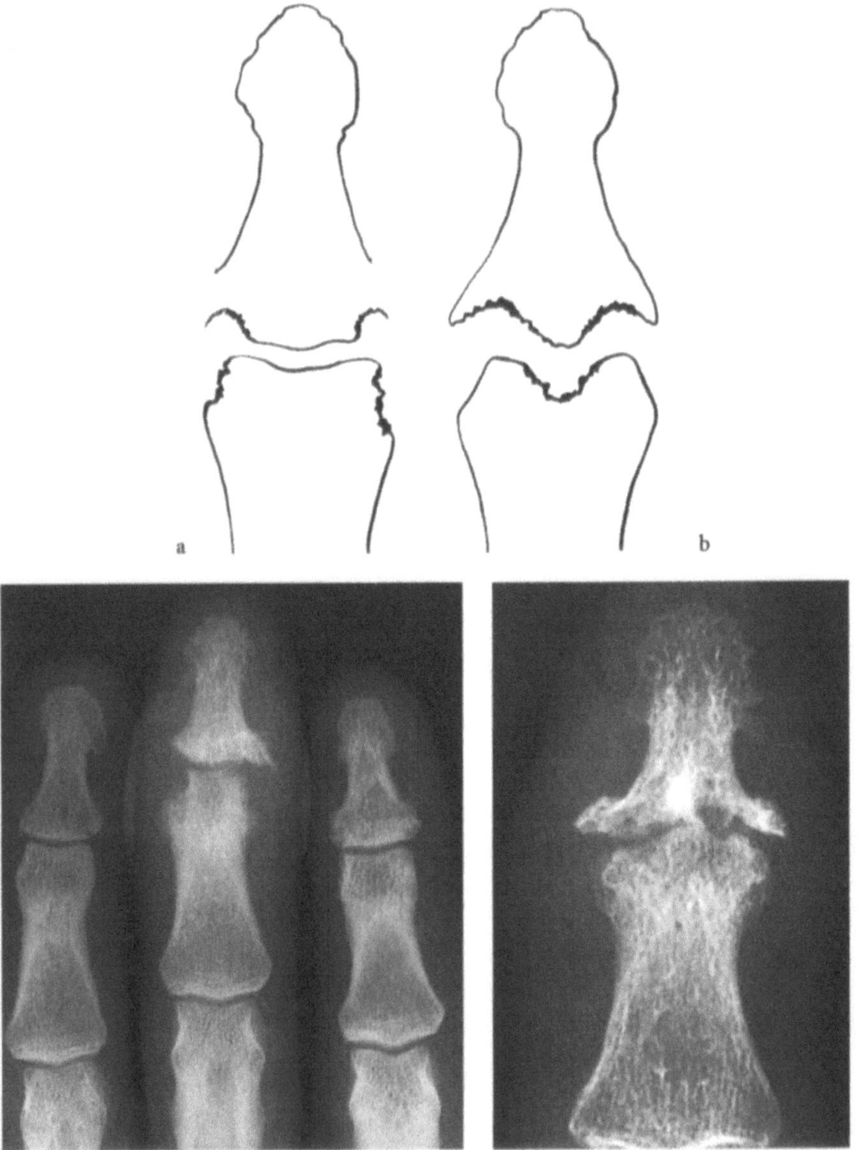

Fig. 16.13 a–d. Distinction between psoriatic arthritis and erosive arthritis is made easy by the work of Dr. WILLIAM MARTEL (from MARTEL et al. 1980). **a** A line drawing demonstrates the site of erosion in a patient with psoriasis where the synovium invades the bare area and is associated with new bone formation. **b** In erosive OA predominant involvement is of the articular surface. Actual examples demonstrate this: **c** psoriatic arthritis and **d** erosive OA

ward as EOA is an articular surface disease whereas psoriasis it is a proliferative erosive lesion of the "bare area" (Fig. 16.13) (MARTEL et al. 1970).

16.2.3
Enthesis Diseases

Enthesis diseases may be purely local, such as rotator-cuff disease or tennis elbow or systemic, as in the case of ankylosing spondylitis. Local lesions occur most frequently at the shoulder (rotator cuff, frozen shoulder or calcific periarthritis), hip (trochanteric bursitis) and elbow ("tennis" elbow). Generalised lesions most frequently affect the insertion of the Achilles tendon, around the pelvis and spinal ligamentous attachments.

16.2.3.1
Is the Enthesis Disease Erosive?

Enthesis erosion characterises a group of disorders including ankylosing spondylitis, Reiter's syndrome, psoriatic spondylitis and Behçet's syndrome (Fig. 16.14). Radiologically the enthesis changes associated with inflammatory bowel diseases such as ulcerative colitis, Crohn's disease and Whipple's disease are indistinguishable from ankylosing spondylitis.

Enthesis disease without erosion is exemplified by Forestier's disease (DISH, diffuse idiopathic skeletal hyperostosis), although many other causes of generalised enthesis ossification are recognised, including fluorosis, hypophosphataemic rickets and other metabolic disorders such as gout and obesity (Fig. 16.15).

Typical sites of enthesis disease include the heel (at the insertion of the Achilles tendon or plantar fascia), the spine (particularly the insertion of the anterior longitudinal ligament and the outer fibres of the annulus fibrosus) and around the pelvis or scapula at muscle origins or attachments.

16.2.3.2
Distinction Between Ankylosing Spondylitis and the Other Erosive Enthesis Diseases

Distinction between causes of the erosive group can be very difficult if not impossible! Discrimination on the basis of the peripheral lesions is not possible, but examination of the sacroiliac joints may help (Fig. 16.16). Essentially sacroiliac joint disease falls into three broad groups:

1. Unilateral sacroiliitis, which strongly suggests infection or old trauma.

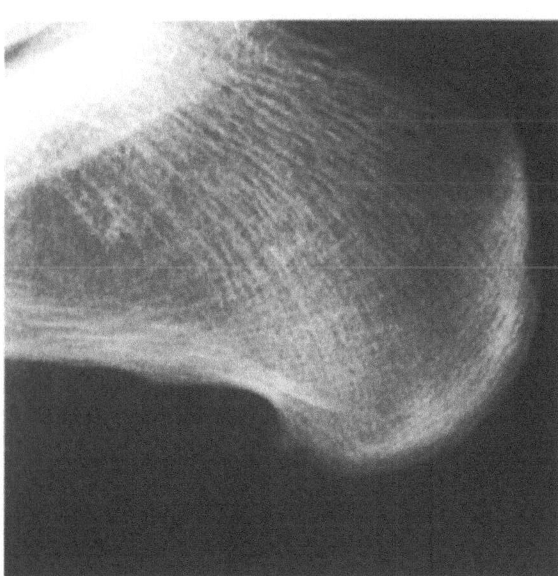

Fig. 16.14. Erosive enthesis disease. A localised view of the os calcis demonstrates ill-defined erosion at the insertion of the plantar fascia with some new bone formation in this patient with ankylosing spondylitis

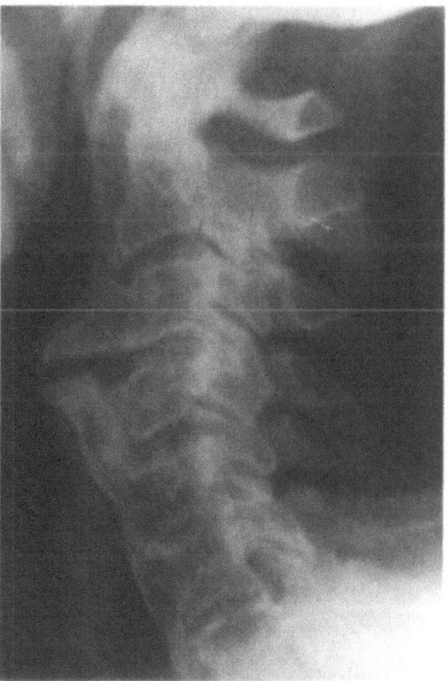

Fig. 16.15. Non-erosive enthesis disease. A lateral cervical spine in a patient with diffuse idiopathic skeletal hyperostosis (DISH) shows florid new bone formation without erosion

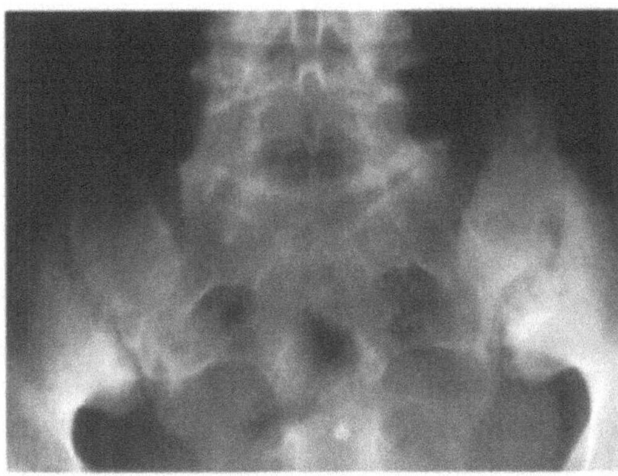
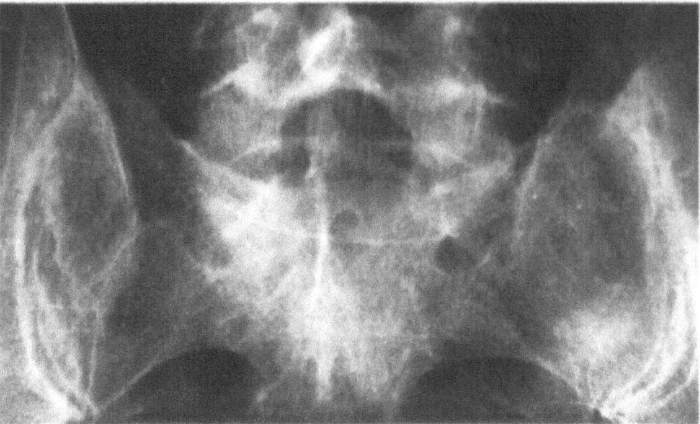

a

b

Fig. 16.16 a,b. Differential diagnosis of sacroiliitis. **a** Typical findings of ankylosing spondylitis, with marked symmetrical erosive disease involving both sacroiliac joints and both com- ponents of the sacroiliac joints. **b** In Behçet's syndrome there is also erosive sacroiliitis, but this is asymmetrical

2. Symmetrical bilateral sacroiliitis, involving both joints and both compartments of the joints, the upper fibrous articulation as well as the lower diarthrodial part. Symmetrical bilateral sacro- iliitis occurs in ankylosing spondylitis (with or without inflammatory bowel disease). Also, severe hyperparathyroidism causes diffuse subchondral erosion.
3. Asymmetrical bilateral disease, which is more typical of psoriasis or Reiter's syndrome. Fre- quently the peripheral lesions will have rather more new bone formation than is the case with ankylosing spondylitis.

Technically OA does not involve the sacroiliac joints as they are not synovial joints. However, "degenerative" changes do occur, and are shown by involvement of the anteroinferior aspect of the joints only. Anterior bridging new bone is also consistent with Forestier's disease or DISH.

16.2.4
Summary of Differential Diagnosis

This section has mentioned only the more common causes of arthritis, but they have been used to illus- trate a simple means of differential diagnosis by using a diagnostic tree. A step-by-step differential diagnosis of arthritis may be summarised as follows:

1. Which part of the joint is involved? Is it synovium, cartilage or bone?
2. If synovium, is joint swelling symmetrical or not? Is the synovium opaque?
3. If synovitis is present, is it erosive?
4. Are the erosions well defined or not?
5. Is new bone formation associated with the erosions?
6. If it is a cartilage disease, is it thick, thin or calcified?
7. What secondary bone changes are there?
8. Is it an enthesopathy?
9. Is it erosive?
10. Are the sacroiliac joints symmetrically involved?

The purpose of this means of differential diagno- sis is to look at pathological processes in a logical fashion. It moves away from pattern recognition, which sees a joint as an entity, towards analysing changes that are occurring within the various tissues of a whole joint organ. This enables understanding of the various manifestations of the diseases them- selves, and acceptance of their variability and overlap. For example, some diseases may involve more than one of the three main tissue sites laid out above. Thus psoriasis may be a disease of both the synovial joints and the enthesis. However, the manifestations of involvement are similar in both tissues. What are the common trigger factors dictat- ing that involvement of more than one tissue site should happen in some patients, but not others? Even broader "inflammatory" overlaps occur. A good example is pustular arthrosteitis (KASPERCZYK and FREYSCHMIDT 1994), or the SAPHO syndrome (KAHN and KAHN 1994). SAPHO is an abbreviation for synovitis, acne, pustulosis, hyperostosis and osteitis. The combination of skin disease (includ- ing psoriasis vulgaris), chronic osteomyelitis and sclerosing bone lesions is diagnostic. Bony changes include asymmetrical sacroiliitis and spinal changes, typical of psoriatic spondylitis, with bizarre involvement of the ribs and sternum known as sternocostoclavicular hyperostosis (KÖHLER et al. 1977).

16.3
The Role of Further Radiological Investigation in Polyarthritis

What purpose do further investigations serve? Are they "just for interest" or a substitute for good clinical skills? Three broad principles should dictate the use of all imaging, for radiologists are clinicians too, not medical photographers!

1. What is the objective, or purpose of the investigation? Is there one?
2. What management decision, if any, hangs on the result of the investigation?
3. Will action will result from the investigation? Will it be of benefit to the patient?

Unless such simple questions can be answered, radiological investigation should not be performed. Before considering the advantages and limitations of each imaging modality, the objectives of investigation need to be restated.

16.3.1
Objectives of Radiological Investigation

With the clinical principles in mind the value of further radiological investigation can be threefold –

1. To establish the diagnosis. In order to investigate a patient properly a sound differential diagnosis is necessary, in order of probability. Naturally this requires that the patient be examined and a good history be taken by a competent clinician. Radiology is not a substitute for clinical acumen or thought! Then, an appropriate radiological investigation should be requested. If the request form is unclear, then the radiologist must decide.

Consider that an adult patient presents with a pain and swelling of the knee for 3 days. Plain film findings are unlikely to identify the causes, save perhaps chondrocalcinosis associated with pseudogout. Joint aspiration is much more likely to be of use if a joint effusion is present. Were symptoms to be measured in weeks then an x-ray may be helpful. Could it be a joint rupture? Ultrasound is a very good means of showing a popliteal cyst, muscle oedema, and patent leg veins. A sudden onset in a middle-aged patient may suggest idiopathic medial femoral condyle necrosis, when a skeletal scintigram or MRI scan would antedate plain film changes by weeks.

Alternatively, in a patient with polyarthritis, radiographs of all the abnormal joints are not needed. Why? Because the diagnosis of typical rheumatoid disease seen in one site is not altered by seeing it at many others. The objective is to make the diagnosis. Hence, images of the hands and the feet (one view of each) will cover most of the polyarticular disorders. Remember that only 30% of rheumatoid patients show erosion in the first year of their disease. However, if early diagnosis is needed (perhaps to confirm the patient actually has an arthropathy), skeletal scintigraphy may be very sensitive to disease.

2. To monitor disease. How often, if at all, a patient is re-examined will depend on clinical circumstance, and the arthropathy from which the patient suffers. However, with a generalised arthropathy, simple x-rays of the hands and feet usually are sufficient to indicate disease status. Joint surveys are not only expensive, arguably valueless and the source of a high radiation burden but also most unpleasant for the patient in pain. Obviously, images of a particularly problematic joint are relevant. Again, consider the purpose of the requested investigation. For example most scoring systems used to assess disease activity [e.g. the Larsen Index (LARSEN et al. 1977)] are usually based on x-rays of the hands. Why take more?

3. To detect complications. Again, the whole spectrum of the radiological imaging orchestra may be needed. To quote a great clinical radiologist (John Roylance, Bristol, UK), the answer to every radiological problem is another film. The skill lies in knowing which film, and when! Selection of the shortest route to answer the clinical question is clearly a matter of experience and an understanding of the benefits and limitations of the imaging modalities. Integration of the modalities is vital, but for brevity the major uses of the various imaging modalities are reviewed below.

16.3.2
Plain Radiography

Although no longer considered exciting in the face of MRI, or whatever, plain films remain the "gold standard" against which other modalities should be measured. Soft tissue changes are shown, although limited to those joints with suitable fat planes adjacent. Bony changes take weeks to show, but spatial resolution is better than with other imaging techniques. Plain films are readily available, cheap (relatively) and carry a low radiation dose. They should be used for:

1. Initial diagnosis: often they are sufficient to diagnose and monitor disease.
2. Pre-operative assessment (with suitable tailoring), e.g. to show the load line or to confirm joint instability. Such films may confirm an element of skeletal dysplasia underlying OA, as at the knee (COOKE et al. 1997).
3. Follow-up of disease, including after surgical intervention.
4. Detection of complications; for example, a single lateral film of the cervical spine in flexion is enough to confirm the presence or absence of atlantoaxial instability in rheumatoid disease.
5. Delineation of synovial masses, joint rupture, venous thrombosis and vasculopathy in polyarteritis nodosa or lupus erythematosus following contrast medium injections.

Fine focus, microfocal techniques are receiving attention once more. Fine detailed examination of lesions (Fig. 16.17) may be helpful in the assessment of patients on treatment regimens (BUCKLAND-WRIGHT 1984). Further, as the fine detail of trabecular pattern can be seen, various methods of texture analysis can be applied, including fractal signature analysis. Differences may be measured between normal bone, juxta-articular osteoporosis and rheumatoid erosion. Similarly, variations of trabecular organisation can be seen in osteoarthritis (LYNCH et al. 1991).

16.3.3
Ultrasound

Ultrasound is relatively cheap, hazard free, widely available and undervalued. However, operator dependence can limit the clinical usefulness. Nonetheless, constantly improving technology, including colour flow and power Doppler, makes this a very valuable adjunct to diagnosis and disease assessment, e.g.:

1. For the demonstration of ganglia, joint cysts, and other transonic masses, and hence first-line investigation of whether soft tissue masses are "cystic" or "solid".
2. To distinguish synovium from joint effusion and to assess masses, both in and adjacent to joints.

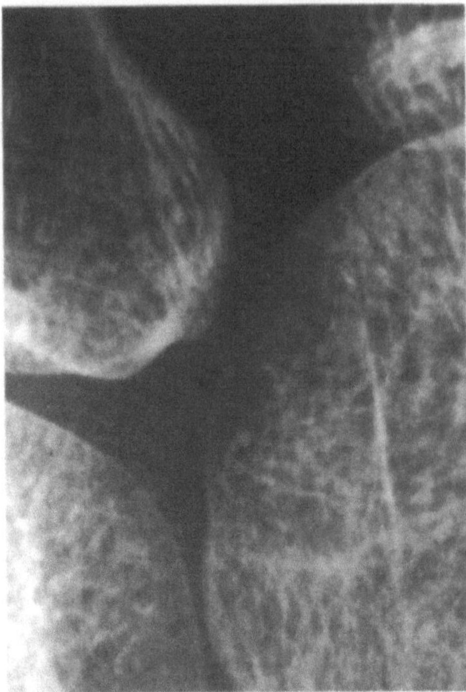

Fig. 16.17. Microfocal radiography. An example of high-resolution radiography demonstrates early erosion of a metacarpal head, not visible on a standard radiograph. (Courtesy of Dr. ULF MEYER, VIENNA)

3. For the differential diagnosis of leg venous thrombosis and knee joint rupture.
4. To evaluate pericardial and cardiac complications of polyarthritis.
5. For the diagnosis of tendinitis, e.g. of the Achilles tendon and the rotator cuff.
6. Potentially, for the assessment of active synovitis and response to therapy by using power Doppler in the suprapatellar region of the knee, prior to and following intra-articular steroid injection (NEWMAN et al. 1996). This role has yet to be fully evaluated.

Other areas where ultrasound is being explored include measurement of cartilage thickness and stiffness when ultrasound probes are applied directly to the surface of cartilage at arthroscopy.

16.3.4
Radionuclide Scintigraphy

The principal radionuclide employed is technetium-99m. This may be bound to a number of pharmaceuticals, of which the commonest is a diphosphonate compound for use in skeletal scintigraphy. It should always be recorded in two phases, blood pool and delayed, since inflammatory lesions will be overlooked if the former is omitted. Technetium-99m scintigraphy of particular value for the following purposes:

1. As a good "screening" investigation. Not only will bony lesions, such as osteoid osteoma, be revealed, but also soft tissue lesions such as tendinitis can be shown (MAURICE and WATT 1989).
2. The detection and demonstration of treatment effect on acute inflammatory lesions such as septic arthritis or disc space infection.
3. The differential diagnosis of symptomatic joint prosthesis. Blood pool images reflect increased vascularity and are as sensitive as a labelled white cell or gallium-67 scan. The delayed, bone phase is sensitive to prosthetic-related problems, but without the blood pool phase such images are not discriminatory.
4. Confirmation of the diagnosis of polyarthritis when x-rays are normal (Fig. 16.18). Scintigraphy is predictive of joint failure in OA of the hand or knee (HUTTON et al. 1986; McCRAE et al. 1992). Skeletal scintigraphy may assist in differential diagnosis. For example four subsets of knee OA have been shown, possibly with prognostic significance (DIEPPE et al. 1993).

5. Demonstration other causes of bone and joint pain, including metastasis and insufficiency fractures.

Technetium-99m may be used with other pharmaceuticals including:

1. With nanocolloid or liposomes to assist in distinguishing between the inflammatory component of synovitis and the secondary bone changes in rheumatoid disease.
2. With human immunoglobulin to obtain a similar differential diagnosis (DE BOIS et al. 1994) (Fig. 16.19). The use of gallium-67 citrate is no longer justified.

16.3.5
Computed Tomography

Mostly CT has been replaced by MRI; however, CT still has a number of important uses, including:

1. Assessment of the lumbar spine in degenerative disc disease, facet joint OA, spondylolysis and spinal stenosis, when MRI may be much more difficult to evaluate.
2. Evaluation of the lung parenchymatous involvement of the rheumatic disorders; however, MRI

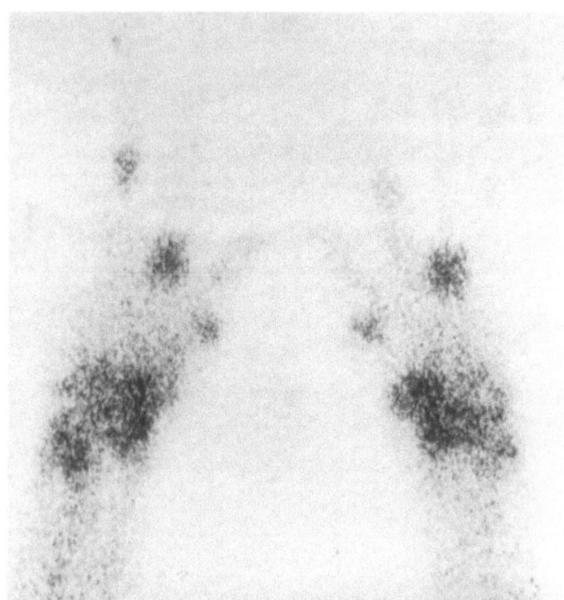
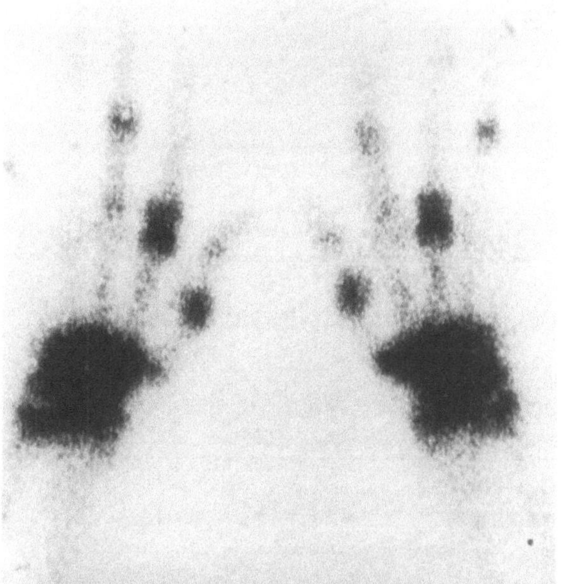

Fig. 16.18 a,b. An example of scintigraphy in the evaluation of joint disease. A patient with rheumatoid arthritis in whom the radiograph was essentially normal. **a** Blood pool and **b** delayed phase bone scan images show obvious involvement of the wrists and small hand joints. The inflammatory nature of the lesion is confirmed by the abnormality on the blood pool image

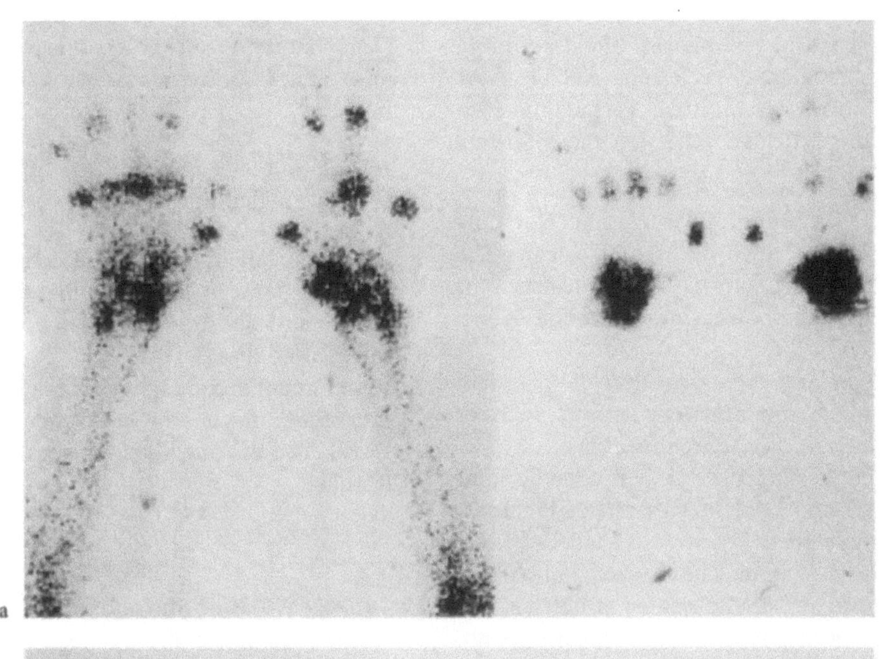
a

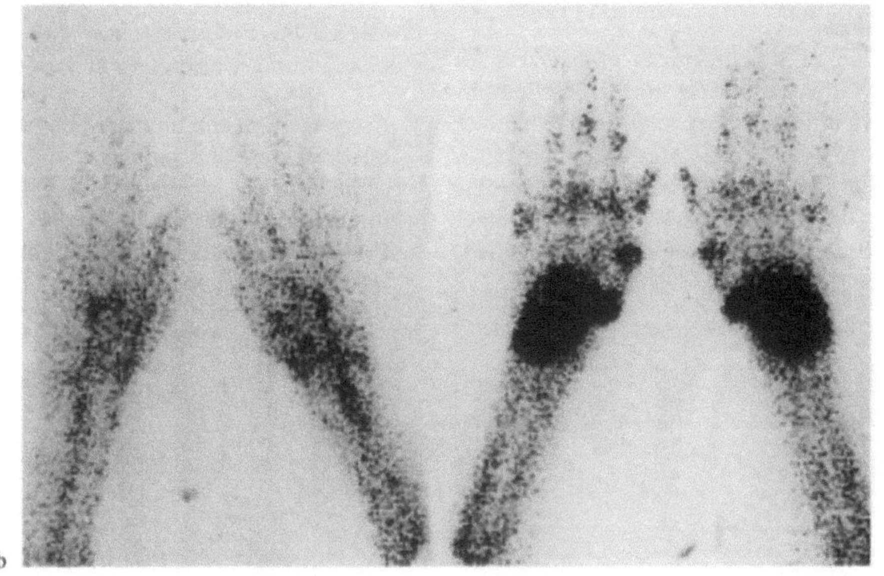
b

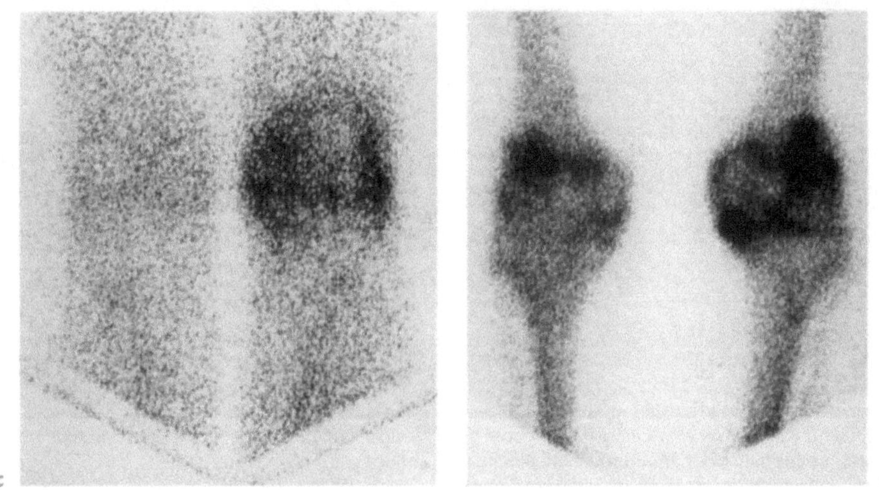
c
d

with breath-hold techniques can also fulfil this function.

3. Assessment of craniocervical complications of rheumatoid disease. Nevertheless, MRI is preferable (Fig. 16.20) as the cord and hind brain are better shown.

4. Evaluation of spondylolysis, when multiple plain x-rays, including oblique views, carry a high radiation burden and the anatomy is better shown by two or three CT scans through the appropriate lamina. Again, MRI may be better for this purpose as it carries no radiation burden and shows discal anatomy too. However, bony structures may be more difficult to evaluate on MRI.

16.3.6
Magnetic Resonance Imaging

Magnetic resonance imaging has become crucial in rheumatological radiology as it combines the virtues of ultrasound, CT and radionuclide scintigraphy as both an anatomical and a dynamic modality. The limitations to its use include patients with pacemakers, claustrophobia, metallic implants and other metallic foreign bodies and cost.

While MRI can duplicate almost all other investigations, availability limits its use to a backup procedure. For example, it is simpler to use ultrasound to assess tendinitis, whereas MRI is the

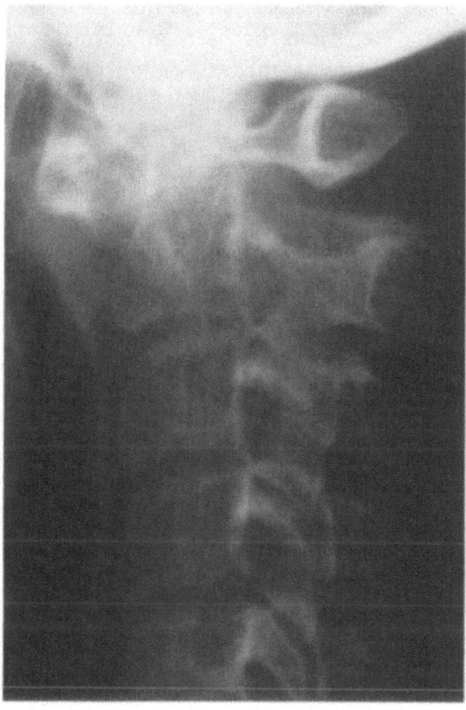
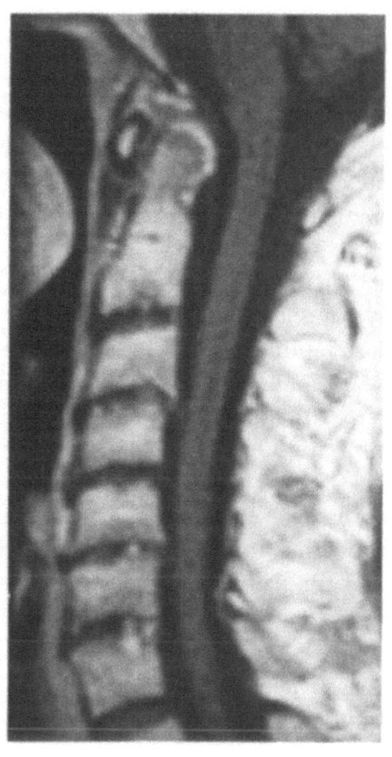

Fig. 16.20 a,b. Craniocervical instability. **a** A radiograph demonstrates obvious cephalic migration of the odontoid peg, horizontal instability and a degree of spinal stenosis between the intact dorsum of the peg and the lamina of C1. **b** In a different case the contrast-enhanced sagittal T1-weighted image demonstrates high signal pannus around a partially eroded peg. Note also high signal in several intervertebral discs, consistent with invasion by rheumatoid pannus

Fig. 16.19 a–d. Other scintigraphic methods for assessing arthritis. Images (**a** and **b**) in which human immunoglobulin (HIG) labelled with technetium-99m are compared with the late phase images of bone scans. In **a** there is evidence of increased activity at the joints on both scans, indicating continued bony and synovial activity. In **b** the HIG is normal, indicating the absence of synovitis, whereas the bone scan still shows increased bone turnover (from DE BOIS et al. 1994). Another comparison is shown between nanocolloid (**c**) and a bone scan with hydroxymethylene diphosphonate (**d**), again in a patient with active synovitis and rheumatoid disease. The nanocolloid scan clearly demonstrates the degree and extent of synovitis, whereas the bone scan demonstrates continued bone turnover

best investigation for craniocervical problems in rheumatoid disease. The roles of MRI can be categorised as follows:

16.3.6.1
Investigation of Choice

Magnetic resonance imaging may be regarded as the investigation of choice for the following purposes:

1. Assessment of craniocervical instability once a plain film or clinical symptoms suggest disease (Fig. 16.20).
2. Assessment of all other spinal problems, although CT may be better for spinal stenosis.
3. Evaluation of soft tissue or synovial masses, popliteal cysts or joint rupture, following ultrasound assessment, if the latter is inconclusive.
4. Identification of the causes of internal joint derangement. Appropriate surface coils will be necessary. Such derangements include meniscal tears, labral tears at the hip, ganglia, carpal instability, triangular fibrocartilage tears and tendon lesions.
5. Evaluation of active synovitis (Fig. 16.21) (KÖNIG et al. 1990) and cartilage disease (Fig. 16.22). A close correlation exists between the rate of enhancement following intravenous contrast medium and inflammatory features such as cellular infiltration and vascular proliferation (TAMAI et al. 1997). Further, active inflammatory pannus shown on MRI in a given joint predicts the likelihood that future bone erosion will be identified in it on plain x-rays (JEVTIC et al. 1996).

16.3.6.2
Investigation Under Evaluation

1. MRI may be useful for assessment of therapy regimens for synovitis (CREAMER et al. 1997). Careful study of hand changes in rheumatoid disease shows that distinct subsets occur, perhaps relevant to appropriate therapy (JEVTIC et al. 1993). Differentiation between types of polyarthritis can be made (JEVTIC et al. 1995).
2. Good demonstration of hyaline cartilage is possible (Fig. 16.23). Possible insights into cartilage degeneration are being explored, as is accuracy of calculation of hyaline cartilage volume (PETERFY et al. 1994).

3. Tracking abnormalities of the patellofemoral joints can be shown by MRI, but the nature and effect of therapy remain unclear.
4. Occasionally an MRI scan can show an obscure cause of disease, particularly on the STIR (short tau inversion recovery) or heavily T2-weighted sequences (Fig. 16.24).

16.3.6.3
Investigation Not Indicated

1. When the diagnosis has already been made by other means.
2. When the scan will not influence patient management.
3. For evaluation of the sacroiliac joints in ankylosing spondylitis. What is normal? Can erosions be seen in normal people even with contrast medium enhancement? It is also unclear whether MRI influences management.

16.3.7
DEXA Scanning

The use of bone absorptiometry techniques has been focused on post-menopausal osteoporosis. However, other important roles are emerging, although still embryonic. These include:

1. Detection and early therapy in algodystrophy
2. Quantification of bone loss in the hands in rheumatoid disease
3. Demonstration of early changes in the knee, comparing density levels in normal people and those with osteoarthritis and rheumatoid disease

16.4
Conclusion

This chapter has laid out a basic radiological differential diagnosis based on disease processes, rather than patterns. Integration of the various "instruments" of the imaging "orchestra" is crucial if cost-effective patient benefit is to be achieved. Always ask, "Why do we want to image this patient, what will it achieve and how will it improve their quality of life?" before embarking upon a course of investigation. Has the most appropriate means of

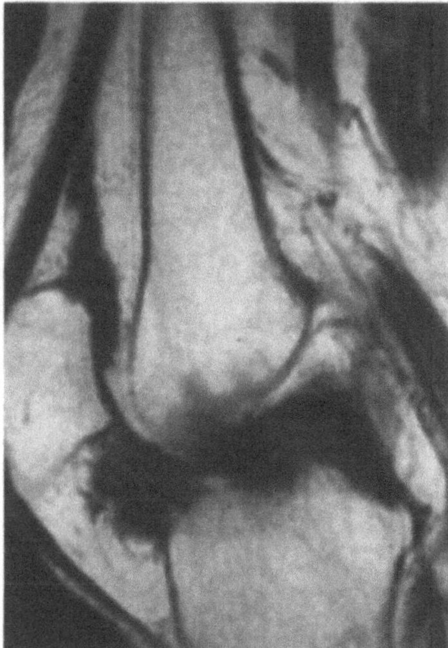

Fig. 16.21 a–c. MRI of synovitis. Normal synovium is not visualised on an MRI scan. **a** Thick, low signal, ill-defined synovium is demonstrated in a patient with haemophilia; the low signal reflects iron deposition. **b** A patient with erosive rheumatoid disease in whom it is difficult to distinguish synovium from joint fluid or to know the nature of the material within the tibial erosion. **c** Following intravenous contrast medium there is clear enhancement of active synovitis

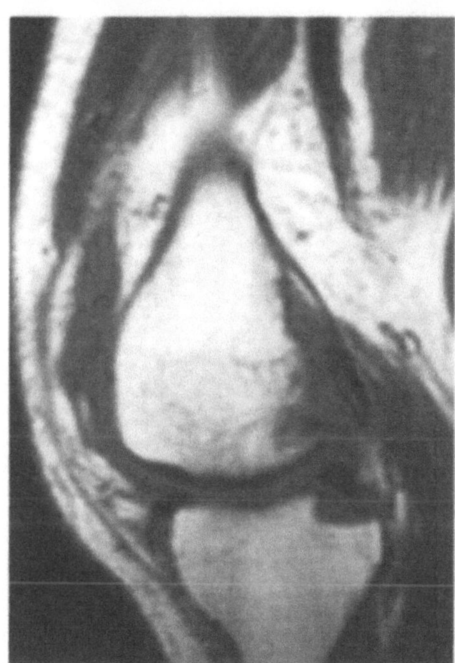

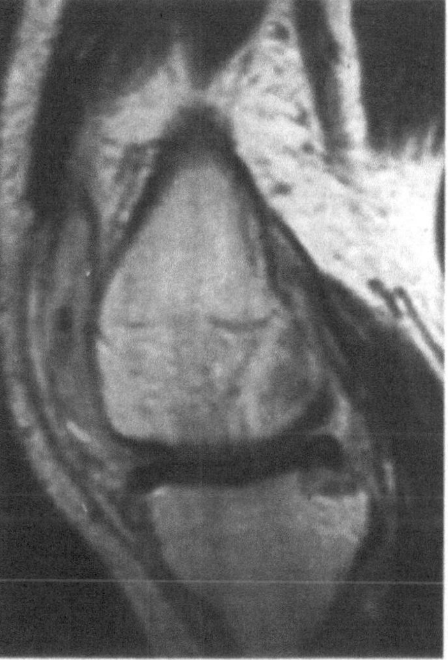

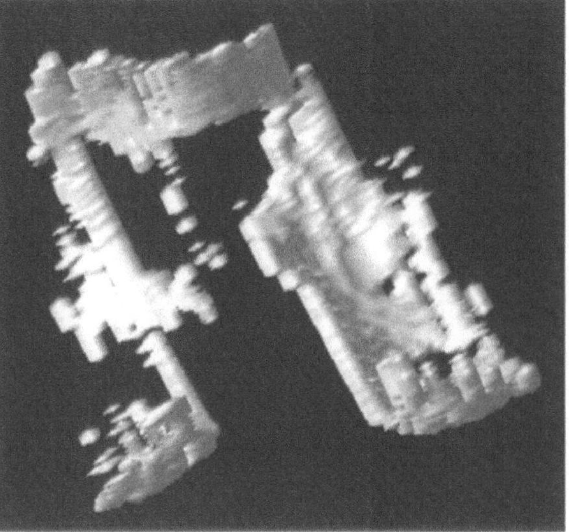

Fig. 16.23. This 3D reconstruction demonstrates obvious hyaline cartilage thinning and areas of total destruction

Fig. 16.22. A 3D reconstruction of normal femoral cartilage derived from a 3D gradient echo data set

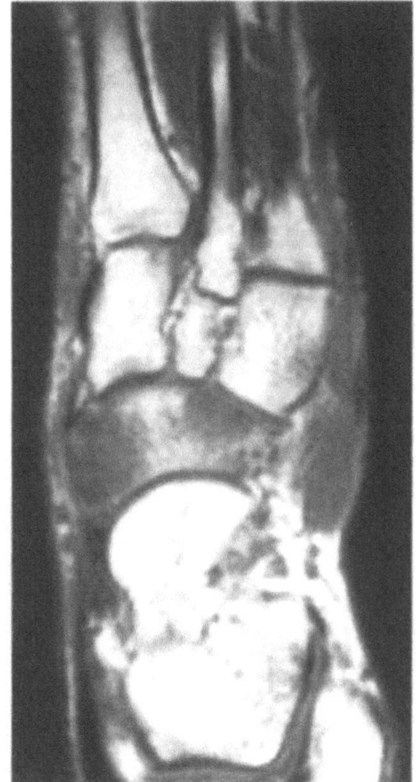

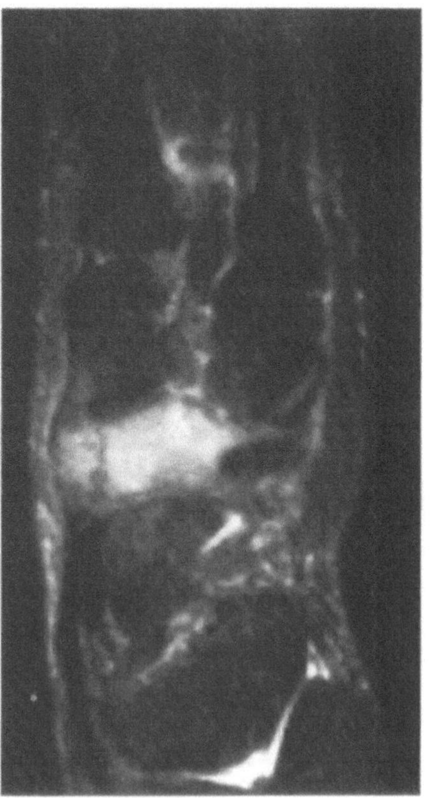

Fig. 16.24 a,b. The value of the STIR sequence. An athletic medical practitioner complained of pain in the forefoot. Low signal is shown in the navicular on a T1-weighted sequence (a). On the STIR sequence (b), however, obvious marrow oedema is present throughout the navicular and a low signal band is demonstrated, confirming an insufficiency fracture

imaging the patient been chosen? X-rays, like drugs, are potentially dangerous; use them wisely and not just for interest's sake!

References

Axford JS, Bomford A, Revell P, et al. (1991) Hip arthropathy in genetic haemochromatosis: radiographic and histologic features. Arthritis Rheum 34:357–361

Booth A, Dieppe PA, Goddard P, Watt I (1989) The radiological manifestations of relapsing polychondritis. Clin Radiol 40:147–149

Brook A, Corbett M (1977) Radiographic changes in early rheumatoid disease. Ann Rheum Dis 36:71–73

Buckland-Wright JC (1984) Microfocal radiographic examination of erosions in the wrist and hand of patients with rheumatoid arthritis. Ann Rheum Dis 43:160–171

Campion GV, McCrae F, Alwan W, et al. (1988) Idiopathic destructive arthritis of the shoulder in the elderly. Semin Arthritis Rheum 17:232–245

Cobby M, Cushnaghan J, Creamer, et al. (1990) Erosive osteoarthritis: is it a separate disease entity? Clin Radiol 42:258–263

Conrozier T, Lequesne MG, Tron AM, Mathieu P, Berdah L, Vignon E (1997) The effects of position on the radiographic joint space in osteoarthritis of the hip. Osteoarthritis Cartilage 5:17–22

Cooke D, Scudamore A, Li J, Wyss U, Bryant T, Costigan P (1997) Axial lower-limb alignment: comparison of knee geometry in normal volunteers and osteoarthritis patients. Osteoarthritis Cartilage 5:39–47

Creamer P, Keen M, Zananiri F, et al. (1997) MRI of the knee: a method of monitoring efficacy of intraarticular therapies. Arthritis Rheum (to be published)

De Bois MHW, Arndt JW, Van der Velde EA, et al. (1994) Joint scintigraphy for quantification of synovitis with ^{99m}Tc-labelled human immunoglobulin G compared with late phase scintigraphy with ^{99m}Tc-labelled diphosphonate. Br J Rheumatol 33:67–73

Dieppe PA, Klippel JH (eds) (1993) Rheumatology. Mosby, St. Louis

Dieppe PA, Alexander GJH, Jones HE, et al. (1982) Pyrophosphate arthropathy: a clinical and radiological study of 105 cases. Ann Rheum Dis 41:371–376

Dieppe PA, Doherty M, MacFarlane DG, et al. (1984) Apatite associated destructive arthritis. Br J Rheumatol 23:84–91

Dieppe PA, Cushnaghan J, Young P, et al. (1993) Prediction of the progression of joint space narrowing in osteoarthritis of the knee by bone scintigraphy. Ann Rheum Dis 52:557–563

Doherty M, Watt I, Dieppe PA (1982) Localised chondrocalcinosis in post-meniscectomy knees. Lancet i:1207–1210

Halverson PB, McCarty DJ, Cheung HS (1984) Milwaukee shoulder syndrome. Semin Arthritis Rheum 14:36–44

Hutton CW, Higgs ER, Jackson PC, et al. (1986) ^{99m}Technetium HMDP bone scanning in generalised nodal osteoarthritis – the 4 hour bone scan image predicts radiographic change Ann Rheum Dis 45:622–626

Jevtic V, Watt I, Rozman B, et al. (1993) Pre contrast and post contrast (Gd-DTPA) magnetic resonance imaging (MRI) of hand joints in patients with rheumatoid arthritis. Clin Radiol 48:176–181

Jevtic V, Watt I, Rozman B, et al. (1995) Distinctive radiological features of small hand joints in rheumatoid arthritis and seronegative spondyloarthritis demonstrated by contrast enhanced (Gd-DTPA) magnetic resonance imaging. Skeletal Radiol 24:351–356

Jevtic V, Watt I, Rozman B, et al. (1996) Prognostic value of contrast enhanced Gd-DTPA MRI for development of bone erosive changes in rheumatoid disease. Br J Rheumatol 35 (Suppl 3):26–30

Kahn MF, Khan MA (1994) The SAPHO syndrome. Baillieres Clin Rheumatol 8:333–362

Kasperczyk A, Freyschmidt J (1994) Pustulotic arthrosteitis: spectrum of bone lesions with palmoplantar pustulosis. Radiology 191:207–211

Köhler H, Uehlinger E, Kutzner J, et al. (1977) Sternocostoclavicular hyperostosis: painful swelling of the sternum, clavicles and upper ribs, report of 2 new cases. Ann Intern Med 87:192–199

König H, Sieper J, Wolf K-J (1990) Rheumatoid arthritis: evaluation of hypervascular and fibrous pannus with dynamic MR imaging enhanced with Gd-DTPA. Radiology 176:473–477

Larsen A, Dale K, Eek M (1977) Radiographic evaluation of rheumatoid arthritis and related conditions by standard reference films. Acta Radiologica (Scand) 18:481–491

Lynch JA, Hawkes DJ, Buckland-Wright JC (1991) Analysis of texture in macroradiographs of osteoarthritic knees using the fractal signature. Phys Med Biol 36:709–722

Martel W, Champion CK, Thompson GR, et al. (1970) A roentgenologically distinctive arthropathy in some patients with the pseudogout syndrome. AJR 109:587–607

Martel W, Stuck KJ, Dworin AM, Hylland RG (1980) Erosive osteoarthritis and psoriatic arthritis: a radiologic comparison in the hand, wrist and foot. AJR 134:125–135

Maurice H, Watt I (1989) ^{99m}Technetium hydroxymethylene diphosphonate (TcHDP) scanning of acute injuries to the lateral ligaments of the ankle. Br J Radiol 62:31–34

McCrae F, Shouls J, Dieppe PA, et al. (1992) Scintigraphic assessment of osteoarthritis of the knee joint. Ann Rheum Dis 51:938–942

Newman JS, Laing TL, McCarthy CJ, Adler RS (1996) Power Doppler sonography of synovitis: assessment of therapeutic response – preliminary observations. Radiology 198:582–584

Peterfy CF, van Dijke C, Janzen DL, et al. (1994) Quantification of articular cartilage in the knee with pulsed saturation transfer subtraction and fat-suppressed MR imaging: optimisation and validation. Radiology 192:485–491

Resnick D (1997) Diagnosis of bone and joint disorders, 4th edn. Saunders, Philadelphia

Resnick D, Shaul SR, Robins JM (1975) Diffuse idiopathic skeletal hyperostosis (DISH): Forestier's disease with extraspinal manifestations. Radiology 115:513–524

Resnick D, Niwayama G, Goergen TG, et al. (1977) Clinical, radiographic and pathologic abnormalities in calcium pyrophosphate deposition disease (CPPD): pseudogout. Radiology 122:1–15

Sutton DA (1992) Textbook of radiology and imaging, 5th edn. Churchill Livingstone, Edinburgh

Tamai K, Yamato M, Yamaguchi T, Ohno W (1997) Dynamic magnetic resonance imaging for the evaluation of synovitis in patients with rheumatoid arthritis. Arthritis Rheum 37:1151–1157

Watt I (1994) Radiology and imaging. In: Doherty M (ed) Osteoarthritis, a colour atlas and text. Wolfe Medical Imaging, London, pp 85–114

17 Bone and Joint Infections

K. Jonsson

CONTENTS

17.1 Introduction

Infections of the musculoskeletal system are common and show an increasing frequency because of the increasing number of immuno-compromised patients. Septic bone and joint infections are serious conditions, and a delay in diagnosis and treatment can have serious consequences. In a child, destruction of a growth plate or joint may be disabling for life. Radiological diagnosis of bone and joint infection may be difficult, mainly because the findings are often nonspecific. The purpose of this chapter is to outline the advantages and disadvantages of different diagnostic tools.

17.2 Osteomyelitis

The term "osteomyelitis" means infection of bone and marrow, most commonly resulting from bacterial infections. The most common organism is *Staphylococcus aureus*, but all kinds of bacteria can cause osteomyelitis, such as *Streptococcus, Escherichia coli, Klebsiella, Salmonella, Haemophilus influenzae*, and *Mycobacterium tuberculosis*. In addition, fungi, parasites, and viruses can cause osteomyelitis.

The term "osteitis" indicates infection of the cortical bone, which may occur in isolation, without involvement of the marrow.

Osteomyelitis is classified as acute, subacute, and chronic. The acute stage starts with abrupt clinical symptoms, which can be local and/or general. Local symptoms are local pain, tenderness, and warmth, while examples of general symptoms are fever, malaise, and nausea. Acute osteomyelitis is often accompanied by laboratory findings such as elevated erythrocyte sedimentation rate (ESR) and increasing levels of C-reactive protein (CRP). If the acute osteomyelitis is not properly treated there is a transition to subacute and chronic stages. Viable organisms may remain in small abscesses and/or in fragments of necrotic bone, which protect them from antibiotic therapy. After several months or years, remaining organisms can flare up with new clinical symptoms. The symptoms of acute osteomyelitis may be vague and neglected and the patient later presents with radiological signs of chronic osteomyelitis, because the acute infection was never diagnosed.

There are several routes of contamination in osteomyelitis:

1. Hematogenous spread
2. Spread from a contiguous source of infection
3. Direct implantation
4. Postoperative infection

Hematogenous osteomyelitis in children often develops without a known focus. In adults the osteomyelitis is preceded by a distant infection or trauma. The infective source may be vague and overlooked, for instance after a *Salmonella* infection. The symptoms of the original infection disappear relatively quickly and the signs of osteomyelitis may appear several months later.

K. Jonsson, MD, PhD, Department of Radiology, University Hospital, S-221 85 Lund, Sweden

The distribution of hematogenous osteomyelitis in tubular bones is largely related to the vascular pattern of the bone. Osteomyelitis has been divided into three categories based on the variation of blood supply of tubular bones (TRUETA 1959):

1. The infantile type, before the age of 1 year
2. The juvenile type, between the age of 1 year and the closure of the growth plates
3. The adult type, occurring after the closure of the growth zone

In the infantile type, vessels from the metaphysis penetrate the growth plate and supply part of the epiphysis. In this way an infectious focus of the metaphysis can rapidly spread to the epiphysis, causing destruction of the growth plate and also septic arthritis.

Between the age of 1 year and the closure of the growth plate no such vascular penetration exists and the infection is limited to the metaphysis. The infection tends to spread laterally and often penetrates the loose periosteum, causing a subperiosteal edema, collection of inflammatory cells, and vascular occlusion. In the juvenile type the infection rarely spreads to the epiphysis or the adjacent joint, but this may happen. In the adult type the cartilage of the growth zone has been reabsorbed and vessels from the metaphysis also supply the epiphysis. An infection in the bone marrow of the metaphysis can spread to the subcortical bone, with an increased risk of spreading into the adjacent joint.

Hematogenous osteomyelitis in adults is more common in the spine, pelvis, or small tubular bones than in the long tubular bones. Because the periosteum is firmly anchored to the cortical bone in adults it is unusual to encounter subperiosteal abscess formation or extensive periostitis.

Spread from a contiguous source may come from a skin or soft tissue infection, for instance in a patient with diabetes. These infections most commonly occur in the hands and feet. In the hands the infection may spread via tendon sheaths, fascial planes, or lymphatics. An infection in one finger may spread to another. The most spectacular example is the spread from the thumb to the fifth finger through tendon sheaths. A skin ulceration due to infection may penetrate to the underlying bone, which is eroded by infection. The skin ulceration or sinus tract from the skin to the bone may be seen on radiographs before the osteomyelitis (Fig. 17.8). Another example of spread from a contiguous source is a tooth infection that spreads to the adjacent bone and causes osteomyelitis.

Infection due to direct implantation is common in penetrating wounds, where foreign bodies are implanted into the soft tissue. This often gives a soft tissue infection which spreads as a continuous source to the underlying bone. Wooden splinters and thorns often cause such infections. A serious cause of direct implantation infection is open fractures, contaminated with dirt or soil. It is often difficult to clean such a contaminated bone from the soil. Infections may also be caused by direct punctures made because of needle biopsy of a lesion or joint puncture, for instance arthrography. If proper hygiene is maintained, infections caused by needle puncture are rare.

An often overlooked but significant injury is human bites (GONZALEZ et al. 1993). The most common cause of this injury is a fist blow to the mouth resulting in ulceration of the dorsum of a metacarpophalangeal joint. Septic arthritis and tenosynovitis are common and have a devastating result if not treated properly. Infections caused by human bites are often more aggressive than those from animal bites.

Postoperative infections are also a kind of inoculation infection, which may be caused by adjacent infected soft tissue or due to improper hygiene in the operating room. Postoperative infections will not be further discussed in this chapter.

When an organism has lodged in the bone marrow, an acute pus-forming inflammation starts. The local edema causes increased intraosseous pressure, leading to a reduction in local blood supply and consequently necrosis and further spread of the infection. Bone trabeculae, matrix, and mineral are absorbed. The inflammatory reaction penetrates the endosteum and spreads under the periosteum, forming subperiosteal abscesses. Finally, the adjacent soft tissues are also infected (Fig. 17.18). Because of vascular occlusion due to the increased pressure, large segments of cortical bone may lose their blood supply and become necrotic. A piece of necrotic bone is called a sequestrum (Fig. 17.5). A sequestrum is surrounded by granulation tissue and often harbors organisms that are not reached by antibiotics. A sequestrum may spontaneously evacuate via a cloaca of the periosteum through a sinus tract to the skin. However, often the sequestrum must be eliminated surgically in order to heal chronic osteomyelitis.

At the site of an infection with cortical destruction and sequestrum formation, an extensive periosteal reaction is often present around the affected bone. This periosteal reaction takes the shape of the original bone and is called an involucrum. The result is that the original cortical bone is seen as a dead sclerotic sequestrum centrally within the newly formed bone (Fig. 17.5).

17.3
Septic Arthritis

Septic arthritis means infection of the structures of a joint. Septic arthritis may be isolated, but may also spread to adjacent bones and cause osteomyelitis. The opposite may also happen, with spread of organisms from osteomyelitis or from infected soft tissues into a joint.

Septic arthritis may develop from the same routes of spread as for osteomyelitis, i.e., hematogenous spread, spread from a contiguous source of infection, direct implantation and postoperative infection. There are several predisposing factors for septic arthritis. Age is one factor; both the newborn and the elderly may have impaired resistance to bacteremia. Aseptic arthritis, such as rheumatoid arthritis, and immunosuppressive therapy in cancer and trans-

plantation also predispose to septic arthritis. The same is true for diabetes and chronic alcoholism (BROWER 1996).

In the neonate the diagnosis of septic arthritis is especially difficult. The signs of these infants may be poor feeding or crying on manipulation. Dislocation or subluxation of a joint, due to effusion, is difficult to evaluate in the neonate. The hip joint is the most difficult joint to examine, but a high degree of suspicion in a child with signs of sepsis should indicate repeated examinations for correct diagnosis.

After the age of 1 year, the symptoms are usually more evident, with pain and gaiting or marked pain on motion of the involved joint. In the adult the symptoms are usually acute pain, swelling and limitation of motion.

A special problem in postoperative infections occurs in total joint replacements or around prosthetic replacements. Pain around a hip or knee prosthesis is most often due to aseptic loosening of the prosthesis. The symptoms may be the same with infection and septic loosening of a prosthesis. Furthermore, the findings at radiography are often

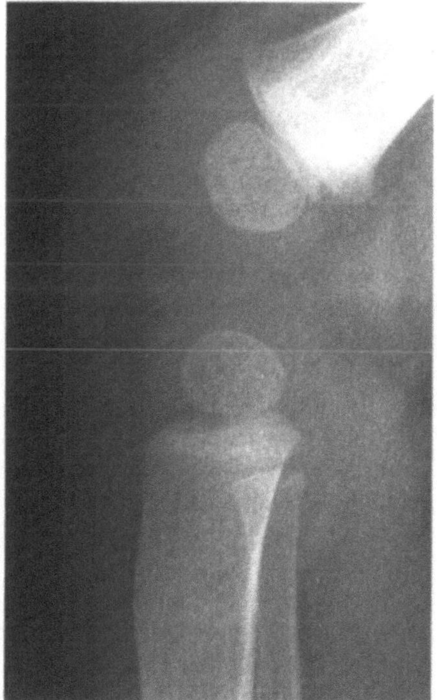
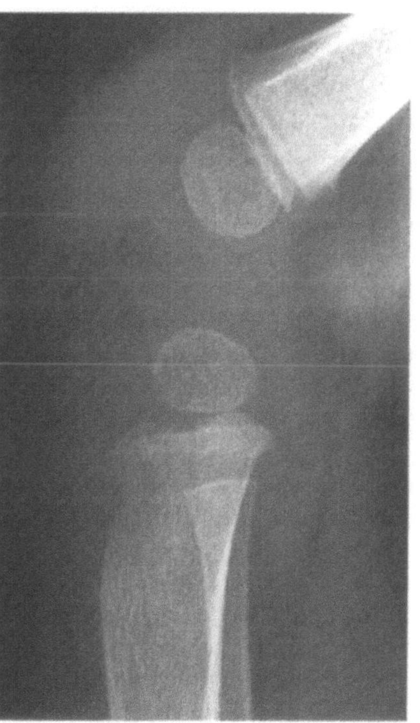

Fig. 17.1 a,b. Nine-month-old boy with local pain over the right knee. **a** Primary examination 3 days after onset of symptoms: no evidence of osteomyelitis. **b** Three days later, i.e., 6 days after onset, there is a considerable destruction in the metaphysis of the tibia

the same with septic and aseptic loosening. In septic loosening there is often an effusion of the joint, but this is difficult to evaluate in the hip. Other methods must be used, which will be discussed later.

17.4
Imaging Methods

Several diagnostic tools are available for use in patients with suspicion of osteomyelitis or septic arthritis. The primary examination should be radiography. Other diagnostic methods are radionuclide investigations, computerized tomography (CT), magnetic resonance imaging (MRI), ultrasonography, fistulography, and fine-needle aspiration biopsy. It is important to know that all these diagnostic tools are more or less nonspecific. The investigation must be judged in combination with radiography and the clinical situation. Only open biopsy and fine-needle aspiration biopsy, where material is sent for culture, are specific in the diagnosis of osteomyelitis or septic arthritis (RESNICK and NIWAYAMA 1995).

17.4.1
Radiography

A patient with hematogenous pyogenic osteomyelitis usually has local swelling and tenderness, which is clinically evident. Radiographs during the first 3–7 days are usually negative with regard to bone destruction (Fig. 17.1a). In children the first radiographic sign is usually a periosteal reaction (Fig. 17.2) followed by relatively rapid development of bone destruction (Fig. 17.1b). Such bone destruction is poorly demarcated and simulates malignant tumor with permeative growth through cortical bone. Occasionally gas is seen within the bone lesions or in the adjacent soft tissues. In subacute and chronic osteomyelitis, Brodie's abscess may develop. These abscesses are usually well demarcated (Fig. 17.3a). The abscess develops when the organisms have reduced virulence or when the host has increased resistance to infections. These changes occur in the metaphysis; they are especially common in the distal tibia, but may be seen in any of the long bones. The abscesses are outlined by inflammatory granulation tissue which is surrounded by sclerotic bone, and the sclerosis is the dominating radiological finding. The central lucency may be difficult to see through the

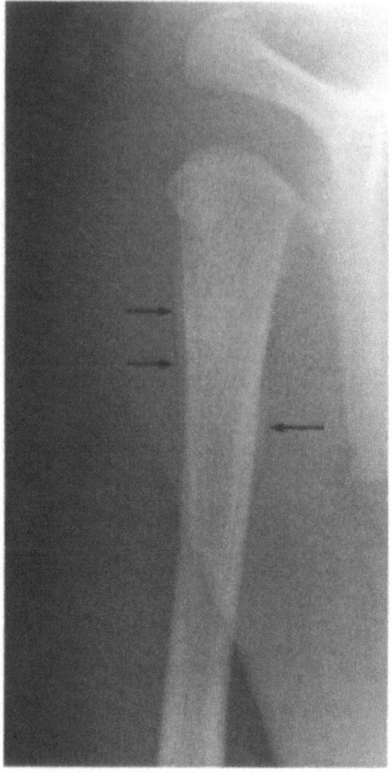

Fig. 17.2. Three-week-old baby. Five days after onset of symptoms there is an extensive periosteal reaction around the proximal humerus (*arrows*). Later destruction developed, which proved to be osteomyelitis

sclerosis. It is connected with the adjacent growth plate of the cortical bone by a tortuous channel (Fig. 17.3b).

Occasionally the sclerotic reaction around an abscess is less pronounced with a geographic lesion (Fig. 17.4).

In osteomyelitis the cortical vessels may be blocked, causing sequestration of a cortical fragment. The sequestrum is usually of higher density than the surrounding bone because of lack of blood supply with no resorption of the fragment. A sequestrum is usually sharply outlined against the viable bone (Fig. 17.5). However, a sequestrum may be quite small and difficult to see on a radiograph, especially in post-traumatic osteomyelitis, and CT is the best mode of diagnosis.

In the subacute and chronic stage osteomyelitis appears dense and sclerotic due to reactive bone formation. The radiographic appearance may resemble osteoid osteoma, fibrous dysplasia, or Ewing's sarcoma. A lucency or nidus may be seen, surrounded by massive sclerosis (Fig. 17.5d). The lesion resembles osteoid osteoma. In osteomyelitis the "ni-

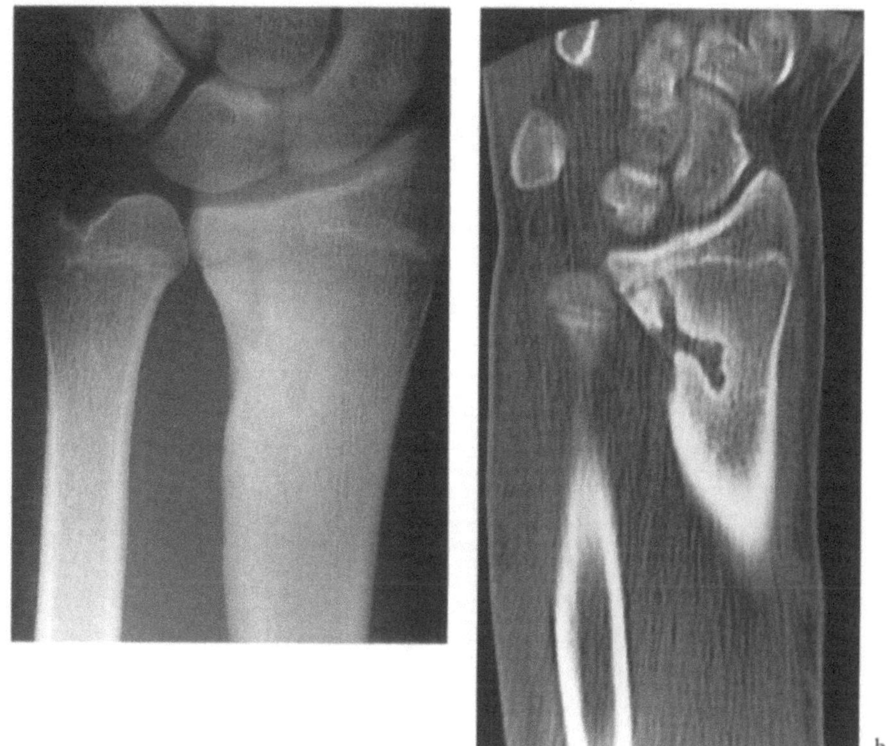

Fig. 17.3 a,b. Brodie's abscess of distal radius. **a** Radiography, where sclerosis dominates the picture. The abscess is vaguely seen through the sclerosis. **b** CT in coronal plane discloses the irregular abscess of the radius, with connection to the adjacent growth plate and the cortical bone

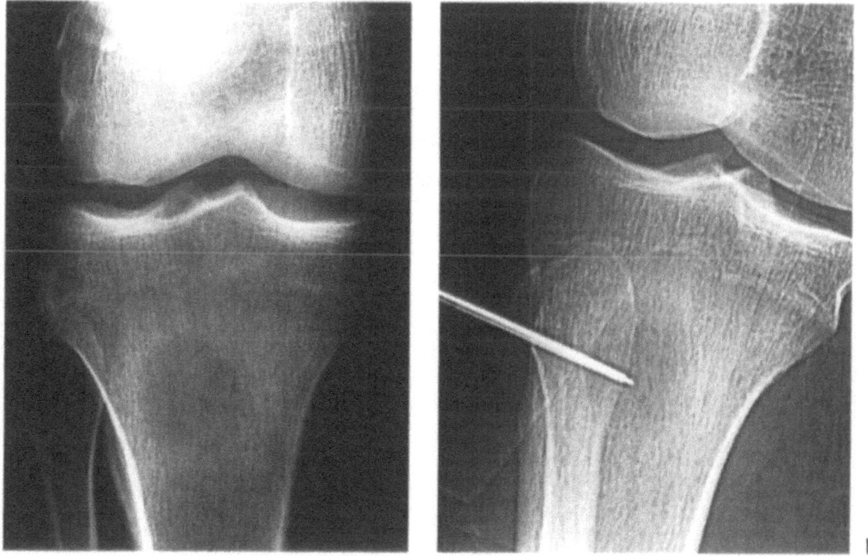

Fig. 17.4 a,b. **a** Abscess of the proximal metaphysis of the right tibia. The lesion appears geographic with a minor sclerotic reaction around the abscess. **b** Puncture of the lesion for aspiration biopsy

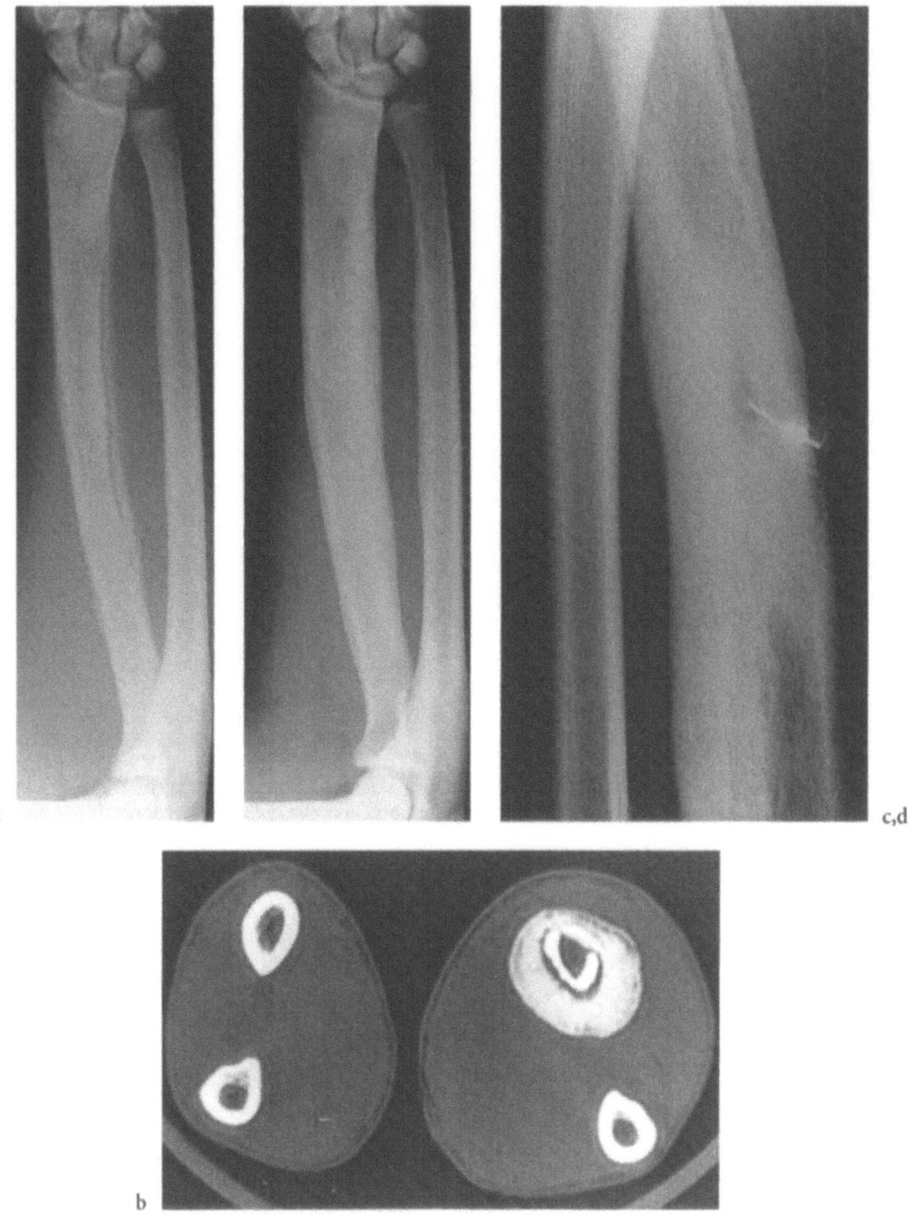

Fig. 17.5 a–d. After trauma to the lower arm this 17-year-old boy developed pain and swelling and received anti-inflammatory drugs and antibiotics for 1 week. The symptoms disappeared, but 3 months later he developed new symptoms with pain and tenderness over the lower arm. **a** Radiograph of the lower arm showed deformity of the midshaft of the radius with a large sequestrum and involucrum formation. Continuous antibiotic therapy improved the clinical situation and the involucrum formation progressed. **b** Four months later the involucrum formation was extensive, as shown in this CT scan. The central sequestrum was surgically removed at that time. **c** Two years later the radius appears healed with a sclerotic bone formation. **d** Six years after the primary treatment the patient had recurrent pain and a small lucent lesion was found within the sclerotic bone. Fine-needle aspiration biopsy from this lucent area disclosed bacteria of the same type as the primary infection

dus" is long, measuring 1.5–2 cm, while the nidus in osteoid osteoma is rounded with a diameter of 5–10 mm. Sometimes flaring of chronic osteomyelitis may be seen on radiographs as a periosteal reaction (Fig. 17.6). Most often there is no indication of activity in a bone with chronic osteomyelitis. Complementary CT or MRI is necessary for correct diagnosis.

A special kind of chronic osteomyelitis in children is chronic recurrent multifocal osteomyelitis (CRMO) (CARR et al. 1993; SUNDARAM et al. 1996). Other names for this condition are condensing

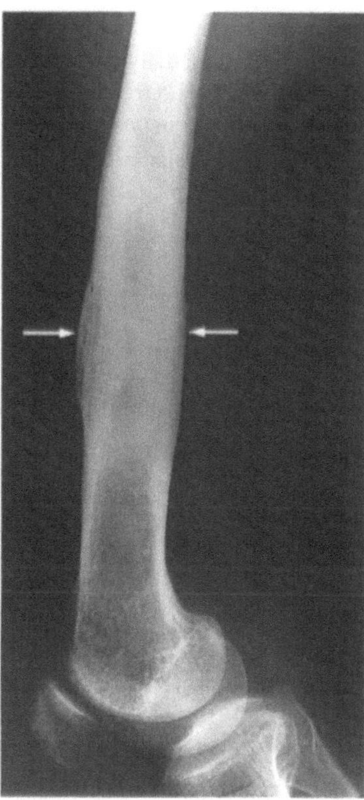

Fig. 17.6. Chronic osteomyelitis in the femur with periosteal reaction indicates flaring (*arrows*)

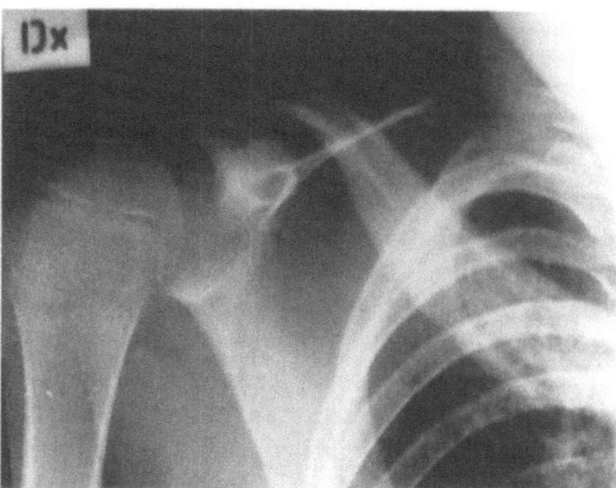

Fig. 17.7. Chronic recurrent multifocal osteomyelitis of the clavicle in a 10-year-old girl

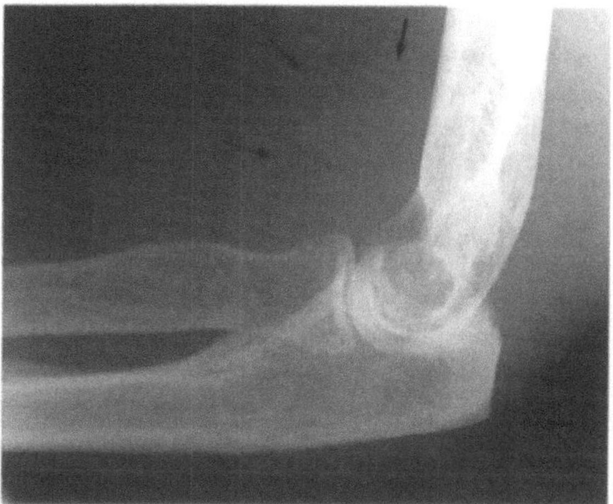

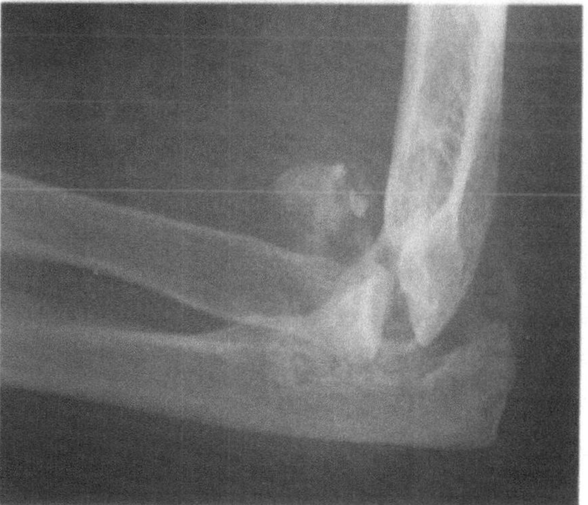

Fig. 17.9 a,b. Tuberculous arthritis of the elbow. Extensive effusion is present within the bone. A fat stripe of the anterior capsule is seen (*arrows*). a No erosion or bone destruction. b Three months later massive bone destruction and fragmentation are seen within the joint

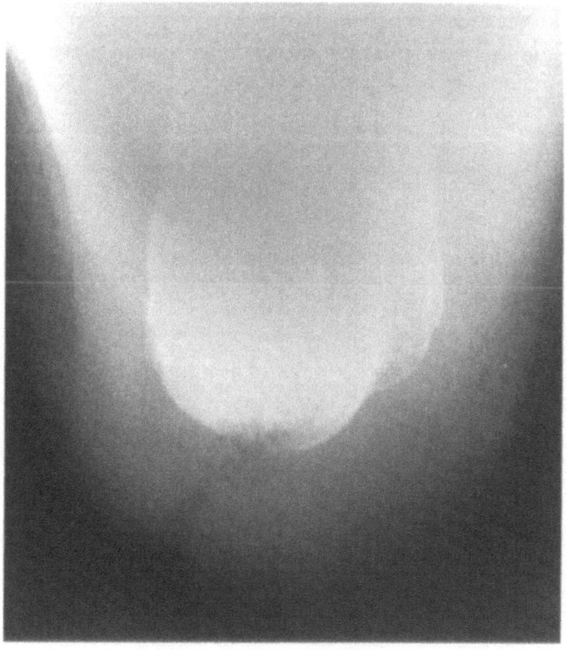

Fig. 17.8. Ulceration of the heel in a patient with diabetes. A sinus tract is seen from the ulceration to the calcaneus, where bone destruction is observed

osteitis of the clavicle in childhood, chronic symmetric plasma cell osteomyelitis, and chronic sclerosing osteomyelitis (Fig. 17.7). Sclerosis and bone formation are the dominating radiographic signs and the involved bone is often enlarged. This condition may simulate Paget's disease or sarcoma. Culture from bone biopsy is usually negative. The patients have pain, tenderness, and swelling, and the disease can remain for a considerable time, but the long-term results are good.

Soft tissue changes may be seen in patients with osteomyelitis. In diabetics an ulceration or sinus tract from soft tissue infection may be seen to the underlying bone (Fig. 17.8). Soft tissue swelling also may be evident on radiographs.

In septic arthritis, joint effusion and soft tissue swelling around the joint are the first radiographic signs of infection. Such effusion is easy to see around the elbow with a fat pad sign (Fig. 17.9) or swelling of the suprapatellar recess of the knee. With progression of the arthritis marginal erosions occur around the joint. This is nonspecific and may occur in any inflammatory disease of the joint. In septic arthritis there is typically a uniform joint space narrowing and loss of the cortical margins of the joint (Fig. 17.10). In deeply located joints such as the sacroiliac joint and the hip joint the early radiographic signs of infection are difficult to detect, and an early diagnosis depends on other imaging methods.

Tuberculous arthritis is most common in the lower extremities. The radiological manifestations are the same as in other types of infection. Soft tissue swelling and periarticular osteopenia may be extensive. The joint space is usually preserved for a long time. Cortical and marginal erosions such as are seen in rheumatoid arthritis may develop (Yao and Sartoris 1995). Rapid destruction of the subcortical bone and fragmentation may, however, occur (Fig. 17.9).

17.4.2
Radionuclide Investigations

17.4.2.1
Technetium-99m Bone Scintigraphy

Radionuclide studies of bone and joints are done with a number of different agents. The most common labelling substance is ^{99m}Tc, and the substances labelled are methylene diphosphonate (MDP), hydroxyethylene diphosphonate (HEDP), or hydroxymethylene diphosphonate (HMDP). These agents are comparable to each other (Vande Streek et al. 1994).They visualize the activity of osteoblasts, i.e., bone formation as a reaction to a destructive process. In osteomyelitis, a bone scan may be positive and show increased activity several days prior to radiographic changes. The finding is nonspecific but more sensitive than radiography. The scan should be done using a three-phase technique (Fig. 17.11). During the first phase, consisting of 2- to 5-s images of the area of suspected osteomyelitis, the first angiographic flow phase is shown. The second phase, the

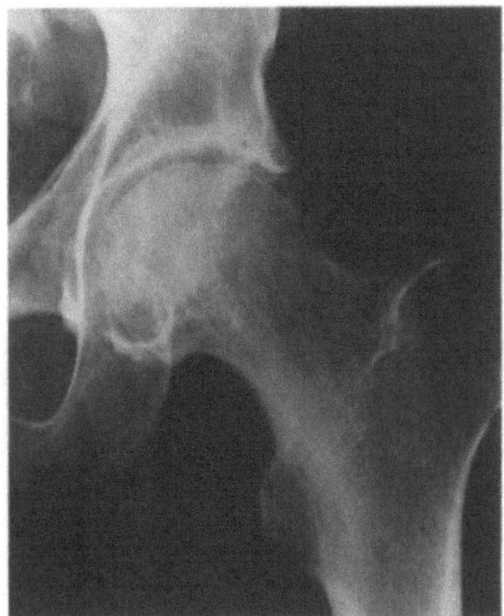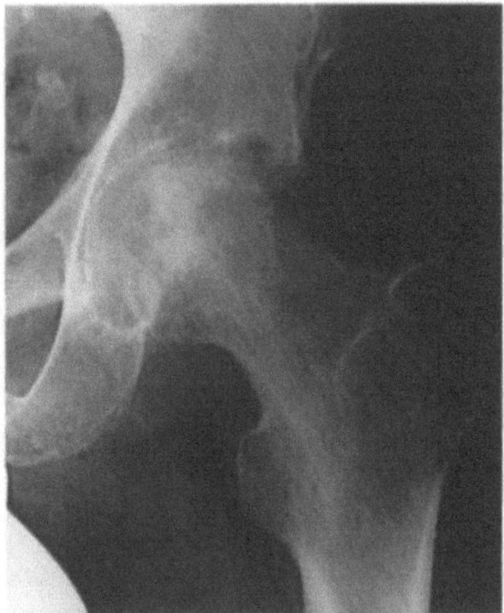

Fig. 17.10 a,b. Patient with agranulocytosis. The patient had hip pain. **a** Initial radiograph shows no evidence of bone destruction. Puncture revealed septic arthritis. **b** Two months later there is a massive destruction of the acetabulum and femoral head

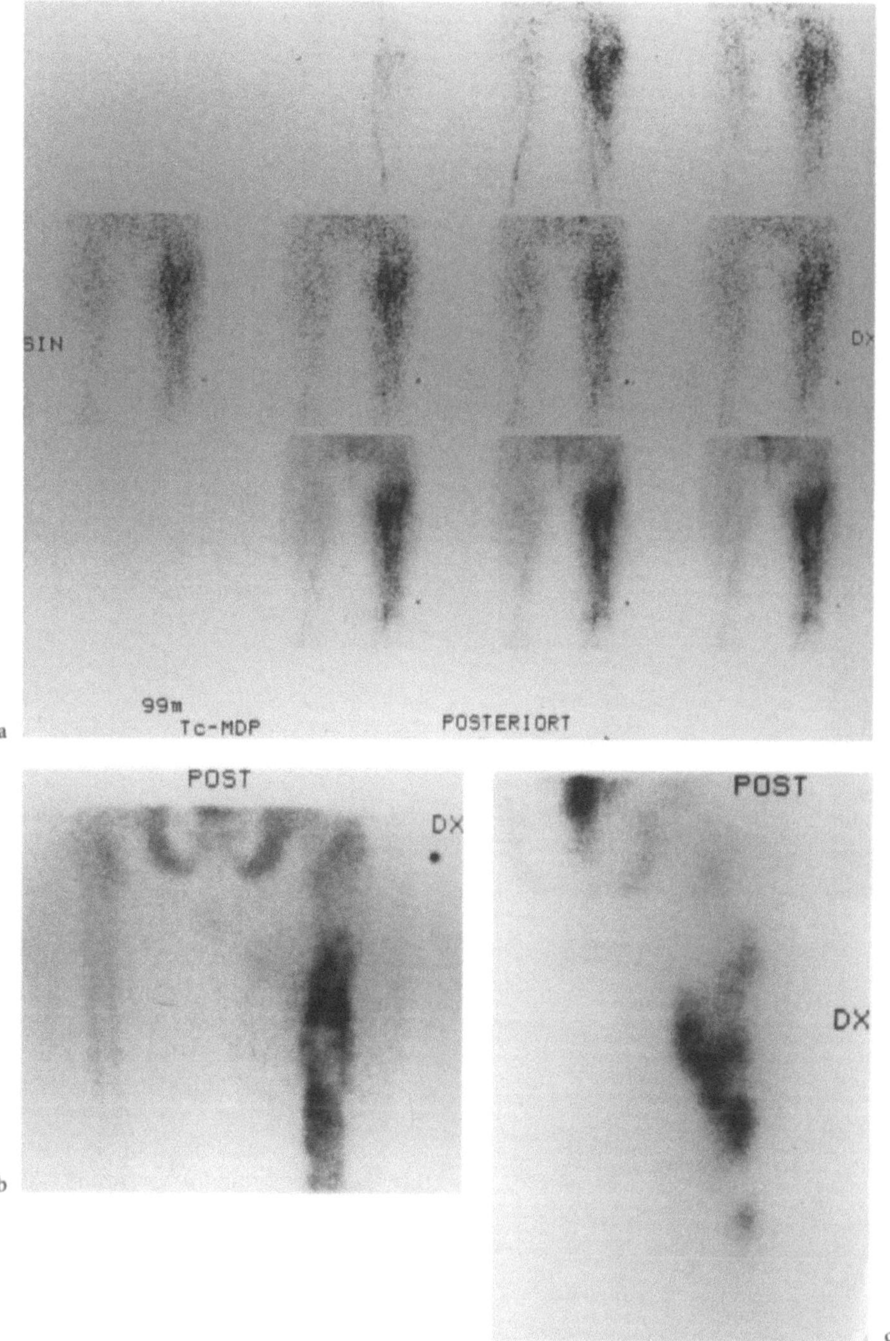

Fig. 17.11 a–c. Chronic osteomyelitis in a patient with open femoral fracture. **a** Three-phase ^{99m}Tc-MDP scintigraphy, angiographic phase. The *upper two rows* represent registration during the first minute and the *bottom row* represents registration after 3, 4, and 5 min, i.e., representing the blood pool phase. **b** Registration after 3 h. In the angiographic phase and the blood pool phase, soft tissue infection is suggested, but the 3-h registration only shows increased activity within bone. Note a central defect in the bone activity, representing a necrotic fragment. **c** ^{67}Ga scintigraphy verifies the soft tissue extension of infection

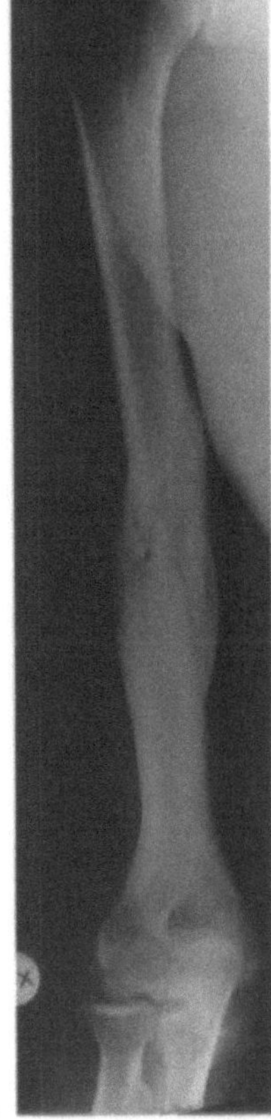

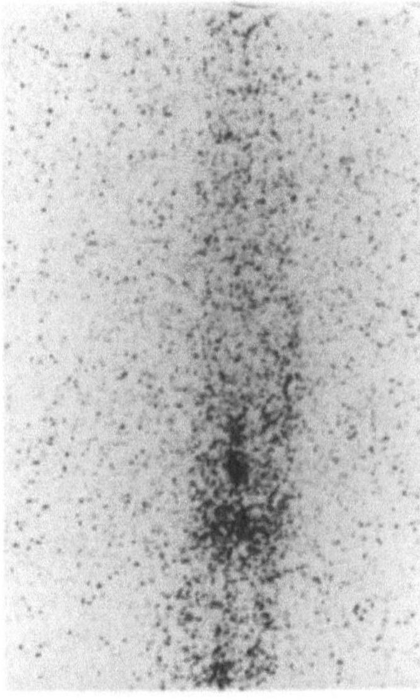

a

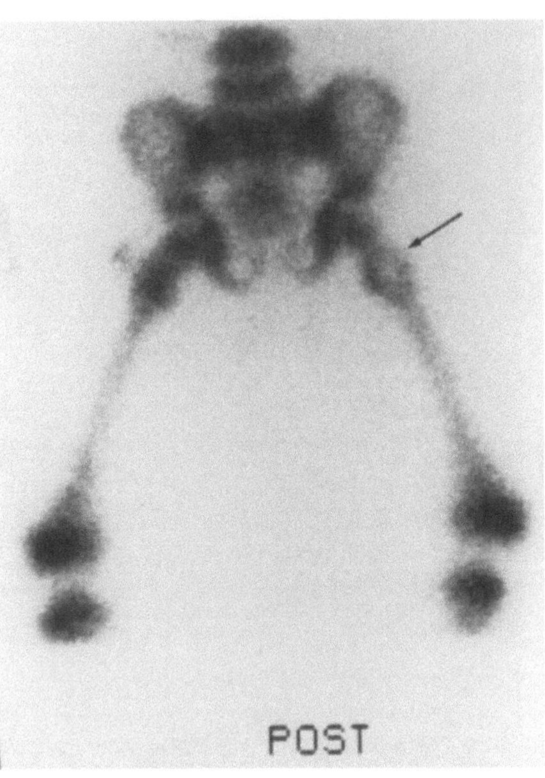

Fig. 17.13. ^{99m}Tc-MDP scintigraphy of a 4-year-old boy 2 days after onset of symptoms. Posterior view revealing a cold spot in the proximal right femur (*arrow*)

c

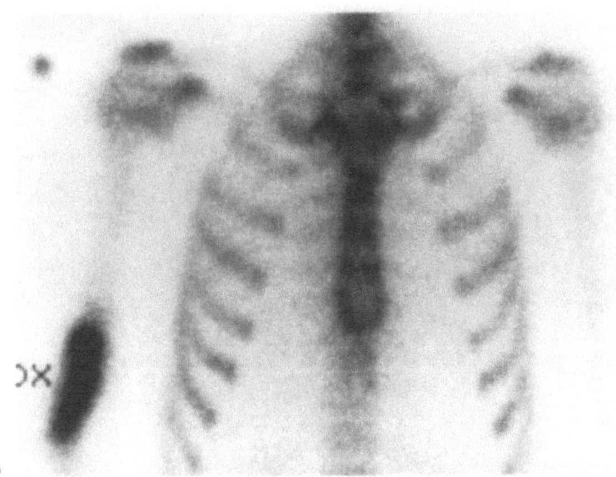

b

Fig. 17.12 a–c. Chronic osteomyelitis in the right humerus located at a healed shaft fracture. **a** No radiographic signs of osteomyelitis. **b** Positive ^{99m}Tc-MDP scintigraphy of the right humerus, 3 h after injection. **c** Positive leukocyte scintigraphy 24 h after injection

blood pool phase, is obtained 5 min after injection. With inflammation, capillaries dilate, causing increased blood pooling. The third phase is a late phase obtained 3 h after injection. In osteomyelitis in-

creased uptake is seen in all three phases, while in soft tissue infection only the first two phases are positive, while the third, delayed, phase is normal or the uptake is diffusely increased (Fig. 17.11b). A 24-h image may be added to the three-phase bone scan. Woven bone, i.e., abnormal bone around osteomyelitis and bone tumors, continues to accumulate MDP, while the accumulation stops in normal cortical bone after 4 h. Thus, there is an increased difference between normal bone and osteomyelitis that is better seen in the 24-h image (SCHAUWECKER 1992).

The three-phase bone scan has a sensitivity and specificity of 95% in adults with normal bone on plain radiographs (SCHAUWECKER 1992). In chronic osteomyelitis, often secondary to derangement of the bone, such as fracture, the bone scan findings are always pathological because of bone remodelling even if there is no active infection (SEABOLD et al. 1989; ROSENTHALL 1992). In patients with radiological findings consistent with chronic osteomyelitis the findings on ^{99m}Tc-MDP scintigraphy are always abnormal. In such cases other studies have to be considered, such as labelled leukocyte scintigraphy or

MRI (Fig. 17.12). In the very early phase of infection in children, usually less than 48 h after the onset, the bone scan may show an area of decreased uptake, a cold spot, due to local edema suppressing the circulation (TUSON et al. 1994) (Fig. 17.13).

17.4.2.2
Gallium-67 Scintigraphy

Gallium accumulates in active inflammation, not only in bone infection and septic arthritis, but also in cellulitis, myositis, tumors, and areas of trauma. After i.v. injection of ^{67}Ga citrate the substance is bound to several plasma proteins. Gallium is transported to inflammatory exudate because it is taken up by leukocytes and bacteria. There is also a leakage of plasma proteins due to increased capillary permeability and hypervascularity. The uptake is registered 48–72 h after injection. Gallium scintigraphy should be compared with conventional bone scintigraphy (Fig. 17.11). Osteomyelitis is diagnosed when there is increased uptake with both ^{99m}Tc and gallium scintigraphy. The uptake of gallium should be equal to or greater than that of ^{99m}Tc (DAVID et al. 1987; BOXEN and BALLINGER 1991.) There are several drawbacks to the use of gallium. The specificity is low, the radiation dose to the patient is high, the interval between administration and scintigraphy is long, and gallium is relatively expensive (FLIVIK et al. 1993).

17.4.2.3
Labelled Leukocytes

Labelling of leukocytes with indium-111 or ^{99m}Tc-hexamethylpropylene amine oxime (HMPAO) is theoretically the ideal technique to detect musculoskeletal infections. However, white cells infiltrate not only septic sites but also sites of aseptic chronic inflammation, osteonecrosis, loosening of prostheses, rheumatoid arthritis, and other noninfectious conditions.

In virgin-bone osteomyelitis a conventional three-phase bone scan is diagnostic and ^{111}In leukocyte scintigraphy is not necessary. In bone that has been violated by surgery, fracture, or other causes of increased bone turnover, conventional bone scintigraphy is not sufficiently specific for infection (Fig. 17.12). In such cases a ^{111}In leukocyte study is a good complementary examination since the leukocytes are not usually incorporated into areas of increased bone turnover (SCHAUWECKER 1989; SEABOLD et al. 1991). In patients with intra-articular or periarticular fracture with post-traumatic arthropathy there may be a false-positive finding on combined ^{99m}Tc-MDP scan and ^{111}In leukocyte scintigraphy, making culture confirmation necessary when a positive bone scan is obtained (SEABOLD et al. 1993). With a leukocyte study it is difficult to determine whether an infection is located in soft tissue or bone. If bone scintigraphy and leukocyte study are combined it is easy to establish this because the bony landmarks are clearly visualized with a bone scan.

A great problem is patients with total hip arthroplasty, since radiographically it may be difficult to differentiate infection from loosening. ^{111}In leukocyte study may be positive in both loosening and infection. The difficult part is that red bone marrow often accumulates around a prosthesis and this gives rise to a positive ^{111}In leukocyte study. Infected bone marrow does not accumulate a marrow-imaging agent such as sulfur colloids. PALESTRO et al. (1990) studied ^{111}In-labeled leukocytes and ^{99m}Tc-sulfur colloid to evaluate infection around hip prostheses. In cases with infection there was incongruity between the two agents, i.e., leukocyte accumulation but absence of uptake of sulfur colloid.

17.4.2.4
Nanocolloid

Technetium-99m nanocolloid scintigraphy is an alternative scintigraphic method in inflammatory diseases. The particle size is small, about 30 nm. A short investigation time is required, and registration is made 30 min after injection. Nanocolloid scintigraphy can be combined with conventional bone scan (Fig. 17.14). FLIVIK et al. (1993) compared the efficacy of nanocolloid scintigraphy with ^{111}In-labelled leukocytes and found no difference between the two agents in terms of sensitivity, specificity, and accuracy. The handling and costs are considerably lower than with ^{111}In labeling. OOI et al. (1993) found less reliable results with nanocolloid than with ^{111}In-labelled leukocytes in orthopedic infections.

In our institution we use the combination of nanocolloid scintigraphy and conventional ^{99m}Tc-MDP scintigraphy in patients with suspected osteomyelitis as the first examination after plain film radiography.

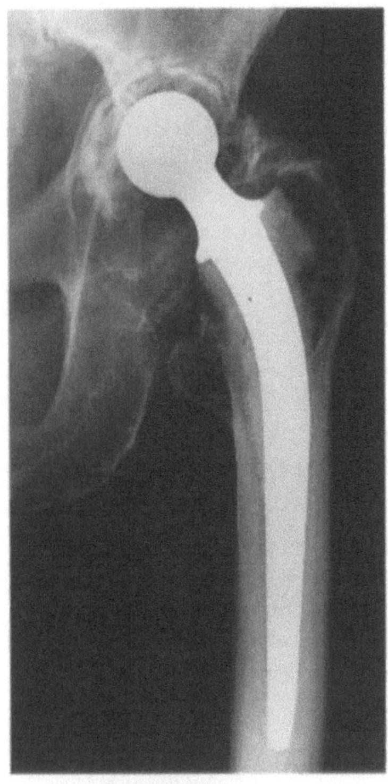

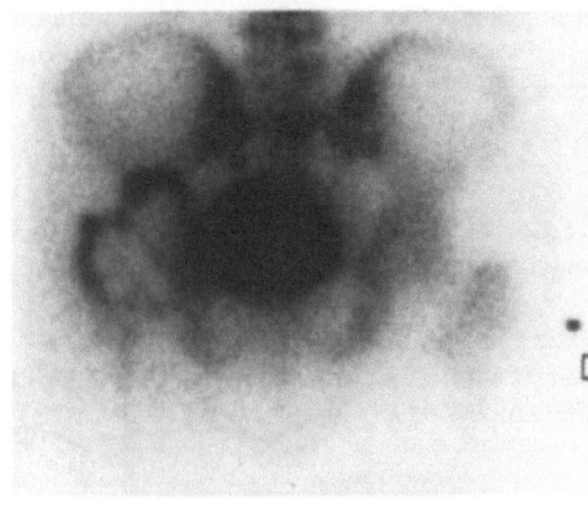

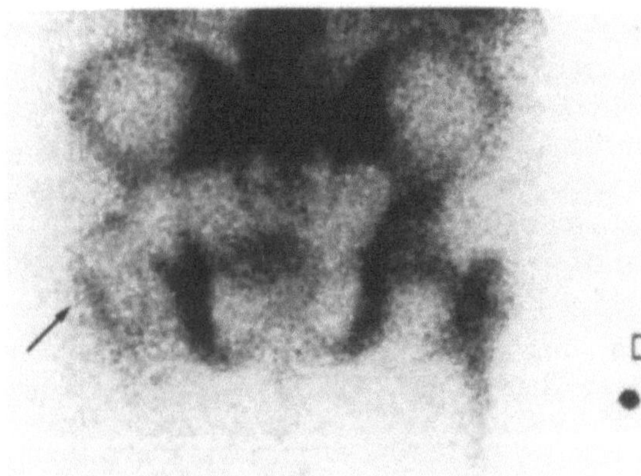

17.4.2.5
Other Radionuclide Methods

Indium-111 labeled human immunoglobulin G scintigraphy is reported to give high sensitivity and specificity for the diagnosis of orthopedic infections (OYEN et al. 1992). Other groups have used monoclonal antibodies, i.e., labeled antigranulocyte antibodies (LIND et al. 1990; BECKER et al. 1996; SCHEIDLER et al. 1994). These methods are still not commonly used.

17.4.3
Computerized Tomography

Computerized tomography is a valuable complement to radiography (Fig. 17.3b), especially in areas with complex anatomy, such as the spine. The extension of bone destruction and soft tissue involvement is easily diagnosed with CT (GOLD et al. 1991; GORDON et al. 1995). A small sequestrum that may be overlooked on radiography is well seen with CT (Fig. 17.5b), as are foreign bodies within the soft tissues. CT is also valuable for monitoring aspiration biopsy.

17.4.4
Fistulography

In a patient with a sinus tract it may be of value to establish whether this sinus tract has a connection with the underlying bone. Such an examination can be done with injection of contrast medium through a device with a cone on the top, occluding the opening of the sinus tract. This method is useful in making a preliminary estimation of the depth and direction of the sinus tract. A more detailed evaluation is usually achieved if the sinus tract is catheterized with a small catheter and the contrast medium is injected as close as possible to the infectious focus (Fig. 17.15). Such an examination can be combined with CT for a more

Fig. 17.14 a–c. Hip arthroplasty. **a** Radiography shows zones around the acetabular component. The picture is nonspecific and might also represent aseptic loosening. **b** ^{99m}Tc-MDP scintigraphy with registration 3 h after infection. Posterior view revealed increased uptake along the acetabulum and the proximal part of the femoral stem. **c** ^{99m}Tc-nanocolloid scintigraphy shows a similar but less extensive area of activity mainly around the proximal part of the femoral component (*arrow*). Needle puncture showed infection

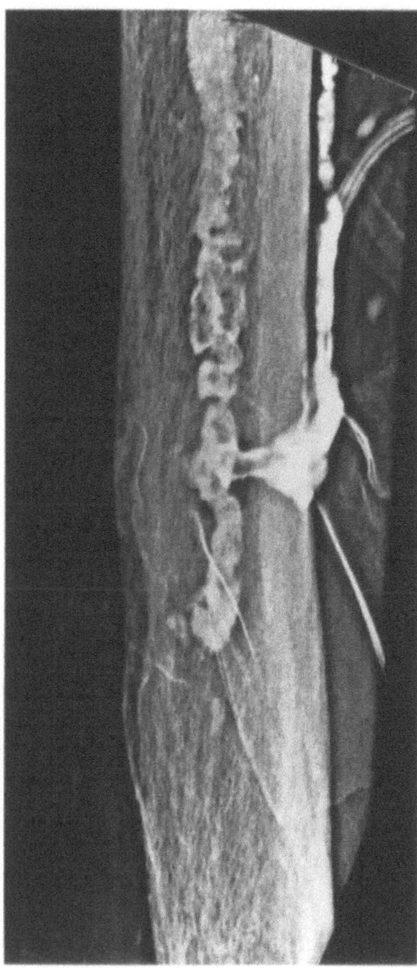

Fig. 17.15. Fistulography where a sinus tract is catheterized with a thin catheter. Contrast medium injection reveals extension of a sinus tract into the medulla

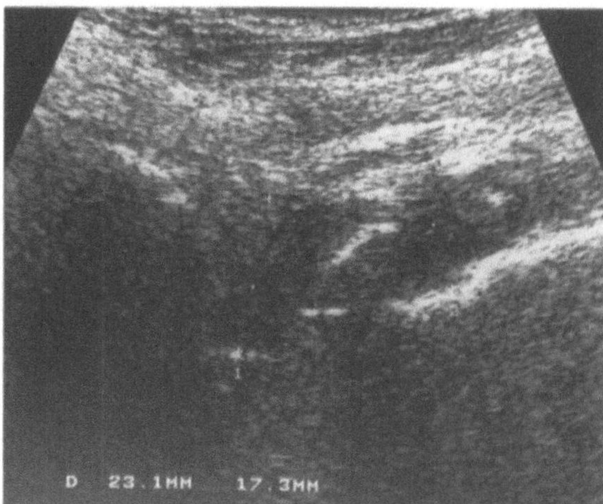

Fig. 17.16. Ultrasonography of an infected hip prosthesis. The shadow marked with "*1*" indicates the neck of the prosthesis. The distance between the calipers indicates marked fluid distension of the joint. Fluid is also seen anterior to the natural femoral neck, indicated by "*2*"

detailed view of the sinus tract, especially close to a joint. A sinus tract around a joint with a prosthesis is difficult to evaluate with CT, because of metallic artifacts. With tract injection and fluoroscopic monitoring of radiographs it is often possible to outline the connection of the tract to the prosthesis, i.e., to establish whether there is infective loosening of the prosthesis.

17.4.5
Ultrasonography

Ultrasonography permits evaluation of fluid collections in deep joints, e.g., the hip joint. Depending on the clinical situation, such a joint effusion indicates aseptic or septic arthritis, which can be further investigated by open biopsy or puncture. Also in patients with total hip arthroplasty, ultrasonography is valuable in demonstrating joint effusion (VAN HOLSBEECK et al. 1994) (Fig. 17.16), which also may be indicative of septic arthritis. In patients with clinically suspected osteomyelitis, ultrasonography can demonstrate the presence of a fluid collection adjacent to the involved bone (ABIRI et al. 1989) and subperiosteal abscesses (ABERNETHY et al. 1993; HOWARD et al. 1993; KAISER and ROSENBORG 1994). Soft tissue abscesses and sinus tracts are also easily diagnosed by means of ultrasonography (Fig. 17.17).

17.4.6
Magnetic Resonance Imaging

In the newborn and the young child, the bone marrow consists almost exclusively of hematopoietic marrow, which has a high water content. In T1-weighted images the signal is low. With increasing age there is a conversion of hematopoietic bone marrow to fatty bone marrow with fat content. In the adult hematopoietic bone marrow persists in the proximal metaphysis of the femora and the humeri and in the bone marrow of the spine and the pelvis. The long tubular bone consists mainly of fatty marrow. This fatty yellow marrow has a high signal intensity on both T1- and T2-weighted images.

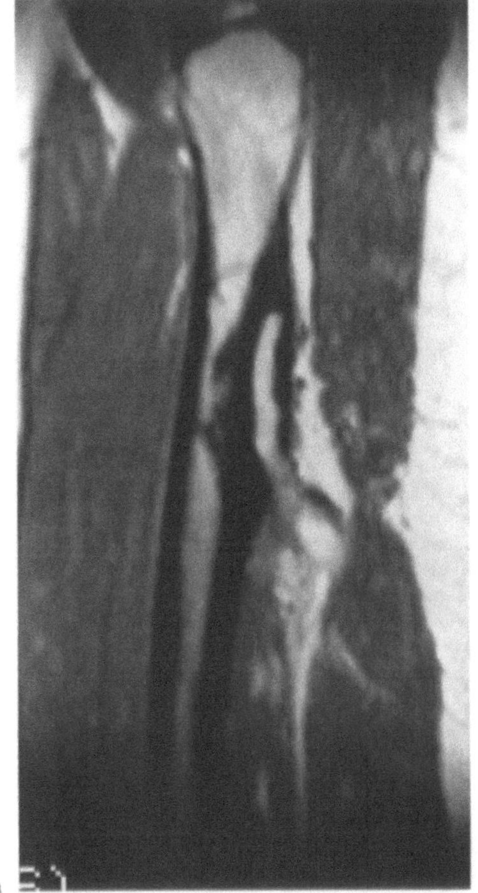

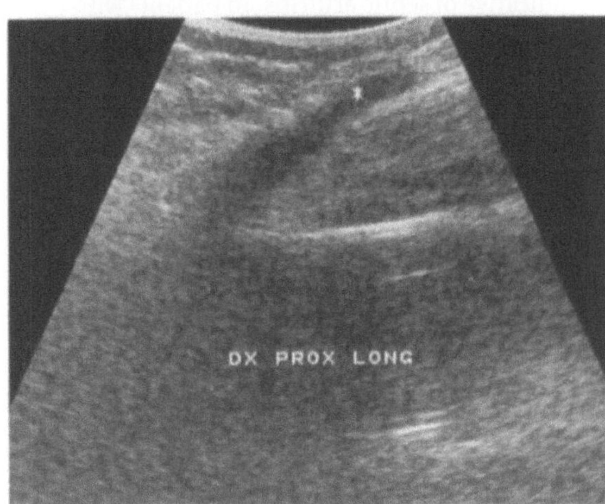

Fig. 17.17 a,b. Chronic osteomyelitis of proximal femur. **a** T2-weighted MRI examination: an abscess of the cortical bone and cloaca with extension to a soft tissue abscess is seen. **b** Ultrasonography in the same patient. The sinus tract and abscess formation are well visualized on ultrasonography (*asterisk*)

In acute osteomyelitis the involved bone marrow is seen as a low-signal-intensity lesion on T1-weighted images, and a high-signal lesion on T2-weighted images. The imaging quality may be increased by the use of short-tau inversion recovery (STIR) images (Fig. 17.18). STIR images are more sensitive to edema and water accumulation than T2-weighted images. The described bone marrow changes are, however, not specific for infection and the findings have to be evaluated in relation to the clinical situation. Improvement in diagnostic accuracy may be achieved by enhancement with gadolinium-DTPA (HOPKINS et al. 1995; DANGMAN et al. 1992) (Fig. 17.19) or by the use of fat-suppressed contrast medium-enhanced MRI (MORRISON et al. 1993).

Several reports have indicated that the sensitivity and specificity of MRI are better than those of various scintigraphic methods (MAZUR et al. 1995; MASON et al. 1989; CROLL et al. 1996). MRI is particularly valuable in diagnosing active infection in patients with chronic osteomyelitis, especially diabetics (CROLL et al. 1996; MORRISON et al. 1995; MASON et al. 1989). MRI is also of great assistance in disclosing a cloaca and sinus tract from an infectious focus (Fig. 17.17).

Although MRI is extremely sensitive there are pitfalls in the diagnosis of osteomyelitis (ERDMAN et al. 1991). In acute osteomyelitis, it may be difficult to differentiate between soft tissue extension of the infection and soft tissue edema (Fig. 17.19). The signal characteristics in both these conditions are the same, although in edema no tissue planes are destroyed and there is no mass effect in edema. The localized soft tissue abscess does produce a mass effect, and there is rim enhancement following the intravenous administration of gadolinium.

In septic arthritis there is typically an inflammatory signal change of the synovium in the affected joint and an effusion in the joint (ERDMAN et al. 1991). Septic arthritis may be complicated by secondary osteomyelitis of the adjacent bone, which on MRI will appear with decreased signal intensity on T1-weighted images and increased signal intensity on T2-weighted images and STIR. However, bone marrow edema in bone adjacent to an aseptic joint shows the same signal changes as in osteomyelitis. On the other hand, osteomyelitis in the metaphysis and epiphysis of a long bone may produce sympathetic effusion of the adjacent bone, which is mistaken for septic arthritis. If the bone marrow changes are seen on only one side of an infected joint or if there are marginal erosions of the bone and peri-

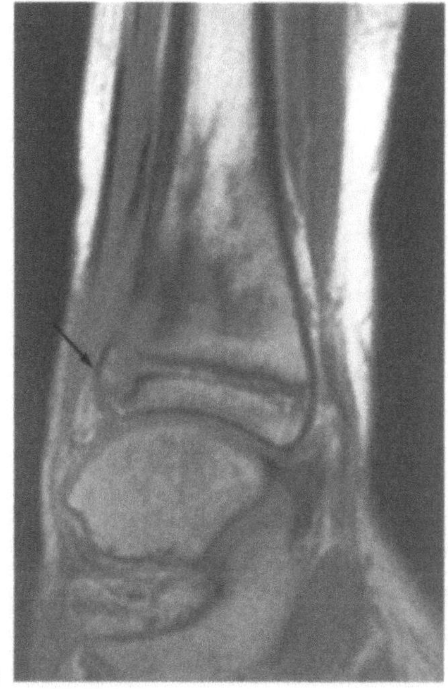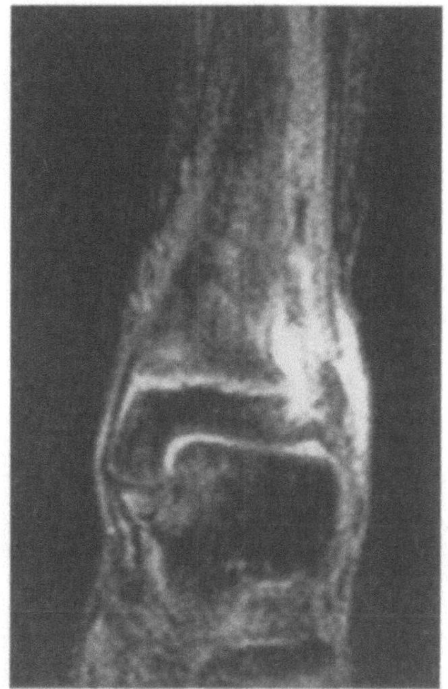

Fig. 17.18 a,b. Infection of distal metaphysis in a 14-year-old boy. a T1-weighted MR image in the sagittal projection: low signal intensity in the distal metaphysis indicates edema. In the more proximal part of the tibia, normal fat signa is evident. There is evidence of extension of infection from the metaphysis to the epiphysis (*arrows*). b STIR sequence in the coronal plane: high signal intensity is present in the metaphysis and epiphysis as well as in the growth plate and adjacent soft tissue. There is also evidence of fluid in the ankle joint

osteal reaction, osteomyelitis rather than edema is likely to be present. Gadolinium enhancement may assist in the differentiation between effusion in septic arthritis and sympathetic effusion by allowing evaluation of the synovial proliferation, which is usually extensive in septic arthritis. This finding does not exclude aspiration of the joint fluid for microscopy and culture.

In chronic osteomyelitis, infectious activity must be suspected if a sequestrum is seen, if there is a periosteal reaction or high signal intensity of the bone marrow on T2- and STIR sequences, or if there is enhancement following gadolinium administration. Again, these signs are nonspecific.

17.5
Needle Aspiration

In the clinical setting an imaging finding may indicate osteomyelitis or adjacent abscess formation, but the imaging techniques per se do not prove a diagnosis. In our institution we frequently employ needle aspiration biopsy for verification of suspected septic arthritis or osteomyelitis. Aspiration of hip joint effusion in a patient with suspected septic arthritis can be guided by fluoroscopy or ultrasonography (TAYLOR and BEGGS 1995). Bone destruction of unknown nature is usually easy to puncture with guidance by either fluoroscopy or CT (WHITE et al. 1995; HOWARD et al. 1994). In such a case, material is used both for cytology and for culture. In osteomyelitis with cortical destruction the lesion is easily punctured with a fine needle with a diameter of 0.7 mm (Fig. 17.5d). If there is intact cortical bone overlying the lesion, a thicker needle with mandrin should be used to penetrate the cortical bone (Fig. 17.4b). In a joint filled with pus it may be difficult to aspirate the pus through a fine needle, and a slightly thicker needle, 0.9 mm or 1.2 mm in diameter, is usually preferable for this purpose.

17.6
Summary

There are many diagnostic tools to verify disease in a patient with suspected osteomyelitis or septic arthri-

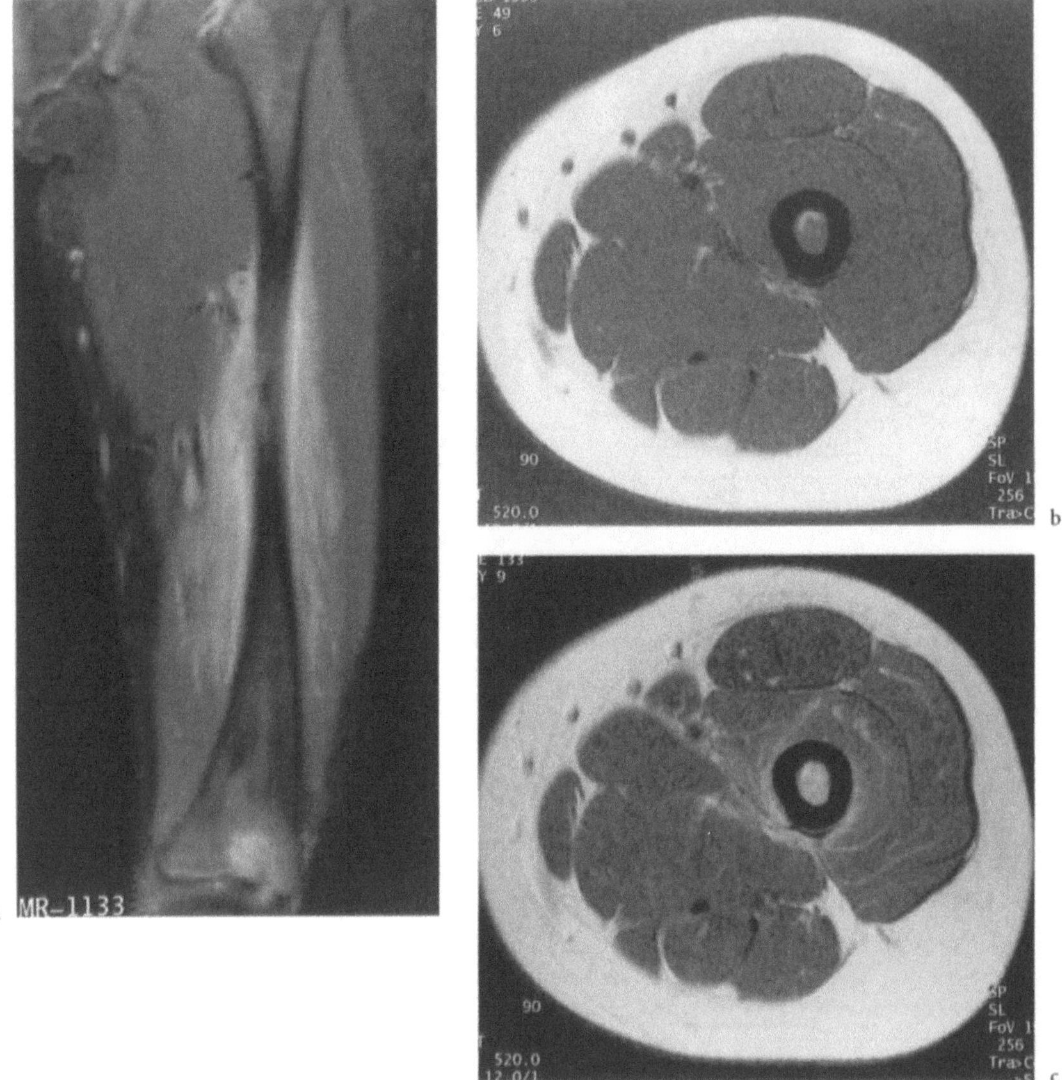

Fig. 17.19 a–d. Osteomyelitis in proximal part of the femur. **a** STIR sequence in the coronal plane. There is a massive edema surrounding the diaphysis of femur. High signal intensity is present in the metaphysis of distal femur. **b** T1-weighted axial view of femur. Low signal intensity within the bone marrow. **c** T1-weighted image after intravenous gadolinium DTPA. Increased signal intensity is seen within the bone marrow and in the edema adjacent to the femur. **d** Radiograph of the femur; lateral view. Cortical destruction (*arrow*) made us believe that this could be an osteosarcoma or Ewing's sarcoma

tis. The choice of diagnostic methods often depends on their availability. The primary examination should be radiography. If this shows clear pathology, needle biopsy or open biopsy should be performed for confirmation. If the radiographic findings are less clear, the next step should be three-phase ^{99m}Tc-MDP scintigraphy or MRI. In a case with radiographic changes that may suggest tumor or osteomyelitis, MRI is mandatory to evaluate extension of the lesion, followed by needle biopsy or open biopsy. In a patient in whom radiography reveals postoperative changes or remnants after previous fracture or chronic osteomyelitis, MRI or leukocyte scintigraphy should be performed to confirm activity. When septic arthritis is suspected, the diagnosis should be confirmed by puncture and aspiration of joint fluid. In a patient with a sinus tract, the extension of the sinus should be confirmed by means of contrast medium injection, i.e., fistulography.

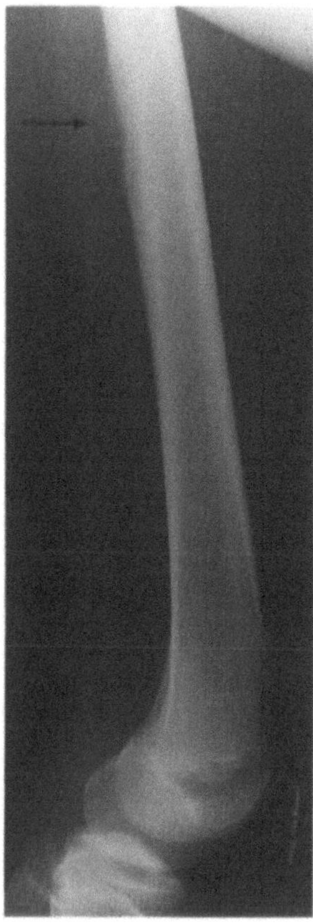

Fig. 17.19d

References

Abernethy LJ, Lee YCP, Orth MC, Cole WG (1993) Ultrasound localization of subperiosteal abscesses in children with late-acute osteomyelitis. J Pediatr Orthop 13:766–768

Abiri MM, Kirpekar M, Ablow RC (1989) Osteomyelitis: detection with US. Radiology 172:509–511

Becker W, Palestro CJ, Winship J, Feld T, Pinsky CM, Wolf F, Goldenberg DM (1996) Rapid imaging of infections with a monoclonal antibody fragment (LeukoScan). Clin Orthop 329:263–272

Boxen I, Ballinger JR (1991) Nuclear medicine detection of inflammation and infection. Curr Opin Radiol 3:840–850

Brower AC (1996) Septic arthritis. Radiol Clin North Am 34:293–309

Carr AJ, Cole WG, Roberton DM, Chow CW (1993) Chronic multifocal osteomyelitis. J Bone Joint Surg [Br] 75:582–591

Croll SC, Nicholas GG, Osborne MA, Wasser TE, Jones S (1996) Role of magnetic resonance imaging in the diagnosis of osteomyelitis in diabetic foot infections. J Vasc Surg 24:266–270

Dangman BC, Hoffer FA, Rand FF, O'Rourke EJ (1992) Osteomyelitis in children: gadolinium-enhanced MR imaging. Radiology 182:743–747

David R, Barron B, Madewell J (1987) Osteomyelitis, acute and chronic. Radiol Clin North Am 25:1171–1201

Erdman WA, Tamburro F, Jayson HT, Weatherall PT, Bond Ferry K, Peshock RM (1991) Osteomyelitis: characteristics and pitfalls of diagnosis with MR imaging. Radiology 180:533–539

Flivik G, Sloth M, Rydholm U, Herrlin K, Lidgren L (1993) Technetium-99m-nanocolloid scintigraphy in orthopedic infections: a comparison with indium-111-labeled leukocytes. J Nucl Med 34:1646–1650

Gold RH, Hawkins RA, Katz RD (1991) Bacterial osteomyelitis: findings on plain radiography, CT, MR and scintigraphy. AJR 157:365–370

Gonzalez MH, Papierski P, Hall RF Jr (1993) Osteomyelitis of the hand after a human bite. J Hand Surg [Am] 18:520–522

Gordon BA, Martinez S, Collins AJ (1995) Pyomyositis: characteristics at CT and MR imaging. Radiology 197:279–286

Hopkins KL, Li KCP, Bergman G (1995) Gadolinium-DTPA-enhanced magnetic resonance imaging of musculoskeletal infectious processes. Skeletal Radiol 24:325–330

Hovi I, Valtonen M, Korhola O, Hekali P (1995) Low-field MR imaging for the assessment of therapy response in musculoskeletal infections. Acta Radiol 36:220–227

Howard CB, Einhorn M, Dagan R, Nyska M (1993) Ultrasound in diagnosis and management of acute haematogenous osteomyelitis in children. J Bone Joint Surg [Br] 75:79–82

Howard CB, Einhorn M, Dagan R, Yagupski P, Porat S (1994) Fine-needle bone biopsy to diagnose osteomyelitis. J Bone Joint Surg [Br] 76:311–314

Kaiser S, Rosenborg M (1994) Early detection of subperiosteal abscesses by ultrasonography. A means for further successful treatment in pediatric osteomyelitis. Pediatr Radiol 24:336–339

Kanlic E, Perry C (1992) Advances in the laboratory and radiographic evaluation of bone and joint infections. Curr Opin Orthop 3:50–53

Lind P, Langsteger W, Koltringer P, Dimai HP, Passi R, Eber O (1990) Immunoscintigraphy of inflammatory processes with a technetium-99m-labelled monoclonal antigranulocyte antibody. J Nucl Med 31:417–423

Mason MD, Zlatkin MB, Esterhai JL, Dalinka MK, Velchik MG, Kressel HY (1989) Chronic complicated osteomyelitis of the lower extremity: evaluation with MR imaging. Radiology 173:355–359

Mazur JM, Ross G, Cummings RJ, Hahn GA Jr, McCluskey WP (1995) Usefulness of magnetic resonance imaging for the diagnosis of acute musculoskeletal infections in children. J Pediatr Orthop 15:144–147

Morrison WB, Schweitzer ME, Bock GW, Mitchell DG, Hume EL, Pathria MN, Resnick D (1993) Diagnosis of osteomyelitis: utility of fat-suppressed contrast-enhanced MR imaging. Radiology 189:251–257

Morrison WB, Schweitzer ME, Wapner KL, Hecht PJ, Gannon FH, Behm WR (1995) Osteomyelitis in feet of diabetics: clinical accuracy, surgical utility, and cost-effectiveness of MR imaging. Radiology 196:557–564

Ooi GC, Belton I, Finlay D (1993) Comparison of technetium 99m nanocolloid and indium 111 leucocytes in the diagnosis of orthopaedic infections. Br J Radiol 66:1025–1030

Oyen WJG, van Horn JR, Claessens RAMJ, Slooff TJJH, van der Meer JWM, Corstens FHM (1992) Diagnosis of bone, joint, and joint prosthesis infections with In-111-labeled nonspecific human immunoglobulin G scintigraphy. Radiology 182:195–199

Palestro CJ, Kim CK, Swyer AJ, Capozzi JD, Solomon RW, Goldsmith SJ (1990) Total-hip arthroplasty: periprosthetic indium-111-labelled leukocyte activity and complementary technetium-99m-sulfur colloid imaging in suspected infection. J Nucl Med 31:1950–1953

Resnick D, Niwayama G (1995) Osteomyelitis, septic arthritis and soft tissue infections; mechanisms and situations. In:

Resnick D (ed) Diagnosis of bone and joint disorders, 3rd edn. Saunders, Philadelphia, pp 2325–2418

Rosenthall L (1992) Radionuclide investigation of osteomyelitis. Curr Opin Radiol 4:62–69

Ruther W, Hotze A, Moller F, Bockisch A, Heitzmann P, Biersack HJ (1990) Diagnosis of bone and joint infection by leucocyte scintigraphy: a comparative study with 99m Tc-HMPAO-labelled leucocytes, 99m Tc-labelled antigranulocyte antibodies and 99m Tc-labelled nanocolloid. Arch Orthop Trauma Surg 110:26–32

Schauwecker DS (1989) Osteomyelitis: diagnosis with In-111-labeled leukocytes. Radiology 171:141–146

Schauwecker DS (1992) The scintigraphic diagnosis of osteomyelitis. AJR 158:9–18

Scheidler J, Leinsinger G, Pfahler M, Kirsch CM (1994) Diagnosis of osteomyelitis. Accuracy and limitations of antigranulocyte antibody imaging compared to three-phase bone scan. Clin Nucl Med 19:731–737

Seabold JE, Nepola JV, Conrad GR, Marsh JL, Montgomery WJ, Bricker JA, Kirchner PT (1989) Detection of osteomyelitis at fracture nonunion sites: comparison of two scintigraphic methods. AJR 152:1021–1027

Seabold JE, Nepola JV, Marsh JL, et al. (1991) Postoperative bone marrow alterations: potential pitfalls in the diagnosis of osteomyelitis with In-111-labeled leukocyte scintigraphy. Radiology 180:741–747

Seabold JE, Ferlic RJ, Marsh JL, Nepola JV (1993) Periarticular bone sites associated with traumatic injury: false-positive findings with In-111-labeled white blood cell and Tc-99m MDP scintigraphy. Radiology 186:845–849

Sundaram M, McDonald D, Engel E, Rotman M, Siegfried E (1996) Chronic recurrent multifocal osteomyelitis: an evolving clinical and radiological spectrum. Skeletal Radiol 25:333–336

Taylor T, Beggs I (1995) Fine needle aspiration in infected hip replacements. Clin Radiol 50:149–152

Trueta J (1959) The three types of acute hematogenous osteomyelitis. A clinical and vascular study. J Bone Joint Surg 41:671–680

Tuson CE, Hoffman EB, Mann MD (1994) Isotope bone scanning for acute osteomyelitis and septic arthritis in children. J Bone Joint Surg [Br] 76:306–310

Vande Streek PR, Carretta RF, Weiland FL (1994) Nuclear medicine approaches to musculoskeletal disease. Current status. Radiol Clin North Am 32:227–253

van Holsbeeck MT, Eyler WR, Sherman LS, et al. (1994) Detection of infection in loosened hip prostheses: efficacy of sonography. AJR 163:381–384

White LM, Schweitzer ME, Deely DM, Gannon F (1995) Study of osteomyelitis: utility of combined histologic and microbiologic evaluation of percutaneous biopsy samples. Radiology 197:840–842

Yao DC, Sartoris DJ (1995) Musculoskeletal tuberculosis. Radiol Clin North Am 33:679–689

18 Joint Prostheses

T.H. Berquist

CONTENTS

18.1
Introduction

Joint arthroplasty procedures have evolved over many years as attempts have been made to improve function and alleviate pain in patients with joint disorders. Joint replacement procedures were developed as an alternative to other techniques such as synovectomy and arthrodesis (Beckenbaugh 1979; Coventry 1996; Peterson 1977).

Early prosthetic designs and techniques improved with the development of newer alloys, more versatile modular designs and better cement techniques. Today, components are available for many joints including the shoulder, elbow, hand, wrist, and foot and ankle (Berquist 1995). Arthroplasty procedures are most commonly used for the hip and knee. Therefore, because of the complexity of image evaluation, our discussion will focus on hip and knee arthroplasty.

T.H. Berquist, MD, FACR, Professor of Diagnostic Radiology, Mayo Medical School; Chair, Department of Diagnostic Radiology, Mayo Clinic Jacksonville, 4500 San Pablo Road, Jacksonville, FL 32224, USA

18.1.1
Indications and Patient Selection

Indications for joint replacement arthroplasty are similar regardless of anatomic region. In most patients, joint replacement is considered because of pain and limitation of activity due to osteoarthritis, rheumatoid arthritis, post-traumatic degenerative disease, avascular necrosis, or congenital deformities (Habermann 1986). Certain neoplasms are also treated with limb salvage procedures using customized arthroplasty components (Berquist 1995; Morrey 1991b). Contraindications to joint replacement procedures include infection, paralysis, neuromuscular dysfunction, and systemic illnesses which preclude surgery (Habermann 1986; Berquist 1995).

18.1.2
Clinical Evaluation

Joint replacement procedures are elective and designed to relieve pain and improve function and/or stability. The success of these procedures is related to factors other than surgical technique, including patient age, weight, activity, occupation, and patient expectations. Most patients are not selected for these procedures until they are beyond 60 years of age because survival of component systems may decrease significantly after 10 years (Morrey 1991b). Age is a good example of how patients may be selected for hip replacement. For example, patients over 65 years of age are considered for joint replacement if pain interferes with activity or sleep and the pain has not responded to a course of conservative therapy over a period of 3–6 months. Younger patients (55–65 years of age) are given a longer trial of conservative therapy. In patients under 55 years of age, activity must be significantly restricted and pain severe to consider them candidates for joint replacement. Osteotomy and even arthrodesis are still considered alternative forms of treatment in younger patients (DeOrio and Blasser 1991).

Although image findings are important, clinical criteria are also important for selecting patients and evaluating surgical results. Different systems for clinical scoring have been developed for the hip and the knee (Berquist 1995).

The Harris scoring system is commonly used for hip assessment (Harris 1969). This scoring system includes four categories: (a) pain (0–44 points), (b) function (0–47 points), (c) deformity (0–4 points) and (d) range of motion (0–5 points). Functional categories include stair climbing, daily activities, walking, and the need for assistance with a cane or crutches.

Several clinical evaluation systems have been used for the knee. The most recent system was developed by The Knee Society (Insall et al. 1989). The knee assessment portion evaluates pain, stability, and range of motion. A well-aligned knee with no pain or instability is given a score of 100 points. Functional categories include walking and stair climbing without assistance, which is also scored at 100 points (Insall et al. 1989). Both hip and knee scoring systems are frequently referred to in orthopedic literature. Therefore, radiologists should be familiar with these systems.

18.2
Preoperative Imaging

Preoperative imaging of the hip and knee is essential to the orthopedic surgeon. Anatomy, bone stock, deformity from congenital disease and previous trauma or surgery, and other factors such as template measurements must be evaluated (Berquist 1995).

Radiologists need to be familiar with features evaluated by surgeons on preoperative images. Also, we must be certain that images are properly obtained to assure accurate measurements.

18.2.1
Preoperative Imaging for Hip Arthroplasty

Most image data required for preoperative evaluation can be obtained from routine radiographs. The standard radiographic series comprises an AP view of the pelvis that includes the upper third of the femur and a lateral view. Magnification markers are included on the film for each view (Berquist 1995). There are numerous features that should be re-

viewed on the AP radiograph (Fig. 18.1). Features and measurements are listed below:

1. Ischial tuberosity line (Fig. 18.1a): A line connecting the ischial tuberosities which is used to evaluate leg length discrepancy and form the acetabular angles.
2. Acetabular angle and femoral head coverage (Fig. 18.1a): The acetabular angle is determined by a line along the acetabular margins and the angle formed with the ischial tuberosity line.
3. Ilioischial or Kohler's line (Fig. 18.1b): This line is drawn from the pelvic border of the sciatic notch to the lateral inferior margin of the obturation foramen; it is used to evaluate acetabular protrusio.
4. Calcar to canal isthmus ratio (Fig. 18.1b): This ratio is calculated by measuring the marrow width at the mid lesser trochanter and at a second level 10 cm below the lesser trochanteric measurement. The ratio of the lower width over the trochanteric width times 100% is normally about 50%.
5. Femoral neck angle (Fig. 18.1c): This angle is formed by lines through the central neck and femoral shaft. The normal angle is 135°.
6. Femoral offset (Fig. 18.1c): This measurement is the distance from the center of the femoral head along a perpendicular line to a line through the central femoral axis.

Other important radiographic features that should be evaluated are bone loss, changes of diffuse idiopathic skeletal hyperostosis (DISH), and prominent osteophytes (Berquist 1995; Pellegrini and Gegoritch 1996).

On occasion, computed tomography (CT) or magnetic resonance imaging (MRI) may be required to evaluate bone loss or soft tissue abnormalities (Berquist 1995). In complex cases, selective injection of the hip with anesthetic is useful to confirm that the source of pain is intra-articular (Berquist 1993).

Orthopedic vendors provide templates for selecting acetabular and femoral component sizes. Templates are overlaid on radiographs to assist with selecting the proper component.

18.2.2
Preoperative Imaging for Knee Arthroplasty

Routine radiographs provide most data necessary for preoperative assessment of patients undergoing evaluation for knee arthroplasty. Our standard radiographic series consists of full-length standing AP, lateral, notch, and merchant views (BERQUIST 1995). The full-length standing view is used to assess joint congruency for the hips, ankles, and knees and to

evaluate tibial and femoral angles and other abnormalities. Measurements obtained from the standing film provide valuable information for the surgeon. Figure 18.2 demonstrates the common measurements obtained by orthopedic surgeons. These measurements include the mechanical axis, femoral axis, the vertical axis, and the femoro-tibial angle (normal 5–7° valgus) (STUART 1991).

Lateral, notch, and merchant views should be evaluated to determine compartment involvement by arthritic diseases, bone loss, and other abnormalities which may impact surgical decisions and implant selection (BERQUIST 1995). CT, MRI, and diagnostic injections may also be useful in selected cases (BERQUIST 1993).

18.3
Component Design and Selection

Preoperative image findings, clinical data (age, weight, activity, and patient expectations), and surgical preference dictate the type of component system that will be used by the surgeon. The number of components available today is too numerous for them to be discussed thoroughly in this chapter. However, it is important for radiologists to understand basic design concepts. Therefore, we will review certain basic concepts of knee and hip components prior to discussing image evaluation of these implants.

Today, most prosthetic systems are modular and designed to be used with or without cement. Component systems use metal alloys, typically titanium or cobalt-chromium-molybdenum (BERQUIST 1995).

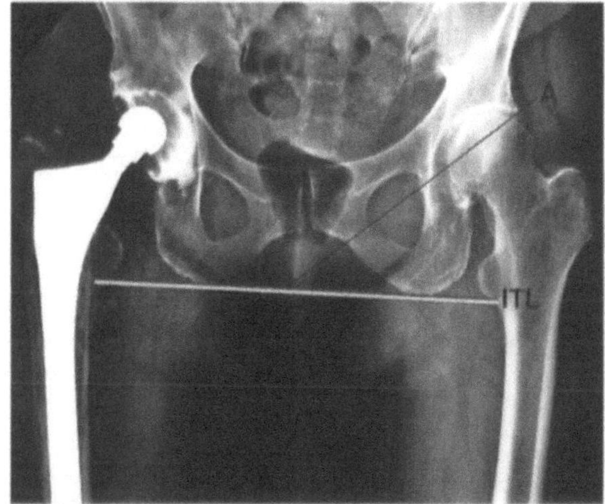

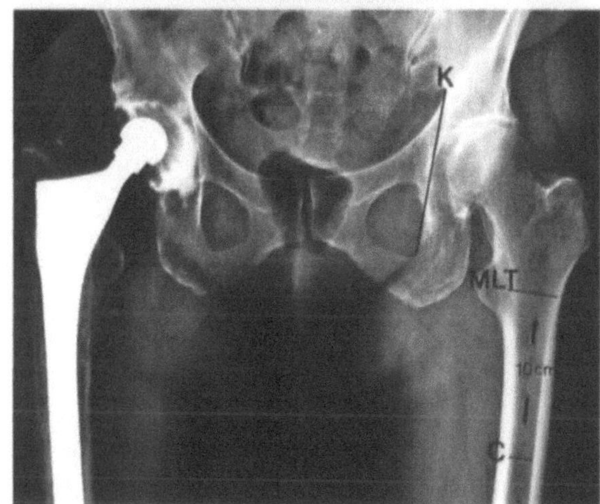

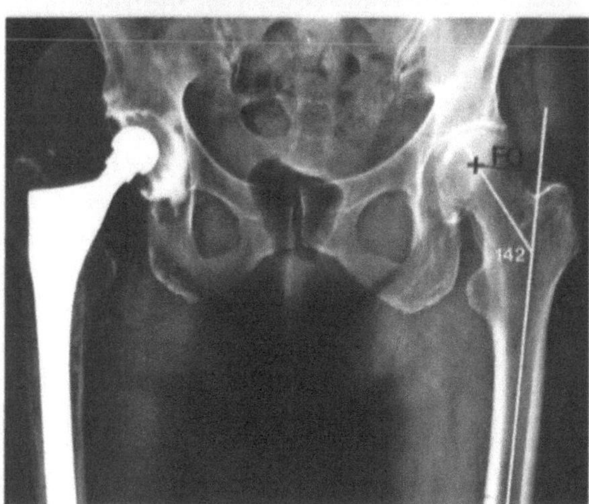

Fig. 18.1 a–c. AP view of the pelvis demonstrating a long stem revision system on the right. Common measurements are demonstrated. **a** The ischial tuberosity line (*ITL*) is used to evaluate leg length discrepancy and to evaluate the acetabular angle. Note the intersection with the lesser trochanters demonstrating no significant leg length discrepancy. The acetabular angle is formed by a line along the acetabular margins connecting to the ischial tuberosity line. In this case the angle measures 40°. The femoral head (*dotted line*) is partially uncovered due to degenerative arthritis and superolateral migration. **b** Kohler's line (*K*) is used to evaluate acetabular protrusio and to assess acetabular migration after surgery. The calcar-canal isthmus ratio – the canal measurement at *C* over the mid lesser trochanteric measurement (*MLT*) ×100% – should be about 50%. **c** The femoral offset (*FO*) is the distance from the center of the femoral head perpendicular to a line along the femoral shaft. The femoral neck angle is formed by lines along the femoral shaft and neck (*white lines*). In this case, the angle is 142°. Normal is 135°

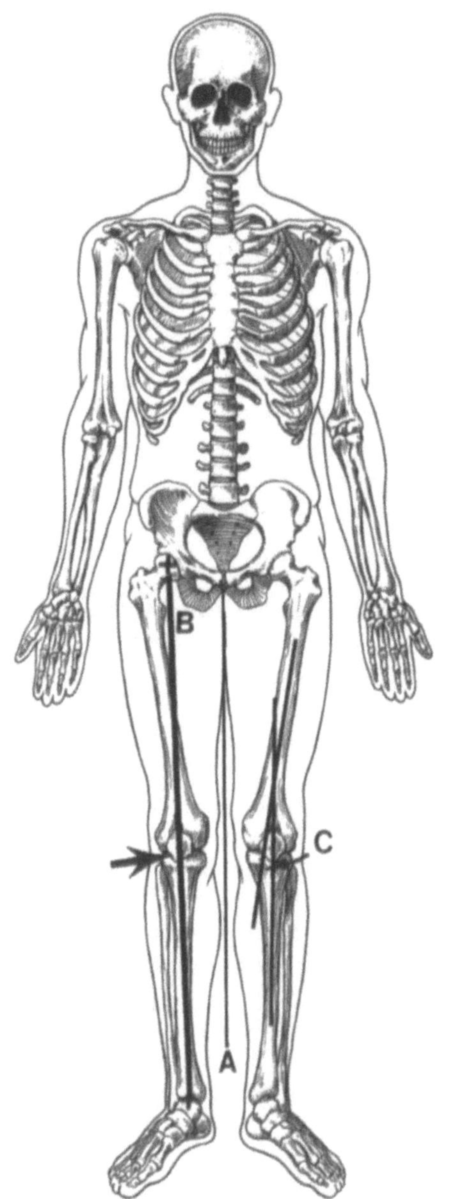

Fig. 18.2. Illustration of skeleton demonstrating measurements evaluated on standing radiographs. *A,* Vertical axis. This is a vertical line extending inferiorly from the pubic symphysis. *B,* Mechanical axis. This is formed by a line from the center of the femoral head (+) to the center of the ankle (+). This line should pass near or through the center of the knee (*arrow*). *C,* Femorotibial angle. This angle is formed by lines through the central axis of the tibia and femur. Normally the angle is 5–7° valgus

Metal components can be designed for press-fit, or they may be porous-coated to allow bone ingrowth. Components may be cemented in place regardless of the coating used. Polyethylene liners are used with metal backed acetabular components in the hip and as inserts for the tibial tray in the knee. All polyethyl-

ene tibial and acetabular components are also available. Ceramic femoral heads are used by some manufacturers (BERQUIST 1995).

18.4
Postoperative Imaging

There are numerous features that must be evaluated after joint replacement arthroplasty. Serial radiographs are still the most useful technique for detection of early abnormalities in patients with joint replacement arthroplasty.

18.4.1
Knee Imaging

Most surgeons prefer the least constrained prosthesis possible for primary reconstruction. Unicompartment replacement can be used for isolated compartment involvement (Fig. 18.3). However, most often total joint replacement is performed. Today, condylar designs (Fig. 18.4) are commonly used when cruciate ligaments are intact. Retaining the posterior cruciate ligament preserves the lever-arm quadriceps function, assists in femorotibial load transfer, and preserves proprioceptive function (RAND 1991). When the cruciate ligaments are deficient, a more constrained posterior-stabilized design may be selected (Fig. 18.5). Deficiency of the collateral ligaments is usually not a significant problem with modern condylar designs (RAND 1991; SCHNEIDER et al. 1986).

Bone loss, significant soft tissue deficiency, fracture deformity, and failure of previous arthroplasty implants must be approached differently. In these situations, revision systems (Fig. 18.5) or hinged systems (Fig. 18.6) may be required (RAND 1991). Most categories of components can be easily identified on routine radiographs (Figs. 18.3–18.6) (BERQUIST 1995).

Numerous radiographic features need to be assessed on postoperative images. Routinely, we obtain standing AP views on full-length films. The same features that are assessed preoperatively are remeasured (see Fig. 18.2). Also, a fluoroscopically positioned series should be obtained to properly align the component interfaces. The tibial and femoral components are aligned in AP and lateral projections. Flexion and extension images are obtained in the lateral projection to assess range of motion. The patellar component is also aligned on the lateral

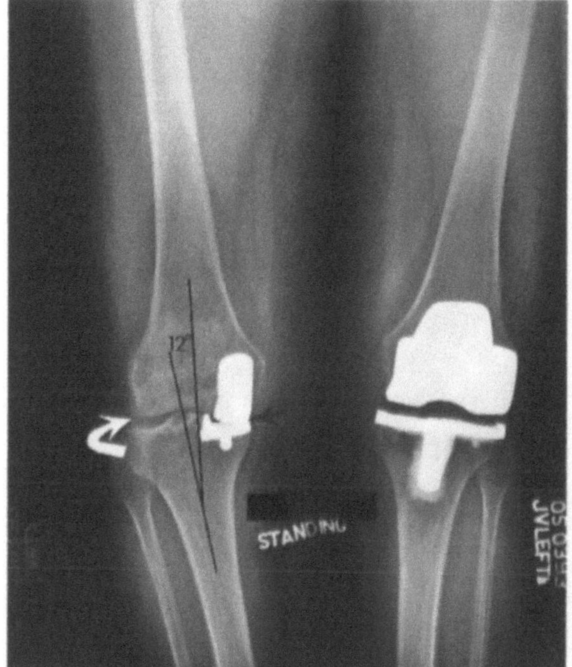

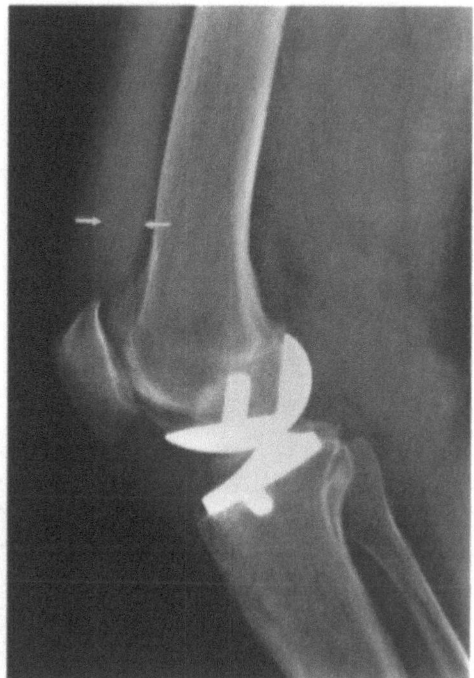

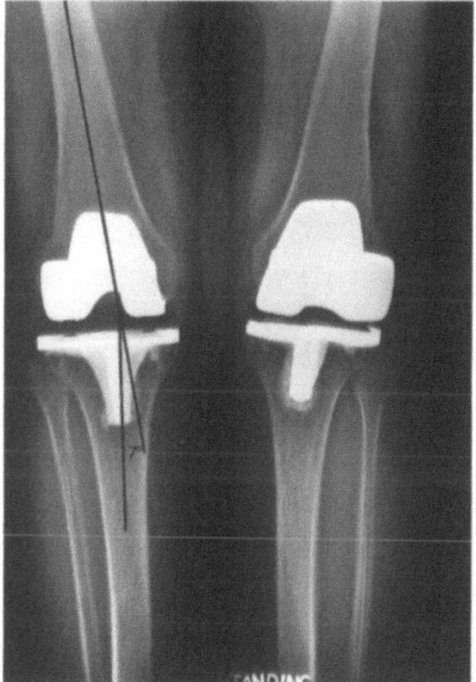

Fig. 18.3. a Standing AP radiographs of the knees demonstrating a medial compartment joint replacement on the right and a total knee replacement with Genesis components on the left. Note the obliteration of the medial joint space (*arrow*) on the right due to polyethylene wear. There is also lateral joint laxity (*curved arrow*). The femorotibial angle is 12° varus instead of the preferred 5–7° valgus. b Lateral radiograph of the right knee showing the uncemented, porous-coated femoral and tibial components. There is a large joint effusion (*arrows*). c Standing AP radiographs of the knee after revision of the right knee with a Johnson & Johnson system. The components are cemented. Note the uniform thickness of the polyethylene insert (*black lines*) on the tibial tray. The femorotibial angle is improved as well (7° valgus)

view. A merchant view is obtained to complete evaluation of the patellar component. These studies serve as a baseline for subtle changes such as lucent lines should complications develop. Stress views can be obtained when instability is suspected (Berquist 1995). Table 18.1 and Figs. 18.7–18.9 summarize features and measurements used postoperatively.

18.4.2
Hip Imaging

Hip components are usually modular with various acetabular components (cemented, press-fit, porous-coated) composed of all polyethylene or metal shells with polyethylene liners. Femoral head

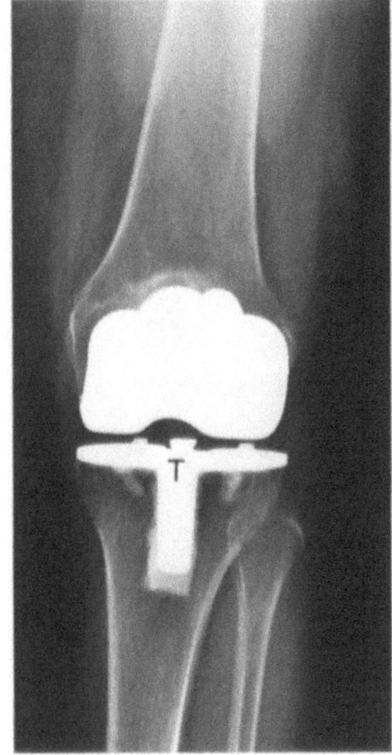 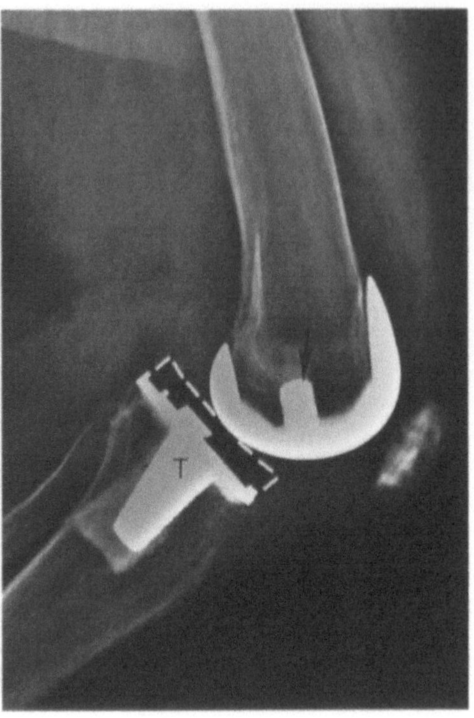

Fig. 18.4. AP (**a**) and lateral (**b**) radiographs demonstrating a cemented, porous coated Howmedica total knee with condylar design. The tibial tray (*T*) is metal with a polyethylene insert (*white lines* in **b**) and the patellar component is all polyethyl-ene. The two pegs (*arrow*) and thin femoral shell indicate a cruciate-sparing system. Compare with the posterior stabi-lized design in Fig. 18.5

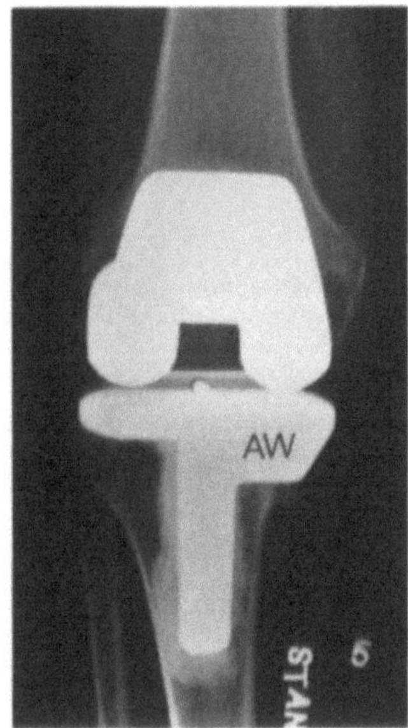 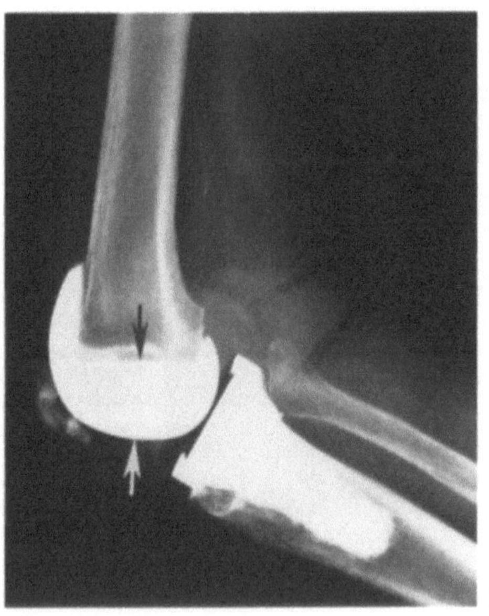

Fig. 18.5. AP (**a**) and lateral (**b**) radiographs of a posterior stabilized system used for revision. Note the tibial stem is longer and there is an augmentation wedge medially (*AW*) because of medial tibial bone loss. Note the thickness of the femoral component (*arrows*) and lack of pegs on the lateral view (**b**). Compare with Fig. 18.4b

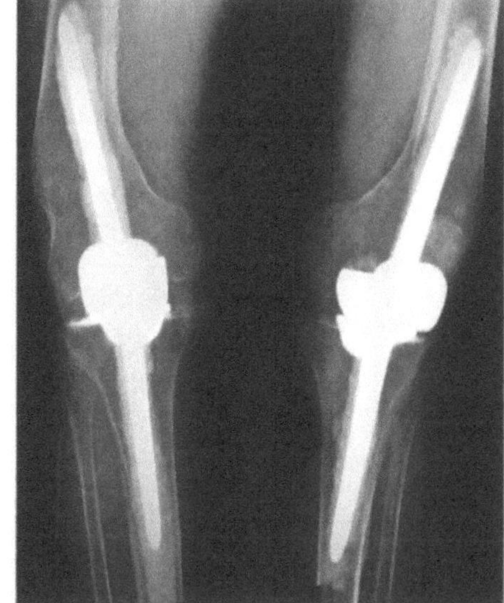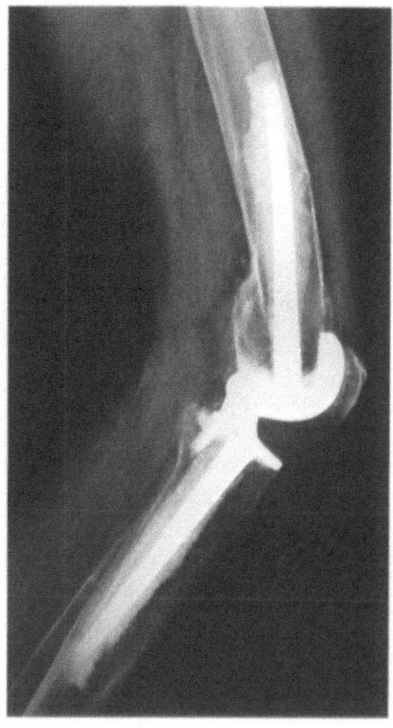

Fig. 18.6. Standing AP (a) and lateral (b) radiographs demonstrating cemented hinged prostheses used in a patient with osteoporosis, an old distal femoral fracture deformity, and muscle atrophy

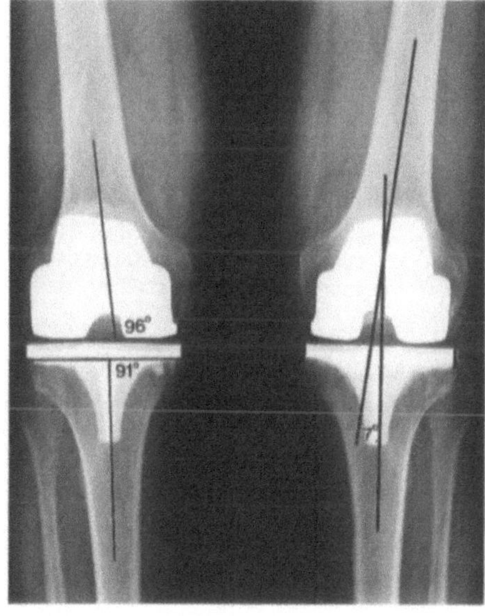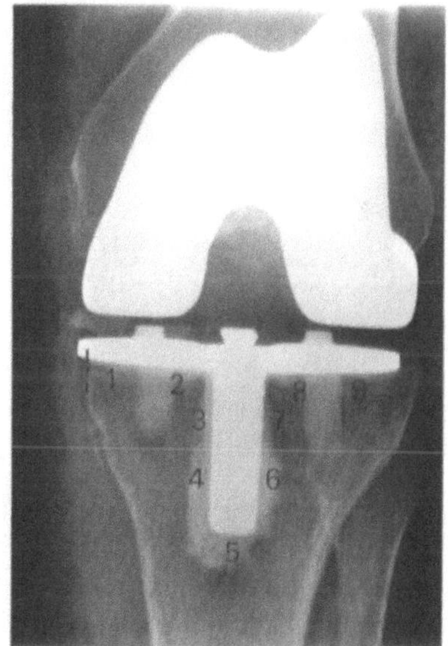

Fig. 18.7. a Standing AP radiograph of the knee with bilateral Johnson & Johnson cemented implants. The normal femoral and tibial component angles are demonstrated on the right and the femorotibial angle of 7° on the left. Note that both tibial trays almost completely cover the bone of the tibia (*vertical black lines*). b Fluoroscopically positioned AP view with tibial tray properly aligned to evaluate the interface with bone and cement. The tibial tray overhangs the medial tibial plateau slightly (*black broken lines*), which can lead to pes anserine bursitis. There are no lucent lines at the bone-cement interfaces in the zones described medial to lateral (*1 through 9*)

Table 18.1. Postoperative evaluation of knee prostheses (BERQUIST 1995; MANASTER 1995)

Radiographic view	Features
Full-length standing AP view (see Figs. 18.2 and 18.7a)	Mechanical axis
	Vertical axis
	Femorotibial angle (5–7° valgus)
Fluoroscopic AP view (Fig. 18.7b)	Tibial tray covers 85%+ of articular surface
	Tibial tray at 90° (±5°) to the tibial shaft
	Femoral component 97–98° to the shaft
	Lucent lines at bone-cement or bone-metal interface <2 mm
Fluoroscopically positioned lateral view (Fig. 18.8)	Tibial tray 90° to the tibial shaft
	Femoral axis should be along a line perpendicular to the femoral shaft
	Patella should be 9–10 mm above the tibial insert
	Zones for metal/bone or cement/bone lucency <2 mm
Merchant view (Fig. 18.9)	Zones for lucency are assessed (<2 mm)
	Patellar symmetry and joint space

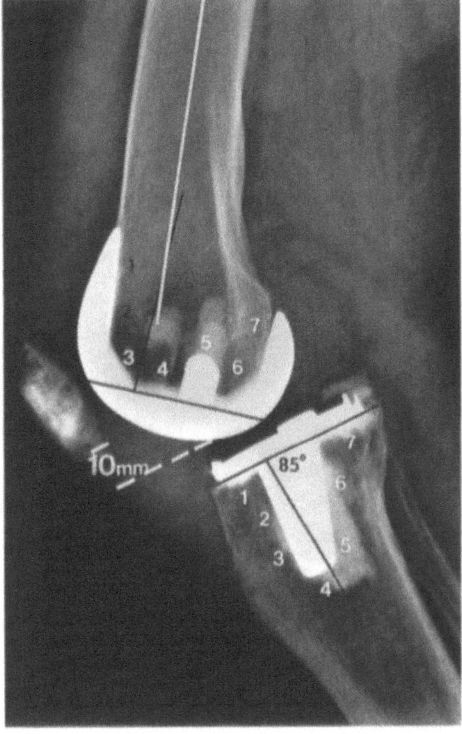

Fig. 18.8. Fluoroscopically positioned lateral view. The tibial tray should be at 90° to the shaft. In this case, it is 85°, or in slight flexion. A line perpendicular to the base of the femoral component (*black line*) should align with the femoral shaft line (*white line*). The inferior patellar margin should be 9–10 mm above the lucent tibial insert (*white dotted lines*). Zones for assessing lucency at the bone-cement interface are numbered. There are no lucent lines in this case

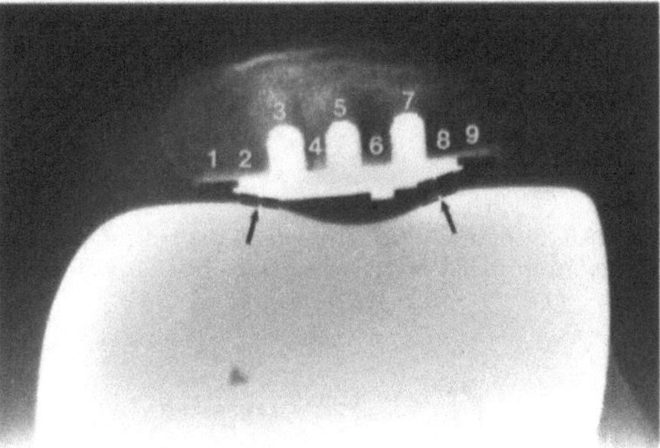

Fig. 18.9. Merchant view in a patient with a porous-coated, metal-backed patellar component. The interface is normal with zones marked for assessment of lucency. Note the patellar asymmetry (*arrows, broken lines*). This may be due to extensor mechanism imbalance or early polyethylene wear

sizes typically vary from 22 to 32 mm in diameter. A larger head reduces the thickness of polyethylene liner that can be used, but theoretically increases stability and range of motion (MORREY 1991a;

BERQUIST 1995). Femoral components have variable length, neck size, and neck angles. Custom components are available for use in patients with bone loss in either the calcar or the trochanteric region. Other variations include collars, stem sleeves, and centralizers. Radiologists should become familiar with components favored by surgeons at their institution or in their practice region.

Postoperative imaging requires serial routine radiographs. Unlike in the knee, fluoroscopic positioning is rarely required for hip implants. AP radiographs of the pelvis and operated femur and a lateral view to include the hip and femur are obtained. The femur included on the AP and lateral views should extend to 4–5 cm below the femoral component or cement plug in the case of cemented components (BERQUIST 1995). Table 18.2 and Figs. 18.10–18.12

Table 18.2. Postoperative evaluation of hip prostheses (BERQUIST 1995)

Radiographic view	Image features
AP view of pelvis (Fig. 18.10)	Acetabular component angle: normal 45°, range 35–55°
	Kohler's line: protrusio measurement
	Medial migration: measures acetabular migration
AP view of the hip and femur (Fig. 18.11)	Femoral component orientation: neutral to slight valgus
	Zones for lucent lines at bone-cement or metal-bone interface
Lateral view (Fig. 18.12)	Acetabular angle neutral to 15° from a vertical line
	Femoral component position

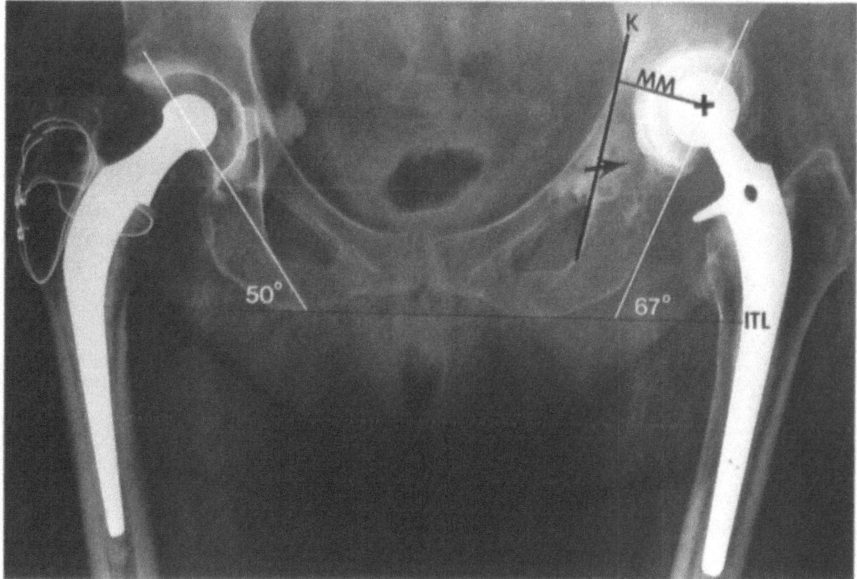

Fig. 18.10. AP view of the pelvis with bilateral cemented total hip arthroplasties. The ischial tuberosity line (*ITL*) shows no leg length discrepancy. The acetabular components are angled 50° on the right (normal 45°, range 35–55°) and 67° on the left due to a loose left acetabular component which has shifted inferiorly (*arrow*). Kohler's line (*K*) shows no acetabular protrusio. Acetabular migration is measured using a line perpendicular (*MM*) to Kohler's line (*K*) to the center of the femoral head (+). Changes are assessed on serial radiographs

summarize image features evaluated on postoperative radiographs.

18.5
Complications of Arthroplasty

Results of hip and knee arthroplasty have improved significantly over the years. Improved results are due to multiple factors including improved cement techniques and better implant designs. Charnley's hip designs used with cement have demonstrated survival rates of nearly 90% over 20 years (CHARNLEY 1974; SALVATI et al. 1981). Similarly, The Knee Society reports 90% good to excellent results in patients with total knee replacements (GILL and MILLS 1991; INSALL et al. 1989).

Most patients with complications following hip or knee arthroplasty present with pain, swelling, reduced function, or instability (BERQUIST 1995; MANASTER 1995; SCHNEIDER et al. 1986; IDUSUYI and MORREY 1996). Table 18.3 summarizes common complications of hip and knee arthroplasties (BERQUIST 1995; MANASTER 1995; IDUSUYI and MORREY 1991; PIRAINO et al. 1990; LYONS et al. 1985; HAYNES et al. 1993).

Clinical findings and appropriate imaging studies provide data necessary to diagnose complications in most cases. Serial radiographs with careful attention to changes in position of components and measurements may be all that is required. However, radionuclide bone scans or direct radioisotope arthrography, subtraction arthrography, and diagnostic injections are also useful (BERQUIST 1995; MANASTER 1995; BERQUIST et al. 1987; MAUS et al. 1987). The next sections will review common complications and appropriate imaging approaches.

18.5.1
Component Loosening

Historically, component loosening has been a major complication in patients with joint replacement arthroplasties (BECKENBAUGH and ILSTRUP 1978; GARCIA-CIMBRELO and MUNUERA 1992; HUE and FITZGERALD 1990; MALONEY et al. 1995). Improved cement techniques and component design have reduced hip component loosening from 57% in the early years to 15%–18% with new metal-backed designs (BERQUIST 1995). Early hinged knee components (see Fig. 18.6) had loosening rates of 20%–30%. However, the use of new condylar designs (see Figs. 18.3, 18.4) has reduced complications due to loosening to a few percent (MORREY 1991b; HUE and FITZGERALD 1990).

Imaging for suspected component loosening can be accomplished with serial radiographs, radionuclide scans, and subtraction arthrography

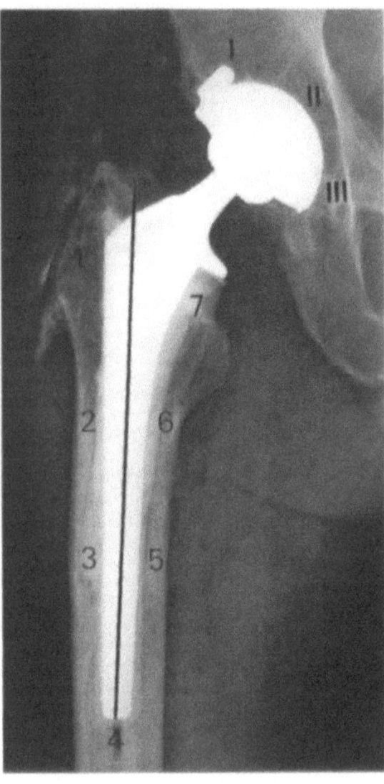

Fig. 18.11. AP radiograph of the hip and femur demonstrating a cemented femoral component and porous-coated, uncemented acetabular component. Ideally, the femur should be included to 5 cm below the component or cement plug. The acetabular interface is evaluated in three zones (*I, II,* and *III*) from lateral to medial. The femoral interface is evaluated in seven zones (1–7) from lateral to medial. The femoral component should be in line with the shaft (*black line*) or in slight valgus position

Table 18.3. Complications of hip and knee arthroplasty (BERQUIST 1995; MANASTER 1995; IDUSUYI and MORREY 1996)

Hip replacement complications	Knee replacement complications
Loosening	Wound healing
Infection	Infection
Dislocation	Extensor mechanism
Pseudobursae	Loosening
Greater trochanteric nonunion	Instability
Fractures	Fracture/dislocation
Osteolysis	Synovitis
Arthropathy	Osteolysis
Polyethylene wear	Polyethylene wear
	Pes anserine bursitis
	Deep venous thrombosis
	Peroneal nerve palsy

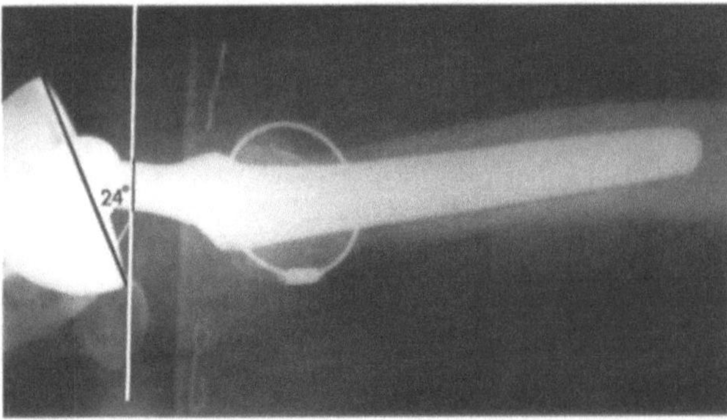

Fig. 18.12. Lateral radiograph of the hip demonstrating an acetabular angle of 24°. The angle is formed by a line along the acetabular cup (*black*) and a vertical line (*white*). The normal angle is neutral to 15°

(BERQUIST 1995; LYONS et al. 1985; MAUS et al. 1987; ROSENTHALL et al. 1985). Serial radiographs (see Figs. 18.7–18.12) provide valuable clues to component loosening. For optimal results, knee components should be fluoroscopically positioned to visualize the bone-metal or bone-cement interfaces (BERQUIST 1995). Radiographic evaluation differs somewhat for cemented and uncemented components; however, accuracy for predicting loosening based on plain film features may be as high as 84% for femoral loosening of hip prostheses (LYONS et al. 1985). Certain features with cemented components are even more accurate (~100%). Loosening is most accurately detected on serial radiographs that demonstrate a change in component position (Figs. 18.13, 18.14), a cement fracture, or lucent zones that have progressed to a width of >2mm (Fig. 18.15) (BERQUIST 1995; LYONS et al. 1985; MANASTER 1995; MALONEY and SMITH 1995). Uncemented components are more difficult to evaluate since components may not be loose even in the presence of lucent lines (KAPLAN et al. 1988; HEEKIN et al. 1993; MANASTER 1995). Progressive widening of lucent zones and shedding of porous-coated beads (Fig. 18.16) or mesh is still useful for predicting loosening

(BERQUIST 1995). Polyethylene wear is also frequently associated with loosening (MANASTER 1995) (Fig. 18.17). Tables 18.4 and 18.5 summarize radiographic features that may indicate component loosening for hip and knee arthroplasties.

Table 18.4. Routine radiographic features for loosening of hip components

Radiographic feature	Cemented components	Uncemented components
Acetabular component		
Position change (Fig. 18.13)	+	+
Cement fracture (Fig. 18.15)	+	N/A
Acetabular fracture or protrusio	+	+
Lucent zone >2mm (zone II most useful) (Fig. 18.16)	+	±
Femoral component		
Varus migration	+	+
Femoral stem fracture	+	+
Cement fracture	+	N/A
Endosteal resorption	+	+
Lucent zones >2mm	+	±
Subsidence	+	+

+, valuable feature; ±, somewhat useful; N/A, does not apply.

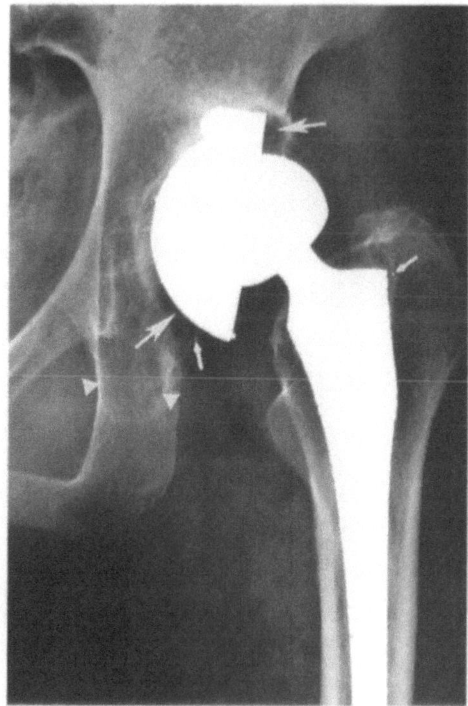
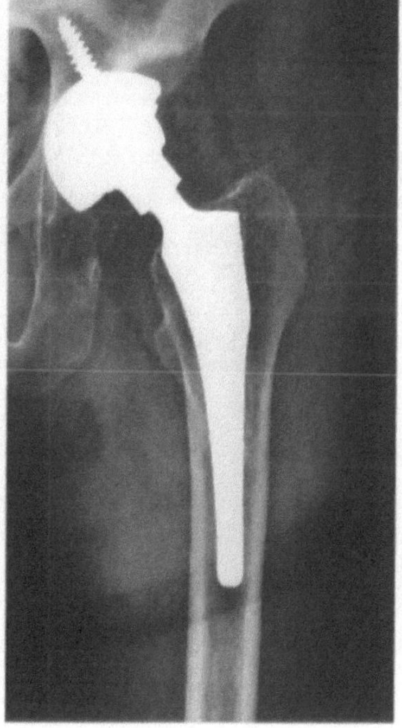

Fig. 18.13. a AP radiograph of the left hip demonstrating a porous-coated, uncemented hip system. The acetabular component has shifted (*large white arrows*) and there is a large area of osteolysis in the ischium (*arrowheads*). *Small white arrows* point to several shedded beads. **b** The acetabular component was revised and fixed with a single cancellous screw. The femoral head was also changed to match the new acetabular component (compare with **a**)

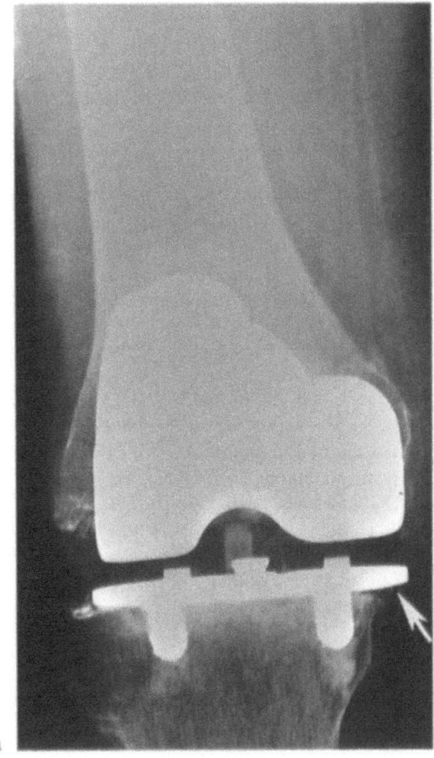

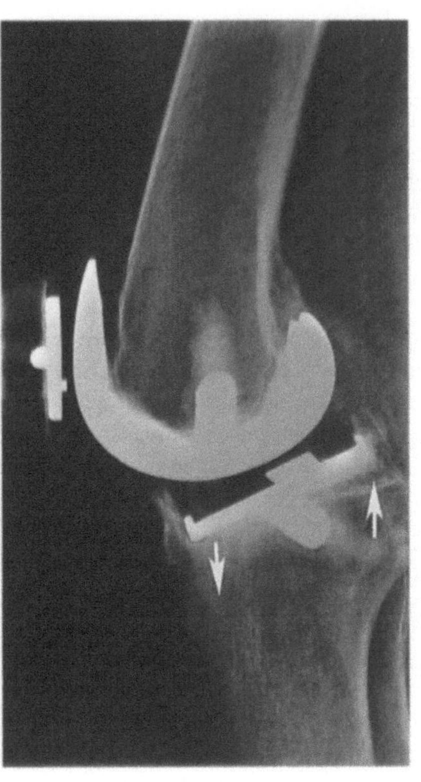

Fig. 18.14. AP (**a**) and lateral (**b**) radiographs of the right knee with a Howmedica PCA system. The tibial tray has moved with anterior depression and posterior elevation on the lateral view (**b**, *arrows*). The AP view (**a**) shows osteolysis under the medial tibial tray and medial overhang (*arrow*), which can lead to pes anserine bursitis

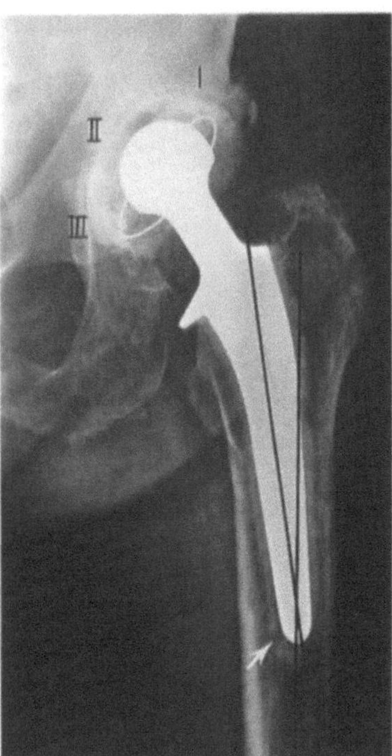

Fig. 18.15. AP radiograph of the left hip in a patient with cemented components. The acetabular component is polyethylene. There is a lucent zone in all three regions (*I–III*) with irregularity in zones I and II due to loosening. There is a cement fracture (*arrow*) at the tip of the femoral component. The component is in varus position (*lines*), indicating loosening of the femoral component

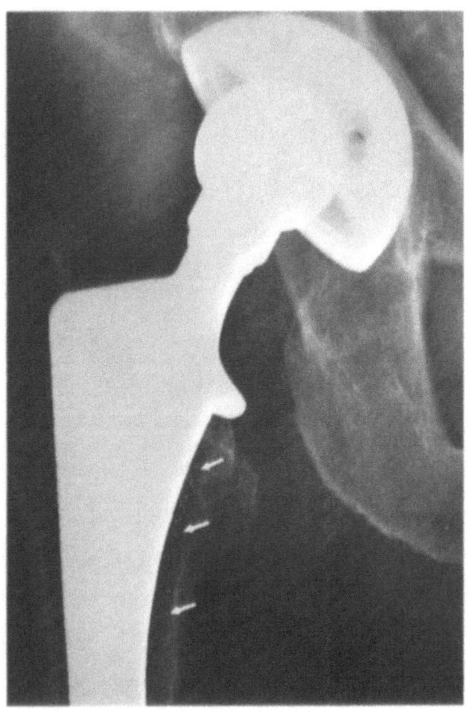

Fig. 18.16. Coned-down AP view of the upper femur demonstrating more than 30 shed beads or particles (*arrows*) in an area of osteolysis about this loose uncemented femoral component

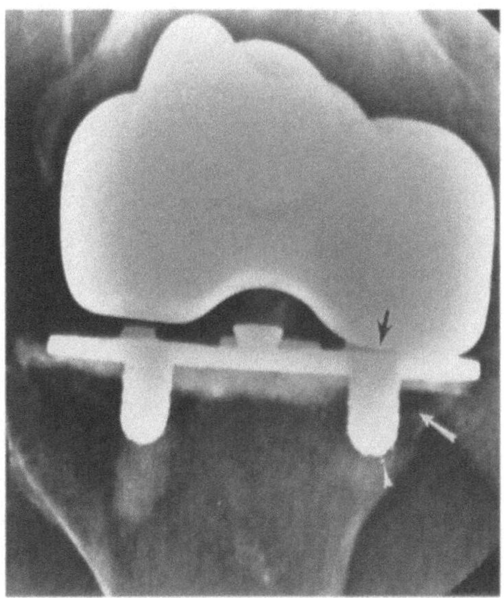

Fig. 18.17. AP fluoroscopically positioned image of the knee with a cemented PCA (Porous-Coated Anatomic) system. The tibial tray overhangs medially; there is osteolysis (*arrow*), bead shedding (*arrowhead*) and asymmetry of the polyethylene (*black arrow*) due to polyethylene wear and loosening of the tibial component

Table 18.5. Radiographic features of loosening of knee components

Radiographic feature	Cemented components	Uncemented components
Component migration (Fig. 18.14)	+	+
Progressive lucent zone >2 mm	+	+
Polyethylene wear[a] (Fig. 18.17)	+	+

+, useful feature; ±, somewhat useful; N/A, does not apply.
[a] May be seen with loosening and/or instability.

Patients with pain and suspected loosening and/or infection may require additional studies to confirm the diagnoses. Subtraction arthrograms, radionuclide scans, or direct radionuclide injection with arthrography and diagnostic anesthetic injections provide valuable information to the orthopedic surgeon (BERQUIST 1993, 1995; LYONS et al. 1985; MAUS et al. 1987; BRAUNSTEIN et al. 1995; LACHIEWICZ et al. 1996).

Selection of imaging techniques differs for the knee and hip. Subtraction arthrograms are frequently used for hip evaluation, but they are less effective in the knee. This technique permits fluid sampling and anesthetic injection to confirm the source of pain and provides valuable information about capsule size, pseudobursae, loosening, and infection. Proper technique requires joint distention which is more easily accomplished in the hip than the knee, where the large suprapatellar bursa decompresses the joint. Subtle loosening (contrast in lucent zones at the metal-cement or bone-cement interfaces) is more difficult to confirm in large joints (BERQUIST 1995; MAUS et al. 1987).

Accuracy of subtraction arthrograms for detection of loosening is superior for cemented than for uncemented components (Fig. 18.18). Sensitivity for femoral component loosening is 96% and specificity is 92% (MAUS et al. 1987). Acetabular component loosening is somewhat more difficult, especially with large pseudocapsules (>20 cc). The two most reliable indicators of acetabular loosening are contrast extending in zone II and a thick irregular area of contrast in any zone. These features improve accuracy for femoral component loosening to 95% (BERQUIST 1995; MAUS et al. 1987).

Radionuclide scans using technetium-99m methylene diphosphonate (MDP) can be helpful for detecting loosening, especially in the knee. However, there is normally some degree of pericomponent increased tracer uptake for up to 2 years after joint replacement (BERQUIST 1995; LYONS et al. 1985; ROSENTHALL et al. 1985).

18.5.2
Infection

Infection after joint replacement may be superficial, related to wound healing, or deep. Most infections are caused by Gram-positive cocci (74%). Patients present with elevated erythrocyte sedimentation rates in 63% of cases (TSUKAYAMA et al. 1996). Deep infection was once a common complication, occurring in 10%–11% of hip replacements and up to 19% of knee arthroplasty patients. Today, with improved technique, the incidence of infection for primary replacement surgery is 0.5%–2% (KAVANAGH et al. 1985; WILSON et al. 1990; RAND 1993). The incidence of infection is high in both the hip and the knee with revision surgery (RAND 1993; KAVANAGH et al. 1985).

Serial radiographs may demonstrate evolving changes to suggest infection. Endosteal scalloping and laminated periosteal new bone formation are associated with infection in 80% of cases (LYONS et al. 1985) (Fig. 18.19). Loosening is frequently associ-

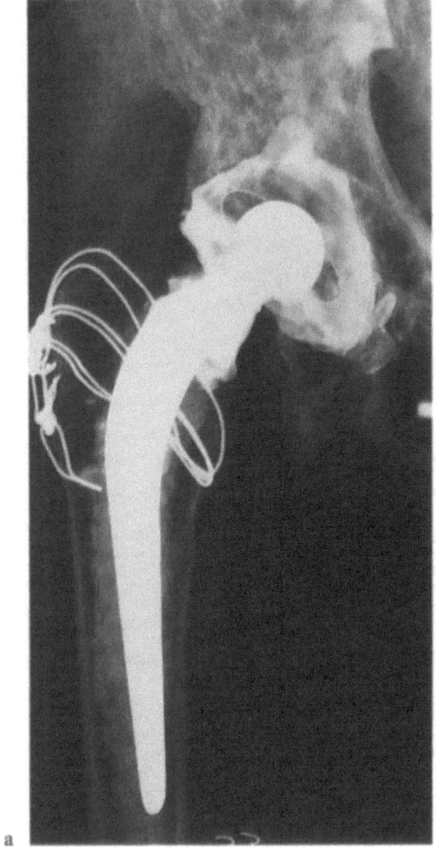
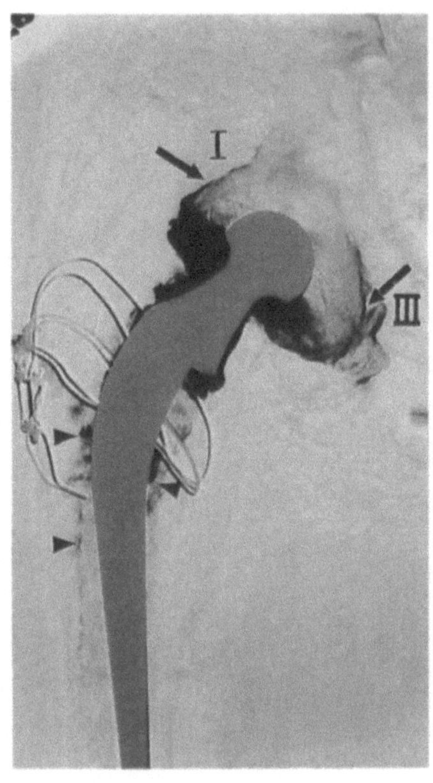

Fig. 18.18 a,b. Patient with a painful cemented right hip arthroplasty and Paget's disease of the ilium. **a** The conventional arthrogram is difficult to interpret due to the opaque cement, which is similar in density to the contrast agent. **b** The subtraction arthrogram allows the contrast dissecting along the bone-cement interfaces of the femoral (*arrowheads*) and acetabular components in zones I and II (*arrows*) to be easily appreciated. Both components were loose

ated with infection so features described in Tables 18.4 and 18.5 may also be seen radiographically (BERQUIST 1995).

Joint aspirations are useful in the hip and knee to obtain fluid samples for culture. Organisms are correctly identified in more than 70% of patients (MAUS et al. 1987). CUCKLER et al. (1991) reported accuracy rates of 83%, with a sensitivity of 67% and a specificity of 92%. Arthrograms are not as useful in the knee as in the hip. However, irregular pseudocapsules and sinus tracts are helpful indicators of infection. Lymphatic filling is not a useful radiographic feature in our experience (LYONS et al. 1985; BERQUIST 1995).

Radionuclide scans are of value for detecting infection, especially in the absence of significant radiographic features. Several techniques have been used. Combined indium-111 labeled white blood cell and technetium-99m MDP scans (Fig. 18.20) are 85%–93% accurate (PALESTRO et al. 1991; MERKEL et al. 1985). Combined leukocyte and sulfur colloid imaging may improve accuracy to 96%–97% (PALESTRO et al. 1991).

18.5.3
Extensor Mechanism Dysfunction

Although loosening and infection are significant complications in patients with either hip or knee replacements, extensor mechanism failure is a more common indication for revision of knee arthroplasties (BERQUIST 1995; MANASTER 1995). The incidence of this complication varies (5%–30%), but it is responsible for up to 50% of knee revision procedures (GRACE and RAND 1988). Abnormalities seen with this complication include patellar fracture, patellar subluxation or dislocation, patellar or quadriceps tendon tears, component loosening, and polyethylene wear (Fig. 18.21). Fluoroscopically

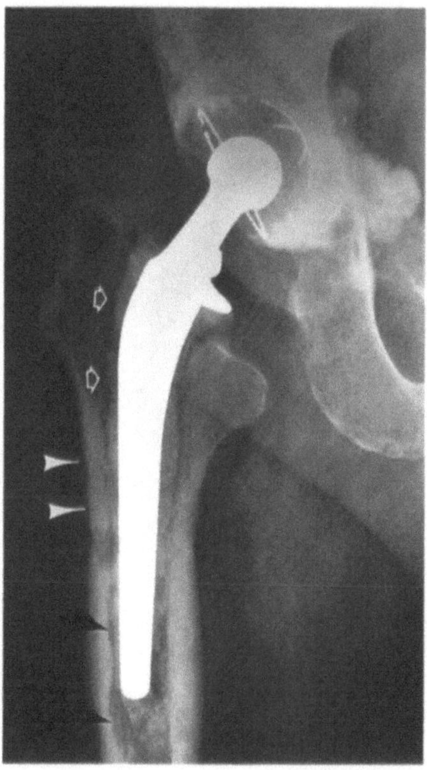

Fig. 18.19. AP radiograph of the right hip and upper femur in a patient with a loose, infected, cemented hip arthroplasty. There is endosteal scalloping (*large black arrows*) and subtle periosteal reaction (*white arrowheads*). Note the thick irregular lucent zones along the upper portion of the femoral component (*white open arrows*)

positioned lateral views, stress views, and merchant views, are usually adequate for radiographic diagnosis (BERQUIST 1995).

18.5.4
Other Complications

There are other complications of joint replacement arthroplasty. They have been summarized in Table 18.3, but some deserve further mention.

Fractures occur most commonly in osteopenic patients, in patients with loose components, in patients with hinged prostheses, or when technical problems occur at the time of implantation. The fractures, as expected, usually occur at the tip of femoral components in the hip or in the supracondylar region with knee prostheses (BERQUIST 1995; SCHNEIDER et al. 1986).

Dislocations of the patella have been noted above. Femorotibial dislocation is unusual though instability after knee replacement may occur in up to 13% of patients (BERQUIST 1995). Hip dislocations occur in about 3% of patients. This complication typically occurs shortly after surgery during the period of initial weight bearing (Fig. 18.22) (BERQUIST 1995).

Osteolysis may occur due to reaction to polyethylene debris, cement, or metal (BERQUIST 1995; MALONEY et al. 1995; TSUKAYAMA et al. 1996;

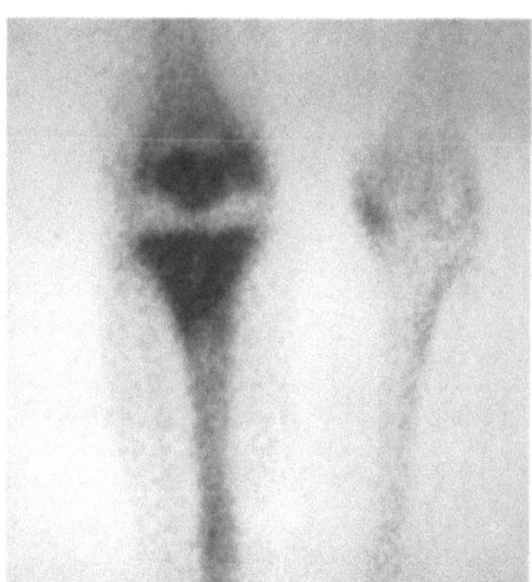

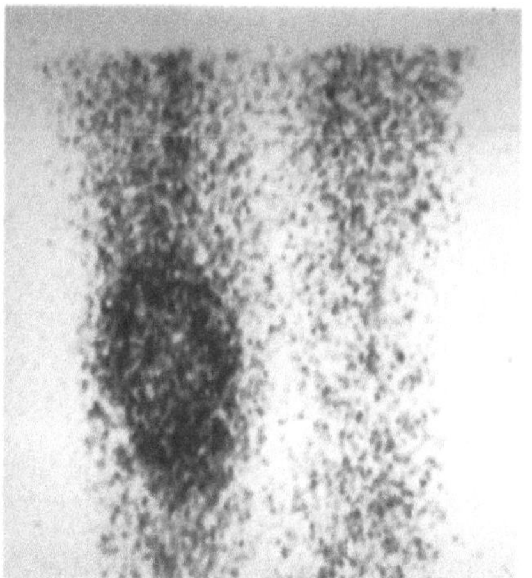

Fig. 18.20 a,b. Patient with infection after knee arthroplasty. a Technetium-99m MDP scan shows increased tracer in the region of the tibial and patellar components with less intense uptake about the femoral component. b Indium-111 leukocyte scan is strongly positive

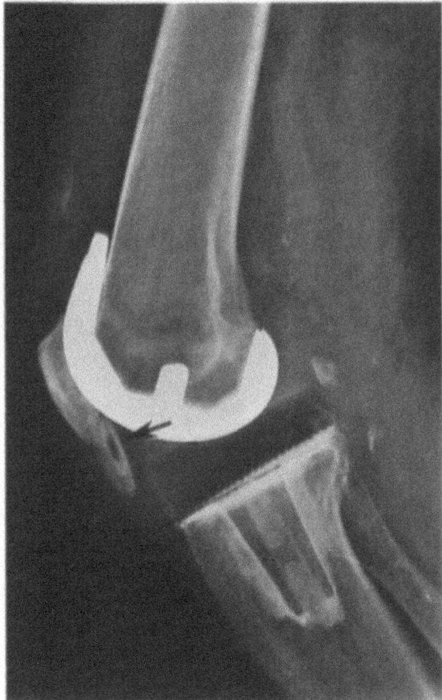

Fig. 18.21. Merchant (**a**) and lateral (**b**) views of the knee demonstrating displacement of the patellar component (*arrow*) and lateral subluxation of the patella

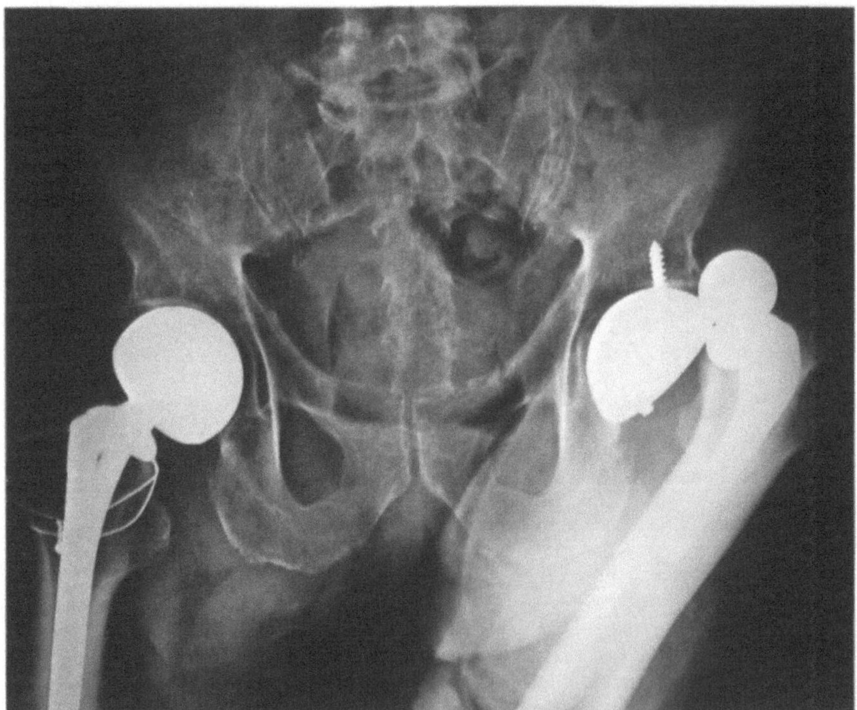

Fig. 18.22. AP radiograph of the pelvis and hips in a patient with a bipolar endoprosthesis on the right side and a recently inserted left total hip arthroplasty. The left hip is dislocated posterosuperiorly

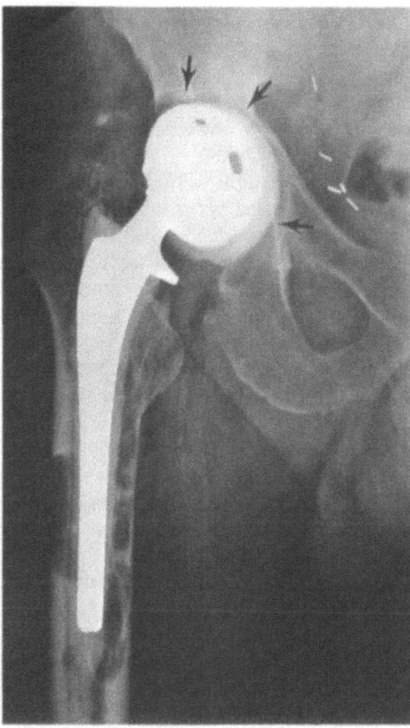

Fig. 18.23. AP radiograph of the pelvis and upper femur in a patient with cemented hip components. There is slightly irregular lucency around the acetabular component (*arrows*) with more aggressive areas of osteolysis about the femoral component. Both components are loose and the acetabular component is vertically oriented

SCHMALZREID et al. 1992; GRIFFITHS et al. 1987). Bone resorption is usually detected with radiographs (Fig. 18.23) as linear or local areas of bone loss (TSUKAYAMA et al. 1996). Large pseudotumor-like areas of resorption may also occur (GRIFFITHS et al. 1987). Loosening is not uncommonly associated with either type of osteolysis (see Fig. 18.16) (BERQUIST 1995; ZICAT et al. 1995).

There are several less frequent complications that are associated with knee replacement. These include peroneal nerve palsy (IDUSUYI and MORREY 1996), deep venous thrombosis, and pes anserine bursitis. The latter is associated with medial overhang of the tibial tray (see Fig. 18.7b) (BERQUIST 1995).

References

Beckenbaugh RD (1979) Total joint arthroplasty: the wrist. Mayo Clin Proc 54:513–515

Beckenbaugh RD, Ilstrup DM (1978) Total hip arthroplasty: a review of three hundred and thirty-three cases with long follow-up. J Bone Joint Surg [Am] 60:306–313

Berquist TH (1993) Diagnostic and therapeutic injections as an aid to musculoskeletal diagnosis. Semin Intervent Radiol 10:326–343

Berquist TH (1995) Imaging atlas of orthopedic appliances and prostheses. Raven Press, New York

Berquist TH, Bender CE, Maus TP, et al. (1987) Pseudobursae: a useful finding in patients with painful hip arthroplasty. AJR 148:103–106

Braunstein EM, Cardinal E, Buckwalter KA, Capello W (1995) Bupivicaine arthrography of the post-arthroplasty hip. Skeletal Radiol 24:519–521

Charnley J (1974) Total hip replacement. JAMA 230:1045

Coventry MB (1996) Historical perspective of hip and arthroplasty. In: Morrey BF (ed) Joint replacement arthroplasty. Churchill Livingstone, New York, pp 491–499

Cuckler JM, Star AM, Alavi A, Noto NB (1991) Diagnosis and management of infected total joint arthroplasty. Orthop Clin North Am 22:523–530

DeOrio JK, Blasser KE (1991) Indications and patient selection. In: Morrey BF (ed) Joint replacement arthroplasty. Churchill Livingstone, New York, pp 547–559

Garcia-Cimbrelo E, Munuera L (1992) Early and late loosening of the acetabular cup after low-friction arthroplasty. J Bone Joint Surg [Am] 74:1119–1128

Gill ES, Mills DM (1991) Long-term follow-up of 1000 consecutive cemented total knee arthroplasties. Clin Orthop 273:66–67

Grace JN, Rand JA (1988) Patellar instability after total knee arthroplasty. Clin Orthop 230:168

Griffiths HA, Burke J, Bonfiglio TA (1987) Granulomatous pseudotumors in total joint replacement. Skeletal Radiol 16:146–152

Habermann ET (1986) Total joint replacement: an overview. Semin Roentgenol 21:7–19

Harris WH (1969) Traumatic arthritis of the hip after dislocation and acetabular fractures: treatment by mold arthroplasty: an end result study using a new method of result evaluation. J Bone Joint Surg [Am] 51:737–755

Haynes DR, Rogers SD, Hay S, et al. (1993) The differences in toxicity and release of bone resorbing mediators induced by titanium and cobalt-chromium-alloy wear particles. J Bone Joint Surg [Am] 75:825–834

Heekin RD, Callaghan JJ, Hopkinson WJ, et al. (1993) The porous-coated anatomic total hip prosthesis inserted without cement: results after five to seven years on a prospective study. J Bone Joint Surg [Am] 75:77–91

Hue FC, Fitzgerald RH Jr (1990) Hinged total knee arthroplasty. J Bone Joint Surg [Am] 65:513–519

Idusuyi OB, Morrey BF (1996) Peroneal nerve palsy after total knee arthroplasty. J Bone Joint Surg [Am] 78:177–184

Insall JN, Dorr LD, Scott RD, Scott SN (1989) Rationale of The Knee Society clinical grading system. Clin Orthop 248:13–14

Kaplan PA, Montesi SA, Jardon OM, Gregory PR (1988) Bone ingrowth prostheses in asymptomatic patients: radiographic features. Radiology 169:221–227

Kavanagh BF, Ilstrup DM, Fitzgerald RH Jr (1985) Revision total hip arthroplasty. J Bone Joint Surg [Am] 67:517–526

Lachiewicz PF, Rogers GD, Thomason HC (1996) Aspiration of the hip joint before revision total hip arthroplasty. J Bone Joint Surg [Am] 78:749–755

Lyons CW, Berquist TH, Lyons JC, et al. (1985) Evaluation of radiographic findings in painful hip arthroplasties. Clin Orthop 195:239–251

Maloney WJ, Smith RL (1995) Periprosthetic osteolysis in total hip arthroplasty: the role of particulate wear debris. J Bone Joint Surg [Am] 77:1448–1461

Maloney WJ, Smith RL, Schmalzried TP, Chiba J, Huene D, Rubash H (1995) Isolation and characterization of wear particles generated in patients who have had failure of a hip arthroplasty without cement. J Bone Joint Surg [Am] 77:1301–1310

Manaster BJ (1995) Total knee arthroplasty: post-operative radiographic findings. AJR 165:899–904

Maus TP, Berquist TH, Bender CE, Rand JA (1987) Arthrographic study of painful hip arthroplasty: refined criteria. Radiology 162:721–727

Merkel KD, Brown ML, Dewanjee MK, Fitzgerald RH Jr (1985) Comparison of indium-labeled leukocyte imaging with sequential technetium gallium scanning in diagnosis of low-grade musculoskeletal sepsis. J Bone Joint Surg [Am] 67:465–476

Morrey BF (1991a) Femoral head size. In: Morrey BF (ed) Joint replacement arthroplasty. Churchill Livingstone, New York, pp 587–599

Morrey BF (1991b) Joint replacement arthroplasty. Churchill Livingstone, New York

Palestro CJ, Sawyer AJ, Kim CK, Goldsmiths J (1991) Infected knee prostheses: diagnosis with indium-111 leukocyte, Tc-99m sulfur colloid, and Tc-99m MDP imaging. Radiology 179:645–648

Pellegrini VD, Gegoritch SJ (1996) Pre-operative irradiation for prevention of heterotopic ossification following total hip arthroplasty. J Bone Joint Surg [Am] 78:870–880

Peterson LFA (1977) Current status of total knee arthroplasty. Arch Surg 112:1099–1104

Piraino D, Richmond D, Freed H, et al. (1990) Total knee replacement: radiographic findings in failure of porous-coated, metal-backed patellar components. AJR 155:555–558

Rand JA (1991) Selection of prostheses. In: Morrey BF (ed) Joint replacement arthroplasty. Churchill Livingstone, New York, pp 981–988

Rand JA (1993) Alternatives to re-implantation for salvage of total knee arthroplasty complicated by infection. J Bone Joint Surg [Am] 75:282–288

Rosenthall L, Addis AE, Hill RO (1985) Combined radionuclide and radiocontrast arthrography for evaluating hip arthroplasty. Nucl Med 10:531–534

Salvati EA, Wilson PD Jr, Jolley MN, et al. (1981) A ten-year follow-up study of our first one hundred consecutive Charnley total hip replacements. J Bone Joint Surg [Am] 63:753–767

Schmalzried TP, Jasty M, Harris WH (1992) Periprosthetic bone loss after total hip arthroplasty. J Bone Joint Surg [Am] 74:849–867

Schneider R, Goldman AB, Insall JN (1986) Knee prosthesis. Semin Roentgenol 21:29–46

Stuart MJ (1991) Indications and patient selection. In: Morrey BF (ed) Joint replacement arthroplasty. Churchill Livingstone, New York, pp 971–979

Tsukayama DT, Estrada R, Gustilo RB (1996) Infection after total hip arthroplasty. J Bone Joint Surg [Am] 78:512–523

Wilson MG, Kelley K, Thornhill TS (1990) Infection as a complication of total knee replacement arthroplasty. J Bone Joint Surg [Am] 72:878–883

Zicat B, Engh CA, Gokan E (1995) Patterns of osteolysis around total hip components inserted with and without cement. J Bone Joint Surg [Am] 77:432–439

19 Musculoskeletal Tumours

A.M. Davies and D. Vanel

CONTENTS

19.1 Introduction

A large variety of tumours and tumour-like lesions arise in the musculoskeletal system. To the unwary observer they present a bewildering spectrum of radiographic appearances that can lead to misinterpretation and suboptimal management. Although primary malignancies are relatively rare, they often pose an intriguing problem for the radiologist, particularly as the pathology is frequently equally challenging. Dramatic progress has been achieved over the past two decades in both the management and the prognosis of musculoskeletal malignancies. Prior to this, early amputation followed by the rapid devel-

opment of metatases was the typical scenario for most patients. The addition of chemotherapy to the surgical management of patients with, for example, conventional osteosarcoma, has improved the 5-year survival from less than 20% to 76% (KROPEI et al. 1991; VETH 1991). Provided wide surgical margins are obtained, no difference is seen between the long-term survival of patients with amputation and that of patients undergoing limb-salvage surgery (SIMON et al. 1986).

During this period of improving outcome the imaging of musculoskeletal tumours has also undergone something of a revolution, with the introduction of newer sophisticated techniques, albeit at a price. The evaluation of the patient with a suspected musculoskeletal tumour requires knowledge of the benefits and limitations of the numerous imaging techniques currently available. Care should be taken to balance the financial costs and invasiveness of each technique against the diagnostic reward. Rarely is it necessary to employ every technique on an individual patient. In this chapter we will discuss the role of imaging in these patients from detection and diagnosis through to the ultimate aim of medical management, a cure. It is beyond the scope of this chapter to describe each tumour entity in detail.

19.2 Detection

The patient with a bone or soft tissue tumour, irrespective of its nature, will typically present with either pain or swelling. The pain may be mild and intermittent initially but later becomes more severe and unremitting, particularly if it is a malignancy. Alternatively, a pathological fracture may be the precipitous presenting feature. Occasionally, a bone tumour, typically a benign lesion, can be an incidental radiographic finding.

Despite newer imaging techniques, the radiograph is the preliminary and single most important imaging investigation. It remains cheap, easily

A.M. DAVIES, MD, MRI Centre, Royal Orthopaedic Hospital, Bristol Road South, Birmingham B31 2AP, UK
D. VANEL, MD, Department of Radiology, Institut Gustave-Roussy, 39, rue Camille Desmoulins, F-94805 Villejuif, France

obtainable and universally available. The radiologist and clinician ignore the value of the radiograph at their peril. Frequently, the diagnosis may be obvious to the trained eye and further imaging is then directed towards staging the lesion. Alternatively, if an abnormality is present on the film and the exact nature is not immediately apparent, certain findings will indicate a differential diagnosis and other forms of imaging can then be employed to assist in making a more definitive radiological diagnosis. If the initial radiograph is normal, however, with persisting and increasing symptoms a repeat radiograph in due course may be indicated.

Early signs of a bone tumour include areas of ill-defined lysis or sclerosis, cortical destruction, periosteal new bone formation and soft tissue swelling. Bone lesions are frequently missed or overlooked on the initial radiograph. In a study performed at one of the authors' institutions (A.M.D.) in approximately 20% cases neither the clinician nor the radiologist at the referring centre detected the bone tumour on the initial radiographs, although evidence was present on retrospective review of the films (GRIMER and SNEATH 1990). A number of features may improve the rate of detection. Attention to good radiographic technique is essential. The fundamental prerequisite for skeletal radiology that at least two views of an area are obtained must be strictly adhered to. Often subtle signs of a lesion may be discernible on one view but not be visualized on the second. It is ironic that the more complex the bony anatomy, the fewer the projections that are obtained. The prime example is the pelvis, where early lesions may be missed due to the curvature of the bones and can be obscured by the overlying soft tissues, bowel gas and vascular calcifications.

Careful scrutiny of the entire film, including the periphery, is necessary and if an abnormality is seen at the edge of the film, further, more extensive views will be required. If the film is overexposed it should be viewed with a bright light or may even need to be repeated; not only should the bone be optimally seen but also the soft tissues for assessment of density changes or displacement of tissue planes.

In the presence of a normal radiograph, referred pain needs to be considered, in which case further radiographs would be indicated. Hip joint pathology presenting with referred pain to the knee is a well-recognized entity in the child.

The pathological process may be well established even in the presence of a normal radiograph. At least 40%–50% of trabecular bone must be destroyed before a discrete area of lucency can be seen on the

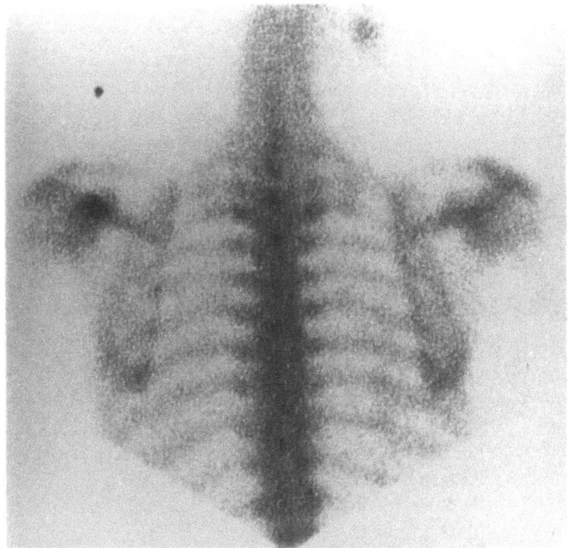

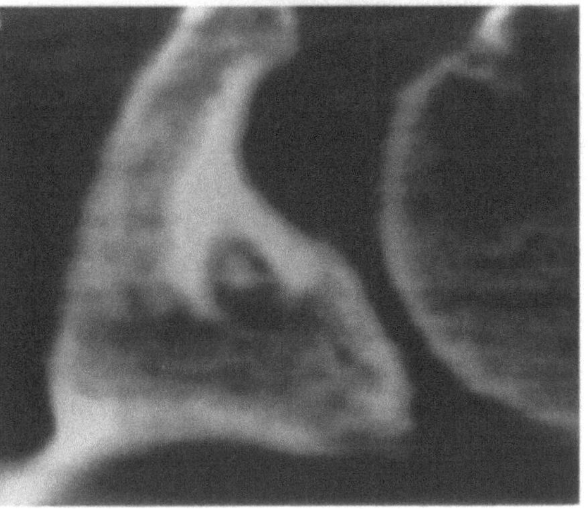

Fig. 19.1. Twenty-two year old male with an osteoid osteoma of the anterior glenoid detected by 3-h bone scintigraphy (a) and confirmed by CT (b)

radiographs. Erosion or destruction of the cortex is more readily apparent.

On occasion radiographically occult lesions can be detected by bone scintigraphy and/or magnetic resonance (MR) imaging (Fig. 19.1). A typical example in this regard is the painful adolescent scoliosis due to an osteoid osteoma, the nature of which is not immediately obvious on the radiograph. An intense focus of increased activity/uptake on bone scintigraphy may suggest the diagnosis. Frequently, however, the high sensitivity of bone scintigraphy will highlight the site of pathology but its lack of specificity means that other imaging will be required to establish the diagnosis (Fig. 19.1). The clinical context should also be considered as certain conditions, such as histiocytosis, myeloma and some malignant round cell tumours, will fail to exhibit increased activity on scintigraphy.

It is well accepted that MR imaging is the most sensitive imaging technique available in the detection of bone marrow disorders, including focal and diffuse neoplasms, marrow packing disorders, myeloproliferative diseases and chronic anaemia (STEINER et al. 1993). A T1-weighted, fat-suppressed T2-weighted or short tau inversion recovery (STIR) sequence will demonstrate optimum contrast between osseous tumour and marrow fat, while fat-suppressed T2-weighted or STIR sequences are preferred for depiction of extraosseous extension and soft tissue tumours. The clinical suspicion of soft tissue masses can be confirmed with ultrasound, computed tomography (CT) or MR imaging. Although ultrasound is cheap and widely available, MR imaging more accurately demonstrates the anatomical location of the mass and would be required for subsequent staging if neoplasia is suspected. MR imaging is preferred to CT due to its superior soft tissue contrast resolution.

19.3
Diagnosis

Once a musculoskeletal tumour has been detected the next objective of imaging is to attempt to characterize the lesion and, in so doing, indicate an appropriate differential diagnosis to the referring clinician. At this stage important maxims that should be appreciated include not over-treating a benign lesion, not under-treating a malignant lesion and not misdirecting the approach to biopsy, which might prejudice subsequent surgical management (MOSER and MADEWELL 1987). In drawing up a differential diagnosis for a particular case the radiologist must first have a knowledge of the different pathologies that may arise in the musculoskeletal system. While understanding of the microscopic features is not required, it is self-evident that an entity will not appear in the differential diagnosis if the radiologist involved is unaware of the existence of a particular condition. Musculoskeletal tumours can best be categorized according to their tissue of origin and then into benign and malignant subtypes (Table 19.1).

Table 19.1. Classification of tumours and tumour-like lesions by tissue of origin

Tissue of origin	Benign lesion	Malignant lesion
Bone forming (osteogenic)	Osteoma Osteoid osteoma Osteoblastoma	Osteosarcoma (and variants)
Cartilage-forming (chondrogenic)	Enchondroma (chondroma) Enchondromatosis (Ollier's disease) Osteochondroma (diaphyseal aclasis) Chondroblastoma Chondromyxoid fibroma	Chondrosarcoma (central) Mesenchymal Clear cell Dedifferentiated Chondrosarcoma (peripheral) Periosteal (juxtacortical)
Fibrous, osteofibrous and fibrohistiocytic (fibrogenic)	Fibrous cortical defect Non-ossifying fibroma Fibrous dysplasia Desmoplastic fibroma Osteofibrous dysplasia	Fibrosarcoma Malignant fibrous histiocytoma
Vascular	Haemangioma Glomus tumour Cystic angiomatosis	Angiosarcoma Haemangioendothelioma Haemangiopericytoma
Haematopoietic, reticuloendothelial and lymphatic	Giant cell tumour Eosinophilic granuloma Lymphangioma	Lymphoma Leukaemia Myeloma (plasmacytoma) Ewing's sarcoma
Neural (neurogenic)	Neurofibroma Neurilemoma	Malignant schwannoma Neuroblastoma Primitive neuroectodermal tumour (PNET)
Notochordal		Chordoma
Fat (lipogenic)	Lipoma	Liposarcoma
Unknown	Simple bone cyst Aneurysmal bone cyst	Adamantinoma

Table 19.2. Peak age incidence of benign and malignant tumours and tumour-like lesions

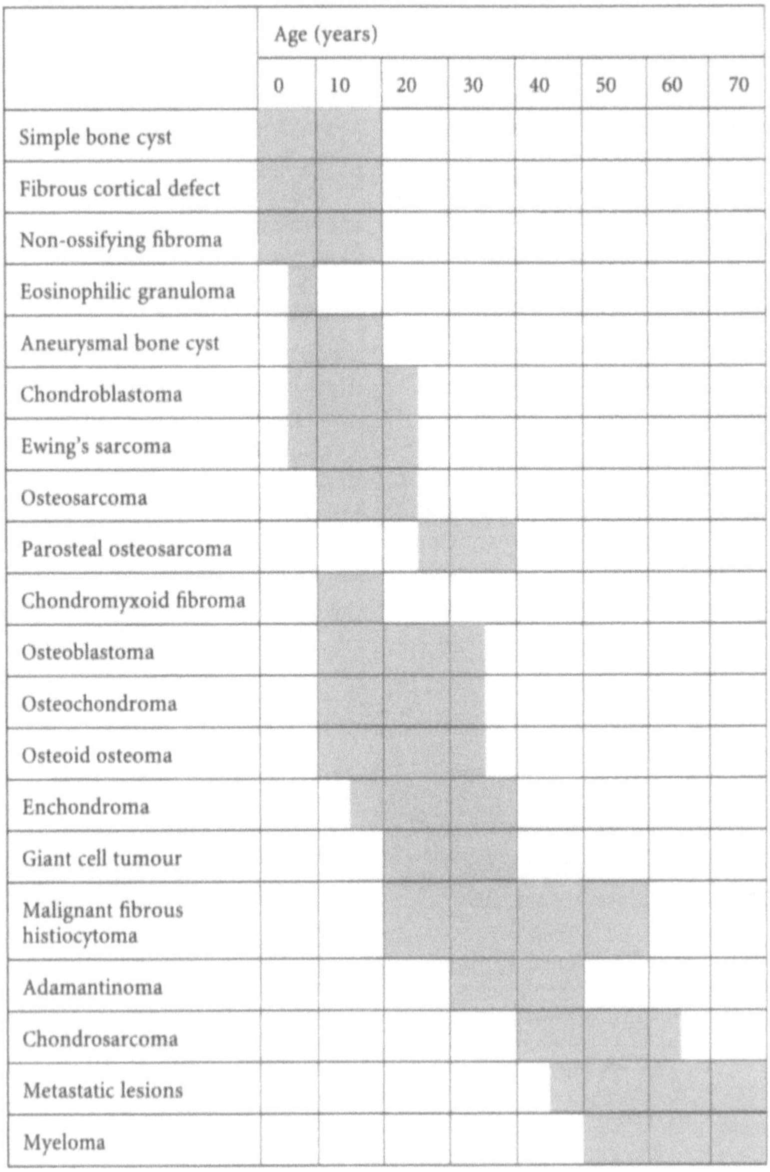

	Age (years)							
	0	10	20	30	40	50	60	70
Simple bone cyst	▓	▓						
Fibrous cortical defect	▓	▓						
Non-ossifying fibroma	▓	▓						
Eosinophilic granuloma		▓						
Aneurysmal bone cyst		▓						
Chondroblastoma		▓						
Ewing's sarcoma		▓						
Osteosarcoma		▓						
Parosteal osteosarcoma			▓					
Chondromyxoid fibroma		▓						
Osteoblastoma		▓	▓					
Osteochondroma		▓						
Osteoid osteoma		▓						
Enchondroma		▓	▓					
Giant cell tumour			▓	▓				
Malignant fibrous histiocytoma			▓	▓				
Adamantinoma			▓	▓				
Chondrosarcoma					▓	▓		
Metastatic lesions					▓	▓	▓	
Myeloma						▓	▓	

Before assessing the imaging the prudent radiologist should establish some basic facts regarding the patient. In recognizing the relevance of certain clinical details the differential diagnosis may then be significantly reduced even before the imaging is taken into account. Important factors to be noted include the following:

1. *Age.* The age of the patient is arguably the single most useful piece of information as it frequently influences the differential diagnosis (Table 19.2). Many musculoskeletal neoplasms exhibit a peak incidence at different ages. For osteosarcoma this is in the second and third decades. Osteosarcoma is, therefore, unlikely to be high in the differential diagnosis in a bone-forming lesion in a middle-aged or elderly patient except in the presence of a pre-existing bone lesion. Metastases and myeloma should always be considered if a bone lesion is identified in a patient over 40 years of age. Similarly, metastatic neuroblastoma should be in the differential diagnosis at 2 years of age or under. Conversely, a tumour arising in adolescence or early adult life is unlikely to be a metastasis.

2. *Gender.* When looking at a large series of patients with a particular type of bone tumour it can be seen that many occur more commonly in boys. In the

individual case this fact does not play a significant role in formulating the differential diagnosis.

3. *Ethnic origin.* Amongst the bone neoplasms, Ewing's sarcoma is unusual in that it is prevalent in Caucasians but is rarely seen in the Afro-Caribbean races. A number of non-neoplastic bone conditions that may on occasion simulate neoplasia also show a racial predisposition, e.g. sickle cell, Gaucher's and Paget's diseases. It is only in isolated cases that the ethnic origin of the patient provides a useful pointer to the diagnosis.

4. *Family history.* There is little evidence of a familial predisposition to the formation of musculoskeletal neoplasms in most instances. The exceptions are certain heriditary bone conditions which may be found in association with malignant change, e.g. diaphyseal aclasis and Ollier's disease.

5. *Previous medical history.* Information that should be noted in all patients, whenever present, is a history of a prior malignancy or a pre-existing bone condition. If such relevant details are forthcoming it is important to establish whether previous imaging exists and, if so, to obtain sight of it for review with the contemporary imaging.

6. *Multiplicity.* It is critical early in the management of a patient to establish whether a lesion is solitary or multiple as this will influence the differential diagnosis. Frequently this question will not be definitively answered until the staging imaging is performed.

In establishing a perspective of the patient as a whole the factors detailed above should be taken in conjunction with one another. For example, age and multiplicity. Multiple bone lesions in the child will suggest a bone dysplasia, histiocytosis, leukaemia or metastatic neuroblastoma, whereas, in the adult, metastastic disease and myeloma are the most likely.

It is at this stage that attention should be turned to the imaging. The radiograph remains the most accurate of all the imaging techniques currently available in determining the differential diagnosis of a bone tumour (KRICUN 1983). The radiologist may attempt a diagnosis from the radiograph in one of two ways. First, the so-called Aunt Minnie (KRICUN 1983) or "pattern recognition" approach. This relies on familiarity with the typical overall appearances of a particular type of tumour. This is all very satisfactory if the tumour is classical in appearance but problems arise if the lesion has atypical features, arises at an unusual site or is mimicked by a differing pathology. The second, preferred approach, that might best be termed "pattern analysis", relies on meticulous recognition of various radiographic signs (LODWICK

1965, 1966). The analysis can be best illustrated by answering a series of five questions. Which bone is affected? Where in that bone is the tumour located? What is the tumour doing to the bone (pattern of destruction)? What form of periosteal reaction, if any, is present? What type of matrix mineralization, if any, is present?

19.3.1
Site in Skeleton

Most bone tumours arise around the knee and in the proximal humerus and as such little diagnostic information can be deduced from noting the affected bone in many cases. There are exceptions. Cartilage tumours of the hands and feet, while common, are almost invariably benign. Both osteofibrous dysplasia and adamantinoma classically involve the diaphysis of the tibia and are extremely rare at any other site. Chordoma characteristically arises from the clivus or sacrum. Although many different tumours may arise in the bony spine, malignant lesions are found predominantly in the anterior part of the vertebra (body), while benign lesions are characteristically found in the posterior elements (neural arch).

19.3.2
Location in Bone

The site of origin of a bone tumour is an important parameter of diagnosis. It reflects the site of greatest cellular activity. During the adolescent growth spurt the most active areas are the metaphyses around the knee and in the proximal humerus. Tumour originating from marrow cells may occur anywhere along the bone. Conventional osteosarcoma will tend to originate in the metaphysis or meta-diaphysis, whereas Ewing's sarcoma originates in the metaphysis or, more distinctively, in the diaphysis (Fig. 19.2). In the child the differential diagnosis of a lesion arising within an epiphysis can be realistically limited to chondroblastoma (Fig. 19.3), epiphyseal abscess (pyogenic or tuberculous) and rarely eosinophilic granuloma. Following skeletal fusion, subarticular lesions, analogous in the adult to the epiphysis, include giant cell tumour (Fig. 19.4), clear cell chondrosarcoma (rare) and intraosseous ganglion.

It can also be helpful to identify the origin of the tumour with respect to the transverse plane of the bone. Is the tumour central, eccentric or cortically

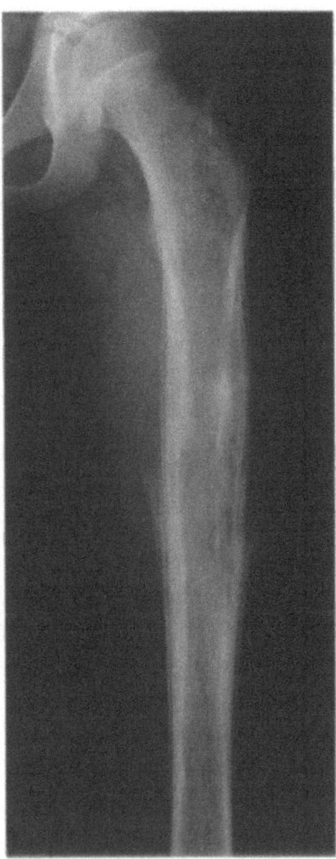

Fig. 19.2. AP radiograph of the femur of a 9-year-old female showing the classic features of a Ewing's sarcoma: there is lamellar periosteal new bone formation with interruption distally (Codman's angles) against a background of permeative bone destruction

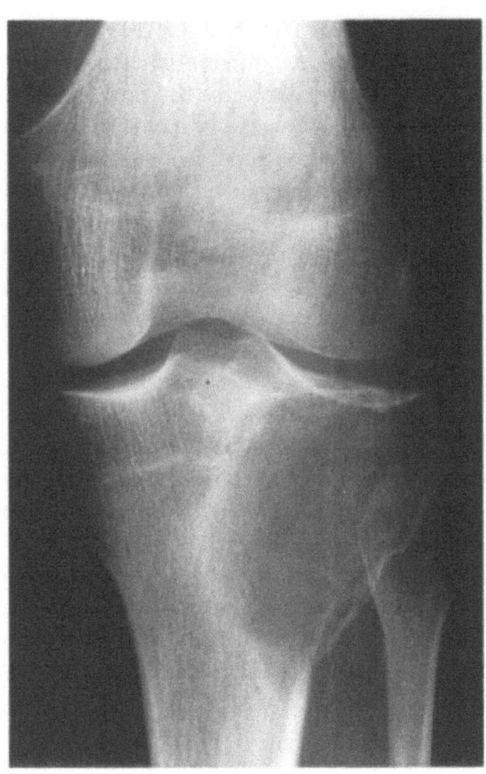

Fig. 19.4. AP radiograph of the knee of a 38-year-old male showing the typical lytic, subarticular, eccentric location of a giant cell tumour

based? For example, a simple bone cyst, fibrous dysplasia and Ewing's sarcoma will tend to be centrally located, whereas giant cell tumour (Fig. 19.4), chondromyxoid fibroma and non-ossifying fibroma (Fig. 19.5) are typically eccentric. Lesions that usually arise in an eccentric position may appear central if the tumour is particularly large or the involved bone is of a small calibre. There are numerous surface lesions of bone which are related to a greater or lesser extent to the cortex (KENAN et al. 1993). A benign example is the periosteal or juxtacortical chondroma. Most of the malignant surface lesions of bone are the rarer forms of osteosarcoma, e.g. periosteal, high-grade surface and parosteal osteosarcoma.

19.3.3
Pattern of Bone Destruction

Analysis of the interface between tumour and host bone is a good indicator of the rate of growth of the lesion. A sharply marginated lesion usually denotes slower growth than a non-marginated lesion. The

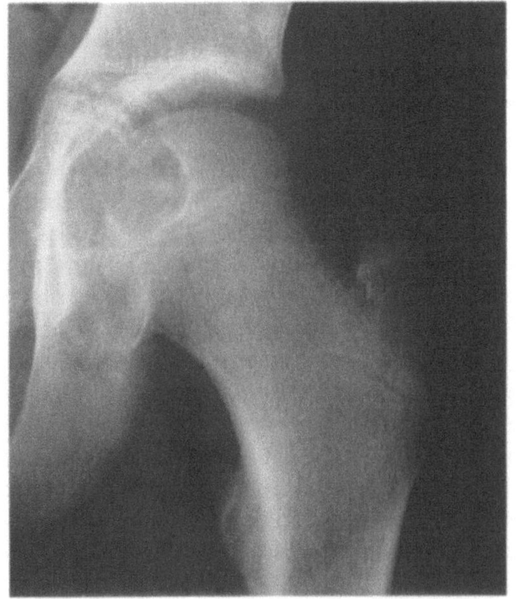

Fig. 19.3. AP radiograph of the hip of a 13-year-old male showing a rounded lytic lesion in the proximal femoral epiphysis typical of a chondroblastoma

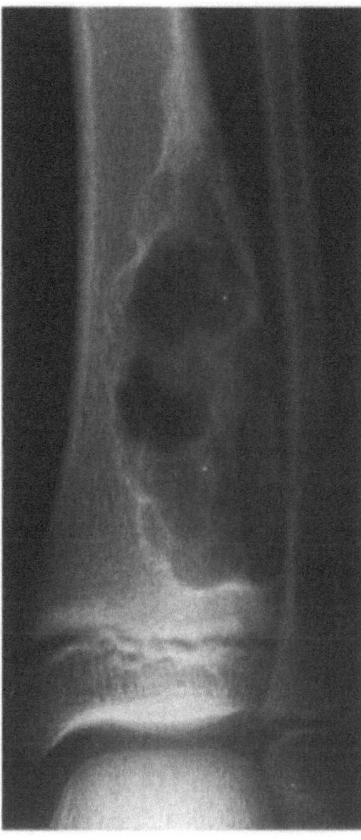

Fig. 19.5. AP radiograph of the distal tibia of a 6-year-old boy showing a lytic, eccentrically located lesion with geographic bone destruction typical of a non-ossifying fibroma

faster the growth, the more aggressive the pattern of destruction and the wider zone of transition between tumour and normal bone. Aggressivity per se does not conclusively indicate malignancy but the malignant tumours tend to be faster growing than their benign counterparts. Classification of the different patterns of bone destruction has not been bettered since Lodwick's seminal work (LODWICK 1965,1966; LODWICK et al. 1980a,b). Each pattern reflects a particular growth rate and thereby a differential diagnosis (KRICUN 1993). It must be noted that any classification is artificial and that a series of bone tumours will exhibit a spectrum of bone destruction from well to ill defined.

19.3.3.1
Geographic Bone Destruction

In this pattern the growth rate is sufficiently indolent that the lesion will appear well marginated with a thin zone of transition (Fig. 19.5). The geographic or type I pattern may be subdivided into IA, IB and IC depending on the appearance of the margin and the

effect on the cortex. Type IA, the slowest growing of all the lesions and thereby the least aggressive, is typified by a sclerotic margin. The thicker the sclerosis, the slower the rate of growth. The vast majority of these lesions will prove to be benign. In type IB the lesion is well defined without the sclerotic margin; while such lesions are still slow growing the rate is slightly greater than type IA. Again, the majority of type IB lesions are benign although some malignancies may on occasion demonstrate this pattern. In type IC the margin is less well defined, indicating a more aggressive pattern. The cortex is also destroyed. Few benign tumours exhibit a type IC pattern. The differential diagnosis in this situation includes giant cell tumour (Fig. 19.4), malignant fibrous histiocytoma and lymphoma of bone.

19.3.3.2
Moth-eaten and Permeative Bone Destruction

Moth-eaten (type II) and permeative (type III) patterns of bone destruction reflect the increasingly aggressive nature of tumours compared with geographic lesions (Fig. 19.2). Again, this is a spectrum of change varying from multiple small areas of lysis to permeation, sometimes almost imperceptible radiographically, characterized by minute tiny cortical defects and a wide, ill-defined zone of transition (Fig. 19.6). The highly aggressive nature of these lesions does not allow the host bone sufficient time to react and produce a response (KRICUN 1993). Typically malignancies, including metastasis, Ewing's sarcoma and osteosarcoma, exhibit a moth-eaten or permeative appearance. Benign tumours in general do not show this pattern of bone destruction.

19.3.4
Periosteal Reaction

The periosteum is the thin layer of soft tissue with osteoblastic properties that lines the outer cortex of bone. It is normally radiolucent but will mineralize when the osteoid-producing cells of the inner cambium layer are stimulated by an adjacent osseous or para-osseous process. The rate of mineralization is partly dependent on the age of the patient. The younger the patient, the more rapid the appearance of radiographic change and vice versa. Periosteal reaction, otherwise known as periosteal new bone formation, may occur in any condition which elevates the periosteum, whether it be blood, pus or

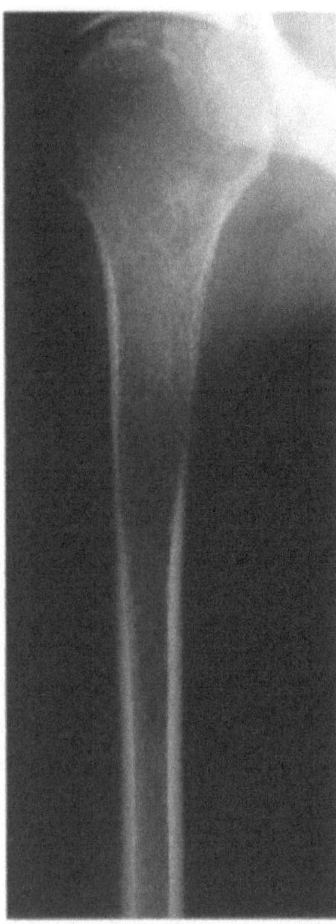

Fig. 19.6. AP radiograph of the humerus of a 22-year-old male showing an extensive Ewing's sarcoma. There is confluent lysis proximally and permeative bone destruction distally with a wide zone of transition

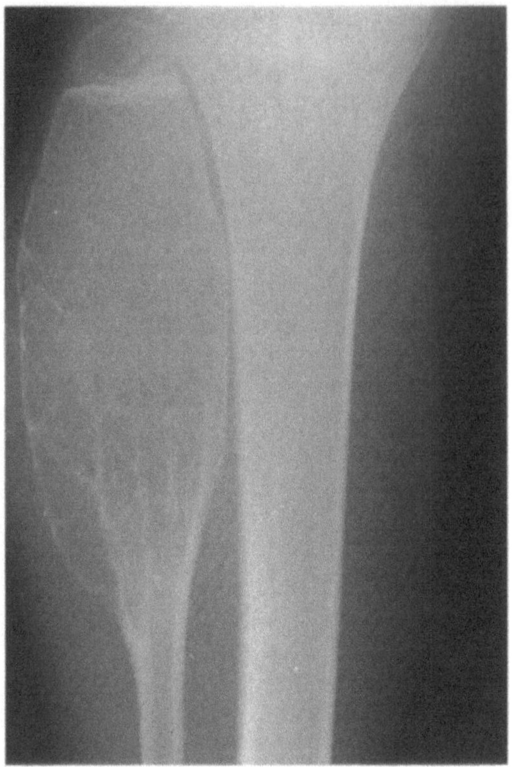

Fig. 19.7. AP radiograph of the proximal fibula of a 13-year-old male showing a typical expansile aneurysmal bone cyst

tumour. The term periostitis, favoured in older texts, is best avoided as it implies an inflammatory aetiology.

The appearance and nature of a periosteal reaction are frequently valuable in narrowing down the differential diagnosis of a bone tumour. A good, albeit complex, classification identifies three broad categories: continuous, discontinuous or interrupted and complex (RAGSDALE et al. 1981; MOSER and MADEWELL 1987).

19.3.4.1
Continuous Periosteal Reaction

A continuous periosteal reaction may be observed with either an intact or a destroyed underlying cortex. In the latter the bone is said to be "expanded" but this is a misnomer as bone cannot be inflated like a balloon (RAGSDALE et al. 1981). Nevertheless the term "cortical expansion" is well entrenched in common usage. It represents a relatively slow process by which endosteal bone resorption is balanced by periosteal new bone formation. In faster growing lesions

the endosteal resorption will exceed periosteal apposition and a thin outer "shell" will be produced (Fig. 19.7). The thickness of this shell is an indicator of the rate of growth of the lesion but is not a good discriminator of benign from malignant lesions. Shells are typically found in benign lesions such as simple bone cyst, aneurysmal bone cyst (Fig. 19.7), chondromyxoid fibroma, fibrous dysplasia and giant cell tumour. They are also well recognized in "expansile" metastases of renal and thyroid origin and plasmacytoma.

Additions to, rather than substitutes for, the original cortex occur with a continuous periosteal reaction with an intact cortex. The periosteal reaction may be solid, a single lamella, lamellated or spiculated. The solid type implies the slow apposition of layers of new bone to the cortex, sometimes termed "cortical thickening" or "cortical hyperostosis" (RAGSDALE et al. 1981). It is seen in chondroma and central chondrosarcoma, and eccentrically in osteoid osteoma. If the solid periosteal reaction is extensive with an undulating quality the differential diagnosis includes chronic osteomyelitis, hypertrophic osteoarthropathy (Fig. 19.8), chronic lymphoedema and varicosities.

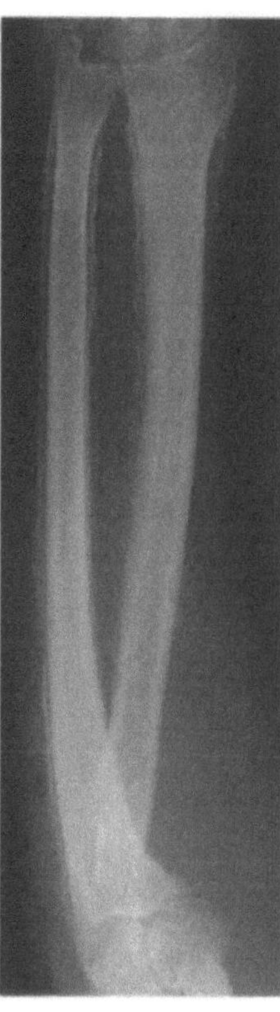

Fig. 19.8. Radiograph of the forearm bones of a 72-year-old male with a history of bronchial carcinoma showing florid solid periosteal new bone due to hypertrophic osteoarthropathy

A single lamella periosteal reaction is formed by a thin radiodense line separated from the cortex by a narrow radiolucent zone. It usually denotes a benign disorder and is frequently seen with traumatic and inflammatory conditions. It should be appreciated that a periosteal reaction is a dynamic process and a single lamella may fill in to produce a solid appearance or go on to have the addition of further lamellae. The lamellated periosteal reaction, otherwise known as onion-skin, is seen in Ewing's sarcoma (Fig. 19.2), osteosarcoma, eosinophilic granuloma of the long bones in children and acute osteomyelitis.

A spiculated periosteal reaction occurs when the mineralization is oriented perpendicular to the cortex and denotes a more rapidly evolving process. It is typical of malignant tumours such as osteosarcoma and Ewing's sarcoma but may be seen in benign tumours such as meningioma, haemangioma of bone and non-neoplastic conditions such as thalassaemia and thyroid acropachy. The location of a spiculated periosteal reaction significantly influences the differential diagnosis.

19.3.4.2
Discontinuous/Interrupted Periosteal Reaction

In the interrupted periosteal reaction the mineralization has been breached in one of two ways. First, the process, usually a tumour, simply occupies the available space, or second, the rate of apposition is exceeded by resorption. Close attention should be paid to the margins of the periosteal reaction. Rapidly growing benign tumours may exhibit a peripheral wedge or buttress with a thin or even non-existent shell. The buttress should be distinguished from the Codman angle, which is the triangular elevation of interupted periosteum with one or more layers of new bone located at the periphery of the lesion (Fig. 19.2) (CODMAN 1926). This pattern is suggestive, if not diagnostic, of malignancy as it may also be seen in osteomyelitis. In malignant bone tumours the site of interruption of a periosteal reaction is usually the area of maximum extraosseous tumour growth.

19.3.4.3
Combined/Complex Periosteal Reaction

More than one pattern of periosteal reaction may be manifest in the same case and reflects the varying rate of growth at different sites in the lesion (Fig. 19.9). The divergent spiculated periosteal reaction, otherwise known as "sun-ray", is a typical example of a complex pattern and is suggestive of osteosarcoma.

19.3.5
Matrix

A number of tumours produce a matrix, the intercellular substance, that can calcify or ossify. The radiodense foci should be differentiated from other causes of calcifications such as fracture callus, sclerotic response adjacent to a tumour, necrotic debris and dystrophic calcification. Radiodense tumour matrix is either osteoid or chondroid. The exception is fibrous dysplasia, where the collagenous matrix may be sufficiently dense to give a ground-glass appearance.

Tumour osteoid is typified by solid (sharp-edged) or cloud to ivory-like (ill-defined edge) patterns (Figs. 19.9, 19.10) (SWEET et al. 1981). Tumour cartilage is variously described as stippled, flocculent, ring-and-arc and popcorn in appearance (Fig. 19.11)

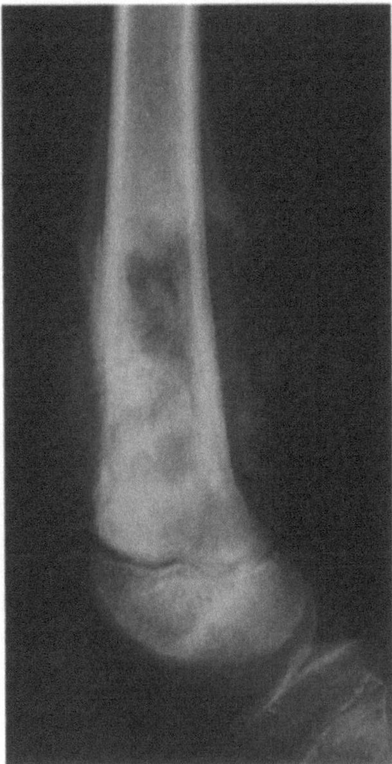

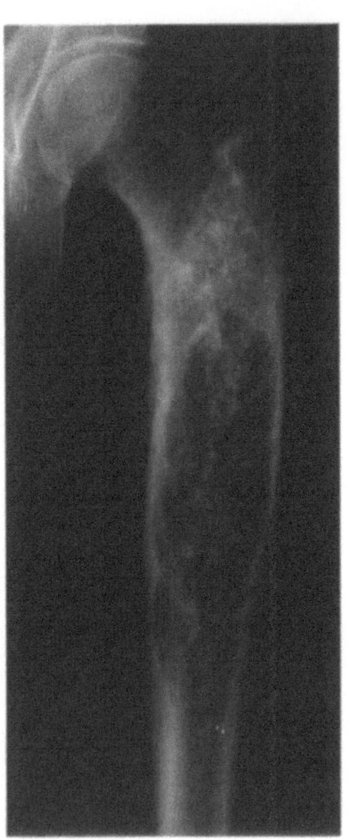

Fig. 19.9. Lateral radiograph of the knee of a 12-year-old female showing an extensive osteosarcoma of the distal femur. There is malignant osteoid mineralization with Codman's angles proximally and a spiculated periosteal reaction distally

Fig. 19.11. AP radiograph of the proximal femur of a 64-year-old female showing an extensive central chondrosarcoma. The tumour is mildly expansile with endosteal scalloping and typical popcorn cartilage mineralization

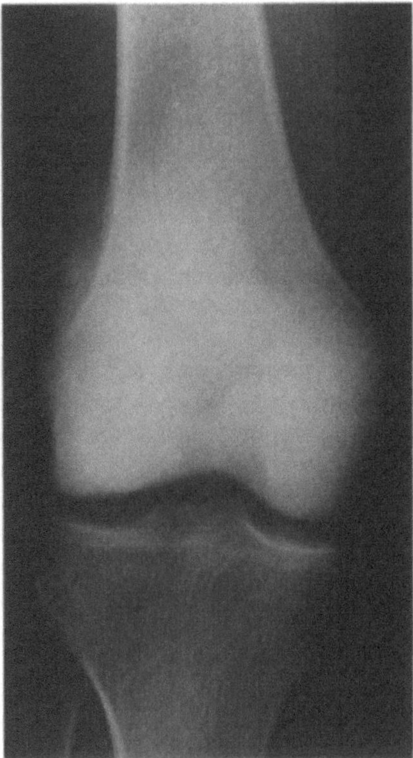

(SWEET et al. 1981). Identifying the pattern of matrix calcification will significantly reduce the differential diagnosis but matrix per se has no influence as to whether the lesion is benign or malignant. The distribution can be helpful. For example, both enchondroma and medullary infarction may show calcification of a similar nature. The distribution in enchondroma is typically central and peripheral in medullary infarction.

19.3.6
Radiographic Diagnosis of Soft Tissue Tumours

The lack of contrast resolution in the soft tissues is a well-recognized limitation of radiography, but the value of the examination should not be underestimated. It may not identify the precise diagnosis, in all but a minority of cases, but can still provide valuable information, e.g. the presence of calcifica-

Fig. 19.10. AP radiograph of the knee of a 20-year-old male showing a sclerotic osteosarcoma of the distal femur indicated by the ivory-like tumour osteoid

tion and bone involvement. Too often the humble radiograph is denigrated as non-contributory because it has failed to identify features that might be termed "positive". The absence of said features, however, can be just as significant. In the presence of a mass the absence, for example, of any bony abnormality immediately indicates that the primary pathology is of soft tissue origin, albeit with a large differential diagnosis. Myositis ossificans, as a more specific example, can be effectively excluded from the differential diagnosis of a mass if there is no radiographic evidence of calcification, in all but the earliest of cases. The majority of soft tissue tumours are of water density similar to that of muscle and are, therefore, only revealed by virtue of their mass effect. This includes displacement or disruption of adjacent fat planes, distortion of the skin contour and involvement of bone.

In a minority of cases, part or all of the tumour may exhibit a radiodensity sufficiently different to that of water for the tumour to be visualized directly. Only fat and gas will give a radiodensity less than that of muscle. Lipomas, the commonest of all the soft tissue tumours, produce a low radiodensity between that of muscle and air. For this reason lipomas are typically well demarcated from the surrounding soft tissues and can be diagnosed with moderate confidence. It should be noted that low-grade liposarcomas may contain variable amounts of lipomatous tissue that will also appear relatively radiolucent.

Increased radiodensity may be seen in the soft tissues due to haemosiderin, calcification or ossification. Calcification or ossification is a feature of a large spectrum of pathologies including congenital, metabolic, endocrine, traumatic and parasitic infections. Soft tissue tumours are one of the less common causes of calcification that the general radiologist can expect to see in his or her routine practice. Analysis of the pattern of calcification within a soft tissue tumour can indicate the tissue type, e.g. phleboliths in haemangioma and "ring-and-arc" calcification in cartilage tumours such as synovial chondromatosis.

Soft tissue sarcomas often present a particular problem to the clinician in that they are rare, with an approximate incidence of only one for every 20 benign soft tissue tumours. The possibility of malignancy is often not considered and the error can be further compounded by the fact that the tumour may have a macroscopically benign appearance. As a result the correct diagnosis may only be made on histological review following inappropriate surgical management.

19.3.7
CT and MR Imaging in Diagnosis

The principal role of CT and MR imaging in the management of the patient with a suspected musculoskeletal sarcoma is in staging (see Sect. 19.4). In selected cases both techniques can be useful in establishing a differential diagnosis. The CT features that should be assessed are similar to those described above when evaluating the radiographs. This reflects the fact that both are radiographic techniques relying on the attenuation of an X-ray source. Cortical breaching, soft tissue extension and faint mineralization are all more readily appreciated on CT scans than on radiographs. This is of particular value in areas of complex anatomy such as the pelvis and spine. Assessment of CT attenuation values will allow distinction between fat-containing and fluid-containing masses, e.g. simple lipoma versus soft tissue sarcoma (Fig. 19.12). Although the physical basis of MR imaging is very different, similar morphological information such as cortical breaching and soft tissue extension can be easily identified. The small signal voids of fine mineralization can easily be missed on MR imaging.

The majority of tumours will have prolonged T1 and T2 relaxation times, thereby showing low to intermediate signal on T1-weighted and high signal on T2-weighted sequences. T1 shortening, with a high signal intensity, will be seen in fat-containing tumours, subacute haemorrhage and Gd-DTPA enhancement. A low signal intensity on T2-weighted images is seen with dense mineralization, hypocellular/fibrous tumours, signal voids from flowing blood, surgical implants and haemosiderin deposition. MR imaging can also be used to measure the thickness of the cartilage cap in differentiating an osteochondroma from a peripheral chondrosarcoma. Care should be taken not to mistake the cartilage cap on MR imaging for an inflamed bursa, which may be found adjacent to osteochondromas. Fluid-fluid levels are well demonstrated by MR imaging in a large number of different musculoskeletal conditions. In the correct clinical and radiographic context they are most commonly seen in aneursymal bone cysts (Fig. 19.13).

19.3.8
Tumour Mimics

There are a large number of disparate bone and soft tissue conditions which can have similar imaging appearances to tumours. What constitutes a tumour

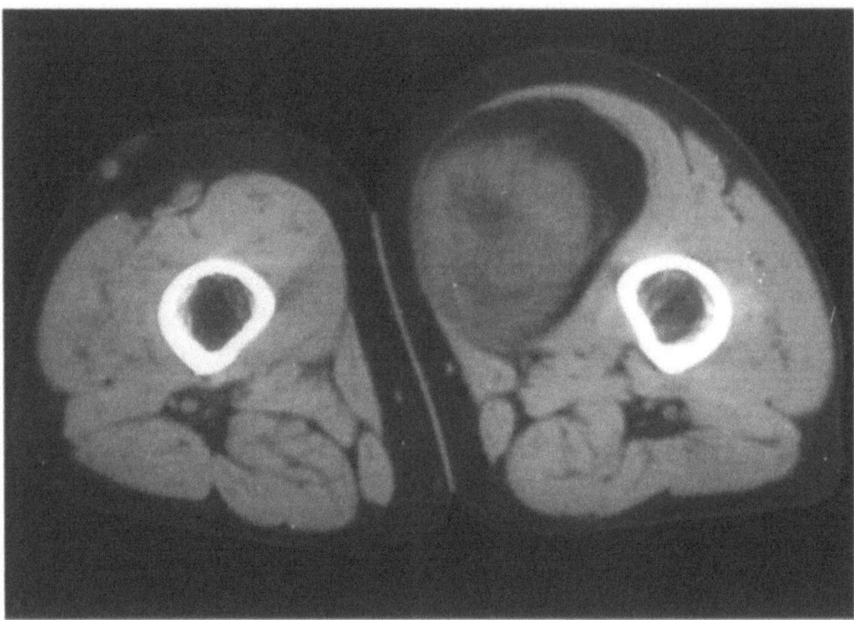

Fig. 19.12. Axial CT of the upper thighs of a 47-year-old male showing a large soft tissue mass in the left adductor compartment. The attenuation of the tumour tissue anterolaterally is that of fat whereas the remainder of the tumour is of water density. The histological diagnosis was a well-differentiated liposarcoma

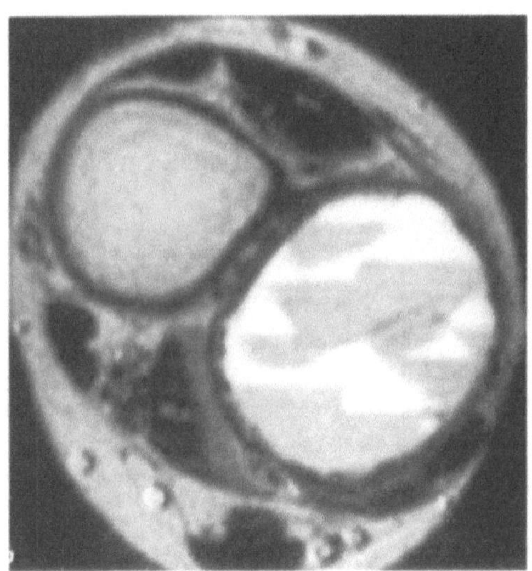

Fig. 19.13. Axial T2-weighted MR image of the distal leg of a 10-year-old male showing marked expansion of the fibula containing numerous fluid-fluid levels typical of an aneurysmal bone cyst

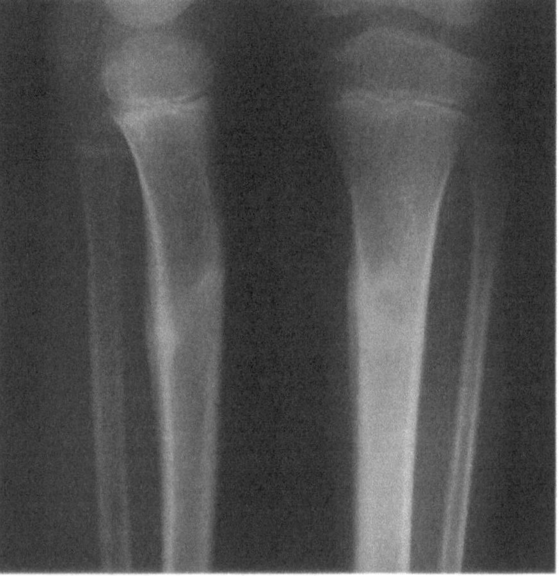

Fig. 19.14. AP and lateral radiographs of the tibia of a 7-year-old female showing the typical periosteal reaction of a proximal tibial stress fracture. This appearance is frequently mistaken for a sarcoma

mimic depends very much on the expertise of the individual reviewing the imaging. The majority can be classified as normal variants, post-traumatic and inflammatory conditions. In the adolescent patient stress fractures and chronic apophyseal avulsion injuries are frequently mistaken for an osseous malignancy (Fig. 19.14). Acute osteomyelitis, at any age, but typically in children, will have an aggressive radiographic appearance, thereby simulating malignancy. In the soft tissues myositis ossificans is probably the non-neoplastic condition most commonly mistaken, both radiographically and histologically, for a malignancy.

There are a number of unrelated non-neoplastic or benign neoplastic lesions which, over the years, have been lumped together and given the term "don't touch me lesions". The radiographic diagnosis of these conditions, to the experienced radiologist, is straightforward and further imaging and biopsy are not indicated in the majority of cases. This category includes fibrous cortical defect, bone island, bone infarct, small foci of fibrous dysplasia, periosteal desmoid, Paget's disease and haemangioma of vertebra.

19.4
Staging

Staging of a suspected bone or soft tissue malignancy is a mandatory part of preoperative evaluation. The purpose of staging is threefold. First, to identify the probable prognosis for the patient. Second, to accurately delineate the tumour to optimize surgical planning. Third, to establish a standard nomenclature by which the extent of a tumour can be categorized. This latter aspect is of limited value to the individual patient but is important when the efficacy of varying drug regimens and different treatment centres is being assessed. A number of staging systems have been advocated, with the Musculoskeletal Tumour Society (ENNEKING et al. 1980; WOLF and ENNEKING 1996) and the American Joint Committee surgical staging systems the most commonly used (RUSSELL et al. 1981). In the Enneking system the tumours are classified as either low grade (I) or high grade (II) and as either intracompartmental (A) or extracompartmental (B). The majority of conventional osteosarcomas will, therefore, be classified as a stage IIB tumour. If metastases are found at presentation the patient is identified as stage III (ENNEKING et al. 1980). The tumour grading is based on histological assessment whereas tumour extent and the presence or absence of metastases relies entirely on imaging.

Staging of the primary tumour requires some form of cross-sectional imaging, be it CT or MR imaging. Because of its improved contrast and multiplanar capability much of the existing literature claims MR imaging to be superior in this respect to CT (AISEN et al. 1986; BLOEM et al. 1988; PETTERSSON et al. 1987; SUNDARAM et al. 1986; ZIMMER et al. 1985). More recent work, however,

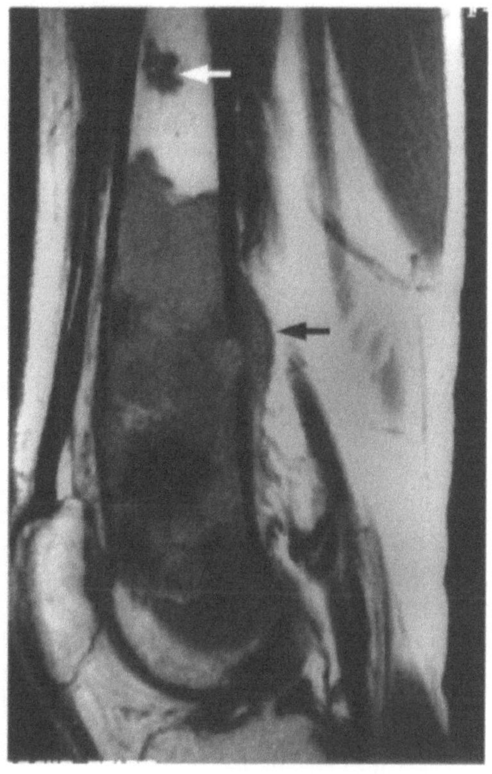

Fig. 19.15. Sagittal T1-weighted MR image of an osteosarcoma of the distal femoral diamctaphysis. There is some early cortical breaching posteriorly (*black arrow*) and a skip metastasis proximally (*white arrow*)

challenges this established belief, concluding that CT and MR imaging are equally accurate in the local staging of malignant bone and soft tissue tumours at certain anatomical sites (PANICEK et al. 1997). Nevertheless, it is likely that most treatment centres will opt to continue using MR imaging, if only because it does not require ionizing radiation.

The minimum requirements of a staging MR examination are a T1-weighted sequence along the line of the affected bone and T2-weighted and/or STIR sequences in orthogonal planes. The T1-weighted images optimise the contrast between the isointense tumour tissue and the high signal intensity marrow fat (Fig. 19.15), whereas the T2-weighted and STIR images highlight extraosseous tumour spread with respect to the surrounding soft tissues. The extent of the tumour in bone and soft tissue should be accurately measured and the relationship to the adjacent joint and neurovascular structures assessed. It is important to include an anatomical reference point on at least one of the sequences in order that measurements can be related to a relevant point. The adjacent joint will usually suffice for this purpose. A large field of view T1-weighted sequence should be included to confirm or exclude skip metastases (Fig. 19.15). Gd-DTPA has little value in the initial staging (SEEGER et al. 1991) although a dynamic contrast-enhanced sequence may be obtained at this stage as a baseline study for the subsequent assessment of tumour response to chemotherapy (see Sect. 19.6). MR angiography can be used to delineate the relationship of the tumour to vessels (SWAN et al. 1995) but should be employed in addition to rather than instead of the conventional MR sequences.

The initial imaging staging also requires frontal and lateral chest radiographs plus a chest CT to exclude occult pulmonary metastases. In patients with osseous tumours bone scintigraphy is necessary to exclude other skeletal lesions, including skip metastases. Those patients treated with neoadjuvant chemotherapy will undergo repeat staging with MR imaging of the primary tumour and CT of the chest prior to definitive surgery to assess the response to chemotherapy and to ensure that the planned surgery remains appropriate.

19.5
Biopsy

With the exception of the "don't touch me lesions", verification of the radiological diagnosis will require a biopsy prior to management decisions. The biopsy should preferably be performed after the appropriate imaging studies as the trauma of the procedure may exaggerate the apparent extent of the tumour. Problems associated with biopsy occur up to 5 times more commonly when it is performed at the referring hospital rather than at the specialist treatment centre (MANKIN et al. 1982, 1996).

Biopsy techniques are covered in further detail in Chap. 7; suffice to say that needle biopsy in experienced hands is a cost-effective and less traumatic alternative to open biopsy (STOKER et al. 1991; SKRZNSKI et al. 1996). The expertise required applies as much to the pathologist interpreting the specimen as to the individual responsible for obtaining it. Close liaison within the team is essential in planning the biopsy and interpreting the results.

19.6
Assessment of Tumour Response to Chemotherapy

The disease-free and overall survival rates for most sarcomas have increased dramatically over the past 20 years, largely as a result of systemic chemotherapy. Neoadjuvant chemotherapy is now routinely used in almost all bone sarcomas with the exception of chondrosarcoma. The role of chemotherapy in soft tissue sarcomas is more contentious although it is frequently employed in many treatment centres. The aim of neoadjuvant chemotherapy is to eradicate potential micrometastases and reduce the size of the primary tumour, thereby facilitating limb-salvage surgery (WINKLER et al. 1988; JURGENS et al. 1988). The histological response of the tumour to chemotherapy is a good prognostic parameter for both osteosarcoma and Ewing's sarcoma but repeated biopsies, monitoring progress prior to definitive surgery, are too invasive to be undertaken routinely. Presurgical imaging is, therefore, of value in assessing tumour response as the results may modify chemotherapy regimens and the timing of surgery.

Radiographic follow-up of patients on chemotherapy may show increased mineralization in osteosarcoma and maturation of periosteal new bone formation in Ewing's sarcoma, indicating a positive response, but the radiographs are of little value in differentiating good and poor responders in individual cases (SMITH et al. 1982; EHARA et al. 1991; HOLSCHER et al. 1996). Angiography and various forms of bone scintigraphy, including positron emission tomography, have all had their advocates but are rarely used routinely. Similarly CT, of proven value in this respect (VANEL et al. 1982; MAIL et al. 1985; SHIRKHODA et al. 1985), has now been superseded by MR imaging.

Magnetic resonance imaging without paramagnetic contrast medium relies on changes in tumour size, margins and signal intensities to predict response to chemotherapy. An increase in tumour volume is usually indicative of a poor histological response in osteosarcoma and may be apparent within 1 month of commencing chemotherapy (HOLSCHER et al. 1995). Conversely, a decrease in tumour volume in osteosarcoma does not allow distinction between good and poor responders (HOLSCHER et al. 1990, 1992). A large reduction in tumour volume is frequently encountered in Ewing's sarcoma but there is considerable overlap between good and poor responders when the reduction is between 25% and 75% (VAN DER WOUDE et al. 1994). Also microscopic residual tumour is often detected even in the presence of almost complete resolution of the soft tissue component of the tumour (MACVICAR et al. 1992).

Considerable research over the past decade has concentrated on the applications of contrast-enhanced MR imaging in monitoring tumour response to chemotherapy. Static contrast-enhanced images following the injection of Gd-DTPA may distinguish necrotic from vascularized tissue but cannot differentiate viable tumour from immature granulation tissue and hyperaemic areas. Dynamic MR imaging following a bolus injection of Gd-DTPA overcomes the problem as viable tumour is characterized by an earlier, more rapid and higher uptake of gadolinium as compared to other tissues (LANG et al. 1995; VAN DER WOUDE 1995a). Alterations in the time-intensity curves measured before and after chemotherapy have been shown to correlate with tumour response (ERLEMANN et al. 1990; FLETCHER et al. 1992). Good responders are typified by a flatter slope when comparing pre- and postchemotherapy curves. Conversely, poor responders show little or no reduction in the slope. Temporal resolution would appear to be important. Some workers claim that a temporal resolution of 3 s is required to differentiate between viable tumour and other vascularized tissues (LANG et al. 1995). Others, however, have claimed that longer intervals, up to 15 s, are adequate (REDDICK et al. 1995; FLETCHER et al. 1996). A variation on the dynamic theme is a parametric technique whereby there is analysis of the contrast-enhanced pattern on consecutive images on a pixel-by-pixel basis. This so-called first-pass technique gives an indication of the vascularization, local blood volume and perfusion of the tumours both qualitatively and quantitatively (VERSTRAETE et al. 1994).

An alternative technique in assessing dynamic studies is to perform a subtraction study comparing precontrast with early postcontrast enhancement images (DE BAERE et al. 1992; VAN DER WOUDE et al. 1995a). Foci of enhancement occurring within 6 s of identification of corresponding arterial enhancement correlate with viable tumour (VAN DER WOUDE et al. 1995a).

Assessment of tumour response using MR spectroscopy, while of research interest, has failed to be introduced into routine patient management. The relatively low cost and non-invasive nature of colour Doppler ultrasound, however, make it a potentially interesting tool for monitoring chemotherapy re-

sponse in the soft tissue component of bone sarcomas (VAN DER WOUDE et al. 1995b).

19.7 Follow-up

Definitive surgery, be it limb salvage or on occasion amputation, is only one phase in the management of the patient with a musculoskeletal sarcoma. Assuming that the patient does not have stage III disease either at presentation or developing during pre-operative chemotherapy, he or she is closely monitored at increasing intervals for evidence of local recurrence, metastatic disease and complications of treatment.

19.7.1 Local Recurrence

Local recurrence of benign bone tumours occurs when inadequate curettage has been performed. This is a relatively common problem with giant cell tumours and is seen occasionally with aneurysmal bone cysts. The clue to the recurrence is the identification of increasing lysis of the surrounding bone and/or bone graft on comparison of serial radiographs (REMEDIOS et al. 1997). A soft tissue mass is usually a late feature unless there was initial packing of the surgical defect with bone cement. In this situation, because of the durability of the cement, the recurrent tumour takes the line of least resistance and spreads early into the soft tissues. MR imaging of curetted bone lesions can give a confusing appearance with variable amounts of fibrous scar, granulation tissue and cystic areas occupying the bony defect. Recurrence, therefore, within the bone can be difficult to identify in the absence of a mass lesion. Fortunately, the time signal intensity enhancement curve for giant cell tumour is usually sufficiently rapid for a dynamic sequence to distinguish recurrence from scar (VAN DER WOUDE et al. 1996).

Local recurrence is a major potential problem with sarcomas, particularly when they are high grade. It is almost inevitable if the original surgical resection was intralesional or marginal. Recurrence may be detected on radiographs as a soft tissue mass with or without bone destruction. Locally recurrent bone sarcoma will usually occur within the soft tissues at the site of initial surgery as the host bone will have been excised and replaced with an endoprosthesis. Detection is more readily achieved if

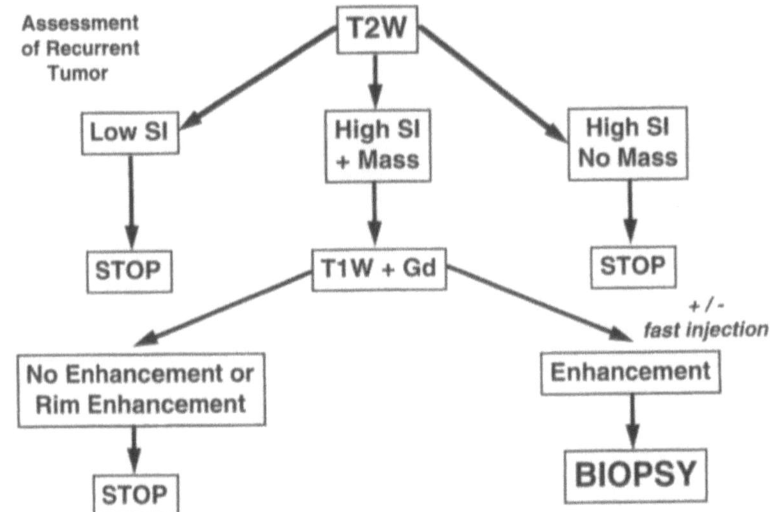

Fig. 19.16. Algorithm for the use of MR imaging in the follow-up of musculoskeletal sarcomas

there is evidence of matrix mineralization. Recurrent tumours with the propensity to mineralize will usually exhibit increased activity on bone scintigraphy but it is rarely used for this purpose.

At the first suggestion of recurrence some form of cross-sectional imaging is indicated, be it CT, MR imaging or, for soft tissue tumours, ultrasound (CHOI et al. 1991). MR imaging and CT have a similar sensitivity in the detection of local recurrent disease when the mass exceeds a volume of 15 cm^3 (REUTHER and MUTSCHLER, 1990). MR imaging is the technique of choice in the detection of early recurrence when local control may still be surgically achievable. While ultrasound does have some attractions, MR imaging will still be required for preoperative evaluation if a recurrence is identified.

The somewhat simplistic principle behind the use of MR imaging to detect recurrence relies on the mass effect and high water content of tumour with respect to the surrounding tissues. An algorithm for the use of MR imaging in the follow-up of sarcomas is given in Fig. 19.16. A T2-weighted or STIR sequence is the most useful in demonstrating a high signal intensity mass (Fig. 19.17) (VANEL et al. 1987, 1994). In the absence of a high signal intensity mass the likelihood of recurrence is extremely remote. Diffuse high signal intensity is frequently seen shortly after surgery or can be prolonged following radiation therapy (BIONDETTI and EHMAN 1992; VANEL et al. 1994; RICHARDSON et al. 1996).

A high signal intensity mass is suggestive but not conclusive for local recurrence as it may be seen in a variety of other postoperative entities (PANICEK

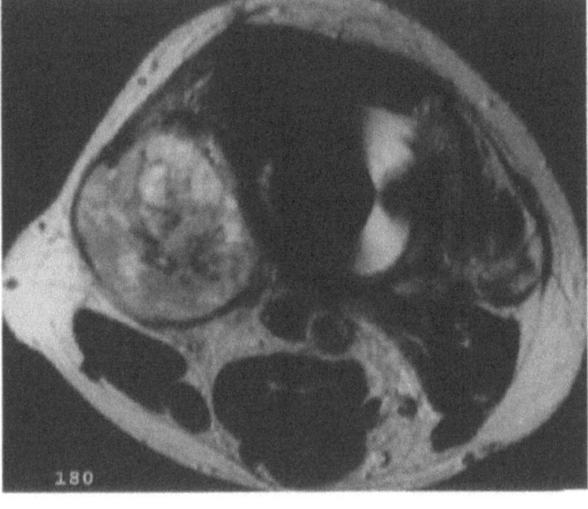

Fig. 19.17. Axial T2-weighted MR image of the distal thigh of a 31-year-old male 4 years after resection of a distal femoral osteosarcoma. There is a moderate metal artefact from the titanium femoral prosthesis. There is a 4 cm in diameter heterogeneous high signal intensity mass due to local recurrence of the tumour lying on the medial aspect of the prosthesis with further ill-defined recurrence laterally

et al. 1995). These include seromas, haematomas, surgical packing material and soft tissue spacers. The use of gadolinium-DTPA will usually permit differentiation between recurrent tumour and non-enhancing non-neoplastic conditions. A dynamic sequence may be required to distinguish tumour from inflammation, particularly shortly after surgery.

There are pitfalls to the stringent application of any algorithm (Fig. 19.16). If the tumour is densely mineralized (e.g. osteosarcoma) or hypocellular (e.g. fibromatosis) there may be mass effect but the predominant signal intensity on the T2-weighted images will be low. If the tumour is relatively hypovascular (e.g. chondrosarcoma) there will be a high signal intensity mass on the T2-weighted images but only rim enhancement simulating seromas. Knowledge of the preoperative imaging characteristics of the primary tumour is clearly valuable.

Optimal scheduling of follow-up MR examinations is contentious. Local recurrence of a musculoskeletal malignancy may occur within weeks of surgery in high-grade sarcomas and up to 15 years later in low-grade sarcomas, such as parosteal osteosarcoma. The cost and resource implications of employing regular, 3- or 6-monthly, follow-up MR examinations in all patients are considerable. There can be little doubt that close follow-up of patients in whom the original surgical resection was intralesional or marginal is worthwhile. There are, however, major problems using MR imaging to distinguish residual disease from surgical trauma early after inappropriate excision (NORIA et al. 1996).

19.7.2
Metastatic Disease

It is generally accepted that it is usually the metastatic disease that will eventually kill the patient with a musculoskeletal sarcoma and rarely the primary tumour itself. It is for this reason that follow-up imaging is concentrated on the site where metastases are most likely to occur, namely the lungs. Chest radiographs are obtained after surgery every 3 months for 2 years, every 6 months for a further 2 years and then annually. Routine follow-up of patients with serial chest CT scans is of doubtful value, particularly in view of the considerable radiation dose involved. CT is indicated if at any stage the chest radiograph is suspicious of early metastatic disease (Fig. 19.18).

Prior to the introduction of chemotherapy virtually all sarcoma metastases arose in the lungs before any other anatomical site. The natural history of osteosarcoma has been modified by chemotherapy in that up to 20% of those who develop metastases will first do so in bone prior to there being any evidence of pulmonary metastases. The prognosis for a patient with osseous osteosarcoma metastases is so

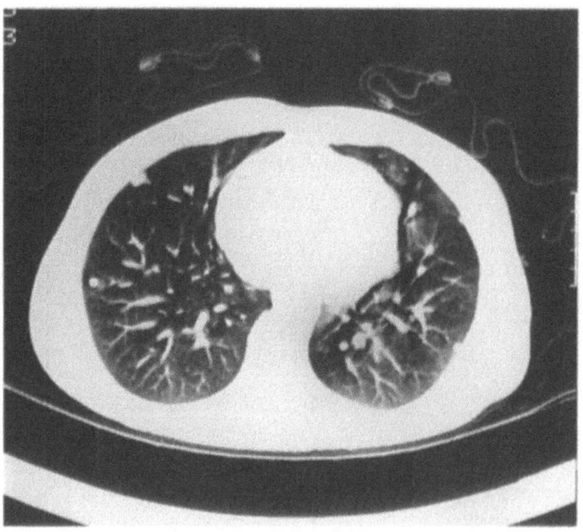

Fig. 19.18. CT chest of a 26-year-old female on follow-up after treatment for a Ewing's sarcoma. There are multiple small peripheral pulmonary metastases which were barely visible on the contemporary chest radiograph

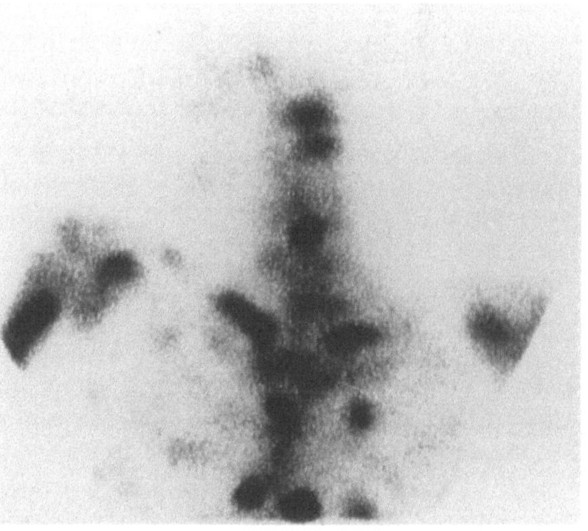

Fig. 19.19. Three-hour bone scintigraphy of the neck and upper thorax showing multiple osseous metastases from a primary osteosarcoma of the femur. Only one of the metastases in the neck was symptomatic at the time of this examination

poor that serial follow-up scintigraphy is unlikely to modify the outcome. Scintigraphy is indicated should a patient on follow-up develop bone pain in order to confirm/exclude and document the extent of metastatic disease (Fig. 19.19). Sarcoma metastases to the other sites, such as the central nervous system and viscera, are uncommon and are usually a late manifestation of the disease.

19.7.3
Complications of Treatment

It should be recognized that the prolonged medical and surgical management of a patient with a musculoskeletal tumour is not without the risk of complications. The structural integrity of the host bone may be sufficiently weakened by curettage of a benign bone lesion that a fracture can occur in the absence of residual tumour. For the patient with a sarcoma complications of treatment can be related to surgery, chemotherapy or radiotherapy where this has been administered. Endoprostheses can become infected or develop mechanical loosening, and the presence of either may be strongly suspected from the radiograph. Neither of these complications is significantly more frequent in patients with a bone tumour than in those who receive implants for other causes despite the prolonged chemotherapy. Rare post-chemotherapy complications that merit mention include methotrexate osteopathy (ECKLUND et al. 1997) and bone infarction following intra-arterial administration which may mimic a skip metastasis on MR imaging (OLLIVIER et al. 1991). Scintigraphy will reveal evidence of bone stress in a remarkably high percentage of patients who have undergone surgery for osteosarcoma, which should not be mistaken for bone metastases (AMI et al. 1987). Well-recognized complications of the use of allografts include infection and fracture. The heterogeneous appearance on MR imaging of allograft revascularization should not be interpreted as recurrent neoplasm or infection (HOEFFNER et al. 1996). In the long-term follow-up of patients who received radiotherapy, pain or functional impairment within the irradiated field should lead to the consideration of bone necrosis or radiation-induced sarcoma.

19.8
Conclusion

With the use of chemotherapy the 5-year survival for most patients with a musculoskeletal sarcoma is in the region of 60%. About 80% of these patients treated in specialist centres will be suitable for some form of limb-salvage surgery and will retain the use of a functioning limb. Imaging, with all its different techniques, has a recognized role in the management of these patients from detection and diagnosis to follow-up after definitive surgery. No imaging investigation should be reported in isolation without knowledge of relevant clinical details and results of

prior investigations. Where possible, the prior investigations themselves should be available for review as the appreciation of the significance of some new observation may well depend on a retrospective review of the previous studies. The importance of the team approach to the management of bone and soft tissue tumours cannot be over-emphasized.

References

Aisen AM, Martel W, Braunstein EM, McMillin KI, Phillips WA, Kling TF (1986) MR imaging and CT evaluation of primary bone and soft tissue tumors. AJR 146:749–756

Ami TB, Treves ST, Tumeh S, Cox-Bryan J, McCarthy C (1987) Stress fractures after surgery for osteosarcoma: scintigraphic assessment. Radiology 164:157–162

Biondetti PR, Ehman RL (1992) Soft-tissue sarcomas: use of textural patterns in skeletal muscle as a diagnostic feature in postoperative MR imaging. Radiology 183:845–848

Bloem JL, Taminiau AHM, Eulderink F, Hermans J, Pauwels EKJ (1988) Radiologic staging of primary bone sarcoma: MR imaging, scintigraphy, angiography, and CT correlated with pathologic examination. Radiology 169:805–810

Choi H, Varma DGK, Fornage BD, Kim EE, Johnston DA (1991) Soft-tissue sarcoma: MR imaging vs sonography for detection of local recurrence after surgery. Radiology 157:353–358

Codman EA (1926) Registry of bone sarcoma. Surg Gynecol Obstet 42:381–393

de Baere T, Vanel D, Shapeero LG, Charpentier A, Terrier P, di Paola M (1992) Osteosarcoma after chemotherapy: evaluation with contrast enhanced subtraction MR imaging. Radiology 185:587–592

Ecklund K, Laor T, Goorin AM, Connolly LP, Jaramillo DJ (1997) Methotrexate osteopathy in patients with osteosarcoma. Radiology 202:543–547

Ehara S, Kattapuram SV, Egglin TK (1991) Ewing's sarcoma: radiographic pattern of healing and bony complications in patients with long-term survival. Cancer 68:1531–1535

Enneking WF, Spanier SS, Goodman MA (1980) A system for the surgical staging of musculoskeletal sarcoma. Clin Orthop 153:106–120

Erlemann R, Sciuk J, Bosse A, Ritter J, Kusnierz-Glaz CR, Peters PE, Wuisman P (1990) Response of osteosarcoma and Ewing's sarcoma to pre-operative chemotherapy: assessment with dynamic and static MR imaging and skeletal scintigraphy. Radiology 175:791–796

Fletcher BD, Hanna SL, Fairclough DL, Gronemeyer SA (1992) Pediatric musculoskeletal tumors: use of dynamic contrast enhanced MR imaging to monitor response to chemotherapy. Radiology 184:243–248

Fletcher BD, Reddick WE, Taylor JS (1996) Dynamic MR imaging of musculoskeletal masses. Radiology 200:869–870

Grimer RJ, Sneath RS (1990) Diagnosing malignant bone tumors: editorial. J Bone Joint Surg [Br] 72:754–756

Hoeffner EG, Ryan JR, Qureshi F, Soulen RL (1996) MR imaging of massive bone allografts with histologic correlation. Skeletal Radiol 25:165–170

Holscher HC, Bloem JL, Nooy MA, Taminiau AHM, Eulderink F, Hermans J (1990) The value of MR imaging in monitoring the effect of chemotherapy in bone sarcomas. AJR 154:763–769

Holscher HC, Bloem JL, Vanel D, Hermans J, Nooy MA, Taminiau AHM, Henry-Amar M (1992) Osteosarcoma: chemotherapy induced changes at MR imaging. Radiology 82:839–844

Holscher HC, Bloem JL, van der Woude HJ, Hermans J, Nooy MA, Taminiau AHM, Hogendoorn PCW (1995) Can MRI predict the histolopathologic response in patients with osteosarcoma after first cycle of chemotherapy? Clin Radiol 50:384–390

Holscher HC, Hermans J, Nooy MA, Taminiau AHM, Hogendoorn PCW, Bloem JL (1996) Can conventional radiographs be used to monitor the effect of neoadjuvant chemotherapy in patients with osteogenic sarcoma? Skeletal Radiol 25:19–24

Jurgens H, Exner U, Gadner H, et al. (1988) Multidisciplinary treatment of primary Ewing's sarcoma of bone; a 6 year experience of a European Cooperative Trial. Cancer 61:23–32

Kenan S, Abdelwahab IF, Klein MJ, Herman G, Lewis MM (1993) Lesions of juxtacortical origin (surface lesions of bone). Skeletal Radiol 22:337–357

Kricun ME (1983) Radiographic evaluation of solitary bone lesions. Orthop Clin North Am 14:39–64

Kricun ME (1993) Imaging of bone tumors. Saunders, Philadelphia, pp 2–45

Kropei D, Schiller C, Ritschl P, Saltzer-Kuntschik M, Kotz R (1991) The management of IIB osteosarcoma. Clin Orthop 270:40–44

Lang P, Honda G, Roberts T, et al. (1995) Musculoskeletal neoplasms: perineoplastic edema versus tumor on dynamic post contrast MR images with spatial mapping of instantaneous enhancement rates. Radiology 197:831–839

Lodwick GS (1965) A probabalistic approach to the diagnosis of bone tumors. Radiol Clin North Am 3:487–497

Lodwick GS (1966) Solitary malignant tumors of bone: the application of predictor variables in diagnosis. Semin Roentgenol 1:293–313

Lodwick GS, Wilson AJ, Farrell C, Virtama P, Dittrich F (1980a) Determining growth rates of focal lesions of bone from radiographs. Radiology 134:577–583

Lodwick GS, Wilson AJ, Farrell C, Virtama P, Dittrich F (1980b) Estimating rate of growth in bone lesions: observer performance and error. Radiology 134:585–590

MacVicar AD, Olliff JFC, Pringle J, Ross-Pinkerton C, Husband JES (1992) Ewing's sarcoma: MR imaging of chemotherapy induced changes with histologic correlation. Radiology 184:859–864

Mail JT, Cohen MD, Mirkin LD, Provisor AJ (1985) Response of osteosarcoma to preoperative high-dose methotrexate chemotherapy: CT evaluation. AJR 144:890–893

Mankin HJ, Lange TA, Spanier SS (1982) The hazards of biopsy in patients with malignant primary bone and soft tissue tumors. J Bone Joint Surg [Am] 64:1121–1127

Mankin HJ, Mankin CJ, Simon MA (1996) The hazards of biopsy revisited. J Bone Joint Surg [Am] 78:656–663

Moser RP, Madewell JE (1987) An approach to primary bone tumors. Radiol Clin North Am 25:1049–1093

Noria S, Davis A, Kandel R, Levesque J, O'Sullivan B, Wunder J, Bell R (1996) Residual disease following unplanned excision of a soft tissue sarcoma of an extremity. J Bone Joint Surg [Am] 78:650–653

Ollivier L, Leclere J, Vanel D, Forest M, Pouillart P, Riche MC, Tomeno B (1991) Femoral infarction following intra-arterial chemotherapy for osteosarcoma of the leg: a possible pitfall in MR imaging. Skeletal Radiol 20:329–332

Panicek DM, Schwartz LH, Heelan RT, Cravelli JF (1995) Non-neoplastic causes of high signal intensity at T2W MR imaging after treatment of musculoskeletal neoplasms. Skeletal Radiol 24:185–190

Panicek DM, Gatsonis CG, Rosenthal DI, et al. (1997) CT and MR imaging in the local staging of primary malignant musculoskeletal neoplasms: report of the radiology diagnostic oncology group. Radiology 202:237–246

Pettersson H, Gillespy IT, Hamlin DJ (1987) Primary musculoskeletal tumors: examination with MR imaging compared with conventional modalities. Radiology 164:237–241

Ragsdale BD, Madewell JE, Sweet DE (1981) Radiologic and pathologic analysis of solitary bone lesions. Part II. Periosteal reaction. Radiol Clin North Am 19:749–783

Reddick WE, Bhargava R, Taylor JS, Meyer WH, Fletcher BD (1995) Dynamic contrast enhanced MR imaging evaluation of osteosarcoma response to neoadjuvant chemotherapy. J Magn Reson Imaging 5:684–694

Remedios D, Safuddin A, Pringle J (1997) Radiological and clinical recurrence of giant cell tumor of bone after the use of cement. J Bone Joint Surg [Br] 79:26–30

Reuther G, Mutschler W (1990) Detection of local recurrent disease in musculoskeletal tumours: MR imaging versus CT. Skeletal Radiol 19:85–90

Richardson ML, Zink-Brody GC, Patten RM, Koh WJ, Conrad EU (1996) MR characterization of post-irradiation soft tissue edema. Skeletal Radiol 25:537–543

Russell WO, Cohen J, Edmonson JH, et al. (1981) Staging system for soft tissue sarcoma. Semin Oncol 8:156–159

Seeger LL, Widoff BE, Bassett LW, Rosen G, Eckardt JJ (1991) Preoperative evaluation of osteosarcoma: value of gadolinium dimeglumine-enhanced MR imaging. AJR 157:347–351

Shirkhoda A, Jaffe N, Wallace S, Ayala AG, Lindell MM, Zornoz AJ (1985) Computed tomography of osteosarcoma after intra-arterial chemotherapy. AJR 144:95–99

Simon MA, Aschliman MA, Thomas N, Mankin HJ (1986) Limb salvage treatment versus amputation of osteosarcoma of the distal end of the femur. J Bone Joint Surg [Am] 68:1331–1337

Skrznski MC, Biermann JS, Montag A, Simon MA (1996) Diagnostic accuracy and charge savings of outpatient core needle biopsy compared with open biopsy of musculoskeletal lesions. J Bone Joint Surg [Am] 78:644–649

Smith J, Heelan RT, Huvos AG (1982) Radiographic changes in primary osteogenic sarcoma following intensive chemotherapy. Radiology 143:355–360

Steiner RM, Mitchell DG, Rao VM, Schweitzer ME (1993) MR imaging of diffuse bone marrow disease. Radiol Clin North Am 31:383–409

Stoker DJ, Cobb JP, Pringle JAS (1991) Needle biopsy of musculoskeletal lesions: a review of 208 procedures. J Bone Joint Surg [Br] 73:498–500

Sundaram M, McGuire MH, Herbold DR, Wolverson MK, Heiberg E (1986) MR imaging in planning limb-salvage surgery for primary malignant tumors of bone. J Bone Joint Surg [Am] 68:809–819

Swan JS, Grist TM, Sproat IA, Heiner JP, Wiersma SR, Heisey DM (1995) Musculoskeletal neoplasms: preoperative evaluation with MR angiography. Radiology 194:519–524

Sweet DE, Madewell JE, Ragsdale BD (1981) Radiologic and pathologic analysis of solitary bone lesions. Part III. Matrix patterns. Radiol Clin North Am 19:785–814

van der Woude HJ, Bloem JL, Holscher HC, et al. (1994) Monitoring the effect of chemotherapy in Ewing's sarcoma of bone with MR imaging. Skeletal Radiol 23:493–500

van der Woude HJ, Bloem JL, Verstraete KL, Taminiau AHM, Nooy MA, Hogendoorn PC (1995a) Osteosarcoma and

Ewing's sarcoma after neoadjuvant chemotherapy: value of dynamic MR imaging in detecting viable tumor before surgery. AJR 165:593–598

van der Woude HJ, Bloem JL, van Oostayen JA, Taminiau AH, Hermans J, Reynierse M, Hogendoorn PC (1995b) Treatment of high grade sarcomas with neoadjuvant chemotherapy: the utility of colour Doppler sonography in predicting histopathologic response. AJR 165:125–133

van der Woude HJ, Verstraete KL, Bloem JL, Hogendoorn PCW, Taminaiu AHM (1996) Giant cell tumor of bone: postsurgical detection of recurrent or residual tumor with fast dynamic contrast enhanced MR imaging (abst.). European Musculoskeletal Oncology Society 1996

Vanel D, Contesso G, Couanet D, Piekarski JD, Sarazin D, Masselot J (1982) Computed tomography in the evaluation of 41 cases of Ewing's sarcoma. Skeletal Radiol 9:8–13

Vanel D, Lacombe MJ, Couanet D, Kalifa C, Spielmann M, Genin J (1987) Musculoskeletal tumor: follow-up with MR imaging after treatment with surgery and radiation therapy. Radiology 164:243–245

Vanel D, Shapeero LG, de Baere T, Gilles R, Tardivon A, Genin J, Guinebretiere JM (1994) MR imaging in the follow-up of malignant and aggressive soft tissue tumors: results of 511 examinations. Radiology 190:263–268

Verstraete KL, de Decour Y, Roels H, Dierich A, Uttendaele D, Kunnen M (1994) Benign and malignant musculoskeletal lesions: dynamic contrast enhanced MR imaging: parametric "first pass" images depict tissue vascularization and perfusion. Radiology 192:835–843

Veth RP (1991) IIB osteosarcoma. Clinical management, local control and survival statistics – The Netherlands. Clin Orthop 270:67–73

Winkler K, Beron G, Delling G, et al. (1988) Neoadjuvant chemotherapy of osteosarcomas: results based on a randomized cooperative trial (COSS 82) with salvage chemotherapy based on histological tumor response. J Clin Oncol 6:329–337

Wolf RE, Enneking WF (1996) The staging and surgery of musculoskleletal neoplasms. Orthop Clin North Am 27:473–481

Zimmer WD, Berquist TH, McLeod RA, et al. (1985) Bone tumors: MR imaging versus computed tomography. Radiology 155:709–718

List of Contributors

Thomas H. Berquist, MD, FACR
Professor of Diagnostic Radiology
Mayo Medical School and
Chair Department of Diagnostic Radiology
Mayo Clinic Jacksonville
4500 San Pablo Road
Jacksonville, FL 32224
USA

H. Bonél, MD
Ludwig-Maximilian-Universität
Klinikum Großhadern
Institut für Radiologische Diagnostik
Marchioninistrasse 15
D-81377 München
Germany

J.A. Bouffard, MD
Division of Musculoskeletal Radiology
Henry Ford Hospital
2799 West Grand Blvd.
Detroit, MI 48202
USA

M. Breitenseher, MD
MR-Einrichtung der Medizinischen Fakultät
Universitätsklinik für Radiodiagnostik
Allgemeines Krankenhaus der Stadt Wien
Lazarettsgasse 14
A-1090 Wien
Austria

Victor N. Cassar-Pullicino, MD
Department of Diagnostic Imaging
The Institute of Orthopaedics
The Robert Jones & Agnes Hunt
Orthopaedic & District Hospital
Oswestry
Shropshire SY10 7AG
United Kingdom

N. Chemla, MD
Department of Radiology B
Hôpital Cochin
27, rue du Faubourg-Saint Jacques
F-75675 Paris
France

Alain Chevrot, MD
Head, Department of Radiology B
Hôpital Cochin
27, rue du Faubourg-Saint Jacques
F-75675 Paris
France

A. Mark Davies, MD
MRI Centre
Royal Orthopaedic Hospital
Bristol Road South
Birmingham B31 2AP
United Kingdom

J.L. Drape, MD
Department of Radiology B
Hôpital Cochin
27, rue du Faubourg-Saint Jacques
F-75675 Paris
France

A.M. Dupont, MD
Department of Radiology B
Hôpital Cochin
27, rue du Faubourg-Saint Jacques
F-75675 Paris
France

Niels Egund, MD
Professor
Røntgenafdeling R
Århus Kommunehospital
Nørrebrogade 44
DK-8000 Århus C
Denmark
Blegdamsvej 9
DK-2100 Copenhagen
Denmark

Harry K. Genant, MD
Professor of Radiology, Medicine, Epidemiology
and Orthopedic Surgery
Chief of the Musculoskeletal Section
Executive Director of the Osteoporosis and
Arthritis Research Group
Department of Radiology, M-392
University of California at San Francisco
San Francisco, CA 94143-0628
USA

F. Gires, MD
Department of Radiology B
Hôpital Cochin
27, rue du Faubourg-Saint Jacques
F-75675 Paris
France

D. Godefroy, MD
Department of Radiology B
Hôpital Cochin
27, rue du Faubourg-Saint Jacques
F-75675 Paris
France

J. Haller, MD
Zentralröntgen
Hanuschkrankenhaus
A-1090 Wien
Austria

Juerg Hodler, MD
Radiology, Balgrist Clinic
University of Zürich
Forchstrasse 340
CH-8008 Zürich
Switzerland

Herwig Imhof, MD, Professor
MR-Einrichtung der Medizinischen Fakultät
Universitätsklinik für Radiodiagnostik
Allgemeines Krankenhaus der Stadt Wien
Lazarettsgasse 14
A-1090 Wien
Austria

Kjell Jonsson, MD, PhD
Department of Radiology
University Hospital
S-221 85 Lund
Sweden

F. Kainberger, MD
MR-Einrichtung der Medizinischen Fakultät
Universitätsklinik für Radiodiagnostik
Allgemeines Krankenhaus der Stadt Wien
Lazarettsgasse 14
A-1090 Wien
Austria

Jeremy J. Kaye, MD
Professor of Radiology
New York Medical College
Chairman
Department of Radiology
Saint Vincents Hospital and Medical Center
153 West 11th Street
New York, NY 10011
USA

Maria Vittoria Maffey, MD
Università degli Studi dell'Aquila
Facoltà di Medicina e Chirurgia
Cattedra di Radiologia
Ospedale Collemaggio
I-67100 L'Aquila
Italy

Carlo Masciocchi, MD
Professor
Università degli Studi dell'Aquila
Facoltà di Medicina e Chirurgia
Cattedra di Radiologia
Ospedale Collemaggio
I-67100 L'Aquila
Italy

Ian W. McCall, MD
Professor, Department of Radiology
Robert Jones and Agnes Hunt
Orthopaedic & District Hospital NHS Trust
Oswestry, Shropshire SY10 7AG
United Kingdom

Eugene G. McNally, MD
Consultant Musculoskeletal Radiologist
Nuffield Orthopaedic Centre
Headington
Oxford OX3 7LD
United Kingdom

A. Minoui, MD
Department of Radiology B
Hôpital Cochin
27, rue du Faubourg-Saint Jacques
F-75675 Paris
France

J. Moutounet, MD
Department of Radiology B
Hôpital Cochin
27, rue du Faubourg-Saint Jacques
F-75675 Paris
France

Willem R. Obermann, MD, PhD
Department of Radiology
University Hospital of Leiden
Albinusdreef 2
2333 ZA Leiden
The Netherlands

E. Pessis, MD
Department of Radiology B
Hôpital Cochin
27, rue du Faubourg-Saint Jacques
F-75675 Paris
France

Holger Pettersson, MD, Professor
Department of Radiology
University Hospital
University of Lund
S-22185 Lund
Sweden

Maximilian Reiser, MD, Professor
Ludwig-Maximilian-Universität
Klinikum Großhadern
Institut für Radiologische Diagnostik
Marchioninistrasse 15
D-81377 München
Germany

L. Sarazin, MD
Department of Radiology B
Hôpital Cochin
27, rue du Faubourg-Saint Jacques
F-75675 Paris
France

Christiaan Schiepers, MD, PhD
Department of Radiological Sciences
Olive View – UCLA Medical Center
14445 Olive View Drive
Sylmar, CA 91342
USA

ERIK R. TJIN A TON, MD
Department of Radiology
University Hospital of Leiden
Albinusdreef 2
2333 ZA Leiden
The Netherlands

S. TRATTNIG, MD
MR-Einrichtung der Medizinischen Fakultät
Universitätsklinik für Radiodiagnostik
Allgemeines Krankenhaus der Stadt Wien
Lazarettsgasse 14
A-1090 Wien
Austria

DANIEL VANEL, MD
Department of Radiology
Institut Gustave Roussy
39, rue Camille Desmoulins
F-94805 Villejuif
France

MARNIX VAN HOLSBEECK, MD
Division of Musculoskeletal Radiology
Henry Ford Hospital
2799 West Grand Blvd.
Detroit, MI 48202
USA

CORNELIUS VAN KUIJK, MD, PhD
Assistant Adjunct Professor of Radiology
Director of Radiographic Laboratory
Osteoporosis and Arthritis Research Group
Department of Radiology, M-392
University of California at San Francisco
San Francisco, CA 94143-0628
USA

IAIN WATT, FRCP, FRCR
Consultant Clinical Radiologist
Department of Clinical Radiology
Bristol Royal Infirmary
Bristol BS2 8HW
United Kingdom

MEDICAL RADIOLOGY
Diagnostic Imaging and Radiation Oncology

Titles in the series already published

RADIATION ONCOLOGY

Interventional Radiation Therapy Techniques - Brachytherapy
Edited by R. Sauer

Radiopathology of Organs and Tissues
Edited by E. Scherer,
C. Streffer, and K.-R. Trott

Concomitant Continuous Infusion Chemotherapy and Radiation
Edited by M. Rotman
and C.J. Rosenthal

Intraoperative Radiotherapy – Clinical Experiences and Results
Edited by F.A. Calvo,
M. Santos, and L.W. Brady

Radiotherapy of Intraocular and Orbital Tumors
Edited by W.E. Alberti
and R.H. Sagerman

Interstitial and Intracavitary Thermoradiotherapy
Edited by M.H. Seegenschmiedt
and R. Sauer

Non-Disseminated Breast Cancer
Controversial Issues
in Management
Edited by G.H. Fletcher
and S.H. Levitt

Current Topics in Clinical Radiobiology of Tumors
Edited by
H.-P. Beck-Bornholdt

Practical Approaches to Cancer Invasion and Metastases
A Compendium of Radiation
Oncologists' Responses
to 40 Histories
Edited by A.R. Kagan with the
Assistance of R.J. Steckel

Radiation Therapy in Pediatric Oncology
Edited by J.R. Cassady

Radiation Therapy Physics
Edited by A.R. Smith

Late Sequelae in Oncology
Edited by J. Dunst and R. Sauer

Mediastinal Tumors. Update 1995
Edited by D.E. Wood
and C.R. Thomas, Jr.

Thermoradiotherapy and Thermochemotherapy
Volume 1:
Biology, Physiology, and Physics
Volume 2:
Clinical Applications
Edited by M.H. Seegenschmiedt,
P. Fessenden, and C.C. Vernon

Carcinoma of the Prostate
Innovations in Management
Edited by Z. Petrovich,
L. Baert, and L.W. Brady

Radiation Oncology of Gynecological Cancers
Edited by H.W. Vahrson

Carcinoma of the Bladder
Innovations in Management
Edited by Z. Petrovich,
L. Baert, and L.W. Brady

Blood Perfusion and Microenvironment of Human Tumors
Implications for Clinical
Radiooncology
Edited by M. Molls and P. Vaupel

Radiation Therapy of Benign Diseases. A Clinical Guide
2nd revised edition
S.E. Order and S.S. Donaldson

MIX
Papier aus verantwortungsvollen Quellen
Paper from responsible sources
FSC® C105338

If you have any concerns about our products,
you can contact us on
ProductSafety@springernature.com

In case Publisher is established outside the EU,
the EU authorized representative is:
Springer Nature Customer Service Center GmbH
Europaplatz 3, 69115 Heidelberg, Germany

Printed by Libri Plureos GmbH
in Hamburg, Germany